INTERMEDIATE ALGEBRA WITH PROBLEM SOLVING

MERVIN L. KEEDY
Purdue University

MARVIN L. BITTINGER
Indiana University—Purdue University at Indianapolis

ADDISON-WESLEY PUBLISHING COMPANY
Reading, Massachusetts • Menlo Park, California
Don Mills, Ontario • Wokingham, England
Amsterdam • Sydney • Singapore • Tokyo • Madrid
Bogotá • Santiago • San Juan

Sponsoring Editor • *Susan Zorn*
Production Supervisor • *Susanah H. Michener*
Editorial and Production Service • *Quadrata, Inc.*
Text Designer • *Vanessa Piñeiro, Piñeiro Design Associates*
Illustrators • *VAP Group Limited, Belanger Associates; Textbook Art Associates*
Art Consultant • *Joseph Vetere*
Manufacturing Supervisor • *Ann DeLacey*
Cover Designer • *Marshall Henrichs*

Photo Credits

1, Peter Vandermark/Stock, Boston **37,** Belanger Associates, with permission of Sozio, Revere, MA **41,** Belanger Associates **49,** Belanger Associates **70,** Mark Antman/Stock, Boston **73,** Daniel S. Brody/Stock, Boston **74,** Belanger Associates **79,** Oldsmobile, General Motors **85,** Belanger Associates **121,** Belanger Associates **153,** Peter Southwick/Stock, Boston **158,** Elizabeth Crews/Stock, Boston **181,** Bohdan Hrynewych/Stock, Boston **206,** Ellis Herwig/Stock, Boston **207,** Dean Abramson/Stock, Boston **209,** Robert V. Eckert/Stock, Boston **218,** Rick Smolan/Stock, Boston **231,** Phyllis Craber Jensen/Stock, Boston **271,** Belanger Associates **282,** Belanger Associates **290,** Frank Siteman/Stock, Boston **297,** Jeff Albertson/Stock, Boston **313,** Belanger Associates **314,** H. Morgan/Stock, Boston **317,** James R. Holland/Stock, Boston **343,** Daniel Brody/Stock, Boston **351,** NASA **359,** Belanger Associates **365,** Oldsmobile, General Motors **379,** Belanger Associates **395,** Frank Siteman/Stock, Boston

Library of Congress Cataloging-in-Publication Data

Keedy, Mervin Laverne.
Intermediate algebra with problem solving.

"Paperback version of the hardback: A problem-solving approach to intermediate algebra, 2nd edition . . . enhanced with . . . worktext features"—Pref.
Includes index.
1. Algebra. 2. Algebra—Programs, exercises, etc. 3. Problem solving. I. Bittinger, Marvin L. II. Keedy, Mervin Laverne. Problem solving approach to intermediate algebra. 2nd ed. III. Title.
QA154.2.K436b 1986 512 86-14155
ISBN 0-201-15650-4(pbk)

ABCDEFGHIJ-MU-89876

PREFACE

This text is a paperbound version of the hardbound *A Problem-Solving Approach to Intermediate Algebra,* 2nd Edition. It has been enhanced with many important worktext features and, as in the hardbound edition, is distinguished by its integrated treatment of problem solving.

MATHEMATICAL CONTENT AND LEVEL

The mathematical content of this text is intermediate algebra. It is intended for use by students who have finished a first course in algebra. Students who have never studied algebra or whose background is otherwise too weak are advised to use the companion volume, *Introductory Algebra with Problem Solving,* by the same authors. A placement test is available to help place students in the proper book. (See the list of supplements that follows this preface.)

WORKTEXT FEATURES

- *Objectives.* The objectives for each section are stated in the margin, and are easy to identify by their symbol i. That same symbol is placed in the text and also in the exercises beside the corresponding material, making the text easy to read and use for self-study and review.
- *Margin exercises.* While working through the material the student is encouraged to do exercises located in the margins. Answers to these exercises are at the back of the book. The margin exercises are designed to get the student *actively involved* in the development of each section of material, and can also serve as a self-study guide. They are very much like the examples in the text and also like the homework exercises, enabling the student to practice as he or she is learning.
- *Exercise sets.* The exercises are carefully indexed to the objectives and identified by the symbols ii. This coding occurs in the basic exercises at the beginning of each exercise set, enabling students to refer to appropriate sections of the text when review is needed.

READABILITY AND TEACHABILITY

The style and format of the book are designed to make the text easy to read and comprehend. Examples to illustrate each skill and concept have been chosen carefully to ensure that they are appropriate and to the point. The exercises in the exercise sets are very much like the examples. Exercises are ample and they are paired and graded, with answers to the odd-numbered exercises at the back of the book. Answers to all exercises are located in the *Student's Guide to Exercises* (see the list of supplements).

The exercises that follow the symbolism

○ ———

are generally different from the examples in the text, and require more than a simple understanding of the material at hand, thus offering a challenge to students. There are also exercises that are designed to be done with a calculator. Those exercises are marked [calculator].

PROBLEM SOLVING

The primary distinguishing feature of this book is its treatment of and emphasis on problem solving. The approach to problem solving is different here, not only because of the increased space devoted to it, but also because of the way students are guided in formulating problems and translating the situation of a problem to mathematical language. Problem-solving techniques are introduced in Chapter 1 and applied throughout the text, with problem-solving practice in most exercise sets.

140 POLYNOMIALS AND POLYNOMIAL FUNCTIONS

9. $3x^8 - 3y^8$	**10.** $9a^4 - a^2b^2$	**11.** $9a^4 - 25a^2b^4$	**12.** $16x^6 - 121x^2y^4$
13. $\frac{1}{25} - x^2$	**14.** $\frac{1}{16} - y^2$	**15.** $0.04x^2 - 0.09y^2$	**16.** $0.01x^2 - 0.04y^2$
17. $a^2 + 2ab + b^2 - 9$	**18.** $x^2 - 2xy + y^2 - 25$	**19.** $r^2 - 2r + 1 - 4s^2$	**20.** $c^2 + 4cd + 4d^2 - 9p^2$
21. $2m^2 + 4mn + 2n^2 - 50b^2$	**22.** $12x^2 + 12x + 3 - 3y^2$	**23.** $9 - (a^2 + 2ab + b^2)$	**24.** $16 - (x^2 - 2xy + y^2)$

Exercises keyed to objectives → **ii** Complete the square.

25. $x^2 + 16x$	**26.** $y^2 - 24y$	**27.** $x^2 - 4.2x$	**28.** $y^2 + 3.6y$

iii Factor by completing the square.

29. $x^2 + 24x + 119$	**30.** $x^2 + 30x + 176$	**31.** $x^2 - 26x + 105$	**32.** $x^2 - 32x + 192$
33. $2x^2 + 56x + 342$	**34.** $3x^2 + 54x + 231$	**35.** $2x^2 - 32ax + 96a^2$	**36.** $3x^2 - 60bx + 192b^2$

Problem-solving practice

37. Susan has four test scores of 78, 92, 79, and 85. What must the fifth test score be in order for her average to be 82?

38. A beaker contains a solution that is 40% ammonia. Another beaker contains a solution that is 60% ammonia. How much of each can be added together in order to obtain 35 liters of a solution that is 55% ammonia? ← **Problem-solving practice**

Challenge exercises integrated into each exercise set → ○ ———

Factor by completing the square.

39. [calculator] $x^2 + 4.482x - 7.403544$

40. [calculator] $5.72x^2 + 35.464x - 1319.2608$ ← **Calculator exercise**

Complete the square.

41. $x^2 + \frac{2}{3}bx$	**42.** $ay^2 + \frac{3}{4}by$	**43.** $ax^2 + bx$	**44.** $ax^2 + bx + c$

Factor. Assume that variables in exponents represent natural numbers.

45. $x^2 - 3x + \frac{5}{4}$	**46.** $x^2 - 5x + \frac{9}{4}$	**47.** $x^2 + 2.6x + 0.69$
48. $x^2 - 3.2x + 1.56$	**49.** $x^2 + \frac{3}{5}x + \frac{8}{100}$	**50.** $x^2 - \frac{7}{2}x + \frac{45}{16}$
51. $x^4y^4 - 8x^2y^2 + 16$	**52.** $-24ab + 16a^2 + 9b^2$	**53.** $\frac{1}{36}x^8 + \frac{4}{18}x^4 + \frac{4}{9}$
54. $0.25x^2 + 0.30x + 0.09$	**55.** $x^{2a} - y^2$	**56.** $x^{4a} - y^{2b}$
57. $4y^{4a} + 12y^{2a} + 10y^{2a} + 30$	**58.** $25y^{2a} - (x^{2b} - 2x^b + 1)$	

Factor by completing the square.

59. $x^{2a} + 14x^a + 45$

60. $x^2 - 2ax - 3a^2$

OBJECTIVE

After finishing Section 4.6, you should be able to:

Objectives for each section clearly marked → **i** **Factor sums and differences of two cubes.**

4.6

SUMS OR DIFFERENCES OF TWO CUBES

i Although a sum of two squares cannot be factored using real-number coefficients, a sum of two cubes can. In this section we develop the patterns for factoring a sum or a difference of two expressions that are cubes. ← **Symbols in text correspond to objectives listed in margin**

Consider the following products:

$$(A + B)(A^2 - AB + B^2) = A(A^2 - AB + B^2) + B(A^2 - AB + B^2)$$
$$= A^3 - A^2B + AB^2 + A^2B - AB^2 + B^3$$
$$= A^3 + B^3$$

and

$$(A - B)(A^2 + AB + B^2) = A(A^2 + AB + B^2) - B(A^2 + AB + B^2)$$
$$= A^3 + A^2B + AB^2 - A^2B - AB^2 - B^3$$
$$= A^3 - B^3.$$

The above equations (reversed) show how we can factor a sum or a difference of two cubes.

$$A^3 + B^3 = (A + B)(A^2 - AB + B^2)$$
$$A^3 - B^3 = (A - B)(A^2 + AB + B^2)$$

Factor.

1. $x^3 - 8$

Margin exercises actively involve students in the development of material

This table of cubes will help in the following examples.

N	0.2	0.1	0	1	2	3	4	5	6	7	8
N^3	0.008	0.001	0	1	8	27	64	125	216	343	512

EXAMPLE 1 Factor: $x^3 - 27$.

$$x^3 - 27 = x^3 - 3^3$$

In one set of parentheses we write the cube root of the first term, x. Then we write the cube root of the second term, -3. This gives us the expression $x - 3$:

$$(x - 3)(\quad).$$

To get the next factor we think of $x - 3$ and do the following:

Square the first term: x^2.
Multiply the terms and then change the sign: $3x$.
Square the second term: 9.

$$(x - 3)(x^2 + 3x + 9).$$

2. $64 - y^3$

Note that we cannot factor $x^2 + 3x + 9$. (It is not a trinomial square nor can it be factored by trial).

Students are encouraged to do margin exercises → DO EXERCISES 1 AND 2.

Factor.

3. $27x^3 + y^3$

4.9

SOLVING PROBLEMS

OBJECTIVE ← **Objectives listed in margin**

After finishing Section 4.9, you should be able to:

i **Solve problems by translating to an equation and then solving the equation using factoring and the principle of zero products.**

i Some problem situations can be translated to polynomial equations, some of which we can now solve. The problem-solving process is the same as for other kinds of equations, except that we can now solve equations by factoring and using the principle of zero products. Let us finish an example that we started earlier, on p. 35.

Students learn to solve sophisticated real-life problems → EXAMPLE 1 There are two square flower gardens on the campus mall. The length of a side of one garden is 4 ft less than the length of a side of the other. The area of the larger garden is 56 ft^2 greater than that of the smaller. Find the dimensions of the garden.

1. The square of a number minus twice the number is 48. Find the number. ← **Margin exercises throughout the text**

1. **Familiarize** We draw a picture and label it.

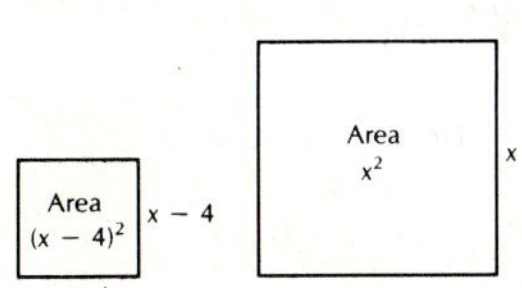

The larger garden has area 56 greater than the smaller garden.

2. **Translate** We use the following statement:

Area of small garden + 56 is area of the large garden.

$$(x - 4)^2 \quad + 56 = \quad x^2.$$

Five steps of problem solving applied throughout the text

Five steps of problem solving applied throughout the text

3. **Carry out** We solve the equation:

$$(x-4)^2+56=x^2$$
$$x^2-8x+16+56=x^2 \quad \text{Squaring a binomial}$$
$$-8x+72=0 \quad \text{Collecting like terms and simplifying}$$
$$x=9.$$

4. **Check** Since we used x to represent the length of a side of the larger garden, it follows that the smaller garden has sides of length $x-4$, or 5. The areas of the gardens are then 5^2, or 25 ft^2, and 9^2, or 81 ft^2. The difference is 56 ft^2, so we do have a solution of the problem.

5. **State** The smaller garden has sides of length 5 ft and the larger garden has sides of length 9 ft.

DO EXERCISE 1.

Students are encouraged to do margin exercises

SKILL MAINTENANCE

Skill maintenance is featured throughout the text.

- The problem-solving practice in the exercise sets covers problem solving from earlier parts of the book, constantly reinforcing the student's thinking skills.
- Each chapter ends with a Summary and Review section designed to give students maximum assistance when they review the material. The chapter objectives are stated in boldface type, followed by exercises testing students' comprehension of them. Answers are given at the back of the book together with section references so the student can restudy material as appropriate.
- A practice test follows the Summary and Review; practice-test answers are in the Instructor's Manual.
- The text also contains cumulative reviews every three to four chapters. These reviews cover all material from Chapter 1 up to that point in the text, and further reinforce mastery of all topics. All answers together with section references are at the back of the book.

Problem-solving practice in exercise sets gives ongoing review of problem-solving technique

Problem-solving practice

21. The perimeter of a rectangular field is 710 m. The length is 115 m greater than the width. Find the dimensions.

22. A student scores a total of 248 points on three tests. The total of the first two scores is 66 more than the score on the third test. The score on the third test is 6 more than that on the second test. Find the scores.

23. Bert has two cats. Their combined age is 12 yr. Their combined weight is 28 lb. If the younger cat weighs three times his age and the older cat weighs twice his age, how old are the two cats?

24. An electrician, a carpenter, and a plumber are hired to work on a house. The electrician earns \$10.50 per hour, the carpenter earns \$9.75 per hour, and the plumber earns \$12 per hour. The first day on the job they earned a total of \$234.75. If the plumber worked two more hours than the carpenter, how many hours did each work?

SUMMARY AND REVIEW: CHAPTER 3

The following contains a summary of what you should be able to do after completing this chapter. The review exercises are for practice. Answers are at the back of the book. If you miss an exercise, restudy the section indicated alongside the answer.

You should be able to:

Solve systems of equations graphically. ← **Objectives summarized in boldface type**

Solve graphically.

1. $4x - y = 10,$ $2x - 2y = 7$

2. $y = 3x + 7,$ $3x + 2y = -4$

Solve systems of equations using the substitution method.

Solve using the substitution method.

3. $7x - 4y = 6,$ $y - 3x = -2$

4. $5x + 5y = -6,$ $2x - 2y = -4$

5. $9x - 6y = 2,$ $-2x = 4y + 5$

Answers at the back of the book with section references

A test for every chapter with answers in the *Instructor's Manual*

TEST: CHAPTER 3

Solve.

1. $x + 3y = -8$, $4x - 3y = 23$
2. $2x + 4y = -6$, $-3x - y = 9$
3. $4x - 6y = 3$, $6x - 4y = -3$
4. $2x + 4y = 18$, $3x + 6y = 26$
5. The perimeter of a rectangle is 96. The length of the rectangle is 6 less than twice the width. Find the dimensions of the rectangle.
6. A fast-food restaurant sold 132 chicken dinners during one day. Two-piece dinners sold for $1.50, and three-piece dinners sold for $2.50. The restaurant's total receipts that day from chicken dinners were $260. How many of each size dinner did they sell?

Answers at the back of the book with section references

Cumulative reviews reinforce mastery of topics

CUMULATIVE REVIEW: CHAPTERS 1–3

Solve.

1. $-14.3 + 29.17 = x$
2. $x + 9.4 = -12.6$
3. $3.9(-11) = x$
4. $-2.4x = -48$
5. $4x + 7 = -14$
6. $-3 + 5x = 2x + 15$
7. $3n - (4n - 2) = 7$
8. $6y - 5(3y - 4) = 10$
9. $14 + 2c = -3(c + 4) - 6$
10. $5x - [4 - 2(6x - 1)] = 12$

Simplify.

11. $x^4 \cdot x^{-6} \cdot x^{13}$
12. $(4x^{-3}y^2)(-10x^4y^{-7})$
13. $(6x^{a+2}y^{b+2})(-2x^{a-2}y^{b+1})$

SUPPLEMENTS

The following supplementary materials are available.

- *Student's Guide to Exercises.* Contains answers for all the exercises in the exercise sets. Contains detailed solutions to all odd-numbered exercises in the exercise sets.
- *Instructor's Manual with Tests and Computer Programs.* Contains six alternate forms of the chapter tests and final exams, a black-line grid master for graphing, and computer programs written in BASIC for each chapter, designed to demonstrate to the student the power and utility of the computer in performing algebraic functions. The *Instructor's Manual* also contains answers to the chapter tests that are in the book.
- *Instructional Software for Algebra.* A highly interactive computer software supplement that covers most of the topics treated in the text. It consists of guided exercises and quizzes, and can be used either to supplement the text or for individualized instruction.
- *Computerized Testing* (TESTGEN II). Contains test questions at two levels of difficulty for every testable objective. The instructor can choose from a testbank of over 2000 questions to create tests using either open-ended or multiple-choice format, or can generate test questions randomly. TESTGEN II will scramble questions within a given test, and also scramble answers within a given multiple-choice question, allowing many permutations of the same test items. High-quality graphics are printed out on the test, and an answer sheet is provided. TESTGEN II is available for both Apple II and IBM PC computers.
- *Printed Test Bank.* The printed version of the computerized test bank questions with solutions. Instructors can choose questions from it to be printed out by TESTGEN II, or can use it to make up their own tests by hand.
- *Videotape Cassettes.* Keyed to objectives in the text, they can supplement the regular lectures or can be used in a number of teaching situations, such as learning labs, or fill-in lectures when instructors are absent. John Jobe of Oklahoma State University speaks to students and works out examples with lucid explanations on 27 thirty-minute tapes. There are also two thirty-minute tapes on problem solving, which cover the entire scope of the material in the text.
- *Audiotapes.* Designed to supplement the text or to be used for individualized instruction.

- *Placement Test.* Statistically validated, this test can be used to place students properly in either this book or other books by the same authors, such as *Algebra and Trigonometry: A Functions Approach,* 4th ed. (1986). More specific placement in this book can be attained using a form of the final exam in the *Instructor's Manual.*

ACKNOWLEDGMENTS

The authors wish to thank many people without whose committed efforts our work could not have been completed successfully. Barbara Dunn, Denis Eckert, Ronnie Khuri, Barbara Johnson, and Karen Parker did a tremendous job checking the manuscript. Randy Becker, Judy Beecher, Virginia Hamilton (Ball State), Judy Penna, Andrew Salvage, and Julie Stephenson were superb in their work on the supplements.

Many instructors provided reviewing information. We thank them for their many suggestions for improvement. They are Don Albers (Menlo College), Harold Hauser (Mt. Hood Community College), Mike Mears (Manatee Junior College), Jane Morrison (Thornton Community College), Harold Oxsen (Diablo Valley College), Richard Semmler (Northern Virginia Community College), and Mary Jo Stieg (Mesa Community College). To the following focus group participants we are likewise grateful: Don Albers (Menlo College), Rosalie O'Mahony (College of San Mateo), Murray Peterson (College of Marin), and Sally Thomas (Orange Coast College).

M.L.K.
M.L.B.

CONTENTS

1

ALGEBRA AND PROBLEM SOLVING 1

1.1 Numbers and Symbolism 2
1.2 Solving Equations 10
1.3 The Nature of Problem Solving 14
1.4 Algebraic Symbolism and Properties of Numbers 21
1.5 Exponential Notation 30
1.6 Scientific Notation 37
1.7 Formulas and Problem Solving 42
Summary and Review 45
Test 47

2

GRAPHS AND FUNCTIONS 49

2.1 Graphs 50
2.2 Graphs of Linear Equations 55
2.3 Another Look at Linear Graphs 61
2.4 Other Equations of Lines 66
2.5 Functions 71
2.6 Linear Functions and Problem Solving 78
Summary and Review 82
Test 83

3

PROBLEM SOLVING AND SYSTEMS OF EQUATIONS 85

3.1 Systems of Equations in Two Variables 86
3.2 Solving Systems of Equations 89
3.3 Solving Problems with Two Equations 94

3.4 Systems of Equations in Three Variables 100
3.5 Solving Problems with Three Equations 105
3.6 Matrices 108
3.7 Inconsistent and Dependent Systems 111
3.8 Business and Economic Applications 114
Summary and Review 117
Test 119
Cumulative Review: Chapters 1–3 119

4
POLYNOMIALS AND POLYNOMIAL FUNCTIONS 121

4.1 Polynomials and Functions 122
4.2 Multiplication 127
4.3 Factoring and Evaluating Polynomials 130
4.4 Factoring Trinomials 133
4.5 Differences of Squares; Completing the Square 137
4.6 Sums or Differences of Two Cubes 140
4.7 Factoring: A General Strategy 143
4.8 Solving Equations by Factoring 145
4.9 Solving Problems 148
Summary and Review 151
Test 152

5
INEQUALITIES AND LINEAR PROGRAMMING 153

5.1 Solving Inequalities 154
5.2 Solving Problems 158
5.3 Compound Inequalities 161
5.4 Absolute Value 164
5.5 Inequalities in Two Variables 168
5.6 Problem Solving Using Linear Programming 173
Summary and Review 178
Test 179

6
RATIONAL EXPRESSIONS, EQUATIONS, AND FUNCTIONS 181

6.1 Rational Expressions: Multiplying and Dividing 182
6.2 Least Common Multiples 189
6.3 Addition and Subtraction 190
6.4 Complex Fractional Expressions 195
6.5 Solving Fractional Equations 199
6.6 Solving Problems 203
6.7 Division of Polynomials 211
6.8 Synthetic Division 214
6.9 Formulas and Problem Solving 217

6.10 Variation and Problem Solving 221
Summary and Review 227
Test 228
Cumulative Review: Chapters 1–6 229

7

EXPONENTIAL AND RADICAL SYMBOLISM; COMPLEX NUMBERS 231

7.1 Radical Expressions 232
7.2 Multiplying and Simplifying with Radical Notation 237
7.3 Dividing and Simplifying with Radical Notation 241
7.4 Addition and Subtraction Involving Radicals 244
7.5 More About Multiplication with Radicals 246
7.6 Rationalizing Numerators or Denominators 247
7.7 Rational Numbers as Exponents 252
7.8 Solving Radical Equations 257
7.9 The Complex Numbers 263
Summary and Review 267
Test 268

8

QUADRATIC EQUATIONS AND FUNCTIONS 271

8.1 Quadratic Equations 272
8.2 The Quadratic Formula 277
8.3 Fractional Equations and Problem Solving 281
8.4 Solutions of Quadratic Equations 285
8.5 Formulas and Problem Solving 288
8.6 Equations Reducible to Quadratic 291
8.7 Other Kinds of Variation 294
8.8 Quadratic Functions 299
8.9 More About Graphing Quadratic Functions 308
8.10 Problem Solving and Quadratic Functions 311
Summary and Review 315
Test 316

9

EXPONENTIAL AND LOGARITHM FUNCTIONS 317

9.1 Exponential Functions 318
9.2 Logarithm Functions 324
9.3 Properties of Logarithm Functions 328
9.4 Logarithm Function Values; Tables and Calculators 332
9.5 Interpolation (Optional) 338
9.6 Exponential and Logarithmic Equations 342
Summary and Review 346
Test 347
Cumulative Review: Chapters 1–9 348

10

EQUATIONS OF SECOND DEGREE AND THEIR GRAPHS 351

10.1 The Distance Formula 352
10.2 Conic Sections: Circles and Ellipses 354
10.3 Conic Sections: Parabolas and Hyperbolas 359
10.4 Systems of Equations 365
10.5 Systems of Second-degree Equations 369
10.6 Quadratic and Rational Inequalities 372
Summary and Review 376
Test 376

11

SEQUENCES, SERIES, AND PROBABILITY 379

11.1 Sequences and Series 380
11.2 Arithmetic Sequences and Series 383
11.3 Geometric Sequences and Series 389
11.4 Probability 395
11.5 The Binomial Theorem 399
Summary and Review 407
Test 408

Cumulative Review: Chapters 1–11 409

Table 1 Powers, Roots, and Reciprocals 411
Table 2 Common Logarithms 413

Answers A-1

Index I-1

Some problems of aerial navigation are solved using algebra.

1 ALGEBRA AND PROBLEM SOLVING

The principal theme of this text is problem solving in algebra. The overall strategy for solving problems is presented in this chapter, and you will find additional and increasing emphasis on problem solving throughout the text.

This chapter begins with a short review of algebraic symbolism and properties of numbers. As you will see, the manipulations of algebra, such as simplifying expressions and solving equations, are based on the properties of numbers.

OBJECTIVES

After finishing Section 1.1, you should be able to:

i **Convert between fractional, decimal, and percent symbolism.**

ii **Find the additive inverse and the reciprocal of a number.**

iii **Add and subtract positive and negative numbers.**

iv **Multiply and divide positive and negative numbers.**

v **Find different fractional names for numbers using the idea of multiplying by 1.**

vi **Write set notation.**

1.1

NUMBERS AND SYMBOLISM

As we have seen, solving equations is important for problem solving. Thus the ability to solve equations is a vital part of algebra. If an equation is complex, a good deal of manipulation of symbols may be required to solve it. For algebraic symbol manipulation, we depend on the properties of numbers. We now consider symbols and properties of numbers. Since most of this material is actually a review of beginning algebra, much of it may be considered optional.

Numbers

The numbers used in elementary algebra are the so-called *real numbers.* There is a real number for every point on a number line.

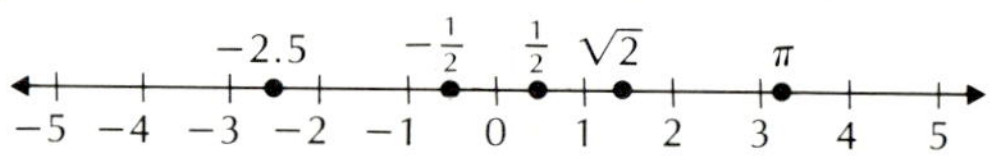

The following are some subsets of the real numbers.

SUBSETS OF THE REAL NUMBERS

Natural numbers: **Those used for counting:**

1, 2, 3, and so on.

Whole numbers: **The set of natural numbers with 0 included:**

0, 1, 2, 3, 4, and so on.

Integers:

0, 1, −1, 2, −2, 3, −3, and so on.

Rational numbers: **Numbers that can be named as a quotient of integers:**

$\frac{2}{3}, -\frac{4}{5}$, **and so on.**

i Notation for Rational Numbers

The rational numbers *can be named* as quotients of integers.

Fractional Notation. *Fractional notation* consists of symbolism such as the following:

$$\frac{2}{3}, \quad \frac{12}{-7}, \quad \frac{-17}{15}, \quad -\frac{9}{7}.$$

Decimal Notation. The rational numbers can also be named using *decimal notation.*

EXAMPLE 1 Find decimal notation for $\frac{5}{8}$.

Since $\frac{5}{8}$ means $5 \div 8$, we divide:

$$\begin{array}{r} 0.625 \\ 8\overline{)5.000} \\ \underline{4\,8} \\ 20 \\ \underline{16} \\ 40 \\ \underline{40} \end{array}$$

Decimal notation for $\frac{5}{8}$ is thus 0.625.

Decimal notation for some numbers repeats.

EXAMPLE 2 Find decimal notation for $\frac{6}{11}$.

We divide:

$$\begin{array}{r} 0.5454\ldots \\ 11\overline{)6.0000} \\ \underline{5\,5} \\ 50 \\ \underline{44} \\ 60 \\ \underline{55} \\ 50 \\ \underline{44} \\ 6 \end{array}$$

Repeating decimal notation can be abbreviated by writing a bar over the repeating part, in this case, $0.\overline{54}$.

DO EXERCISES 1 AND 2.

EXAMPLE 3 Find fractional notation for 3.6.

$$3.6 = \frac{36}{10}$$

The last decimal place is *tenths* so the denominator is 10.

EXAMPLE 4 Find fractional notation for 0.75.

$$0.75 = \frac{75}{100}$$

The denominator is 100 because the last decimal place is *hundredths.*

DO EXERCISES 3 AND 4.

Percent Notation. We can also name rational numbers using percent notation. Let us recall what percent means.

Find decimal notation.

1. $\frac{7}{8}$

2. $\frac{7}{11}$

Find fractional notation.

3. 5.9

4. 0.125

Find decimal notation and fractional notation.

5. 3.5%

6. 38.4%

Find decimal notation and percent notation.

7. $\frac{3}{8}$

8. $\frac{17}{20}$

9. Find percent notation for 0.0035.

10. Find decimal notation for 235%.

The symbol % means "× 0.01" or "$\times \frac{1}{100}$."

We use this definition to change to or from percent notation.

EXAMPLE 5 Find decimal notation and fractional notation for 67.9%.

$67.9\% = 67.9 \times 0.01$ Replacing % by × 0.01

$= 0.679$ Multiplying. This is decimal notation.

$= \frac{679}{1000}$ Changing to fractional notation

EXAMPLE 6 Find decimal notation and percent notation for $\frac{5}{8}$.

$$8\overline{)5.000} = 0.625$$

Decimal notation is 0.625.

$\frac{5}{8} = 0.625 \times 100 \times 0.01$ Multiplying by 100 × 0.01, or 1

$= (0.625 \times 100) \times 0.01$

$= 62.5 \times 0.01$

$= 62.5\%$ Replacing × 0.01 by %

Percents can be less than 1% or greater than 100%.

EXAMPLES Find percent notation.

7. $0.005 = 0.005 \times 100 \times 0.01$
$= (0.005 \times 100) \times 0.01$
$= 0.5 \times 0.01$
$= 0.5\%$

8. $1.35 = 1.35 \times 100 \times 0.01$
$= 135 \times 0.01$
$= 135\%$

DO EXERCISES 5–10.

ii Inverses and Reciprocals

Two numbers whose sum is 0 are called *additive inverses*, or simply *inverses* of each other.

EXAMPLES Find the additive inverse.

9. 18 The additive inverse is −18 because $18 + (-18) = 0$.

10. −12 The inverse is 12, because $-12 + 12 = 0$.

11. 0 The inverse is 0, because $0 + 0 = 0$.

We often just say "inverse," rather than "additive inverse." A symbol such as −8 can be read "negative 8," but it can also be read "the inverse of 8," because negative

8 is the inverse of 8. On the other hand, it is better not to read a symbol like $-x$ as "negative x," because we do not know whether it represents a positive number, a negative number, or 0.

When we replace a number by its additive inverse, we sometimes say that we are "changing the sign."

Two numbers whose product is 1 are called *reciprocals* of each other. They are also sometimes called *multiplicative inverses.*

EXAMPLES Find the reciprocal.

12. 3 The reciprocal is $\frac{1}{3}$ because $3 \cdot \frac{1}{3} = 1$.

13. $\frac{7}{8}$ The reciprocal is $\frac{8}{7}$ because $\frac{7}{8} \cdot \frac{8}{7} = 1$.

14. $-\frac{2}{3}$ The reciprocal is $-\frac{3}{2}$ because $-\frac{2}{3} \cdot (-\frac{3}{2}) = 1$.

Note that the reciprocal of a negative number is negative.

DO EXERCISES 11–18.

iii Addition and Subtraction: Rules of Signs

Following are the rules of signs for addition of positive and negative numbers. Recall that the absolute value of a number is its distance from 0.

RULES OF SIGNS: ADDITION

Two positive numbers: **The sum is positive.**

Two negative numbers: **The sum is negative.**

One positive number and one negative number:

a) If the positive number has the larger absolute value, the sum is positive.

b) If the negative number has the larger absolute value, the sum is negative.

c) If the numbers have the same absolute value, the sum is 0.

EXAMPLES Add.

15. $-5 + (-9) = -14$

16. $-5 + 7 = 2$

17. $1 + (-5) = -4$

18. $-15 + 15 = 0$

DO EXERCISES 19–28.

To subtract, we can add an inverse. (We sometimes say that we "change the sign of the subtrahend and then add.")

EXAMPLES Subtract.

19. $3 - 8 = 3 + (-8)$ Changing the sign and changing the subtraction to an addition

$= -5$

Find the additive inverse.

11. 23

12. -41

Find the reciprocal.

13. $\frac{2}{3}$

14. $-\frac{4}{5}$

15. Find $-a$ when a is 9.

16. Find $-a$ when a is -23.

17. Find $\frac{1}{a}$ when a is $\frac{2}{3}$.

18. Find $\frac{1}{a}$ when a is $-\frac{5}{9}$.

Add.

19. $-7 + (-11)$

20. $-8 + (-9)$

21. $20 + (-9)$

22. $-8.9 + (-9.7)$

23. $-\frac{6}{5} + \left(-\frac{23}{5}\right)$

24. $-\frac{2}{5} + \frac{3}{10}$

25. $-\frac{2}{3} + \left(-\frac{5}{6}\right)$

26. $-7 + 7$

27. $16 + (-16)$

28. $0 + 0$

Subtract.

29. $-2-3$

30. $5-8$

31. $-2-(-5)$

32. $8-(-9)$

33. $-10-6$

34. $-23.7-5.9$

35. $-\frac{11}{12}-\left(-\frac{23}{12}\right)$

36. $-\frac{3}{5}-\frac{3}{10}$

37. $\frac{2}{3}-\left(-\frac{5}{6}\right)$

38. **a)** $17-23$

b) $-17-23$

c) $-17-(-23)$

Multiply.

39. $-4.7(-9.1)$

40. $\frac{3}{4}\cdot\frac{7}{7}$

41. $\frac{4}{3}\cdot\frac{-2}{-2}$

42. $\frac{-9}{2}\cdot\frac{-1}{-1}$

43. $\frac{14}{-5}\cdot\frac{-1}{-1}$

44. $\frac{-17}{-25}\cdot\frac{-1}{-1}$

Divide.

45. $\frac{-28}{-14}$

46. $\frac{125}{-5}$

47. $\frac{-75}{25}$

48. $\frac{-4.2}{-21}$

20. $8-(-6)=8+6$
$=14$

21. $-12-17=-12+(-17)$
$=-29$

DO EXERCISES 29–38.

iv Multiplication and Division: Rules of Signs

Following are the rules of signs for multiplication or division of positive and negative numbers.

RULES OF SIGNS: MULTIPLICATION AND DIVISION

Two positive numbers: **The answer is positive.**

Two negative numbers: **The answer is positive.**

One positive number and one negative number: **The answer is negative.**

EXAMPLES Multiply or divide.

22. $3\times(-4)=-12$

23. $(-7)\cdot(-3)=21$

24. $\frac{-8}{2}=-4$

25. $\frac{8}{-2}=-4$

26. $\frac{-8}{-2}=4$

27. $-\frac{-8}{-2}=-\left(\frac{-8}{-2}\right)=-4$

28. $\frac{-7}{3}\cdot\frac{-8}{-5}=\frac{(-7)(-8)}{3(-5)}$ Multiplying numerators and multiplying denominators

$=\frac{56}{-15}$

$=-\frac{56}{15}$

Note that

$$\frac{-8}{2}=\frac{8}{-2}=-\frac{8}{2}=-\frac{-8}{-2}.$$

This can be generalized.

For any number *a* and any nonzero number *b*,

$$\frac{-a}{b}=\frac{a}{-b}=-\frac{a}{b}=-\frac{-a}{-b}.$$

DO EXERCISES 39–48.

To divide, we can multiply by a reciprocal. (We sometimes say that we "invert the divisor and multiply.")

EXAMPLES Divide.

29. $\frac{1}{4} \div \frac{3}{5} = \frac{1}{4} \cdot \frac{5}{3}$ Replacing the divisor by its reciprocal and changing the division to a multiplication

$= \frac{5}{12}$

30. $\frac{2}{3} \div (-\frac{4}{9}) = \frac{2}{3} \cdot (-\frac{9}{4})$

$= -\frac{18}{12}$, or $-\frac{3}{2}$

DO EXERCISES 49 AND 50.

v Multiplying by 1

When a number is multiplied by 1, the result is the original number. We can use that simple fact with fractional notation to do several things.

EXAMPLE 31 Find fractional notation for $\frac{3}{5}$ having a denominator of 20.

We multiply by 1, using $\frac{4}{4}$ as a name for 1:

$$\frac{3}{5} \cdot \frac{4}{4} = \frac{12}{20}.$$

EXAMPLE 32 Find fractional notation for $\dfrac{-7}{8}$ having a denominator of -16.

We multiply by 1, using $\dfrac{-2}{-2}$ as a name for 1:

$$\frac{-7}{8} \cdot \frac{-2}{-2} = \frac{14}{-16}.$$

DO EXERCISES 51–54.

EXAMPLE 33 Simplify: $-\frac{18}{12}$.

We first factor numerator and denominator:

$$-\frac{18}{12} = -\frac{3 \cdot 6}{2 \cdot 6}.$$

Next, we factor the fraction, with one factor equal to 1:

$$-\frac{3 \cdot 6}{2 \cdot 6} = -\frac{3}{2} \cdot \frac{6}{6}.$$

Then we leave off the factor of 1 to get $-\frac{3}{2}$.

DO EXERCISES 55–60.

Set Notation

To name a set, we use braces { }. For example, to name the set consisting of the first three natural numbers, we name those numbers inside braces, as follows:

$$\{1, 2, 3\}.$$

Divide by multiplying by a reciprocal.

49. $-\dfrac{3}{4} \div \dfrac{7}{8}$

50. $-\dfrac{12}{5} \div \left(-\dfrac{7}{15}\right)$

51. Find fractional notation for $\frac{9}{5}$ having a denominator of 25.

52. Find fractional notation for $\frac{3}{7}$ having a denominator of 42.

53. Find fractional notation for $\frac{7}{8}$ having a denominator of -64.

54. Find fractional notation for $\frac{-12}{7}$ having a denominator of -49.

Simplify.

55. $\dfrac{8}{6}$

56. $\dfrac{27}{6}$

57. $\dfrac{-26}{39}$

58. $\dfrac{-15}{70}$

59. $\dfrac{18}{-36}$

60. $\dfrac{22}{-55}$

61. Write set notation for the set of integers between -3 and 7.

62. Write set builder notation for the set of all odd numbers from 5 to 11, inclusive.

EXAMPLE 34 Name the set consisting of the first four even natural numbers.

$$\{2, 4, 6, 8\}$$

We can also name a set by specifying conditions under which a number is in the set, as follows. The symbolism is read as indicated, and the notation is known as *set builder notation.*

$$\{x \mid x \text{ is an odd number less than } 5\}$$

The set of all x | such that | x is an odd number less than 5

EXAMPLE 35 Write set builder notation for the set of all even numbers between 5 and 25.

$$\{x \mid x \text{ is an even number between 5 and 25}\}$$

DO EXERCISES 61 AND 62.

EXERCISE SET 1.1

i Find decimal notation.

1. $\frac{3}{8}$ **2.** $\frac{1}{8}$ **3.** $\frac{5}{3}$ **4.** $\frac{7}{6}$

Find fractional notation.

5. 2.7 **6.** 13.91 **7.** 0.145 **8.** 0.0213

Find percent notation.

9. $\frac{4}{5}$ **10.** $\frac{7}{25}$ **11.** $\frac{3}{8}$ **12.** $\frac{1}{8}$ **13.** $\frac{12}{5}$

14. $\frac{14}{5}$ **15.** $\frac{3}{80}$ **16.** $\frac{3}{160}$ **17.** 0.412 **18.** 0.325

19. 0.001 **20.** 0.0025 **21.** 1.25 **22.** 2.34

Find decimal and fractional notation.

23. 56.7% **24.** 47.5% **25.** 6.85% **26.** 8.23%

27. 120% **28.** 180% **29.** 0.5% **30.** 0.1%

ii Find the additive inverse.

31. 7 **32.** -8 **33.** -2.8 **34.** 0

Change the sign.

35. 10 **36.** -8 **37.** 0 **38.** $-2x$

Find the reciprocal.

39. $\frac{3}{4}$ **40.** $\frac{9}{10}$ **41.** $-\frac{7}{8}$ **42.** $-\frac{5}{6}$

43. 15 **44.** -75 **45.** 4.4 **46.** -5.5

iii Add.

47. $-12 + (-16)$ **48.** $-11 + (-18)$ **49.** $-8 + (-8)$ **50.** $-6 + (-6)$
51. $8 + (-3)$ **52.** $9 + (-4)$ **53.** $11 + (-7)$ **54.** $12 + (-8)$
55. $-16 + 9$ **56.** $-23 + 8$ **57.** $-24 + 0$ **58.** $-34 + 0$
59. $-8.4 + 9.6$ **60.** $-6.3 + 8.2$ **61.** $-2.62 + (-6.24)$ **62.** $-5.83 + (-7.43)$
63. $-\frac{2}{7} + \frac{3}{7}$ **64.** $-\frac{5}{6} + \frac{1}{6}$ **65.** $-\frac{11}{12} + \left(-\frac{5}{12}\right)$ **66.** $-\frac{3}{8} + \left(-\frac{7}{8}\right)$
67. $\frac{2}{5} + \left(-\frac{3}{10}\right)$ **68.** $-\frac{3}{4} + \frac{1}{8}$

Subtract.

69. $5 - 7$ **70.** $9 - 12$ **71.** $-5 - 7$ **72.** $-9 - 12$
73. $-6 - (-11)$ **74.** $-7 - (-12)$ **75.** $10 - (-5)$ **76.** $28 - (-16)$
77. $15.8 - 27.4$ **78.** $17.2 - 34.9$ **79.** $-18.01 - 11.24$ **80.** $-19.04 - 15.76$
81. $-\frac{21}{4} - \left(-\frac{7}{4}\right)$ **82.** $-\frac{16}{5} - \left(-\frac{3}{5}\right)$ **83.** $-\frac{1}{2} - \left(-\frac{1}{12}\right)$ **84.** $-\frac{3}{4} - \left(-\frac{3}{2}\right)$

iv Multiply.

85. $3(-7)$ **86.** $5(-8)$ **87.** $-2 \cdot 4$ **88.** $-5 \cdot 9$
89. $-8(-2)$ **90.** $-7(-3)$ **91.** $-9 \cdot 14$ **92.** $-8 \cdot 17$
93. $-6(-5.7)$ **94.** $-7(-6.1)$ **95.** $-4.2(-6.3)$ **96.** $-7.4(-9.6)$
97. $-3\left(-\frac{2}{3}\right)$ **98.** $-5\left(-\frac{3}{5}\right)$ **99.** $-3(-4)(5)$ **100.** $-6(-8)(9)$
101. $-\frac{3}{5} \cdot \frac{4}{7}$ **102.** $-\frac{5}{4} \cdot \frac{11}{3}$ **103.** $\frac{9}{11} \cdot \left(-\frac{11}{9}\right)$ **104.** $-\frac{13}{7} \cdot \left(-\frac{5}{2}\right)$
105. $-\frac{2}{3} \cdot \left(-\frac{2}{3}\right) \cdot \left(-\frac{2}{3}\right)$ **106.** $-\frac{4}{5} \cdot \left(-\frac{4}{5}\right) \cdot \left(-\frac{4}{5}\right)$

Divide.

107. $\frac{-8}{4}$ **108.** $\frac{-16}{2}$ **109.** $\frac{56}{-8}$ **110.** $\frac{63}{-7}$
111. $\frac{-77}{-11}$ **112.** $\frac{-48}{-6}$ **113.** $\frac{-5.4}{-18}$ **114.** $\frac{-8.4}{-12}$
115. $\frac{2}{7} \div \left(-\frac{11}{3}\right)$ **116.** $\frac{3}{5} \div \left(-\frac{6}{7}\right)$ **117.** $-\frac{10}{3} \div \left(-\frac{2}{15}\right)$ **118.** $-\frac{12}{5} \div \left(-\frac{3}{10}\right)$
119. $18.6 \div (-3.1)$ **120.** $39.9 \div (-13.3)$ **121.** $(-75.5) \div (-15.1)$

v By multiplying by 1, do the following.

122. Find the fractional notation for $\frac{7}{12}$ having a denominator of -60.

123. Find fractional notation for $\frac{14}{-9}$ having a denominator of 81.

Simplify by factoring and removing a factor of 1.

124. $\frac{-13}{39}$ **125.** $\frac{65}{-25}$ **126.** $-\frac{461}{369}$ **127.** $\frac{99}{-66}$

vi Use set-builder notation to name the sets.

128. The set consisting of all the odd numbers between 8 and 30.

129. The set consisting of all multiples of 4 between 34 and 49.

130. $\{0, 1, 2, 3, 4\}$

131. $\{-3, -2, -1, 0, 1, 2\}$

OBJECTIVES

After finishing Section 1.2, you should be able to:

i **Determine whether a number is a solution of an equation.**

ii **Solve simple equations using the addition and multiplication principles.**

iii **Solve certain equations that contain parentheses.**

1. Consider the replacements 5, 8, 4, 0, and 2. Which of them make the equation $2x + 3 = 11$ false?

2. Find the replacement that makes $3x + 2 = 11$ true.

1.2

SOLVING EQUATIONS

Solving equations is essential for problem solving in algebra. In this section you will review and practice the solving of simple equations.

By an *equation* we mean a number sentence with the verb "=." Such a sentence *says* that the symbols on either side of the equals sign represent, or *name,* the same thing. For example:

$1 + 10 = 11$	is an equation. It *says* that $1 + 10$ and 11 are two names of the same number. The equation is true.
$7 - 8 = 9 - 13$	is an equation. It *says* that $7 - 8$ and $9 - 13$ are two names of the same number. The equation is false.
$x - 9 = 3$	is an equation. It *says* that $x - 9$ and 3 represent the same number. The equation is neither true nor false, because we do not know what x represents.

i Solutions

If an equation contains a variable it may be neither true nor false. Some replacements may make it true and some may make it false.

The replacements making an equation true are called its *solutions.* When we find them, *no matter how,* we say that we have *solved* the equation.

EXAMPLE 1 Determine whether 5 is a solution of the equation $2x + 3 = 13$.

We substitute 5 for the variable and see whether we get a true sentence.

$2x + 3$	$= 13$
$2 \cdot 5 + 3$	13
$10 + 3$	
13	

Substituting 5 for x and then simplifying the left side

Since both sides of the equation represent 13, the number 5 is a solution.

DO EXERCISES 1 AND 2.

ii The Addition and Multiplication Principles

You are already familiar with adding or multiplying on both sides of an equation by the same number. Let's see why that can be done.

Consider the equation $a = b$. It says that a and b represent the *same number.* Suppose $a = b$ is true and then add a number c to a. We will get the same answer if we add c to b, because a and b are the same number. The same is true if we multiply by c.

THE ADDITION AND MULTIPLICATION PRINCIPLES

If $a = b$ is true, then:

a) $a + c = b + c$ **is true for any number** c;

b) $a \cdot c = b \cdot c$ **is true for any number** c.

EXAMPLE 2 Solve: $y - 4.7 = 13.9$.

$$\begin{aligned} y - 4.7 &= 13.9 \\ y - 4.7 + 4.7 &= 13.9 + 4.7 \quad \text{Using the addition principle; adding 4.7} \\ y + 0 &= 13.9 + 4.7 \\ y &= 18.6 \end{aligned}$$

Check:

$y - 4.7 = 13.9$	
$18.6 - 4.7$	13.9
13.9	

Substituting 18.6 for y and calculating

The solution is 18.6.

In Example 2, why did we add 4.7 on both sides? Because we wanted the variable y alone on one side of the equation. When we added 4.7, we got 0 and that left y alone on the left. Then the other side shows us what calculations to make to find the answer.

EXAMPLE 3 Solve: $4x = 9$.

$$\begin{aligned} 4x &= 9 \\ \tfrac{1}{4} \cdot 4x &= \tfrac{1}{4} \cdot 9 \quad \text{Using the multiplication principle, we multiply by } \tfrac{1}{4}\text{, the reciprocal of 4.} \\ 1 \cdot x &= \tfrac{9}{4} \\ x &= \tfrac{9}{4} \end{aligned}$$

Simplifying

Check:

$4x = 9$	
$4 \cdot \frac{9}{4}$	9
9	

The solution is $\frac{9}{4}$.

In Example 3, why did we choose to multiply by $\frac{1}{4}$? Because we wanted x alone on one side of the equation. We multiplied by the reciprocal of 4. Then we got $1 \cdot x$, which simplified to x. This eliminated the 4 on the left.

DO EXERCISES 3–6.

EXAMPLE 4 Solve: $3x - 4 = 13$.

$$\begin{aligned} 3x - 4 &= 13 \\ 3x - 4 + 4 &= 13 + 4 \quad \text{Using the addition principle, we add 4.} \\ 3x &= 17 \quad \text{Simplifying} \\ \tfrac{1}{3} \cdot 3x &= \tfrac{1}{3} \cdot 17 \quad \text{Using the multiplication principle, we multiply by } \tfrac{1}{3}. \\ x &= \tfrac{17}{3} \end{aligned}$$

Check:

$3x - 4 = 13$	
$3 \cdot \frac{17}{3} - 4$	13
$17 - 4$	
13	

The solution is $\frac{17}{3}$, or $5\frac{2}{3}$.

In Example 4, we used the addition principle first. That usually makes the work simpler.

Solve using the addition principle.

3. $x + 9 = 2$

4. $x + \dfrac{1}{4} = -\dfrac{3}{4}$

5. $13 = -25 + y$

6. $y - 61.4 = 78.9$

Solve using the multiplication principle.

7. $8x = 10$

8. $-\frac{3}{7}y = 21$

9. $-4x = -\frac{6}{7}$

10. $-12.6 = 4.2y$

Solve.

11. $-4 + 9x = 8$

Clear of fractions.

12. $\frac{2}{3} - \frac{5}{6}y = \frac{1}{3}$

Clear of decimals.

13. $6.3x - 9.1 = 3$

Solve.

14. $\frac{5}{2}x + \frac{9}{2}x = 21$

15. $1.4x - 0.9x + 0.7 = -2.2$

16. $-4x + 2 + 5x = 3x - 15$

There are various ways to solve equations. With the method we are developing, the idea is to find an equation so simple that we can *see* what the solution is. Then we check to see if that solution is also a solution of the original equation.

DO EXERCISES 7–10.

When there are fractions or decimals in an equation, it sometimes helps to eliminate them at the outset. We do that by using the multiplication principle. The process is called *clearing* of decimals or fractions.

EXAMPLE 5 Clear the equation $\frac{3}{4}x + \frac{1}{2} = \frac{3}{2}$ of fractions.

We multiply on both sides by the least common multiple of the denominators—in this case, 4.

$$4(\tfrac{3}{4}x + \tfrac{1}{2}) = 4 \cdot \tfrac{3}{2} \qquad \text{Multiplying by 4 on both sides}$$

$$4 \cdot \tfrac{3}{4}x + 4 \cdot \tfrac{1}{2} = 4 \cdot \tfrac{3}{2} \qquad \text{Carrying out the multiplication}$$

$$3x + 2 = 6 \qquad \text{Simplifying}$$

EXAMPLE 6 Clear the equation $12.4 - 5.12x = 3.14$ of decimals.

We multiply on both sides by a power of ten to get rid of the decimal points.

$$100(12.4 - 5.12x) = 100(3.14) \qquad \text{Multiplying on both sides by 100}$$

$$100 \times 12.4 - 100 \times 5.12x = 100 \times 3.14 \qquad \text{Carrying out the multiplication}$$

$$1240 - 512x = 314 \qquad \text{Simplifying}$$

When there are like terms in an equation, we combine them on each side. Then if there are still like terms on opposite sides, we get them on the same side by using the addition principle.

EXAMPLE 7 Solve: $8x + 6 - 2x = -4x - 14$.

$$8x + 6 - 2x = -4x - 14$$

$$6x + 6 = -4x - 14 \qquad \text{Combining like terms on the left}$$

$$6x + 4x + 6 = -14 \qquad \text{Adding } 4x \text{ on both sides}$$

$$10x + 6 = -14 \qquad \text{Combining like terms on the left}$$

$$10x = -20 \qquad \text{Adding } -6 \text{ on both sides}$$

$$\tfrac{1}{10} \cdot 10x = \tfrac{1}{10} \cdot (-20) \qquad \text{Multiplying on both sides by } \tfrac{1}{10}$$

$$x = -2 \qquad \text{Simplifying}$$

Check:

$8x + 6 - 2x$	$= -4x - 14$
$8(-2) + 6 - 2(-2)$	$-4(-2) - 14$
$-16 + 6 + 4$	$8 - 14$
-6	-6

The solution is -2.

Here is a summary of the method of solving equations, as used in the preceding examples:

1. **Clear of fractions or decimals if that is needed.**
2. **Collect like terms on both sides of the equation, if necessary.**
3. **Get all like terms with letters on one side and all other terms on the other side.**
4. **Collect like terms again, if necessary.**
5. **Use the multiplication principle to get the letter alone on one side.**

DO EXERCISES 11–16.

iii Equations Containing Parentheses

Equations containing parentheses can often be solved by first multiplying to remove parentheses and then proceeding as before.

EXAMPLE 8 Solve: $3(7 - 2x) = 14 - 8(x - 1)$.

$$3(7 - 2x) = 14 - 8(x - 1)$$
$$21 - 6x = 14 - 8x + 8 \quad \text{Multiplying to remove parentheses}$$
$$21 - 6x = 22 - 8x \quad \text{Combining like terms}$$
$$8x - 6x = 22 + (-21) \quad \text{Adding } -21 \text{ and also } 8x$$
$$2x = 1 \quad \text{Combining like terms}$$
$$x = \tfrac{1}{2} \quad \text{Multiplying by } \tfrac{1}{2}$$

Check:

$3(7 - 2x)$	$= 14 - 8(x - 1)$
$3(7 - 2 \cdot \frac{1}{2})$	$14 - 8(\frac{1}{2} - 1)$
$3(7 - 1)$	$14 - 8(-\frac{1}{2})$
$3 \cdot 6$	$14 + 4$
18	18

The solution is $\frac{1}{2}$.

DO EXERCISES 17 AND 18.

Special Cases

There are equations with no solution.

EXAMPLE 9 Solve: $-8x + 5 = 14 - 8x$.

$$-8x + 5 = 14 - 8x$$
$$5 = 14 \quad \text{Adding } 8x\text{, we get a false equation.}$$

No matter what number we try for x we get a false sentence. Thus the equation has no solution.

There are equations for which any real number is a solution.

EXAMPLE 10 Solve: $-8x + 5 = 5 - 8x$.

$$-8x + 5 = 5 - 8x$$
$$5 = 5 \quad \text{Adding } 8x\text{, we get a true equation.}$$

Replacing x by any real number gives a true sentence. Thus any real number is a solution.

DO EXERCISES 19 AND 20.

Solve.

17. $30 + 7(x - 1) = 3(2x + 7)$

18. $3(y - 1) - 1 = 2 - 5(y + 5)$

Try to solve.

19. $4 + 7x = 7x + 9$

20. $3 + 9x = 9x + 3$

EXERCISE SET 1.2

Solve. Don't forget to check.

1. $4x - 12 = 60$	**2.** $4x - 6 = 70$	**3.** $5y + 3 = 28$	**4.** $7t + 11 = 74$
5. $2y - 11 = 37$	**6.** $3x - 13 = 29$	**7.** $-4x - 7 = -35$	**8.** $-9y + 8 = -91$
9. $5x + 2x = 56$	**10.** $3x + 7x = 120$	**11.** $9y - 7y = 42$	**12.** $8t - 3t = 65$

13. $-6y - 10y = -32$ **14.** $-9y - 5y = 28$ **15.** $7y - 1 = 23 - 5y$ **16.** $15x + 20 = 8x - 22$

17. $5 - 4a = a - 13$ **18.** $8 - 5x = x - 16$ **19.** $3m - 7 = -7 - 4m - m$

20. $5x - 8 = -8 + 3x - x$ **21.** $5r - 2 + 3r = 2r + 6 - 4r$ **22.** $5m - 17 - 2m = 6m - 1 - m$

23. $4x - 2x - 2 = 2x$ **24.** $2x + 4 + x = 4 + 3x$ **25.** $\frac{1}{4} + \frac{3}{8}y = \frac{3}{4}$

26. $\frac{1}{5} + \frac{3}{10}x = \frac{4}{5}$ **27.** $-\frac{5}{2}x + \frac{1}{2} = -18$ **28.** $0.9y - 0.7 = 4.2$

29. $0.8t - 0.3t = 6.5$ **30.** $1.4x + 5.02 = 0.4x$ **31.** $2(x + 6) = 8x$

32. $3(y + 5) = 8y$ **33.** $80 = 10(3t + 2)$ **34.** $27 = 9(5y - 2)$

35. $180(n - 2) = 900$ **36.** $210(x - 3) = 840$ **37.** $5y - (2y - 10) = 25$

38. $8x - (3x - 5) = 40$ **39.** $0.7(3x + 6) = 1.1 - (x + 2)$ **40.** $0.9(2x + 8) = 20 - (x + 5)$

41. $\frac{1}{8}(16y + 8) - 17 = -\frac{1}{4}(8y - 16)$ **42.** $\frac{1}{6}(12t + 48) - 20 = -\frac{1}{8}(24t - 144)$ **43.** $a + (a - 3) = (a + 2) - (a + 1)$

44. $0.8 - 4(b - 1) = 0.2 + 3(4 - b)$

Solve. Remember to check. (With a calculator, you need not clear of decimals.)

45. $0.0008x = 0.0000056$ **46.** $43.008z = 1.201135$

47. $4.23x - 17.898 = -1.65x - 42.454$ **48.** $-0.00458y + 1.7787 = 13.002y - 1.005$

An identity is an equation that is true for all sensible replacements. Determine which of the following are identities.

49. $2(x - 3) + 5 = 3(x - 2) + 5$ **50.** $3(x - 4) = 3x - 4$

51. $5(x + 3) = 5x + 15$ **52.** $\dfrac{6y + 4}{2} = 6y + 2$

OBJECTIVES

After finishing Section 1.3, you should be able to:

i **Familiarize yourself with a simple problem situation.**

ii **Perform the other steps in solving a simple problem.**

iii **Use the five-step problem-solving strategy to solve simple problems with algebra.**

1.3 THE NATURE OF PROBLEM SOLVING

In this section we look at problem solving. Although we are interested mainly in the use of algebra to solve problems, much of what we say here applies to solving all kinds of problems.

What do we mean by a *problem*? By *solving a problem*? It is common to speak of an "addition problem," meaning a specific addition to be done, such as

$$\begin{array}{r} 876 \\ 915 \\ +\ 784 \\ \hline \end{array}$$

In this book, we do not use the word "problem" in that way. We mean instead some question to which we need or want to find an answer. Perhaps we can best illustrate with some examples of what we will call *problems*.

EXAMPLES OF PROBLEMS

1. **What is the least expensive way to get a degree at this university?**
2. **What automobile, for my purposes, is the best one to buy?**
3. **If I drive to Auburn for lunch, will I be able to get back in time to keep my doctor's appointment?**
4. **An airplane is on submarine patrol over the Atlantic Ocean, flying out to sea, perpendicular to the shoreline. When the wind speed is 40 mph, at what point must the airplane turn and head for home, so as not to run out of fuel over the ocean?**

These problems are all different, but there are some similarities. We cannot give rules for problem solving, but there is a general *strategy* that can be used. Some problems can be solved using algebra and some cannot, but the overall strategy can be used in any case.

The Five-step Strategy

Since you have already studied algebra, you have had some experience with problem solving. The following steps make up a strategy that you may already have used in algebra. They also constitute a good strategy for problem solving in general, although in some cases some of the steps may not apply.

THE FIVE STEPS IN PROBLEM SOLVING IN ALGEBRA

1. ***Familiarize*** **yourself with the problem situation.**
2. ***Translate*** **to mathematical language.**
3. ***Carry out*** **some mathematical manipulation.**
4. ***Check*** **your possible answer in the original problem.**
5. ***State*** **the answer clearly.**

i The First Step: Familiarization

Of the five steps, probably the most important is the first one: becoming familiar with the problem situation. Here are some hints for familiarization.

HINTS FOR FAMILIARIZATION

1. **If a problem is presented to you in words, read it carefully.**
2. **Make a list of the information given and the question to be answered.**
3. **Find further information.**
4. **Make a table of the information given and the information you have collected.**
5. **Make a drawing and label it with known information. Also indicate unknown information.**
6. **Make some guesses or estimates of the answer.**

EXAMPLE 1 How might you familiarize yourself with the situation of Problem 1: "What is the least expensive way to get a degree at this university?"

You will obviously need to find further information in order to solve this problem. You might:

a) Talk to various students, especially upperclassmen. Ask them questions about tuition, costs of books and materials, living expenses, and so on.

b) Get a university catalog and study it.

c) Talk to counselors.

When you collect the information, it is a good idea to list it in a table. The nature of the table depends on what you find out. Here is an example.

			Living expenses		
	Tuition	Books and supplies	Dorm	Apartment	Share room
Business school					
Home economics					
Agriculture					
Liberal arts					

EXAMPLE 2 How might you familiarize yourself with the situation of Problem 4: "At what point must the airplane turn and head for home, so as not to run out of fuel over the ocean?"

First, read the problem very carefully, probably more than once. Then make a list of information known and information needed. In this case, a drawing can be very helpful.

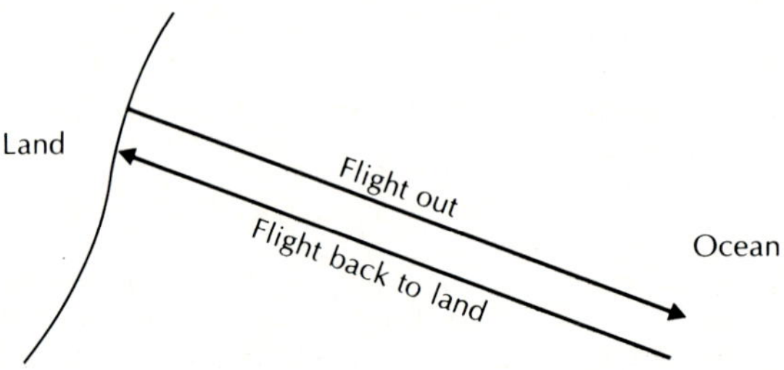

The wind speed is 40 mph, but we do not know the direction. There are other things we do not know. We will either have to find the information or make an estimate of it. Let us suppose that you have found, or estimated, the needed data. You should then make a table, as shown here.

Wind speed	40 mph
Wind direction	From the east
Direction of flight	Outward easterly
Speed of airplane	150 mph
Number of hours of fuel	10

In this case, another drawing might help. Note how we list pertinent information in the figure.

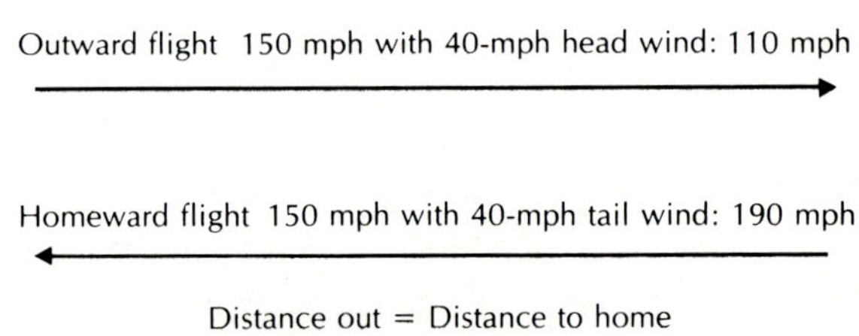

There is a relation between distance, speed, and time. Although the problem doesn't mention it, it will be needed. You might need to look it up.

$$\text{Speed} = \frac{\text{Distance}}{\text{Time}}, \quad \text{or} \quad \text{Distance} = \text{Speed} \times \text{Time}.$$

Using a table, again, we can lay out the information in a useful way.

	Distance	Speed	Time
Outward	Unknown	110 mph	Unknown
Homeward	Unknown, but same as distance outward	190 mph	Unknown, but total of outward and homeward is 10 hr

At this point, you might try to guess an answer. The average of the outward and homeward speeds is 150 mph. Half the fuel will be used in 5 hours. If you fly 150 mph for 5 hr, you will travel 750 miles. Could 750 miles be the answer? If you check it out, you will see that it is not, because that kind of a trip will require more than 10 hours.

After doing this much, you should be reasonably familiar with the situation.

DO EXERCISE 1.

Familiarize Familiarize yourself with the following problem situation. Do not try to solve the problem. Instead list as many things as you can to help you become familiar with the situation.

1. *The problem.* One angle of a triangle is three times as great as a second angle. The third angle is half as large as the second angle. Find the measures of the angles.

The familiarization process varies, not only from problem to problem, but from person to person. You may go about it differently from a fellow student. The important thing is that you be thorough.

ii The Other Steps

Remember that the second step in problem solving is to translate the situation to mathematical language. That might consist of making a graph in some cases. In other cases it might consist of noting that certain numbers are to be added. In algebra, it often consists of *writing an equation.*

The next step is to carry out some sort of mathematical manipulation. For some problems that might consist of doing an addition, subtraction, division, or the like. For problems solvable using algebra, that step often consists of solving an equation. Then we check and state the answer.

We will illustrate Steps 2–5 of the problem-solving process using some simple situations that can be solved using algebra. As you will see, these steps are not really new to you.

EXAMPLE 3 The number 64 is 0.8 of what number?

Let us assume that you have familiarized yourself with the situation, that you have read and reread the question, and that you know what it means. The problem is so explicitly stated that we can translate right away:

Translate

64 is 0.8 of what number?

$64 = 0.8 \cdot x$

"is" translates to $=$; "of" translates to $\cdot$; "what number" translates to a variable x

Next, we carry out some mathematical manipulation. In this case, we solve the equation.

Carry out We solve the equation:

$$64 = 0.8x.$$

We divide on both sides by 0.8:

$$\frac{64}{0.8} = \frac{0.8}{0.8}x.$$

Now we simplify:

$$\frac{64}{0.8} = 1 \cdot x = x$$

$$80 = x. \quad \text{Dividing 64 by 0.8}$$

The next step is to check in the original problem. Is 80 the answer?

Check We must find out whether 0.8 of 80 is 64. We multiply:

$$\begin{array}{r} 80 \\ \times\ 0.8 \\ \hline 64 \end{array}$$

The number checks.

The last step is to state the answer clearly.

State The answer is 80, that is, 80 is the number such that 0.8 of it is 64.

EXAMPLE 4 The number 92 is 4 more than half of what number?

We assume that you have familiarized yourself with the problem. We go on to Step 2.

Translate 92 is 4 more than half of what number?

$$92 = 4 \quad + \quad \tfrac{1}{2} \quad \cdot \quad x$$

Carry out We solve the equation $92 = 4 + \frac{1}{2}x$:

$$\begin{aligned} 92 - 4 &= \tfrac{1}{2}x && \text{Subtracting 4 on both sides} \\ 88 &= \tfrac{1}{2}x \\ 88 \cdot 2 &= 2 \cdot \tfrac{1}{2}x && \text{Multiplying by 2 on both sides} \\ 176 &= 1 \cdot x = x. \end{aligned}$$

Check We check to see whether 92 is 4 more than half of 176. Half of 176 is 88. Four more than 88 is 92, so the possible answer checks.

State The answer to the problem is the number 176.

DO EXERCISES 2–9.

Finding Information

In the familiarization process, as illustrated in Example 2, we needed some information other than mathematics itself. We needed the definition of speed from the field of physics. No amount of pure mathematics alone could have allowed us to find an answer, or even to do a complete familiarization.

In many problem-solving situations, you will find that you need some information that comes from a subject other than mathematics. You could not possibly solve all problems using mathematics by itself, so you should expect it necessary for many problems to find and use some fact, formula, law, or principle from another field, such as physics, chemistry, economics, or biology, in order to find a solution.

Problem Solving

As your ability to solve equations increases, so too will your ability to solve problems. Here is another example to illustrate the five-step problem-solving process.

EXAMPLE 5 Three numbers are such that the second is 6 less than 3 times the first and the third is 2 more than $\frac{2}{3}$ the first. The sum of the three numbers is 150. Find the largest number.

Translate Translate the following to equations.

2. 35 is $\frac{3}{5}$ of what number?

3. 87 more than half of what number is 115?

Carry out

4. Solve the equation obtained in Exercise 2.

5. Solve the equation obtained in Exercise 3.

Check

6. Determine whether your answer to Exercise 4 is an answer to Exercise 2.

7. Determine whether your answer to Exercise 5 is an answer to Exercise 3.

State

8. State the answer to Exercise 2.

9. State the answer to Exercise 3.

We proceed according to the five-step process.

1. **Familiarize** We note that there are three numbers involved, and we want to find the largest one. We list the other information in a table.

First number	
Second number	6 less than 3 times the first
Third number	2 more than $\frac{2}{3}$ the first

$$\text{First} + \text{Second} + \text{Third} = 150$$

2. **Translate** We need to assign a letter to something that is unknown. But there are three unknown numbers. Which should we start with? From the table, it looks as though it would be the first number. We shall use x to represent the first number. (We often say, "Let x = first number.")

We can now name the second and third numbers using x. (We often say that we name them "in terms of x.") We'll go back to the table and add another column:

First number		x
Second number	6 less than 3 times the first	$3x - 6$
Third number	2 more than $\frac{2}{3}$ the first	$\frac{2}{3}x + 2$

From the first step, we know that the sum of the three numbers is 150. Substituting, we obtain an equation in terms of x:

$$\underbrace{\text{First}} + \underbrace{\text{second}} + \underbrace{\text{third}} = \underbrace{150}$$

$$x + (3x - 6) + (\tfrac{2}{3}x + 2) = 150$$

3. **Carry out** We solve the equation:

$x + 3x - 6 + \frac{2}{3}x + 2 = 150$ Leaving off unnecessary parentheses

$(4 + \frac{2}{3})x - 4 = 150$ Collecting like terms

$\frac{14}{3}x - 4 = 150$

$\frac{14}{3}x = 154$ Adding 4 on both sides

$x = \frac{3}{14} \cdot 154$ Multiplying on both sides by $\frac{3}{14}$

$x = 33.$

We could check to see whether 33 is a solution of the equation, but we can skip that step, because we shall check later, in the original problem.

Going back to the table, we can find the other two numbers:

Second: $3x - 6 = 3 \cdot 33 - 6 = 93;$

Third: $\frac{2}{3}x + 2 = \frac{2}{3} \cdot 33 + 2 = 24.$

4. **Check** We go back to the original problem. We have three numbers: 33, 93, and 24. Is the second one 6 less than 3 times the first?

$$3 \times 33 - 6 = 99 - 6 = 93$$

The answer is *yes*.

10. Use the five-step process to solve the following problem. A certain number is three times a second number. A third number is twice the second number. The sum of the first number, twice the second number, and three times the third number is 165. Find the numbers.

Is the third 2 more than $\frac{2}{3}$ the first?

$$\frac{2}{3} \times 33 + 2 = 22 + 2 = 24$$

The answer is *yes*.

Is the sum of the three numbers 150?

$$33 + 93 + 24 = 150$$

The answer is *yes*. The numbers do check.

5. **State** The problem says to find the largest number, so the answer is: "The largest number is 93."

DO EXERCISE 10.

EXERCISE SET 1.3

i For each problem, familiarize yourself with the situation. Then translate to mathematical language. You need not actually solve the problem, merely carry out the first two steps.

1. A 12-ft piece of rope is to be cut into two pieces, one piece 4 ft longer than the other. How should the rope be cut?

2. A piece of wire 10 m long is to be cut into two pieces, one of them $\frac{2}{3}$ as long as the other. How should the wire be cut?

3. One angle of a triangle is three times as great as a second angle. The third angle measures 12° less than twice the second angle. Find the measures of the angles.

4. One angle of a triangle is four times as great as a second angle. The third angle measures 5° more than twice the second angle. Find the measures of the angles.

5. Find three consecutive odd integers such that the sum of the first, two times the second, and three times the third is 70.

6. Find two consecutive even integers such that two times the first plus three times the second is 76.

7. A piece of wire 100 cm long is to be cut into two pieces and those pieces are each to be bent to make a square. The length of a side of one square is to be 2 cm greater than the length of a side of the other. How should the wire be cut?

8. A piece of wire 100 cm long is to be cut into two pieces and those pieces are each to be bent to make a square. The area of one square is to be 144 cm^2 greater than that of the other. How should the wire be cut?

9. Three numbers are such that the second is six less than three times the first and the third is two more than two-thirds of the second. The sum of the three numbers is 172. Find the largest number.

10. An appliance store is having a sale on 13 TV sets. They are displayed in order of increasing price from left to right. The price of each set differs by $20 from either set next to it. For the price of the set at the extreme right, a customer can buy both the second and seventh sets. What is the price of the least expensive set?

11. A student's scores on five tests are 93, 89, 72, 80, and 96. What must the score be on the next test so that the average will be 88?

12. The changes in population of a city for three consecutive years are, respectively, 20% increase, 30% increase, and 20% decrease. What is the percent of total change for those three years?

ii, **iii** Solve each problem. Carry out all five problem-solving steps.

13. The number 38.2 is less than some number by 12.1. What is the number?

14. The number 173.5 is greater than a certain number by 16.8. What is the number?

15. The number 128 is 0.4 of what number?

16. The number 456 is $\frac{1}{3}$ of what number?

17. One number is greater than another by 12. The sum of the numbers is 114. What is the larger number?

18. One number is less than another by 65. The sum of the numbers is 92. What is the smaller number?

19. One number is twice another number. The sum of the numbers is 495. What are the numbers?

20. One number is five times another. The sum of the numbers is 472. What are the numbers?

21. Solve the problem of Exercise 10.

22. Solve the problem of Exercise 11.

23. Solve the problem of Exercise 12.

24. Find three consecutive odd integers such that the sum of the first, two times the second, and three times the third is 70.

25. One angle of a triangle is three times as great as a second angle. The third angle measures 12° less than twice the second angle. Find the measures of the angles.

26. A piece of wire 100 cm long is to be cut into two pieces and those pieces are each to be bent to make a square. The length of a side of one square is to be 2 cm greater than the length of a side of the other. How should the wire be cut?

27. The height and sides of a triangle are four consecutive integers. The height is the first integer, and the base is the fourth integer. The perimeter of the triangle is 42 in. Find the area of the triangle.

28. The salary of an employee is reduced $n\%$ during a time when a company was having financial difficulty. By what percent would the company then have to raise the salary in order to bring it back to where it was before the reduction?

1.4

ALGEBRAIC SYMBOLISM AND PROPERTIES OF NUMBERS

OBJECTIVES

After finishing Section 1.4, you should be able to:

- **i** **Use the identity properties of 0 and 1 to write equivalent expressions.**
- **ii** **Use the commutative and associative properties to write equivalent expressions.**
- **iii** **Use the distributive law to obtain equivalent expressions, collecting like terms and simplifying.**
- **iv** **Multiply by -1 to obtain expressions for additive inverses.**
- **v** **Remove certain parentheses to write equivalent, simpler expressions.**
- **vi** **Carry out computations given by computer symbolism, performing them in the proper order.**

In solving equations we manipulate algebraic symbols in various ways, such as collecting like terms. For example, instead of

$$3y + 2y,$$

we might write

$$5y$$

with the firm belief that the two expressions represent the same number, no matter what y may represent. In that sense the expressions $3y + 2y$ and $5y$ are equivalent. We define equivalent expressions in general as follows.

If two expressions represent the same number for *any* sensible replacement(s) of the variable(s), the expressions are said to be *equivalent*.

By a nonsensible replacement for an expression, we mean a number which when substituted in the expression does not give us a number. For example, the number 1 is not sensible for the expression

$$\frac{4 + x}{x - 1}$$

because it gives a denominator of 0, and we cannot divide by 0.

i The Identity Properties

How can we tell whether two expressions are equivalent? The answer is that we depend on the fundamental properties of numbers. We state two such properties now and show how they can be used.

THE IDENTITY PROPERTIES OF 1 AND 0

For any number a,

$$1 \cdot a = a.$$

For any number a,

$$a + 0 = a.$$

Recall that 1 can be named $\frac{b}{b}$ for any number b, except 0. Also note that 0 can be named $(b - b)$ for any number b.

EXAMPLE 1 Write two expressions equivalent to

$$\frac{x+2}{2}$$

by multiplying by 1.

a) Let us choose the name $\frac{8}{8}$ for 1. Then we multiply:

$$\frac{x+2}{2} = \frac{x+2}{2} \cdot \frac{8}{8} \qquad \text{Multiplying by 1}$$

$$= \frac{(x+2)8}{2 \cdot 8} \qquad \text{Multiplying numerators and denominators}$$

$$= \frac{8x+16}{16}.$$

The expression

$$\frac{8x+16}{16}$$

is equivalent to

$$\frac{x+2}{2}.$$

The two expressions will name the same number for any number x.

b) Let us choose the name $\frac{y}{y}$ for 1. Then we have:

$$\frac{x+2}{2} = \frac{x+2}{2} \cdot \frac{y}{y} \qquad \text{Multiplying by 1}$$

$$= \frac{(x+2)y}{2y} \qquad \text{Multiplying numerators and denominators}$$

$$= \frac{xy+2y}{2y}.$$

The expressions

$$\frac{x+2}{2} \quad \text{and} \quad \frac{xy+2y}{2y}$$

are equivalent. They will name the same number for any replacements for x and y, except that y cannot be 0.

EXAMPLE 2 Obtain an expression equivalent to

$$\frac{6xy}{2x},$$

but simpler. Use the identity property of 1.

We do the reverse of what we did in Example 1. We factor:

$$\frac{6xy}{2x} = \frac{2 \cdot 3 \cdot xy}{2x} \qquad \text{Factoring the numerator}$$

$$= \frac{2x}{2x} \cdot \frac{3y}{1} \qquad \text{Factoring the fractional expression}$$

$$= \frac{3y}{1} = 3y. \qquad \text{“Removing” the factor } \frac{2x}{2x}$$

The expressions

$$\frac{6xy}{2x} \quad \text{and} \quad 3y$$

are equivalent. They will name the same number for all replacements for x and y, except that x cannot be 0.

DO EXERCISES 1 AND 2.

EXAMPLE 3 Write an expression equivalent to $4x - 2$ by adding 0.

Let us choose the name $6y - 6y$ for 0. Then we have:

$$\begin{aligned} 4x - 2 &= 4x - 2 + (6y - 6y) && \text{Adding 0} \\ &= 4x - 2 + 6y - 6y \\ &= 4x + 6y - 2 - 6y && \text{Rearranging terms} \end{aligned}$$

The expressions $4x - 2$ and $4x + 6y - 2 - 6y$ are equivalent. They will name the same number for any replacements for x and y. This time there are no nonsensible replacements.

ii The Commutative and Associative Laws

When we rearranged terms in Example 3, we changed the order in which we wrote the terms. How do we know that we can do that and still have an equivalent expression? The answer to the question depends on the commutative and associative laws.

The commutative laws have to do with changing order.

THE COMMUTATIVE LAWS

Addition. **For any numbers a and b,**

$$a + b = b + a.$$

Multiplication. **For any numbers a and b,**

$$a \cdot b = b \cdot a.$$

The associative laws have to do with changing grouping.

THE ASSOCIATIVE LAWS

Addition. **For any numbers a, b, and c,**

$$a + (b + c) = (a + b) + c.$$

Multiplication. **For any numbers a, b, and c,**

$$a \cdot (b \cdot c) = (a \cdot b) \cdot c.$$

EXAMPLE 4 Write an expression equivalent to $3x + 4y$, using the commutative law of addition.

$$3x + 4y = 4y + 3x \qquad \text{Using the commutative law to change order}$$

The expressions are equivalent. They will name the same number for any numbers x and y.

1. Write two expressions equivalent to

$$\frac{2x - 3}{x + 1}$$

by multiplying by 1.

a) Multiply by $\frac{7}{7}$.

b) Multiply by $\frac{2y}{2y}$.

2. Obtain an expression equivalent to

$$\frac{10yz}{5z},$$

but simpler, by "removing" a factor of 1.

3. a) Write an expression equivalent to $5y + 7z$ using the commutative law of addition.

   b) Write an expression equivalent to $a \cdot b$ using the commutative law of multiplication.

4. Write an expression equivalent to $(5 + 2y) + 7z$ using the associative law of addition.

5. Write an expression equivalent to $2x + (y + z)$ using both the commutative and associative laws of addition.

EXAMPLE 5 Write an expression equivalent to $(3x + 7y) + 9z$, using the associative law of addition.

$$(3x + 7y) + 9z = 3x + (7y + 9z)$$

The expressions are equivalent. They will name the same number for any numbers x, y, and z.

EXAMPLE 6 Write an expression equivalent to

$$\frac{5}{x} \cdot (2yz),$$

using the commutative and associative laws of multiplication.

$$\frac{5}{x} \cdot (2yz) = \left(\frac{5}{x} \cdot 2y\right) \cdot z \qquad \text{Using the associative law}$$

$$= \left(2y \cdot \frac{5}{x}\right) \cdot z \qquad \text{Using the commutative law in the parentheses}$$

Although you don't probably think about it, you use the associative law of multiplication when you multiply a term such as $8x$ by a number such as 5, to get $40x$:

$$5(8x) = (5 \cdot 8)x = 40x.$$

DO EXERCISES 3–5.

iii The Distributive Laws

Let's look at some other laws that enable us to write equivalent expressions. We say that multiplication is distributive over addition, meaning that when we multiply a number by a sum we can either add first or multiply first. Multiplication is also distributive over subtraction.

THE DISTRIBUTIVE LAWS

Multiplication over addition. **For any numbers a, b, and c,**

$$a(b + c) = ab + ac.$$

Multiplication over subtraction. **For any numbers a, b, and c,**

$$a(b - c) = ab - ac.$$

An Agreement About Parentheses. Parentheses indicate that a certain computation is to be done before another. For example, in

$$a(b + c),$$

the addition is to be done first. In an expression such as $ab + ac$ or $ab - ac$, the multiplications are to be done first. We make an agreement about omitting the parentheses in such cases; that is,

$(ab) + (ac)$ can be shortened to $ab + ac$, and

$(ab) - (ac)$ can be shortened to $ab - ac$.

Equivalent Expressions. The distributive laws allow us to multiply algebraic expressions, factor them, or collect like terms, knowing that we will obtain equivalent expressions.

EXAMPLE 7 Obtain an expression equivalent to $x(5y + 4)$ by multiplying.

We use a distributive law to get

$$x(5y + 4) = 5xy + 4x.$$

The expressions $x(5y + 4)$ and $5xy + 4x$ are equivalent. They will name the same number for any replacements for x and y.

When we do the opposite of what we did in Example 7, we say that we are *factoring* an expression.

EXAMPLE 8 Obtain an expression equivalent to $3xy - 6x$ by factoring.

We use a distributive law to get

$$3xy - 6x = 3x(y - 2).$$

The expressions $3xy - 6x$ and $3x(y - 2)$ are equivalent. They will name the same number for any replacements for x and y.

When we collect like terms, we also depend on the distributive laws.

EXAMPLE 9 Obtain an expression equivalent to $8xy + 3xy - 4xy$ by collecting like terms.

We use both distributive laws to get

$$\begin{aligned} 8xy + 3xy - 4xy &= (8 + 3 - 4)xy \\ &= 7xy. \end{aligned}$$

The expressions $8xy + 3xy - 4xy$ and $7xy$ are equivalent. They will name the same number for any replacements for x and y.

DO EXERCISES 6–8.

6. Obtain an expression equivalent to $2x(3y + z)$ by multiplying.

7. Obtain an expression equivalent to $4yz + 12y$ by factoring.

8. Obtain an expression equivalent to $3xy + 5yz - 2xy + 4yz + 7z$.

iv Multiplying by −1

When we multiply a number by -1, we get its additive inverse. For example,

$$-1 \cdot 8 = -8 \qquad \text{(the additive inverse of 8)}$$

and

$$-1(-5) = 5 \qquad \text{(the additive inverse of } -5\text{)}.$$

> **For any number a,**
>
> $$-1 \cdot a = -a$$
>
> **(negative 1 times a is the additive inverse of a.) In other words, multiplying by -1 changes the sign.**

Using this fact, we can obtain equivalent expressions.

EXAMPLE 10 Obtain an expression equivalent to $-(a - b)$, using the property of -1.

$$\begin{aligned} -(a - b) &= -1 \cdot (a - b) && \text{Replacing } - \text{ by } -1\,\cdot \\ &= -1 \cdot a - (-1) \cdot b && \text{Using a distributive law} \\ &= -a - (-b) && \text{Replacing } -1 \cdot a \text{ by } -a \text{ and } (-1) \cdot b \text{ by } -b \\ &= -a + b, \quad \text{or} \quad b - a \end{aligned}$$

9. Obtain an expression equivalent to $-(2x + y - 3)$ using the property of -1.

10. Find an expression for the additive inverse.

a) $x - y$

b) $2x - 5y$

c) $3y - 5z$

11. Obtain an equivalent but simpler expression by first removing parentheses.

$$5xy - 2x(y + 1)$$

12. Simplify.

$$9x - \{2[5(x - 3) - 3(x + 2)] - 4\}$$

The expressions $-(a - b)$ and $b - a$ are equivalent. They will name the same number for all replacements for a and b.

Example 10 illustrates something worth remembering, because it gives us a useful shortcut, that is,

The inverse of an expression $a - b$ is $b - a$.

DO EXERCISES 9 AND 10.

v Removing Certain Parentheses

We are often faced with altering an expression in which parentheses are preceded by a subtraction sign. To simplify the expression, we may wish to remove the parentheses. That can be done, provided that we change the sign of every term inside the parentheses.

EXAMPLES Obtain equivalent but simpler expressions, by first removing parentheses.

11. $6x - (4x + 2) = 6x - 4x - 2$ Removing parentheses and changing the sign of every term inside

$= 2x - 2$ Collecting like terms

The expressions $6x - (4x + 2)$ and $2x - 2$ are equivalent. They will name the same number for every number x.

12. $x - 3(x + y) = x + [-3(x + y)]$ Changing the subtraction to an addition

$= x + [-3x - 3y]$ Using a distributive law

$= -2x - 3y$ Collecting like terms

The expressions $x - 3(x + y)$ and $-2x - 3y$ are equivalent. They will name the same number for all replacements for x and y.

In expressions like the one in Example 12 we ordinarily do not write out all of the steps, but get to the answer as quickly as possible.

DO EXERCISE 11.

Parentheses Within Parentheses

When parentheses occur within parentheses, we use different symbols, such as [ ] (brackets) and { } (braces). You should compute what is within the innermost ones first, and then work from the inside out.

EXAMPLE 13 Simplify: $6y - \{4[3(y - 2) - 4(y + 2)] - 3\}$.

$6y - \{4[3(y - 2) - 4(y + 2)] - 3\}$

$= 6y - \{4[3y - 6 - 4y - 8] - 3\}$ Multiplying to remove the innermost parentheses

$= 6y - \{4[-y - 14] - 3\}$ Collecting like terms in the inner parentheses

$= 6y - \{-4y - 56 - 3\}$ Multiplying to remove the inner parentheses

$= 6y - \{-4y - 59\}$ Collecting like terms

$= 6y + 4y + 59$ Removing parentheses

$= 10y + 59$ Collecting like terms

The original expression and $10y + 59$ are equivalent. They will name the same number for every number y.

DO EXERCISE 12.

vi Computer Language and Order of Operations

When several operations occur together, in what order should they be done? We have agreements about such calculations, and we also use parentheses. In hand-written algebra, little, if any, difficulty arises in knowing in which order operations are to be performed. With computers, however, there may be some problem, because the symbolism is different. The order in which a computer performs operations is the same as it is in ordinary algebra, but it is more important to give some thought to order of operations when using a computer language. The computer symbols we shall use to illustrate are from the computer language known as BASIC.

BASIC notation	Algebra notation
A + B	$a + b$
A − B	$a - b$
A*B	ab or $a \cdot b$
A/B	$\frac{a}{b}$ or $a \div b$
A^B	a^b
( )	( )

We now state the rules for order of operations.

RULES FOR ORDER OF OPERATIONS

1. **Do all calculations within parentheses before operations outside.**
2. **Evaluate all exponential expressions.**
3. **Do all multiplications and divisions.**
4. **Do all additions and subtractions.**

Computers actually carry out all multiplications and divisions in order, from left to right. If you are computing by hand, it can be helpful to think of each division as a multiplication by a reciprocal. Then, by the commutative and associative laws, you know that you can do the multiplications in any order that you prefer. A similar statement holds with respect to the order of additions and subtractions. Computers, however, follow the rules above.

Let us apply these rules to sentences we see in arithmetic or algebra and then with sentences in BASIC language.

EXAMPLE 14 Calculate: $4000 \cdot (1 + 0.12)^3$.

1. Do all calculations within parentheses:

$$4000 \cdot (1 + 0.12)^3 = 4000 \cdot (1.12)^3. \qquad \text{Adding 1 and 0.12}$$

2. Evaluate all exponential expressions:

$$4000 \cdot (1.12)^3 = 4000 \cdot 1.404928. \qquad \text{Calculating } 1.12^3$$

3. Do all multiplications and divisions:

$$4000 \cdot 1.404928 = 5619.712.$$

Now let us apply the rules to the BASIC language.

EXAMPLE 15 Calculate: `6.4 + 15 * 23 − (18 + 29)^2`

We do calculations within parentheses:

```
6.4 + 15 * 23 − (18 + 29)^2 = 6.4 + 15 * 23 − 47^2
```

Do the following calculations in the order in which a computer would perform them.

13. 5 * 3^2/4 * 5 − (2^3 + 4)/2

14. 4^2 * 3/2 * 3 + (3 * 2)^2/4

Write BASIC notation.

15. $\dfrac{(x+y)^2}{3-xy}$

16. $(a-b^2)\left(a^2-2b^3+\dfrac{4}{a^2}\right)$

Now we do the exponentiation:

```
6.4 + 15 * 23 - 47^2 = 6.4 + 15 * 23 - 2209
```

Next, we do the multiplication, followed by the addition and subtraction:

```
6.4 + 15 * 23 - 2209 = 6.4 + 345 - 2209
                     = 351.4 - 2209
                     = -1857.6
```

EXAMPLE 16 Calculate: `5 + 3 * 8^2/(5 + 1) - (4^2 - 11)^3`

We do all calculations within parentheses:

```
5 + 3 * 8^2/6 - (4^2 - 11)^3
```

In the parentheses remaining, we have an exponentiation and a subtraction. The calculation in parentheses is as follows.

```
4^2 - 11 = 16 - 11     Doing the exponentiation
         = 5           Doing the subtraction
```

Since the computations in parentheses are now completed, we have

```
5 + 3 * 8^2/6 - 5^3
```

We now do the two remaining exponentiations to get

```
5 + 3 * 64/6 - 125
```

Next we do the multiplications and divisions in order, from left to right:

```
5 + 32 - 125
```

We end by doing the additions and subtractions in order, from left to right:

```
-88
```

DO EXERCISES 13 AND 14.

To program a computer to perform a task, it is important to be able to convert from algebraic symbolism to BASIC notation.

EXAMPLES Write BASIC notation.

17. $(a+b)(c-d)$ `(A + B) * (C - D)`

18. $\left(\dfrac{ab}{a^2-b^2}\right)^3$ `(A * B/(A^2 - B^2))^3`

19. $(a-b)\left(a^3-2a^2-\dfrac{c}{d^3}\right)$ `(A - B) * (A^3 - 2 * A^2 - C/D^3)`

DO EXERCISES 15 AND 16.

EXERCISE SET 1.4

i Write an equivalent expression by multiplying by 1. Use the given name for 1.

1. $\dfrac{x+3}{5}$; $\dfrac{6}{6}$

2. $\dfrac{x-4}{x+3}$; $\dfrac{y}{y}$

3. $\dfrac{2x-3}{3x+2}$; $\dfrac{2y}{2y}$

4. $\dfrac{5y-2x}{2x+4y}$; $\dfrac{3z}{3z}$

Write an equivalent expression by simplifying using the property of 1.

5. $\frac{4y}{4x}$ 6. $\frac{8xy}{16y}$ 7. $\frac{5y}{-2xy}$ 8. $\frac{3x(y+1)}{2(y+1)}$

ii Write an equivalent expression using a commutative law.

9. $5y + 2x$ 10. $4y + x$ 11. $2x - 3y$ 12. $5y - 4x$ 13. $\frac{x}{2} \cdot \frac{y}{3}$ 14. $3 \cdot \frac{y}{4}$

Write an equivalent expression using an associative law.

15. $2 \cdot (8x)$ 16. $x \cdot (3y)$ 17. $x + (2y + 5)$ 18. $(3y + 4) + 10$

iii Obtain an equivalent expression by multiplying.

19. $3(a + 1)$ 20. $8(x + 1)$ 21. $4(x - y)$ 22. $9(a - b)$
23. $-5(2a + 3b)$ 24. $-2(3c + 5d)$ 25. $2a(b - c + d)$ 26. $5x(y - z + w)$
27. $2\pi r(h + 1)$ 28. $P(1 + rt)$ 29. $\frac{1}{2}h(a + b)$ 30. $\pi r(1 + s)$

Obtain an equivalent expression by factoring.

31. $8x + 8y$ 32. $7a + 7b$ 33. $9p - 9$ 34. $12x - 12$
35. $7x - 21$ 36. $6y - 36$ 37. $xy + x$ 38. $ab + a$
39. $2x - 2y + 2z$ 40. $3x + 3y - 3z$ 41. $3x + 6y - 3$ 42. $4a + 8b - 4$

Obtain an equivalent expression by collecting like terms.

43. $4a + 5a$ 44. $9x + 3x$ 45. $8b - 11b$ 46. $9c - 12c$
47. $14y + y$ 48. $13x + x$ 49. $12a - a$ 50. $15x - x$
51. $t - 9t$ 52. $x - 6x$ 53. $5x - 3x + 8x$
54. $3x - 11x + 2x$ 55. $5x - 8y + 3x$ 56. $9a - 10b + 4a$
57. $7c + 8d - 5c + 2d$ 58. $12a + 3b - 5a + 6b$ 59. $4x - 7 + 18x + 25$
60. $13p + 5 - 4p + 7$ 61. $13x + 14y - 11x - 47y$ 62. $17a + 17b - 12a - 38b$

iv, v Obtain an equivalent but simpler expression, by first removing parentheses.

63. $a - (2a + 5)$ 64. $x - (5x + 9)$ 65. $4m - (3m - 1)$
66. $5a - (4a - 3)$ 67. $3d - 7 - (5 - 2d)$ 68. $8x - 9 - (7 - 5x)$
69. $-2(x + 3) - 5(x - 4)$ 70. $-9(y + 7) - 6(y - 3)$ 71. $5x - 7(2x - 3)$
72. $8y - 4(5y - 6)$ 73. $9a - [7 - 5(7a - 3)]$ 74. $12b - [9 - 7(5b - 6)]$
75. $5\{-2 + 3[4 - 2(3 + 5)]\}$ 76. $7\{-7 + 8[5 - 3(4 + 6)]\}$
77. $2y + \{7[3(2y - 5) - (8y + 7)] + 9\}$ 78. $7b - \{6[4(3b - 7) - (9b + 10)] + 11\}$

vi Calculate.

79. `5 * (6 + 2)^2/(5 - 1)` 80. `23 - 7 * 2/(5 - 3) + 4^2`
81. `3^4 * 5 - 7 + 3^(2 + 1)` 82. `4 * (3 - (2 + 4^2) + 5)^2`

Write BASIC notation.

83. $a^3 + b^3$ 84. $a^2 - 2ab + b^2$ 85. $\left(\frac{a + b^2}{a^2 b}\right)^3$
86. $b^2 - 4ac$ 87. $\frac{a^3}{b} + \frac{a^2}{b^3}$ 88. $c - d + \left(\frac{c + d}{a - b}\right)^2$

Problem solving

89. The changes in population of a city for three consecutive years are, respectively, 30% increase, 20% increase, and 20% decrease. What is the percent of total change for those three years?

90. Three numbers are such that the second is 6 less than 3 times the first and the third is 2 more than $\frac{2}{3}$ the second. The sum of the three numbers is 172. Find the largest number.

Evaluate each expression when $a = -1$, $b = 7$, $c = -6$, and $d = \frac{1}{2}$.

91. $\dfrac{d(ab - c)}{2a}$

92. $\dfrac{ad + bd}{b - a} - \dfrac{ac}{d}$

93. Remove parentheses and simplify:

$$(0.0079x - 0.000843y) - (-0.00793x - 0.000853y).$$

Simplify.

94. $-\{-[-(-9)]\}$

95. $\frac{2}{3}[2(x + y) + 4(x + 4y)]$

96. $-4[3(x - y - z) - 3(2x + y - 5z)]$

97. $[-(7a - b) - (a + 5b)] - [2(a + \frac{1}{2}b) + 3(7a - \frac{5}{3}b)]$

98. $0.01\{0.1(x - 2y) - [0.001(3x + y) - (0.2x - 0.1y)]\} - (x - y)$

1.5

EXPONENTIAL NOTATION

OBJECTIVES

After finishing Section 1.5, you should be able to:

- **i** **Write expressions with or without exponents.**
- **ii** **Use exponential notation in multiplication or division.**
- **iii** **Use exponential notation in raising powers to powers.**
- **iv** **Use exponential notation in translating problem situations to mathematical language.**

i Exponential notation is a kind of shorthand. It enables us to write compactly what might otherwise take a lot of space. Let us begin by recalling the definition for whole-number exponents.

> **EXPONENTIAL NOTATION**
>
> **A symbol a^n, where n is an integer greater than 1, means**
>
> $$\underbrace{a \cdot a \cdot a \cdot \ldots \cdot a \cdot a}_{n \text{ factors.}}$$
>
> **In a^n, a is called the *base* and n is the *exponent.***
> **A symbol a^1 means a.**
> **A symbol a^0 means 1, provided that $a \neq 0$.**

We now define exponential notation for negative-integer exponents. To see how the definition might be made, look for a pattern in the following.

$$10^2 = 100$$
$$10^1 = 10$$
$$10^0 = 1$$
$$10^{-1} = ?$$
$$10^{-2} = ?$$

In order for the pattern to continue, 10^{-1} would have to be $\frac{1}{10}$ and 10^{-2} would have to be $\frac{1}{100}$. We make the definition accordingly. In fact, we are going to generalize a bit.

> **EXPONENTIAL NOTATION, INTEGER EXPONENTS**
>
> **A symbol a^{-n}, where n is any integer, means**
>
> $$\frac{1}{a^n}.$$
>
> **In other words, a^n and a^{-n} are reciprocals.**

We should note that a symbol a^{-n} might be meaningless if $a = 0$.

EXAMPLES Write equivalent expressions without negative exponents.

1. 7^{-3} By the definition, 7^{-3} is the reciprocal of 7^3, so the answer is

$$\frac{1}{7^3}.$$

2. $\frac{1}{5^{-3}}$ By the definition, 5^{-3} is the reciprocal of 5^3, so the answer is 5^3.

3. $(3x)^{-4}$ The answer is

$$\frac{1}{(3x)^4}.$$

4. $\frac{y^{-4}x^2}{z^{-2}}$ We use the definition, obtaining

$$\frac{\frac{1}{y^4}x^2}{\frac{1}{z^2}}.$$

If we wish, we can simplify the answer to

$$\frac{x^2z^2}{y^4}.$$

DO EXERCISES 1 AND 2.

EXAMPLES Write equivalent expressions with negative exponents.

5. 5^7 By the definition, 5^7 is the reciprocal of 5^{-7}, so the answer is

$$\frac{1}{5^{-7}}.$$

6. $\frac{1}{8^{-5}}$ By the definition, 8^{-5} is the reciprocal of 8^5, so the answer is 8^5.

7. $5x^3$ By the definition, x^3 and x^{-3} are reciprocals, so the answer is

$$5 \cdot \frac{1}{x^{-3}} \quad \text{or} \quad \frac{5}{x^{-3}}.$$

8. $(-12)^4$ By the definition, $(-12)^4$ and $(-12)^{-4}$ are reciprocals, so the answer is

$$\frac{1}{(-12)^{-4}}.$$

DO EXERCISES 3 AND 4.

ii Properties of Exponents

For whole-number exponents, we can often simplify a product by adding exponents; for example,

$$a^3 \cdot a^2 = a^5.$$

Rewrite without negative exponents.

1. 10^{-4}

2. $(-4)^{-3}$

Rewrite with negative exponents.

3. $\frac{1}{4^3}$

4. $\frac{1}{(-5)^4}$

Multiply and simplify.

5. $8^{-3}8^{7}$

6. $y^{7}y^{-2}$

7. $(9x^{4})(-2x^{7})$

8. $(-3x^{-4})(25x^{-10})$

9. $(5x^{-3}y^{4})(-2x^{-9}y^{-2})$

10. $(4x^{-2}y^{4})(15x^{2}y^{-3})$

We now consider cases in which one or both exponents are negative integers:

$$\begin{aligned} b^5 \cdot b^{-2} &= b \cdot b \cdot b \cdot b \cdot b \cdot \frac{1}{b \cdot b} && \text{By the definition of exponents} \\ &= \frac{b \cdot b}{b \cdot b} \cdot b \cdot b \cdot b \\ &= b^3. \end{aligned}$$

Adding exponents again gives the correct result. Let us now consider a case in which both exponents are negative:

$$\begin{aligned} c^{-3} \cdot c^{-2} &= \frac{1}{c \cdot c \cdot c} \cdot \frac{1}{c \cdot c} && \text{By the definition of exponents} \\ &= \frac{1}{c \cdot c \cdot c \cdot c \cdot c} && \text{Multiplying} \\ &= c^{-5}. && \text{Simplifying} \end{aligned}$$

Again, adding exponents gives the correct result. That is always the case, even if exponents are negative or zero.

In multiplication with exponential notation, we can add exponents if the bases are the same, that is,

$$a^m a^n = a^{m+n}.$$

Equivalent expressions are obtained when we use the above property.

EXAMPLES Multiply and simplify. When letters occur in exponents, assume that they represent integers.

9. $x^4 \cdot x^3 = x^{4+3} = x^7$

10. $4^5 \cdot 4^{-3} = 4^{5+(-3)} = 4^2 = 16$

11. $(-2)^{-3}(-2)^7 = (-2)^{-3+7} = (-2)^4 = 16$

12. $$\begin{aligned} (8x^4y^{-2})(-3x^{-3}y) &= 8 \cdot (-3) \cdot x^4 \cdot x^{-3} \cdot y^{-2} \cdot y^1 \\ &= -24x^{4-3}y^{-2+1} \\ &= -24xy^{-1}, \quad \text{or} \quad -\frac{24x}{y}, \quad \text{or} \quad \frac{-24x}{y} \end{aligned}$$

13. $$\begin{aligned} (4x^ay^b)(2x^{a+1}y^{2-b}) &= 4 \cdot 2x^{a+(a+1)}y^{b+(2-b)} \\ &= 8x^{2a+1}y^2 \end{aligned}$$

There is no standard way of simplifying in cases like Example 12. In some cases it is better to have all exponents positive, but in some it is not. We give answers in two ways. You should also be able to give answers in two ways.

DO EXERCISES 5–10.

We now consider a division:

$$\frac{8^5}{8^3} \text{ means } \frac{8 \cdot 8 \cdot 8 \cdot 8 \cdot 8}{8 \cdot 8 \cdot 8}.$$ This simplifies to $8 \cdot 8$, or 8^2.

We can obtain the result by subtracting exponents. This is always the case, even if exponents are negative or zero.

In division with exponential notation, we can subtract exponents if the bases are the same, that is,

$$\frac{a^m}{a^n} = a^{m-n}.$$

Again, using this property, we obtain equivalent expressions.

EXAMPLES Divide and simplify. When letters occur in exponents, assume that they represent integers.

14. $\dfrac{5^7}{5^3} = 5^{7-3} = 5^4$ Subtracting exponents

15. $\dfrac{5^7}{5^{-3}} = 5^{7-(-3)} = 5^{7+3}$ Subtracting exponents (adding an inverse)

$= 5^{10}$

16. $\dfrac{9^{-2}}{9^5} = 9^{-2-5} = 9^{-7}$, or $\dfrac{1}{9^7}$

17. $\dfrac{7^{-4}}{7^{-5}} = 7^{-4-(-5)}$

$= 7^{-4+5} = 7^1 = 7$

18. $\dfrac{16x^4y^7}{-8x^3y^9} = \dfrac{16}{-8} \cdot \dfrac{x^4}{x^3} \cdot \dfrac{y^7}{y^9}$

$= -2xy^{-2}$, or $-\dfrac{2x}{y^2}$, or $\dfrac{-2x}{y^2}$

19. $\dfrac{14x^7y^{-3}}{4x^5y^{-5}} = \dfrac{14}{4} \cdot \dfrac{x^7}{x^5} \cdot \dfrac{y^{-3}}{y^{-5}}$

$= \frac{7}{2}x^2y^2$

20. $\dfrac{25x^{3a-5}y^{a+b-1}}{-5x^{a+5}y^{4+b}} = \dfrac{25}{-5}x^{(3a-5)-(a+5)}y^{(a+b-1)-(4+b)}$

$= -5x^{2a-10}y^{a-5}$

In cases such as Examples 14–20 above, it may help to think as follows: After writing the base, write the top exponent. Then write a subtraction sign. Then write the bottom exponent. Then do the subtraction. For example,

$$\frac{x^{-3}}{x^{-5}} = x^{-3-(-5)}$$

Writing the base and the top exponent — Writing a subtraction sign — Writing the bottom exponent

DO EXERCISES 11–15.

We do not define 0^0. Let's see why. We know that 0^0 is equal to 0^{1-1}. But 0^{1-1} is also equal to $0/0$. We have already seen that we must leave $0/0$ undefined, so we also leave 0^0 undefined.

Divide and simplify.

11. $\dfrac{4^8}{4^5}$

12. $\dfrac{5^4}{5^{-2}}$

13. $\dfrac{10^{-8}}{10^{-2}}$

14. $\dfrac{42y^7x^6}{-21y^{-3}x^{10}}$

15. $\dfrac{33a^5b^{-2}}{22a^2b^{-4}}$

Simplify.

16. $(3^7)^6$

17. $(8^2)^{-7}$

Simplify.

18. $(2xy)^3$

19. $(4x^{-2}y^7)^2$

20. $(-2x^4y^2)^5$

21. $(10x^{-4}y^7z^{-2})^3$

22. $(3x^{2a}y^{b+1})^{2a}$

iii Raising Powers to Powers

Consider the expression $(5^2)^4$. This means

$$5^2 \cdot 5^2 \cdot 5^2 \cdot 5^2, \quad \text{or} \quad 5^8.$$

We can obtain the result by multiplying the exponents.

Consider $(8^{-2})^3$. This means

$$\frac{1}{8^2} \cdot \frac{1}{8^2} \cdot \frac{1}{8^2}, \quad \text{or} \quad \frac{1}{8^6}, \quad \text{which is} \quad 8^{-6}.$$

Again, we could obtain the result by multiplying the exponents. This works, in general, whether the exponents are positive, negative, or zero.

To raise a power to a power, we multiply exponents:

$$(a^m)^n = a^{m \cdot n}.$$

EXAMPLES Simplify.

21. $(3^5)^7 = 3^{5 \cdot 7} = 3^{35}$

22. $(x^{-5})^4 = x^{-5 \cdot 4} = x^{-20}$

DO EXERCISES 16 AND 17.

We consider cases in which there are several factors inside the parentheses. Consider $(5^2 \cdot 7^3)^4$. This means

$$(5^2 \cdot 7^3) \cdot (5^2 \cdot 7^3) \cdot (5^2 \cdot 7^3) \cdot (5^2 \cdot 7^3), \quad \text{or} \quad 5^8 \cdot 7^{12}.$$

Again, we can obtain the result by multiplying the exponents. This is true, in general, for positive, negative, or zero exponents.

To raise an expression with several factors to a power, we raise each factor to the power, multiplying exponents:

$$(a^m b^n)^p = a^{m \cdot p} \cdot b^{n \cdot p}.$$

EXAMPLES Simplify. When letters occur in exponents, assume that they represent integers.

23. $(3x^2y^{-2})^3 = 3^3(x^2)^3(y^{-2})^3 = 3^3x^6y^{-6}$, or $27x^6y^{-6}$, or $\dfrac{27x^6}{y^6}$

24. $(5x^3y^{-5}z^2)^4 = 5^4(x^3)^4(y^{-5})^4(z^2)^4$, or $625x^{12}y^{-20}z^8$, or $\dfrac{625x^{12}z^8}{y^{20}}$

25. $(4x^{a+1}y^{b-1})^{2c} = 4^{2c}(x^{a+1})^{2c}(y^{b-1})^{2c}$
$= 4^{2c}x^{2ac+2c}y^{2bc-2c}$

DO EXERCISES 18–22.

We now consider raising a quotient to a power. Let's take as an example

$$\left(\frac{3^2}{5^3}\right)^4.$$

This means

$$\left(\frac{3^2}{5^3}\right)\cdot\left(\frac{3^2}{5^3}\right)\cdot\left(\frac{3^2}{5^3}\right)\cdot\left(\frac{3^2}{5^3}\right), \quad \text{or} \quad \frac{3^8}{5^{12}}.$$

Once more, we see that we can obtain the result by multiplying the exponents. This is true, in general, for positive, negative, or zero exponents.

To raise a quotient to a power, raise both numerator and denominator to the power, multiplying exponents:

$$\left(\frac{a^m}{b^n}\right)^p = \frac{a^{m\cdot p}}{b^{n\cdot p}}.$$

EXAMPLES Simplify.

26. $\left(\frac{x^2}{y^{-3}}\right)^{-5} = \frac{x^{2\cdot(-5)}}{y^{-3\cdot(-5)}} = \frac{x^{-10}}{y^{15}}$, or $\frac{1}{x^{10}y^{15}}$

27. $\left(\frac{2x^3y^{-2}}{3y^4}\right)^5 = \frac{(2x^3y^{-2})^5}{(3y^4)^5} = \frac{2^5x^{3\cdot 5}y^{-2\cdot 5}}{3^5y^{4\cdot 5}}$

$= \frac{32x^{15}y^{-10}}{243y^{20}} = \frac{32x^{15}}{243y^{30}}$, or $\frac{32}{243}x^{15}y^{-30}$

DO EXERCISES 23–25.

Simplify.

23. $\left(\frac{x^{-3}}{y^4}\right)^{-3}$

24. $\left(\frac{3x^2y^{-3}}{2y^5}\right)^2$

25. $\left(\frac{2x^{-2}y^3}{5y^{-2}}\right)^{2a}$

iv Translating with Exponential Notation

The following example shows how we can use exponential notation in problem solving. We will not solve the problem now, but will show how to translate it to mathematical language.

EXAMPLE 28 There are two square flower gardens on the campus mall. The length of a side of one garden is 4 ft less than the length of a side of the other. The area of the larger garden is 56 ft^2 greater than that of the smaller. (*Note:* ft^2 is exponential notation for *square feet*). Find the dimensions of the garden.

1. **Familiarize** We draw a picture and label it.

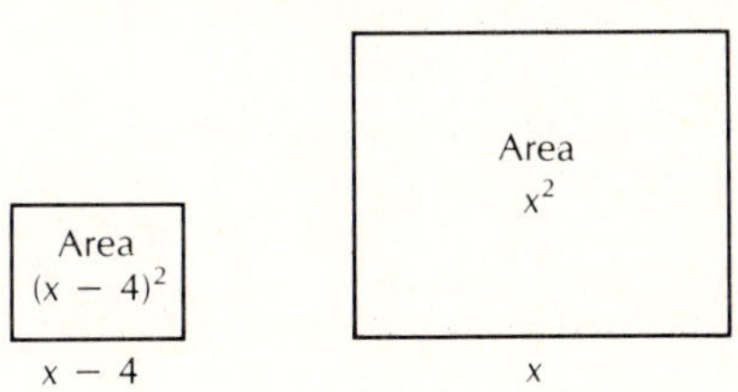

The larger garden has area 56 greater than the smaller garden.

2. **Translate** We use the statement:

Area of small garden + 56 is area of large garden.

$$(x-4)^2 + 56 \stackrel{?}{=} x^2$$

As you can see, for problem solving we will need to find out how to solve equations in which exponential notation appears. We shall do that in Chapter 4.

DO EXERCISE 26.

Translate the following problem situation to mathematical language. You need not solve the problem.

26. A cube has edges of length *a* cm. For what size cube is the total surface area, in square centimeters, the same number as its volume, in cubic centimeters?

EXERCISE SET 1.5

i Write equivalent expressions without negative exponents.

1. 6^{-3}
2. 8^{-4}
3. 9^{-5}
4. 16^{-2}
5. $(-11)^{-1}$
6. $(-4)^{-3}$
7. $(5x)^{-3}$
8. $(4xy)^{-5}$
9. x^2y^{-3}
10. $2a^2b^{-5}$
11. x^2y^{-2}
12. $a^2b^{-3}c^4d^{-5}$
13. $\dfrac{x^3}{y^{-2}}$
14. $\dfrac{y^4z^3}{x^{-1}}$
15. $\dfrac{y^{-5}}{x^2}$
16. $\dfrac{z^{-4}}{3x^5}$
17. $\dfrac{y^{-5}}{x^{-3}}$
18. $\dfrac{(7x)^{-4}}{y^{-7}}$
19. $\dfrac{x^{-2}y^7}{z^{-4}}$
20. $\dfrac{y^4z^{-3}}{x^{-2}}$

Write equivalent expressions with negative exponents.

21. $\dfrac{1}{3^4}$
22. $\dfrac{1}{9^2}$
23. $\dfrac{1}{(-16)^2}$
24. $\dfrac{1}{(-8)^6}$
25. 6^4
26. 8^5
27. $6x^2$
28. $-4y^5$
29. $\dfrac{1}{(5y)^3}$
30. $\dfrac{1}{(5x)^5}$
31. $\dfrac{1}{3y^4}$
32. $\dfrac{1}{4b^3}$

ii In Exercises 33–98, assume that all letters that appear in exponents represent integers.

Multiply and simplify.

33. $5^6 \cdot 5^3$
34. $6^2 \cdot 6^6$
35. $8^{-6} \cdot 8^2$
36. $9^{-5} \cdot 9^3$
37. $8^{-2} \cdot 8^{-4}$
38. $9^{-1} \cdot 9^{-6}$
39. $b^2 \cdot b^{-5}$
40. $a^4 \cdot a^{-3}$
41. $a^{-3} \cdot a^4 \cdot a^2$
42. $x^{-8} \cdot x^5 \cdot x^3$
43. $(2x)^3(3x)^2$
44. $(9y)^2(2y)^3$
45. $(14m^2n^3)(-2m^3n^2)$
46. $(6x^5y^{-2})(-3x^2y^3)$
47. $(-2x^{-3})(7x^{-8})$
48. $(6x^{-4}y^3)(-4x^{-8}y^{-2})$
49. $5^{a+1} \cdot 5^{2a-1}$
50. $7^{a+b} \cdot 7^{a+1}$
51. $(3x^ay^b)(2x^{a-1}y^{3-b})$
52. $(6x^{5a+1}y^{-a})(2x^{2-a}y^{a+3})$

Divide and simplify.

53. $\dfrac{6^8}{6^3}$
54. $\dfrac{7^9}{7^4}$
55. $\dfrac{4^3}{4^{-2}}$
56. $\dfrac{5^8}{5^{-3}}$
57. $\dfrac{10^{-3}}{10^6}$
58. $\dfrac{12^{-4}}{12^8}$
59. $\dfrac{9^{-4}}{9^{-6}}$
60. $\dfrac{2^{-7}}{2^{-5}}$
61. $\dfrac{a^3}{a^{-2}}$
62. $\dfrac{y^4}{y^{-5}}$
63. $\dfrac{9a^2}{(-3a)^2}$
64. $\dfrac{24a^5b^3}{-8a^4b}$
65. $\dfrac{-24x^6y^7}{18x^{-3}y^9}$
66. $\dfrac{14a^4b^{-3}}{-8a^8b^{-5}}$
67. $\dfrac{-18x^{-2}y^3}{-12x^{-5}y^5}$
68. $\dfrac{-14a^{14}b^{-5}}{-18a^{-2}b^{-10}}$
69. $\dfrac{10^{2a+1}}{10^{a+1}}$
70. $\dfrac{11^{b+2}}{11^{3b-3}}$
71. $\dfrac{9a^{x-2}}{3a^{2x+2}}$
72. $\dfrac{-12x^{a+1}}{4x^{2-a}}$
73. $\dfrac{45x^{2a+4}y^{b+1}}{-9x^{a+3}y^{2+b}}$
74. $\dfrac{-28x^{b+5}y^{4+c}}{7x^{b-5}y^{c-4}}$

iii Simplify.

75. $(4^3)^2$
76. $(5^4)^5$
77. $(8^4)^{-3}$
78. $(9^3)^{-4}$
79. $(6^{-4})^{-3}$
80. $(7^{-8})^{-5}$
81. $(3x^2y^2)^3$
82. $(2a^3b^4)^5$
83. $(-2x^3y^{-4})^{-2}$
84. $(-3a^2b^{-5})^{-3}$
85. $(-6a^{-2}b^3c)^{-2}$
86. $(-8x^{-4}y^5z^2)^{-4}$
87. $\left(\dfrac{4^{-3}}{3^4}\right)^3$
88. $\left(\dfrac{5^2}{4^{-3}}\right)^{-3}$
89. $\left(\dfrac{2x^3y^{-2}}{3y^{-3}}\right)^3$
90. $\left(\dfrac{-4x^4y^{-2}}{5x^{-1}y^4}\right)^{-4}$
91. $(7^a)^{2b}$
92. $(3^{a+2})^a$
93. $(12^{3-a})^{2b}$
94. $(x^{a-1})^{3b}$
95. $(5x^{a-1}y^{b+1})^{2c}$
96. $(4x^{3a}y^{2b})^{5c}$
97. $\dfrac{4x^{2a+3}y^{2b-1}}{2x^{a+1}y^{b+1}}$
98. $\dfrac{25x^{a+b}y^{b-a}}{-5x^{a-b}y^{b+a}}$

iv Translate each problem situation to mathematical language. You need not solve.

99. A farmer owns two square pens. The length of a side of one is 50 ft greater than the length of a side of the other. The sum of the areas of the pens is 15,700 ft^2. What is the area of the smaller pen?

100. On a large estate there are two circular flower gardens. The radius of one is twice that of the other. The difference in their areas is 147 ft^2. Find the sum of the areas of the gardens.

$x^2+(x+50)^2=15{,}700$

Simplify.

101. $(-2)^0 - (-2)^3 - (-2)^{-1} + (-2)^4 - (-2)^{-2}$

102. $2(6^1 \cdot 6^{-1} - 6^{-1} \cdot 6^0)$

103. $\dfrac{(-8)^{-2} \cdot (8 - 8^0)}{2^{-6}}$

104. $\left[\dfrac{1}{(-3)^{-2}} - (-3)^1\right] \cdot \left[(-3)^2 + (-3)^{-2}\right]$

105. $\dfrac{(2^{-2})^a \cdot (2^b)^{-a}}{(2^{-2})^{-b}(2^b)^{-2a}}$

106. $\{[(8^{-a})^{-2}]^b\}^{-c} \cdot [(8^0)^a]^c$

107. $\left[\dfrac{(-3x^{-2a}y^{5b})^{-2}}{(2x^{4a}y^{-8b})^{-3}}\right]^2$

108. $\left[\left(\dfrac{a^{-2c}}{b^{7c}}\right)^{-3}\left(\dfrac{a^{4c}}{b^{-3c}}\right)^2\right]^{-a}$

1.6

SCIENTIFIC NOTATION

OBJECTIVES

After finishing Section 1.6, you should be able to:

- **i** **Convert between scientific notation and ordinary decimal notation.**
- **ii** **Use scientific notation in multiplication and division.**
- **iii** **Use scientific notation in approximating and estimating.**
- **iv** **Use scientific notation in problem solving.**

There are many kinds of symbolism, or *notation,* for numbers. You are already familiar with fractional notation, decimal notation, and percent notation. In this section, we study another notation, one that is especially useful for work with very large numbers or very small numbers. It is thus important for problem solving. It is also useful in estimating.

The following are examples of scientific notation:

6.4×10^{13} means 64000000000000;

4.6×10^{-6} means 0.0000046;

3.75×10^{-2} means 0.0375.

Scientific notation **for a number consists of exponential notation for a power of 10, and, if needed, decimal notation for a number *a* between 1 and 10 and a multiplication sign. The following are both scientific notation:**

$$a \times 10^b; \qquad 10^b.$$

i Conversions

We can convert to scientific notation by multiplying by 1, choosing a name like $10^b \cdot 10^{-b}$ for the number 1.

EXAMPLE 1 Light travels about 9,460,000,000,000 kilometers in one year. Write scientific notation for the number.

We want to move the decimal point 12 places, between the 9 and the 4, so we choose $10^{-12} \times 10^{12}$ as a name for 1. Then we multiply.

$$9{,}460{,}000{,}000{,}000 \times 10^{-12} \times 10^{12} \qquad \text{Multiplying by 1}$$

$$= 9.46 \times 10^{12} \qquad \text{The } 10^{-12} \text{ moved the decimal point 12 places to the left and we have scientific notation.}$$

DO EXERCISES 1 AND 2.

1. Convert 460,000,000,000 to scientific notation.

2. The distance from the earth to the sun is about 93,000,000 miles. Write scientific notation for this number.

EXAMPLE 2 Write scientific notation for 0.0000000000156.

We want to move the decimal point 11 places. We choose $10^{11} \times 10^{-11}$ as a name for 1, and then multiply.

$$0.0000000000156 \times 10^{11} \times 10^{-11} \qquad \text{Multiplying by 1}$$

$$= 1.56 \times 10^{-11} \qquad \text{The } 10^{11} \text{ moved the decimal point 11 places to the right and we have scientific notation.}$$

You should try to make conversions to scientific notation mentally as much as possible.

DO EXERCISES 3 AND 4.

3. Convert 0.00000001235 to scientific notation.

4. The mass of a hydrogen atom is about 0.0000000000000000000000017 gram. Write scientific notation for this number.

EXAMPLES Convert mentally to decimal notation.

3. $7.893 \times 10^5 = 789{,}300$ Moving the decimal point 5 places to the right
4. $4.7 \times 10^{-8} = 0.000000047$ Moving the decimal point 8 places to the left

DO EXERCISES 5 AND 6.

Convert to decimal notation.

5. 7.893×10^{11}

6. 5.67×10^{-5}

ii Multiplying and Dividing

Multiplying and dividing with scientific notation is easy because we can use the properties of exponents.

EXAMPLE 5 Multiply and write scientific notation for the answer.

$$(3.1 \times 10^5)(4.5 \times 10^{-3}).$$

We apply the commutative and associative laws to get

$$(3.1 \times 4.5)(10^5 \times 10^{-3}) = 13.95 \times 10^2.$$

To find scientific notation for the result, we convert 13.95 to scientific notation and then simplify:

$$13.95 \times 10^2 = (1.395 \times 10^1) \times 10^2 = 1.395 \times 10^3.$$

DO EXERCISES 7 AND 8.

Multiply and write scientific notation for the answer.

7. $(9.1 \times 10^{-17})(8.2 \times 10^3)$

8. $(1.12 \times 10^{-8})(5 \times 10^{-7})$

EXAMPLE 6 Divide and write scientific notation for the answer.

$$\frac{6.4 \times 10^{-7}}{8.0 \times 10^6}.$$

We solve as follows:

This shows two divisions.

$$\frac{6.4 \times 10^{-7}}{8.0 \times 10^{6}} = \frac{6.4}{8.0} \times \frac{10^{-7}}{10^{6}} \quad \text{Factoring}$$

$$= 0.8 \times 10^{-13} \quad \text{Doing the divisions separately}$$

$$= (8.0 \times 10^{-1}) \times 10^{-13} \quad \text{Converting 0.8 to scientific notation}$$

$$= 8.0 \times 10^{-14}.$$

DO EXERCISES 9 AND 10.

In the preceding example, you may be tempted to quit when you have obtained 0.8×10^{-13}. That would not be an incorrect answer, except that it is not yet scientific notation, because 0.8 is not a number between 1 and 10.

iii Estimating and Approximating

Scientific notation is helpful in estimating or approximating.

EXAMPLE 7 Estimate: $\dfrac{780{,}000{,}000 \times 0.00071}{0.000005}$.

We first convert to scientific notation:

$$\frac{(7.8 \times 10^{8})(7.1 \times 10^{-4})}{5 \times 10^{-6}}.$$

Next, we group as follows:

$$\frac{7.8 \times 7.1}{5} \times \frac{(10^{8})(10^{-4})}{10^{-6}}.$$

We round the numbers that are between 1 and 10 and calculate:

$$\frac{8 \times 7}{5} = \frac{56}{5} \approx \frac{55}{5} = 11.$$

This answer will vary, depending on how you round. Next, we consider the powers of 10:

$$\frac{10^{8} \times 10^{-4}}{10^{-6}} = \frac{10^{4}}{10^{-6}} = 10^{10}.$$

We now have an approximation of 11×10^{10}, or 1.1×10^{11}.

DO EXERCISE 11.

With some practice, you will be able to do this kind of estimating mentally. It is important that you acquire this skill to use in problem solving.

It is helpful to remember the following:

$10^{3} = 1000$ (thousand);

$10^{6} = 1{,}000{,}000$ (million);

$10^{9} = 1{,}000{,}000{,}000$ (billion);

$10^{12} = 1{,}000{,}000{,}000{,}000$ (trillion).

Divide and write scientific notation for the answer.

9. $\dfrac{4.2 \times 10^{5}}{2.1 \times 10^{2}}$

10. $\dfrac{1.1 \times 10^{-4}}{2.0 \times 10^{-7}}$

11. Estimate. Answers will vary depending on how you round.

$$\frac{830{,}000{,}000 \times 0.0000012}{3{,}100{,}000}$$

12. A sheet of plastic has a thickness of 0.0003 mm. The sheet, which is 1.34 m wide, is rolled into a roll. The length of the plastic on the roll is 103.4 m. Use scientific notation to find the volume of the plastic sheet.

SOME EXAMPLES USING SCIENTIFIC NOTATION

The distance from the planet Pluto to the sun is 3.664×10^9 miles.

The mass of an electron is 9.11×10^{-28} grams.

The population of the United States was 2.04×10^8.

The wavelength of a certain red light is 6.6×10^{-5} cm.

The gross national product (GNP) one year was \$$9.32 \times 10^{11}$.

An electron has a charge of 4.8×10^{-10} electrostatic units.

iv Scientific Notation in Problems

The following example shows how scientific notation can be useful in problem solving.

EXAMPLE 8 A certain kind of wire will be used to construct 350 km of transmission line. The wire has a diameter of 1.2 cm. What is the volume of wire needed for one wire 350 km long?

1. **Familiarize** Drawing a picture, we see that we have a cylinder (a very *long* one). Its length is 350 km and the base has a diameter of 1.2 cm.

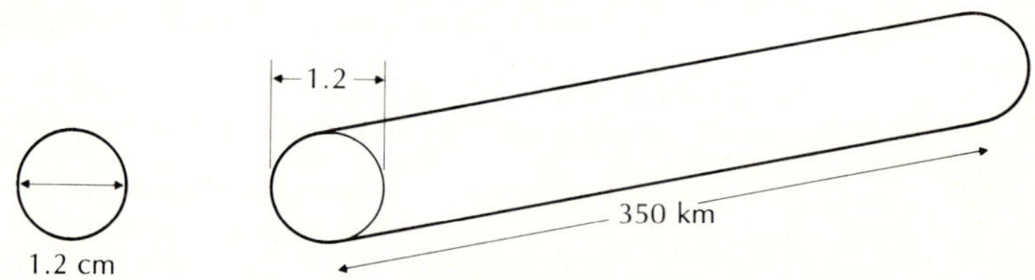

At this stage, you may need to look up a formula for the volume of a cylinder. It is

$$V = \pi r^2 h,$$

where r is the radius of the base and h is the height (in this case, the length of the wire).

2. **Translate** We will use the volume formula, but it will help to make the units the same. Let's put everything into meters:

Length: 350 km = 350,000 m, or 3.5×10^5 m;

Diameter: 1.2 cm = 0.012 m, or 1.2×10^{-2} m.

The radius, which we will need in the formula, is half the diameter:

Radius: 0.6×10^{-2} m, or 6×10^{-3} m.

We now substitute into the formula:

$$V = \pi(6 \times 10^{-3})^2 \cdot 3.5 \times 10^5.$$

3. **Carry out** We do the calculation. We shall use 3.14 for π.

$$\begin{aligned} V &= 3.14 \times (6 \times 10^{-3})^2 \times 3.5 \times 10^5 \\ &= 3.14 \times 6^2 \times 10^{-6} \times 3.5 \times 10^5 \\ &= (3.14 \times 6^2 \times 3.5) \times (10^{-6} \times 10^5) \\ &= 395.64 \times 10^{-1} \\ &\approx 3.96 \times 10^1, \quad \text{or} \quad 39.6 \end{aligned}$$

($\approx$ means "is approximately")

4. **Check** In this case, about all we can do is recheck the calculations and the translation.

5. **State** The answer is that the volume of wire is about 39.6 m^3 (cubic meters).

DO EXERCISE 12.

EXERCISE SET 1.6

i Convert to scientific notation.

1. 47,000,000,000
2. 2,600,000,000,000
3. 863,000,000,000,000,000
4. 957,000,000,000,000,000
5. 0.000000016
6. 0.000000263
7. 0.00000000007
8. 0.00000000009
9. 407,000,000,000
10. 3,090,000,000,000

Convert to decimal notation.

11. 4×10^{-4}
12. 5×10^{-5}
13. 6.73×10^{8}
14. 9.24×10^{7}
15. 8.923×10^{-10}
16. 7.034×10^{-2}
17. 9.03×10^{10}
18. 1.01×10^{12}

ii Multiply and write scientific notation for the answer.

19. $(2.3 \times 10^{6})(4.2 \times 10^{-11})$
20. $(6.5 \times 10^{3})(5.2 \times 10^{-8})$
21. $(2.34 \times 10^{-8})(5.7 \times 10^{-4})$
22. $(3.26 \times 10^{-6})(8.2 \times 10^{-6})$
23. $(3.2 \times 10^{6})(2.6 \times 10^{4})$
24. $(3.11 \times 10^{3})(1.01 \times 10^{13})$
25. $(3.01 \times 10^{-5})(6.5 \times 10^{7})$
26. $(4.08 \times 10^{-10})(7.7 \times 10^{5})$

Divide and write scientific notation for the answer.

27. $\dfrac{8.5 \times 10^{8}}{3.4 \times 10^{5}}$
28. $\dfrac{5.1 \times 10^{6}}{3.4 \times 10^{3}}$
29. $\dfrac{4.0 \times 10^{-6}}{8.0 \times 10^{-3}}$
30. $\dfrac{7.5 \times 10^{-9}}{2.5 \times 10^{-4}}$
31. $\dfrac{12.6 \times 10^{8}}{4.2 \times 10^{-3}}$
32. $\dfrac{3.2 \times 10^{-7}}{8.0 \times 10^{8}}$
33. $\dfrac{2.42 \times 10^{5}}{1.21 \times 10^{-5}}$
34. $\dfrac{9.36 \times 10^{-11}}{3.12 \times 10^{11}}$

iii Estimate.

35. $\dfrac{(6.1 \times 10^{4})(7.2 \times 10^{-6})}{9.8 \times 10^{-4}}$
36. $\dfrac{(8.05 \times 10^{-11})(5.9 \times 10^{7})}{3.1 \times 10^{14}}$
37. $\dfrac{780{,}000{,}000 \times 0.00071}{0.000005}$
38. $\dfrac{830{,}000{,}000 \times 0.12}{3{,}100{,}000}$
39. $\dfrac{43{,}000{,}000 \times 0.095}{63{,}000}$
40. $\dfrac{0.0073 \times 0.84}{0.000006}$
41. $\dfrac{40{,}000 \times 0.29}{0.057 \times 160{,}000}$
42. $\dfrac{80{,}000 \times 0.64}{0.032 \times 450{,}000}$

iv ***Problem solving***

43. A certain thin plastic sheet is used in many applications of building and landscaping. The sheet is packaged in rolls 1 m wide and 30 m long. The thickness of the sheet is 0.8 mm. Find the volume of plastic in a roll.

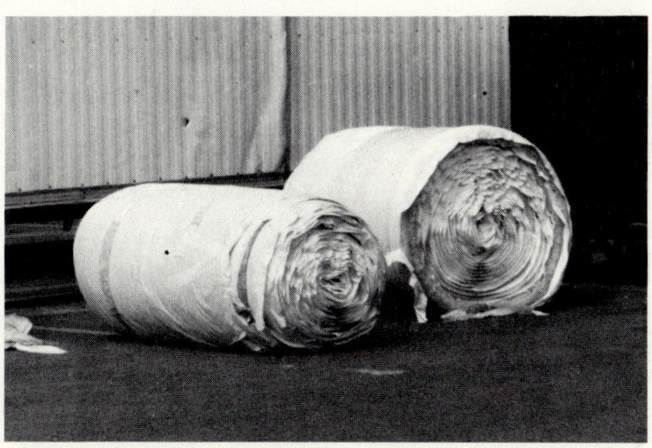

44. The distance that light travels in 100 years is approximately 5.8×10^{14} miles. How far does light travel in 13 weeks?

45. The average distance of the earth from the sun is about 9.3×10^{7} miles. About how far does the earth travel in a yearly orbit about the sun?

46. A *mil* is one thousandth of a dollar. The taxation rate in a certain school district is 5 mils for every dollar of assessed valuation. The assessed valuation for the district is 13.4 million dollars. How much tax revenue will be raised?

47. Compare $8 \cdot 10^{-90}$ and $9 \cdot 10^{-91}$. Which is the larger value? How much larger? Write scientific notation for the difference.

48. Find 8^{-x} when $8^{3x} = 27$.

49. Evaluate: $(4096)^{0.05}(4096)^{0.2}$.

50. What is the unit's digit in $(513)^{127}$?

51. Given that $a = 4 \cdot 3^b$ and $c = 2 - 3^{-b}$, find c in terms of a.

Simplify.

52. $[7y(7-8)^{-2} - 8y(8-7)^{-2}]^{-(-2)^2}$

53. $\dfrac{3^{q+3} - 3^2(3^q)}{3(3^{q+4})}$

54. $[3^{-(3s+1)} - 3^{-(3s-1)} + 3^{-3s}]^{-2^2}$

OBJECTIVES

After finishing Section 1.7, you should be able to:

i **Solve a formula for a specified letter.**

ii **Use formulas in solving problems.**

1. A formula for the area of a triangle is $A = \frac{1}{2}bh$. Solve for b.

1.7

FORMULAS AND PROBLEM SOLVING

You have already used formulas in solving problems. In fact, in the familiarization and translating steps, you often need to recall or look up a formula.

A formula is a kind of recipe, or rule, for doing a certain kind of calculation, and is often stated by an equation. For example, the formula

$$C = \pi d$$

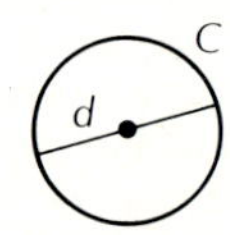

tells us that to find the circumference of a circle we should multiply the diameter d by the number π.

i Solving Formulas

Suppose that we know the circumference of a circle and wish to find its diameter. To do so, we can get d alone on one side of the equation. We "solve" for d. In solving problems it is often necessary to do that.

EXAMPLE 1 Solve the formula $C = \pi d$ for d.

$$C = \pi d \quad \text{We want this letter alone.}$$

$$\frac{1}{\pi} \cdot C = \frac{1}{\pi} \cdot \pi \cdot d \quad \text{Multiplying by } \frac{1}{\pi}$$

$$\frac{C}{\pi} = d$$

The new formula says that to find the diameter, we should divide the circumference by π.

When we solve a formula, we do the same things we would do to solve any equation. The idea is to get a certain letter alone on one side of the equals sign.

In most formulas there are several letters. Some of those letters are variables because they can represent various numbers. Other letters represent only one number. Such letters are called constants. We find that the letters are variables in some applications and constants in others. In the formula $C = \pi d$, the Greek letter π is always a constant.

DO EXERCISE 1.

EXAMPLE 2 Solve for b: $A = \frac{5}{2}(b - 20)$.

$$A = \tfrac{5}{2}(b - 20)$$

$$\tfrac{2}{5}A = b - 20 \qquad \text{Multiplying by } \tfrac{2}{5}$$

$$\tfrac{2}{5}A + 20 = b \qquad \text{Adding 20}$$

DO EXERCISE 2.

EXAMPLE 3 Solve for r: $H = 2r + 3m$.

$$H = 2r + 3m \qquad \text{We want this letter alone.}$$

$$H - 3m = 2r \qquad \text{Adding } -3m$$

$$\frac{H - 3m}{2} = r \qquad \text{Multiplying by } \tfrac{1}{2}$$

Compare Example 3 with an ordinary equation with numbers.

EXAMPLE 3A Solve for x: $11 = 2x + 3 \cdot 5$.

$$11 = 2x + 3 \cdot 5$$

$$11 - 3 \cdot 5 = 2x \qquad \text{Adding } -3 \cdot 5$$

$$\frac{11 - 3 \cdot 5}{2} = x \qquad \text{Multiplying by } \tfrac{1}{2}$$

In this case we can simplify further, whereas in Example 3 we could not.

DO EXERCISE 3.

EXAMPLE 4 The formula $A = P + Prt$ tells how much a principal P, in dollars, will be worth when invested at simple interest at a rate r in t years. Solve the formula for P.

$$A = P + Prt \qquad \text{We want this letter alone.}$$

$$A = P(1 + rt) \qquad \text{Factoring (or collecting like terms)}$$

$$A \cdot \frac{1}{1 + rt} = P(1 + rt) \cdot \frac{1}{1 + rt} \qquad \text{Multiplying by } \frac{1}{1 + rt}$$

$$\frac{A}{1 + rt} = P \qquad \text{Simplifying}$$

Compare Example 4 with an equation in which there are numbers.

EXAMPLE 4A Solve for x: $7 = x + 4 \cdot 5 \cdot x$.

$$7 = x(1 + 4 \cdot 5) \qquad \text{Factoring, or collecting like terms}$$

$$7 \cdot \frac{1}{1 + 4 \cdot 5} = x(1 + 4 \cdot 5) \cdot \frac{1}{1 + 4 \cdot 5} \qquad \text{Multiplying by } \frac{1}{1 + 4 \cdot 5}$$

$$\frac{7}{1 + 4 \cdot 5} = x \qquad \text{Simplifying}$$

2. Solve for c: $P = \frac{3}{5}(c + 10)$.

3. Solve for m: $H = 2r + 3m$.

4. Solve for Q: $T = Q + Qiy$.

In this case we can simplify further, to $\frac{7}{21} = \frac{1}{3} = x$, whereas in Example 4 we could not simplify further.

DO EXERCISE 4.

ii Formulas in Translating

The next example illustrates the use of formulas in translating problem situations to mathematical language and in carrying out the mathematical manipulations (steps 2 and 3 of the problem-solving process).

EXAMPLE 5 You deposit \$75 in an account that pays 8% simple interest. How long will it take for the money to earn \$3 interest?

2. Translate If necessary, look up the simple interest formula

$$I = Prt,$$

where I is the amount of interest, P is the principal, r is the interest rate, and t is the time, in years.

What we want to know is the time, so we solve the formula for t:

$$I = Prt$$

$$\frac{1}{Pr} \cdot I = \frac{1}{Pr} \cdot Prt \qquad \text{Multiplying by } \frac{1}{Pr}$$

$$\frac{I}{Pr} = t. \qquad \text{Simplifying}$$

We now have a formula that gives t. It says to divide I by the product Pr.

3. Carry out We now put the numbers in the formula and perform the indicated calculation:

$$t = \frac{I}{Pr}$$

$$= \frac{3}{75 \cdot 8\%} \qquad \text{Substituting}$$

$$= \frac{3}{75 \times 0.08} \qquad \text{Using the definition of \%}$$

$$= 0.5.$$

The possible solution to the problem (we still must check) is that 0.5 or $\frac{1}{2}$ year will be required.

5. The area of a trapezoid is 88 cm^2. The height of the figure is 8 cm and one of the bases has length 10. What is the length of the other base? (*Hint:* if necessary, look up the formula for the area of a trapezoid.)

DO EXERCISE 5.

EXERCISE SET 1.7

i Solve.

1. $A = lw$, for l (an area formula)
2. $A = lw$, for w
3. $W = EI$, for I (an electricity formula)
4. $W = EI$, for E
5. $F = ma$, for m (a physics formula)
6. $F = ma$, for a
7. $I = Prt$, for P (an interest formula)
8. $I = Prt$, for r

9. $E = mc^2$, for m (a relativity formula)

10. $E = mc^2$, for c^2

11. $P = 2l + 2w$, for l (a perimeter formula)

12. $P = 2l + 2w$, for w

13. $c^2 = a^2 + b^2$, for a^2 (a geometry formula)

14. $c^2 = a^2 + b^2$, for b^2

15. $A = \pi r^2$, for r^2 (a circle formula)

16. $A = \pi r^2$, for π

17. $W = \frac{11}{2}(h - 40)$, for h

18. $C = \frac{5}{9}(F - 32)$, for F (a temperature formula)

19. $V = \frac{4}{3}\pi r^3$, for r^3 (a volume formula)

20. $V = \frac{4}{3}\pi r^3$, for π

21. $A = \frac{1}{2}h(a + b)$, for h (an area formula)

22. $A = \frac{1}{2}h(a + b)$, for b

23. $F = \dfrac{mv^2}{r}$, for m (a physics formula)

24. $F = \dfrac{mv^2}{r}$, for v^2

ii ***Problem solving***

25. You are going to buy certificates of deposit at a bank. They pay 7% simple interest. You want your money to earn \$110 in a year. How much will you have to invest?

26. You have \$250 to invest for 6 months, and you expect it to earn at least \$8 in that time. What rate of simple interest will your money have to earn?

27. A garden is being constructed in the shape of a trapezoid.

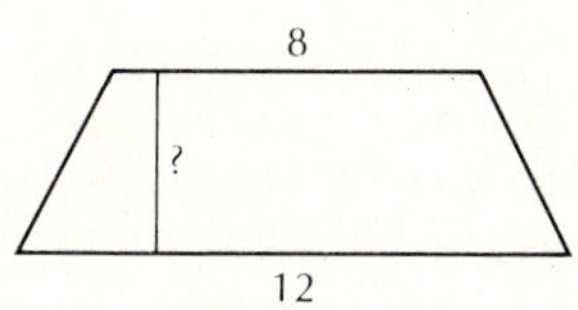

The dimensions are as shown in the figure. The unknown dimension is to be such that the area of the garden is 90 ft^2. Find that unknown dimension. (You may have to look up a formula for the area of a trapezoid.)

28. A rectangular garden is being constructed. There is 76 ft of fencing available, so the perimeter must be 76 ft at most. The width of the garden is to be 13 ft. What should the length be, in order to use just 76 ft of fence? (See the formula in Exercise 11.)

Solve.

29. $s = v_it + \frac{1}{2}at^2$, for a

30. $A = \pi rs + \pi r^2$, for s

31. $\dfrac{P_1V_1}{T_1} = \dfrac{P_2V_2}{T_2}$, for V_1

32. $\dfrac{P_1V_1}{T_1} = \dfrac{P_2V_2}{T_2}$, for T_2

33. In Exercise 7, you solved the formula $I = Prt$ for P. Now use it to find how much principal would be needed in order to earn \$6 in two thirds of a year at 12% simple interest.

34. In Exercise 8, you solved the formula $I = Prt$ for r. Now use it to find what rate of interest would be required for a principal of \$120 to earn \$6.60 in half a year.

35. In Exercise 13 you solved for a^2. How might you solve for a?

36. In Exercise 19 you solved for r^3. How might you solve for r?

SUMMARY AND REVIEW: CHAPTER 1

The following contains a summary of what you should be able to do after completing this chapter. The review exercises are for practice. Answers are at the back of the book. If you miss an exercise, restudy the section indicated alongside the answer.

You should be able to:

Perform the indicated operations.

Add.

1. $-14 + (-12)$

2. $-6.2 + 8.4$

3. $2.93 + (-4.56)$

Subtract.

4. $36.2 - 73.4$

5. $-12.4 - 15.1$

6. $-\dfrac{4}{5} - \left(-\dfrac{1}{3}\right)$

Multiply.

7. $-4.1(2.7)$

8. $4.5(-12)$

9. $-\dfrac{3}{5}\left(-\dfrac{7}{10}\right)$

Divide.

10. $\dfrac{-16.8}{-2.1}$

11. $\dfrac{-7.2}{-12}$

12. $\dfrac{4}{9} \div \left(-\dfrac{5}{18}\right)$

Multiply by a quantity in parentheses.

Multiply.

13. $-7(4a + 3b)$

14. $3x(-2y - 5z)$

Simplify an algebraic expression by collecting like terms.

Collect like terms.

15. $6x - 12x + 9x$

16. $41x + 17y - 10x - 23y$

Solve equations.

Solve.

17. $12x + 15 = 9x - 39$

18. $6x - (4x - 7) = 35$

Translate problems to mathematical language.

Translate and carry out the first two steps.

19. One angle of a triangle is three times as great as a second angle. The third angle is 9° less than five times the second angle. Find the measures of the angles.

20. A student's scores on four tests are 85, 91, 75, and 83. What must the score be on the fifth test so that the average will be 84?

Solve problems using the five problem-solving steps.

Solve.

21. The number 47.9 is less than some number by 14.5. What is the number?

22. One number is three times another number. The sum of the numbers is 164. What are the numbers?

23. Find three consecutive even integers such that the sum of twice the first, three times the second, and four times the third is 220.

Use the properties of numbers and algebraic symbolism to simplify algebraic expressions.

Simplify.

24. $5n - (2n + 1)$

25. $4y - 7(9y - 3)$

26. $-7(x + 2) - 4(x + 3)$

27. $7c - [4 - 3(6c - 9)]$

Use the properties of exponents to multiply, divide, and simplify algebraic expressions.

Multiply and simplify.

28. $x^7 \cdot x^{-2} \cdot x^{-5}$

29. $(7x^3y^{-6})(-5x^{-2}y^{-4})$

30. $(7x^{a+1}y^{b+1})(-3x^{a+2}y^{b+3})$

Divide and simplify.

31. $\dfrac{y^5}{y^{-6}}$

32. $\dfrac{-25a^{12}b^{-7}}{-35a^{-4}b^{-3}}$

33. $\dfrac{8b^{2x-1}}{-2b^{x+4}}$

Simplify.

34. $(-8a^3b^{-6})^{-2}$

35. $\left(\dfrac{4x^4y^{-5}}{3x^{-3}}\right)^3$

36. $(5x^{2a}y^{b-1})^{4c}$

Use scientific notation to multiply and divide large and small numbers.

Multiply.

37. $(4.5 \times 10^3)(2.4 \times 10^{-6})$

38. $(2.18 \times 10^8)(1.21 \times 10^{-10})$

Divide.

39. $\dfrac{2.4 \times 10^{-5}}{4.8 \times 10^6}$

40. $\dfrac{2.8 \times 10^{-3}}{7.0 \times 10^{-5}}$

Estimate.

41. $\dfrac{(2.03 \times 10^{-9})(5.5 \times 10^4)}{2.5 \times 10^{12}}$

42. $\dfrac{21{,}000{,}000 \times 0.027}{44{,}000}$

Solve problems using scientific notation.

Solve.

43. The average distance from the planet Jupiter to the sun is 4.8×10^8 miles. About how far does Jupiter travel in one orbit around the sun?

Solve a formula for a specified letter.

Solve.

44. $C = 2\pi r$, for π

45. $\frac{P_1V_1}{T_1} = \frac{P_2V_2}{T_2}$, for V_2

46. $A = \frac{s^2\sqrt{3}}{4}$, for s (area of an equilateral triangle)

Simplify.

47. $(3^{-1} + 2^{-1})^{-2}$

Solve.

48. $\frac{1}{z} = \frac{1}{x} + \frac{1}{y}$, for x

49. Solve the equation $a(y + b) = by + c$ for y.

TEST: CHAPTER 1

1. A student's scores on five tests are 94, 80, 76, 91, and 75. What must the score be on the sixth test so that the average will be 85?

2. One number is 27.4 less than another number. The sum of the numbers is 83.6. What are the numbers?

Add.

3. $-25 + (-16)$

4. $-10.5 + 6.8$

5. $6.21 + (-8.32)$

Subtract.

6. $29.5 - 43.7$

7. $-17.8 - 25.4$

8. $-\frac{7}{3} - \left(-\frac{3}{4}\right)$

Multiply.

9. $-6.4(5.3)$

10. $8.4(-10.1)$

11. $-\frac{2}{7}\left(-\frac{5}{14}\right)$

Divide.

12. $\frac{-42.6}{-7.1}$

13. $\frac{-9.1}{-13}$

14. $\frac{2}{5} \div \left(-\frac{3}{10}\right)$

Collect like terms.

15. $4y - 15y + 19y$

16. $35a + 16b - 24a - 43b$

Multiply.

17. $-11(3c - 4d)$

18. $4x(-6y - 7z)$

Solve.

19. $13x - 7 = 41x + 49$

20. $9y - (5y - 3) = 33$

21. Find three consecutive odd integers such that the sum of four times the first, three times the second, and two times the third is 167.

Simplify.

22. $-5(x - 4) - 3(x + 7)$

23. $6b - [7 - 2(9b - 1)]$

24. $(12x^{-4}y^{-7})(-6x^{-6}y)$

25. $\frac{-27a^{x+1}}{3a^{x-2}}$

26. $(-6x^2y^{-4})^{-2}$

27. $\left(\frac{2x^3y^{-6}}{-4y^{-2}}\right)^2$

28. $(4x^{3a}y^{b+1})^{2c}$

Multiply.

29. $(2.9 \times 10^8)(6.1 \times 10^{-4})$

30. $(9.05 \times 10^{-3})(2.22 \times 10^{-5})$

Divide.

31. $\frac{3.6 \times 10^7}{7.2 \times 10^{-3}}$

32. $\frac{1.8 \times 10^{-4}}{4.8 \times 10^{-7}}$

33. Estimate: $\frac{43{,}000{,}000 \times 0.064}{27{,}000}$.

Solve.

34. The average distance from the planet Venus to the sun is 6.7×10^7 miles. About how far does Venus travel in one orbit around the sun?

35. $A = \frac{1}{2}h(b_1 + b_2)$, for b_1 (area of a trapezoid)

---

36. Simplify: $\frac{(-16x^{x-1}y^{y-2})(2x^{x+1}y^{y+1})}{(-7x^{x+2}y^{y+2})(8x^{x-2}y^{y-1})}$.

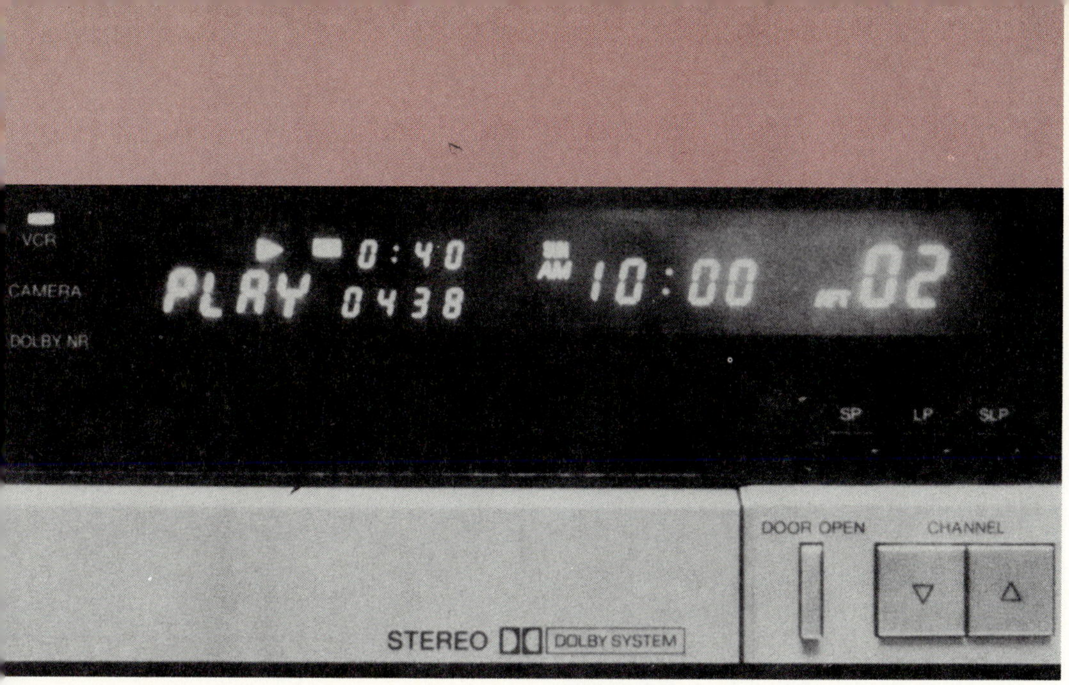

What will the counter read at the end of $1\frac{1}{2}$ hr? Functions and algebra can be used to solve such problems.

2

GRAPHS AND FUNCTIONS

Graphs are useful because they allow us to see relationships pictorially. For example, a graph of an equation in two variables helps us to see how those variables are related. Graphs are useful in problem solving, and in other ways as well. In this chapter you will take a good look at graphs of equations.

A certain kind of relationship between two variables is known as a *function*. That kind of relation is highly important in mathematics in general, and in problem solving in particular. You will learn what we mean by a function and then begin to use functions in solving problems.

OBJECTIVES

After finishing Section 2.1, you should be able to:

i **Plot points on a plane, given their coordinates.**

ii **Determine whether an ordered pair of numbers is a solution of an equation with two variables.**

iii **Graph equations of the type $y = mx$.**

iv **Graph equations of the type $y = x^2 + c$.**

Plot these points on the graph below.

1. (6, 4)
2. (4, 6)
3. (−3, 5)
4. (−4, −3)
5. (4, −2)
6. (0, 3)
7. (3, 0)
8. (0, −4)
9. (−4, 0)

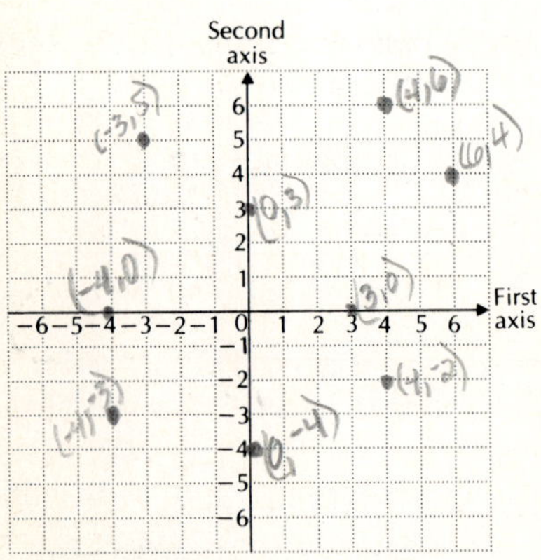

2.1

GRAPHS

For some problems, when we translate to mathematical language we may draw a graph. A graph can show in a pictorial way some information that allows us to find an answer to a problem. In this section we begin to learn how to construct graphs.

Points and Ordered Pairs

On a number line each point is the graph of a number. On a plane each point is the graph of a number pair. We use two perpendicular number lines, called *axes.* The point at which they cross is called the *origin* and is labeled 0. The arrows show the positive directions.

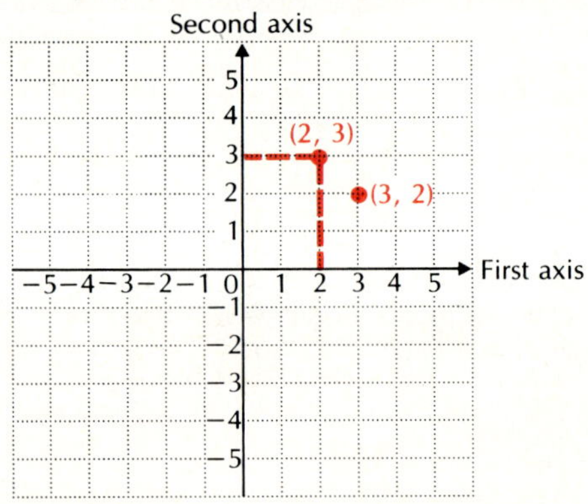

i **Plotting Points.** Note that (2, 3) and (3, 2) give different points. These are called *ordered pairs* of numbers since it makes a difference which number comes first.

EXAMPLE 1 Plot the point (−4, 3).

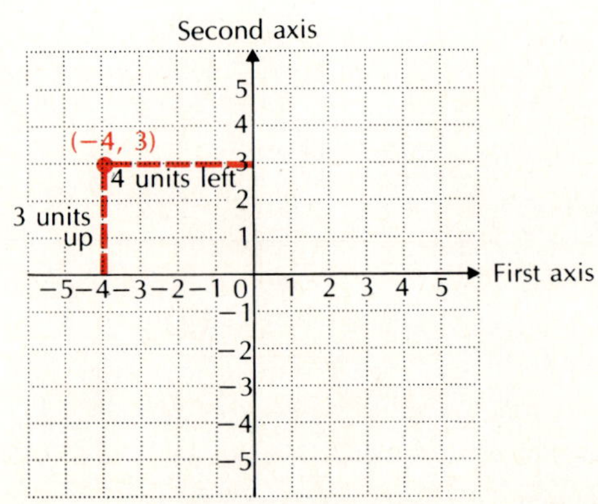

The first number, −4, tells us the distance in the first, or horizontal, direction. We move 4 units *left.* The second number tells us the distance in the second, or vertical, direction. We move 3 units *up.*

DO EXERCISES 1–9.

The numbers in an ordered pair are called *coordinates.* In $(-4, 3)$, the *first coordinate* is -4 and the *second coordinate** is 3.

Quadrants. The axes divide the plane into four regions, as shown, called *quadrants.* In region I (the *first* quadrant) both coordinates of a point are positive. In region II (the *second* quadrant) the first coordinate is negative and the second coordinate is positive.

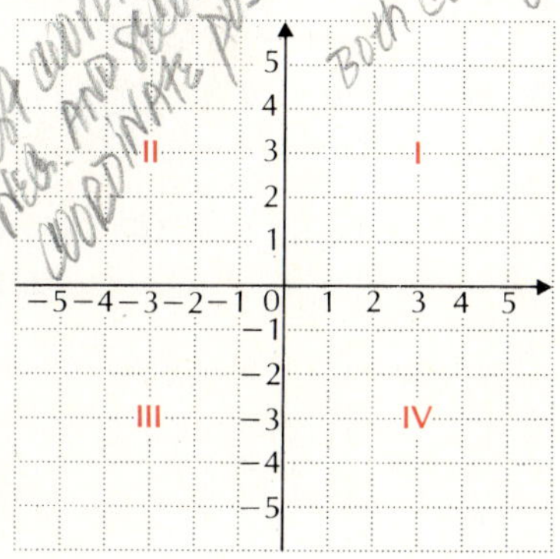

DO EXERCISES 10 AND 11.

10. What can you say about the coordinates of a point in the third quadrant?

11. What can you say about the coordinates of a point in the fourth quadrant?

ii Solutions of Equations

If an equation has two variables, its solutions are pairs of numbers. We usually take the variables in alphabetical order, hence we get *ordered* pairs of numbers for solutions.

EXAMPLE 2 Determine whether the following ordered pairs are solutions of the equation $y = 3x - 1$: $(-1, -4)$, $(7, 5)$.

y	$= 3x - 1$
-4	$3(-1) - 1$
	$-3 - 1$
	-4

We substitute -1 for x and -4 for y (alphabetical order of variables).

The equation becomes true: $(-1, -4)$ is a solution.

y	$= 3x - 1$
5	$3 \cdot 7 - 1$
	$21 - 1$
	20

We substitute.

The equation becomes false: $(7, 5)$ is not a solution.

DO EXERCISES 12–14.

12. Determine whether $(9, -2)$ is a solution of the equation $y = 4x + 17$.

13. Determine whether $(0, 23)$ is a solution of $y = 4x + 17$.

14. Determine whether $(2, -2)$ is a solution of $2y + 6x = 8$.

* The first coordinate is sometimes called the *abscissa* and the second coordinate is called the *ordinate.*

iii Graphing Equations of the Type $y = mx$

To *graph* an equation means to make a drawing of its solutions, that is, its *solution set.* If an equation has a graph that is a line, we can graph it by plotting a few points and then drawing a line through them.

EXAMPLE 3 Graph the equation $y = x$.

We will use alphabetical order. Thus the first axis is the x-axis and the second axis is the y-axis.

Next, we find some ordered pairs that are solutions of the equation. In this case, it is easy. Here are a few:

$$(0, 0), \quad (1, 1), \quad (5, 5), \quad (-1, -1), \quad (-6, -6).$$

Now we plot these points. We can see that if we were to plot a million solutions, the dots that we draw would merge into a solid line. We see the pattern, so we can draw the line with a ruler. The line is the graph of the equation $y = x$. We label the line $y = x$ on the graph paper.

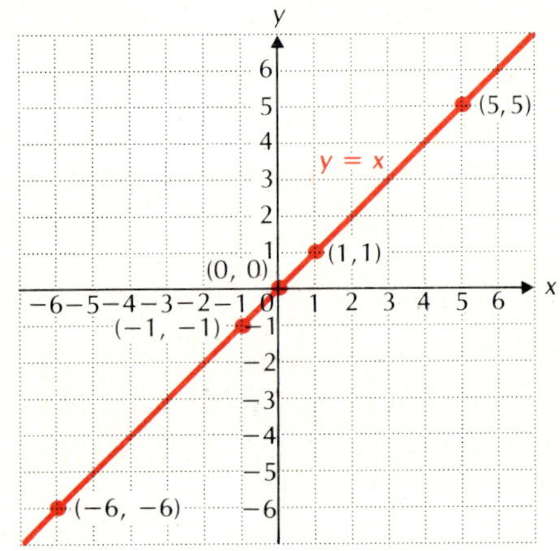

EXAMPLE 4 Graph the equation $y = 2x$.

We find some ordered pairs that are solutions. This time we shall keep them in a table. To find an ordered pair, we can choose *any* number for x and then determine y. For example, if we choose 3 for x, then $y = 2 \cdot 3$ (substituting into the equation of the line), or 6. We make some negative choices for x, as well as positive ones. If a number takes us off the graph paper, we generally do not use it.

x	y
0	0
1	2
3	6
−2	−4
−3	−6

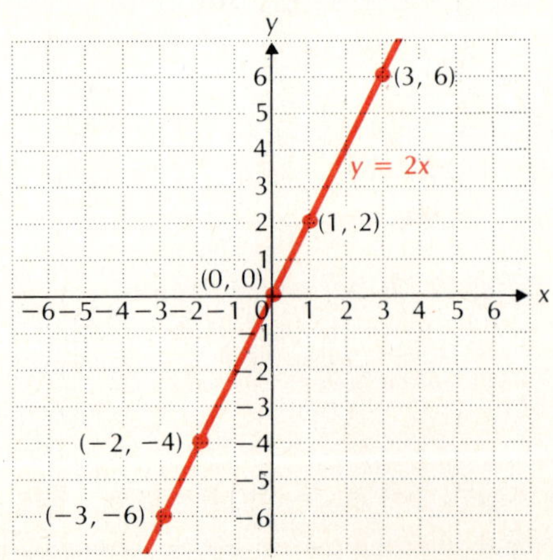

Next, we plot these points. If we had enough of them they would make a solid line. We can draw the line with a ruler, and we label it $y = 2x$.

EXAMPLE 5 Graph the equation $y = -5x$.

To find an ordered pair, we choose any number for x and then determine y. For example, if we choose 3 for x, we get $y = -5(3)$, or -15.

When we choose -2 for x, we get $y = -5(-2)$, or 10. We find several ordered pairs, plot them, and draw the line.

x	y
3	−15
−2	10
0	0
1	−5
−1	5

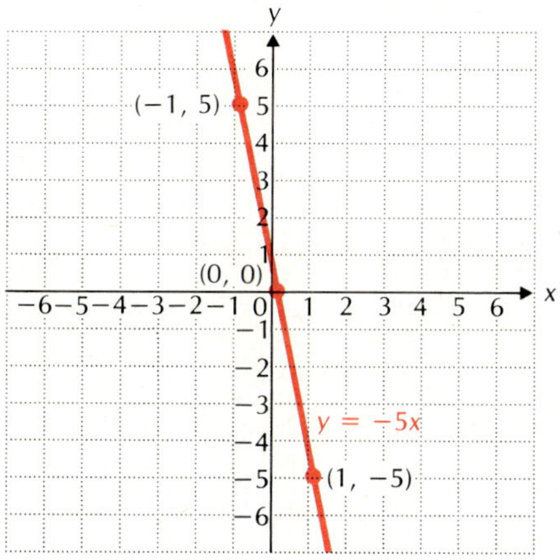

DO EXERCISES 15–17.

iv Graphing Equations of the Type $y = x^2 + c$

There are many equations whose graphs are not straight lines. Let's look at an equation having another type of graph.

EXAMPLE 6 Graph $y = x^2 - 5$.

We select numbers for x and find the corresponding values for y. The table gives us the ordered pairs $(0, -5)$, $(-1, -4)$, and so on.

x	y
0	−5
−1	−4
1	−4
−2	−1
2	−1
−3	4
3	4

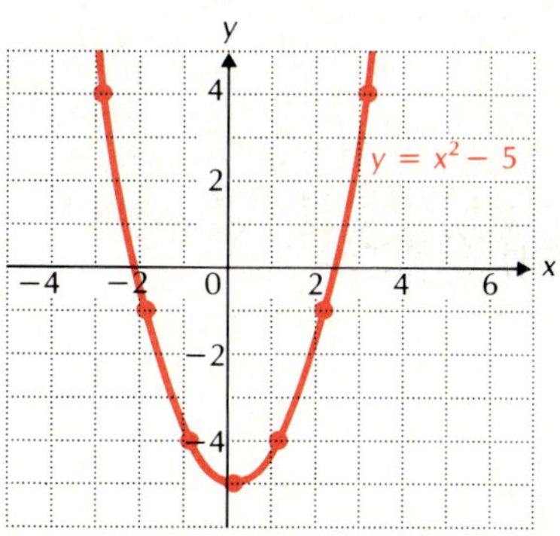

Next we plot these points. We note that as the absolute value of x increases, $x^2 - 5$ also increases. Thus the graph is a curve that rises gradually on either side of the y-axis, as shown in the figure.

You can always use a calculator to find as many values as desired. This can be especially helpful when you are uncertain about the shape of a graph.

15. Graph the equation $y = \frac{1}{2}x$.

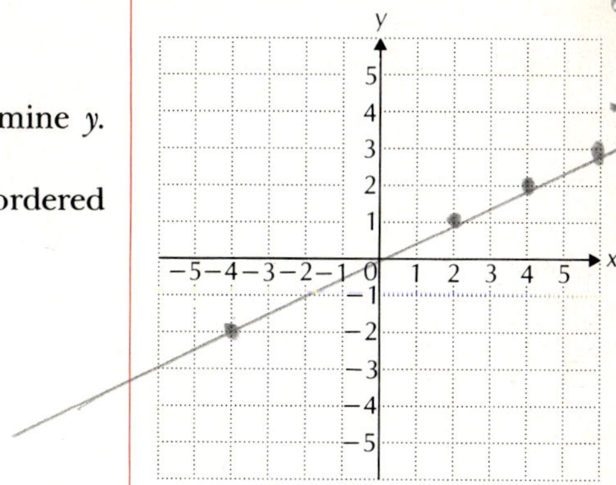

16. Graph the equation $y = 3x$.

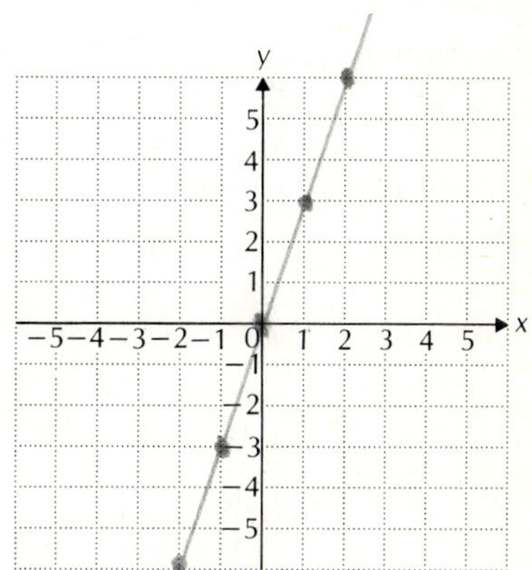

17. Graph the equation $y = -2x$.

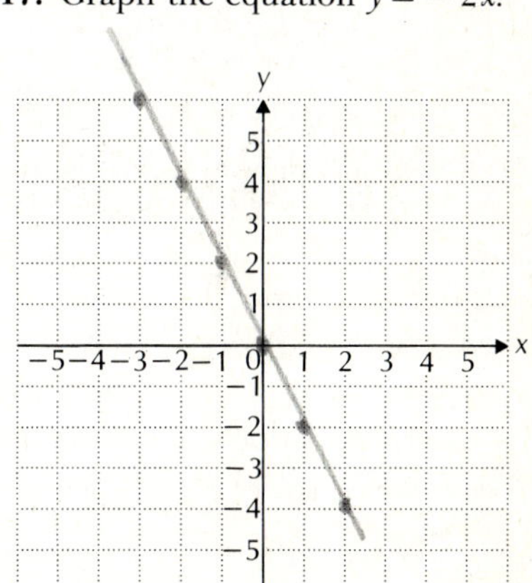

EXERCISE SET 2.1

i Plot the following points.

1. $A(5, 3)$, $B(2, 4)$, $C(0, 2)$, $D(0, -6)$, $E(3, 0)$, $F(-2, 0)$, $G(1, -3)$, $H(-5, 3)$, $J(-4, 4)$

2. $A(3, 5)$, $B(1, 5)$, $C(0, 4)$, $D(0, -4)$, $E(5, 0)$, $F(-5, 0)$, $G(1, -5)$, $H(-7, 4)$, $J(-5, 5)$

3. $A(3, 0)$, $B(4, 2)$, $C(5, 4)$, $D(6, 6)$, $E(3, -4)$, $F(3, -3)$, $G(3, -2)$, $H(3, -1)$

4. $A(1, 1)$, $B(2, 3)$, $C(3, 5)$, $D(4, 7)$, $E(-2, 1)$, $F(-2, 2)$, $G(-2, 3)$, $H(-2, 4)$, $J(-2, 5)$, $K(-2, 6)$

5. Plot the points $M(2, 3)$, $N(5, -3)$, $P(-2, -3)$. Draw $\overline{MN}$, $\overline{NP}$, and $\overline{MP}$. ($\overline{MN}$ means the line segment from M to N). What kind of geometric figure is formed? What is its area?

6. Plot the points $Q(-4, 3)$, $R(5, 3)$, $S(2, -1)$, and $T(-7, -1)$. Draw $\overline{QR}$, $\overline{RS}$, $\overline{ST}$, and $\overline{TQ}$. What kind of figure is formed? What is its area?

ii Determine whether the ordered pair is a solution of the indicated equation.

7. $(1, -1)$; $y = 2x - 3$

8. $(2, 5)$; $y = 3x - 1$

9. $(3, 4)$; $3s + t = 4$

10. $(2, 3)$; $2p + q = 5$

11. $(3, 5)$; $4x - y = 7$

12. $(2, 7)$; $5x - y = 3$

13. $(0, \frac{3}{5})$; $2a + 5b = 3$

14. $(0, \frac{3}{2})$; $3f + 4g = 6$

15. $(2, -1)$; $4r + 3s = 5$

16. $(2, -4)$; $5w + 2z = 2$

17. $(3, 2)$; $3x - 2y = -4$

18. $(1, 2)$; $2x - 5y = -6$

iii, **iv** Graph each of the following.

19. $y = -2x$

20. $y = -\frac{1}{2}x$

21. $y = 3x$

22. $y = 4x$

23. $y = -\frac{3}{2}x$

24. $y = -\frac{3}{4}x$

25. $y = \frac{1}{5}x$

26. $y = x^2$

27. $y = -x^2$

28. $y = x^2 + 2$

29. $y = x^2 - 2$

30. $x = y^2 + 2$

Problem-solving practice

31. A garden is being constructed in the shape of a right triangle. One side is 12 ft long. How tall should the triangle be in order to make the area of the garden 156 ft^2?

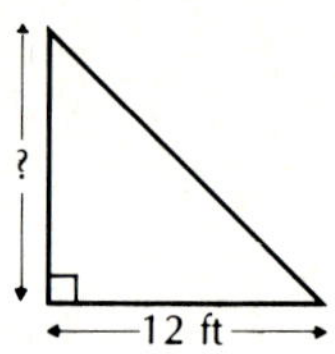

32. What rate of interest is required in order that a principal of \$320 earn \$17.60 in half a year?

---

33. If $(-10, -2)$, $(-3, 4)$, and $(6, 4)$ are the coordinates of three of the four vertices of a parallelogram, what are the coordinates of the fourth vertex?

34. Graph $y = 6x$, $y = 3x$, $y = \frac{1}{2}x$, $y = -6x$, $y = -3x$, and $y = -\frac{1}{2}x$ using the same set of axes, and compare the slants of the lines. What does the number in front of the x tell you about the slant of the line?

35. Which of the following equations have $(-\frac{1}{3}, \frac{1}{4})$ as a solution?

a) $-\frac{3}{2}x - 3y = -\frac{1}{4}$

b) $8y - 15x = \frac{7}{2}$

c) $0.16y = -0.09x + 0.1$

d) $2(-y + 2) - \frac{1}{4}(3x - 1) = 4$

36. Use a calculator. Find and graph at least 20 points for the following equation:

$$y = \frac{1}{3}x^3 - x + \frac{2}{3}.$$

37. Graph $y = 2x$, $y = 2x - 3$, and $y = 2x + 3$ using the same set of axes. What does the sign of the number added to $2x$ tell you about the location of the line with respect to the graph of $y = 2x$?

38. If $(-1, 1)$ and $(4, -4)$ are the endpoints of a diagonal of a square, what are the coordinates of the other two vertices of the square?

39. Plot the following points: $A(-5, 2)$, $B(-3, 2)$, $C(-2, \frac{1}{2})$, $D(-1, 2)$, $E(1, 2)$, $F(-1, -1)$, $G(1, -4)$, $H(-1, -4)$, $I(-2, -2\frac{1}{2})$, $J(-3, -4)$, $K(-5, -4)$, $L(-3, -1)$. Draw $\overline{AB}$, $\overline{BC}$, $\overline{CD}$, $\overline{DE}$, $\overline{EF}$, $\overline{FG}$, $\overline{GH}$, $\overline{HI}$, $\overline{IJ}$, $\overline{JK}$, $\overline{KL}$, and $\overline{LA}$, in order to find an "unknown."

2.2

GRAPHS OF LINEAR EQUATIONS

Graphs can have different kinds of shapes. For now we are interested in those equations whose graphs are straight lines. Such graphs, and their equations, are called *linear.*

i Graphs of Equations of the Type $y = mx + b$

We know that the graph of any equation $y = mx$ is a straight line through the origin. What will happen if we add a number b on the right-hand side to get an equation $y = mx + b$?

EXAMPLE 1 Graph $y = x + 2$ and compare it with the graph of $y = x$.

We first make a table of values.

x	y (or $x + 2$)
0	2
1	3
5	7
−1	1
−2	0
−7	−5

Then we graph and compare. The graph of $y = x + 2$ is a line. It looks just like the graph of $y = x$, but it is moved up 2 units.

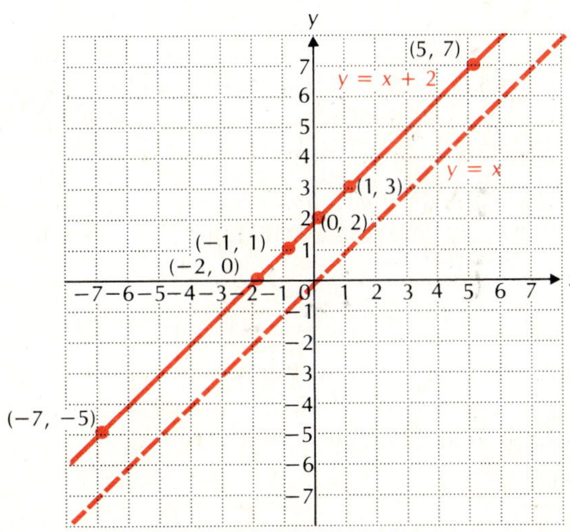

DO EXERCISES 1 AND 2.

OBJECTIVES

After finishing Section 2.2, you should be able to:

i Given an equation $y = mx + b$, determine the slope and y-intercept and graph the equation.

ii Graph linear equations with a missing variable.

iii Determine which equations are linear.

1. Graph $y = x + 3$ and compare with the graph of $y = x$.

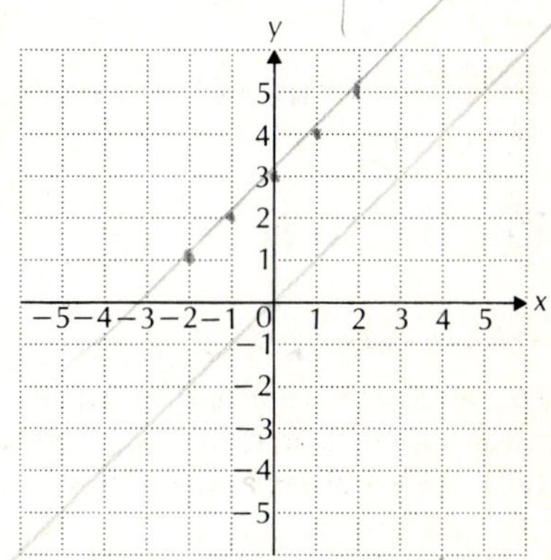

2. Graph $y = x - 1$ and compare with the graph of $y = x$.

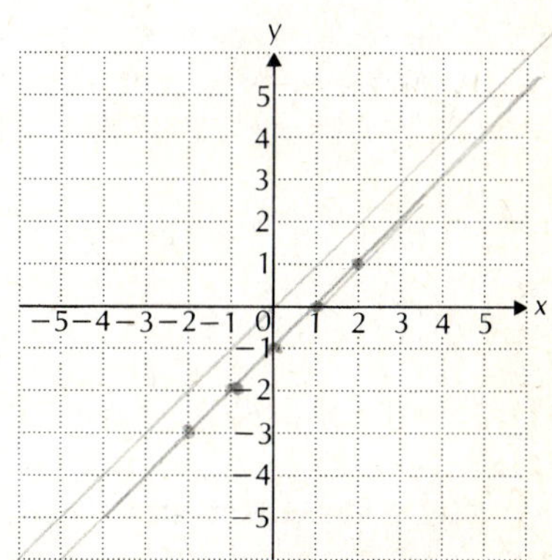

3. Graph $y = 2x + 1$ and compare with the graph of $y = 2x$.

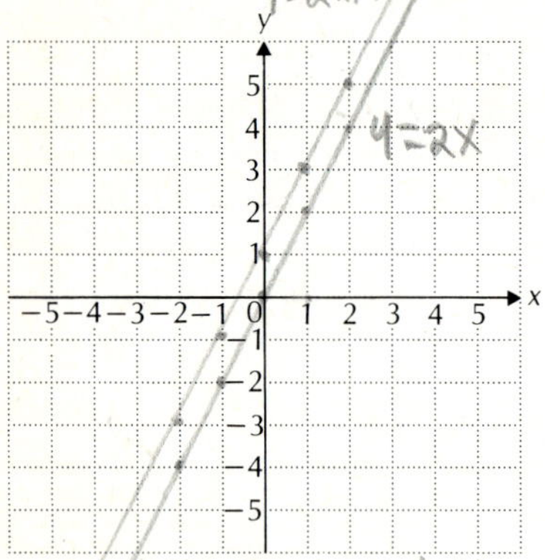

4. Graph $y = 2x - 4$ and compare with the graph of $y = 2x$.

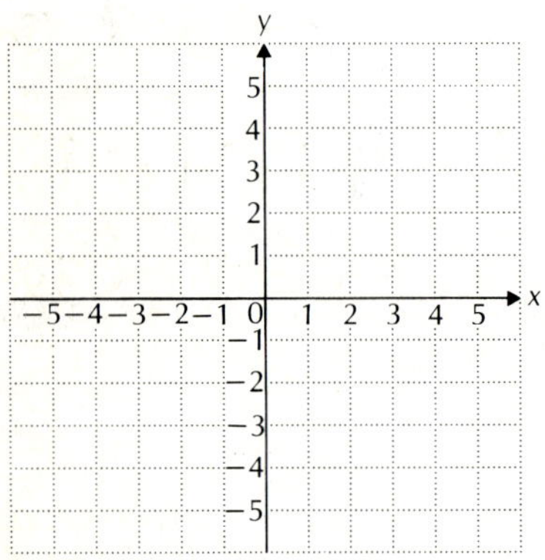

Find the y-intercept.

5. $y = -5.4x + 7$

6. $y = 12.5x - 17$

EXAMPLE 2 Graph $y = 2x - 3$ and compare it with the graph of $y = 2x$.

We first make a table of values. Then we graph and compare.

x	y (or $2x - 3$)
0	-3
1	-1
3	3
-2	-7
2	1

The graph of $y = 2x - 3$ is a line. It looks just like the graph of $y = 2x$, but it is moved down 3 units.

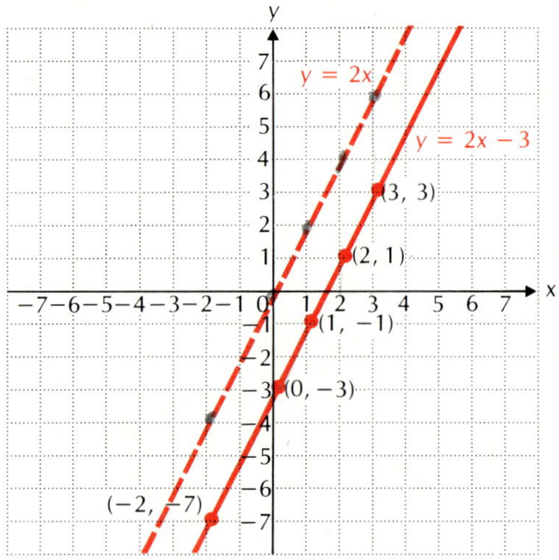

DO EXERCISES 3 AND 4.

Intercepts. The graph of an equation $y = mx$ is a line through the origin. Any equation $y = mx + b$ also has a graph that is a line. It is parallel to $y = mx$ but moved up or down. The graph of $y = mx + b$ goes through the point $(0, b)$. This point is called the *y-intercept*. We also refer to the number b as the y-intercept.

EXAMPLE 3 Find the y-intercept: $y = -5x + 4$.

The y-intercept is $(0, 4)$, or simply 4.

EXAMPLE 4 Find the y-intercept: $y = 5.3x - 12$.

The y-intercept is $(0, -12)$, or simply -12.

DO EXERCISES 5 AND 6.

We now look at the number m in the equation $y = mx + b$. The number m is called the *slope*, and it shows how the line slants. The greater the slope, the steeper the line slants, from left to right. If the slope is negative, the line slants downward from left to right.

EXAMPLE 5 Determine the slope and y-intercept for $y = \frac{2}{3}x + 4$ and draw its graph.

The *y-intercept* is the point (0, 4), so we know one point on the graph. The *slope* is the number $\frac{2}{3}$. It tells us that from *any* point of the graph, the line slants *up* 2 units for every 3 *over* to the right. We start at the y-intercept. We go *up* 2 units and *over* 3. This gives us the point (3, 6) and we can now draw the graph.

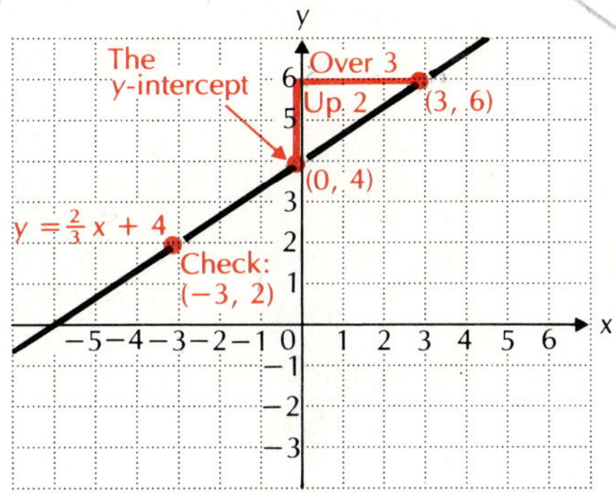

Important: As a check we choose some other value for x, say -3, and determine y (in this case, 2). We plot that point and see if it is on the line. If it is not, there has been some error.

DO EXERCISES 7 AND 8.

In Example 5 the slope is $\frac{2}{3}$. We can find some other names for $\frac{2}{3}$ by multiplying by 1 and then find other points of the graph.

$$\frac{2}{3} = \frac{2}{3} \cdot \frac{2}{2} = \frac{4}{6}$$ Thus from any point of the graph we can find another point by going *up* 4 units and *over* 6.

$$\frac{2}{3} = \frac{2}{3} \cdot \frac{3}{3} = \frac{6}{9}$$ Thus we can find a point by going *up* 6 and *over* 9, from any point of the graph.

When numbers are negative, we reverse directions.

$$\frac{2}{3} = \frac{2}{3} \cdot \frac{-1}{-1} = \frac{-2}{-3}$$ Thus from any point of the graph we can find another point by going *down* 2 units and then *to the left* 3 units.

If the slope of a line is negative, it slants downward from left to right.

EXAMPLE 6 Graph: $y = -\frac{1}{2}x + 5$.

The y-intercept is (0, 5). The slope is $-\frac{1}{2}$, or $\frac{-1}{2}$. From the y-intercept we go *down* 1 unit and to the right 2 units. That gives us the point (2, 4). We can now draw the graph.

Determine the slope and y-intercept and then graph. Be sure to find a third point as a check.

7. $y = \frac{3}{4}x - 2$

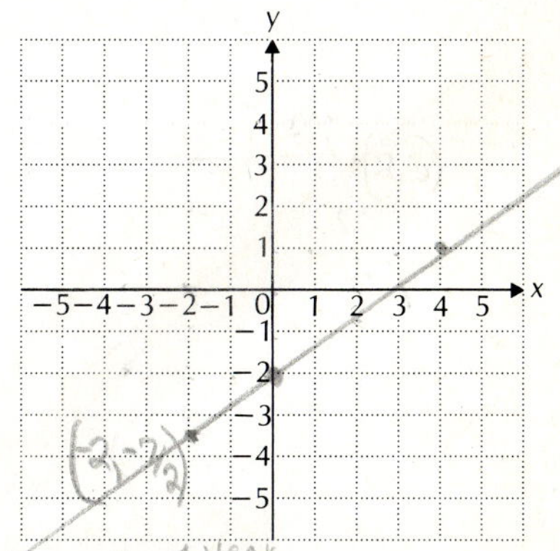

8. $y = \frac{3}{2}x + 1$

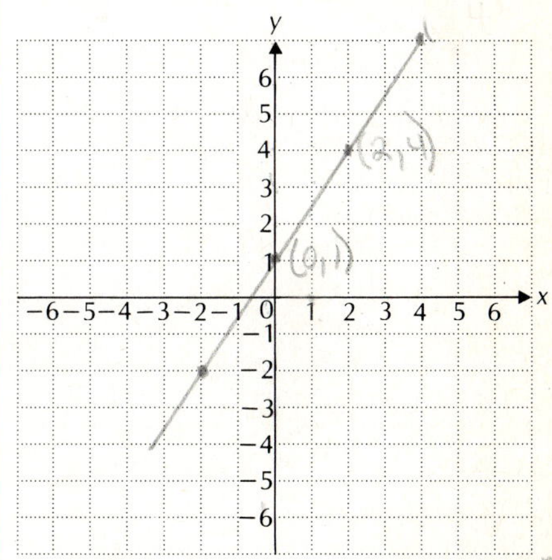

Determine the slope and y-intercept and then graph. Be sure to find a third point as a check.

9. $y = -\frac{3}{5}x + 5$

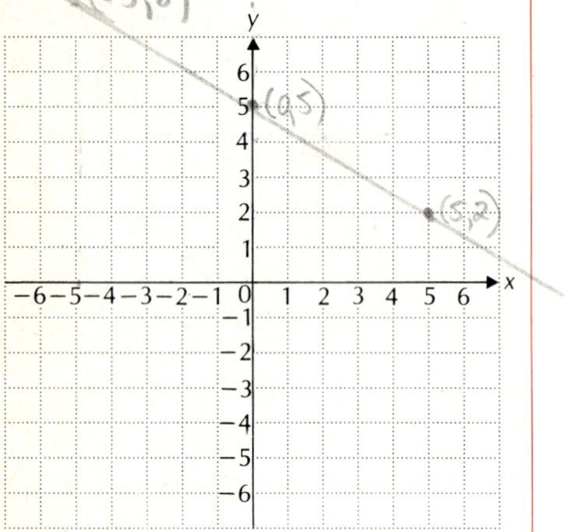

10. $y = -\frac{5}{3}x - 4$

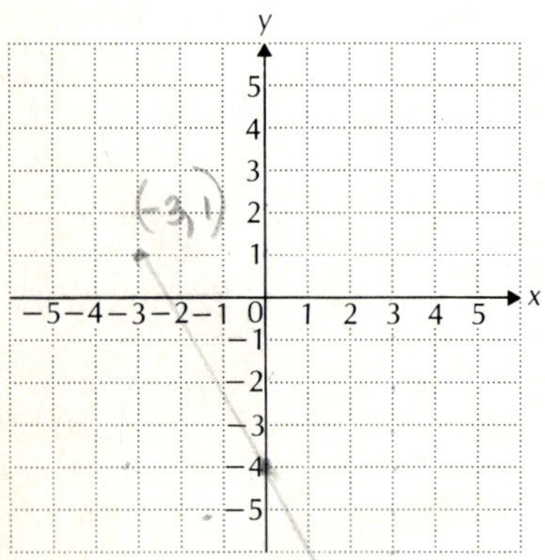

As a check, we rename the slope and find another point:

$$\frac{-1}{2} = \frac{-1}{2} \cdot \frac{-3}{-3} = \frac{3}{-6}.$$

Thus we can go *up* 3 units and then *to the left* 6 units. This gives the point $(-6, 8)$.

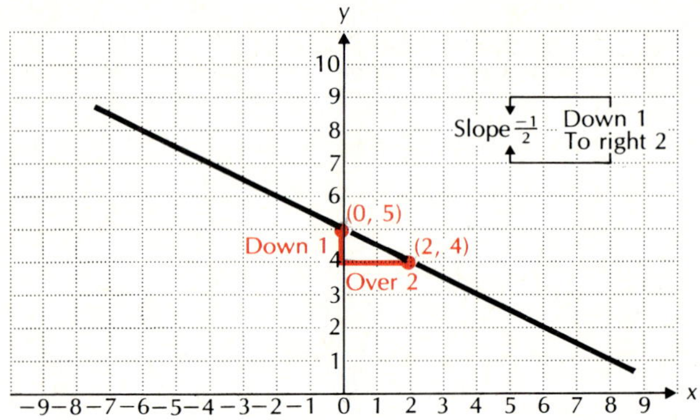

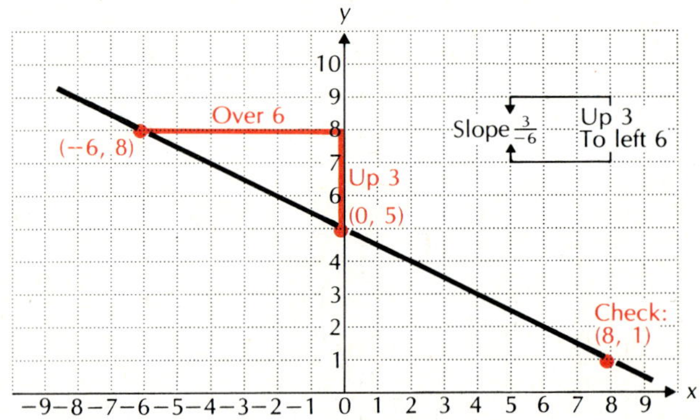

We can also check by choosing a number for x and finding y. If we choose 8 for x, we find that y is 1. Thus the point (8, 1) should be on the graph.

DO EXERCISES 9 AND 10.

All the equations we have been graphing so far have graphs that are straight lines. Therefore, they are called *linear* equations. We know that any equation $y = mx + b$ is linear, but there are other equations that have straight lines for graphs.

The *slope-intercept* equation of a line is $y = mx + b$.
Any equation $y = mx + b$ has a graph that is a straight line.
It goes through the y-intercept $(0, b)$ and has slope m.

ii Graphs That Are Parallel to an Axis

Some equations have graphs that are parallel to one of the axes. These equations have a missing variable. In the following example, x is missing.

EXAMPLE 7 Graph: $y = 3$.

Since x is missing, any number for x will do. Thus all ordered pairs $(x, 3)$ are solutions. The graph is a line parallel to the x-axis.

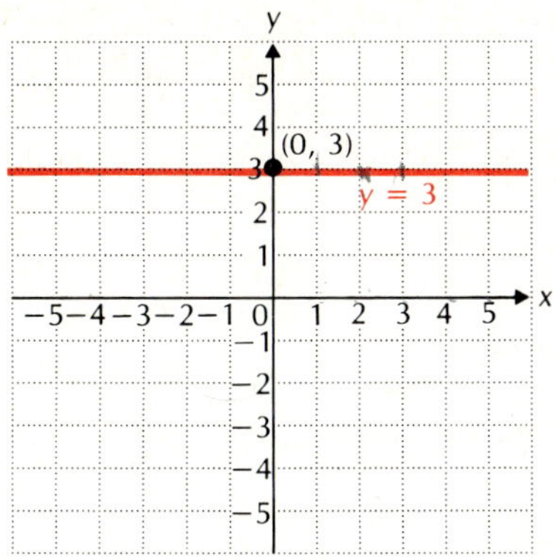

DO EXERCISES 11 AND 12.

In the following example, y is missing and the graph turns out to be parallel to the y-axis.

EXAMPLE 8 Graph: $x = -2$.

Since y is missing, any number for y will do. Thus we know that all ordered pairs $(-2, y)$ are solutions. The graph is a line parallel to the y-axis.

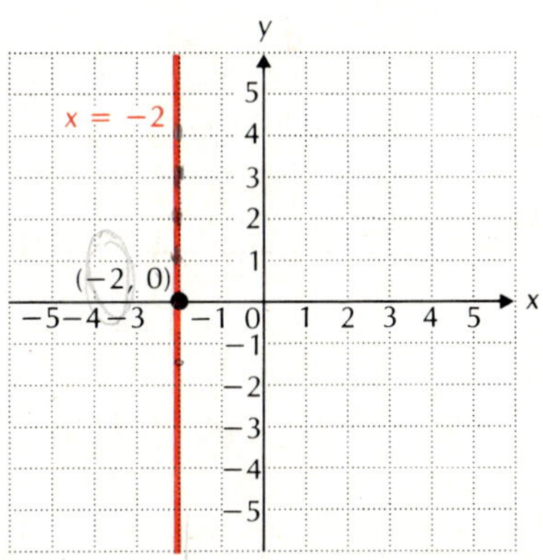

The graph of an equation $x = a$, where a is a constant, is a vertical line through the point $(a, 0)$.

The graph of an equation $y = b$, where b is a constant, is a horizontal line through the point $(0, b)$.

DO EXERCISE 13.

iii Which Equations Are Linear?

A *linear* equation is one having a straight line for its graph. From the examples in this section, we know which equations are linear.

A *linear equation* is one that is equivalent to an equation $y = mx + b$ or an equation $x = c$, where m, b, and c are constants.

Graph.

11. $y = 4$

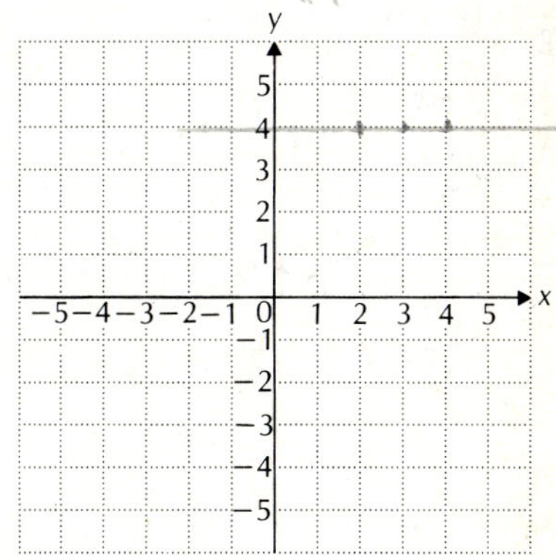

12. $3y + 6 = 0$
(*Hint:* Solve for y.)

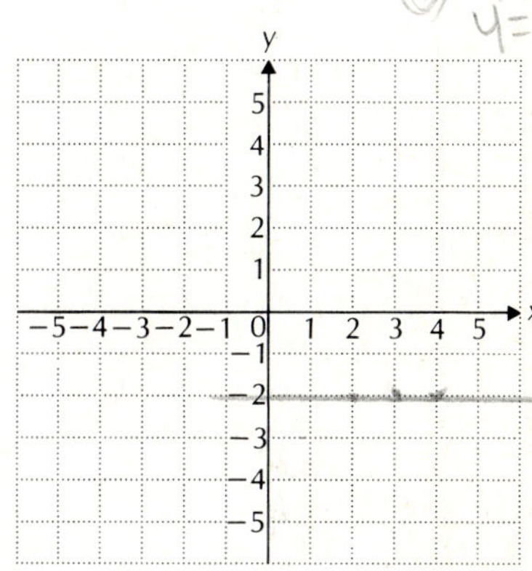

Graph.

13. $x = -5$

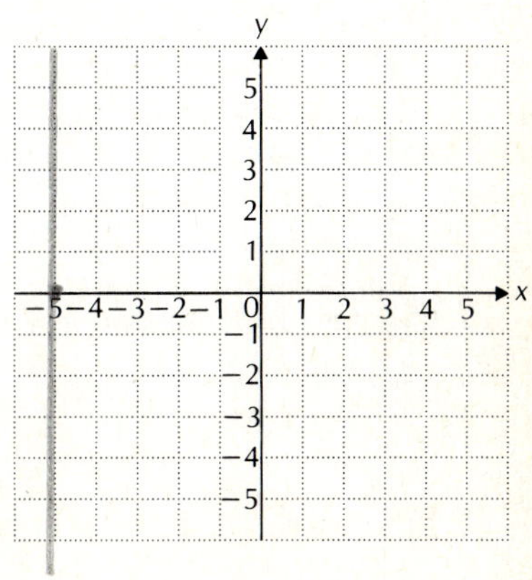

Keep in mind that the constants m, b, and c can be 0. Therefore an equation $y = mx$ is linear. Also, the equation $y = 0$ is linear. (It has a slope of 0 and its y-intercept is also 0.)

EXAMPLE 9 Show that the equation $2x - 3y = 6$ is linear. Determine the slope and the y-intercept.

To show that the equation is equivalent to an equation $y = mx + b$, we solve for y:

$$2x = 6 + 3y \quad \text{Adding } 3y$$
$$2x - 6 = 3y \quad \text{Adding } -6$$
$$\tfrac{1}{3}(2x - 6) = \tfrac{1}{3} \cdot 3y \quad \text{Multiplying by } \tfrac{1}{3}$$
$$\tfrac{2}{3}x - 2 = y$$
$$y = \tfrac{2}{3}x - 2 \quad \text{Reversing the equation}$$

Since we have an equation $y = mx + b$, we know that the graph is a straight line. The slope is $\frac{2}{3}$ and the y-intercept is -2.

EXAMPLES Which of these equations are linear?

10. $5x + 10 = 0$

Since there is no y in the equation, we solve for x to see if we get an equation $x = c$:

$$5x = -10 \quad \text{Adding } -10$$
$$\tfrac{1}{5} \cdot 5x = \tfrac{1}{5}(-10) \quad \text{Multiplying by } \tfrac{1}{5}$$
$$x = -2.$$

Since we have an equation $x = c$, we know that the equation is linear. The graph is parallel to the y-axis, 2 units to the left of that axis.

11. $2x^2 = 3y - 4$

We solve for y:

$$2x^2 + 4 = 3y \quad \text{Adding 4}$$
$$\tfrac{1}{3}(2x^2 + 4) = \tfrac{1}{3} \cdot 3y \quad \text{Multiplying by } \tfrac{1}{3}$$
$$\tfrac{2}{3}x^2 + \tfrac{4}{3} = y, \quad \text{or} \quad y = \tfrac{2}{3}x^2 + \tfrac{4}{3}.$$

The equation is not linear because it has an x^2-term.

DO EXERCISES 14–16.

Which equations are linear?

14. $2y - 4x = 6$

15. $2y - \dfrac{5}{y} = 3y$

16. $4x - 5 = y^2 + 2$

EXERCISE SET 2.2

i Determine the slope and y-intercept.

1. $y = 4x + 5$ **2.** $y = 5x + 3$ **3.** $y = -2x - 6$ **4.** $y = -5x + 7$ **5.** $y = -\frac{6}{16}x - 0.2$

6. $y = \frac{15}{7}x + 2.2$ **7.** $3x + 2y = 6$ **8.** $5y - 4x = 20$ **9.** $4x = 2y + 16$ **10.** $5y + 25 = -50$

Determine the slope and y-intercept. Then draw a graph. Be sure to use a third point as a check.

11. $y = \frac{5}{2}x + 1$ **12.** $y = \frac{2}{5}x + 4$ **13.** $y = -\frac{5}{2}x + 4$ **14.** $y = -\frac{2}{5}x + 3$ **15.** $y = 2x - 5$

16. $y = -2x + 4$ **17.** $y = \frac{1}{3}x + 6$ **18.** $y = -3x + 6$ **19.** $y = -0.25x + 2$ **20.** $y = 1.5x - 3$

ii Graph.

21. $3y = 9$ **22.** $6y + 24 = 0$ **23.** $3x = -15$

24. $2x = 10$ **25.** $4y + 3x = 12 + 3x$ **26.** $6x - 4y + 12 = -4y$

iii Determine which equations are linear.

27. $3x + 5y + 15 = 0$ 28. $5x - 3y = 15$ 29. $3x - 12 = 0$ 30. $16 + 4y = 0$ 31. $2x + 4y^2 = 19$

32. $3y = 5x^2 + 4$ 33. $5x - 4xy = 12$ 34. $3y = 7xy - 5$ 35. $\frac{3y}{4x} = 5y + 2$ 36. $6y - \frac{4}{y} = 0$

In Exercises 37–46, assume that a, b, and c are constants and that x and y are variables. Determine the slope and the y-intercept.

37. $ay = -5x + b$ 38. $bx + 2 = 4y - 9a$ 39. $ax + by = c$ 40. $ax + by = c - ay$

Determine which equations are linear.

41. $ax + 3y = b - c$ 42. $by = cx - dy + 2$ 43. $a^2x = by + 5$

44. $axy + cy = 2x$ 45. $\frac{x}{a} - by = 17$ 46. $axy + \frac{b}{x} = -axy + 2$

Problem-solving practice

47. The second angle of a triangle is twice as large as the first. The third angle is 44° greater than the first. Find the measures of the angles.

48. A piece of wire 32.8 ft long is to be cut into two pieces and those pieces are each to be bent to make a square. The length of a side of one square is to be 2.2 ft greater than the length of a side of the other. How should the wire be cut?

2.3

ANOTHER LOOK AT LINEAR GRAPHS

OBJECTIVES

After finishing Section 2.3, you should be able to:

i Graph linear equations using intercepts.

ii Given two points of a line, find its slope.

iii Identify lines with zero slope and lines that do not have a slope.

i Graphing Using Intercepts

Once you have determined that an equation is linear, you can draw its graph rather quickly—more quickly than with the methods we considered in Section 2.2. The more rapid procedure consists of plotting the x-intercept and the y-intercept and then drawing a straight line through those points. To do that, it is not necessary to solve the equation for either y or x.

EXAMPLE 1 Graph: $3x + 2y = 12$.

This equation is linear. (We could show that it is by solving for y.) To find the y-intercept, we let $x = 0$ and see what y must be. If we just cover up the $3x$ or ignore it, that amounts to letting x be 0. So we cover up $3x$ and see that $2y = 12$. Then y must be 6. The y-intercept is (0, 6).

To find the x-intercept, we can cover up the y-term and look at the rest of the equation. We have

$$3x = 12, \quad \text{or} \quad x = 4.$$

The x-intercept is (4, 0).

We plot these points and draw the line. A third point should be used as a check.

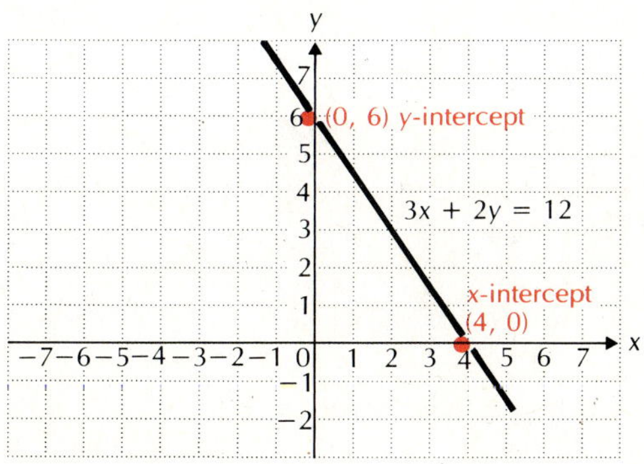

Graph using intercepts.

1. $4y - 12 = -6x$

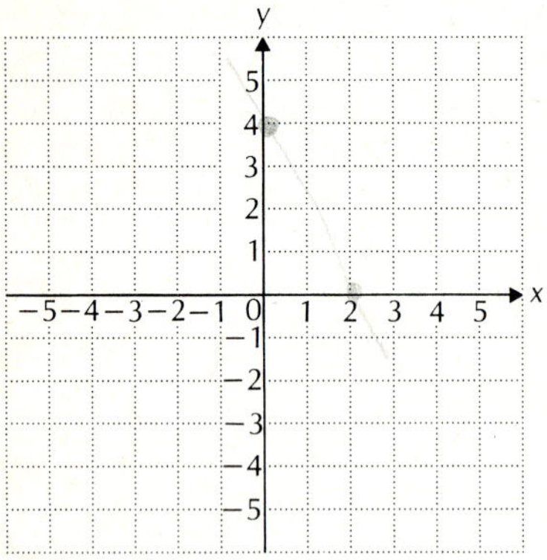

Graph.

2. $2y + x = 0$

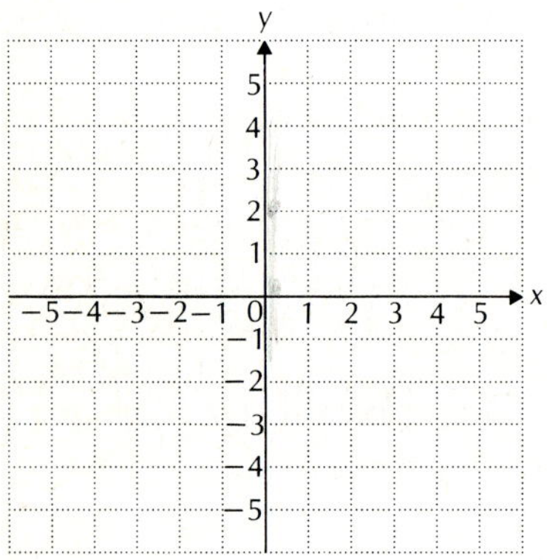

3. $8x + 4y = 1$

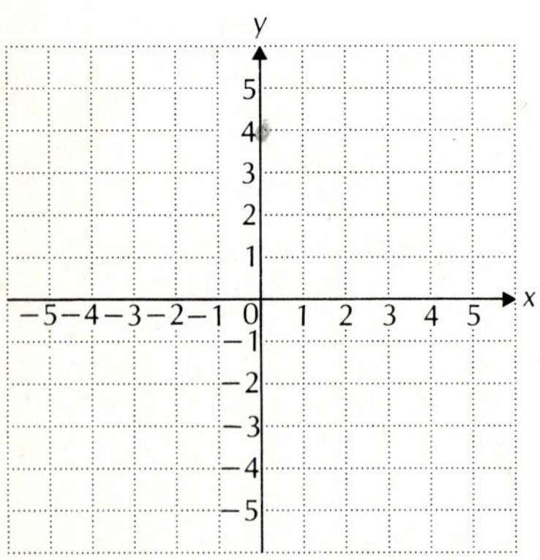

4. Two points of a line are (1, 2) and (6, 9). Find the slope of the line.

An x-intercept is a point $(a, 0)$. To find a, let $y = 0$ in the equation.
A y-intercept is a point $(0, b)$. To find b, let $x = 0$ in the equation.

DO EXERCISE 1.

We summarize the procedure.

TO GRAPH LINEAR EQUATIONS

1. **Is there a variable missing? If so, solve for the other one. The graph will be a line parallel to an axis.**
2. **If no variable is missing, find the intercepts. Graph using the intercepts if that is feasible.**
3. **If the intercepts are too close together, choose another point farther from the origin.**
4. **In any case, use a third point as a check.**

DO EXERCISES 2 AND 3.

ii Finding the Slope

If we know the coordinates of *any* two points of a line, we can find its slope. We divide the vertical distance (the rise) by the horizontal distance (the run). That is, slope = rise/run.

EXAMPLE 2 Two points of a line are (1, 3) and (5, 6). Find the slope of the line.

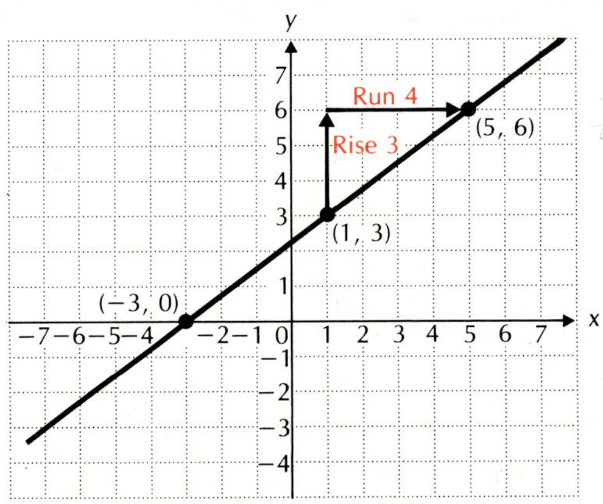

We graph the points and draw the line. Then we mark the rise (the distance *up*) and the run (the distance *over*) by drawing a right triangle. The rise is 3 (from 3 to 6) and the run is 4 (from 1 to 5).

$$\text{Slope} = \frac{\text{rise}}{\text{run}} = \frac{3}{4}$$

DO EXERCISE 4.

We can choose *any* two points of a line to find its slope.

EXAMPLE 3 Find the slope of the line in Example 2 by using a different pair of points.

We can use *any* two points. Let's choose $(-3, 0)$ and $(5, 6)$. The rise is 6 (from 0 to 6) and the run is 8 (from -3 to 5).

$$\text{Slope} = \frac{\text{rise}}{\text{run}} = \frac{6}{8} = \frac{3}{4}$$

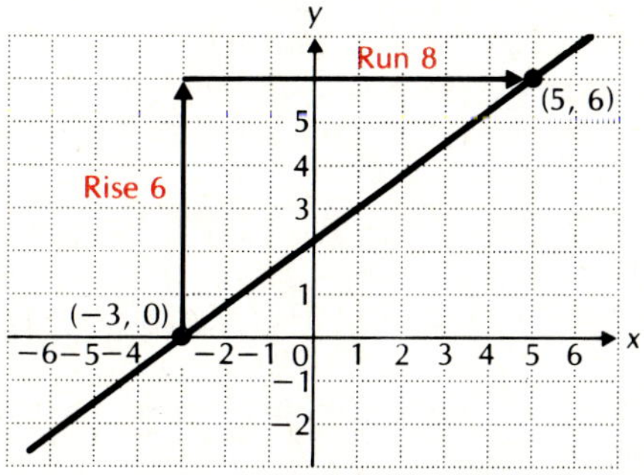

DO EXERCISE 5.

5. For the line of Examples 2 and 3, find the slope using the points $(-3, 0)$ and $(1, 3)$.

EXAMPLE 4 Two points of a line are $(-5, -3)$ and $(6, 2)$. Find the slope of the line.

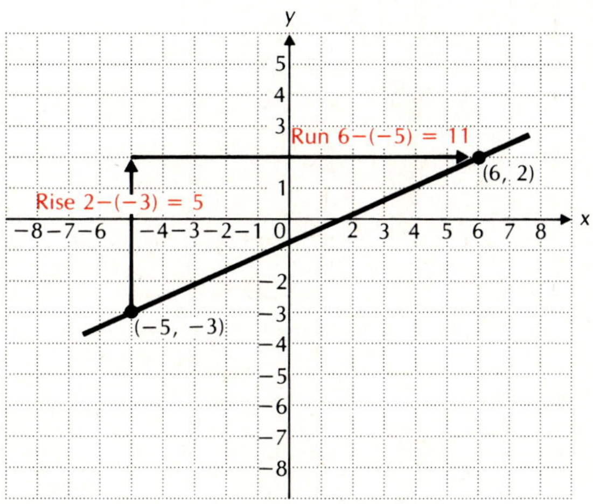

We graph the points and draw the line. We mark the rise and the run, which we find by subtracting:

$$\text{Rise} = 2 - (-3) = 5 \quad (\text{from } -3 \text{ to } 2);$$

$$\text{Run} = 6 - (-5) = 11 \quad (\text{from } -5 \text{ to } 6);$$

$$\text{Slope} = \frac{\text{rise}}{\text{run}} = \frac{5}{11}.$$

DO EXERCISE 6.

6. Two points of a line are $(-5, 3)$ and $(3, -10)$. Find the slope of the line.

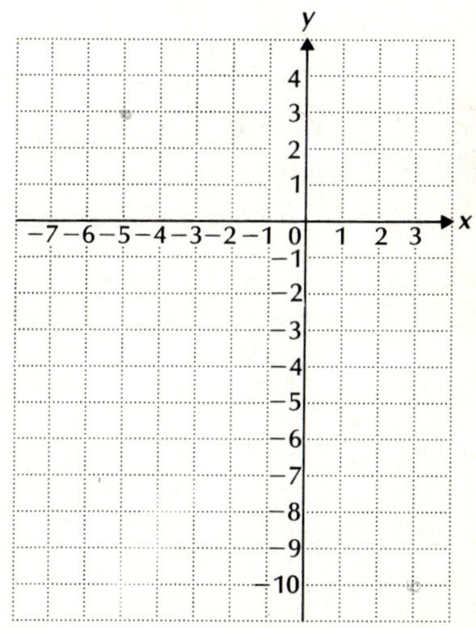

In Example 4, what will happen if we subtract coordinates in the opposite order? For the rise we get $-3 - 2$, or -5. For the run we get $-5 - 6 = -11$. For the slope we get

$$\text{Slope} = \frac{\text{rise}}{\text{run}} = \frac{-5}{-11} = \frac{5}{11}.$$

Find the slope of the line through each pair of points.

7. (4, −3) and (−8, −7)

8. (4.2, −6.8) and (−8.3, 2.1)

This is the same answer we got before. We can subtract either way, so long as we subtract in the same order both times.

> **If (x_1, y_1) and (x_2, y_2) are two points of a nonvertical line, the slope m of the line can be calculated as follows:**
>
> $$m = \frac{y_2 - y_1}{x_2 - x_1} \quad \text{or} \quad \frac{y_1 - y_2}{x_1 - x_2}.$$

Note: For a nonvertical line, x_1 and x_2 are different. (For a vertical line, $x_1 = x_2$.) Thus we do not need to worry about encountering a zero denominator as we calculate slope.

EXAMPLES Find the slope of the line through each pair of points.

5. (5, −2) and (−7, −5)

$$\text{Slope} = \frac{y_2 - y_1}{x_2 - x_1} = \frac{-5 - (-2)}{-7 - 5} = \frac{-5 + 2}{-12}$$

$$= \frac{-3}{-12} = \frac{1}{4}, \quad \text{or} \quad 0.25$$

We could also have subtracted in the opposite order, as follows:

$$\frac{y_1 - y_2}{x_1 - x_2} = \frac{-2 - (-5)}{5 - (-7)} = \frac{3}{12} = \frac{1}{4}.$$

6. (−4.2, 5.9) and (8.4, −2.2)

$$\text{Slope} = \frac{y_2 - y_1}{x_2 - x_1} = \frac{-2.2 - 5.9}{8.4 - (-4.2)} = \frac{-8.1}{12.6}$$

$$\approx -0.643$$ $\approx$ means "is approximately equal to."

DO EXERCISES 7 AND 8.

iii Zero Slope and Lines Without Slope

If two points have the same second coordinate, what about the slope of the line joining them? Since $y_2 = y_1$, we have

$$\frac{y_2 - y_1}{x_2 - x_1} = \frac{0}{x_2 - x_1} = 0.$$

> **Every horizontal line has a slope of 0.**

Suppose two different points are on a vertical line. Then they have the same first coordinate. Since $x_2 = x_1$, we have

$$\frac{y_2 - y_1}{x_2 - x_1} = \frac{y_2 - y_1}{0}.$$

Since we cannot divide by 0, this is undefined.

Vertical lines do not have a slope.

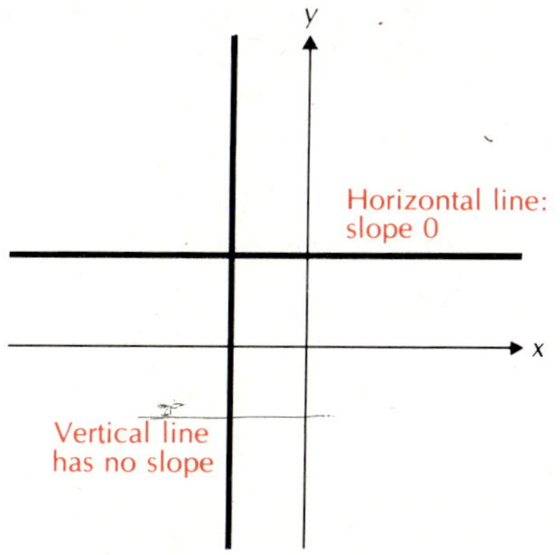

DO EXERCISES 9–11.

Find the slope, if it exists, of the line through each pair of points.

9. $(6, -4)$ and $(-10, -4)$

10. $(8, -5)$ and $(8, 13)$

11. Which equations have zero slope and which have no slope?

a) $3y = 6$

b) $5x = 13$

c) $5y + x = 5y + 4$

d) $y + 2 = 3x - 12$

e) $3x - 7 = 3y + 3x$

EXERCISE SET 2.3

i Find the intercepts. Then graph, using the intercepts and a third point as a check.

1. $x - 2 = y$ **2.** $x - 4 = y$ **3.** $3x - 1 = y$

4. $3x - 4 = y$ **5.** $5x - 4y = 20$ **6.** $3x + 5y = 15$

7. $y = -5 - 5x$ **8.** $y = -2 - 2x$ **9.** $5y = -15 + 3x$

10. $7x = 3y - 21$ **11.** $6x - 7 + 3y = 9x - 2y + 8$ **12.** $7x - 8 + 4y = 8y + 4x + 4$

13. $1.4y - 3.5x = -9.8$ **14.** $3.6x - 2.1y = 22.68$

ii For each pair of points, find the slope, if it exists, of the line containing them.

15. $(3, 8)$ and $(9, -4)$ **16.** $(17, -12)$ and $(-9, -15)$ **17.** $(-8, -7)$ and $(-9, -12)$

18. $(14, 3)$ and $(2, 12)$ **19.** $(-16.3, 12.4)$ and $(-5.2, 8.7)$ **20.** $(14.4, -7.8)$ and $(-12.5, -17.6)$

21. $(3.2, -12.8)$ and $(3.2, 2.4)$ **22.** $(-1.5, 7.6)$ and $(-1.5, 8.8)$

iii Which equations have zero slope and which have no slope?

23. $3x = 12 + y$ **24.** $5y - 12 = 3x$ **25.** $5x - 6 = 15$ **26.** $-12 = 4x - 7$

27. $5y = 6$ **28.** $19 = -6y$ **29.** $y - 6 = 14$ **30.** $3y - 5 = 8$

31. $12 - 4x = 9 + x$ **32.** $15 + 7x = 3x - 5$ **33.** $2y - 4 = 35 + x$ **34.** $2x - 17 + y = 0$

35. $3y + x = 3y + 2$ **36.** $x - 4y = 12 - 4y$ **37.** $3y - 2x = 5 + 9y - 2x$ **38.** $17y + 4x + 3 = 7 + 4x$

Problem-solving practice

39. The fare for a taxi ride from Johnson Street to Elm Street is \$5.20. If the rate of the taxi is \$1.00 for the first $\frac{1}{2}$ mile and 30¢ for each additional $\frac{1}{4}$ mile, how far is it from Johnson Street to Elm Street?

40. Seven more than three times a number is the same as thirteen less than five times the number. What is the number?

○

41. Find the x-intercept of $y = mx + b$, assuming that $m \neq 0$.

42. Consider the equation $Ax + By = C$, where A, B, and C are constants. Allow that the constants may be 0, except that A and B are not *both* 0.

a) Show that the equation is linear.
b) Find the intercepts.
c) Find the slope.

43. Suppose that two linear equations have the same y-intercept, but that equation A has an x-intercept that is half the x-intercept of equation B. How do the slopes compare?

44. Determine a so that the slope of the line through this pair of points has the value m.

$$(-2, 3a), (4, -a); \qquad m = -\tfrac{5}{12}$$

45. Find the slope of the line that contains the pair of points.

a) $(5b, -6c)$ $(b, -c)$
b) (b, d) $(b, d + e)$
c) $(c + f, a + d)$ $(c - f, -a - d)$

46. A line contains the points $(-100, 4)$ and $(0, 0)$. List four more points of the line.

47. Find the slopes of the sides of the triangle formed by joining the midpoints of the triangle whose vertices are $A(-8, -2)$, $B(-2, 6)$, and $C(4, -10)$. *The midpoint formula:* If the endpoints of a segment are (x_1, y_1) and (x_2, y_2), then the coordinates of the midpoint are

$$\left(\frac{x_1 + x_2}{2}, \frac{y_1 + y_2}{2}\right).$$

OBJECTIVES

After finishing Section 2.4, you should be able to:

i Given the slope of a line and a point on the line, find an equation of the line.

ii Given two points of a line, find an equation of the line.

iii Given a line and a point, find an equation of the line parallel to the given line and containing the point, and an equation of the line perpendicular to the given line and containing the point.

2.4 OTHER EQUATIONS OF LINES

If we know the slope of a line and also its y-intercept, we can find an equation of the line and draw its graph. There are other conditions that determine a line. Since we can draw a line if we know two of its points, there must also be a way to find an equation of that line if we know two of its points. In this section, we consider conditions that determine lines and allow us to find an equation.

i Point–Slope Equations

Suppose we know one point on a line and its slope. We can find an equation of the line. Suppose the known point is (x_1, y_1) and the slope is m. Then consider any other point (x, y) of the line.

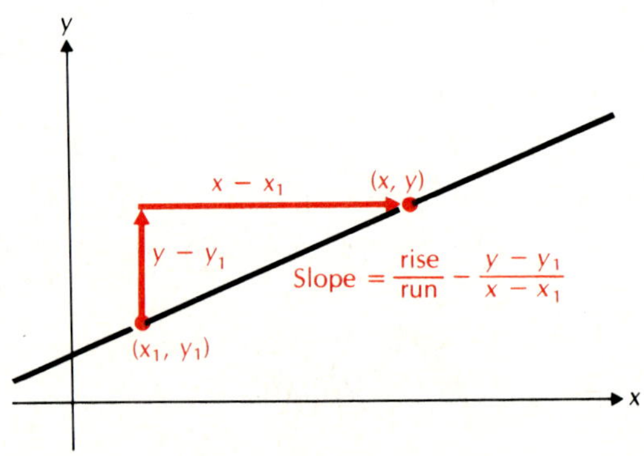

From this drawing we see that the slope of the line is

$$\frac{y - y_1}{x - x_1}.$$

The slope is also m. This gives us the equation

$$\frac{y - y_1}{x - x_1} = m. \tag{1}$$

Multiplying by $x - x_1$ gives us the equation we are looking for.

The *point-slope equation* of a line is

$$y - y_1 = m(x - x_1).$$

1. Find an equation of the line containing the point $(1, -2)$, with slope 3.

Note that Equation (1) is not true when $x = x_1$ and $y = y_1$, but the point–slope equation is. Thus Equation (1) is not an equation of the entire line.

Equation (1) is easy to remember. You can easily get the point–slope equation from it by multiplying by $x - x_1$.

EXAMPLE 1 Find an equation of the line containing the point $(\frac{1}{2}, -1)$ with a slope of 5.

We substitute in the equation $y - y_1 = m(x - x_1)$:

$$y - (-1) = 5(x - \tfrac{1}{2}) \quad \text{Substituting}$$
$$y + 1 = 5x - \tfrac{5}{2} \quad \text{Simplifying}$$
$$y = 5x - \tfrac{7}{2}.$$

EXAMPLE 2 Find an equation of the line with x-intercept $(4, 0)$ and slope 2.

$$y - 0 = 2(x - 4) \quad \text{Substituting}$$
$$y = 2x - 8 \quad \text{Simplifying}$$

Vertical lines do not have point–slope equations, because they do not have slopes. All other lines do.

DO EXERCISES 1 AND 2.

2. Find an equation of the line containing the point $(-5, \frac{3}{4})$, with slope $-\frac{4}{3}$.

ii Two-point Equations

Suppose we know two points of a nonvertical line. Call them (x_1, y_1) and (x_2, y_2). We can find the slope m:

$$m = \frac{y_2 - y_1}{x_2 - x_1}.$$

We substitute in the point–slope equation, and get the two-point equation. From this we can write an equation of a nonvertical line whenever we know two points on the line.

The *two-point equation* of a nonvertical line is

$$y - y_1 = \frac{y_2 - y_1}{x_2 - x_1}(x - x_1).$$

EXAMPLE 3 Find an equation of the line containing the points $(2, 3)$ and $(1, -4)$.

Let us take $(2, 3)$ to be (x_1, y_1) and $(1, -4)$ to be (x_2, y_2). We substitute:

$$y - 3 = \frac{-4 - 3}{1 - 2}(x - 2)$$

$$y - 3 = 7(x - 2). \quad \text{Simplifying}$$

3. Find an equation of the line containing the points $(4, -3)$ and $(1, 2)$.

4. Find an equation of the line containing the points $(-3, -5)$ and $(-4, 12)$.

If we wish, we can simplify further to get the slope–intercept equation:

$$y - 3 = 7x - 14$$
$$y = 7x - 11.$$

We can also take the points in the opposite order. If we take $(1, -4)$ to be (x_1, y_1) and $(2, 3)$ to be (x_2, y_2), we get

$$y - (-4) = \frac{3 - (-4)}{2 - 1}(x - 1)$$
$$y + 4 = 7(x - 1).$$

Simplifying further to get the slope–intercept equation, we obtain

$$y + 4 = 7x - 7$$
$$y = 7x - 11.$$

Note that in Example 3, what we did amounted to finding the slope and then substituting it into the *point-slope* equation. If you wish to think of it that way, you need memorize only that formula and not the *two-point* formula above.

DO EXERCISES 3 AND 4.

iii Parallel and Perpendicular Lines

If two lines are vertical they are parallel. How can we tell whether nonvertical lines are parallel? The answer is simple: We look at their slopes.

Two nonvertical lines are parallel if they have the same slope.

If one line is vertical and another is horizontal, they are perpendicular. Otherwise, how can we tell whether two lines are perpendicular?

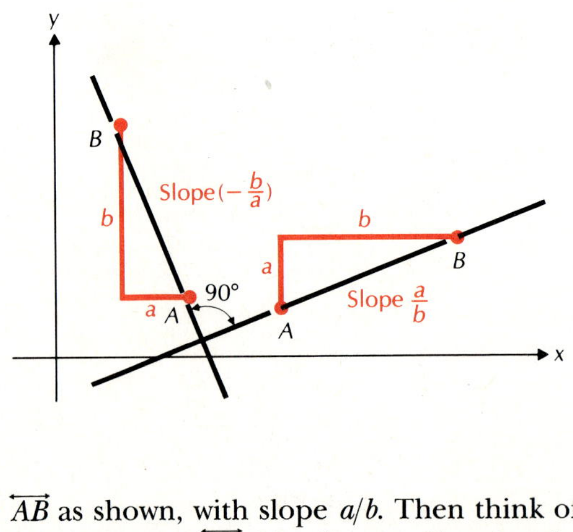

Consider a line $\overleftrightarrow{AB}$ as shown, with slope a/b. Then think of rotating the figure 90° to get a line perpendicular to $\overleftrightarrow{AB}$. For the new line the rise and run are interchanged, but the run is now negative. Thus the slope of the new line is $-b/a$. Let us multiply the slopes:

$$\frac{a}{b}\left(-\frac{b}{a}\right) = -1.$$

This is the condition under which lines will be perpendicular.

Two lines are perpendicular if the product of their slopes is -1. (If one line has slope m, the slope of a line perpendicular to it is $-1/m$. That is, we take the reciprocal and change the sign.) The lines are also perpendicular if one of them is vertical and the other is horizontal.

EXAMPLE 4 Given the line $8y = 7x - 24$ and the point $(-1, 2)$.

a) Find an equation of a line containing the point and parallel to the line.

We find the slope:

$$y = \tfrac{7}{8}x - 3. \qquad \text{Solving for } y$$

Now we use the point–slope equation:

$$y - 2 = \tfrac{7}{8}[x - (-1)] \quad \text{or} \quad y = \tfrac{7}{8}x + \tfrac{23}{8}. \qquad \text{Substituting } \tfrac{7}{8} \text{ for the slope and } (-1, 2) \text{ for the point}$$

b) Find an equation of the line containing the point and perpendicular to the line.

To find the slope we take the reciprocal of $\frac{7}{8}$ and then change the sign. The slope is $-\frac{8}{7}$.

$$y - 2 = -\tfrac{8}{7}[x - (-1)] \quad \text{or} \quad y = -\tfrac{8}{7}x + \tfrac{6}{7} \qquad \text{Substituting } -\tfrac{8}{7} \text{ for the slope and } (-1, 2) \text{ for the point}$$

DO EXERCISES 5 AND 6.

5. Find an equation of the line parallel to the line $8x = 7y - 24$ and containing the point $(2, -1)$.

6. Find an equation of the line perpendicular to the line $2x - 4y = 12$ and containing the point $(5, 4)$.

EXERCISE SET 2.4

i Find an equation of the line having the specified slope and containing the point indicated.

1. $m = 4$, $(3, 2)$ **2.** $m = 5$, $(5, 4)$ **3.** $m = -2$, $(4, 7)$ **4.** $m = -3$, $(7, 3)$

5. $m = 3$, $(-2, -4)$ **6.** $m = 1$, $(-5, -7)$ **7.** $m = -2$, $(8, 0)$ **8.** $m = -3$, $(-2, 0)$

9. $m = 0$, $(0, -7)$ **10.** $m = 0$, $(0, 4)$

ii Find an equation of the line containing each pair of points.

11. $(1, 4)$ and $(5, 6)$ **12.** $(2, 6)$ and $(4, 1)$ **13.** $(-1, -1)$ and $(2, 2)$

14. $(-3, -3)$ and $(6, 6)$ **15.** $(-2, 0)$ and $(0, 5)$ **16.** $(6, 0)$ and $(0, -3)$

17. $(3, 5)$ and $(-5, 3)$ **18.** $(4, 6)$ and $(-6, 4)$ **19.** $(0, 0)$ and $(5, 2)$

20. $(0, 0)$ and $(7, 3)$ **21.** $(-4, -7)$ and $(-2, -1)$ **22.** $(-2, -3)$ and $(-4, -6)$

iii Without graphing, tell whether the graphs of each pair of equations are parallel.

23. $x + 6 = y$, $y - x = -2$ **24.** $2x - 7 = y$, $y - 2x = 8$ **25.** $y + 3 = 5x$, $3x - y = -2$

26. $y + 8 = -6x$, $-2x + y = 5$ **27.** $y = 3x + 9$, $2y = 6x - 2$ **28.** $y = -7x - 9$, $-3y = 21x + 7$

Write an equation of the line containing the specified point and parallel to the indicated line.

29. $(3, 7)$, $x + 2y = 6$ **30.** $(0, 3)$, $3x - y = 7$ **31.** $(2, -1)$, $5x - 7y = 8$

32. $(-4, -5)$, $2x + y = -3$ **33.** $(-6, 2)$, $3x - 9y = 2$ **34.** $(-7, 0)$, $5x + 2y = 6$

Without graphing, tell whether the graphs of each pair of equations are perpendicular.

35. $y = 4x - 5,$ $4y = 8 - x$

36. $2x - 5y = -3,$ $2x + 5y = 4$

37. $x + 2y = 5,$ $2x + 4y = 8$

38. $y = -x + 7,$ $y = x + 3$

Write an equation of the line containing the specified point and perpendicular to the indicated line.

39. $(2, 5),\quad 2x + y = -3$

40. $(4, 0),\quad x - 3y = 0$

41. $(3, -2),\quad 3x + 4y = 5$

42. $(-3, -5),\quad 5x - 2y = 4$

43. $(0, 9),\quad 2x + 5y = 7$

44. $(-3, -4),\quad -3x + 6y = 2$

Problem-solving practice

45. The price of a radio, including 5% sales tax, is \$36.75. Find the price of the radio before the tax was added.

46. A basketball team increases its score by seven points in each of three consecutive games. If the team scored a total of 228 points in all three games, what was its score in the first game?

47. A real estate salesman earned \$720 in commission during one month. If he receives commission at a rate of 2% of his total sales over \$100,000, how much were his total sales for the month?

48. There are four more nickels than dimes in a piggy bank. If the total amount of money in the bank is \$2.45, how many nickels are in the bank?

---

49. (*An application: Road grade*). The steepness of a road is given by a number, usually expressed in percent, known as the *grade*. For example, if a road has a 3% grade, it rises 3 ft for every 100 ft of horizontal distance.

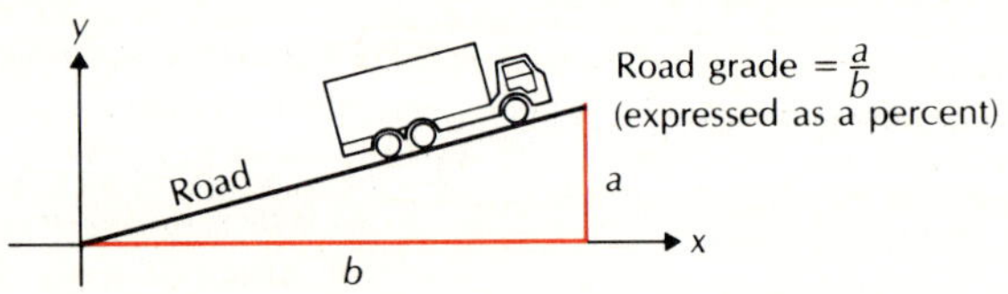

Find the grade in each case. Then find an equation of the road surface, as shown.

a)

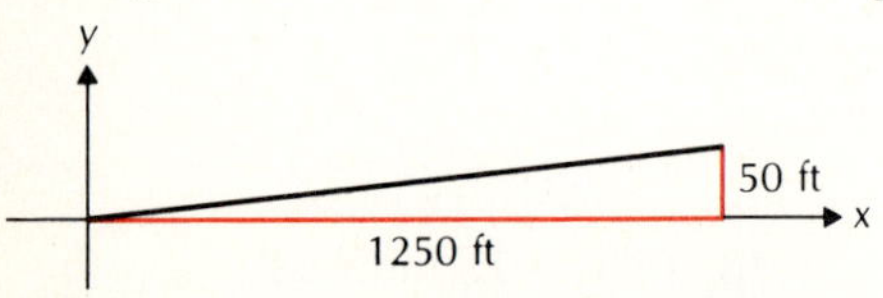

b)

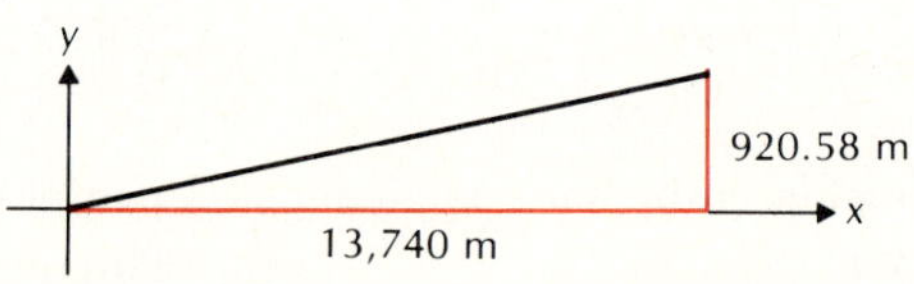

50. Use graph paper. Plot the points $A(0, 0)$, $B(8, 2)$, $C(11, 6)$, and $D(3, 4)$. Draw $\overline{AB}$, $\overline{BC}$, $\overline{CD}$, and $\overline{DA}$. Find the slopes of these four segments. Compare the slopes of $\overline{AB}$ and $\overline{CD}$. Compare the slopes of $\overline{BC}$ and $\overline{DA}$.

51. Find an equation of the line containing $(4, -2)$ and parallel to the line containing $(-1, 4)$ and $(2, -3)$.

52. Find an equation of the line containing $(-1, 3)$ and perpendicular to the line containing $(3, -5)$ and $(-2, 7)$.

53. Use slopes to show that the triangle with vertices $(-2, 7)$, $(6, 9)$, and $(3, 4)$ is a right triangle.

54. Find an equation of the line containing each pair of points.

a) $(-0.2, 0.7), (-0.7, -0.3)$

b) $(\frac{1}{11}, \frac{1}{2}), (-\frac{10}{11}, -2)$

55. Which pairs of the following four equations represent perpendicular lines?

(1) $7y - 3x = 21$

(2) $-3x - 7y = 12$

(3) $7y + 3x = 21$

(4) $3y + 7x = 12$

56. Find a number a for which the graphs of $5y = ax + 5$ and $\frac{1}{4}y = \frac{1}{10}x - 1$ will be parallel.

57. Find a number k for which the graphs of $x + 7y = 70$ and $y + 3 = kx$ will be perpendicular.

58. Write an equation of the line that has y-intercept $(0, \frac{5}{7})$ and is parallel to the graph of $6x - 3y = 1$.

59. Write an equation of the line that has x-intercept $(-1.2, 0)$ and is perpendicular to the graph of $6x - 3y = 1$.

60. Write an equation of the line that has x-intercept $(-3, 0)$ and y-intercept $(0, \frac{2}{5})$.

2.5 FUNCTIONS

OBJECTIVES

After finishing Section 2.5, you should be able to:

- **i** **Given a correspondence, determine whether it is a function.**
- **ii** **Given a function defined by a formula, find function values (outputs) for specified inputs.**
- **iii** **Find and use function values from a table.**
- **iv** **Find function values when a function is defined by a formula.**
- **v** **Graph functions and use the graph to find function values.**
- **vi** **Recognize graphs of functions.**

i In this section, we consider a mathematical concept that is highly useful in problem solving—that is, a *function*. A function is a certain kind of correspondence, from one set to another. For example:

To each person in a math class	there corresponds	his or her age.
To each item in a store	there corresponds	its price.
To each positive real number	there corresponds	its positive square root.

In each example, the first set is called the *domain*. The second set is called the *range*. Given a member of the domain, there is *just one* member of the range to which it corresponds. This kind of correspondence is called a *function*.

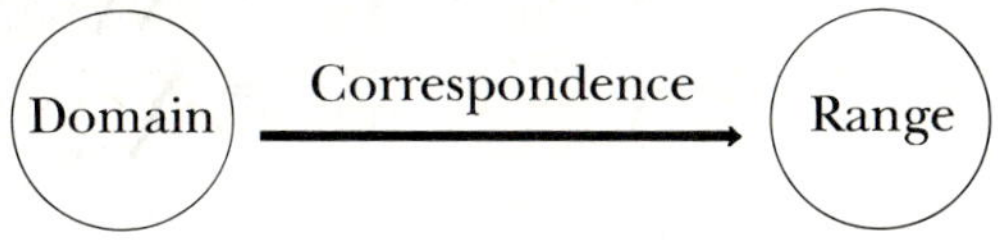

EXAMPLES Which of the following correspondences are functions?

1.

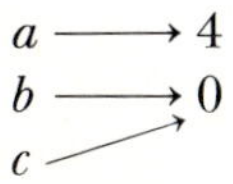

The correspondence is a function because for each member of the domain, there corresponds just one member of the range.

2.

The correspondence is not a function because a member of the domain (cheese pizza) corresponds to more than one member of the range.

A *function* is a correspondence between a first set, called the *domain*, and a second set, called the *range*, such that to any member of the domain, there corresponds *exactly one* member of the range.

1. Which of the following are functions?

	Domain	Correspondence	Range
a)	A family	Each person's age	A set of positive numbers
b)	A math class	Each person's final grade	A set of letters
c)	The people in a city	Each person's brothers	A set of males

2. In this correspondence, which is the domain and which is the range?

The items in a store correspond to their prices.

EXAMPLES Which of the following are functions?

	Domain	*Correspondence*	*Range*
3.	A family	Each person's weight	A set of positive numbers

This correspondence is a function, because each person has *only one* weight.

4.	The natural numbers	The square of each number	A set of natural numbers

This correspondence is a function, because each natural number has *only one* square.

5.	The set of positive real numbers	The square roots of each number	The set of all real numbers

This correspondence is not a function, because every positive real number has two square roots.

DO EXERCISES 1 AND 2.

ii Notation for Functions

To understand function notation, it helps to imagine a *function machine*. Think of putting a member of the domain (an *input*) into the machine. The machine knows the correspondence and gives you a member of the range (the *output*).

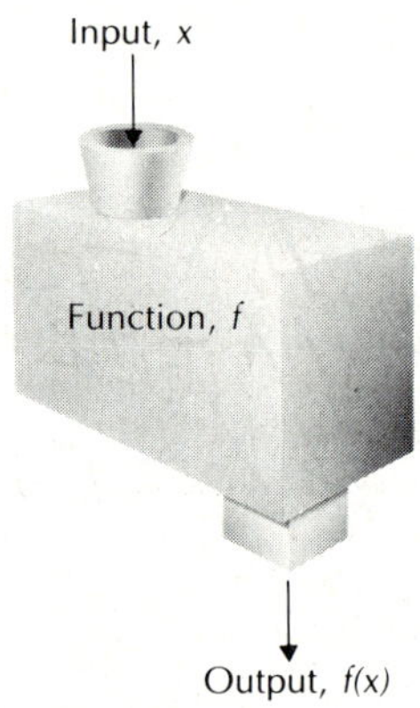

The function has been named f. We call the input x, and its output $f(x)$. This is read "f of x," or "f at x," or "the value of f at x." Note that $f(x)$ does *not* mean "f times x."

Some functions can be described by formulas or equations. For example, $f(x) = 2x + 3$ describes the function that takes an input x, multiplies it by 2, and then adds 3.

$$f(x) = 2x + 3$$

Input (x); Double ($2x$); Add 3 ($+3$)

To find the output $f(4)$, we take the input, 4, double it, and add 3 to get 11. That is, we substitute 4 into the formula for $f(x)$:

$$f(4) = 2 \cdot 4 + 3$$
$$= 11.$$

EXAMPLES Find these function values.

6. For $f(x) = 3x + 2$, find $f(5)$.

$$f(5) = 3 \cdot 5 + 2 = 17$$

7. For $g(x) = 5x^2 - 4$, find $g(3)$.

$$g(3) = 5(3)^2 - 4 = 41$$

8. For $A(x) = 3x^2 + 2x$, find $A(-2)$.

$$A(-2) = 3(-2)^2 + 2(-2) = 8$$

DO EXERCISES 3–5.

iii Functions and Tables

Tables are often used in problem solving and furnish examples of functions. To use tables in problem solving, we generally look up some function value in them.

EXAMPLE 9 The cost of replacing a defective tire is a function of tread depth, and can be found from the following table.

	Tread depth, in millimeters								
	9	8	7	6	5	4	3	2	1
Percent charged	No charge	20	30	40	55	70	80	90	100

Find the cost of replacing a tire whose original price was \$64.50 and whose tread depth is 4 mm.

2. **Translate** The problem translates to:

$$\text{Cost} = ?\% \times \text{Original cost}$$

From the table, we know that the percent charged is 70%, so we have:

$$\text{Cost} = 70\% \times \$64.50.$$

3. **Carry out** We calculate:

$$\text{Cost} = 0.70 \times 64.50 \quad \text{Converting to decimal notation}$$
$$= \$45.15.$$

DO EXERCISE 6.

Find these function values.

3. $f(x) = 2x + 5$; find $f(6)$.

4. $P(x) = 2x^3 - x + 1$; find $P(2)$.

5. $r(x) = 2x^2 - x - 1$; find $r(-1)$.

6. The number of computers sold in a month by the *Personal Computer Haven* is related to the profit, as shown in the following table.

Number sold	Profit in \$
15	−500
20	−100
25	0
30	300
40	600
45	800
50	1000

Find the profit per computer sold in a month when the store sold 40 computers.

7. The volume V of a sphere is a function of the radius r:

$$V(r) = \frac{4}{3}\pi r^3.$$

Find the volume of a sphere with a radius of 5 in.

iv Functions and Formulas

Functions are often described by formulas, and are used extensively in solving problems. You are probably already familiar with this, except that you may not have seen function notation in that connection. For example, a formula for finding the area of a circle is

$$A = \pi r^2.$$

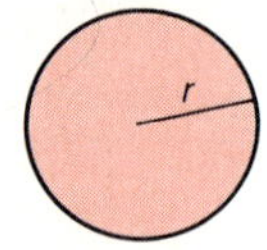

The area A is thus a function of r. To emphasize that fact, we often write

$$A(r) = \pi r^2.$$

EXAMPLE 10 Find the area of a circle having a radius of 12 cm.

We use the area function $A(r) = \pi r^2$:

$$\begin{aligned} A(12) &= \pi(12)^2 \quad \text{Substituting for } r \\ &= \pi \times 144 \\ &\approx 452 \text{ cm}^2. \end{aligned}$$

DO EXERCISE 7.

v Functions and Graphs

Functions are often described by graphs. To use a graph in problem solving, we do about what we would do with a table, except that we find a function value by reading a graph instead of a table. In what follows we first draw a graph and then estimate a function value from it.

EXAMPLE 11 A person buys a video cassette recorder on which there is a revolution counter. There is also a booklet with a table that relates the counter reading and the time for which the tape has run.

Time of tape, in hours	Counter reading
0	000
1	300
2	500
3	675
4	800

a) Use the data in the table to draw a graph of the counter reading as a function of the time.

b) Use the graph to estimate the counter reading at the end of a $1\frac{1}{2}$-hr program.

We solve as follows.

a) We draw a graph first. We plot the points and connect them with a smooth curve.

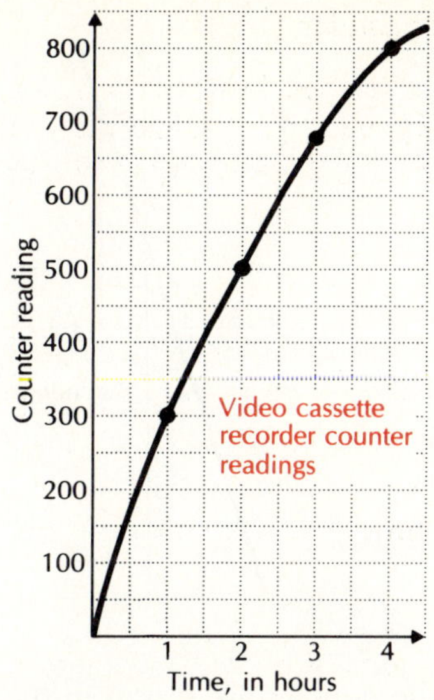

b) We then find the number $1\frac{1}{2}$ on the first (horizontal) axis. Then we move along the vertical line to the curve, and from there to the left, to the second (vertical) axis. There, we read an approximate function value. The counter reading is about 410.

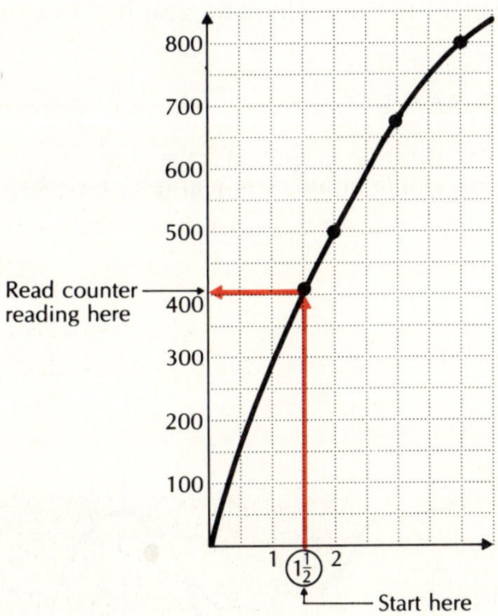

DO EXERCISES 8 AND 9.

From the preceding examples, we can see why a certain terminology is used with functions, a terminology that distinguishes the variables. The variable that is used for the inputs is called the *independent* variable and the variable used for the outputs is called the *dependent* variable.

vi Recognizing Graphs of Functions. Consider the function $f(x) = 3x + 2$. If you think of it as $y = 3x + 2$, you realize that the graph is a line, and that you already know how to graph it. The function is one of a class called *linear functions* (functions having straight-line graphs).

8. The following table relates the cost of shipping a crate of machinery parts to the weight of the contents.

Weight	Cost in $
1 T	350
2 T	450
3 T	550
4 T	660
5 T	760

Use the data in the table to draw a graph of the cost of shipping as a function of the weight of the contents.

9. Use the graph of Exercise 8 to find the cost of shipping a crate weighing each of the following.

a) 2.3 T
b) 0.5 T

Which of the following are graphs of functions?

10.

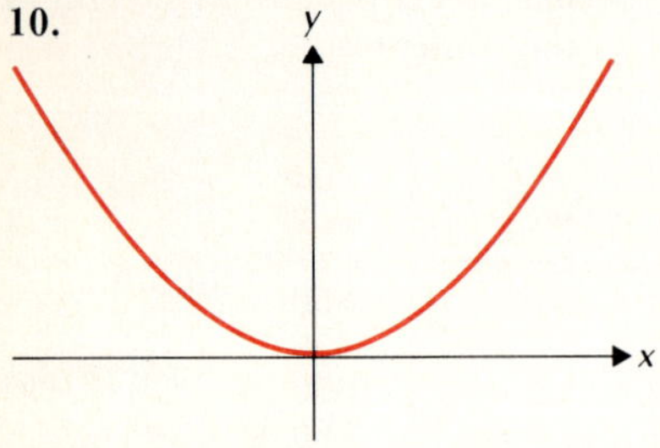

11.

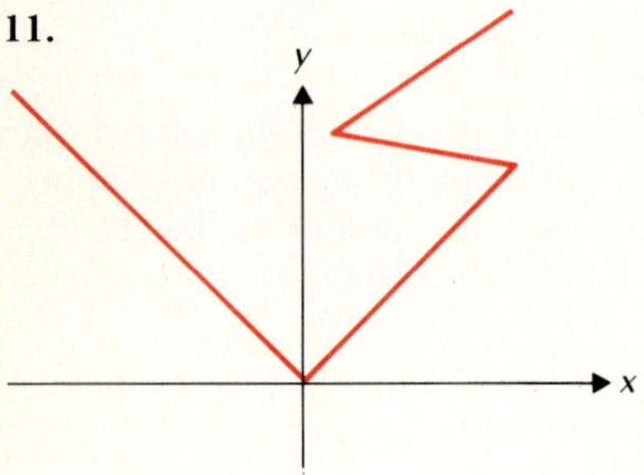

12.

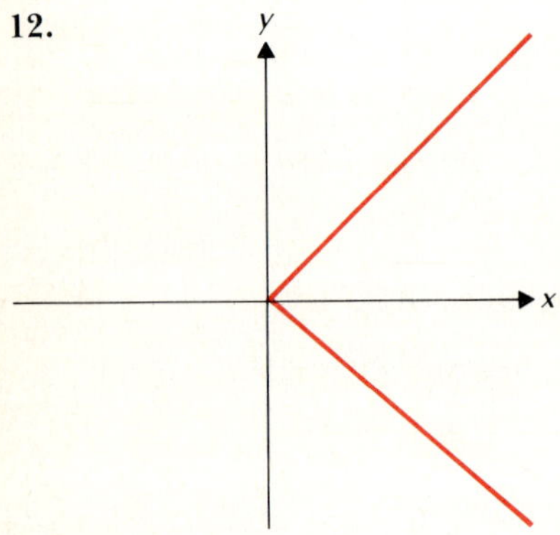

13.

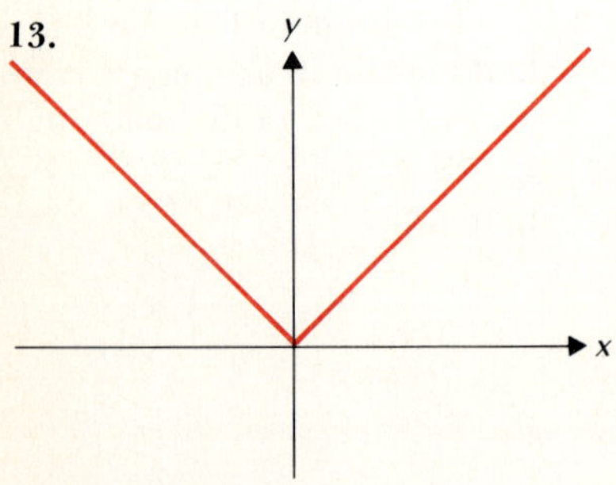

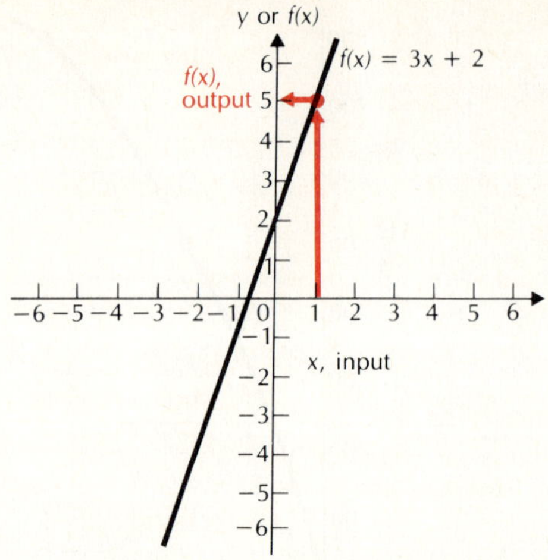

The members of the domain are shown in the figure. They are represented by the points on the horizontal axis. The members of the range are represented by the points on the vertical axis. To find a function value, we locate the input on the horizontal axis, move as shown, and then find the output on the vertical axis.

Some graphs are graphs of functions and some are not. (For example, see the graphs below.) If a vertical line drawn anywhere on the graph can intersect the curve at more than one point, then there will be more than one number in the range corresponding to a member of the domain. The graph is not the graph of a function.

THE VERTICAL LINE TEST

If it is possible for a vertical line to intersect a graph more than once, the graph is not a graph of a function.

DO EXERCISES 10–13.

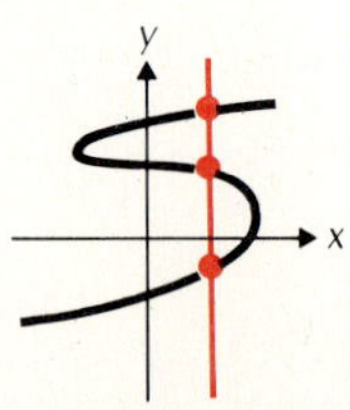

Not a function

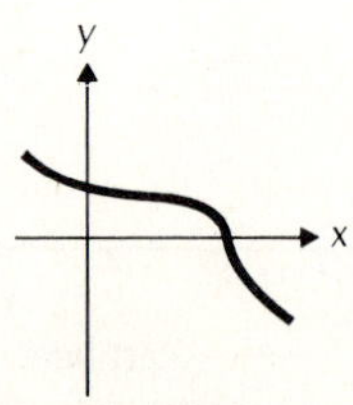

A function

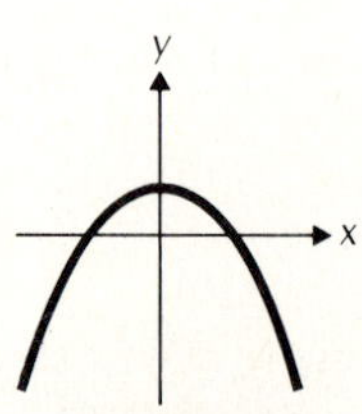

A function

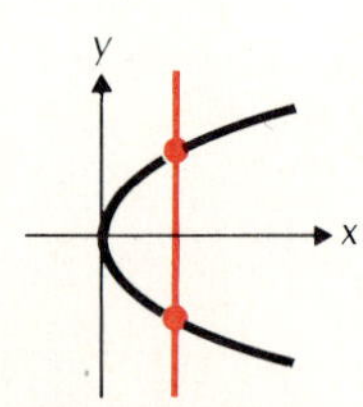

Not a function

EXERCISE SET 2.5

i Which of the following correspondences are functions?

1.

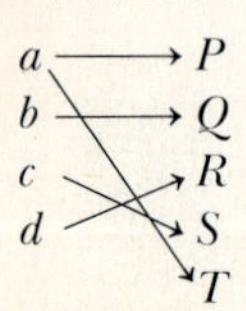

2.

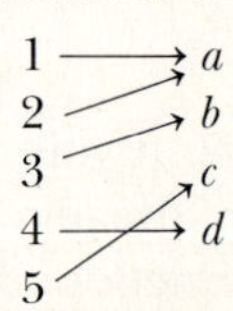

3.

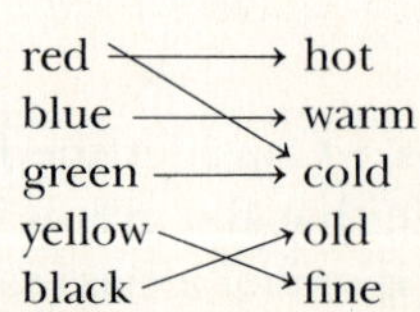

4.

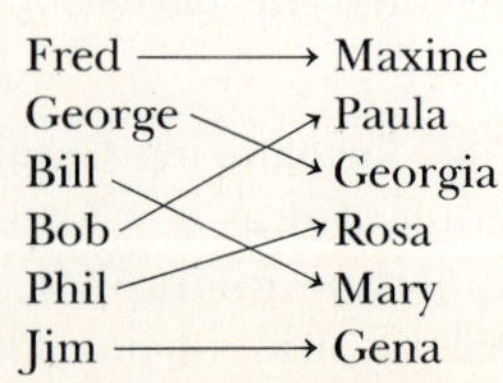

Which of the following are functions?

	Domain	*Correspondence*	*Range*
5.	A math class	Each person's seat number	A set of numbers
6.	A set of numbers	Square each number and then add 4	A set of numbers
7.	A set of geometric figures	Find the area of each figure	A set of numbers
8.	A set of lines	The slope	A set of numbers

ii Find the function values.

9. $g(x) = x + 1$

a) $g(0)$ **b)** $g(-4)$
c) $g(-7)$ **d)** $g(8)$

10. $h(x) = x - 4$

a) $h(4)$ **b)** $h(8)$
c) $h(-3)$ **d)** $h(-4)$

11. $f(x) = 5x^2 + 4x$

a) $f(0)$ **b)** $f(-1)$
c) $f(3)$ **d)** $f(t)$

12. $g(x) = 3x^2 - 2x$

a) $g(0)$ **b)** $g(-1)$
c) $g(3)$ **d)** $g(t)$

13. $f(x) = 3x^2 + 2x - 1$

a) $f(2)$ **b)** $f(3)$
c) $f(-3)$ **d)** $f(1)$

14. $h(x) = 4x^2 - x + 2$

a) $h(3)$ **b)** $h(0)$
c) $h(-1)$ **d)** $h(-2)$

15. $f(x) = \dfrac{x^2 - x - 2}{2x^2 - 5x - 3}$

a) $f(0)$ **b)** $f(4)$
c) $f(-1)$ **d)** $f(3)$

16. $s(x) = \dfrac{3x - 4}{2x + 5}$

a) $s(10)$ **b)** $s(2)$
c) $s(-\frac{5}{2})$ **d)** $s(-1)$
e) $s(1)$

Consider a function g described by $g(x) = -2x - 4$ and a function h described by $h(x) = 3x^2$. Find the following.

17. $g(5) + h(-2)$ **18.** $g(-4) \cdot h(4)$ **19.** $2g(-3) - 5h(12)$ **20.** $g(a + 4) - g(a)$

iii

21. Using the table in Example 9, find the cost of replacing a tire with a tread depth of 7 mm, whose original price was $78.50.

22. Using the table in Example 9, find the cost of replacing a tire with a tread depth of 3 mm, whose original price was $72.40.

iv The function A described by $A(s) = s^2 \dfrac{\sqrt{3}}{4}$ gives the area of an equilateral triangle with side s. Find the function values.

23. $A(4)$ **24.** $A(6)$

The function V described by $V(r) = \frac{4}{3}\pi r^3$ gives the volume of a sphere with radius r. Find the function values.

25. $V(3)$ **26.** $V(6)$

v For Exercises 27 and 28, use this graph, which shows the cost of manufacturing screwdrivers as a function of the number produced.

27. Find the cost of producing 1200 screwdrivers.

28. Find the cost of producing 5100 screwdrivers.

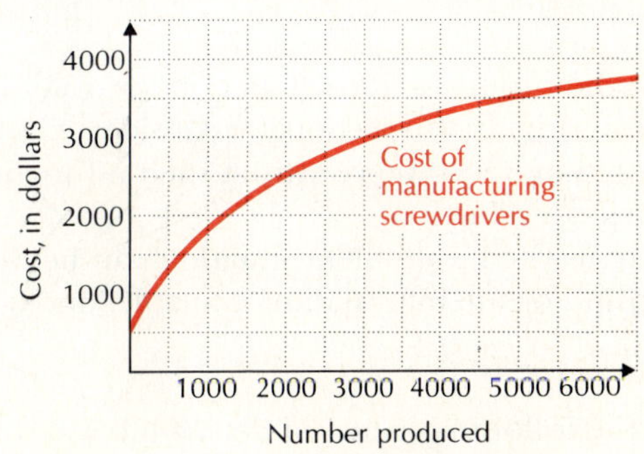

vi Which of the following are graphs of functions?

29.

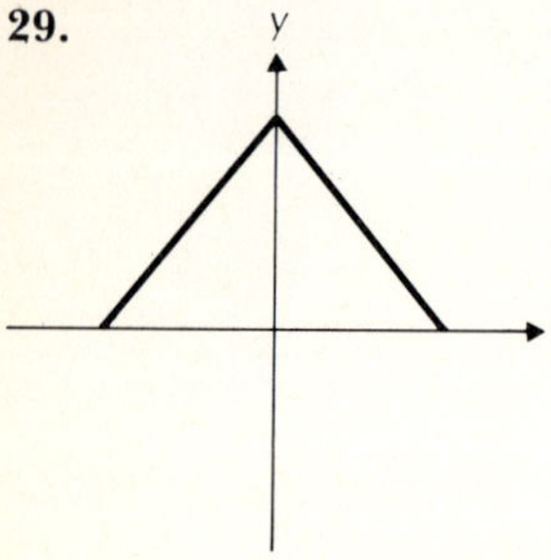

30.

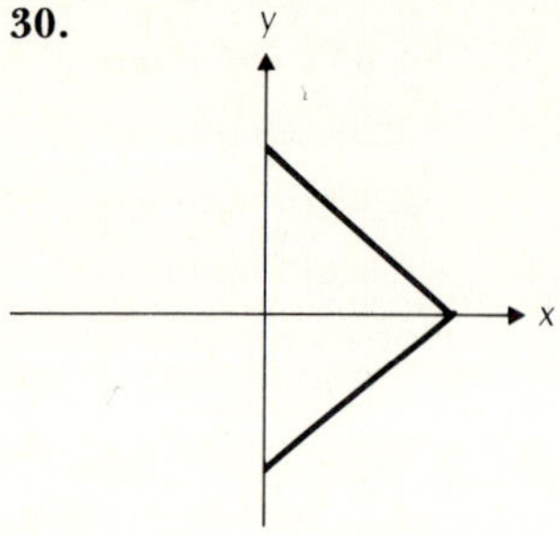

31.

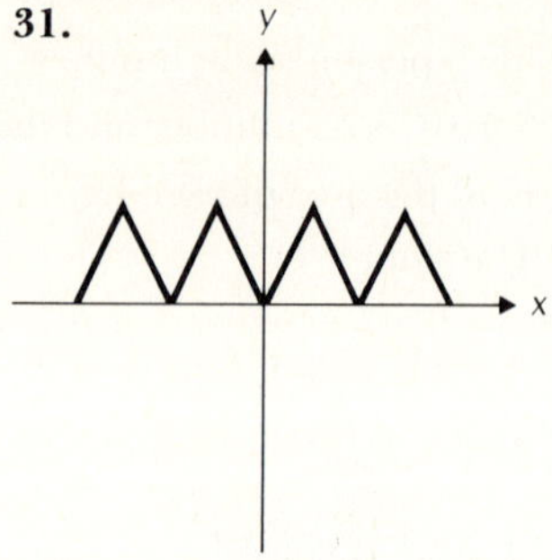

32.

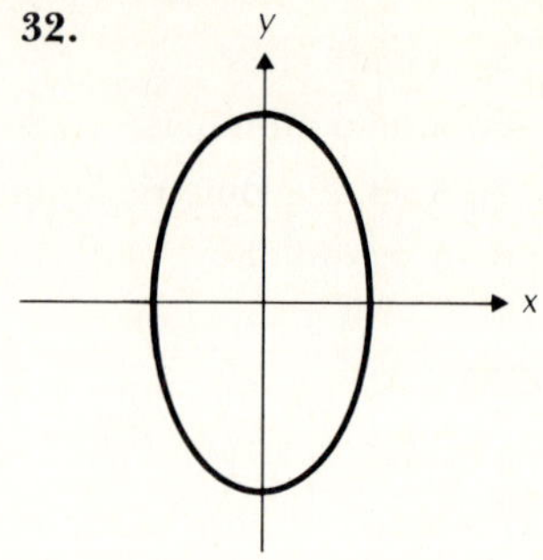

Problem-solving practice

33. Find three consecutive even integers such that the sum of the first, two times the second, and three times the third is 124.

34. The changes in the salary of a vice president of a corporation for three consecutive years are, respectively, 10% increase, 15% increase, and 5% decrease. What is the percent of total change for those three years?

For each of these functions, use a calculator to find the indicated function values.

35. ▦ $f(x) = 4.3x^2 - 1.4x$

a) $f(1.034)$ **b)** $f(-3.441)$
c) $f(27.35)$ **d)** $f(-16.31)$

36. ▦ $g(x) = 2.2x^3 + 3.5$

a) $g(17.3)$ **b)** $g(-64.2)$
c) $g(0.095)$ **d)** $g(-6.33)$

○

Which of the following are functions?

	Domain	*Correspondence*	*Range*
37.	A set of pairs of numbers (a, b)	Take the larger of a and b.	A set of numbers
38.	A set of ordered pairs of numbers (a, b)	Subtract a from b. Then take the larger of a and $b - a$.	A set of numbers

39. For a function g, you know that $g(-1) = -7$ and that $g(3) = 8$. Suppose that the graph of the function is a straight line. Find an equation, or formula, for g.

40. You know that for a linear function f, $f(x - 1) = 5x$. What is $f(6)$?

The *composite* of the functions f and g is the function h where $h(x) = f(g(x))$. To find $h(x)$, begin by finding $g(x)$. Then use this value to evaluate f. Let $f(x) = 2x + 1$ and $g(x) = 1 - x^2$.

41. Show that $f(g(0)) = f(1) = 3$.
42. Show that $g(f(0)) = g(1) = 0$.
43. Find $g(g(2))$.
44. Find $f(f(-3))$.
45. Find $g(f(3))$.
46. Find $f(g(3))$.

2.6

LINEAR FUNCTIONS AND PROBLEM SOLVING

i We have seen how to use a linear function, or other function, to find function values in problem solving. In this section, we consider how to construct a linear function that fits a situation. Once we construct such a function, then we can use it as before to find function values.

How do we know when a problem situation can be described by a linear function? We plot a graph of the information available, and see whether a straight line fits it.

EXAMPLE 1 Plot the following data and determine whether a linear function gives an approximate fit.

OBJECTIVES

After finishing Section 2.6, you should be able to:

i **Find a linear equation that fits two data points and use the equation to make predictions.**

Crickets are known to chirp faster when the temperature is higher. Here are some measurements made at different temperatures.

Temperature °C	6	8	10	15	20
Number of chirps per min	11	29	47	75	107

We make a graph with a T (temperature) axis and an N (number per minute) axis and plot these data. We see that they do lie approximately on a straight line, so a linear function fits this situation.

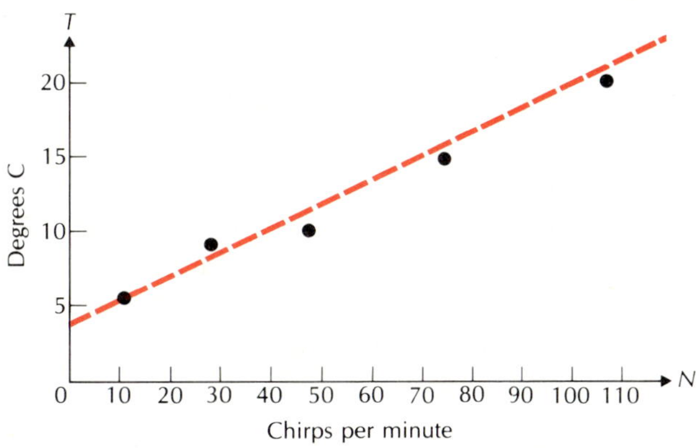

EXAMPLE 2 Wind friction, or *air resistance*, increases with speed. Here are some measurements made in a wind tunnel. Plot the data and determine whether a linear function will give an approximate fit.

Velocity, in km/h	10	21	34	40	45	52
Force of resistance, in kg	3	4.2	6.2	7.1	15.1	29.0

We make a graph with a V (velocity) axis and an F (force) axis and plot the data. They do not lie on a straight line, even approximately. Therefore, we cannot use a linear function in this situation.

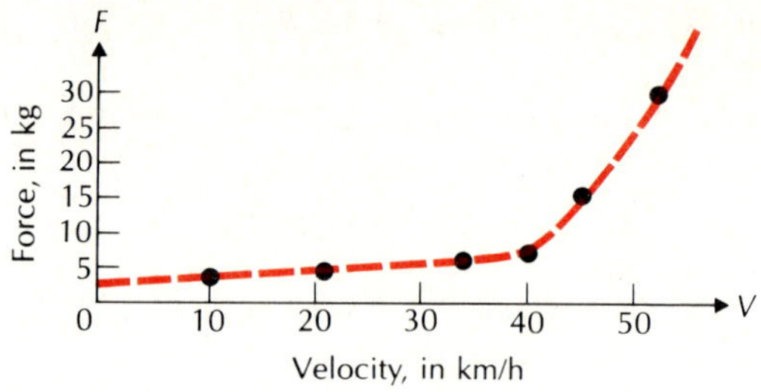

When we have a situation that a linear function fits, we can find a function and use it to solve problems and make predictions.

EXAMPLE 3 When crickets chirp 40 times per minute, the temperature is 10°C. When they chirp 112 times per minute, the temperature is 20°C. We know that a linear function fits this situation.

a) Find a linear function that fits the data.

b) From your function, find the temperature when crickets chirp 76 times per minute; 100 times per minute.

a) **Translate** To find the function, we use the two known ordered pairs (40, 10°) and (112, 20°). We call these *data points.* We find the two-point equation, using these data points:

$$T - T_1 = \frac{T_2 - T_1}{N_2 - N_1}(N - N_1)$$

$$T - 10 = \frac{20 - 10}{112 - 40}(N - 40) \qquad \text{Substituting}$$

$$T - 10 = \frac{10}{72}(N - 40) = \frac{5}{36}N - \frac{5}{36}\cdot 40$$

$$T = \frac{5}{36}N + \frac{40}{9}, \quad \text{or} \quad \frac{5N + 160}{36} \qquad \text{Solving for } T$$

We now have a linear function:

$$T(N) = \frac{5N + 160}{36}.$$

b) **Carry out** To find the temperature when $N = 76$, we find $T(76)$:

$$T(76) = \frac{5 \cdot 76 + 160}{36} = 15.$$

To find the temperature when $N = 100$, we find $T(100)$:

$$T(100) = \frac{5 \cdot 100 + 160}{36} \approx 18.3.$$

State When the number of chirps per minute is 76, the temperature is 15°. When the number of chirps per minute is 100, the temperature is about 18.3°.

In a problem situation, the domain and range are often determined by physical conditions. For the function of Example 3, any temperature for which the formula gives a negative number of chirps per minute would be meaningless. Also, imagine what would happen to a cricket at $-40°$ or at 200°!

DO EXERCISE 1.

1. (*Records in the 100-meter dash*). It has been found that running records change with time according to linear equations. In 1920 the record for the 100-meter dash was 10.43 seconds. In 1970 it was 9.93 seconds.
   a) Fit a linear equation to the data points. (Use R for the record and t for the number of years since 1920.)
   b) Use your equation to predict the record in 1982; in 1990.
   c) In what year will the record be 9.0 seconds?

EXERCISE SET 2.6

i

1. (*Life expectancy of females in the United States.*) In 1950, the life expectancy of females was 72 years. In 1970, it was 75 years. Let E represent the life expectancy and t the number of years since 1950 ($t = 0$ gives 1950 and $t = 10$ gives 1960).

a) Fit a linear function to the data points. [They are (0, 72) and (20, 75).]
b) Use the function of (a) to predict the life expectancy of females in 1989; in 1992.

2. (*Life expectancy of males in the United States.*) In 1950, the life expectancy of males was 65 years. In 1970, it was 68 years. Let E represent life expectancy and t the number of years since 1950.

a) Fit a linear function to the data points.
b) Use the function of (a) to predict the life expectancy of males in 1988; in 1995.

3. (*Natural gas demand.*) In 1950, natural gas demand in the United States was 20 quadrillion joules. In 1960, the demand was 22 quadrillion joules. Let D represent the demand for natural gas t years after 1950.

a) Fit a linear function to the data points.
b) Use the function of (a) to predict the natural gas demand in 1987; in 2000.

4. (*Records in the 1500-meter run.*) In 1930, the record for the 1500-meter run was 3.85 minutes. In 1950, it was 3.70 minutes. Let R represent the record in the 1500-meter run and t the number of years since 1930.

a) Fit a linear function to the data points.
b) Use the function of (a) to predict the record in 1984; in 1989.
c) When will the record be 3.3 minutes?

5. (*Records in the 400-meter run.*) In 1930, the record for the 400-meter run was 46.8 seconds. In 1970, it was 43.8 seconds. Let R represent the record in the 400-meter run and t the number of years since 1930.

a) Fit a linear function to the data points.
b) Use the function of (a) to predict the record in 1990; in 2000.
c) When will the record be 40 seconds?

6. (*The cost of a taxi ride.*) The cost of a taxi ride for 2 miles in Eastridge is \$2.50. For 3 miles the cost is \$2.75.

a) Fit a linear function to the data points.
b) Use the function to find the cost of a 7-mile ride.

7. (*The cost of renting a car.*) If you rent a car for one day and drive it 100 miles, the cost is \$40. If you drive it 150 miles, the cost is \$55.

a) Fit a linear function to the data points.
b) Use the function to find how much it will cost to rent the car for one day if you drive it 200 miles.

○

Solve each problem, assuming that a linear function fits the situation.

8. The value of a copying machine is \$5200 when it is purchased. After 2 years its value is \$4225. Find its value after 8 years.

9. Water freezes at 32° Fahrenheit and at 0° Celsius. Water boils at 212°F and at 100°C. What Celsius temperature corresponds to a room temperature of 70° Fahrenheit?

10. A business determines that when it sells 7000 units of a product it will take in \$22,000. For the sale of 8000 units it will take in \$25,000. How much will it take in for the sale of 10,000 units?

11. A piece of copper pipe has a length of 100 cm at 18°C. At 20°C the length of the pipe changes to 100.00356 cm. Find the length of the pipe at 40°C and at 0°C.

12. For a linear function f, $f(-1) = 3$ and $f(2) = 4$.

a) Find an equation for f.
b) Find $f(3)$.
c) Find a such that $f(a) = 100$.

13. (*Sales commissions*). A person applying for a sales position is offered alternative salary plans.

Plan A: A base salary of \$600 per month plus a commission of 4% of the gross sales for the month

Plan B: A base salary of \$700 per month plus a commission of 6% of the gross sales for the month in excess of \$10,000

a) For each plan formulate a function that expresses monthly earnings as a function of gross sales, x.
b) For what gross sales values is Plan B preferable?

14. (*Anthropology: Estimating heights*). An anthropologist can use certain linear functions to estimate the height of a male or female, given the length of certain bones. A *humerus* is the bone from the elbow to the shoulder. Let x = the length of the humerus in centimeters. Then the height, in cm, of a male with a humerus of length x is given by $M(x) = 2.89x + 70.64$. The height, in cm, of a female with a humerus of length x is given by $F(x) = 2.75x + 71.48$. A 45-cm humerus was uncovered in some ruins.

a) Assuming it was from a male, how tall was he?
b) Assuming it was from a female, how tall was she?

15. (*Advertising*). An automobile dealer discovers that when \$1000 is spent on radio advertising, weekly sales increase by \$101,000. When \$1250 is spent on radio advertising, sales increase by \$126,000. Suppose that the sales increase is a linear function of the amount of radio advertising. Then calculate the increase in sales when the amount spent on radio advertising is \$1500; when it is \$2000.

16. (*Publishing*). A publisher pays royalties to authors on the basis of the number of words in a manuscript. A 40,000-word manuscript earns 9.5% of the retail price of the book. A 45,000-word manuscript earns 10.4% of the retail price. Suppose that a linear function fits the situation. Then calculate the royalty percentage on a 60,000-word manuscript.

SUMMARY AND REVIEW: CHAPTER 2

The following contains a summary of what you should be able to do after completing this chapter. The review exercises are for practice. Answers are at the back of the book. If you miss an exercise, restudy the section indicated alongside the answer.

You should be able to:

Determine whether an ordered pair is a solution of an equation.

Determine whether the ordered pair is a solution.

1. $(3, 7)$, $4p - q = 5$

2. $(-2, 4)$, $x - 3y = 10$

3. $(0, -\frac{1}{2})$, $3a + 4b = 2$

4. $(8, -2)$, $3c - 2d = 28$

Graph an equation.

Graph.

5. $y = -3x + 2$

6. $y = -x^2 + 1$

Find the slope and *y*-intercept of a linear equation.

Find the slope and *y*-intercept.

7. $y = -4x - 9$

8. $-6y + 2x = 7$

Graph a linear equation.

Graph.

9. $8x + 32 = 0$

Determine whether an equation is linear.

Which of these are linear equations?

10. $2x - 7 = 0$

11. $3x - 8y = 7$

12. $2a + 7b^2 = 3$

13. $2p - \dfrac{7}{q} = 1$

Graph a linear equation using intercepts.

Graph, using intercepts.

14. $-2x + 4y = 7$

Find the slope (if it exists) of a line, given two points.

Find the slope of the line.

15. Containing the points $(4, 5)$ and $(-3, 1)$

16. Containing the points $(-16.4, 2.8)$ and $(-16.4, 3.5)$

Find an equation of a line given a point and the slope, or two points.

Find an equation of the line.

17. With slope -2 and containing $(-3, 4)$

18. Containing $(2, 5)$ and $(-4, 0)$

Determine without graphing whether pairs of equations are parallel or perpendicular.

Determine whether the lines are parallel or perpendicular.

19. $y + 5 = -x$,
$x - y = 2$

20. $3x - 5 = 7y$,
$7y - 3x = 7$

21. $4y + x = 3$,
$2x + 8y = 5$

Find an equation of a line, containing a given point, that is perpendicular or parallel to a given line.

Find an equation of the line.

22. Containing the point $(2, -5)$ and parallel to the line $3x - 5y = 9$

23. Containing the point $(2, -5)$ and perpendicular to the line $3x - 5y = 9$

Find function values.

Find the function values.

24. $g(x) = 2x - 5$

a) $g(0)$ **b)** $g(2)$ **c)** $g(-3)$ **d)** $g(7)$

25. $h(x) = x^2 - 2x + 1$

a) $h(0)$ **b)** $h(-1)$ **c)** $h(2)$ **d)** $h(-3)$

Use linear functions in problem solving.

Solve.

26. The cost of a taxi ride for 2.5 miles is \$3.00. For 3.0 miles, the cost is \$3.75.

a) Fit a linear function to the data points.
b) Use the linear function to find the cost of a 5-mile ride.

---

27. Find the slope and y-intercept of the graph representing Fahrenheit temperatures as a function of Celsius temperatures.

28. Determine the value of a so that the lines $3x - 4y = 12$ and $ax + 6y = -9$ are parallel.

TEST: CHAPTER 2

Determine whether the ordered pair is a solution of the equation.

1. $(2, 5)$, $3y - 4z = -14$

2. $(-6, 8)$, $2s - t = -4$

3. $(0, -5)$, $x - 4y = -20$

4. $(1, -4)$, $-2p + 5q = 18$

Graph.

5. $y = -5x + 4$

6. $y = -2x^2 + 3$

Find the slope and y-intercept.

7. $y = 3x - 5$

8. $-3y + 4x = 9$

9. $-2bx + ay = c$

10. Graph: $3x - 18 = 0$.

11. Which of these are linear equations?

a) $8x - 7 = 0$ **b)** $4b - 9a^2 = 2$ **c)** $2x - 5y = 3$ **d)** $1 - \frac{1}{x} = 4$

12. Graph: $-5x + 2y = -12$.

Find the slope, if it exists, of the line containing the following points.

13. $(-2, -2)$ and $(6, 3)$

14. $(-3.1, 5.2)$ and $(-4.4, 5.2)$

15. Which of these equations has zero slope and which has no slope?

a) $2y = 7$ **b)** $3x - 4y = 6x - 4y$

16. Find an equation of the line with slope of 4 and containing $(-2, -4)$.

17. Find an equation of the line containing $(3, -1)$ and $(4, -2)$.

Determine without graphing whether the pairs of lines are parallel or perpendicular.

18. $4y + 2 = 3x$,
$-3x + 4y = -12$

19. $y = -2x + 5$,
$2y - x = 6$

Find an equation of the line.

20. Containing $(-3, 2)$ and parallel to the line $2x - 5y = 8$

21. Containing $(-3, 2)$ and perpendicular to the line $2x - 5y = 8$

22. Find the following function values, given $g(x) = -3x - 4$ and $h(x) = x^2 + 1$.

a) $g(0)$ **b)** $h(-2)$ **c)** $g(-3) + h(5)$ **d)** $3g(1) - 2h(-2)$

23. (*The cost of renting a car.*) If you rent a car for one day and drive it 250 miles, the cost is \$100. If you drive it 300 miles, the cost is \$125.

a) Fit a linear function to the data points.

b) Use the function to find how much it will cost to rent the car for one day and drive it 500 miles.

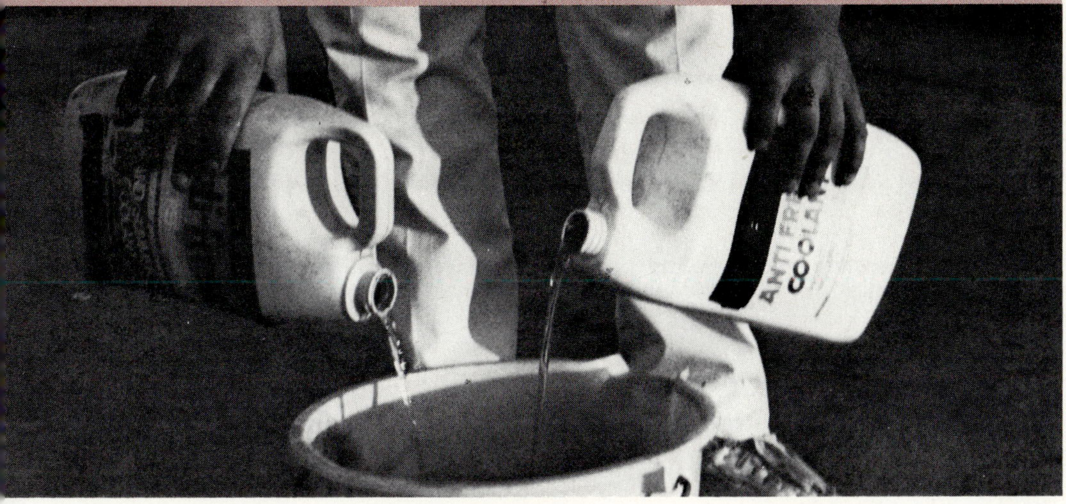

3

PROBLEM SOLVING AND SYSTEMS OF EQUATIONS

How should solutions be mixed to obtain 65% antifreeze? Systems of equations can be used to solve such problems.

As you have probably already noticed, the most difficult and time-consuming part of solving a problem is usually the translation of the situation to mathematical language. When you know how to translate using more than one equation, the translation is often much easier. In this chapter you will learn how to do such translations. That is, you will translate to a system of equations. You will learn to solve systems of equations and then apply that skill to the solving of problems.

OBJECTIVES

After finishing Section 3.1, you should be able to:

i **Translate a problem situation to mathematical language using two equations.**

ii **Determine whether an ordered pair is a solution of a system of two equations.**

iii **Solve systems of two linear equations graphically.**

Translate the following problem situation to mathematical language using two equations.

1. Jones's salary is \$100 more than twice Smith's salary. Jones's salary is \$900 less than three times Smith's salary.

2. Determine whether (20, 40) is a solution of the system

$$2x = y$$
$$4y + x = 180.$$

3. Determine whether $(-2, 3)$ is a solution of the system

$$2a - 5b = 7$$
$$5b + 3a = -4.$$

3.1

SYSTEMS OF EQUATIONS IN TWO VARIABLES

i Translating

Let us see how to use more than one equation, as we translate a problem to mathematical language.

EXAMPLE 1 Translate the following problem situation to mathematical language, using two equations.

Eight times a certain number added to five times a second number is 184. The first number is three more than the second number.

2. **Translate** There are two statements in the problem. We translate the first one.

$$\underbrace{\text{8 times a certain number}}_{8x}\ \underbrace{\text{added to}}_{+}\ \underbrace{\text{5 times a second number}}_{5y}\ \underbrace{\text{is}}_{=}\ \underbrace{184.}_{184}$$

Here x represents the first number and y represents the second. Now we translate the second statement, remembering to use the same variables in the same way.

$$\underbrace{\text{The first number}}_{x}\ \underbrace{\text{is}}_{=}\ \underbrace{3}_{3}\ \underbrace{\text{more than}}_{+}\ \underbrace{\text{the second number.}}_{y}$$

We now have two equations:

$$8x + 5y = 184,$$
$$x = 3 + y.$$

If we can find numbers x and y that make *both* equations true, we may have an answer to the problem.

DO EXERCISE 1.

ii Systems of Equations and Solutions

A pair of equations such as those in Example 1 is known as a *system of equations.* A *solution* of such a system is an ordered pair of numbers that makes *both* equations true.

EXAMPLE 2 Determine whether (4, 7) is a solution of the system

$$x + y = 11,$$
$$3x - y = 5.$$

We use alphabetical order of the variables. Thus we replace x by 4 and y by 7:

$x + y = 11$	
$4 + 7$	11
11	

$3x - y = 5$	
$3 \cdot 4 - 7$	5
$12 - 7$	
5	

The pair (4, 7) makes both equations true, so it is a solution of the *system.* We sometimes describe such a solution by saying that $x = 4$ and $y = 7$.

DO EXERCISES 2 AND 3.

iii Graphing Systems of Equations

Recall that the graph of an equation is a picture of its solution set. If the graph of an equation is a line, then every point on that line corresponds to an ordered pair that is a solution of the equation. If we graph a *system* of two linear equations, the point at which the lines intersect will be a solution of *both* equations.

EXAMPLE 3 Solve this system graphically:

$$y - x = 1$$
$$y + x = 3.$$

All ordered pairs from line A are solutions of the first equation. All ordered pairs from line B are solutions of the second equation. The point of intersection has coordinates that make *both* equations true. The solution seems to be the point (1, 2). Graphing is not perfectly accurate, so solving by graphing may give only approximate answers. The pair (1, 2) does check, so it is the solution.

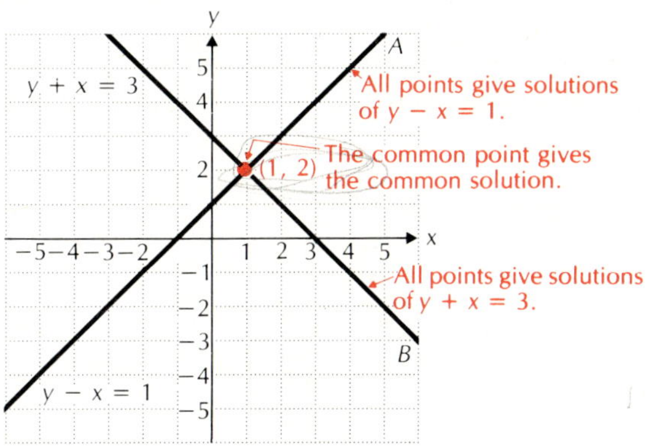

DO EXERCISES 4 AND 5.

Some systems of equations have no solution. That happens when the graphs are parallel lines. Such equations are called *inconsistent,* because they are never *both* true at once. Some equations have only one solution, as in Example 3. Some have an infinite (unending) set of solutions. That happens when the equations have exactly the same graph. Such equations in two variables are called *dependent.* The three possibilities are shown in the following graphs.

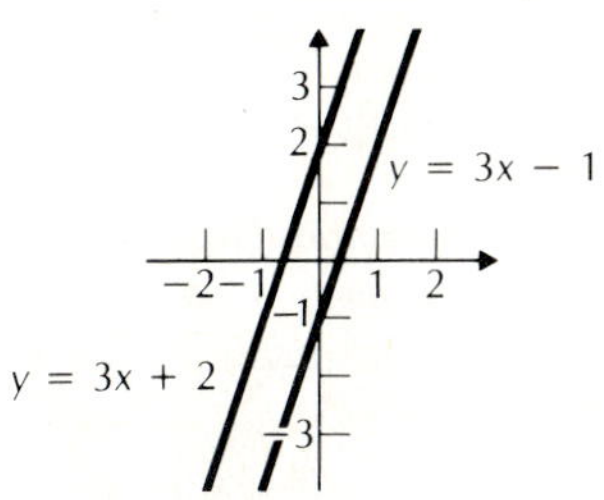

No solution.
Graphs are parallel.
Equations are *inconsistent* because there is no solution.

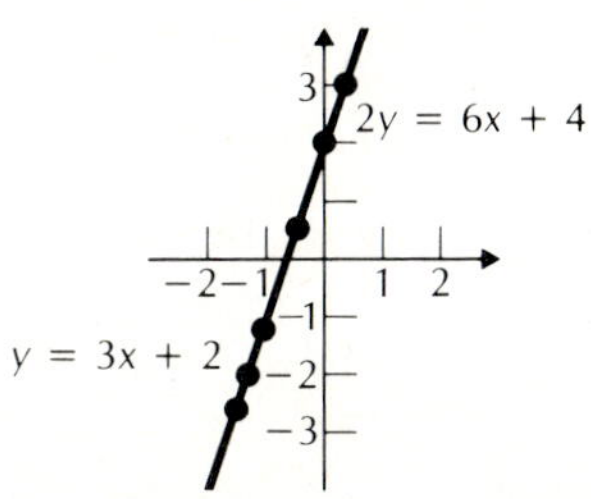

Many solutions.
Equations have the same graph.
Equations are *dependent* (but they are consistent because there is at least one solution).

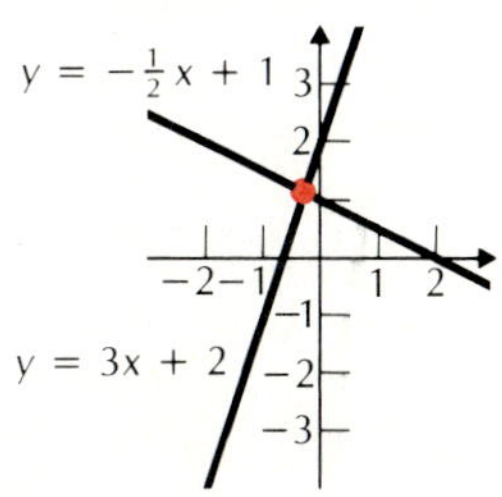

One solution.
Graphs intersect.
Equations are *consistent* (and not dependent).

Solve graphically.

4. $-2x + y = 1$
$3x + y = 1$

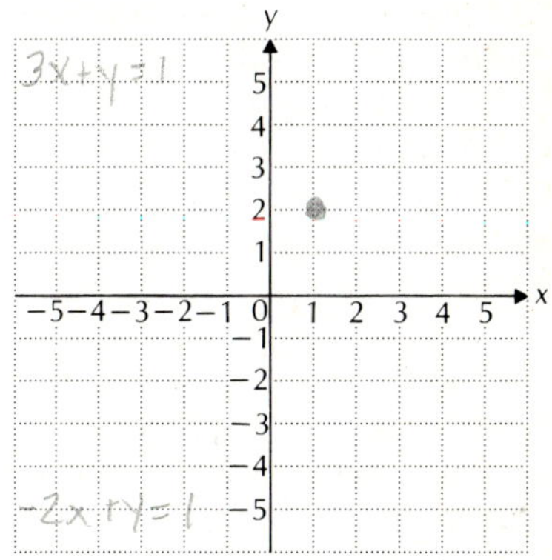

5. $x = 2y$
$4y + x = 6$

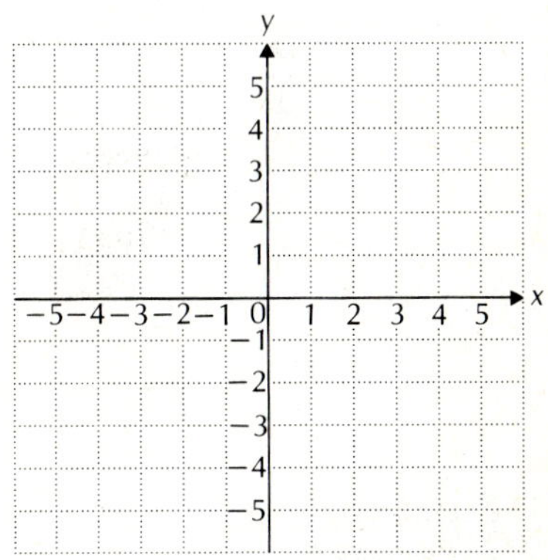

EXERCISE SET 3.1

i Translate these problem situations to systems of equations.

1. The sum of two numbers is -42. The first number minus the second is 52. What are the numbers?

2. The difference between two numbers is 11. Twice the smaller plus three times the larger is 123. What are the numbers?

3. The perimeter of a rectangular field is 628 m. The length of the field is 6 m greater than its width. What are the dimensions of the field?

4. One week a business sold 40 scarves. White ones sold for \$4.95 and printed ones sold for \$7.95. In all, \$282 worth of scarves were sold. How many of each kind were sold?

5. Bill is nine years younger than his sister. The sum of their ages is 43. How old are Bill and his sister?

6. A basketball player scored 18 times during one game. He scored a total of 30 points, two for each field goal and one for each free throw. How many field goals did he make? How many free throws?

7. At a barbecue there were 250 dinners served. Children's plates were \$1.50 each and adult's plates were \$2.00. If the total amount of money collected for dinners at the barbecue was \$441, how many of each type of plate was served?

8. Alan and Sarah drove a total of 16 hours on their trip. Alan drove at an average speed of 50 miles per hour, while Sarah drove at an average speed of 55 miles per hour. If their trip was 940 miles long, how many hours did each of them drive?

9. An airplane has a total of 152 seats. The number of coach-class seats is five more than six times the number of first-class seats. How many of each type of seat are there on the plane?

10. A disc jockey must play 12 commercial spots during one hour of a radio show. Each commercial is either 30 seconds or 60 seconds long. If the total commercial time during that hour is 10 minutes, how many 30-second commercials were played that hour? How many 60-second commercials?

ii Determine whether the ordered pair is a solution of the system of equations. Remember to use alphabetical order of variables.

11. $(1, 2)$; $4x - y = 2$, $10x - 3y = 4$

12. $(-1, -2)$; $2x + y = -4$, $x - y = 1$

13. $(2, 5)$; $y = 3x - 1$, $2x + y = 4$

14. $(-1, -2)$; $x + 3y = -7$, $3x - 2y = 12$

15. $(1, 5)$; $x + y = 6$, $y = 2x + 3$

16. $(5, 2)$; $a + b = 7$, $2a - 8 = b$

17. $(2, -7)$; $3a + b = -1$, $2a - 3b = -8$

18. $(2, 1)$; $3p + 2q = 5$, $4p + 5q = 2$

19. $(3, 1)$; $3x + 4y = 13$, $5x - 4y = 11$

20. $(5, 4)$; $-a + 3b = 7$, $2a + b = 16$

21. $(-2, 3)$; $6x - y = -15$, $y = 4x + 11$

22. $(4, -2)$; $-3x - 2y = -8$, $y = 2x - 5$

iii Solve each system graphically. Be sure to check.

23. $x + y = 4$, $x - y = 2$

24. $x - y = 3$, $x + y = 5$

25. $2x - y = 4$, $5x - y = 13$

26. $3x + y = 5$, $x - 2y = 4$

27. $4x - y = 9$, $x - 3y = 16$

28. $2y = 6 - x$, $3x - 2y = 6$

29. $a = 1 + b$, $b = -2a + 5$

30. $x = y - 1$, $2x = 3y$

31. $2u + v = 3$, $2u = v + 7$

32. $2b + a = 11$, $a - b = 5$

33. $y = -\frac{1}{3}x - 1$, $4x - 3y = 18$

34. $y = -\frac{1}{4}x + 1$, $2y = x - 4$

35. $6x - 2y = 2$, $9x - 3y = 1$

36. $y - x = 5$, $2x - 2y = 10$

37. $2x - y = 1$, $x + y = 2$

38. $x = -3$, $y = 2$

39. $x + 2y = 2$, $3x - 2y = -10$

40. $a + 2b = 4$, $b - a = 3$

○

Translate to systems of equations.

41. Burl is twice as old as his son. Ten years ago Burl was three times as old as his son. How old are they now?

42. Mike and Barbara have been mathematics professors at a state university. Together they have 46 years of service. Two years ago Mike had taught 2.5 times as many years as Barbara. How long has each taught at the university?

43. A piece of posterboard has a perimeter of 156 in. If you cut 6 in. off the width, the length becomes four times the width. What are the dimensions of the original piece of posterboard?

44. A baseball team has played 104 games. Before winning their last eight games in a row, they had won three times as many games as they had lost. What is their present win-loss record?

45. John has 58 coins, consisting of dimes and quarters, in his bank. His grandmother gave him 7 quarters to put in his bank, which meant he had four times as many quarters as dimes. How many quarters and how many dimes does John have in his bank?

3.2

SOLVING SYSTEMS OF EQUATIONS

Graphing is not an efficient or accurate method for solving systems of equations. We now consider more efficient methods.

i The Substitution Method

EXAMPLE 1 Solve this system:

$$x + y = 5$$
$$x = y + 1.$$

The second equation says that x and $y + 1$ name the same thing. In the first equation we can substitute $y + 1$ for x:

$$x + y = 5$$
$$(y + 1) + y = 5 \qquad \text{Substituting } y + 1 \text{ for } x$$

We solve this last equation, using methods learned earlier:

$$(y + 1) + y = 5$$
$$2y + 1 = 5 \qquad \text{Removing parentheses and collecting like terms}$$
$$2y = 4 \qquad \text{Adding } -1 \text{ on both sides}$$
$$y = 2. \qquad \text{Multiplying on both sides by } \tfrac{1}{2}$$

We return to the original pair of equations, and substitute 2 for y in either of them. Calculation will be easier if we choose the second one:

$$x = y + 1$$
$$x = 2 + 1$$
$$x = 3.$$

We obtain the ordered pair (3, 2). We check to be sure it is a solution.

Check:

$x + y = 5$	
$3 + 2$	5
5	

$x = y + 1$	
3	$2 + 1$
	3

Since (3, 2) checks, it is the solution.

DO EXERCISES 1 AND 2.

Suppose neither equation of a pair has a variable alone on one side. We then solve one equation for one of the variables.

EXAMPLE 2 Solve this system:

$$2x + y = 6$$
$$3x + 4y = 4.$$

First we solve one equation for one variable. We solve the first equation for y because that will make the work simpler:

$$y = 6 - 2x.$$

OBJECTIVES

After finishing Section 3.2, you should be able to:

i **Solve systems of equations in two variables by the substitution method.**

ii **Solve systems of equations in two variables by the addition method.**

Solve by the substitution method.

1. $x + y = 6$
$y = x + 2$

2. $y = 7 - x$
$2x - y = 8$

(*Caution:* Use parentheses when you substitute. Then be very careful about removing them.)

Solve. Use the substitution method.

3. $2y + x = 1$
$y - 2x = 8$

4. $8x + 5y = 184$
$x - y = -3$

Then we substitute $6 - 2x$ for y in the second equation and solve for x:

$$\begin{aligned} 3x + 4(6 - 2x) &= 4 && \text{Substituting } 6 - 2x \text{ for } y \\ 3x + 24 - 8x &= 4 && \text{Multiplying to remove parentheses} \\ 3x - 8x &= 4 - 24 \\ -5x &= -20 \\ x &= 4. \end{aligned}$$

Now we can substitute 4 for x in either equation and solve for y. We use the first equation:

$$\begin{aligned} 2x + y &= 6 \\ 2 \cdot 4 + y &= 6 \\ y &= -2. \end{aligned}$$

We obtain $(4, -2)$. This checks, so it is the solution.

DO EXERCISES 3 AND 4.

ii The Addition Method

The *addition method* for solving systems of equations makes use of the *addition principle.* Some systems are much easier to solve using the addition principle.

Consider this system:

$$\begin{aligned} 2x - 3y &= 0 \\ -4x + 3y &= -1. \end{aligned}$$

The first equation says that $2x - 3y$ and 0 represent the same number. We use the addition principle with the second equation. We add the "same thing" on both sides. On the left we call it $2x - 3y$. On the right we call it 0:

$$-4x + 3y + (2x - 3y) = -1 + 0$$

or

$$-2x = -1.$$

Using this method we try to get an equation with only one variable. We often say that we have *eliminated* a variable. The addition method is sometimes called the *method of elimination.*

EXAMPLE 3 Solve this system:

$$\begin{aligned} 2x - 3y &= 0 \\ -4x + 3y &= -1. \end{aligned}$$

We have

$$\begin{array}{rcrl} 2x - 3y &=& 0 & \\ -4x + 3y &=& -1 & \\ \hline -2x + \ \ 0 &=& -1 & \text{Adding} \end{array}$$

We have made one variable "disappear." Now we solve this equation for x:

$$\begin{aligned} -2x &= -1 \\ x &= \tfrac{1}{2}. \end{aligned}$$

We substitute $\frac{1}{2}$ for x in the first equation and solve for y:

$$2 \cdot \tfrac{1}{2} - 3y = 0 \quad \text{Substituting}$$
$$\left.\begin{aligned} 1 - 3y &= 0 \\ y &= \tfrac{1}{3} \end{aligned}\right\} \quad \text{Solving for } y$$

We obtain $(\frac{1}{2}, \frac{1}{3})$. This checks, so it is the solution.

DO EXERCISES 5 AND 6.

Note in Example 3 that a term in one equation and a term in the other were additive inverses of each other. Thus their sum was 0. That enabled us to eliminate a variable.

In order to eliminate a variable we sometimes must multiply one or both of the equations by something before adding.

EXAMPLE 4 Solve this system:

$$3x + 3y = 15$$
$$2x + 6y = 22.$$

If we add, we do not eliminate a variable. However, if the $3y$ in the first equation were $-6y$, we could. So we multiply by -2 on both sides of the first equation.

$$\begin{aligned} -6x - 6y &= -30 \quad \text{Multiplying by } -2 \text{ on both sides} \\ 2x + 6y &= 22 \\ \hline -4x + 0 &= -8 \quad \text{Adding} \\ x &= 2 \quad \text{Solving for } x \end{aligned}$$

Then

$$\begin{aligned} 2 \cdot 2 + 6y &= 22 \quad \text{Substituting 2 for } x \text{ in second equation} \\ 4 + 6y &= 22 \\ y &= 3 \quad \text{Solving for } y \end{aligned}$$

We obtain (2, 3), or $x = 2$, $y = 3$. This checks, so it is the solution.

DO EXERCISE 7.

Sometimes we must multiply twice to make two terms become additive inverses.

EXAMPLE 5 Solve this system:

$$2x + 3y = 12$$
$$5x + 7y = 29.$$

From first equation: $10x + 15y = 60$ Multiplying by 5

From second equation: $-10x - 14y = -58$ Multiplying by -2

$$\begin{aligned} 10x + 15y &= 60 \\ -10x - 14y &= -58 \\ \hline 0 + y &= 2 \quad \text{Adding} \\ y &= 2 \quad \text{Solving for } y \end{aligned}$$

Solve. Use the addition method.

5. $5x + 3y = 17$
$-5x + 2y = 3$

6. $-3a + 2b = 0$
$3a - 4b = -1$

Solve. Use the addition method.

7. $2y + 3x = 12$
$-4y + 5x = -2$

8. Solve. Use the addition method.

$$3x + 5y = 30$$
$$5x + 3y = 34$$

9. Clear of decimals.

$$1.32x + 9.12y = 0.1$$
$$4.2x - 0.6y = 1.2$$

(*Hint:* Multiply the first equation by 100 and the second one by 10.)

10. Clear of fractions.

$$\frac{3}{5}x + \frac{2}{3}y = \frac{1}{3}$$
$$\frac{3}{4}x - \frac{1}{3}y = \frac{1}{4}$$

We then have

$$2x + 3 \cdot 2 = 12 \quad \text{Substituting 2 for } y \text{ in first equation}$$
$$2x = 6$$
$$x = 3. \quad \text{Solving for } x$$

We obtain (3, 2), or $x = 3$, $y = 2$. This checks, so it is the solution.

DO EXERCISE 8.

When decimals or fractions occur, we *clear* before solving.

EXAMPLE 6 Clear of decimals:

$$0.5x + 0.3y = 1.9$$
$$1.2x - 0.3y = 12.4.$$

We solve as follows:

$$5x + 3y = 19 \quad \text{Multiplying on both sides by 10}$$
$$12x - 3y = 124. \quad \text{Multiplying on both sides by 10}$$

We can now finish solving as in Example 3.

DO EXERCISE 9.

EXAMPLE 7 Clear of fractions:

$$\tfrac{1}{2}x + \tfrac{2}{3}y = 1$$
$$\tfrac{3}{4}x - \tfrac{1}{3}y = 2.$$

We solve as follows:

$$6(\tfrac{1}{2}x + \tfrac{2}{3}y) = 6 \cdot 1 \quad \text{Multiplying on both sides by 6}$$
$$12(\tfrac{3}{4}x - \tfrac{1}{3}y) = 12 \cdot 2 \quad \text{Multiplying by 12}$$

We have multiplied each equation by the LCM of the denominators. Now we simplify.

First equation:

$$6 \cdot \tfrac{1}{2}x + 6 \cdot \tfrac{2}{3}y = 6$$
$$3x + 4y = 6$$

Second equation:

$$12 \cdot \tfrac{3}{4}x - 12 \cdot \tfrac{1}{3}y = 24$$
$$9x - 4y = 24$$

We can now finish solving as in Example 3.

DO EXERCISE 10.

There are some preliminary things that we might do to make the steps of the addition method simpler. One is to first write the equations in the form $Ax + By = C$. Another is to interchange two equations before beginning. For example, if we have the system

$$5x + y = -2$$
$$x + 7y = 3$$

we may prefer to write the second equation first. This accomplishes two things. One is that the x-coefficient is 1 in the first equation. After y has been found, this will make solving for x easier. The other is that this makes the x-coefficient in the second equation a multiple of the first.

The method illustrated in Examples 3–7 establishes a step-by-step procedure, or *algorithm*, which could even be done mechanically by a computer.

DO EXERCISES 11–13.

Solve.

11. $4x + 3y = -6$
$-4x + 2y = 16$

12. $9x - 2y = -4$
$3x + 4y = 1$

13. $0.2x + 0.3y = 0.1$
$0.3x - 0.1y = 0.7$

Which Method Is Better?

The substitution method is usually better when a variable has a coefficient of 1, as in the following examples:

$$8x + 4y = 180 \qquad x + y = 561$$
$$-x + y = 3; \qquad x = 2y.$$

Both methods work. When in doubt, use the addition method.

EXERCISE SET 3.2

i Solve using the substitution method.

1. $3x + 5y = 3$, $x = 8 - 4y$

2. $2x - 3y = 13$, $y = 5 - 4x$

3. $9x - 2y = 3$, $3x - 6 = y$

4. $x = 3y - 3$, $x + 2y = 9$

5. $5m + n = 8$, $3m - 4n = 14$

6. $4x + y = 1$, $x - 2y = 16$

7. $4x + 12y = 4$, $-5x + y = 11$

8. $-3b + a = 7$, $5a + 6b = 14$

9. $3x - y = 1$, $2x + 2y = 2$

10. $5p + 7q = 1$, $4p - 2q = 16$

11. $3x - y = 7$, $2x + 2y = 5$

12. $5x + 3y = 4$, $x - 4y = 3$

ii Solve using the addition method.

13. $x + 3y = 7$, $-x + 4y = 7$

14. $x + y = 9$, $2x - y = -3$

15. $2x + y = 6$, $x - y = 3$

16. $x - 2y = 6$, $-x + 3y = -4$

17. $9x + 3y = -3$, $2x - 3y = -8$

18. $6x - 3y = 18$, $6x + 3y = -12$

19. $5x + 3y = 19$, $2x - 5y = 11$

20. $3x + 2y = 3$, $9x - 8y = -2$

21. $5r - 3s = 24$, $3r + 5s = 28$

22. $5x - 7y = -16$, $2x + 8y = 26$

23. $0.3x - 0.2y = 4$, $0.2x + 0.3y = 1$

24. $0.7x - 0.3y = 0.5$, $-0.4x + 0.7y = 1.3$

25. $\frac{1}{2}x + \frac{1}{3}y = 4$, $\frac{1}{4}x + \frac{1}{3}y = 3$

26. $\frac{2}{3}x + \frac{1}{7}y = -11$, $\frac{1}{7}x - \frac{1}{3}y = -10$

27. $3p - 3q = 7$, $3p + 5q = 17$

28. $3x + 4y = 4$, $x - 2y = 2$

29. $4x + 5y = 20$, $3x - y = 18$

30. $\frac{1}{3}x + \frac{4}{5}y = 7$, $\frac{1}{6}x - \frac{2}{5}y = -4$

Solve.

31. $5x - 9y = 7$, $7y - 3x = -5$

32. $a - 2b = 16$, $b + 3 = 3a$

33. $3(a - b) = 15$, $4a = b + 1$

34. $1.3x - 0.2y = 12$, $0.4x + 17y = 89$

35. $x - \frac{1}{10}y = 100$, $y - \frac{1}{10}x = -100$

36. $\frac{1}{8}x + \frac{3}{5}y = \frac{19}{2}$, $-\frac{3}{10}x - \frac{7}{20}y = -1$

37. $0.05x + 0.25y = 22$, $0.15x + 0.05y = 24$

38. $10x + y = 306$, $10y + x = 90$

○

Solve.

39. ▦ $3.5x - 2.1y = 106.2$, $4.1x + 16.7y = -106.28$

40. $\dfrac{x + y}{2} - \dfrac{x - y}{5} = 1$, $\dfrac{x - y}{2} + \dfrac{x + y}{6} = -2$

41. Solve for x and y in terms of a and b:

$$5x + 2y = a$$
$$x - y = b$$

42. Determine a and b for which $(-4, -3)$ will be a solution of the system.

$$ax + by = -26$$
$$bx - ay = 7$$

43. The points $(0, -3)$ and $(-\frac{3}{2}, 6)$ are two of the solutions of the equation $px - qy = -1$. Find p and q.

Each of the following is a system of equations that is *not* linear. But each is *reducible to linear*, because an appropriate substitution (say u for $1/x$ and v for $1/y$) yields a linear system. Solve for the new variable and then solve for the original variable.

44. $\frac{1}{x} - \frac{3}{y} = 2$

$\frac{6}{x} + \frac{5}{y} = -34$

45. $\frac{2}{x} + \frac{1}{y} = 0$

$\frac{5}{x} + \frac{2}{y} = -5$

OBJECTIVES

After finishing Section 3.3, you should be able to:

i **Solve problems by translating them to systems of two equations and solving the system.**

3.3

SOLVING PROBLEMS WITH TWO EQUATIONS

i You are now in a much better position with respect to solving problems, because you can use systems of equations. That usually makes the translating step easier.

EXAMPLE 1 One day Glovers, Inc., sold 20 pairs of gloves. Cloth gloves sold for \$4.95 per pair and pigskin gloves sold for \$7.50 per pair. The company took in \$137.25. How many of each kind were sold?

1. **Familiarize** There are many ways to go about familiarizing ourselves with a problem situation. This time, let's make a guess and do some calculations. We guess:

They sold 12 pairs of cloth gloves and
11 pairs of pigskin gloves.

Does the guess check? Does it make sense?

The total number of pairs of gloves sold was supposed to be 20, so our guess cannot be right. Let's try another guess:

12 pairs of cloth gloves and
8 pairs of pigskin gloves.

Now the total number sold is 20, so our guess is right in that respect.

How much money was taken in? The problem says that cloth gloves sold for \$4.95, so they took in

$$12(\$4.95)$$

from the cloth gloves. They took in

$$8(\$7.50)$$

from the pigskin gloves. This makes the total taken in

$$12(\$4.95) + 8(\$7.50) = \$59.40 + \$60.00 = \$119.40.$$

Our guess is not the answer to the problem, because the total taken in, according to the problem, was \$137.25. However, we are more familiar with the problem situation.

Let's list the information in a table. That usually helps a great deal in the familiarization process.

Kind of glove	Price	Number sold	Money taken in
Cloth	\$4.95	x	$4.95x$
Pigskin	\$7.50	y	$7.50y$
	Total	20	137.25

2. **Translate** We shall let $x =$ the number of pairs of cloth gloves sold and $y =$ the number of pairs of pigskin gloves sold. The total number sold was 20, so we have

$$x + y = 20.$$

The amount taken in for cloth gloves is then $4.95x$ and the amount taken in for pigskin gloves is $7.50y$. These amounts are in dollars. The total was \$137.25, so we have

$$4.95x + 7.50y = 137.25.$$

We can multiply on both sides by 100 in order to clear of decimals. Thus we have the translation, a system of equations:

$$x + y = 20$$
$$495x + 750y = 13{,}725.$$

3. **Carry out** We solve the system of equations, using either the substitution method or the addition method. In this case the substitution method is simpler. When we solve, we obtain

$$x = 5 \quad \text{and} \quad y = 15.$$

4. **Check** We check the possible answer in the original problem. Remember that x is the number of cloth gloves and y is the number of pigskin gloves. Thus:

Money from cloth gloves: $\$4.95x = 4.95 \times 5 = \$\ 24.75$

Money from pigskin gloves: $\$7.50y = 7.50 \times 15 = \$112.50$

Total $\$137.25$

Number of gloves: $x + y = 5 + 15 = 20$

The numbers check.

5. **State** The answer is that Glovers sold 5 pairs of cloth gloves and 15 pairs of pigskin gloves.

DO EXERCISE 1.

EXAMPLE 2 (*A mixture problem.*) Solution A is 2% alcohol. Solution B is 6% alcohol. A service station owner wants to mix the two to get 60 liters of a solution that is 3.2% alcohol. How many liters of each should the owner use?

1. **Familiarize** We keep information in a table.

	Amount of solution	Percent of alcohol	Amount of alcohol in solution
A	x liters	2%	$2\%x$
B	y liters	6%	$6\%y$
Mixture	60 liters	3.2%	0.032×60, or 1.92 liters

Note that we have used x for the number of liters of A and y for the number of liters of B. To get the amount of alcohol, we multiply by the percentages, of course.

2. **Translate** If we add x and y in the first column we get 60, and this gives us one equation:

$$x + y = 60.$$

1. A store sold 30 sweatshirts. They sold white ones for \$8.95 and red ones for \$9.50. They took in \$272.90. How many of each color did they sell?

2. A gardener has two kinds of solutions containing weedkiller and water. One is 5% weedkiller and the other is 15% weedkiller. The gardener needs 100 liters of a 12% solution and wants to make it by mixing. How much of each solution should be used?

If we add the amounts of alcohol in the third column we get 1.92, and this gives us another equation:

$$2\%x + 6\%y = 1.92.$$

After changing percents to decimals and clearing, we have this system:

$$x + y = 60$$
$$2x + 6y = 192.$$

3. **Carry out** We solve the system

$$x + y = 60$$
$$2x + 6y = 192$$

to get

$$-2x - 2y = -120 \quad \text{Multiplying in the first equation by } -2$$
$$2x + 6y = 192$$
$$0 + 4y = 72 \quad \text{Adding}$$
$$y = 18 \quad \text{Solving for } y$$

$$x + 18 = 60 \quad \text{Substituting into the first equation of the system}$$
$$x = 42$$

4. **Check** Remember, $x =$ the number of liters of 2% solution and $y =$ the number of liters of 6% solution.

Total number of liters of mixture: $x + y = 42 + 18 = 60$

Amount of alcohol: $2\% \times 42 + 6\% \times 18 = 0.02 \times 42 + 0.06 \times 18 = 1.92$ liters

Percentage of alcohol in mixture: $\frac{1.92}{60} = 0.032$, or 3.2%

The numbers check in the original problem.

5. **State** The answer is that the owner should use 42 L of 2% alcohol and 18 L of 6% alcohol.

DO EXERCISE 2.

EXAMPLE 3 (*An interest problem.*) Two investments are made totaling \$4800. In the first year they yield \$412 in simple interest. Part of the money is invested at 8% and the rest at 9%. Find the amount invested at each rate of interest.

1. **Familiarize** Keeping information in a table will help. The columns in the table come from the formula for simple interest: $I = Prt$.

	Principal, P	Rate of interest, r	Time, t	Interest, I
First investment	x	8%	1 yr	$8\%x$
Second investment	y	9%	1 yr	$9\%y$
Total	\$4800			\$412

2. **Translate** Note that we have used x and y for the numbers of dollars invested. They total \$4800, and this gives us one equation:

$$x + y = 4800.$$

Look at the last column. The interest, or *yield,* totals \$412. This gives us a second equation:

$$8\%x + 9\%y = 412, \quad \text{or} \quad 0.08x + 0.09y = 412.$$

After we multiply on both sides to clear of decimals, we have

$$8x + 9y = 41{,}200.$$

3. **Carry out** We solve the system

$$x + y = 4800$$
$$8x + 9y = 41{,}200$$

We find that $x = 2000$ and $y = 2800$.

5. **State** The numbers check in the problem, so we know that \$2000 is invested at 8% and \$2800 at 9%.

DO EXERCISE 3.

EXAMPLE 4 The perimeter of a rectangular field is 194 yards. The length is two more than four times the width. Find the length and the width.

1. **Familiarize** We make a drawing and label it. We recall, or look up, the definition of *perimeter*: $P = 2l + 2w$.

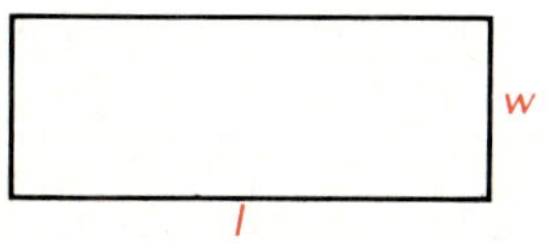

2. **Translate** We have used l for length and w for width. We now translate as follows:

The perimeter is 194

$$2l + 2w \quad = 194.$$

We translate the second statement.

The length is 2 more than four times the width.

$$l \quad = 2 \quad + \quad 4 \quad \cdot \quad w$$

We now have the system of equations

$$2l + 2w = 194$$
$$l = 2 + 4w.$$

3. **Carry out** In this case, it is probably easier to use the substitution method, substituting $2 + 4w$ for l in the first equation:

$2(2 + 4w) + 2w = 194$	Substituting
$4 + 8w + 2w = 194$	Multiplying to remove parentheses on the left
$4 + 10w = 194$	Collecting like terms
$10w = 190$	Adding -4
$w = \frac{190}{10}$, or 19.	Multiplying by $\frac{1}{10}$

3. \$3700 is invested for a year at simple interest, yielding \$297. Part of the money is invested at 7% and the rest at 9%. How much was invested at each rate?

4. The perimeter of a rectangle is 90 ft. The length is 20 ft greater than the width. Find the length and the width.

Then we substitute 19 for w in the second equation:

$$\begin{aligned} l &= 2 + 4 \cdot 19 \\ &= 2 + 76 \\ &= 78. \end{aligned}$$

5. **State** The numbers 78 and 19 check in the original problem, so we know that the length is 78 yards and the width is 19 yards.

DO EXERCISE 4.

EXAMPLE 5 (*A motion problem.*) A train leaves Sioux City traveling east at 30 km/h. Two hours later another train leaves Sioux City traveling in the same direction on a parallel track at 45 km/h. At what point will the faster train overtake the slower one?

1. **Familiarize** We make a drawing.

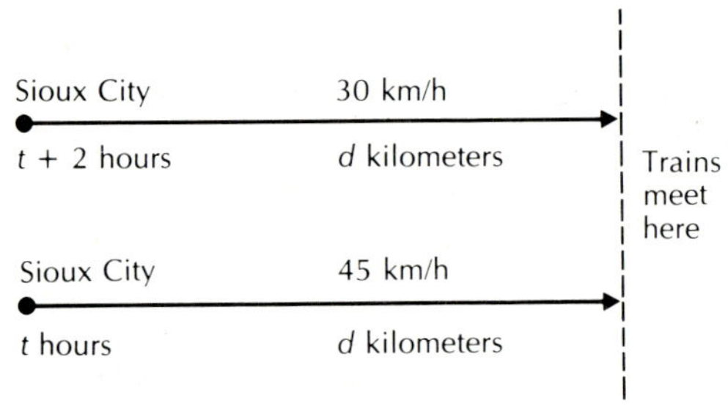

We know that distance, speed, and time are related by the definition of speed:

$$\text{Speed} = \frac{\text{Distance}}{\text{Time}}, \quad \text{or} \quad r = \frac{d}{t}.$$

We often solve for d, to get $d = rt$. From the drawing we see that the distances are the same. We have already used the letter d for distance and t for the time traveled by the faster train. Then $t + 2$ is the time traveled by the slower one.

We list the information in a table, the columns of the table being determined by the formula $d = rt$.

	Distance	Rate	Time
Slow train	d	30	$t + 2$
Fast train	d	45	t

2. **Translate** Using $d = rt$ in each row of the table, we get an equation. Thus we have a system of equations:

$$d = 30(t + 2)$$
$$d = 45t.$$

3. **Carry out** We solve the system:

$$\begin{aligned} 30(t + 2) &= 45t \qquad \text{Using substitution} \\ 30t + 60 &= 45t \\ 60 &= 15t \\ 4 &= t. \end{aligned}$$

Thus the time for the faster train should be 4 hours and the time for the slower train 6 hours.

4. **Check** At 45 km/h the faster train will travel $45 \cdot 4$, or 180 km, in 4 hours. At 30 km/h the slower train will travel $30 \cdot 6$, or 180 km, in 6 hours. The numbers check.

5. **State** The answer is that the trains will meet at a point 180 km east of Sioux City.

DO EXERCISE 5.

5. A train leaves Barstow traveling east at 35 km/h. One hour later a faster train leaves Barstow, also traveling east on a parallel track at 40 km/h. How far from Barstow will the faster train catch the slower one?

EXERCISE SET 3.3

i

1. The sum of a certain number and a second number is -42. The first number minus the second is 52. Find the numbers.

2. The sum of two numbers is -63. The first number minus the second is -41. Find the numbers.

3. The difference between two numbers is 16. Three times the larger number is nine times the smaller. What are the numbers?

4. The difference between two numbers is 11. Twice the smaller number plus three times the larger number is 123. What are the numbers?

5. Soybean meal is 16% protein; corn meal is 9% protein. How many pounds of each should be mixed together to get a 350-pound mixture that is 12% protein?

6. A chemist has one solution that is 25% acid and a second that is 50% acid. How many liters of each should be mixed together to get 10 liters of a solution that is 40% acid?

7. One canned juice drink is 15% orange juice; another is 5% orange juice. How many liters of each should be mixed together to get 10 liters that is 10% orange juice?

8. Antifreeze A is 18% alcohol. Antifreeze B is 10% alcohol. How many liters of each should be mixed to get 20 liters of a mixture that is 15% alcohol?

9. Two investments are made totaling \$8800. For a certain year these investments yield \$1326 in simple interest. Part of the \$8800 is invested at 14% and part at 16%. Find the amount invested at each rate.

10. Two investments are made totaling \$15,000. For a certain year these investments yield \$1432 in simple interest. Part of the \$15,000 is invested at 9% and part at 10%. Find the amount invested at each rate.

11. \$1150 is invested, part of it at 12% and part of it at 11%. The total yield is \$133.75. How much was invested at each rate?

12. \$27,000 is invested, part of it at 10% and part of it at 12%. The total yield is \$2990. How much was invested at each rate?

13. A train leaves a station and travels north at 75 km/h. Two hours later a second train leaves on a parallel track and travels north at 125 km/h. How far from the station will they meet?

14. Two cars leave town traveling in opposite directions. One travels 80 km/h and the other 96 km/h. In how many hours will they be 528 km apart?

15. Two motorcycles travel toward each other from Chicago and Indianapolis, which are about 350 km apart, at rates of 110 and 90 km/h. They started at the same time. In how many hours will they meet?

16. Two planes travel toward each other from cities that are 780 km apart at rates of 190 and 200 km/h. They started at the same time. In how many hours will they meet?

17. One day a store sold 30 sweatshirts. White ones cost \$9.95 and yellow ones cost \$10.50. In all, \$310.60 worth of sweatshirts were sold. How many of each color were sold?

18. One week a business sold 40 scarves. White ones cost \$4.95 and printed ones cost \$7.95. In all, \$282 worth of scarves were sold. How many of each kind were sold?

19. One day a store sold 45 pens, one kind at \$8.50 and another kind at \$9.75. In all, \$398.75 was taken in. How many of each kind were sold?

20. At a club play, 117 tickets were sold. Adult's tickets cost \$1.25 and children's tickets cost \$0.75. In all, \$129.75 was taken in. How many of each kind of ticket were sold?

21. Carlos is 8 years older than his sister Maria. Four years ago Maria was $\frac{2}{3}$ as old as Carlos. How old are they now?

22. Paula is 12 years older than her brother Bob. Four years from now Bob will be $\frac{2}{3}$ as old as Paula. How old are they now?

23. The perimeter of a rectangular field is 628 m. The length of the field exceeds its width by 6 m. Find the dimensions.

24. The perimeter of a lot is 190 m. The width is one fourth of the length. Find the dimensions.

25. The perimeter of a rectangle is 86 cm. The length is 19 cm greater than the width. Find the length and the width.

26. The perimeter of a rectangle is 384 m. The length is 82 m greater than the width. Find the length and the width.

27. James and Joan are mathematics teachers. They have a total of 46 years of teaching. Two years ago James had taught 2.5 times as many years as Joan. How long has each taught?

28. Nancy jogs and walks to school each day. She averages 4 km/h walking and 8 km/h jogging. The distance from home to school is 6 km and she makes the trip in 1 hour. How far does she jog in a trip?

29. The ten's digit of a two-digit positive integer is 2 more than three times the unit's digit. If the digits are interchanged, the new number is 13 less than half the given number. Find the given integer. (*Hint:* Let $x =$ the ten's place digit and $y =$ the unit's place digit; then $10x + y$ is the number.)

30. A limited edition of a book published by a historical society was offered for sale to its membership. The cost was one book for \$12 or two books for \$20. The society sold 880 books and the total amount of money taken in was \$9840. How many members ordered two books?

31. The measure of one of two supplementary angles is 8° more than three times the other. Find the measure of the larger of the two angles.

32. The numerator of a fraction is 12 more than the denominator. The sum of the numerator and the denominator is 5 more than three times the denominator. What is the reciprocal of the fraction?

33. An automobile radiator contains 16 liters of antifreeze and water. This mixture is 30% antifreeze. How much of this mixture should be drained and replaced with pure antifreeze so that there will be 50% antifreeze?

34. A train leaves Union Station for Central Station, 216 km away, at 9 A.M. One hour later, a train leaves Central Station for Union Station. They meet at noon. If the second train had started at 9 A.M. and the first train at 10:30 A.M., they would still have met at noon. Find the speed of each train.

35. An automobile gets 18 miles per gallon (mpg) in city driving and 24 mpg in highway driving. The car is driven 465 miles on a full tank of 23 gallons of gasoline. How many miles were driven in the city and how many were driven on the highway?

36. Phil and Phyllis are siblings. Phyllis has twice as many brothers as she has sisters. Phil has the same number of brothers as sisters. How many girls and how many boys are in the family?

OBJECTIVES

After finishing Section 3.4, you should be able to:

i **Determine whether an ordered triple is a solution of a system of three equations in three variables.**

ii **Solve systems of three equations in three variables.**

1. Consider the system

$$\begin{aligned} 4x + 2y + 5z &= 6 \\ 2x - y + z &= 5 \\ x + 2y - z &= 0. \end{aligned}$$

a) Determine whether (1, 2, 3) is a solution.

b) Determine whether (2, −1, 0) is a solution.

3.4

SYSTEMS OF EQUATIONS IN THREE VARIABLES

Some problem situations naturally give rise to a translation to two equations. Others more naturally give rise to a translation to three equations, or more. In this section we consider how to solve systems of three linear equations. Later, we will use such systems to solve problems.

i Identifying Solutions

A solution of a system of three equations in three variables is an ordered triple that makes *all three* equations true.

EXAMPLE 1 Determine whether $(\frac{3}{2}, -4, 3)$ is a solution of the following system.

$$\begin{aligned} 4x - 2y - 3z &= 5 \\ -8x - y + z &= -5 \\ 2x + y + 2z &= 5 \end{aligned}$$

We substitute $(\frac{3}{2}, -4, 3)$ into the three equations, using alphabetical order.

$$\begin{array}{r|l} \multicolumn{2}{c}{4x - 2y - 3z = 5} \\ \hline 4 \cdot \frac{3}{2} - 2(-4) - 3 \cdot 3 & 5 \\ 6 + 8 - 9 & \\ 5 & \end{array} \qquad \begin{array}{r|l} \multicolumn{2}{c}{-8x - y + z = -5} \\ \hline -8 \cdot \frac{3}{2} - (-4) + 3 & -5 \\ -12 + 4 + 3 & \\ -5 & \end{array}$$

$$\begin{array}{r|l} \multicolumn{2}{c}{2x + y + 2z = 5} \\ \hline 2 \cdot \frac{3}{2} + (-4) + 2 \cdot 3 & 5 \\ 3 - 4 + 6 & \\ 5 & \end{array}$$

The triple makes all three equations true, so it is a solution.

DO EXERCISE 1.

ii Solving

Graphical methods for solving linear equations in three variables are unsatisfactory, because a three-dimensional coordinate system is required. The substitution method becomes cumbersome for most systems of more than two equations. Therefore, we will use the addition method. It is essentially the same for systems of three equations as for systems of two equations.

EXAMPLE 2 Solve the following system of equations.

$$\begin{aligned} x + y + z &= 4 \quad &① \\ x - 2y - z &= 1 \quad &② \\ 2x - y - 2z &= -1 \quad &③ \end{aligned}$$

These numbers indicate the equations in the first, second, and third positions, respectively.

We begin by multiplying ① by -1, and adding it to ② to eliminate x from the second equation. Equations ① and ③ are unchanged by this step.

$$\begin{aligned} x + y + z &= 4, \quad &① \\ -3y - 2z &= -3, \quad &② \\ 2x - y - 2z &= -1. \quad &③ \end{aligned}$$

$$\begin{aligned} -x - y - z &= -4 \quad &\text{(Multiplying ① by } -1) \\ x - 2y - z &= 1 \quad &② \\ \hline -3y - 2z &= -3 \quad &\text{(Adding)} \end{aligned}$$

To eliminate x from the third equation we multiply ① by -2 and add it to ③:

$$\begin{aligned} x + y + z &= 4, \quad &① \\ -3y - 2z &= -3, \quad &② \\ -3y - 4z &= -9. \quad &③ \end{aligned}$$

$$\begin{aligned} -2x - 2y - 2z &= -8 \quad &\text{(Multiplying ① by } -2) \\ 2x - y - 2z &= -1 \quad &③ \\ \hline -3y - 4z &= -9 \quad &\text{(Adding)} \end{aligned}$$

Next we eliminate y from the third equation by multiplying ② by -1 and adding the result to ③:

$$\begin{aligned} x + y + z &= 4 \quad &① \\ -3y - 2z &= -3 \quad &② \\ -2z &= -6 \quad &③ \end{aligned}$$

$$\begin{aligned} 3y + 2z &= 3 \quad &\text{(Multiplying ② by } -1) \\ -3y - 4z &= -9 \quad &③ \\ \hline -2z &= -6 \quad &\text{(Adding)} \end{aligned}$$

When the system is in this triangular form, we can easily solve for the three variables.

First we solve ③ for z:

$$\begin{aligned} -2z &= -6 \\ z &= 3. \quad &\text{(Multiplying ③ by } -\tfrac{1}{2}) \end{aligned}$$

Next we substitute 3 for z in ② and solve for y:

$$\begin{aligned} -3y - 2(3) &= -3 \\ -3y - 6 &= -3 \\ -3y &= 3 \\ y &= -1. \end{aligned}$$

Finally, we substitute -1 for y and 3 for z in ①:

$$\begin{aligned} x + (-1) + 3 &= 4 \\ x + 2 &= 4 \\ x &= 2. \end{aligned}$$

2. Solve. Don't forget to check.

$$\begin{aligned} 4x - y + z &= 6 \\ -3x + 2y - z &= -3 \\ 2x + y + 2z &= 3 \end{aligned}$$

We have obtained the triple $(2, -1, 3)$. We check, in *all three* equations:

$x + y + z = 4$	
$2 + (-1) + 3$	4
4	

$x - 2y - z = 1$	
$2 - 2(-1) - 3$	1
1	

$2x - y - 2z = -1$	
$2 \cdot 2 - (-1) - 2 \cdot 3$	-1
-1	

The solution is $(2, -1, 3)$.

In Example 2, we obtained a system of equations equivalent to the original, but in a *triangular form.* After that, finding the solution was not difficult. The following is a description of the procedure. It should be followed for systems of three or more variables, except in very simple cases.

TO SOLVE A SYSTEM OF THREE LINEAR EQUATIONS

1. **Make each x-coefficient a multiple of the first. Do so by multiplying or interchanging equations, or both.**
2. **Eliminate the x-terms from the second and third equations.**
3. **Make the y-coefficient of the third equation a multiple of the second. Do so by multiplying or interchanging equations.**
4. **Eliminate the y-term from the third equation.**
5. **Solve the third equation for z. Then substitute upward to find y and x.**

DO EXERCISE 2.

EXAMPLE 3 Solve the following system of equations.

$$\begin{aligned} 2x - 4y + 6z &= 22 \quad ① \\ 4x + 2y - 3z &= 4 \quad ② \\ 3x + 3y - z &= 4 \quad ③ \end{aligned}$$

We begin by multiplying ③ by 2, to make each x-coefficient a multiple of the first:

$$\begin{aligned} 2x - 4y + 6z &= 22, \\ 4x + 2y - 3z &= 4, \\ 6x + 6y - 2z &= 8. \end{aligned}$$

Next, we multiply ① by -2 and add it to ②. We also multiply ① by -3 and add it to ③:

$$\begin{aligned} 2x - 4y + 6z &= 22 \\ 10y - 15z &= -40 \\ 18y - 20z &= -58. \end{aligned}$$

Now we multiply ③ by -5 to make the y-coefficient a multiple of the y-coefficient in ②:

$$\begin{aligned} 2x - 4y + 6z &= 22 \\ 10y - 15z &= -40 \\ -90y + 100z &= 290. \end{aligned}$$

Next, we multiply ② by 9 and add it to ③:

$$\begin{aligned} 2x - 4y + 6z &= 22 \\ 10y - 15z &= -40 \\ -35z &= -70. \end{aligned}$$

Now we solve ③ for z:

$$\begin{aligned} -35z &= -70 \\ z &= 2. \end{aligned}$$

Next we substitute 2 for z in ②, and solve for y:

$$\begin{aligned} 10y - 15(2) &= -40 \\ 10y - 30 &= -40 \\ y &= -1. \end{aligned}$$

Finally, we substitute -1 for y and 2 for z in ①:

$$\begin{aligned} 2x - 4(-1) + 6(2) &= 22 \\ 2x + 4 + 12 &= 22 \\ x &= 3. \end{aligned}$$

We have obtained the triple $(3, -1, 2)$. It checks, but the check is left to the reader.

DO EXERCISE 3.

EXAMPLE 4 Solve:

$$\begin{aligned} x + y + z &= 180 \quad ① \\ x \qquad - z &= -70 \quad ② \\ 2y - z &= 0. \quad ③ \end{aligned}$$

Note that there is no y in equation (2).

We use equations ① and ③ and eliminate y.

$$\begin{aligned} x + y + z &= 180 \quad ① \\ 2y - z &= 0 \quad ③ \end{aligned}$$

$$\begin{aligned} -2x - 2y - 2z &= -360 \quad \text{Multiplying by } -2 \\ 2y - z &= 0 \\ \hline -2x \qquad - 3z &= -360 \quad ④ \end{aligned}$$

Now we solve the resulting system of equations ② and ④.

$$\begin{aligned} x - z &= -70 \quad ② \\ -2x - 3z &= -360 \quad ④ \end{aligned}$$

$$\begin{aligned} 2x - 2z &= -140 \\ -2x - 3z &= -360 \\ \hline -5z &= -500 \\ z &= 100 \end{aligned}$$

Continuing as we did in the previous examples, we get the solution $(30, 50, 100)$.

DO EXERCISE 4.

3. Solve. Don't forget to check.

$$\begin{aligned} 2x + y - 4z &= 0 \\ x - y + 2z &= 5 \\ 3x + 2y + 2z &= 3 \end{aligned}$$

4. Solve. Don't forget to check.

$$\begin{aligned} x + y + z &= 100 \\ x - y \qquad &= -10 \\ x \qquad - z &= -30 \end{aligned}$$

EXERCISE SET 3.4

i

1. Determine whether (30, 50, 100) is a solution of the system.

$$x + y + z = 180$$
$$x - z = -70$$
$$2y - z = 0$$

2. Determine whether (2, −1, −2) is a solution of the system.

$$x + y - 2z = 5$$
$$2x - y - z = 7$$
$$-x - 2y + 3z = 6$$

ii Solve.

3. $x + y + z = 6$
$2x - y + 3z = 9$
$-x + 2y + 2z = 9$

4. $2x - y + z = 10$
$4x + 2y - 3z = 10$
$x - 3y + 2z = 8$

5. $2x - y - 3z = -1$
$2x - y + z = -9$
$x + 2y - 4z = 17$

6. $x - y + z = 6$
$2x + 3y + 2z = 2$
$3x + 5y + 4z = 4$

7. $2x - 3y + z = 5$
$x + 3y + 8z = 22$
$3x - y + 2z = 12$

8. $6x - 4y + 5z = 31$
$5x + 2y + 2z = 13$
$x + y + z = 2$

9. $3a - 2b + 7c = 13$
$a + 8b - 6c = -47$
$7a - 9b - 9c = -3$

10. $x + y + z = 0$
$2x + 3y + 2z = -3$
$-x + 2y - 3z = -1$

11. $2x + 3y + z = 17$
$x - 3y + 2z = -8$
$5x - 2y + 3z = 5$

12. $2x + y - 3z = -4$
$4x - 2y + z = 9$
$3x + 5y - 2z = 5$

13. $2x + y + z = -2$
$2x - y + 3z = 6$
$3x - 5y + 4z = 7$

14. $2x + y + 2z = 11$
$3x + 2y + 2z = 8$
$x + 4y + 3z = 0$

15. $x - y + z = 4$
$5x + 2y - 3z = 2$
$3x - 7y + 4z = 8$

16. $2x + y + 2z = 3$
$x + 6y + 3z = 4$
$3x - 2y + z = 0$

17. $4x - y - z = 4$
$2x + y + z = -1$
$6x - 3y - 2z = 3$

18. $a + 2b + c = 1$
$7a + 3b - c = -2$
$a + 5b + 3c = 2$

19. $2r + 3s + 12t = 4$
$4r - 6s + 6t = 1$
$r + s + t = 1$

20. $10x + 6y + z = 7$
$5x - 9y - 2z = 3$
$15x - 12y + 2z = -5$

21. $4a + 9b = 8$
$8a + 6c = -1$
$6b + 6c = -1$

22. $3p + 2r = 11$
$q - 7r = 4$
$p - 6q = 1$

Solve.

23. $\frac{x+2}{3} - \frac{y+4}{2} + \frac{z+1}{6} = 0$
$\frac{x-4}{3} + \frac{y+1}{4} - \frac{z-2}{2} = -1$
$\frac{x+1}{2} + \frac{y}{2} + \frac{z-1}{4} = \frac{3}{4}$

24. $0.2x + 0.3y + 1.1z = 1.6$
$0.5x - 0.2y + 0.4z = 0.7$
$-1.2x + y - 0.7z = -0.9$

25. $w + x + y + z = 2$
$w + 2x + 2y + 4z = 1$
$w - x + y + z = 6$
$w - 3x - y + z = 2$

26. $w + x - y + z = 0$
$w - 2x - 2y - z = -5$
$w - 3x - y + z = 4$
$2w - x - y + 3z = 7$

For Exercises 27 and 28, let u represent $1/x$, v represent $1/y$, and w represent $1/z$. Solve for u, v, and w, and then solve for x, y, and z.

27. $\frac{2}{x} - \frac{1}{y} - \frac{3}{z} = -1$
$\frac{2}{x} - \frac{1}{y} + \frac{1}{z} = -9$
$\frac{1}{x} + \frac{2}{y} - \frac{4}{z} = 17$

28. $\frac{2}{x} + \frac{2}{y} - \frac{3}{z} = 3$
$\frac{1}{x} - \frac{2}{y} - \frac{3}{z} = 9$
$\frac{7}{x} - \frac{2}{y} + \frac{9}{z} = -39$

29. Determine a, b, and c such that (2, 3, −4) is a solution of the system.

$$ax + by + cz = -11$$
$$bx - cy + az = -19$$
$$ax + cy - bz = 9$$

In each case three solutions of an equation are given. Find the equation using a system of equations.

30. $Ax + By + Cz = 12$;
$(1, \frac{3}{4}, 3)$, $(\frac{4}{3}, 1, 2)$, and $(2, 1, 1)$

31. $z = b - mx - ny$;
$(1, 1, 2)$, $(3, 2, -6)$, and $(\frac{3}{2}, 1, 1)$

3.5

SOLVING PROBLEMS WITH THREE EQUATIONS

i Solving systems of three or more equations is important in many applications. Systems of equations arise very often in the use of statistics, for example, in such fields as the social sciences. They also come up in problems of business, science, and engineering, for example.

EXAMPLE 1 The sum of three numbers is 4. The first number minus twice the second minus the third is 1. Twice the first number minus the second minus twice the third is -1. Find the numbers.

1. **Familiarize** In this case there are three obvious statements in the problem. The translation looks as though it can be made directly from these statements, as soon as we decide what letters to assign to the unknown numbers.

2. **Translate** Let us call the three numbers x, y, and z. Then we can translate directly, from the words of the problem, as follows.

$$\underbrace{\text{The sum of the three numbers}}_{x+y+z}\ \underbrace{\text{is}}_{=}\ \underbrace{4}_{4}$$

$$\underbrace{\text{The first number}}_{x}\ \underbrace{\text{minus}}_{-}\ \underbrace{\text{twice the second}}_{2y}\ \underbrace{\text{minus}}_{-}\ \underbrace{\text{the third}}_{z}\ \underbrace{\text{is}}_{=}\ \underbrace{1.}_{1}$$

$$\underbrace{\text{Twice the first number}}_{2x}\ \underbrace{\text{minus}}_{-}\ \underbrace{\text{the second}}_{y}\ \underbrace{\text{minus}}_{-}\ \underbrace{\text{twice the third}}_{2z}\ \underbrace{\text{is}}_{=}\ \underbrace{-1.}_{-1}$$

We now have a system of three equations:

$$\begin{aligned} x + y + z &= 4 \\ x - 2y - z &= 1 \\ 2x - y - 2z &= -1. \end{aligned}$$

3. **Carry out** We need to solve the system of equations. Note that we have already solved the system, as Example 2 in Section 3.4. We obtained $(2, -1, 3)$ for the solution.

4. **Check** We go to the original problem. The first statement says that the sum of the three numbers is 4. That checks.

   The second statement says that the first number minus twice the second minus the third is 1. We calculate:

$$\begin{aligned} x - 2y - z &= 1 \\ 2 - 2(-1) - 3 &= 1 \\ 2 + 2 - 3 &= 1 \\ 1 &= 1. \end{aligned}$$

   That checks.

   We leave the check of the third statement to the reader.

5. **State** The answer is that the three numbers are 2, -1, and 3.

DO EXERCISE 1.

OBJECTIVE

After finishing Section 3.5, you should be able to:

i **Solve problems by translating to a system of three equations in three variables.**

1. The sum of three numbers is 3. The first number minus twice the second minus the third is 4. Twice the first number plus the other two is 5. Find the numbers.

2. One angle of a triangle is twice as large as a second angle. The remaining angle is 20° greater than the first angle. Find the measures of the angles.

EXAMPLE 2 In a triangle the largest angle is 70° greater than the smallest angle. The largest angle is twice as large as the remaining angle. Find the measure of each angle.

1. **Familiarize** The first thing to do in a case like this is to make a drawing. Since we don't know the size of any angle, we have used x, y, and z for the measures of the angles. Here we use x for the smallest angle, z for the largest angle, and y for the remaining angle.

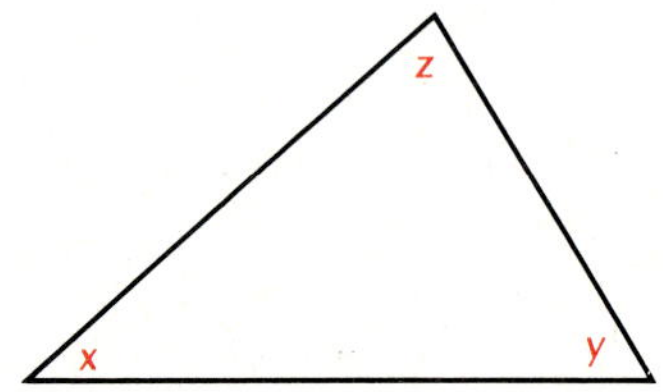

A geometric fact will be needed here, the fact that the measures of the angles of a triangle add up to 180°.

2. **Translate** This geometric fact about triangles gives us one equation:

$$x + y + z = 180.$$

There are two statements in the problem that we can translate almost directly.

The largest angle is 70° greater than the smallest angle.

$$z = 70 + x$$

The largest angle is twice as large as the remaining angle.

$$z = 2y$$

We now have a system of three equations:

$$\begin{aligned} x + y + z &= 180 \\ x + 70 &= z \\ 2y &= z; \end{aligned} \qquad \text{or} \qquad \begin{aligned} x + y + z &= 180 \\ x - z &= -70 \\ 2y - z &= 0. \end{aligned}$$

3. **Carry out** We solve the system. The details are left to the reader, but the solution is (30, 50, 100).

4. **Check** The sum of the numbers is 180, so that checks.

The largest angle measures 100° and the smallest measures 30°. The largest angle is 70° greater than the smallest.

The remaining angle measures 50°. The largest angle measures 100°, so it is twice as large. We do have an answer to the problem.

5. **State** The measures of the angles of the triangle are 30°, 50°, and 100°.

DO EXERCISE 2.

EXAMPLE 3 In a factory there are three machines, A, B, and C. When all three are running, they produce 222 suitcases per day. If A and B work, but C does not, they produce 159 suitcases per day. If B and C work, but A does not, they produce 147 suitcase per day. What is the daily production of each machine?

1. **Familiarize** Keeping information in a table will be of great help in this problem.

Machines working	A	B	C	A and B	B and C	A, B, and C
Daily production	x	y	z	159	147	222

2. **Translate** From the table, it is apparent that we have decided to use x, y, and z for the daily productions of the individual machines. Also from the table, we can readily obtain three equations:

$$x + y + z = 222 \quad \text{(All three machines work)}$$
$$x + y = 159 \quad \text{(A and B work)}$$
$$y + z = 147. \quad \text{(B and C work)}$$

3. **Carry out** We solve the system of equations. The details are left to the reader, but the solution is (75, 84, 63).

4. **Check** The numbers in the triple are supposed to be the daily numbers of suitcases produced. We add the three numbers, and obtain 222. That checks, since the problem states that the total number produced by all three machines is 222.

When A and B are working, the numbers are 75 and 84. They add up to 159. That checks.

When B and C are working, the numbers are 84 and 63. They add up to 147. That checks.

5. **State** The answer is that machine A produces 75 suitcases per day, machine B produces 84 suitcases per day, and machine C produces 63 suitcases per day.

DO EXERCISE 3.

3. There are three machines, A, B, and C, in a factory. When all three work, they produce 287 screws per hour. When only A and C work, they produce 197 screws per hour. When only A and B work, they produce 202 screws per hour. How many screws per hour can each produce alone?

EXERCISE SET 3.5

i Solve.

1. The sum of three numbers is 105. The third is 11 less than 10 times the second. Twice the first is 7 more than 3 times the second. Find the numbers.

2. The sum of three numbers is 57. The second is 3 more than the first. The third is 6 more than the first. Find the numbers.

3. The sum of three numbers is 5. The first number minus the second plus the third is 1. The first minus the third is 3 more than the second. Find the numbers.

4. The sum of three numbers is 26. Twice the first minus the second is 2 less than the third. The third is the second minus three times the first. Find the numbers.

5. In triangle ABC, the measure of angle B is 2° more than three times the measure of angle A. The measure of angle C is 8° more than the measure of angle A. Find the angle measures.

6. In triangle ABC, the measure of angle B is three times the measure of angle A. The measure of angle C is 30° greater than the measure of angle A. Find the angle measures.

7. In triangle ABC, the measure of angle B is twice the measure of angle A. The measure of angle C is 80° more than that of angle A. Find the angle measures.

8. In triangle ABC, the measure of angle B is three times that of angle A. The measure of angle C is 20° more than that of angle A. Find the angle measures.

9. Gina sells magazines part time. On Thursday, Friday, and Saturday, she sold $66 worth. On Thursday she sold $3 more than on Friday. On Saturday she sold $6 more than on Thursday. How much did she take in each day?

10. Pat picked strawberries on three days. He picked a total of 87 quarts. On Tuesday he picked 15 quarts more than on Monday. On Wednesday he picked 3 quarts fewer than on Tuesday. How many quarts did he pick each day?

11. Linda has a total of 225 on three tests. The sum of the scores on the first and second tests exceeds her third score by 61. Her first score exceeds her second by 6. Find the three scores.

12. Fred, Jane, and Mary made a total bowling score of 575. Fred's score was 15 more than Jane's. Mary's was 20 more than Jane's. Find the scores.

13. In a factory there are three polishing machines, A, B, and C. When all three of them are working, 5700 lenses can be polished in one week. When only A and B are working, 3400 lenses can be polished in one week. When only B and C are working, 4200 lenses can be polished in one week. How many lenses can be polished in a week by each machine?

14. Sawmills A, B, and C can produce 7400 board-feet of lumber per day. Mills A and B together can produce 4700 board-feet, while mills B and C together can produce 5200 board-feet. How many board-feet can each mill produce by itself?

15. When three pumps, A, B, and C, are running together, they can pump 3700 gallons per hour. When only A and B are running, 2200 gallons per hour can be pumped. When only A and C are running, 2400 gallons per hour can be pumped. What is the pumping capacity of each pump?

16. Three welders, A, B, and C, can weld 37 linear feet per hour when working together. Welders A and B together can weld 22 linear feet per hour, while A and C together can weld 25 linear feet per hour. How many linear feet per hour can each weld alone?

17. Tammy's age is the sum of the ages of Carmen and Dennis. Carmen's age is 2 more than the sum of the ages of Dennis and Mark. Dennis's age is four times Mark's age. The sum of all four age is 42. How old is Tammy?

18. Find a three-digit positive integer such that the sum of all three digits is 14, the ten's digit is 2 more than the unit's digit, and if the digits are reversed the number is unchanged.

19. Find the sum of the angle measures at the tips of the star in this figure.

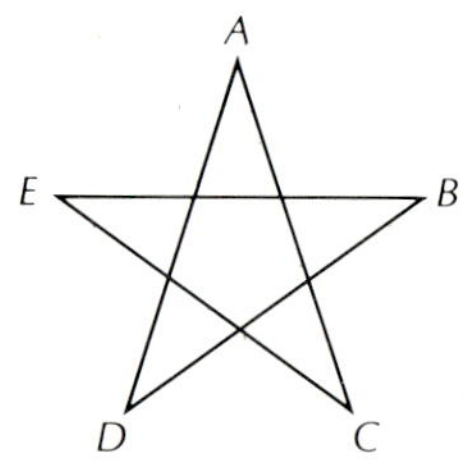

20. A theater audience of 100 people consists of men, women, and children. The ticket prices are \$10 for men, \$3 for women, and 50¢ for children. The total amount of money taken in is \$100. How many men, women, and children are in attendance? Does there seem to be some information missing? Do some more careful reasoning.

21. Find the year in which the first U.S. transcontinental railroad was completed. The following are some facts about the number. The sum of the digits in the year is 24. The one's digit is one more than the hundred's digit. Both the ten's and one's digits are multiples of three.

22. Hal gives Tom as many raffle tickets as Tom has and Gary as many as Gary has. In like manner, Tom then gives Hal and Gary as many tickets as each then has. Similarly, Gary gives Hal and Tom as many tickets as each then has. If each finally has 40 tickets, with how many tickets does Tom begin?

23. At a county fair, adults' tickets sold for \$5.50, senior citizens' tickets sold for \$4.00, and children's tickets sold for \$1.50. On the opening day the number of children's and senior citizens' tickets sold was 30 more than half the number of adults' tickets sold. The number of senior citizens' tickets sold was 5 more than 4 times the number of children's tickets. How many of each type of ticket were sold if the total receipts from the ticket sales were \$14,970?

24. The graph of $y = ax^2 + bx + c$ contains the three points $(4, 2)$, $(2, 0)$, and $(1, 2)$. Find a, b, and c.

25. Debbie bought three record albums. She paid full price for the first, and then received a 10% discount on the second and a 20% discount on the third. The second album's discount price was twice the price of the first album. The third album's original price was twice the discount price of the second. Debbie paid a total of \$27.90 for the three albums. What did each cost originally?

OBJECTIVE

After finishing Section 3.6, you should be able to:

i **Solve systems of two or three linear equations using matrices.**

3.6

MATRICES

i In solving systems of equations, we perform computations with the constants. The variables play no important role in the process. We can simplify writing by omitting the variables. For example, the system

$$\begin{aligned} 3x + 4y &= 5, \\ x - 2y &= 1 \end{aligned} \qquad \text{simplifies to} \qquad \begin{matrix} 3 & 4 & 5 \\ 1 & -2 & 1 \end{matrix}$$

if we leave off the variables and omit the operation and equals signs.

In the above example we have written a rectangular array of numbers. Such an array is called a *matrix* (plural, *matrices*). We ordinarily write brackets around matrices. The following are examples of matrices:

$$\begin{bmatrix} 4 & 1 & 3 & 5 \\ 1 & 0 & 1 & 2 \\ 6 & 3 & -2 & 0 \end{bmatrix}, \quad \begin{bmatrix} 6 & 2 & 1 & 4 & 7 \\ 1 & 2 & 1 & 3 & 1 \\ 4 & 0 & -2 & 0 & -3 \end{bmatrix}, \quad \begin{bmatrix} 1 & 2 \\ 145 & 0 \\ -7 & 9 \\ 8 & 1 \\ 0 & 0 \end{bmatrix}.$$

The *rows* of a matrix are horizontal, and the *columns* are vertical.

$$\begin{bmatrix} 5 & -2 & 2 \\ 1 & 0 & 1 \\ 0 & 1 & 2 \end{bmatrix} \begin{matrix} \leftarrow \text{row 1} \\ \leftarrow \text{row 2} \\ \leftarrow \text{row 3} \end{matrix}$$

column 1 column 2 column 3

Let us now use matrices to solve systems of linear equations.

EXAMPLE 1 Solve:

$$\begin{aligned} 2x - y + 4z &= -3, \\ x \quad - 4z &= 5, \\ 6x - y + 2z &= 10. \end{aligned}$$

We first write a matrix, using only the constants. Note that where there are missing terms we must write 0's.

$$\begin{bmatrix} 2 & -1 & 4 & -3 \\ 1 & 0 & -4 & 5 \\ 6 & -1 & 2 & 10 \end{bmatrix}$$

We do exactly the same calculations using the matrix that we would do if we wrote the entire equations. The first step, if possible, is to interchange the rows so that each number in the first column below the first number is a multiple of that number. We do this by interchanging rows 1 and 2.

$$\begin{bmatrix} 1 & 0 & -4 & 5 \\ 2 & -1 & 4 & -3 \\ 6 & -1 & 2 & 10 \end{bmatrix}$$

This corresponds to interchanging equation ① with equation ②.

Next we multiply the first row by -2 and add it to the second row.

$$\begin{bmatrix} 1 & 0 & -4 & 5 \\ 0 & -1 & 12 & -13 \\ 6 & -1 & 2 & 10 \end{bmatrix}$$

This corresponds to multiplying equation ① by -2 and adding it to equation ②.

Now we multiply the first row by -6 and add it to the third row.

$$\begin{bmatrix} 1 & 0 & -4 & 5 \\ 0 & -1 & 12 & -13 \\ 0 & -1 & 26 & -20 \end{bmatrix}$$

This corresponds to multiplying equation ① by -6 and adding it to equation ③.

Next we multiply row 2 by -1 and add it to the third row.

$$\begin{bmatrix} 1 & 0 & -4 & 5 \\ 0 & -1 & 12 & -13 \\ 0 & 0 & 14 & -7 \end{bmatrix}$$

This corresponds to multiplying equation ② by -1 and adding it to equation ③.

Solve using matrices.

1. $5x - 2y = -44$
   $2x + 5y = -6$

2. $x - 2y + 3z = 4$
   $2x - y + z = -1$
   $4x + y + z = 1$

If we now put the variables back, we have the following:

$$\begin{aligned} x - 4z &= 5 \\ -y + 12z &= -13 \\ 14z &= -7. \end{aligned}$$

We solve ③ for z and get $z = -\frac{1}{2}$. Next we substitute $-\frac{1}{2}$ for z in ② and solve for y. $-y + 12(-\frac{1}{2}) = -13$, so $y = 7$. Since there is no y-term in ① we need only substitute $-\frac{1}{2}$ for z in ① and solve for x: $x - 4(-\frac{1}{2}) = 5$, so $x = 3$. The solution is $(3, 7, -\frac{1}{2})$.

Note that in the preceding example our goal was to get the matrix in *triangular form*,

$$\begin{bmatrix} a & b & c & d \\ 0 & e & f & g \\ 0 & 0 & h & k \end{bmatrix},$$

where there are just 0's below the *main diagonal*, formed by a, e, and h. Then we put the variables back and complete the solution.

All the operations used in the preceding correspond to operations with the equations and they produce equivalent systems of equations. We call the matrices *row-equivalent*, and the operations that produce them *row-equivalent operations*. Each of the following row-equivalent operations produces equivalent matrices:

a) Interchanging any two rows of a matrix;

b) Multiplying each element of a row by the same nonzero constant;

c) Multiplying each element of a row by a nonzero number and adding the result to another row.

DO EXERCISES 1 AND 2.

EXERCISE SET 3.6

i Solve using matrices.

1. $4x + 2y = 11,$
$3x - y = 2$

2. $3x - 3y = 11,$
$9x - 2y = 5$

3. $x + 2y - 3z = 9,$
$2x - y + 2z = -8,$
$3x - y - 4z = 3$

4. $x - y + 2z = 0,$
$x - 2y + 3z = -1,$
$2x - 2y + z = -3$

5. $5x - 3y = -2,$
$4x + 2y = 5$

6. $3x + 4y = 7,$
$-5x + 2y = 10$

7. $4x - y - 3z = 1,$
$8x + y - z = 5,$
$2x + y + 2z = 5$

8. $3x + 2y + 2z = 3,$
$x + 2y - z = 5,$
$2x - 4y + z = 0$

9. $p + q + r = 1,$
$p - 2q - 3r = 3,$
$4p + 5q + 6r = 4$

10. $m + n + t = 6,$
$m - n - t = 0,$
$m + 2n + t = 5$

11. $2x + 2y - 2z - 2w = -10,$
$x + y + z + w = -5,$
$x - y + 4z + 3w = -2,$
$3x - 2y + 2z + w = -6$

12. $2x - 3y + z - w = -8,$
$x + y - z - w = -4,$
$x + y + z + w = 22,$
$x - y - z - w = -14$

Problem solving

13. A collection of 34 coins consists of dimes and nickels. The total value is \$1.90. How many dimes and how many nickels are there?

14. A collection of 43 coins consists of dimes and quarters. The total value is \$7.60. How many dimes and how many quarters are there?

15. A collection of 22 coins consists of nickels, dimes, and quarters. The total value is \$2.90. There are six more nickels than there are dimes. How many of each type of coin are there?

16. A collection of 18 coins consists of nickels, dimes, and quarters. The total value is \$2.55. There are two more quarters than there are dimes. How many of each type of coin are there?

17. A tobacco dealer has two kinds of tobacco. One is worth \$4.05 per lb and the other is worth \$2.70 per lb. The dealer wants to blend the two tobaccos to get a 15-lb mixture worth \$3.15 per lb. How much of each kind of tobacco should be used?

18. A grocer mixes candy worth \$0.80 per lb with nuts worth \$0.70 per lb to get a 20-lb mixture worth \$0.77 per lb. How many pounds of candy and how many pounds of nuts should be used?

3.7

INCONSISTENT AND DEPENDENT SYSTEMS

Some systems of equations have no solution. Others may have more than one solution. In this section, we consider such systems.

i Inconsistent Equations

If a system of equations has a solution, we say that it is *consistent.* If a system does not have a solution, we say that it is *inconsistent.*

EXAMPLE 1 Determine whether this system is consistent or inconsistent:

$$\begin{aligned} x - 3y &= 1 \\ -2x + 6y &= 5. \end{aligned}$$

We attempt to find a solution:

$$\begin{aligned} 2x - 6y &= 2 && \text{Multiplying the first equation by 2} \\ -2x + 6y &= 5 \\ \hline 0 &= 7 && \text{Adding} \end{aligned}$$

The last equation says that $0 \cdot x + 0 \cdot y = 7$. There are no numbers x and y for which this is true, so there is no solution. The system is inconsistent.

Whenever we obtain a statement such as $0 = 7$, which is clearly false, we know that the system we are trying to solve is inconsistent.

DO EXERCISE 1.

EXAMPLE 2 Determine whether this system is consistent or inconsistent:

$$\begin{aligned} y + 3z &= 4, && (1) \\ -x - y + 2z &= 0, && (2) \\ x + 2y + z &= 1. && (3) \end{aligned}$$

We attempt to find a solution. If we add equations (2) and (3), we get

$$y + 3z = 1.$$

OBJECTIVES

After finishing Section 3.7, you should be able to:

i **Determine whether a system of two or three linear equations is consistent or inconsistent.**

ii **Determine whether a system of two or three linear equations is dependent or independent.**

1. Determine whether these systems are consistent or inconsistent.

a) $3x - y = 2$
$6x - 2y = 3$

b) $x + 4y = 2$
$2x - y = 1$

2. Determine whether these systems are consistent or inconsistent.

a) $$\begin{aligned} x + 2y + z &= 1 \\ 3x + 3y + z &= 2 \\ 2x + y \quad &= 2 \end{aligned}$$

b) $$\begin{aligned} x \quad + z &= 1 \\ y + z &= 1 \\ x + y \quad &= 1 \end{aligned}$$

3. Determine whether these systems are dependent or independent.

a) $$\begin{aligned} 3x - 2y &= 1 \\ -6x + 4y &= -2 \end{aligned}$$

b) $$\begin{aligned} x - y &= 2 \\ x + y &= 4 \end{aligned}$$

If we multiply this by -1 and then add the result to equation (1), we get

$$0 = 3.$$

The system is inconsistent.

DO EXERCISE 2.

ii Dependent Equations

Consider this system:

$$\begin{aligned} 2x + 3y &= 1 \\ 4x + 6y &= 2. \end{aligned}$$

If we multiply the first equation by 2, we get the second equation, so the two equations have the same solutions (are equivalent). The system of two equations is equivalent to only one of them. We call them *dependent.*

If a system of *n* linear equations is equivalent to a system of fewer than *n* of them, we say that the system is *dependent.* If such is not the case, we call the system *independent.*

EXAMPLE 3 Determine whether this system is dependent or independent:

$$\begin{aligned} 2x + 3y &= 1 \\ 4x + 6y &= 2. \end{aligned}$$

We attempt to solve:

$$\begin{aligned} -4x - 6y &= -2 \qquad \text{Multiplying the first equation by } -2 \\ 4x + 6y &= 2 \\ \hline 0 &= 0 \qquad \text{Adding} \end{aligned}$$

The last equation says that $0 \cdot x + 0 \cdot y = 0$. This is true for all numbers x and y. The system is dependent.

Whenever we get an obviously true sentence like this, we know that the system we are trying to solve is dependent.

DO EXERCISE 3.

EXAMPLE 4 Determine whether this system is dependent or independent:

$$\begin{aligned} x + 2y + \ z &= 1 \\ x - \ y + \ z &= 1 \\ 2x + \ y + 2z &= 2. \end{aligned}$$

We attempt to solve:

$$\begin{aligned} x + 2y + z &= 1 \\ -3y \quad &= 0 \qquad \text{Multiplying the first equation by } -1 \text{ and adding it to the second} \\ -3y \quad &= 0. \qquad \text{Multiplying the first equation by } -2 \text{ and adding it to the third} \end{aligned}$$

It is now obvious that our system of three equations is equivalent to a system of two equations. The system is dependent.

When we attempt to solve a system and obtain two equations that are identical, or obviously equivalent, we know that the system is dependent. Or, if we obtain a statement that is obviously true, such as $3 = 3$, we know that the system is dependent.

DO EXERCISE 4.

4. Determine whether these systems are dependent or independent.

a) $x + y + 2z = 1$
$x - y + z = 1$
$2x + 3z = 2$

b) $x + y + 2z = 1$
$x - y + z = 1$
$x + 2y + z = 2$

EXERCISE SET 3.7

i Determine whether these systems are consistent or inconsistent.

1. $x + 2y = 6$
$2x = 8 - 4y$

2. $y - 2x = 1$
$2x - 3 = y$

3. $y - x = 4$
$x + 2y = 2$

4. $y + x = 5$
$y = x - 3$

5. $x - 3 = y$
$2x - 2y = 6$

6. $3y = x - 2$
$3x = 6 + 9y$

7. $x + z = 0$
$x + y + 2z = 3$
$y + z = 2$

8. $x + y = 0$
$x + z = 1$
$2x + y + z = 2$

9. $x + z = 0$
$x + y = 1$
$y + z = 1$

10. $2x + y = 1$
$x + 2y + z = 0$
$x + z = 1$

ii Determine whether these systems are dependent or independent.

11. $x - 3 = y$
$2x - 2y = 6$

12. $3y = x - 2$
$3x = 6 + 9y$

13. $y - x = 4$
$x + 2y = 2$

14. $y + x = 5$
$y = x - 3$

15. $2x + 3y = 1$
$x + 1.5y = 0.5$

16. $15x + 6y = 20$
$7.5x - 10 = -3y$

17. $x + z = 0$
$x + y = 1$
$y + z = 1$

18. $2x + y = 1$
$x + 2y + z = 0$
$x + z = 1$

19. $x + y + z = 1$
$-x + 2y + z = 2$
$2x - y = -1$

20. $y + z = 1$
$x + y + z = 1$
$x + 2y + 2z = 2$

Problem-solving practice

21. The difference of two numbers is 4. The second number is 6 less than the first number. What are the numbers?

22. The graph of the equation $y = ax^2 + bx$ contains the points $(2, 6)$ and $(1, 3)$. Find a and b.

23. Two years ago, Roland was 5 times as old as Jennifer. Ten years from now, Roland's age will be 6 more than twice Jennifer's age. How old are they now?

24. The perimeter of a rectangle is 192. The width is 12 less than the length, and the sum of the length and 3 times the width is 140. Find the dimensions of the rectangle.

25. A hockey team won 15 more games than it tied, lost 10 more games than it won, and lost 7 more games than it tied. How many games did it win?

26. The sum of three numbers is 71. The first number is 12 more than twice the third number. The sum of the second number and 3 times the third number is 59. What are the numbers?

Solve. If a system has more than one solution, list three of them.

27. $9x - 3y = 15$
$6x - 2y = 10$

28. $2s - 3t = 9$
$4s - 6t = 9$

29. $x + 2y - z = -8$
$2x - y + z = 4$
$8x + y + z = 2$

30. $2x + y + z = 0$
$x + y - z = 0$
$x + 2y + 2z = 0$

31. $2x + 4y + 8z = 5$
$x + 2y + 4z = 13$
$4x + 8y + 16z = 10$

32. $x + y + z = 3$
$5x + 5y + 5z = 15$
$2x + 2y + 2z = 6$

33. Classify each of the systems in Exercises 27, 29, and 31 as consistent or inconsistent, dependent or independent.

34. Classify each of the systems in Exercises 28, 30, and 32 as consistent or inconsistent, dependent or independent.

Determine the constant k such that each system is dependent.

35. $6x - 9y = -3$
$-4x + 6y = k$

36. $8x - 16y = 20$
$10x - 20y = k$

Consider the following dependent systems. For each system, find an ordered pair in terms of y that describes the entire solution set of the system.

37. $2x + 3y = 1$
$4x + 6y = 2$

38. $-6x + 4y = 10$
$3x - 2y = -5$

OBJECTIVES

After finishing Section 3.8, you should be able to:

i Given information on revenue and cost, find the profit and the break-even point.

ii Given supply and demand curves, find the equilibrium point.

3.8

BUSINESS AND ECONOMIC APPLICATIONS

i Break-even Analysis

When a company manufactures a product, it invests money (this is its *cost*). When it sells the product, it takes in money (called *revenue*). It hopes to make a profit. Profit is computed as follows:

$$\text{Profit} = \text{Revenue} - \text{Cost}$$

or

$$P = R - C.$$

If R is greater than C, the company makes a profit. If C is greater than R, the company has a loss. If $R = C$, the company breaks even.

There are two kinds of costs. First, there are costs like rent, insurance, buying machinery, and so on. These costs, which must be paid whether a product is produced or not, are called *fixed costs.* Then, when a product is being produced there are costs for labor, materials, marketing, and so on. These are called *variable costs.* They vary according to the amount being produced. Adding these together, we find the *total cost* of producing a product.

EXAMPLE 1 Snowy Electronics, Inc., is planning to make a new kind of radio. Fixed costs the first year will be \$90,000, and it will cost \$15 to produce each radio (variable costs). Each radio will sell for \$26. Determine the profit and break-even point.

a) We know that $P = R - C$, and that

$$C = (\text{Fixed costs}) \text{ plus } (\text{Variable costs}) \quad \text{or}$$
$$C = 90{,}000 + 15x,$$

where x is the number of radios produced. We know that

$$R = 26x. \quad \text{(\$26 times the number of radios sold. We are assuming that all radios produced are sold.)}$$

So we have

$$P = R - C$$
$$P = 26x - [90{,}000 + 15x], \quad \text{or}$$
$$P = 11x - 90{,}000.$$

b) The break-even point occurs when revenue equals total cost. So we set $R = C$ and solve:

$$26x = 90{,}000 + 15x$$
$$11x = 90{,}000$$
$$x = 8181.8.$$

The firm will need to produce and sell 8182 radios in order to break even (8181 radios produces a slight loss, and 8182 a slight profit).

It is interesting to graph the equations in a problem like that in Example 1. We begin as follows:

$$C = (\text{Fixed costs}) \text{ plus } (\text{Variable costs})$$
$$C = 90{,}000 + 15x \qquad (1)$$
$$R = 26x \qquad (2)$$

C and R are both in dollars.

Equation (1) has an intercept on the \$-axis of 90,000 and has a slope of 15. Equation (2) has a graph that goes through the origin, with a slope of 26. Profit occurs where the revenue is greater than the cost. Loss occurs where revenue is less than cost. The break-even point occurs where the graphs cross.

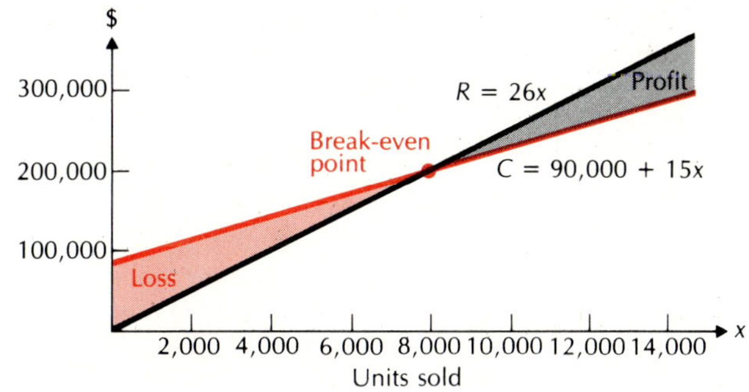

DO EXERCISE 1.

1. Suppose, in Example 1, that

$$C = 80{,}000 + 20x$$

and

$$R = 36x.$$

a) Find the profit.

b) Find the break-even point.

ii Supply and Demand

Demand Varies with Price. As the price of coffee varied over a period of years, the amount sold varied. The table and graph both show how the amount *demanded* by consumers went down as the price went up.

Demand schedule

Price, *P*, per kg	Quantity, *Q*, in millions of kg
\$1	25
2	20
3	15
4	10
5	5

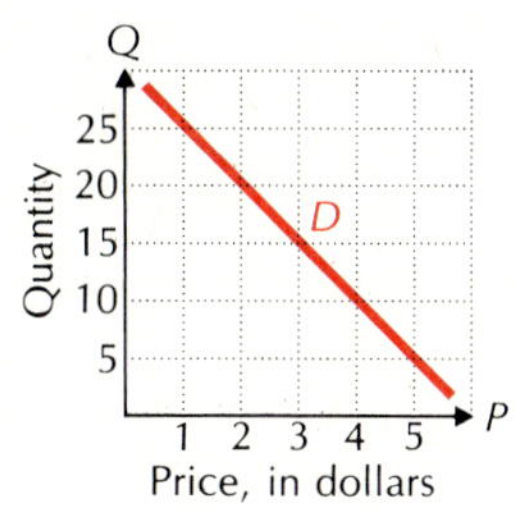

Supply Varies with Price. As the price of coffee varied, we see that the amount available varied. The amount of coffee on the market (the *supply*) went up as the price went up.

Supply schedule

Price, *P*, per kg	Quantity, *Q*, in millions of kg
\$2.00	5
2.50	10
3.00	15
3.50	20
4.00	25

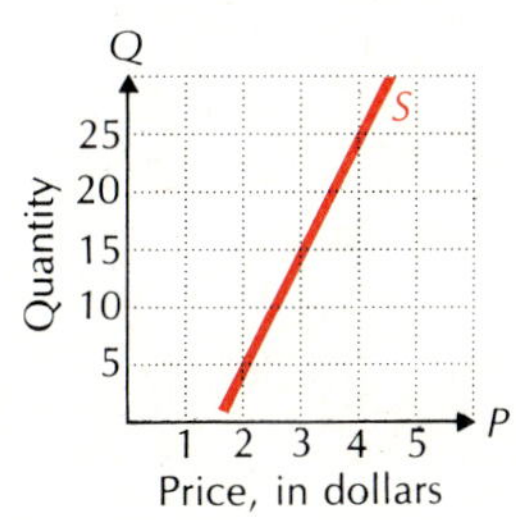

The Equilibrium Point. Let us look at the graphs together. We see that as supply increases, demand decreases. As supply decreases, demand increases. The point of intersection is called the *equilibrium point.* At that price, the amount that the seller will supply is the same amount that the consumer will buy.

2. Demand and supply are given by

$$D = 1000 - 50P$$

and

$$S = 300 + 4P.$$

Find the equilibrium point.

Any ordered pair of coordinates from this graph is (price, quantity), because the *first* axis is the *price* axis and the *second* axis is the *quantity* axis.

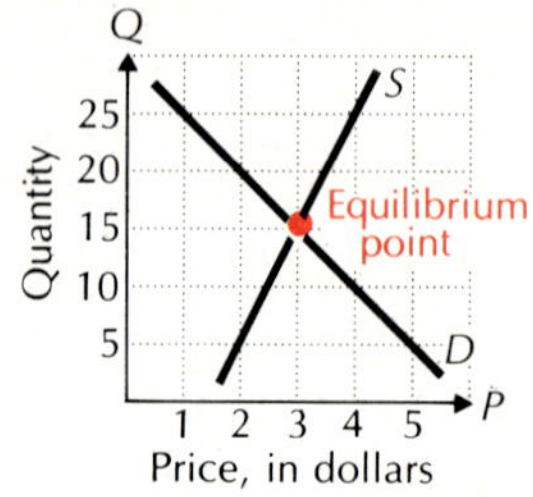

EXAMPLE 2 Find the equilibrium point when demand D and supply S are given by

$$D = 1000 - 60P$$

and

$$S = 200 + 4P.$$

a) We first set $D = S$ and solve:

$$\begin{aligned} 1000 - 60P &= 200 + 4P \\ 1000 - 64P &= 200 \quad \text{Adding } -4P \\ -64P &= -800 \quad \text{Adding } -1000 \\ P &= \frac{-800}{-64} = 12.5. \end{aligned}$$

Thus the equilibrium price, P_E, is \$12.50 per unit.

b) To find the equilibrium quantity we substitute P_E into either D or S. We use S:

$$\begin{aligned} 200 + 4P &= 200 + 4(12.5) \\ &= 200 + 50 = 250. \end{aligned}$$

Thus the equilibrium quantity is 250 units, and the equilibrium point is (\$12.50, 250).

DO EXERCISE 2.

EXERCISE SET 3.8

i For each of the following, find (a) the profit and (b) the break-even point.

1. $C = 45x + 600{,}000$, $R = 65x$

2. $C = 25x + 360{,}000$, $R = 70x$

3. $C = 10x + 120{,}000$, $R = 60x$

4. $C = 30x + 49{,}500$, $R = 85x$

5. $C = 20x + 10{,}000$, $R = 100x$

6. $C = 40x + 22{,}500$, $R = 85x$

7. $C = 15x + 75{,}000$, $R = 55x$

8. $C = 22x + 16{,}000$, $R = 40x$

9. $C = 50x + 195{,}000$, $R = 125x$

10. $C = 34x + 928{,}000$, $R = 128x$

ii Find the equilibrium point.

11. $D = 2000 - 60P$, $S = 460 + 94P$

12. $D = 1000 - 10P$, $S = 250 + 5P$

13. $D = 760 - 13P$, $S = 430 + 2P$

14. $D = 800 - 43P$, $S = 210 + 16P$

15. $D = 7500 - 25P$, $S = 6000 + 5P$

16. $D = 8800 - 30P$, $S = 7000 + 15P$

17. $D = 1600 - 53P$, $S = 320 + 75P$

18. $D = 5500 - 40P$, $S = 1000 + 85P$

19. $D = 4750 - 62P$, $S = 3350 + 38P$

20. $D = 6190 - 18P$, $S = 5045 + 7P$

Problem-solving practice

21. The perimeter of a rectangular field is 710 m. The length is 115 m greater than the width. Find the dimensions.

22. A student scores a total of 248 points on three tests. The total of the first two scores is 66 more than the score on the third test. The score on the third test is 6 more than that on the second test. Find the scores.

23. Bert has two cats. Their combined age is 12 yr. Their combined weight is 28 lb. If the younger cat weighs three times his age and the older cat weighs twice his age, how old are the two cats?

24. An electrician, a carpenter, and a plumber are hired to work on a house. The electrician earns $10.50 per hour, the carpenter earns $9.75 per hour, and the plumber earns $12 per hour. The first day on the job they earned a total of $234.75. If the plumber worked two more hours than the carpenter, how many hours did each work?

---

25. An electronics company is planning to introduce a new line of computers. For the first year the fixed costs for setting up the production line are $125,000. The variable costs for producing each computer are $750. The sales department projects that 400 computers will be sold during the first year. The revenue from each computer is $1050.

a) Find the total cost C of producing x computers.
b) Find the total revenue R from the sale of x computers.
c) Find the total profit P.
d) What profit or loss will the company realize if expected sales of 400 computers occur?
e) Find the break-even point.

26. A manufacturer is planning a new type of lamp. For the first year, the fixed costs for setting up the production line are $22,500. The variable costs for producing each lamp are estimated to be $40. The sales department projects that 3000 lamps can be sold during the first year. The revenue from each lamp is to be $85.

a) Find the total cost C of producing x lamps.
b) Find the total revenue R from the sale of x lamps.
c) Find total profit P.
d) What profit or loss will the company realize if expected sales of 3000 lamps occur?
e) Find the break-even point.

27. A clothing firm is planning a new line of sport coats. For the first year, the fixed costs for setting up the production line are $10,000. The variable costs for producing each coat are $20. The sales department projects that 2000 coats can be sold during the first year. The revenue from each coat is to be $100.

a) Find the total cost C of producing x coats.
b) Find the total revenue R from the sale of x coats.
c) Find the total profit P.
d) What profit or loss will the company realize if expected sales of 2000 coats occur?
e) Find the break-even point.

28. A custom-printing company is planning on adding painter's caps to its product line. For the first year, the fixed costs for setting up the production line are $16,400. The variable costs for producing a dozen caps are $6.00. The sales department projects that 3000 dozen caps can be sold during the first year. The revenue on each dozen caps will be $18.00.

a) Find the total cost C of producing x dozen caps.
b) Find the total revenue R from the sale of x dozen caps.
c) Find the total profit P.
d) What profit or loss will the company realize if expected sales of 3000 dozen caps occur?
e) Find the break-even point.

SUMMARY AND REVIEW: CHAPTER 3

The following contains a summary of what you should be able to do after completing this chapter. The review exercises are for practice. Answers are at the back of the book. If you miss an exercise, restudy the section indicated alongside the answer.

You should be able to:

Solve systems of equations graphically.

Solve graphically.

1. $4x - y = 10,$
$2x - 2y = 7$

2. $y = 3x + 7,$
$3x + 2y = -4$

Solve systems of equations using the substitution method.

Solve using the substitution method.

3. $7x - 4y = 6,$
$y - 3x = -2$

4. $5x + 5y = -6,$
$2x - 2y = -4$

5. $9x - 6y = 2,$
$-2x = 4y + 5$

Solve systems of equations using the addition method.

Solve using the addition method.

6. $8x - 2y = 10,$
$-3x - 4y = -17$

7. $4x - 7y = 18,$
$9x + 14y = 40$

8. $3x - 5y = -4,$
$5x - 3y = 4$

Solve problems by translating to a system of two linear equations and then solving the system.

Solve.

9. Jimmy has \$20.00 to spend at the store. He can spend all the money for two record albums and a poster, or he can buy one record album and two posters and have \$1.00 left over. What is the price of a record album? What is the price of a poster?

10. One day a store sold a total of 48 pens and pencils. Each pen costs 35¢ and each pencil costs 13¢. If the total sales from pens and pencils that day was \$11.74, how many of each were sold?

11. A beaker of alcohol contains a solution that is 30% alcohol. In another beaker is a solution that is 50% alcohol. How much of each can be added to obtain 40 L of a solution that is 45% alcohol?

Solve systems of linear equations containing three variables.

Solve.

12. $x + 2y + z = 10$
$2x - y + z = 8$
$3x + y + 4z = 2$

13. $3x + 2y + z = 3$
$6x - 4y - 2z = -34$
$-x + 3y - 3z = 14$

14. $2x - 5y - 2z = -4$
$7x + 2y - 5z = -6$
$-2x + 3y + 2z = 4$

Solve problems by translating into a system of three equations and then solving the system.

Solve.

15. In triangle ABC, the measure of angle A is four times the measure of angle C and the measure of angle B is 45° more than the measure of angle C. What are the measures of the angles of the triangle?

16. Find the three-digit number in which the sum of the digits is 11, the ten's digit is three less than the sum of the hundred's and unit's digits, and the unit's digit is five less than the hundred's digit.

17. Lynn has \$194 in her purse, consisting of \$20, \$5, and \$1 bills. The number of \$1 bills is one less than the total number of \$20 and \$5 bills. If she has 39 bills in her purse, how many of each denomination does she have?

Solve systems of equations using matrices.

Solve using matrices.

18. $x + y + z = 6$
$2x - y + 3z = 9$
$-x + 2y + 2z = 9$

19. $2x + y + z = -2$
$2x - y + 3z = 6$
$3x - 5y + 4z = 7$

Determine whether systems of equations are consistent or inconsistent.

Determine whether the following are consistent or inconsistent.

20. $-5x + 5y = -6$
$2x - 2y = 4$

21. $3x + y = 2$
$x + 3y + z = 0$
$x + z = 2$

Determine whether systems of equations are dependent or independent.

Determine whether the following are dependent or independent.

22. $3x + 4y = 6$
$1.5x - 3 = -2y$

23. $x + y + 2z = 1$
$x - y + z = 1$
$x + 2y + z = 2$

Solve problems involving profit and loss.

Solve.

24. A furniture company is planning to produce a new type of bed. For the first year, the fixed costs for setting up the production line are \$35,000. The variable costs for producing each bed are \$175. The sales department projects that 1200 beds can be sold during the first year. The revenue from each bed is \$225.

a) Find the total cost C of producing x beds.
b) Find the total revenue R from the sale of x beds.
c) Find the total profit P.
d) What profit or loss will the company realize if the expected sales of 1200 beds occur?
e) Find the break-even point.

Solve the system.

25. $xy = 1$
$x + 2y = 3$

26. $x^2 + y^2 = 25$
$2x + y = 10$

TEST: CHAPTER 3

Solve.

1. $x + 3y = -8$
$4x - 3y = 23$

2. $2x + 4y = -6$
$-3x - y = 9$

3. $4x - 6y = 3$
$6x - 4y = -3$

4. $2x + 4y = 18$
$3x + 6y = 26$

5. The perimeter of a rectangle is 96. The length of the rectangle is 6 less than twice the width. Find the dimensions of the rectangle.

6. A fast-food restaurant sold 132 chicken dinners during one day. Two-piece dinners sold for \$1.50, and three-piece dinners sold for \$2.50. The restaurant's total receipts that day from chicken dinners were \$260. How many of each size dinner did they sell?

Solve.

7. $-3x + y - 2z = 8$
$-x + 2y - z = 5$
$2x + y + z = -3$

8. $6x + 2y - 4z = 15$
$-3x - 4y + 2z = -6$
$4x - 6y + 3z = 8$

Solve using matrices.

9. $2x + y + z = 2$
$x - z = 0$
$2y + 3z = 1$

10. $x + 3y - 3z = 12$
$3x - y + 4z = 0$
$-x + 2y - z = 1$

11. A hockey team played 81 games in a season. They won 1 less game than 3 times the number of ties, and they lost 8 fewer games than they won. How many games did they win?

12. A sporting goods manufacturer is planning a new type of tennis racket. For the first year the fixed costs for setting up production lines are \$40,000. The variable costs for producing each racket are \$45. The sales department predicts that 1500 rackets can be sold during the first year. The revenue from each racket is \$80.

a) Find the total cost C of producing x tennis rackets.
b) Find the total revenue R from the sale of x tennis rackets.
c) Find the total profit P.
d) What profit or loss will the company realize if the expected sales of 1500 rackets occur?
e) Find the break-even point.

CUMULATIVE REVIEW: CHAPTERS 1–3

Solve.

1. $-14.3 + 29.17 = x$

2. $x + 9.4 = -12.6$

3. $3.9(-11) = x$

4. $-2.4x = -48$

5. $4x + 7 = -14$

6. $-3 + 5x = 2x + 15$

7. $3n - (4n - 2) = 7$

8. $6y - 5(3y - 4) = 10$

9. $14 + 2c = -3(c + 4) - 6$

10. $5x - [4 - 2(6x - 1)] = 12$

Simplify.

11. $x^4 \cdot x^{-6} \cdot x^{13}$

12. $(4x^{-3}y^2)(-10x^4y^{-7})$

13. $(6x^{a+2}y^{b+2})(-2x^{a-2}y^{b+1})$

14. $\dfrac{y^4}{y^{-6}}$

15. $\dfrac{-10a^7b^{-11}}{25a^{-4}b^{22}}$

16. $\left(\dfrac{3x^4y^{-2}}{4x^{-5}}\right)^4$

Multiply.

17. $(2.6 \times 10^4)(4.3 \times 10^{-5})$

18. $(1.95 \times 10^{-3})(5.73 \times 10^8)$

Divide.

19. $\dfrac{3.2 \times 10^{-3}}{8.0 \times 10^{-6}}$

20. $\dfrac{2.42 \times 10^5}{6.05 \times 10^{-2}}$

21. Solve for b_1: $A = \frac{1}{2}h(b_1 + b_2)$.

Determine whether the ordered pair is a solution of the equation.

22. $(-3, 4)$; $5a - 2b = -23$

23. $(2, -1)$; $-x - 5y = 4$

Graph.

24. $y = -2x + 3$

25. $y = x^2 - 1$

26. $4x + 16 = 0$

27. $-3x + 2y = 6$

28. Find the slope and y-intercept of $-4y + 9x = 12$.

29. Find the slope, if it exists, of the line containing the points $(2, 7)$ and $(-1, 3)$.

30. Find an equation of the line with slope -3 and containing the point $(2, -11)$.

31. Find an equation of the line containing the points $(-6, 3)$ and $(4, 2)$.

Given $g(x) = 4x - 3$ and $h(x) = -2x^2 + 1$, find the following function values.

32. $g(-1)$

33. $h(4)$

34. $-g(0)$

35. $g(2) + h(-2)$

36. $3g(-3) - 2h(-1)$

37. $g(a) - h(2a)$

Solve.

38. $3x + y = 4,$
$6x - y = 5$

39. $4x + 4y = 4,$
$5x - 3y = -19$

40. $6x - 10y = -22,$
$-11x - 15y = 27$

41. $x + y + z = -5,$
$2x + 3y - 2z = 8,$
$x - y + 4z = -21$

42. $2x + 5y - 3z = -11,$
$-5x + 3y - 2z = -7,$
$3x - 2y + 5z = 12$

Problem solving

43. Find three consecutive odd numbers such that the sum of 4 times the first number and 5 times the third number is 47.

44. A student's scores on four tests are 83, 92, 100, and 85. What must the score be on the fifth test so that the average will be 90?

45. The perimeter of a rectangle is 32 cm. If 5 times the width equals 3 times the length, what are the dimensions of the rectangle?

46. The total of Elizabeth's rent, electric bill, and phone bill for March was \$340. The electric bill was $\frac{1}{4}$ of the rent and the rent was \$200 more than the phone bill. What was the amount of the electric bill?

How many balls are in the stack? A polynomial function can give the answer.

4 POLYNOMIALS AND POLYNOMIAL FUNCTIONS

A *polynomial expression* is a certain kind of symbol with one or more terms. We will be more specific about that in this chapter. These kinds of expressions give rise to certain functions, called *polynomial functions*, and to certain equations, called *polynomial equations*.

In this chapter you will become familiar with polynomials in their several aspects, learning to manipulate them and then use them in problem solving.

OBJECTIVES

After finishing Section 4.1, you should be able to:

i **Determine the degree of a polynomial and the degree of each term; identify terms, coefficients, monomials, binomials, and trinomials; and arrange polynomials in ascending or descending order.**

ii **Find function values for polynomial functions.**

iii **Combine similar terms.**

iv **Add polynomials.**

v **Rename the additive inverse of a polynomial without using parentheses.**

vi **Subtract polynomials.**

1. Identify the terms and the leading term.

$$-92x^5 - 8x^4 + y^2 + 5$$

2. Identify the coefficient of each term and the leading coefficient.

$$5x^3y - 4xy^2 - 2x^3 + pq - y - 5$$

3. Determine the degree of each term and the degree of the polynomial.

$$2y - 4 - 5x + 7x^2y^3z^2 + 5xy^2$$

Identify the leading term and the leading coefficient.

4.1

POLYNOMIALS AND FUNCTIONS

A *polynomial* is a type of algebraic expression. A certain kind of function related to polynomial expressions is called a *polynomial function.*

i Polynomial Expressions

Expressions like these are called *polynomials in one variable:*

$$5x^2, \quad 8a, \quad 2, \quad 2x + 3, \quad -7x + 5, \quad 2y^2 + 5y - 3,$$
$$5a^4 - 3a^2 + \tfrac{1}{4}a - 8, \quad b^6 + 3b^5 - 8b + 7b^4 + \tfrac{1}{2}.$$

Expressions like these are called *polynomials in several variables:*

$$5x - xy^2 + 7y + 2, \quad 9xy^2z - 4x^3z + (-14x^4y^2) + 9, \quad 15x^3y^2.$$

The polynomial $5x^3y - 7xy^2 + 2$ has three *terms:*

$$5x^3y, \quad -7xy^2, \quad \text{and} \quad 2.$$

The *coefficients* of the terms are 5, -7, and 2.

The *degree of a term* is the sum of the exponents of the variables, if there are variables. The degree of a constant term is 0, except that we do not assign any degree to a zero term. The *degree of a polynomial* is the same as its term of highest degree.

The *leading term* of a polynomial is the term of highest degree. Its coefficient is called the *leading coefficient.*

EXAMPLE 1 Determine the degree of each term and the degree of the polynomial. Then find the leading term and the leading coefficient.

$$6x^2 + 8x^2y^3 - 17xy - 24xy^2z^4 + 2y + 3$$

Term	$6x^2$	$8x^2y^3$	$-17xy$	$-24xy^2z^4$	$2y$	3
Degree	2	5	2	7	1	0
Leading term	$-24xy^2z^4$					
Leading coefficient	-24					

The degree of the polynomial is 7.

DO EXERCISES 1–3.

A polynomial with a single term is called a *monomial.* A polynomial of two terms is a *binomial,* and one with three terms is a *trinomial.* We usually arrange polynomials in one variable so that the exponents decrease (descending order) or so that they increase (ascending order).

A polynomial of degree 0 or 1 is called *linear.* Polynomials in one variable are called *quadratic* if they are of degree 2, and are called *cubic* if they are of degree 3.

We can also arrange the terms of polynomials in several variables in ascending or descending order, with respect to one of the variables.

EXAMPLE 2 Arrange in descending powers of x: $y^4 + 2 - 5x^2 + 3x^3y + 7xy$.

$$3x^3y - 5x^2 + 7xy + y^4 + 2$$

DO EXERCISES 4–6.

ii Polynomial Functions

A polynomial function is a function like the following:

$$P(x) = 5x^7 + 3x^5 - 4x^2 - 5,$$

where the expression on the right is a polynomial. To find values of a polynomial function, we simply substitute into the polynomial expression.

EXAMPLE 3 For the polynomial function $P(x) = 2x - 3$, find the following function values: (a) $P(0)$; (b) $P(10)$; (c) $P(-10)$.

a) $P(0) = 2 \cdot 0 - 3 = -3$

b) $P(10) = 2 \cdot 10 - 3 = 17$

c) $P(-10) = 2(-10) - 3 = -23$

DO EXERCISE 7.

For polynomial functions of greater degree, there is a more efficient procedure for finding function values. We will consider it later in the chapter.

Polynomial functions are useful in many problem-solving situations, as shown in the following.

EXAMPLE 4 The number of oranges in a pile like the one shown is approximated by the polynomial function

$$N(x) = \tfrac{1}{6}x^3 + \tfrac{1}{2}x^2 + \tfrac{1}{3}x,$$

where x is the number of layers and $N(x)$ is the number of oranges. Find the approximate number of oranges in a pile with 9 layers.

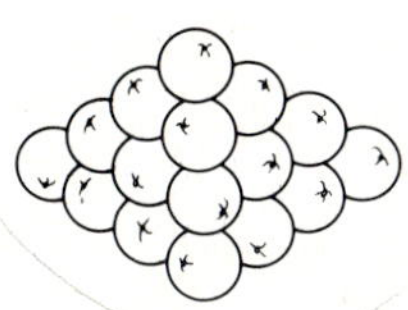

1. **Familiarize** We will find the number by substituting for x in the polynomial. Let's see how the numbers run by making a few substitutions:

$$N(0) = \tfrac{1}{6}(0)^3 + \tfrac{1}{2}(0)^2 + \tfrac{1}{3}(0) = 0$$

(if there are no layers at all, then there are no oranges);

$$N(1) = \tfrac{1}{6}(1)^3 + \tfrac{1}{2}(1)^2 + \tfrac{1}{3}(1) = 1$$

(if there is 1 layer, there will be just 1 orange);

$$\begin{aligned} N(2) &= \tfrac{1}{6}(2)^3 + \tfrac{1}{2}(2)^2 + \tfrac{1}{3}(2) \\ &= \tfrac{1}{6} \cdot 8 + \tfrac{1}{2} \cdot 4 + \tfrac{1}{3} \cdot 2 \\ &= 4 \end{aligned}$$

(if there are 2 layers, there will be about 4 oranges).

Consider the following polynomials.

a) $3x^2 - 2$

b) $5x^3 + 9x - 3$

c) $4x^2$

d) $-7y$

e) -3

f) $8x^3 - 2x^2$

g) $-4y^2 - 5y - 5$

h) $5 - 3x$

4. Identify the monomials, the binomials, and the trinomials.

5. Identify the linear polynomials, the quadratic polynomials, and the cubic polynomials.

6. Arrange

$$5x^4y - 3y^2 + 3x^2y^3 + x^3 - 5$$

in descending powers of y.

7. For the polynomial function $P(x) = x^2 - 2x + 5$, find the following function values.

a) $P(0)$

b) $P(4)$

c) $P(-2)$

8. The cost, in cents per kilometer, of operating an automobile at speed s (in km/h) is approximated by the polynomial $0.002s^2 - 0.21s + 15$. How much does it cost to operate at 80 km/h?

2. **Translate** We actually have the translation. It is the given polynomial function.

3. **Carry out** We substitute 9 for x and calculate:

$$\begin{aligned} N(9) &= \tfrac{1}{6}(9)^3 + \tfrac{1}{2}(9)^2 + \tfrac{1}{3}(9) \\ &= \tfrac{1}{6} \cdot 729 + \tfrac{1}{2} \cdot 81 + 3 \\ &= 121.5 + 40.5 + 3 \\ &= 165. \end{aligned}$$

4. **Check** In this case, the best kind of check is to redo the calculation and to make sure that we have not written anything incorrectly.

5. **State** The answer is that when there are 9 layers, there will be about 165 oranges.

DO EXERCISE 8.

Example 4 illustrates the fact that in many problems an answer may not be exact. The function of that problem gave us a pretty good answer, but because oranges are not exactly the same size and shape, the actual number might be a little different from our answer.

iii Calculations with Polynomials

Combine similar terms.

9. $3y - 4x + 6xy^2 - 2xy^2$

10. $3xy^3 + 2x^3y + 5xy^3 - 3x^2y - 6x^2y$

If two terms of a polynomial have the same letters raised to the same powers, the terms are called *similar*, or *like terms*. Similar terms can be "combined" or "collected" using the distributive laws.

EXAMPLES Combine similar terms.

5. $3x^2 - 4y + 2x^2 = 3x^2 + 2x^2 - 4y$ Rearranging using the commutative and associative laws

$= (3 + 2)x^2 - 4y$ Using a distributive law

$= 5x^2 - 4y$

6. $4x^3 + 5x - 4x^2 - 2x^3 + 5x^2 = 2x^3 + x^2 + 5x$

7. $3x^2y + 5xy^2 - 3x^2y - xy^2 = 4xy^2$

DO EXERCISES 9 AND 10.

iv Addition. The sum of two polynomials can be found by writing a plus sign between them and then combining similar terms. Ordinarily this can be done mentally.

EXAMPLE 8 Add: $-3x^3 + 2x - 4$ and $4x^3 + 3x^2 + 2$.

$$(-3x^3 + 2x - 4) + (4x^3 + 3x^2 + 2) = x^3 + 3x^2 + 2x - 2$$

Using columns is often helpful. To do so we write the polynomials one under the other, writing like terms under one another and leaving spaces for missing terms. Let us do the addition in Example 8 using columns.

$$\begin{array}{rrrr} -3x^3 & & +\,2x & -\,4 \\ 4x^3 & +\,3x^2 & & +\,2 \\ \hline x^3 & +\,3x^2 & +\,2x & -\,2 \end{array}$$

EXAMPLE 9 Add: $4ax^2 + 4bx - 5$ and $3ax^2 + 5bx + 8$.

$$\begin{array}{rrr} 4ax^2 & +\,4bx & -\,5 \\ 3ax^2 & +\,5bx & +\,8 \\ \hline 7ax^2 & +\,9bx & +\,3 \end{array}$$

Although using columns is helpful for complicated examples, you should attempt to write only the answer whenever you can.

EXAMPLE 10 Add: $13x^3y + 3x^2y - 5y$ and $x^3y + 4x^2y - 3xy + 3y$.

$$(13x^3y + 3x^2y - 5y) + (x^3y + 4x^2y - 3xy + 3y) = 14x^3y + 7x^2y - 3xy - 2y$$

DO EXERCISES 11–13.

v Additive Inverses. If the sum of two polynomials is 0, they are called *additive inverses* of each other. For example, $(3x - 2) + (-3x + 2) = 0$, so the additive inverse of $(3x - 2)$ is $(-3x + 2)$. We can say the same thing using algebraic symbolism, as follows:

The additive inverse of $(3x - 2)$ is $-3x + 2$.

$$- \quad (3x - 2) = -3x + 2$$

The additive inverse of a polynomial $P(x)$ can be symbolized by $-P(x)$ or by replacing each term by its additive inverse (changing the sign of every term). The two expressions will, of course, be equivalent.

EXAMPLE 11 Write two equivalent expressions for the additive inverse of $7xy^2 - 6xy - 4y + 3$.

a) $-(7xy^2 - 6xy - 4y + 3)$ Writing an inverse sign in front

b) $-7xy^2 + 6xy + 4y - 3$ Changing the sign of every term

DO EXERCISES 14–16.

vi Subtraction. To subtract one polynomial from another, we add the inverse of the subtrahend.

EXAMPLE 12 Subtract: $(-9x^5 + 2x^2 + 4) - (2x^5 + 4x^3 - 3x^2)$.

$$\begin{aligned}(-9x^5 + 2x^2 + 4) &- (2x^5 + 4x^3 - 3x^2)\\ &= (-9x^5 + 2x^2 + 4) + (-2x^5 - 4x^3 + 3x^2) \quad \text{Adding the inverse of the subtrahend}\\ &= -11x^5 - 4x^3 + 5x^2 + 4\end{aligned}$$

After some practice, you will find that you can skip some steps, by mentally "changing the sign of each term" and then collecting like terms. Eventually, all you will write is the answer.

DO EXERCISES 17–19.

We can also use columns for subtraction. We change signs mentally in the subtrahend and then add.

EXAMPLE 13 Subtract: $(4x^2y - 6x^3y^2 + x^2y^2) - (4x^2y + x^3y^2 + 3x^2y^3)$.

$$\begin{array}{r} 4x^2y - 6x^3y^2 \qquad\qquad + x^2y^2 \\ 4x^2y + x^3y^2 + 3x^2y^3 \\ \hline -7x^3y^2 - 3x^2y^3 + x^2y^2 \end{array}$$

As with addition, you should avoid the use of columns as much as possible. After sufficient experience, you should write only the answer.

DO EXERCISES 20–22.

Add.

11. $3x^3 + 4x^2 - 7x - 2$ and $-7x^3 - 2x^2 + 3x + 4$

12. $7y^5 - 5$ and $3y^5 - 4y^2 + 10$

13. $5p^2q^4 - 2p^2q^2 - 3q$ and $-6p^2q^2 + 3q + 5$

Write two equivalent expressions for the additive inverse.

14. $4x^3 - 5x^2 + \frac{1}{4}x - 10$

15. $8xy^2 - 4x^3y^2 - 9x - \frac{1}{5}$

16. $-9y^5 - 8y^4 + \frac{1}{2}y^3 - y^2 + y - 1$

Subtract.

17. $(6x^2 + 4) - (3x^2 - 1)$

18. $(9y^3 - 2y - 4) - (-5y^3 - 8)$

19. $(-3p^2 + 5p - 4) - (-4p^2 + 11p - 2)$

Subtract.

20. $(2y^5 - y^4 + 3y^3 - y^2 - y - 7) - (-y^5 + 2y^4 - 2y^3 + y^2 - y - 4)$

21. $(4p^4q - 5p^3q^2 + p^2q^3 + 2q^4) - (-5p^4q + 5p^3q^2 - 3p^2q^3 - 7q^4)$

22. $(\frac{3}{2}y^3 - \frac{1}{2}y^2 + 0.3) - (\frac{1}{2}y^3 + \frac{1}{2}y^2 - \frac{4}{3}y + 0.2)$

EXERCISE SET 4.1

i Determine the degree of each term and the degree of the polynomial. Then find the leading term and the leading coefficient.

1. $-11x^4 - x^3 + x^2 + 3x - 9$
2. $t^3 - 3t^2 + t + 1$
3. $y^3 + 2y^7 + x^2y^4 - 8$
4. $u^2 + 3v^5 - u^3v^4 - 7$
5. $a^5 + 4a^2b^4 + 6ab + 4a - 3$
6. $8p^6 + 2p^4t^4 - 7p^3t + 5p^2 - 14$

Arrange in descending powers of y.

7. $x^2y^2 + x^3y - xy^3 + 1$
8. $x^3y - x^2y^2 + xy^3 + 6$
9. $3ay - 5a^5y^5 - 4a^2y^3$
10. $by^6 - 4b^7y^2 + 7ay^5 - 4y - 5b$

Arrange in ascending powers of x.

11. $-9x^3y + 3xy^3 + x^2y^2 + 2x^4$
12. $5x^2y^2 - 9xy + 8x^3y^2 - 5x^4$
13. $4ax - 7ab + 4x^6 - 7ax^2$
14. $5xy^8 - 3ax^5 + 4ax^3 - 12a + 5x^5$

ii For each polynomial function, find the specified function value.

15. Find $P(4)$ and $P(0)$: $P(x) = 4x^2 - 3x + 2$.
16. Find $Q(3)$ and $Q(-1)$: $Q(x) = -5x^3 + 7x^2 - 12$.
17. Find $P(-2)$ and $P(\frac{1}{3})$: $P(y) = 8y^3 - 12y - 5$.
18. Find $Q(-3)$ and $Q(0)$: $Q(y) = 9y^3 + 8y^2 - 4y - 9$.
19. Find $P(a)$: $P(x) = -4x^3 + 2x^2 - 7x + 5$.
20. Find $Q(-b)$: $Q(x) = 6x^3 - 12x^2 - 8x - 3$.
21. Find $P(2a)$: $P(x) = x^3 - 4x^2 + 3x - 7$.
22. Find $Q\left(\frac{c}{2}\right)$: $Q(x) = -7x^3 + 10x^2 + 6$.
23. Find $P(x + a)$: $P(x) = -3x^3 + 6x^2 - 4$.
24. Find $Q(2x + 1)$: $Q(x) = 6x^3 + 8x - 11$.

iii Collect like terms.

25. $6x^2 - 7x^2 + 3x^2$
26. $-2y^2 - 7y^2 + 5y^2$
27. $5x - 4y - 2x + 5y$
28. $4a - 9b - 6a + 3b$
29. $5a + 7 - 4 + 2a - 6a + 3$
30. $9x + 12 - 8 - 7x + 5x + 10$
31. $3a^2b + 4b^2 - 9a^2b - 6b^2$
32. $5x^2y^2 + 4x^3 - 8x^2y^2 - 12x^3$
33. $8x^2 - 3xy + 12y^2 + x^2 - y^2 + 5xy + 4y^2$
34. $a^2 - 2ab + b^2 + 9a^2 + 5ab - 4b^2 + a^2$
35. $4x^2y - 3y + 2xy^2 - 5x^2y + 7y + 7xy^2$
36. $3xy^2 + 4xy - 7xy^2 + 7xy + x^2y$

iv Add.

37. $3x^2 + 5y^2 + 6$ and $2x^2 - 3y^2 - 1$
38. $9y^2 + 8y - 4$ and $12y^2 - 5y + 8$
39. $2a + 3b - c$ and $4a - 2b + 2c$
40. $5x - 4y + 2z$ and $9x + 12y - 8z$
41. $a^2 - 3b^2 + 4c^2$ and $-5a^2 + 2b^2 - c^2$
42. $x^2 - 5y^2 - 9z^2$ and $-6x^2 + 9y^2 - 2z^2$
43. $x^2 + 2x - 3xy - 7$ and $-3x^2 - x + 2xy + 6$
44. $3a^2 - 2b + ab + 6$ and $-a^2 + 5b - 5ab - 2$
45. $7x^2y - 3xy^2 + 4xy$ and $-2x^2y - xy^2 + xy$
46. $7ab - 3ac + 5bc$ and $13ab - 15ac - 8bc$
47. $2r^2 + 12r - 11$ and $6r^2 - 2r + 4$ and $r^2 - r - 2$
48. $5x^2 + 19x - 23$ and $-7x^2 - 11x + 12$ and $-x^2 - 9x + 8$
49. $3 + 4a - 6a^2$ and $-4 - 2a + 3a^2$
50. $-12 - 6x + 4x^2$ and $-15 + x - 7x^2$
51. $4xy + 3xy^2 - x^2y$ and $-9xy + 2x^2y - 3x^2y$
52. $3x - 7x^3 + 4x^5$ and $-4x - 4x^3 - 2x^5$

v Name the additive inverse of each polynomial in two ways.

53. $5x^3 - 7x^2 + 3x - 6$
54. $-8y^4 - 18y^3 + 4y - 9$
55. $-12y^5 + 4ay^4 - 7by^2$
56. $7ax^3y^2 - 8by^4 - 7abx - 12ay$

vi Subtract.

57. $(8x - 4) - (-5x + 2)$
58. $(9y + 3) - (-4y - 2)$
59. $(-3x^2 + 2x + 9) - (x^2 + 5x - 4)$
60. $(-9y^2 + 4y + 8) - (4y^2 + 2y - 3)$
61. $(5a - 2b + c) - (3a + 2b - 2c)$
62. $(8x - 4y + z) - (4x + 6y - 3z)$
63. $(3x^2 - 2x - x^3) - (5x^2 - 8x - x^3)$
64. $(8y^2 - 3y - 4y^3) - (3y^2 - 9y - 7y^3)$
65. $(5a^2 + 4ab - 3b^2) - (9a^2 - 4ab + 2b^2)$
66. $(9y^2 - 14yz - 8z^2) - (12y^2 - 8yz + 4z^2)$
67. $(2 - 3y) - (6y - 7)$
68. $(4 + 8x^2 - 5x) - (-2x^2 + 3x - 2)$
69. $(6ab - 4a^2b + 6ab^2) - (3ab^2 - 10ab - 12a^2b)$
70. $(10x - 4x^2 - 3) - (-9x^2 + 4 - 7x)$
71. $(0.09y^4 - 0.052y^3 + 0.93) - (0.03y^4 - 0.084y^3 + 0.94y^2)$
72. $(\frac{5}{8}x^4 - \frac{1}{4}x^2 - \frac{1}{2}) - (-\frac{3}{8}x^4 + \frac{3}{4}x^2 + \frac{1}{2})$

73. Add.

$$15.36x^2 - 4.039y^2 + 1.002x \quad \text{and}$$
$$12.58x^2 + 13.144y^2 + 7.148x$$

74. Subtract.

$$(8035x - 1735) - (-4111x + 3892)$$

75. If there are n teams in a league and each team plays each other once in a season, we can find the total number of games played by a polynomial function $f(n) = \frac{1}{2}(n^2 - n)$. Find the number of games played when there are 8 teams; when there are 20 teams.

76. The cost, in cents per kilometer, of operating an automobile at speed s is approximated by the polynomial function $C(s) = 0.002s^2 - 0.21s + 15$. How much does it cost to operate at 50 km/h? at 80 km/h?

For the polynomial functions $P(x)$ and $Q(x)$, find each of the following.

$$P(x) = 13x^5 - 22x^4 - 36x^3 + 40x^2 - 16x + 75$$
$$Q(x) = 42x^5 - 37x^4 + 50x^3 - 28x^2 + 34x + 100$$

77. $P(x) + Q(x)$ **78.** $P(x) - Q(x)$ **79.** $Q(x) - P(x)$ **80.** $4[P(x)] + 3[Q(x)]$

81. Find a polynomial function that gives the surface area of a box like this one, with an open top and dimensions as shown.

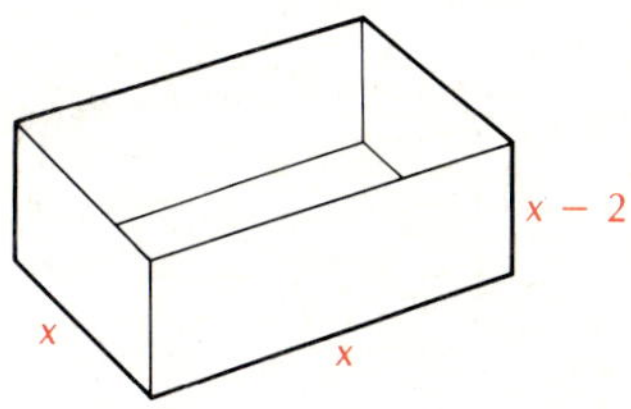

Add. Assume that the exponents are natural numbers.

82. $(2x^{2a} + 4x^a + 3) + (6x^{2a} + 3x^a + 4)$

83. $(47x^{4a} + 3x^{3a} + 22x^{2a} + x^a + 1) + (37x^{3a} + 8x^{2a} + 3)$

Subtract. Assume that the exponents are natural numbers.

84. $(3x^{6a} - 5x^{5a} + 4x^{3a} + 8) - (2x^{6a} + 4x^{4a} + 3x^{3a} + 2x^{2a})$

85. $(2x^{5b} + 4x^{4b} + 3x^{3b} + 8) - (x^{5b} + 2x^{3b} + 6x^{2b} + 9x^b + 8)$

4.2 MULTIPLICATION

To multiply polynomials, we use the distributive laws in order to be sure that we obtain equivalent expressions. For example, suppose we multiply two polynomials $P(x)$ and $Q(x)$ and obtain a polynomial $R(x)$:

$$P(x) \cdot Q(x) = R(x).$$

If we substitute a number for the variable in $P(x) \cdot Q(x)$, we will get the same value as if we made that substitution in the polynomial $R(x)$.

i Multiplication of Any Two Polynomials

When we apply the distributive laws to the multiplication of polynomials, it becomes apparent that we multiply each term of one polynomial by every term of the other and then add the results.

EXAMPLE 1 Multiply $4x^4y - 7x^2y + 3y$ by $2y - 3x^2y$.

$$\begin{array}{rl} 4x^4y - 7x^2y + 3y & \\ 2y - 3x^2y & \\ \hline 8x^4y^2 - 14x^2y^2 + 6y^2 & \text{Multiplying by } 2y \\ -12x^6y^2 + 21x^4y^2 - 9x^2y^2 & \text{Multiplying by } -3x^2y \\ \hline -12x^6y^2 + 29x^4y^2 - 23x^2y^2 + 6y^2 & \text{Adding} \end{array}$$

DO EXERCISES 1–3.

OBJECTIVES

After finishing Section 4.2, you should be able to:

- **i** **Multiply any two polynomials.**
- **ii** **Use a rule to multiply two binomials.**
- **iii** **Use a rule to square a binomial.**
- **iv** **Use a rule to multiply a sum and difference of the same two expressions.**
- **v** **Cube a binomial.**

Multiply.

1. $p - 3$ and $p^3 + 4p^2 - 5$

2. $2x^3 + 4x - 5$ and $x - 4$

3. $-4x^3 - 2x + 1$ and $-2x^2 - 3x + 6$

ii Products of Two Binomials

To multiply any two polynomials, we multiply each term of one by every term of the other. For two binomials we can generally multiply mentally, by using the following rule: Multiply the first terms, then the outside terms, then the inside terms, then the last terms (FOIL):

$$(A + B)(C + D) = \underset{F}{AC} + \underset{O}{AD} + \underset{I}{BC} + \underset{L}{BD}.$$

EXAMPLES Multiply.

2. $(3xy + 2x)(x^2 + 2xy^2) = \underset{F}{3x^3y} + \underset{O}{6x^2y^3} + \underset{I}{2x^3} + \underset{L}{4x^2y^2}$
3. $(2x - 3)(y + 2) = 2xy + 4x - 3y - 6$
4. $(2x + 3y)(x - 4y) = 2x^2 - 8xy + 3xy - 12y^2 = 2x^2 - 5xy - 12y^2$

DO EXERCISES 4–6.

Multiply.

4. $(y - 4)(y + 10)$

5. $(p + 5q)(2p - 3q)$

6. $(x^2y + 2x)(xy^2 + y^2)$

iii Squares of Binomials

A special case of a product of two binomials is the *square* of a binomial, that is, the product of a binomial and itself. The pattern, or rule, to remember and use is as follows. (You can obtain the formulas for yourself by doing the actual calculations.)

$$(A + B)^2 = A^2 + 2AB + B^2;$$
$$(A - B)^2 = A^2 - 2AB + B^2$$

The square of a sum or difference of two expressions is the square of the first, plus or minus twice the product of the expressions, plus the square of the second.

It can be helpful to memorize the words of the rule and say them while calculating.

EXAMPLES Multiply.

$$(A - B)^2 = A^2 - 2\ A\ \ B + B^2$$

5. $(y - 5)^2 = y^2 - 2(y)(5) + 5^2$
6. $(2x + 3y)^2 = (2x)^2 + 2(2x)(3y) + (3y)^2$
 $= 4x^2 + 12xy + 9y^2$
7. $(3x^2 - 5xy^2)^2 = (3x^2)^2 - 2(3x^2)(5xy^2) + (5xy^2)^2$
 $= 9x^4 - 30x^3y^2 + 25x^2y^4$

DO EXERCISES 7–9.

Multiply.

7. $(a - b)^2$

8. $(x + 8)^2$

9. $(3x - 7)^2$

iv Products of Sums and Differences

Another special case of a product of two binomials is the product of a sum and a difference. The pattern, or rule, to remember and use follows. (Again, you can obtain the rule for yourself by doing the actual calculation.)

$$(A + B)(A - B) = A^2 - B^2$$

The product of the sum and difference of two expressions is the square of the first expression minus the square of the second.

EXAMPLES Multiply.

$$(A + B)(A - B) = A^2 - B^2$$

8. $(y + 5)(y - 5) = y^2 - 5^2$
$= y^2 - 25$

9. $(2xy^2 + 3x)(2xy^2 - 3x) = (2xy^2)^2 - (3x)^2$
$= 4x^2y^4 - 9x^2$

10. $(5y + 4 + 3x)(5y + 4 - 3x) = (5y + 4)^2 - (3x)^2$
$= 25y^2 + 40y + 16 - 9x^2$

11. $(3xy^2 + 4y)(-3xy^2 + 4y) = -(3xy^2)^2 + (4y)^2$
$= 16y^2 - 9x^2y^4$

DO EXERCISES 10–13.

v Cubing Binomials

Remembering the special rules for multiplying can save a great deal of time. Another special case of a polynomial product is the cube of a binomial. The pattern, or rule, to be remembered is as follows. (Again, you can obtain the rule for yourself by doing the actual multiplying.)

$$(A + B)^3 = A^3 + 3A^2B + 3AB^2 + B^3$$

EXAMPLES Multiply.

12. $(x + 2)^3 = x^3 + 3x^2(2) + 3x(2)^2 + 2^3$
$= x^3 + 6x^2 + 12x + 8$

13. $(x - 2)^3 = [x + (-2)]^3$
$= x^3 + 3x^2(-2) + 3x(-2)^2 + (-2)^3$
$= x^3 - 6x^2 + 12x - 8$

14. $(5m^2 - 4n^3)^3 = (5m^2)^3 + 3(5m^2)^2(-4n^3) + 3(5m^2)(-4n^3)^2 + (-4n^3)^3$
$= 125m^6 - 300m^4n^3 + 240m^2n^6 - 64n^9$

DO EXERCISES 14–17.

Multiply.

10. $(x + 8)(x - 8)$

11. $(4y - 7)(4y + 7)$

Multiply.

12. $(7x^2y + 2y)(7x^2y - 2y)$

13. $(2x + 3 - 5y)(2x + 3 + 5y)$

Multiply.

14. $(x + 1)^3$

15. $(x - 1)^3$

16. $(t^2 - 3b)^3$

17. $(2a^2 - 5b^2)^3$

EXERCISE SET 4.2

i Multiply.

1. $2y^2$ and $5y$
2. $-3x^2$ and $2xy$
3. $5x$ and $-4x^2y$
4. $-3ab^2$ and $2a^2b^2$
5. $2x^3y^2$ and $-5x^2y^4$
6. $7a^2bc^4$ and $-8ab^3c^2$
7. $2x$ and $3 - x$
8. $4a$ and $a^2 - 5a$
9. $3ab$ and $a + b$
10. $2xy$ and $2x - 3y$
11. $5cd$ and $3c^2d - 5cd^2$
12. a^2 and $2a^2 - 5a^3$
13. $2x + 3$ and $3x - 4$
14. $2a - 3b$ and $4a - b$
15. $s + 3t$ and $s - 3t$
16. $y + 4$ and $y - 4$
17. $x - y$ and $x - y$
18. $a + 2b$ and $a + 2b$
19. $y + 8x$ and $2y - 7x$
20. $x + y$ and $x - 2y$
21. $a^2 - 2b^2$ and $a^2 - 3b^2$
22. $2m^2 - n^2$ and $3m^2 - 5n^2$
23. $x - 4$ and $x^2 + 4x + 16$
24. $y + 3$ and $y^2 - 3y + 9$
25. $x + y$ and $x^2 - xy + y^2$
26. $a - b$ and $a^2 + ab + b^2$
27. $a^2 + a - 1$ and $a^2 + 4a - 5$
28. $x^2 - 2x + 1$ and $x^2 + x + 2$
29. $4a^2b - 2ab + 3b^2$ and $ab - 2b + a$
30. $2x^2 + y^2 - 2xy$ and $x^2 - 2y^2 - xy$

ii, **iii** Multiply.

31. $(a + 2)(a + 3)$
32. $(x + 5)(x + 8)$
33. $(y + 3)(y - 2)$
34. $(y - 4)(y + 7)$
35. $(2a + \frac{1}{3})^2$
36. $(3c - \frac{1}{2})^2$
37. $(x - 2y)^2$
38. $(2s + 3t)^2$
39. $(b - \frac{1}{3})(b - \frac{1}{2})$
40. $(x - \frac{1}{2})(x - \frac{1}{4})$

41. $(2x + 9)(x + 2)$ **42.** $(3b + 2)(2b - 5)$ **43.** $(2x - 3y)(2x + y)$ **44.** $(2a - 3b)(2a - b)$ **45.** $(x + 3)^2$

46. $(y - 7)^2$ **47.** $(2x^2 - 3y^2)^2$ **48.** $(3s^2 + 4t^2)^2$ **49.** $(a^2b^2 + 1)^2$ **50.** $(x^2y - xy^2)^2$

iv Multiply.

51. $(c + 2)(c - 2)$ **52.** $(x - 3)(x + 3)$ **53.** $(2a + 1)(2a - 1)$ **54.** $(3 - 2x)(3 + 2x)$

55. $(3m - 2n)(3m + 2n)$ **56.** $(3x + 5y)(3x - 5y)$ **57.** $(x^2 + yz)(x^2 - yz)$ **58.** $(2a^2 + 5ab)(2a^2 - 5ab)$

v Multiply.

59. $(y + 5)^3$ **60.** $(t - 7)^3$ **61.** $(m^2 - 2n)^3$ **62.** $(3t^2 + 4)^3$

63. $(a^2 + 2y)^3$ **64.** $(2b - y^2)^3$ **65.** $(3a - 2b^2)^3$ **66.** $(2a^2 + 3b)^3$

Problem-solving practice

67. Jim worked a total of 17 days last month at his father's restaurant. He earned \$25 a day when he worked during the week and \$35 a day when he worked during the weekend. Last month Jim earned \$485. How many weekdays did he work?

68. The perimeter of a triangle is 174. The lengths of the three sides are consecutive even numbers. What are the lengths of the sides of the triangle?

Multiply. Assume that variables in exponents represent natural numbers.

69. $(6y)^2(-\frac{1}{3}x^2y^3)^3$ **70.** $[(a^{2n})^{2n}]^4$ **71.** $(-r^6s^2)^3\left(-\frac{r^2}{6}\right)^2(9s^4)^2$

72. $(z^{n^2})^{n^3}(z^{4n^3})^{n^2}$ **73.** $(-\frac{3}{8}xy^2)(-\frac{7}{9}x^3y)(-\frac{8}{7}x^2y)^2$ **74.** $(a^xb^{2y})(\frac{1}{2}a^{3x}b)^2$

75. $(-8s^3t)(2s^5 - 3s^3t^4 + st^7 - t^{10})$ **76.** $y^3z^n(y^{3n}z^3 - 4yz^{2n})$ **77.** $[(2x - 1)^2 - 1]^2$

78. $[x + y + 1][x^2 - x(y + 1) + (y + 1)^2]$ **79.** $[(a + b)(a - b)][5 - (a + b)][5 + (a + b)]$

80. $(y - 1)^6(y + 1)^6$ **81.** $(r^2 + s^2)^2(r^2 + 2rs + s^2)(r^2 - 2rs + s^2)$

82. $(3x^5 - \frac{5}{11})^2$ **83.** $(a - b + c - d)(a + b + c + d)$

84. $(\frac{2}{3}x + \frac{1}{3}y + 1)(\frac{2}{3}x - \frac{1}{3}y - 1)$ **85.** $[2(y - 3) - 6(x + 4)][5(y - 3) - 4(x + 4)]$

86. $(x - \frac{1}{7})(x^2 + \frac{1}{7}x + \frac{1}{49})$ **87.** $(4x^2 + 2xy + y^2)(4x^2 - 2xy + y^2)$

88. $(x^2 - 7x + 12)(x^2 + 7x + 12)$ **89.** $(x^a + y^b)(x^a - y^b)(x^{2a} + y^{2b})$

90. $[1 + \frac{1}{5}(x + 1)^2]^2$ **91.** $[a - (b - 1)][(b - 1)^2 + a(b - 1) + a^2]$

92. $(x - 1)(x^2 + x + 1)(x^3 + 1)$ **93.** $[(\frac{1}{3}x^3 - \frac{2}{3}y^2)(\frac{1}{3}x^3 + \frac{2}{3}y^2)]^2$

94. $10(0.1x^4y - 0.01xy^4)^2$ **95.** $(x^{a-b})^{a+b}$

96. $(M^{x+y})^{x+y}$

The amount to which \$1000 will grow in 3 years, when interest is compounded annually, is given by the polynomial function

$$A(r) = 1000r^3 + 3000r^2 + 3000r + 1000,$$

where r is the rate of interest.

97. Find the amount to which \$1000 will grow in 3 years at 12%.

98. Find the amount to which \$1000 will grow in 3 years at 15%.

OBJECTIVES

After finishing Section 4.3, you should be able to:

i **Factor polynomials whose terms have a common factor.**

ii **Factor certain polynomials with four terms by grouping.**

iii **Evaluate polynomials after first rewriting them in nested form.**

4.3 FACTORING AND EVALUATING POLYNOMIALS

Factoring is the reverse of multiplication. To *factor* an expression means to write an equivalent expression that is a product. What we do is based on the distributive laws, so that any substitution made will produce the same value in the original expression and in the factored expression.

i Terms with Common Factor

When factoring, you should look first for factors common to all the terms of an expression.

EXAMPLE 1 Factor out a common factor: $4y^2 - 8$.

$$4y^2 - 8 = 4 \cdot y^2 - 4 \cdot 2 \quad \text{Noting that 4 is a common factor}$$
$$= 4(y^2 - 2)$$

In some cases there is more than one common factor. In Example 2, for instance, 5 is a common factor, and x^3 is also a common factor. You should try to write the answer directly.

EXAMPLES Factor out a common factor.

2. $5x^4 - 20x^3 = 5x^3(x - 4)$
3. $12x^2y - 20x^3y = 4x^2y(3 - 5x)$
4. $10p^6q^2 - 4p^5q^3 + 2p^4q^4 = 2p^4q^2(5p^2 - 2pq + q^2)$

DO EXERCISES 1–4.

ii Factoring by Grouping

In expressions of four or more terms, there may be a common binomial factor. We proceed as in the following examples.

EXAMPLE 5 Factor: $y^2 + 3y + 4y + 12$.

$$y^2 + 3y + 4y + 12 = (y^2 + 3y) + (4y + 12) \quad \text{Grouping}$$
$$= y(y + 3) + 4(y + 3) \quad \text{Factoring out common factors}$$
$$= (y + 4)(y + 3) \quad \text{Factoring out } (y + 3)$$

EXAMPLE 6 Factor: $4x^3 - 15 + 20x^2 - 3x$.

$$4x^3 - 15 + 20x^2 - 3x = 4x^3 + 20x^2 - 3x - 15$$
$$= 4x^2(x + 5) - 3(x + 5)$$
$$= (4x^2 - 3)(x + 5)$$

EXAMPLE 7 Factor: $ax^2 + ay + bx^2 + by$.

$$ax^2 + ay + bx^2 + by = a(x^2 + y) + b(x^2 + y)$$
$$= (a + b)(x^2 + y)$$

DO EXERCISES 5–8.

iii Finding Polynomial Function Values

We have already evaluated polynomial functions by substitution and by direct calculation. For polynomials of higher degree, it is helpful to do some factoring before computing, especially if a lot of evaluating is to be done. The method we are about to describe is called *nested evaluation*.

EXAMPLE 8 For the polynomial function $P(x) = x^4 + 2x^3 - 4x^2 + 5x - 1$, find $P(5)$.

We first find the nested form of $P(x)$ by factoring parts of $P(x)$:

$$P(x) = x(x^3 + 2x^2 - 4x + 5) - 1 \quad \text{Factoring out an } x \text{ from the first four terms}$$
$$= x(x(x^2 + 2x - 4) + 5) - 1 \quad \text{Factoring out an } x \text{ from the polynomial that was in parentheses in step (1)}$$
$$= x(x(x(x + 2) - 4) + 5) - 1. \quad \text{Factoring out an } x \text{ from the polynomial that was in parentheses in step (2)}$$

Factor out a common factor.

1. $3x^2 - 6$
2. $4x^5 - 8x^3$
3. $9y^4 - 15y^3 + 3y^2$
4. $6x^2y - 21x^3y^2 + 3x^2y^3$
5. $(p + q)(x + 2) + (p + q)(x + y)$
6. $(y+3)(y-21)+(y+3)(y+10)$
7. $x^2 + 5x - 4x - 20$
8. $5y^3 + 2y^2 - 10y - 4$

9. Write nested form and then find $P(6)$.

$P(x) = x^5 - 2x^4 + 2x^3 - 5x^2 - 4x + 6$

We now have nested form. To find $P(5)$ we substitute 5 for x and then calculate, working from the inside out. Note that the only operations we need to do are multiplication and addition or subtraction:

$$P(5) = 5(5(5(5 + 2) - 4) + 5) - 1 \quad \text{Substituting 5}$$
$$= 5(5(5 \cdot 7 - 4) + 5) - 1 \quad \text{Doing the calculation in the innermost parentheses}$$
$$= 5(5 \cdot 31 + 5) - 1 \quad \text{Doing the next innermost calculation}$$
$$= 5 \cdot 160 - 1 \quad \text{Doing the next innermost calculation}$$
$$= 799.$$

DO EXERCISE 9.

It is important when using nested evaluation to write a zero in place of any missing term.

EXAMPLE 9 For the polynomial function $Q(x) = 3x^5 + 2x^4 - 3x^2 + 2x + 1$, find nested form.

10. Write nested form and then find $P(-4)$.

$P(x) = x^5 + 5x^3 + 4x^2 + 12$

We first write a 0 for the missing third-degree term and then proceed as in Example 8:

$$Q(x) = 3x^5 + 2x^4 + 0x^3 - 3x^2 + 2x + 1$$
$$= x(3x^4 + 2x^3 + 0x^2 - 3x + 2) + 1$$
$$= x(x(3x^3 + 2x^2 + 0x - 3) + 2) + 1.$$

We continue to factor in this way until we obtain

$$Q(x) = x(x(x(x(3x + 2) + 0) - 3) + 2) + 1.$$

We now have nested form.

You should note that when you evaluate a polynomial function using nested form, the only operations you need to do are addition or subtraction, and multiplication. You do not need to raise numbers to powers. If you have a calculator with a power key, raising numbers to powers is not difficult, but you may still find the nested form faster and more convenient, if you have a lot of evaluating to do.

DO EXERCISE 10.

EXERCISE SET 4.3

i Factor.

1. $4a^2 + 2a$
2. $6y^2 + 3y$
3. $y^2 - 5y$
4. $x^2 + 9x$
5. $y^3 + 9y^2$
6. $x^3 + 8x^2$
7. $6x^2 - 3x^4$
8. $8y^2 + 4y^4$
9. $4x^2y - 12xy^2$
10. $5x^2y^3 + 15x^3y^2$
11. $3y^2 - 3y - 9$
12. $5x^2 - 5x + 15$
13. $4ab - 6ac + 12ad$
14. $8xy + 10xz - 14xw$
15. $10a^4 + 15a^2 - 25a - 30$
16. $12t^5 - 20t^4 + 8t^2 - 16$

ii Factor.

17. $a(b - 2) + c(b - 2)$
18. $a(x^2 - 3) - 2(x^2 - 3)$
19. $(x - 2)(x + 5) + (x - 2)(x + 8)$
20. $(m - 4)(m + 3) + (m - 4)(m - 3)$
21. $a^2(x - y) + a^2(x - y)$
22. $3x^2(x - 6) + 3x^2(x - 6)$
23. $ac + ad + bc + bd$
24. $xy + xz + wy + wz$
25. $b^3 - b^2 + 2b - 2$
26. $y^3 - y^2 + 3y - 3$
27. $y^2 - 8y - y + 8$
28. $t^2 + 6t - 2t - 12$
29. $24x^3 - 36x^2 + 72x$
30. $10a^4 + 15a^2 - 25a - 30$
31. $x^6 + x^5 - x^3 + x^2$
32. $y^4 - y^3 + y^2 + y$
33. $2y^4 + 6y^2 + 5y^2 + 15$
34. $2xy + x^2y - 6 - 3x$

iii Find nested form.

35. $P(x) = x^4 - 3x^3 - 5x^2 + 4x + 2$

36. $P(x) = x^4 + 5x^3 + 7x^2 - 2x - 4$

37. $P(x) = 5x^5 - 3x^3 + 4x^2 - 5x + 1$

38. $Q(x) = 4x^5 + 2x^4 - 5x^2 - 7$

Find nested form and then find the indicated function values.

39. $P(x) = 2x^4 - 3x^3 + 5x^2 + 6x - 4$; $P(5)$, $P(-2)$, $P(10)$

40. $Q(x) = 2x^5 - 3x^4 + 5x^2 - 7x + 12$; $Q(-4)$, $Q(7)$, $Q(11)$

Problem-solving practice

41. The classified department of a newspaper charges \$4.00 for the first 15 words of an ad, and 25¢ for each additional word. Sam wants to spend at least \$5.00 but no more than \$10.00 on an ad to sell his car. What size ads (in words) will fit into Sam's budget?

42. One angle of a triangle is four times another angle. The third angle is 75°. What are the measures of the first two angles?

Factor. Assume that all exponents are natural numbers.

43. $4y^{4a} + 12y^{2a} + 10y^{2a} + 30$

44. $2x^{4p} + 6x^{2p} + 5x^{2p} + 15$

45. $4x^{a+b} + 7x^{a-b}$

46. $7y^{2a+b} - 5y^{a+b} + 3y^{a+2b}$

The number of diagonals of a polygon having n sides is given by the polynomial function $P(n) = \frac{1}{2}n^2 - \frac{3}{2}n$.

47. Find the number of diagonals in a polygon with 12 sides.

48. Find the number of diagonals in a polygon with 117 sides.

49. For the function $P(x) = 0.1x^4 - x^2 + 0.4$:

a) Find nested form.

b) Calculate and plot points; then draw a graph of the function for values of x from -4 to 4.

4.4

FACTORING TRINOMIALS

i Trinomial Squares

Some trinomials are squares of binomials. Here is an example:

$$x^2 + 6x + 9 = (x + 3)^2.$$

Such trinomials are called *trinomial squares*. They can, of course, be factored. It is important to be able to recognize trinomial squares.

For a trinomial to be a square, the following must hold.

a) **Two of the terms must be squares, such as A^2 and B^2.**

b) **There must be no minus sign before A^2 or B^2.**

c) **If we multiply A and B (the square roots of these expressions) and double the result, we get the remaining term, $2AB$, or its additive inverse, $-2AB$.**

EXAMPLES Determine whether these are trinomial squares.

1. $x^2 + 10x + 25$

   a) Two terms are squares: x^2 and 25.

   b) There is no minus sign before either x^2 or 25.

   c) If we multiply the square roots, x and 5, and double, we get $10x$, the remaining term.

   Thus the trinomial is a square.

OBJECTIVES

After finishing Section 4.4, you should be able to:

i **Factor trinomial squares.**

ii **Factor trinomials of the type $x^2 + ax + c$.**

iii **Factor trinomials of the type $ax^2 + bx + c$.**

1. Which of the following are trinomial squares?
   a) $x^2 + 6x + 9$
   b) $x^2 - 8x + 16$
   c) $x^2 + 6x + 11$
   d) $4x^2 - 20x + 25$
   e) $16x^2 - 20x + 25$
   f) $5x^2 + 14x + 16$
   g) $x^2 + 8x - 16$
   h) $x^2 - 8x - 16$

2. $4x + 16 + 3x^2$

a) Only one term, 16, is a square ($3x^2$ is not a square). Therefore the trinomial is not a square.

DO EXERCISE 1.

To factor trinomial squares, we use the same equations that we used to multiply out squares of binomials. They are as follows.

$$A^2 + 2AB + B^2 = (A + B)^2;$$
$$A^2 - 2AB + B^2 = (A - B)^2$$

EXAMPLE 3 Factor: $x^2 - 10x + 25$.

$x^2 - 10x + 25 = (x - 5)^2$ — We find the square terms and write their square roots with a minus sign between them.

Factor.

2. $x^2 + 14x + 49$

EXAMPLE 4 Factor: $16y^2 + 49 + 56y$.

$16y^2 + 49 + 56y = (4y + 7)^2$ — We find the square terms and write their square roots with a plus sign between them.

EXAMPLE 5 Factor: $-20xy + 4y^2 + 25x^2$.

$$-20xy + 4y^2 + 25x^2 = (2y - 5x)^2$$

3. $9y^2 - 30y + 25$

EXAMPLE 6 Factor: $2x^2 - 12x + 18$.

$$2x^2 - 12x + 18 = 2(x^2 - 6x + 9) \quad \text{Removing the common factor}$$
$$= 2(x - 3)^2 \quad \text{Factoring the trinomial square}$$

DO EXERCISES 2–5.

4. $16x^2 + 72xy + 81y^2$

In factoring, we must always remember to look *first* for a factor common to all the terms.

EXAMPLE 7 Factor: $-4y^2 - 144y^8 + 48y^5$.

$$-4y^2 - 144y^8 + 48y^5 = -4y^2(1 + 36y^6 - 12y^3) \quad \text{Removing the common factor}$$
$$= -4y^2(1 - 12y^3 + 36y^6) \quad \text{Changing order}$$
$$= -4y^2(1 - 6y^3)^2 \quad \text{Factoring the trinomial square}$$

5. $16x^4 - 40x^2y^3 + 25y^6$

DO EXERCISE 6.

ii Trinomials $x^2 + bx + c$

The trinomials that we consider now are not squares and they have a leading coefficient of 1. Consider this product:

Factor.

6. $-8a^2 + 24ab - 18b^2$

$$(x + 3)(x + 5) = \overset{F}{x^2} + \overset{O}{5x} + \overset{I}{3x} + \overset{L}{15}$$
$$= x^2 + 8x + 15.$$

Note that the coefficient 8 is the sum of 3 and 5, and that 15 is the product of 3 and 5. In general,

$$(x + a)(x + b) = x^2 + (a + b)x + ab.$$

To factor we can use the preceding equation in reverse, as follows:

$$x^2 + (a + b)x + ab = (x + a)(x + b).$$

Finding a and b is a trial-and-error process. Practice helps.

EXAMPLE 8 Factor: $x^2 - 3x - 10$.

We look for pairs of integers whose product is -10 and whose sum is -3.

Pairs of factors	Sum of factors
$-2,\ 5$	3
$2,\ -5$	-3
$10,\ -1$	9
$-10,\ 1$	-9

Thus the desired integers are 2 and -5. Then

$$x^2 - 3x - 10 = (x + 2)(x - 5).$$

We can check by multiplying.

DO EXERCISES 7 AND 8.

Factor. Check by multiplying.

7. $x^2 + 5x - 14$

8. $y^2 - y - 2$

iii Trinomials $ax^2 + bx + c$

We now consider trinomials that are not squares and for which the leading coefficient is not 1. We present two methods for factoring such trinomials.

Method 1. To use this method, we begin by multiplying the leading coefficient and the constant term. We then factor that product in a certain way, as in the following example.

EXAMPLE 9 Factor: $8x^2 + 22x + 15$.

a) We multiply the leading coefficient by the constant term:

$$8 \cdot 15 = 120.$$

b) We factor 120 to obtain factors whose *sum* is the coefficient of the first-degree term, 22:

$$120 = 10 \cdot 12, \quad \text{and} \quad 10 + 12 = 22, \text{ the desired coefficient.}$$

c) We split the middle term and then factor by grouping:

$$\begin{aligned} 8x^2 + 22x + 15 &= 8x^2 + 10x + 12x + 15 && \text{Splitting the middle term} \\ &= 2x(4x + 5) + 3(4x + 5) && \text{Factoring by grouping} \\ &= (2x + 3)(4x + 5) \end{aligned}$$

Method 2. To use this method, we factor both the leading coefficient and the constant term by trial, until we obtain the correct middle term.

EXAMPLE 9A Factor: $8x^2 + 22x + 15$.

We look first for a common factor. There is none other than 1. We now look for numbers whose product is 8. They are 1 and 8, 2 and 4, -1 and -8, and -2 and -4. Since the leading term is positive, we can deal only with the positive factors. We have these possibilities:

$$(2x + \quad)(4x + \quad) \quad \text{or} \quad (x + \quad)(8x + \quad).$$

Factor.

9. $3x^2 + 5x + 2$

10. $4x^2 - 3 + 4x$

Factor.

11. $24y^2 - 46y + 10$

12. $16x^2 - 12 + 16x$

13. $p^2q^2 + 7pq + 12$

14. $2x^4y^6 - 3x^2y^3 - 20$

Now we look for numbers whose product is 15, the constant term. The possibilities are 1 and 15 and 3 and 5. We try them in the blank spaces until we have the correct middle term. The correct answer is

$$(2x + 3)(4x + 5).$$

DO EXERCISES 9 AND 10.

EXAMPLE 10 Factor: $12x^2 - 22x - 14$.

We first look for a common factor. The number 2 is a common factor so we factor it out:

$$2(6x^2 - 11x - 7).$$

We now factor the expression in parentheses, using either Method 1 or Method 2, obtaining

$$(2x + 1)(3x - 7).$$

We must not forget to include the common factor 2. The answer is

$$2(2x + 1)(3x - 7).$$

EXAMPLE 11 Factor: $x^2y^2 + 5xy + 4$.

In this case, we can treat xy as if it were a single variable:

$$\begin{aligned} x^2y^2 + 5xy + 4 &= (xy)^2 + 5xy + 4 \\ &= (xy + 4)(xy + 1). \end{aligned}$$

DO EXERCISES 11–14.

EXERCISE SET 4.4

i, **ii**, **iii** Factor:

1. $y^2 - 6y + 9$
2. $x^2 - 8x + 16$
3. $x^2 + 14x + 49$
4. $x^2 + 16x + 64$
5. $x^2 + 1 + 2x$
6. $x^2 + 1 - 2x$
7. $2a^2 + 8a + 8$
8. $4a^2 - 16a + 16$
9. $y^2 + 36 - 12y$
10. $y^2 + 36 + 12y$
11. $-18y^2 + y^3 + 81y$
12. $24a^2 + a^3 + 144a$
13. $12a^2 + 36a + 27$
14. $20y^2 + 100y + 125$
15. $2x^2 - 40x + 200$
16. $32x^2 + 48x + 18$
17. $1 - 8d + 16d^2$
18. $64 + 25y^2 - 80y$
19. $x^2 + 9x + 20$
20. $y^2 + 8y + 15$
21. $y^2 - 8y + 16$
22. $a^2 - 10a + 25$
23. $x^2 - 27 - 6x$
24. $t^2 - 15 - 2t$
25. $m^2 - 3m - 28$
26. $x^2 - 2x - 8$
27. $14x + x^2 + 45$
28. $12y + y^2 + 32$
29. $y^2 + 2y - 63$
30. $x^2 + 3x - 40$
31. $t^2 - 11t + 28$
32. $y^2 - 14y + 45$
33. $3x + x^2 - 10$
34. $x + x^2 - 6$
35. $x^2 + 5x + 6$
36. $y^2 + 8y + 7$
37. $32 + 4y - y^2$
38. $56 + x - x^2$
39. $15 + t^2 + 8t$
40. $3b^2 + 8b + 4$
41. $9x^2 + 15x + 4$
42. $6y^2 - y - 2$
43. $3a^2 - a - 4$
44. $-7a + 6a^2 - 10$
45. $-35z + 12z^2 - 3$
46. $9a^2 + 6a - 8$
47. $4t^2 + 4t - 15$
48. $3x^2 - 16x - 12$
49. $6x^2 - 5x - 25$
50. $6x^2 - 15 - x$
51. $10y^2 - 12 - 7y$
52. $3a^2 - 10a + 8$
53. $12a^2 - 7a + 1$
54. $35y^2 + 34y + 8$
55. $9a^2 + 18a + 8$
56. $2t + 5t^2 - 3$
57. $4x + 15x^2 - 3$
58. $8x^2 - 16 - 28x$
59. $18x^2 - 24 - 6x$
60. $3x^3 - 5x^2 - 2x$
61. $18y^3 - 3y^2 - 10y$
62. $24x^2 - 2 - 47x$
63. $15y^2 - 10 - 19y$
64. $21x^2 + 37x + 12$
65. $10y^2 + 23y + 12$
66. $17x + 40x^2 - 12$
67. $2y + 24y^2 - 15$
68. $12a^2 - 17a + 6$
69. $20a^2 - 23a + 6$
70. $10a^2 - 15 - 19a$

Factor.

71. $x^4 + 11x^2 - 80$
72. $y^4 + 5y^2 - 84$
73. $x^2 - \frac{4}{25} + \frac{3}{5}x$
74. $y^2 - \frac{8}{49} + \frac{2}{7}y$
75. $y^2 + 0.4y - 0.05$
76. $t^2 + 0.6t - 0.27$
77. $2x^2 + xy - 6y^2$
78. $2m^2 + mn - 10n^2$
79. $-6xy + 8x^2 - 9y^2$
80. $-7ts + 2t^2 - 4s^2$
81. $7a^2b^2 + 6 + 13ab$
82. $9x^2y^2 - 4 + 5xy$

Factor. Assume that variables in exponents represent positive integers.

83. $3x^2 + 12x - 495$
84. $-225x + x^3$
85. $4y^3 - 96y^2 + 576y$
86. $3xy^2 - 150xy + 1875x$
87. $7x^2 - 4375y^2$
88. $12x^2 - 72xy + 108y^2$
89. $15t^3 - 60t^2 - 315t$
90. $216x + 78x^2 + 6x^3$
91. $x^{2a} + 5x^a - 24$
92. $x^2 + ax + bx + ab$
93. $4x^{2a} - 4x^a - 3$
94. $bdx^2 + adx + bcx + ac$
95. $\frac{1}{4}p^2 - \frac{2}{5}p + \frac{4}{25}$
96. $\frac{4}{27}r^2 + \frac{5}{9}rs + \frac{1}{12}s^2 - \frac{1}{3}rs$

97. The area of a ring, as shown, is given by the polynomial $\pi R^2 - \pi r^2$. Factoring the polynomial, we have $\pi(R^2 - r^2)$. Find the area of a ring for which $R = 10$ cm and $r = 4$ cm. Use 3.14 for π.

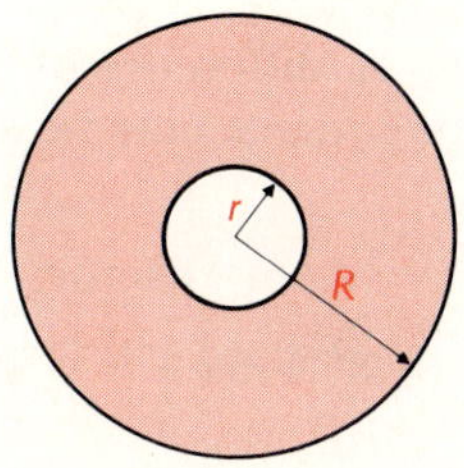

4.5

DIFFERENCES OF SQUARES; COMPLETING THE SQUARE

OBJECTIVES

After finishing Section 4.5, you should be able to:

i **Factor differences of two squares.**
ii **Complete the square.**
iii **Factor by completing the square.**

i To factor a difference of two expressions that are squares, we can use a pattern that we used earlier, for multiplying a sum and a difference.

$$A^2 - B^2 = (A + B)(A - B)$$

To factor a difference of two squares, write the square root of the first expression *plus* the square root of the second, times the square root of the first *minus* the square root of the second.

EXAMPLE 1 Factor: $x^2 - 9$.

$$x^2 - 9 = x^2 - 3^2 = (x + 3)(x - 3)$$

EXAMPLE 2 Factor: $25y^6 - 49x^2$.

$$\begin{array}{ccccccccc} A^2 & - & B^2 & = (& A & + B)(& A & - & B) \\ \downarrow & & \downarrow & & \downarrow & \downarrow & \downarrow & & \downarrow \\ 25y^6 & - & 49x^2 & = (& 5y^3 & + 7x)(& 5y^3 & - & 7x) \end{array}$$

EXAMPLE 3 Factor: $5 - 5x^2y^6$.

$$5 - 5x^2y^6 = 5(1 - x^2y^6)$$
$$= 5(1 + xy^3)(1 - xy^3) \quad \text{Factoring the difference of squares}$$

EXAMPLE 4 Factor: $2x^4 - 8y^4$.

$$2x^4 - 8y^4 = 2(x^4 - 4y^4) \quad \text{Removing the common factor}$$
$$= 2(x^2 + 2y^2)(x^2 - 2y^2) \quad \text{Factoring the difference of squares}$$

Factor.

1. $y^2 - 4$

2. $49x^4 - 25y^{10}$

3. $25x^2y^2 - 4a^2$

4. $9x^2 - 16y^6$

5. $20x^2 - 5y^2$

6. $81x^4y^2 - 16y^2$

Factor.

7. $x^2 - 10x + 25 - 4y^2$

EXAMPLE 5 Factor: $16x^4y - 81y$.

$$16x^4y - 81y = y(16x^4 - 81) \quad \text{Removing the common factor}$$
$$= y(4x^2 + 9)(4x^2 - 9) \quad \text{Factoring the difference of squares}$$
$$= y(4x^2 + 9)(2x + 3)(2x - 3) \quad \text{Factoring } 4x^2 - 9\text{, which is a difference of squares}$$

In Example 5 it may be tempting to try to factor $(4x^2 + 9)$. Note that it is a sum of two expressions that are squares. Such expressions cannot be factored using real numbers.

DO EXERCISES 1–6.

A difference of squares can have more than two terms. For example, one of the squares may be a trinomial. We can factor by a kind of grouping.

EXAMPLE 6 Factor: $x^2 + 6x + 9 - y^2$.

$$x^2 + 6x + 9 - y^2 = (x^2 + 6x + 9) - y^2 \quad \text{Grouping as a trinomial minus } y^2 \text{ to show a difference of squares}$$
$$= (x + 3)^2 - y^2$$
$$= (x + 3 + y)(x + 3 - y)$$

DO EXERCISE 7.

ii Completing the Square

We look for a pattern in a trinomial that is the square of a binomial. Let us square the binomial $x + 5$.

$$(x + 5)^2 = x^2 + 10x + 25$$

This coefficient is $2 \cdot 5$. (10)
This coefficient is 5^2. (25)

If we knew only the first two terms, $x^2 + 10x$, we could find the third term by taking half of 10 and squaring the result.

EXAMPLE 7 What must be added to $x^2 + 12x$ in order to make it a trinomial square?

We take half the coefficient of x and square it.

$$x^2 + 12x$$

→ Half of 12 is 6, and $6^2 = 36$. We add 36.

Thus $x^2 + 12x + 36$ is a trinomial square. It is equivalent to $(x + 6)^2$.

The process illustrated in Example 7 is called *completing the square.* Completing the square can be used in factoring as well as a number of other ways.

EXAMPLE 8 Complete the square on $x^2 - 8ax$.

We take half of $-8a$ (the coefficient of x) and get $-4a$.

We square $-4a$: $(-4a)^2 = 16a^2$.

We add the result to obtain $x^2 - 8ax + 16a^2$.

This is a trinomial square because it is equivalent to $(x - 4a)^2$.

EXAMPLE 9 Complete the square on $y^2 + \frac{3}{4}y$.

We take half of $\frac{3}{4}$ (the coefficient of y) and get $\frac{1}{2} \cdot \frac{3}{4} = \frac{3}{8}$.

We square $\frac{3}{8}$: $(\frac{3}{8})^2 = \frac{9}{64}$.

We add the result to obtain $y^2 + \frac{3}{4}y + \frac{9}{64}$.

DO EXERCISES 8–11.

iii Factoring by Completing the Square

Completing the square can be used to factor trinomials.

EXAMPLE 10 Factor by completing the square: $x^2 - 22x + 112$.

We first complete the square on $x^2 - 22x$. We take half of -22, which is -11. We then square -11 to get $(-11)^2 = 121$.

Now we add 0 to the original trinomial, naming it $121 - 121$:

$$\begin{aligned} x^2 - 22x + 112 &= x^2 - 22x + 112 + (121 - 121) && \text{Adding } 121 - 121 \\ &= (x^2 - 22x + 121) + 112 - 121 && \text{Grouping} \\ &= (x^2 - 22x + 121) - 9 \\ &= (x - 11)^2 - 3^2 \\ &= (x - 11 + 3)(x - 11 - 3), \quad \text{or } (x - 8)(x - 14). \end{aligned}$$

EXAMPLE 11 Factor by completing the square: $2x^2 + 80x + 768$.

$$2x^2 + 80x + 768 = 2(x^2 + 40x + 384) \qquad \text{Removing the common factor 2}$$

To complete the square we take half of 40 to obtain 20. We then square 20 to get 400.

We add 0, naming it $400 - 400$:

$$\begin{aligned} 2(x^2 + 40x + 384) &= 2(x^2 + 40x + 400 - 400 + 384) && \text{Adding } 400 - 400 \\ &= 2(x^2 + 40x + 400 - 16) && \text{Grouping} \\ &= 2[(x + 20)^2 - 4^2] \\ &= 2(x + 20 + 4)(x + 20 - 4), \quad \text{or } 2(x + 24)(x + 16). \end{aligned}$$

EXAMPLE 12 Factor by completing the square: $x^2 - 6.2x + 8.61$.

To complete the square we take half of -6.2, obtaining -3.1. We then square -3.1 to get $(-3.1)^2 = 9.61$.

Next we add 0, naming it $9.61 - 9.61$:

$$\begin{aligned} x^2 - 6.2x + 8.61 &= x^2 - 6.2x + 8.61 + 9.61 - 9.61 && \text{Adding } 9.61 - 9.61 \\ &= (x^2 - 6.2x + 9.61) + 8.61 - 9.61 && \text{Grouping} \\ &= (x - 3.1)^2 - 1 \\ &= (x - 3.1 + 1)(x - 3.1 - 1), \quad \text{or } (x - 2.1)(x - 4.1) \end{aligned}$$

DO EXERCISES 12–14.

Complete the square.

8. $x^2 + 14x$

9. $y^2 - 10by$

10. $x^2 - \frac{2}{5}x$

11. $x^2 + 4.2x$

Factor by completing the square.

12. $x^2 + 26x + 120$

13. $2x^2 - 48x + 126$

14. $x^2 + 8.4x + 13.64$

EXERCISE SET 4.5

i Factor.

1. $x^2 - 16$ **2.** $y^2 - 9$ **3.** $6x^2 - 6y^2$ **4.** $8x^2 - 8y^2$

5. $4xy^4 - 4xz^4$ **6.** $25ab^4 - 25az^4$ **7.** $4a^3 - 49a$ **8.** $9x^3 - 25x$

9. $3x^8 - 3y^8$ **10.** $9a^4 - a^2b^2$ **11.** $9a^4 - 25a^2b^4$ **12.** $16x^6 - 121x^2y^4$

13. $\frac{1}{25} - x^2$ **14.** $\frac{1}{16} - y^2$ **15.** $0.04x^2 - 0.09y^2$ **16.** $0.01x^2 - 0.04y^2$

17. $a^2 + 2ab + b^2 - 9$ **18.** $x^2 - 2xy + y^2 - 25$ **19.** $r^2 - 2r + 1 - 4s^2$ **20.** $c^2 + 4cd + 4d^2 - 9p^2$

21. $2m^2 + 4mn + 2n^2 - 50b^2$ **22.** $12x^2 + 12x + 3 - 3y^2$ **23.** $9 - (a^2 + 2ab + b^2)$ **24.** $16 - (x^2 - 2xy + y^2)$

ii Complete the square.

25. $x^2 + 16x$ **26.** $y^2 - 24y$ **27.** $x^2 - 4.2x$ **28.** $y^2 + 3.6y$

iii Factor by completing the square.

29. $x^2 + 24x + 119$ **30.** $x^2 + 30x + 176$ **31.** $x^2 - 26x + 105$ **32.** $x^2 - 32x + 192$

33. $2x^2 + 56x + 342$ **34.** $3x^2 + 54x + 231$ **35.** $2x^2 - 32ax + 96a^2$ **36.** $3x^2 - 60bx + 192b^2$

Problem-solving practice

37. Susan has four test scores of 78, 92, 79, and 85. What must the fifth test score be in order for her average to be 82?

38. A beaker contains a solution that is 40% ammonia. Another beaker contains a solution that is 60% ammonia. How much of each can be added together in order to obtain 35 liters of a solution that is 55% ammonia?

Factor by completing the square.

39. (calculator) $x^2 + 4.482x - 7.403544$ **40.** (calculator) $5.72x^2 + 35.464x - 1319.2608$

Complete the square.

41. $x^2 + \frac{2}{3}bx$ **42.** $ay^2 + \frac{3}{4}by$ **43.** $ax^2 + bx$ **44.** $ax^2 + bx + c$

Factor. Assume that variables in exponents represent natural numbers.

45. $x^2 - 3x + \frac{5}{4}$ **46.** $x^2 - 5x + \frac{9}{4}$ **47.** $x^2 + 2.6x + 0.69$

48. $x^2 - 3.2x + 1.56$ **49.** $x^2 + \frac{3}{5}x + \frac{8}{100}$ **50.** $x^2 - \frac{7}{2}x + \frac{45}{16}$

51. $x^4y^4 - 8x^2y^2 + 16$ **52.** $-24ab + 16a^2 + 9b^2$ **53.** $\frac{1}{36}x^8 + \frac{4}{18}x^4 + \frac{4}{9}$

54. $0.25x^2 + 0.30x + 0.09$ **55.** $x^{2a} - y^2$ **56.** $x^{4a} - y^{2b}$

57. $4y^{4a} + 12y^{2a} + 10y^{2a} + 30$ **58.** $25y^{2a} - (x^{2b} - 2x^b + 1)$

Factor by completing the square.

59. $x^{2a} + 14x^a + 45$ **60.** $x^2 - 2ax - 3a^2$

OBJECTIVE

After finishing Section 4.6, you should be able to:

i **Factor sums and differences of two cubes.**

4.6 SUMS OR DIFFERENCES OF TWO CUBES

i Although a sum of two squares cannot be factored using real-number coefficients, a sum of two cubes can. In this section we develop the patterns for factoring a sum or a difference of two expressions that are cubes.

Consider the following products:

$$\begin{aligned}(A + B)(A^2 - AB + B^2) &= A(A^2 - AB + B^2) + B(A^2 - AB + B^2)\\ &= A^3 - A^2B + AB^2 + A^2B - AB^2 + B^3\\ &= A^3 + B^3\end{aligned}$$

and

$$\begin{aligned}(A - B)(A^2 + AB + B^2) &= A(A^2 + AB + B^2) - B(A^2 + AB + B^2)\\ &= A^3 + A^2B + AB^2 - A^2B - AB^2 - B^3\\ &= A^3 - B^3.\end{aligned}$$

The above equations (reversed) show how we can factor a sum or a difference of two cubes.

$$A^3 + B^3 = (A + B)(A^2 - AB + B^2)$$
$$A^3 - B^3 = (A - B)(A^2 + AB + B^2)$$

This table of cubes will help in the following examples.

N	0.2	0.1	0	1	2	3	4	5	6	7	8
N^3	0.008	0.001	0	1	8	27	64	125	216	343	512

EXAMPLE 1 Factor: $x^3 - 27$.

$$x^3 - 27 = x^3 - 3^3$$

In one set of parentheses we write the cube root of the first term, x. Then we write the cube root of the second term, -3. This gives us the expression $x - 3$:

$$(x - 3)(\qquad).$$

To get the next factor we think of $x - 3$ and do the following:

Square the first term: x^2.
Multiply the terms and then change the sign: $3x$.
Square the second term: 9.

$$(x - 3)(x^2 + 3x + 9).$$

Note that we cannot factor $x^2 + 3x + 9$. (It is not a trinomial square nor can it be factored by trial).

DO EXERCISES 1 AND 2.

EXAMPLE 2 Factor: $125x^3 + y^3$.

$$125x^3 + y^3 = (5x)^3 + y^3$$

In one set of parentheses we write the cube root of the first term, then a plus sign, and then the cube root of the second term:

$$(5x + y)(\qquad).$$

To get the next factor, we think of $5x + y$ and do the following:

Square the first term: $(5x)^2$, or $25x^2$.
Multiply the terms and then change the sign: $-5xy$.
Square the second term: y^2.

$$(5x + y)(25x^2 - 5xy + y^2).$$

DO EXERCISES 3 AND 4.

EXAMPLE 3 Factor: $128y^7 - 250x^6y$.

We first look for a common factor:

$$2y(64y^6 - 125x^6) = 2y[(4y^2)^3 - (5x^2)^3]$$
$$= 2y(4y^2 - 5x^2)(16y^4 + 20x^2y^2 + 25x^4).$$

EXAMPLE 4 Factor: $64a^6 - 729b^6$.

$(8a^3 - 27b^3)(8a^3 + 27b^3)$ Factoring a difference of squares

Factor.

1. $x^3 - 8$

2. $64 - y^3$

Factor.

3. $27x^3 + y^3$

4. $8y^3 + z^3$

Factor.

5. $16x^7y + 54xy^7$

6. $729x^6 - 64y^6$

7. $x^3 - 0.027$

Each factor is a sum or difference of cubes. We factor:

$$(2a - 3b)(4a^2 + 6ab + 9b^2)(2a + 3b)(4a^2 - 6ab + 9b^2).$$

EXAMPLE 5 Factor: $y^3 + 0.008$.

We note that $0.008 = (0.2)^3$. Thus we have a sum of two cubes, and we factor as follows:

$$(y + 0.2)(y^2 - 0.2y + 0.04).$$

Remember the following about factoring sums or differences of squares and cubes:

Sum of cubes:	$A^3 + B^3 = (A + B)(A^2 - AB + B^2)$
Difference of cubes:	$A^3 - B^3 = (A - B)(A^2 + AB + B^2)$
Difference of squares:	$A^2 - B^2 = (A + B)(A - B)$
Sum of squares:	$A^2 + B^2$ **cannot be factored**

DO EXERCISES 5–7.

EXERCISE SET 4.6

i Factor.

1. $x^3 + 8$
2. $c^3 + 27$
3. $y^3 - 64$
4. $z^3 - 1$
5. $w^3 + 1$
6. $x^3 + 125$
7. $8a^3 + 1$
8. $27x^3 + 1$
9. $y^3 - 8$
10. $p^3 - 27$
11. $8 - 27b^3$
12. $64 - 125x^3$
13. $64y^3 + 1$
14. $125x^3 + 1$
15. $8x^3 + 27$
16. $27y^3 + 64$
17. $a^3 - b^3$
18. $x^3 - y^3$
19. $a^3 + \frac{1}{8}$
20. $b^3 + \frac{1}{27}$
21. $2y^3 - 128$
22. $3z^3 - 3$
23. $24a^3 + 3$
24. $54x^3 + 2$
25. $rs^3 + 64r$
26. $ab^3 + 125a$
27. $5x^3 - 40z^3$
28. $2y^3 - 54z^3$
29. $x^3 + 0.001$
30. $y^3 + 0.125$
31. $64x^6 - 8t^6$
32. $125c^6 - 8d^6$

Problem-solving practice

33. The length of a rectangle is 7 ft less than its width. If the length is increased by 2 ft, the perimeter is then 66 ft. What is the area of the original rectangle?

34. Tickets for the afternoon show at the movies cost \$2.00 for adults and 50¢ for children. At the evening show it costs \$4.00 for adults and \$1.50 for children. On a certain day the afternoon show grossed \$300, and the evening show grossed \$650. If the attendance of both adults and children was the same at both shows, how many children attended the evening show?

○

Factor. Assume that variables in exponents represent natural numbers.

35. $x^{6a} + y^{3b}$
36. $a^3x^3 - b^3y^3$
37. $3x^{3a} + 24y^{3b}$
38. $\frac{8}{27}x^3 + \frac{1}{64}y^3$
39. $\frac{1}{24}x^3y^3 + \frac{1}{3}z^3$
40. $\frac{1}{16}x^{3a} + \frac{1}{2}y^{6a}z^{9b}$

41. A 4×4 ft sandbox is placed on a square lawn that is x ft on a side. Find a polynomial function giving the area of the part of the lawn not covered.

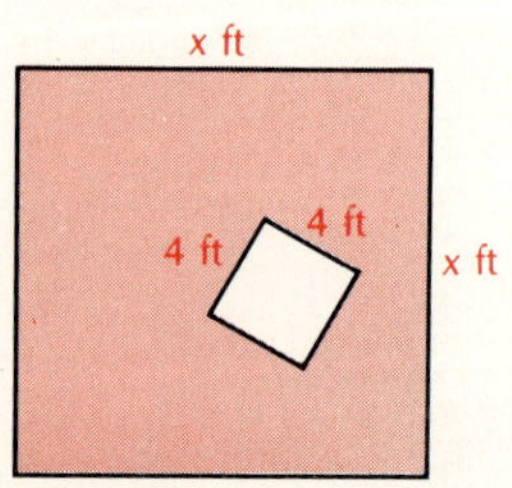

42. Find a polynomial function for the area of the color region.

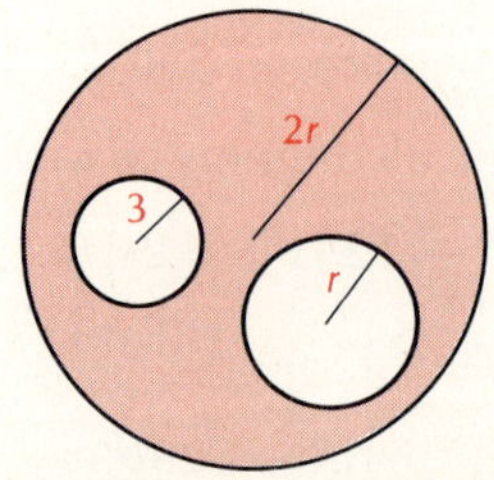

4.7

FACTORING: A GENERAL STRATEGY

OBJECTIVE

After finishing Section 4.7, you should be able to:

i Factor polynomials using any of the methods considered in this chapter.

i Factoring is an important algebraic skill. Once you know what kind of expression you have to factor, the factoring can be done without too much difficulty. The following table outlines a strategy for factoring.

> **A. Always look first for a common factor.**
>
> **B. Then look at the number of terms.**
>
> ***Two terms:*** **Try factoring as a difference of squares first. Next, try factoring as a sum or difference of cubes. Do *not* try to factor a *sum* of squares.**
>
> ***Three terms:*** **Determine whether the trinomial is a square. If so, you know how to factor. If not, try trial and error.**
>
> ***Four or more terms:*** **Try factoring by grouping and removing a common binomial factor. Next, try grouping into a difference of squares, one of which is a trinomial.**
>
> **C. Always *factor completely.* If a factor with more than one term can be factored, you should factor it.**

EXAMPLE 1 Factor: $10a^2x - 40b^2x$.

A. We look first for a common factor:

$$10x(a^2 - 4b^2). \qquad \text{Factoring out the largest common factor}$$

B. The factor $a^2 - 4b^2$ has only two terms. It is a difference of squares. We factor it:

$$10x(a + 2b)(a - 2b).$$

C. Have we factored completely? Yes, because no factor with more than one term can be factored further.

EXAMPLE 2 Factor: $x^6 - y^6$.

A. We look for a common factor. There isn't one (other than 1 or -1).

B. There are only two terms. It is a difference of squares: $(x^3)^2 - (y^3)^2$. We factor it:

$$(x^3 + y^3)(x^3 - y^3).$$

One factor is a sum of two cubes, and the other factor is a difference of two cubes. We factor them:

$$(x + y)(x^2 - xy + y^2)(x - y)(x^2 + xy + y^2).$$

C. We have factored completely because no factors can be factored further.

In Example 2, had we thought of factoring first as a difference of two cubes, we would have had

$$\begin{aligned}(x^2)^3 - (y^2)^3 &= (x^2 - y^2)(x^4 + x^2y^2 + y^4)\\ &= (x + y)(x - y)(x^4 + x^2y^2 + y^4).\end{aligned}$$

In this case, we might have missed some factors; $x^4 + x^2y^2 + y^4$ can be factored, but we might not have noticed.

DO EXERCISES 1–3.

Factor completely.

1. $3y^3 - 12x^2y$

2. $7a^3 - 7$

3. $64x^6 - 729y^6$

Factor.

4. $3x - 6 - bx^2 + 2bx$

5. $5y^4 + 20x^6$

6. $6x^2 - 3x - 18$

7. $a^3 - ab^2 - a^2b + b^3$

8. $3x^2 + 18ax + 27a^2$

9. $2x^2 - 20x + 50 - 18b^2$

EXAMPLE 3 Factor: $10x^6 + 40y^2$.

A. We remove the largest common factor: $10(x^6 + 4y^2)$.

B. In the parentheses there are two terms, a sum of squares, which cannot be factored.

EXAMPLE 4 Factor: $2x^2 + 50a^2 - 20ax$.

A. We remove the largest common factor: $2(x^2 + 25a^2 - 10ax)$.

B. In the parentheses there are three terms. The trinomial is a square. We factor it:

$$2(x - 5a)^2.$$

C. No factor with more than one term can be factored further.

EXAMPLE 5 Factor: $6x^2 - 20x - 16$.

A. We remove the largest common factor: $2(3x^2 - 10x - 8)$.

B. In the parentheses there are three terms. The trinomial is not a square. We factor by trial:

$$2(x - 4)(3x + 2).$$

C. We cannot factor further.

EXAMPLE 6 Factor: $3x + 12 + ax^2 + 4ax$.

A. There is no common factor (other than 1 or -1).

B. There are four terms. We try grouping to remove a common binomial factor:

$3(x + 4) + ax(x + 4)$ Factoring two grouped binomials

$(x + 4)(3 + ax)$. Removing the common binomial factor

C. No factor with more than one term can be factored further.

EXAMPLE 7 Factor: $y^2 - 9a^2 + 12y + 36$.

A. There is no common factor (other than 1 or -1).

B. There are four terms. We try grouping to remove a common binomial factor, but that is not possible. We try grouping as a difference of squares:

$(y^2 + 12y + 36) - 9a^2$ Grouping

$(y + 6 + 3a)(y + 6 - 3a)$. Factoring the difference of squares

C. No factor with more than one term can be factored further.

EXAMPLE 8 Factor: $x^3 - xy^2 + x^2y - y^3$.

A. There is no common factor (other than 1 or -1).

B. There are four terms. We try grouping to remove a common binomial factor:

$x(x^2 - y^2) + y(x^2 - y^2)$ Factoring two grouped binomials

$(x + y)(x^2 - y^2)$. Removing the common binomial factor

C. A factor can be factored further:

$(x + y)(x + y)(x - y)$. Factoring a difference of squares

No factor with more than one term can be factored further, so we have factored completely.

DO EXERCISES 4–9.

EXERCISE SET 4.7

i Factor completely.

1. $x^2 - 144$
2. $y^2 - 81$
3. $2x^2 + 11x + 12$
4. $8a^2 + 18a - 5$
5. $3x^4 - 12$
6. $2xy^2 - 50x$
7. $a^2 + 25 + 10a$
8. $p^2 + 64 + 16p$
9. $2x^2 - 10x - 132$
10. $3y^2 - 15y - 252$
11. $9x^2 - 25y^2$
12. $16a^2 - 81b^2$
13. $m^6 - 1$
14. $64t^6 - 1$
15. $x^2 + 6x - y^2 + 9$
16. $t^2 + 10t - p^2 + 25$
17. $250x^3 - 128y^3$
18. $27a^3 - 343b^3$
19. $8m^3 + m^6 - 20$
20. $-37x^2 + x^4 + 36$
21. $ac + cd - ab - bd$
22. $xw - yw + xz - yz$
23. $4c^2 - 4cd + d^2$
24. $70b^2 - 3ab - a^2$
25. $-7x^2 + 2x^3 + 4x - 14$
26. $9m^2 + 3m^3 + 8m + 24$
27. $2x^3 + 6x^2 - 8x - 24$
28. $3x^3 + 6x^2 - 27x - 54$
29. $16x^3 + 54y^3$
30. $250a^3 + 54b^3$
31. $36y^2 - 35 + 12y$
32. $2b - 28a^2b + 10ab$
33. $a^8 - b^8$
34. $2x^4 - 32$
35. $a^3b - 16ab^3$
36. $x^3y - 25xy^3$

Problem-solving practice

37. There are 75 questions on a college entrance examination. Two points are awarded for each correct answer and one-half point is deducted for each wrong answer. A score of 100 indicates how many correct and how many wrong answers?

38. A pentagon with all five sides congruent has the same perimeter as an octagon with all eight sides congruent. One side of the pentagon is 2 less than 3 times one side of the octagon. Find the perimeters.

○

Factor. Assume that variables in exponents represent natural numbers.

39. $20x^2 - 4x - 39$
40. $30y^4 - 97xy^2 + 60x^2$
41. $72a^2 + 284a - 160$
42. $3x^2y^2z + 25xyz^2 + 28z^3$
43. $7x^3 - \frac{7}{8}$
44. $-16 + 17(5 - y^2) - (5 - y^2)^2$
45. $[(c - d)^3 - d^3]^2$
46. $(x - p)^2 - p^2$
47. $x^4 - 50x^2 + 49$
48. $(y - 1)^4 - (y - 1)^2$
49. $s^6 - 729t^6$
50. $x^6 - 2x^5 + x^4 - x^2 + 2x - 1$
51. $27x^{6s} + 64y^{3t}$
52. $5c^{100} - 80d^{100}$
53. $4x^2 + 4xy + y^2 - r^2 + 6rs - 9s^2$
54. $(y + x)^3 - x^3$
55. $c^4d^4 - a^{16}$
56. $(1 - x)^3 + (x - 1)^6$
57. $c^{2w+1} + 2c^{w+1} + c$
58. $24x^{2a} - 6$
59. $y^9 - y$
60. $1 - \dfrac{x^{27}}{1000}$
61. $3a^2 + 3b^2 - 3c^2 - 3d^2 + 6ab - 6cd$
62. $3(x + 1)^2 + 9(x + 1) - 12$
63. $(a + 2)^3 - (a - 2)^3$
64. $8(a - 3)^2 - 64(a - 3) + 128$
65. Suppose that $\left(x + \dfrac{2}{x}\right)^2 = 6$. Then find $x^3 + \dfrac{8}{x^3}$.

4.8

SOLVING EQUATIONS BY FACTORING

OBJECTIVE

After finishing Section 4.8, you should be able to:

i Solve equations by factoring and using the principle of zero products.

i An equation with a polynomial on one side and 0 on the other,

$$P(x) = 0$$

is called a *polynomial equation.* Any equation equivalent to such an equation is also considered to be a polynomial equation. Polynomial equations occur a great deal in applied problems, so the ability to solve them is an important skill. One way of solving a polynomial equation involves factoring the polynomial and using a property of the real numbers.

The Principle of Zero Products

When we multiply two or more numbers, the product will be 0 if one of the factors is 0. Conversely, if a product is 0, then at least one of the factors must be 0. This property of real numbers gives us another principle for solving equations.

Solve.

1. $x^2 + 8 = 6x$

THE PRINCIPLE OF ZERO PRODUCTS

For any real numbers a and b:

If $a \cdot b = 0$, then $a = 0$ or $b = 0$.

If $a = 0$ or $b = 0$, then $a \cdot b = 0$.

To use this principle in solving equations, we make sure that there is 0 on one side of the equation and a factorization on the other side.

EXAMPLE 1 Solve: $x^2 - 3x = 28$.

In order to use the principle of zero products, we must have 0 on one side of the equation, so we add -28 on both sides:

$$x^2 - 3x - 28 = 0. \quad \text{Getting 0 on one side}$$

We need a factorization on the other side, so we factor the polynomial:

$$(x - 7)(x + 4) = 0. \quad \text{Factoring}$$

We now set each factor equal to 0 (this is a use of the principle of zero products):

$$x - 7 = 0 \quad \text{or} \quad x + 4 = 0. \quad \text{Using the principle of zero products}$$

We now have two simple linear equations. We solve them separately:

$$x = 7 \quad \text{or} \quad x = -4.$$

We have two possible solutions.

Check:

$x^2 - 3x - 28 = 0$	
$7^2 - 3(7) - 28$	0
$49 - 21 - 28$	
0	

$x^2 - 3x - 28 = 0$	
$(-4)^2 - 3(-4) - 28$	0
$16 + 12 - 28$	
0	

The numbers 7 and -4 are both solutions.

To solve using the principle of zero products:

1. **Obtain a 0 on one side of the equation.**
2. **Factor the other side.**
3. **Set each factor equal to 0.**
4. **Solve the resulting equations.**

DO EXERCISE 1.

When you solve an equation using the principle of zero products, you may wish to check by substitution as in Example 1. Such a check will detect errors in solving. The procedure, however, does not produce any possible solutions that are not actual solutions, unless errors have been made.

It is important in using the principle of zero products that there be a 0 on one side of the equation. If neither side of an equation is 0, the procedure will not work at all.

EXAMPLE 2 Solve: $3y^2 + 7y + 2 = 0$.

$$3y^2 + 7y + 2 = 0$$

$$(3y + 1)(y + 2) = 0 \quad \text{Factoring}$$

$$3y + 1 = 0 \quad \text{or} \quad y + 2 = 0 \quad \text{Using the principle of zero products}$$

$$y = -\tfrac{1}{3} \quad \text{or} \quad y = -2$$

The solutions are $-\frac{1}{3}$ and -2.

DO EXERCISE 2.

EXAMPLE 3 Solve: $5b^2 = 10b$.

$$5b^2 = 10b$$

$$5b^2 - 10b = 0 \quad \text{Getting 0 on one side}$$

$$5b(b - 2) = 0 \quad \text{Factoring}$$

$$5b = 0 \quad \text{or} \quad b - 2 = 0 \quad \text{Using the principle of zero products}$$

$$b = 0 \quad \text{or} \quad b = 2$$

The solutions are 0 and 2.

DO EXERCISE 3.

EXAMPLE 4 Solve: $x^2 - 6x + 9 = 0$.

$$x^2 - 6x + 9 = 0$$

$$(x - 3)(x - 3) = 0 \quad \text{Factoring}$$

$$x - 3 = 0 \quad \text{or} \quad x - 3 = 0 \quad \text{Using the principle of zero products}$$

$$x = 3 \quad \text{or} \quad x = 3$$

There is only one solution, 3.

DO EXERCISE 4.

EXAMPLE 5 Solve: $3x^3 - 9x^2 = 30x$.

$$3x^3 - 9x^2 = 30x$$

$$3x^3 - 9x^2 - 30x = 0 \quad \text{Getting 0 on one side}$$

$$3x(x^2 - 3x - 10) = 0 \quad \text{Factoring out a common factor}$$

$$3x(x + 2)(x - 5) = 0 \quad \text{Factoring the trinomial}$$

$$3x = 0 \quad \text{or} \quad x + 2 = 0 \quad \text{or} \quad x - 5 = 0 \quad \text{Using the principle of zero products}$$

$$x = 0 \quad \text{or} \quad x = -2 \quad \text{or} \quad x = 5$$

The solutions are 0, -2, and 5.

DO EXERCISE 5.

Solve.

2. $5y + 2y^2 = 3$

Solve.

3. $8b^2 = 16b$

Solve.

4. $25 + x^2 = -10x$

Solve.

5. $x^3 + x^2 = 6x$

EXERCISE SET 4.8

i Solve.

1. $x^2 + 3x = 28$

2. $y^2 - 4y = 45$

3. $y^2 + 16 = 8y$

4. $r^2 + 1 = 2r$

5. $x^2 - 12x + 36 = 0$

6. $y^2 + 16y + 64 = 0$

7. $9x + x^2 + 20 = 0$

8. $8y + y^2 + 15 = 0$

9. $x^2 + 8x = 0$ **10.** $t^2 + 9t = 0$ **11.** $x^2 - 9 = 0$ **12.** $p^2 - 16 = 0$

13. $z^2 = 36$ **14.** $y^2 = 81$ **15.** $x^2 + 14x + 45 = 0$ **16.** $y^2 + 12y + 32 = 0$

17. $y^2 + 2y = 63$ **18.** $a^2 + 3a = 40$ **19.** $p^2 - 11p = -28$ **20.** $x^2 - 14x = -45$

21. $32 + 4x - x^2 = 0$ **22.** $27 + 12t + t^2 = 0$ **23.** $3b^2 + 8b + 4 = 0$ **24.** $9y^2 + 15y + 4 = 0$

25. $8y^2 - 10y + 3 = 0$ **26.** $4x^2 + 11x + 6 = 0$ **27.** $6z - z^2 = 0$ **28.** $8y - y^2 = 0$

29. $12z^2 + z = 6$ **30.** $6x^2 - 7x = 10$ **31.** $5x^2 - 20 = 0$ **32.** $6y^2 - 54 = 0$

33. $2x^2 - 15x = -7$ **34.** $x^2 - 9x = -8$ **35.** $21r^2 + r - 10 = 0$ **36.** $12a^2 - 5a - 28 = 0$

37. $15y^2 = 3y$ **38.** $18x^2 = 9x$ **39.** $x^2 - \frac{1}{25} = 0$ **40.** $y^2 - \frac{1}{64} = 0$

41. $2x^3 - 2x^2 = 12x$ **42.** $50y + 5y^3 = 35y^2$ **43.** $2x^3 = 128x$ **44.** $147y = 3y^3$

Problem-solving practice

45. At noon, two cars start from the same location going in opposite directions at different speeds. After seven hours they are 651 miles apart. If one car is traveling 15 miles per hour slower than the other car, what are their respective speeds?

46. At the beginning of the month an appliance store had 150 televisions in stock. During the month they sold 45% of their color televisions and 60% of their black-and-white televisions. If they sold a total of 78 televisions, how many of each type did they sell?

Solve.

47. $x(x + 8) = 16(x - 1)$ **48.** $m(m + 9) = 4(2m + 5)$ **49.** $(a - 5)^2 = 36$

50. $(x - 6)^2 = 81$ **51.** $(3x^2 - 7x - 20)(x - 5) = 0$ **52.** $(8x + 11)(12x^2 - 5x - 2) = 0$

53. $(x + 1)^3 = (x - 1)^3 + 26$ **54.** $(x - 2)^3 = x^3 - 2$

OBJECTIVE

After finishing Section 4.9, you should be able to:

i **Solve problems by translating to an equation and then solving the equation using factoring and the principle of zero products.**

4.9

SOLVING PROBLEMS

i Some problem situations can be translated to polynomial equations, some of which we can now solve. The problem-solving process is the same as for other kinds of equations, except that we can now solve equations by factoring and using the principle of zero products. Let us finish an example that we started earlier, on p. 35.

EXAMPLE 1 There are two square flower gardens on the campus mall. The length of a side of one garden is 4 ft less than the length of a side of the other. The area of the larger garden is 56 ft^2 greater than that of the smaller. Find the dimensions of the garden.

1. **Familiarize** We draw a picture and label it.

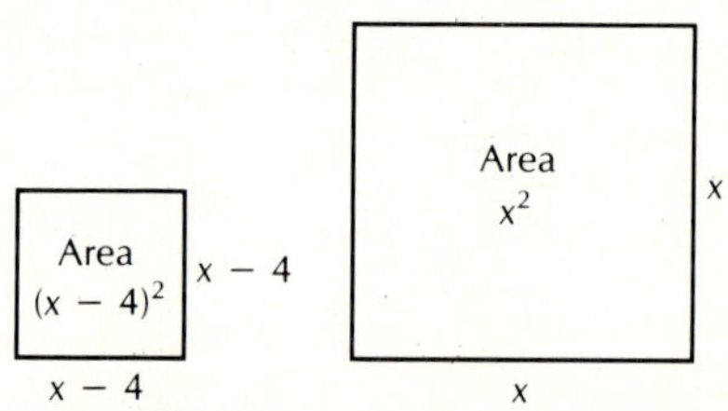

The larger garden has area 56 greater than the smaller garden.

2. **Translate** We use the following statement:

Area of small garden + 56 is area of the large garden.

$$(x - 4)^2 \quad + 56 = \quad x^2.$$

3. **Carry out** We solve the equation:

$$(x-4)^2 + 56 = x^2$$
$$x^2 - 8x + 16 + 56 = x^2 \quad \text{Squaring a binomial}$$
$$-8x + 72 = 0 \quad \text{Collecting like terms and simplifying}$$
$$x = 9.$$

4. **Check** Since we used x to represent the length of a side of the larger garden, it follows that the smaller garden has sides of length $x - 4$, or 5. The areas of the gardens are then 5^2, or 25 ft^2, and 9^2, or 81 ft^2. The difference is 56 ft^2, so we do have a solution of the problem.

5. **State** The smaller garden has sides of length 5 ft and the larger garden has sides of length 9 ft.

DO EXERCISE 1.

Example 1 is interesting in that the equation turned out to be linear, even though it didn't seem as though it would. In the following example, that does not happen and we will need to use the principle of zero products.

EXAMPLE 2 The width of a rectangular swimming pool is 2 m less than the length. The area is 15 m^2. Find the dimensions of the pool.

1. **Familiarize** We make a drawing and label the length and width.

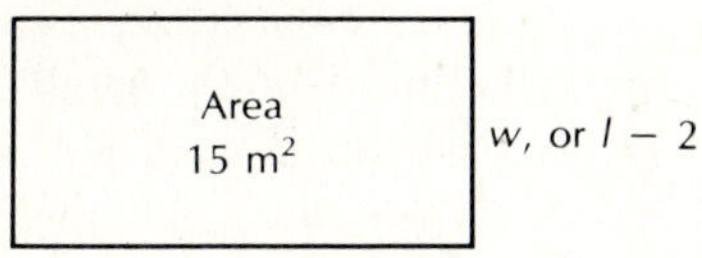

We recall the formula for the area of a rectangle:

$$A = l \cdot w.$$

2. **Translate** We can now translate to a system of two equations:

$$15 = l \cdot w,$$
$$w = l - 2.$$

3. **Carry out** We solve the system of equations:

$$15 = l \cdot w,$$
$$w = l - 2.$$

Substituting from the second equation into the first, we obtain

$$15 = l(l-2).$$

We now have an equation in one variable, which we can solve:

$$15 = l^2 - 2l \quad \text{Multiplying on the right}$$
$$0 = l^2 - 2l - 15 \quad \text{Getting 0 on one side}$$
$$0 = (l-5)(l+3) \quad \text{Factoring}$$
$$l - 5 = 0 \quad \text{or} \quad l + 3 = 0 \quad \text{Using the principle of zero products}$$
$$l = 5 \quad \text{or} \quad l = -3.$$

1. The square of a number minus twice the number is 48. Find the number.

2. The width of a rectangle is 5 cm less than the length. The area is 24 cm^2. Find the dimensions.

4. **Check** We check the possible solutions in the original problem. The number -3 is not a solution because length cannot be negative in this problem. If the length is 5 and the width is 2 m less, or 3, then the area will be $3 \cdot 5$, or 15 m^2. We have a solution.

5. **State** The answer is that the pool has length 5 m and width 3 m.

DO EXERCISE 2.

EXERCISE SET 4.9

i Problem Solving

1. If 4 times the square of a number is 21 more than 8 times the number, what is the number?

2. If 4 times the square of a number is 45 more than 8 times the number, what is the number?

3. The square of a number plus the number is 132. What is the number?

4. The square of a number plus the number is 156. What is the number?

5. The length of the top of a table is 5 cm more than the width. Find the length and width if the area is 84 cm^2.

6. The length of the top of a workbench is 4 cm greater than the width. The area is 96 cm^2. Find the length and the width.

7. Sam Stratton is planning a garden 25 m longer than it is wide. The garden will have an area of 7500 m^2. What will its dimensions be?

8. A flower bed is to be 3 m longer than it is wide. The flower bed will have an area of 108 m^2. What will its dimensions be?

9. The sum of the squares of two consecutive odd positive integers is 202. Find the integers.

10. The sum of the squares of two consecutive odd positive integers is 394. Find the integers.

11. If the sides of a square are lengthened by 4 cm, the area becomes 49 cm^2. Find the length of a side of the original square.

12. If the sides of a square are lengthened by 6 m, the area becomes 144 m^2. Find the length of a side of the original square.

13. The base of a triangle is 9 cm greater than the height. The area is 56 cm^2. Find the height and base.

14. The base of a triangle is 5 cm less than the height. The area is 18 cm^2. Find the height and base.

15. The perimeter of a square is 4 more than its area. Find the length of a side.

16. The area of a square is 12 more than its perimeter. Find the length of a side.

17. Three consecutive even integers are such that the square of the first plus the square of the third is 136. Find the three integers.

18. Three consecutive even integers are such that the square of the third is 76 more than the square of the second. Find the three integers.

19. Find three consecutive integers such that the product of the first and third minus the second is one more than 10 times the third.

20. Find three consecutive integers such that four times the square of the third less three times the square of the first minus 41 is twice the square of the second.

○

21. The sum of two numbers is 17, and the sum of their squares is 205. Find the numbers.

22. A rectangular piece of tin is twice as long as it is wide. Squares 2 cm on a side are cut out of each corner and the ends are turned up to make a box whose volume is 480 cm^3. What are the dimensions of the piece of tin?

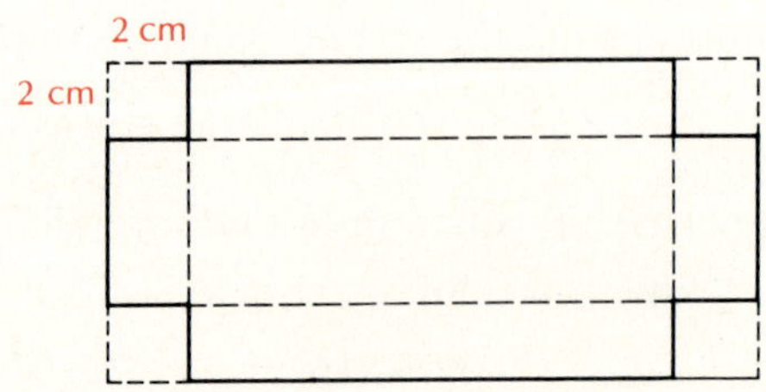

Solve for x.

23. $(x + 6)(x - 3) + b = (x - 5)(x - 2) - 3b$

24. $x^2 - 3(x - b) + c = (x - b)^2$

25. The top and base of a fish tank are rectangles whose length is 10 in. more than the width. If the depth and the width of the tank total 50 in. and the combined area of the top and base is 400 in^2 less than the total area of the four sides, what are the dimensions of the tank?

27. The hypotenuse of a right triangle is 3 cm longer than one of its legs and 6 cm more than the other leg. What is the area of the triangle?

26. A rectangular swimming pool with dimensions of 11 m and 8 m is built in a rectangular backyard. The area of the backyard is 1120 m^2. If the strip of yard surrounding the pool is of uniform width, how wide is the strip?

SUMMARY AND REVIEW: CHAPTER 4

The following contains a summary of what you should be able to do after completing this chapter. The review exercises are for practice. Answers are at the back of the book. If you miss an exercise, restudy the section indicated alongside the answer.

You should be able to:

Add polynomials.

Add.

1. $-6x^3 - 4x^2 + 3x + 1$ and $5x^3 + 2x + 6x^2 + 1$

2. $3x^4 + 3x^3 - 8x + 9$ and $-6x^4 + 4x + 7 + 3x$

3. $4x^3 - 2x^2 - 7x + 5$ and $8x^2 - 3x^3 - 9 + 6x$

4. $-9xy^2 - xy + 6x^2y$ and $-5x^2y - xy + 4xy^2$ and $12x^2y - 3xy^2 + 6xy$

Subtract polynomials.

Subtract.

5. $(3x - 5) - (-6x + 2)$

6. $(4a - b + 3c) - (6a - 7b - 4c)$

7. $(9p^2 - 4p + 4) - (-7p^2 + 4p + 4)$

8. $(6x^2 - 4xy + y^2) - (2y^2 + 3xy - 2y^2)$

Multiply polynomials.

Multiply.

9. $(3x^2y)(-6xy^3)$

10. $(4ab + 3c)(2ab - c)$

11. $(x^2 - 3y - 4)(2x + 5y)$

12. $(5x^2 - 7x + 3)(4x^2 + 2x - 9)$

13. $(6m - n)^2$

14. $(x + 2)^3$

Factor (a) polynomials when the terms have a common factor, (b) differences of two squares, (c) trinomial squares, (d) trinomials of the type $ax^2 + bx + c$, $a \neq 1$, (e) polynomials with four terms by grouping, and (f) polynomials by completing the square. Apply the general strategy for factoring.

Factor.

15. $6x^2 + 5x$

16. $9y^4 - 3y^2$

17. $15x^4 - 18x^3 + 21x^2 - 9x$

18. $a^2 - 12a + 27$

19. $3m^2 + 14m + 8$

20. $25x^2 + 20x + 4$

21. $4y^2 - 16$

22. $a^4 - 81$

23. Factor $3y^3 + 6y^2 - 5y - 10$ by grouping.

24. Factor $3x^2 + 36x - 135$ by completing the square.

25. $27x^3 - 8$

26. $0.064b^3 - 0.125c^3$

27. $y^5 - y$

28. $2y^8 - 8z^8$

Solve equations by factoring.

Solve.

29. $x^2 - 20x = -100$

30. $6b^2 - 13b + 6 = 0$

31. $8y^2 + 5 = 14y$

Solve problems involving equations that can be factored.

32. The area of a square is 5 more than 4 times the length of a side. What is the length of a side of the square?

33. The sum of the squares of three consecutive odd numbers is 83. Find the numbers.

34. A photograph is 3 in. longer than it is wide. When a 2-in. border is placed around the photograph, the total area of the photograph and the border is 108 in^2. Find the dimensions of the photograph.

35. Factor:

$$128x^6 - 2y^6.$$

36. Simplify:

$$\frac{2y^2 - 11y + 12}{6y^2 - 6y - 12} \div \frac{3y^2 - 14y + 8}{2y^2 - 6y + 4}.$$

37. The sum of a number and its reciprocal is $\frac{65}{28}$. What is the number?

TEST: CHAPTER 4

Add.

1. $-6x^3 + 3x^2 - 4y$ and $3x^3 - 2y - 7y^2$

2. $4a^3 - 2a^2 + 6a - 5$ and $3a^3 - 3a + 2 - 4a^2$

3. $5m^3 - 4m^2n - 6mn^2 - 3n^3$ and $9mn^2 - 4n^3 + 2m^3 + 6m^2n$

Subtract.

4. $(9a - 4b) - (3a + 4b)$

5. $(4x^2 - 3x + 7) - (-3x^2 + 4x - 6)$

6. $(2r^3 - 4r^2 + 6r - 7) - (5r^3 - 3r^2 - 4r + 6)$

Multiply.

7. $(-4x^2y)(-16xy^2)$

8. $(6a - 5b)(2a + b)$

9. $(x - y)(x^2 - xy - y^2)$

10. $(3m^2 + 4m - 2)(-m^2 - 3m + 5)$

11. $(4y - 9)^2$

12. $(x - 2y)^3$

Factor.

13. $9x^2 + 7x$

14. $24y^3 + 16y^2$

15. $14m^3 - 21m^2 - 7m + 35$

16. $p^2 - 12p - 28$

17. $12m^2 + 20m + 3$

18. $9y^2 - 25$

19. $3r^4 - 3$

20. $15x^4 - 10y^3$

21. $(z + 1)^2 - b^2$

22. $x^8 - y^8$

Solve.

23. $x^2 - 18 = 3x$

24. $5y^2 - 125 = 0$

25. $2x^2 + 21 = -17x$

26. The area of a square is 96 more than the perimeter. Find the length of a side of the square.

27. Factor: $a^2 - 2ab + b^2 - c^2$.

How can a company make the most profit? Inequalities can be a help in solving such problems.

5 INEQUALITIES AND LINEAR PROGRAMMING

Inequalities are mathematical sentences containing verbs such as $<$ (is less than). Like equations, inequalities are important in solving problems. They are solved using principles that are similar to the equation-solving principles.

In this chapter you will solve inequalities and solve problems that can be translated to inequalities. You will also be introduced to problem solving that involves systems of inequalities.

OBJECTIVES

After finishing Section 5.1, you should be able to:

i **Solve and graph inequalities using the addition principle.**

ii **Solve and graph inequalities using the multiplication principle.**

iii **Solve and graph inequalities using both principles.**

Insert the proper symbol, $<$, $>$, or $=$, to make a true sentence.

1. 1 8

2. 5 -2

3. -3 -8

Which of the following are true?

4. $-5 \leqslant -5$

5. $8 \geqslant -1$

6. $-5 \leqslant -10$

Determine whether the given number is a solution of the inequality.

7. $3 - x < 2$, 4

8. $3x + 2 > -1$, -2

9. $3x + 2 \leqslant 4x - 3$, 5

5.1

SOLVING INEQUALITIES

i The order of the real numbers is often shown on a number line.

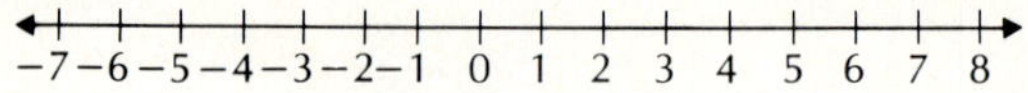

If a number occurs to the left of another, we say that the first number *is less than* the second, or the second *is greater than* the first. We use the symbol $<$ to mean *is less than* and $>$ to mean *is greater than.*

EXAMPLES Insert the proper symbol: $<$, $>$, or $=$.

1. 1 5 Since 1 occurs to the left of 5, $1 < 5$.

2. 5 1 Since $1 < 5$, we also know that $5 > 1$.

3. -3 7 Since -3 occurs to the left of 7, $-3 < 7$.

4. 7 -3 Since $-3 < 7$, we also know that $7 > -3$.

5. -9 -2 Since -9 occurs to the left of -2, $-9 < -2$.

6. -2 -9 Since $-9 < -2$, we also know that $-2 > -9$.

DO EXERCISES 1–3.

The symbol $\leqslant$ means *is less than or equal to,* and the symbol $\geqslant$ means *is greater than or equal to.*

EXAMPLES Which of the following are true?

7. $-1 \leqslant 7$ True, because -1 is less than 7 ($-1 < 7$ is true).

8. $3 \leqslant 3$ True, because $3 = 3$ is true.

9. $-5 \geqslant 7$ False, because $-5 > 7$ and $-5 = 7$ are not true.

DO EXERCISES 4–6.

An *inequality* is any sentence having one of the verbs $<$, $>$, $\leqslant$, or $\geqslant$. As for equations, a *solution* of an inequality is any number making it a true sentence. The set of all solutions is called the *solution set.* When we have found the set of all solutions of an inequality, *no matter how* we might find them, we say that we have *solved* the inequality.

EXAMPLES Determine whether the specified number is a solution of the inequality.

10. $x + 3 < 6$; 5
We substitute and get $5 + 3 < 6$, or $8 < 6$, a false sentence. Therefore, 5 is not a solution.

11. $2x - 3 > -3$; 1
We substitute and get $2(1) - 3 > -3$, or $-1 > -3$, a true sentence. Therefore, 1 is a solution.

12. $4x - 1 \leqslant 3x + 2$; 3
We substitute and get $4(3) - 1 \leqslant 3(3) + 2$, or $11 \leqslant 11$, a true sentence. Therefore, 3 is a solution.

DO EXERCISES 7–9.

A graph of an inequality is a drawing showing all of the inequality's solutions on a number line. The graph is a picture of the solution set.

EXAMPLE 13 Graph $x < 2$ on a number line.

The solutions consist of all numbers less than 2, so we shade all numbers less than 2. Note that 2 is not a solution. We indicate this by using an open circle at 2.

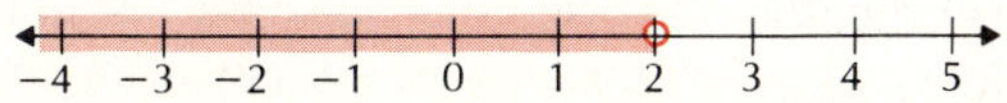

The solution set graphed in Example 13 can be named as follows:

$$\{x \mid x < 2\}.$$

DO EXERCISE 10.

EXAMPLE 14 Graph $x \leqslant 2$ on a number line.

We draw a picture of the set $\{x \mid x \leqslant 2\}$. This time the solution set consists of 2 and also all the numbers less than 2. We shade all numbers less than 2 and use a solid circle at 2 to indicate that it is also a solution.

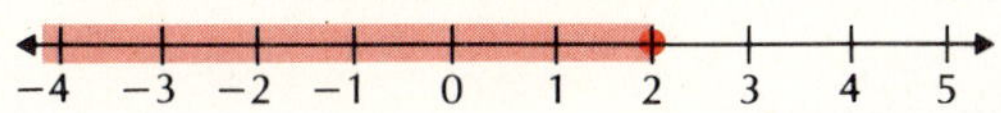

DO EXERCISE 11.

The Addition Principle

There is an addition principle for inequalities, similar to the one for solving equations.

THE ADDITION PRINCIPLE FOR INEQUALITIES

If $a < b$ is true, then $a + c < b + c$ is true for any number c. Similar statements hold for $>$, $\leqslant$, and $\geqslant$.

To solve inequalities using the addition principle, we do almost exactly what we do when solving equations. We transform the inequality into a simpler one by adding the same number on both sides of the inequality.

EXAMPLE 15 Solve $x + 3 > 6$. Then graph.

$$x + 3 > 6$$
$$x + 3 + (-3) > 6 + (-3) \quad \text{Using the addition principle; adding } -3$$
$$x > 3$$

The solution set is $\{x \mid x > 3\}$. The graph is as follows.

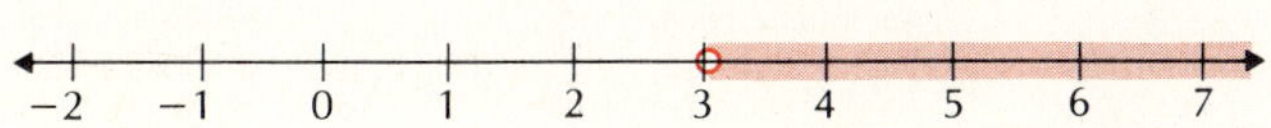

DO EXERCISES 12–14.

10. Graph $x < 3$.

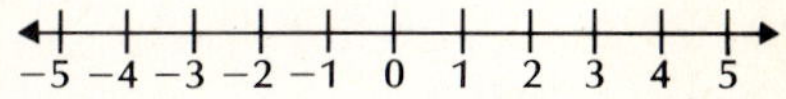

11. Graph $x \geqslant -2$.

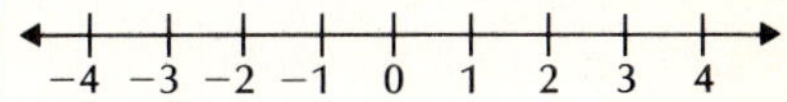

Solve. Then graph.

12. $x + 6 > 9$

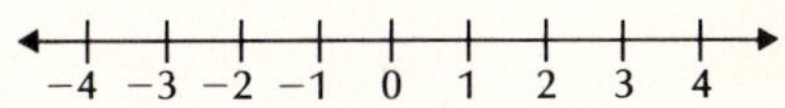

13. $x + 4 \leqslant 7$

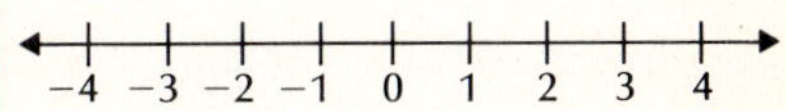

14. $3x - 1 \geqslant 2x - 3$

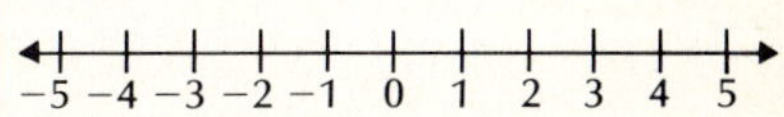

Checking

In Example 15, there was an unlimited number of solutions. We cannot check them by substituting into the original inequality, as we do for equations, since there are too many of them. We do not really need to check, however. Let us see why. Consider the first and last inequalities in Example 15:

$$x + 3 > 6, \qquad x > 3.$$

Any number that makes the first one true must make the last one true. We know this by the addition principle. Now the question is, will any number that makes the last one true also be a solution of the first one? Let us use the addition principle again.

$$x > 3$$

$$x + 3 > 3 + 3 \qquad \text{Using the addition principle; adding 3}$$

$$x + 3 > 6$$

Now we know that any number that makes $x > 3$ true will also make $x + 3 > 6$ true. Whenever we use the addition principle, the first and last inequalities will have the same solutions, as long as we have not added an expression with nonsensible replacements. That is, the first and last inequalities will be equivalent.

ii The Multiplication Principle

The multiplication principle for inequalities is somewhat different from the principle for equations.

Consider this true inequality:

$$4 < 9.$$

If we multiply both numbers by 2, we get another true inequality:

$$8 < 18.$$

If we multiply both numbers by -3, we get a false inequality:

$$-12 < -27.$$

However, if we now reverse the inequality symbol, we get a true inequality:

$$-12 > -27.$$

THE MULTIPLICATION PRINCIPLE FOR INEQUALITIES

If $a < b$ is true, then $ac < bc$ is true for any positive number c and $ac > bc$ is true for any negative number c. Similar statements hold for $>$, $\leqslant$, and $\geqslant$.

The important thing to remember is that if you multiply by a negative number, you must reverse the inequality symbol.

When we solve an inequality using the multiplication principle, we can multiply by any number except zero. We do not need to check any solutions, provided we do not multiply by an expression containing a variable.

EXAMPLES Solve.

16. $3y < \frac{3}{4}$

$$\tfrac{1}{3} \cdot 3y < \tfrac{1}{3} \cdot \tfrac{3}{4} \qquad \text{Multiplying by } \tfrac{1}{3}$$

$$y < \tfrac{1}{4}$$

Any number less than $\frac{1}{4}$ is a solution. The solution set is $\{y \mid y < \frac{1}{4}\}$.

17. $-4x < \frac{4}{5}$

$$-\frac{1}{4}\cdot(-4x) > -\frac{1}{4}\cdot\frac{4}{5} \qquad \text{Multiplying by } -\tfrac{1}{4} \text{ and reversing the inequality sign}$$
$$x > -\frac{1}{5}$$

Any number greater than $-\frac{1}{5}$ is a solution. The solution set is $\{x \mid x > -\frac{1}{5}\}$.

18. $-5x \geqslant -80$

$$-\frac{1}{5}\cdot(-5x) \leqslant -\frac{1}{5}\cdot(-80) \qquad \text{Multiplying by } -\tfrac{1}{5} \text{ and reversing} \geqslant$$
$$x \leqslant 16$$

The solution set is $\{x \mid x \leqslant 16\}$.

DO EXERCISES 15–17.

Solve.

15. $5y \leqslant \frac{3}{2}$

16. $-2y > \frac{5}{6}$

17. $-\frac{1}{3}x \leqslant -4$

iii Using the Principles Together

We use the addition and multiplication principles together in solving inequalities in much the same way as for equations.

EXAMPLE 19 Solve: $16 - 7y \geqslant 10y - 4$.

$$16 - 7y \geqslant 10y - 4$$
$$-16 + 16 - 7y \geqslant -16 + 10y - 4 \qquad \text{Adding } -16$$
$$-7y \geqslant 10y - 20$$
$$-10y - 7y \geqslant -10y + 10y - 20 \qquad \text{Adding } -10y$$
$$-17y \geqslant -20$$
$$-\frac{1}{17}\cdot(-17y) \leqslant -\frac{1}{17}\cdot(-20) \qquad \text{Multiplying by } -\tfrac{1}{17} \text{ and reversing the inequality sign}$$
$$y \leqslant \frac{20}{17}$$

The solution set is $\{y \mid y \leqslant \frac{20}{17}\}$.

DO EXERCISES 18–20.

Solve.

18. $6 - 5y \geqslant 7$

19. $3x + 5x < 4$

20. $17 - 5y \leqslant 8y - 5$

EXERCISE SET 5.1

Find the solution set. In Exercises 1–20, also graph.

i

1. $x \leqslant 4$ **2.** $y \leqslant -1$ **3.** $x > 5$ **4.** $x > 3$

5. $x + 8 > 3$ **6.** $x + 5 > 2$ **7.** $y + 3 < 9$ **8.** $y + 4 < 10$

9. $a + 9 \leqslant -12$ **10.** $a + 7 \leqslant -13$ **11.** $t + 14 \geqslant 9$ **12.** $x - 9 \leqslant 10$

13. $y - 8 > -14$ **14.** $y - 9 > -18$ **15.** $x - 11 \leqslant -2$ **16.** $y - 18 \leqslant -4$

ii

17. $8x \geqslant 24$ **18.** $9t < -81$ **19.** $0.3x < -18$ **20.** $0.5x < 25$

21. $-9x \geqslant -8.1$ **22.** $-8y \leqslant 3.2$ **23.** $-\frac{3}{4}x \geqslant -\frac{5}{8}$ **24.** $-\frac{5}{6}y \leqslant -\frac{3}{4}$

iii

25. $2x + 7 < 19$ **26.** $5y + 13 > 28$ **27.** $5y + 2y \leqslant -21$ **28.** $-9x + 3x \geqslant -24$

29. $2y - 7 < 5y - 9$ **30.** $8x - 9 < 3x - 11$ **31.** $0.4x + 5 \leqslant 1.2x - 4$ **32.** $0.2y + 1 > 2.4y - 10$

33. $3x - \frac{1}{8} \leqslant \frac{3}{8} + 2x$ **34.** $2x - 3 < \frac{13}{4}x + 10 - 1.25x$

Problem-solving practice

35. John earned between \$36 and \$63 each weekend that he mowed lawns. He received \$4.50 for each lawn he mowed. Set up one inequality to figure out the least number of lawns and the most number of lawns he mowed during any one weekend.

36. The scores on a math test ranged from a low of 42 to a high of 96. If 3 points were awarded for each correct answer and 6 points added to each test for a homework assignment that every student turned in, what was the range of correct answers on the test?

Solve. Assume that a, b, c, d, and m are positive constants.

37. $4(3y - 2) \geqslant 9(2y + 5)$

38. $4m + 5 \geqslant 14(m - 2)$

39. $3(2 - 5x) + 2x < 2(4 + 2x)$

40. $2(0.5 - 3y) + y > (4y - 0.2)8$

41. $5[3m - (m + 4)] > -2(m - 4)$

42. $[8x - 3(3x + 2)] - 5 \geqslant 3(x + 4) - 2x$

43. $3ax + 2x \geqslant 5ax - 4$

44. $6by - 4y \leqslant 7by + 10$

45. $a(by - 2) \geqslant b(2y + 5)$

46. $c(6x - 4) < d(3 + 2x)$

47. $c(2 - 5x) + dx > m(4 + 2x)$

48. $a(3 - 4x) + cx < d(5x + 2)$

49. For what values of k is x positive, if we assume that x satisfies $kx + 14 \geqslant 5$?

50. For what values of k is $y < 1$, if we assume that y satisfies $ky < 5 - 3y$?

51. Determine whether the following statements are true or false. If false, give a counterexample.

a) For any real numbers a, b, c, and d, if $a < b$ and $c < d$, then $a - c < b - d$.

b) For any real numbers x and y, if $x < y$, then $x^2 < y^2$.

52. Determine whether the inequalities

$$x < 3 \quad \text{and} \quad x + \frac{1}{x} < 3 + \frac{1}{x}$$

are equivalent. Give reasons to support your answer.

53. Determine whether the inequalities

$$x < 3 \quad \text{and} \quad 0 \cdot x < 0 \cdot 3$$

are equivalent. Give reasons to support your answer.

OBJECTIVE

After finishing Section 5.2, you should be able to:

i **Solve problems by translating to inequalities.**

5.2

SOLVING PROBLEMS

i Certain problems can be translated quite naturally to inequalities rather than equations. The five-step problem-solving process can still be used. The translating step, of course, involves inequalities and the following step consists of solving an inequality.

EXAMPLE 1 In a business course there will be three tests. You must get a total score of 270 or more for an A. You get scores of 91 and 86 on the first two tests. What scores on the last test will give you an A?

1. **Familiarize** Making a table can help in the familiarization process.

Test 1	91
Test 2	86
Test 3	x
Total	270 or better

Note that we have used x to represent the score on the third test.

2. **Translate** Looking at the table, we can rather easily see how to write an inequality:

$$91 + 86 + x \geqslant 270.$$

3. **Carry out** We solve the inequality, obtaining

$$177 + x \geqslant 270 \quad \text{Collecting like terms}$$
$$x \geqslant 93. \quad \text{Adding } -177$$

4. **Check** If you get 93 on the third test, your total score will be 91 + 86 + 93, which is 270. Any higher score will also give you an A.

5. **State** A score of 93 or better will give you an A.

DO EXERCISE 1.

EXAMPLE 2 On your new job you can be paid in one of two ways.

Plan A: A salary of \$600 per month, plus a commission of 4% of sales.

Plan B: A salary of \$800 per month, plus a commission of 6% of sales in excess of \$10,000.

For what amount of sales is Plan A better than Plan B, if we assume that sales are always more than \$10,000?

1. **Familiarize** Listing information in a table will again be helpful.

Plan A: Monthly income	Plan B: Monthly income
\$600 salary 4% of sales *Total:* 600 + 4% of sales	\$800 salary 6% of sales over \$10,000 *Total:* \$800 + 6% of sales over \$10,000

2. **Translate** We write an inequality, stating that the income from Plan A is greater than the income from Plan B.

$$\underset{600 + 4\%S}{\textit{Income from Plan A}} > \underset{800 + 6\%(S - 10{,}000)}{\textit{Income from Plan B}}$$

Note that we have used S to represent the total amount of sales for the month.

3. **Carry out** We solve the inequality:

$$600 + 4\%S > 800 + 6\%S - 6\% \times 10{,}000$$
$$600 + 0.04S > 800 + 0.06S - 600 \quad \text{Changing to decimal notation}$$
$$600 + 0.04S > 200 + 0.06S \quad \text{Collecting like terms}$$
$$400 > 0.02S \quad \text{Adding } -200 - 0.04S$$
$$20{,}000 > S \quad \text{Multiplying by } 1/0.02$$

4. **Check** For $S = 20{,}000$, the income from Plan A is

$$600 + 4\% \cdot 20{,}000, \quad \text{or} \quad \$1400.$$

The income from Plan B is

$$800 + 6\% \cdot 10{,}000, \quad \text{or} \quad \$1400.$$

1. In a psychology course, there will be five tests. To get a B, 400 points are needed. You get scores of 91, 86, 73, and 79 on the first four tests. What scores on the last test will give you at least a B?

2. A painter can be paid in two ways.

Plan A: \$500 plus \$4 per hour.
Plane B: Straight \$9 per hour.

Suppose the job takes n hours. For what values of n is plan A better for the painter?

We calculate for some amount of sales less than \$20,000 and for some amount greater than \$20,000. Suppose sales are \$15,000. Then

Plan A gives $600 + 4\% \cdot 15{,}000$, or \$1200;
Plan B gives $800 + 6\% \cdot 5000$, or \$1100.

For sales of \$25,000,

Plan A gives $600 + 4\% \cdot 25{,}000$, or \$1600;
Plan B gives $800 + 6\% \cdot 15{,}000$, or \$1700.

For these values of sales the solution of the inequality gives correct results. We cannot check all possible values of S in this manner so we shall stop here.

5. **State** For sales under \$20,000, Plan A is better.

DO EXERCISE 2.

EXERCISE SET 5.2

i Solve.

1. A car rents for \$30 per day, plus 20¢ per mile. You are on a daily budget of \$96. What mileages will allow you to stay within the budget?

2. A car can be rented for \$35 per day with unlimited mileage, or for \$28 per day, plus 19¢ per mile. For what daily mileages would the unlimited mileage plan save you money?

3. You are taking a history course. There will be 4 tests. You have scores of 89, 92, and 95 on the first three. You must make a total of 360 to get an A. What scores on the last test will give you an A?

4. You are taking a science course. There will be 5 tests, each worth 100 points. You have scores of 94, 90, and 89 on the first three. You must make a total of 450 to get an A. What scores on your fourth test will still keep you eligible for an A?

5. You are going to invest \$25,000, part at 14% and part at 16%. What is the most that can be invested at 14% in order to make at least \$3600 interest per year?

6. You are going to invest \$20,000, part at 12% and part at 16%. What is the most that can be invested at 12% in order to make at least \$3000 interest per year?

7. In planning for a college dance, you find that one band will play for \$250 plus 50% of the total ticket sales. Another band will play for a flat fee of \$550. In order for the first band to produce more profit for the school than the other band, what is the highest price you can charge per ticket, assuming that 300 people will attend?

8. A bank offers two checking account plans. Plan A charges a base service charge of \$2.00 per month, plus 15¢ per check. Plan B charges a base service charge of \$4.00 per month and 9¢ per check. For what numbers of checks per month would Plan B be better than Plan A?

9. A medical insurance company offers two plans. With Plan A, you pay the first \$100 of your medical bills and they pay 80% of the rest. With Plan B, you pay the first \$250 of your medical bills and they pay 90% of the rest. For what total amounts of medical bills will Plan B save you money?

10. You can spend \$3.50 at the laundromat washing your clothes, or you can have them do it for 40¢ per pound of clothes. For what weights of clothes will it save you money to wash your clothes yourself?

11. On your new job, you can be paid in one of two ways.

Plan A: A salary of \$500 per month, plus a commission of 4% gross sales

Plan B: A salary of \$750 per month plus a commission of 5% of gross sales over \$8000

For what gross sales is Plan B better than Plan A, assuming that gross sales are always more than \$8000?

12. On your new job you can be paid in one of two ways.

Plan A: A salary of \$25,000 per year

Plan B: A salary of \$1500 per month, plus a commission of 6% on gross sales

For what gross sales is Plan A better than Plan B?

13. A mason can be paid in one of two ways.

Plan A: \$500 plus \$5.00 per hour

Plan B: Straight \$8.00 per hour

Suppose the job takes n hours. For what values of n is Plan A better for the mason than Plan B?

14. A mason can be paid in one of two ways.

Plan A: \$300 plus \$6.00 per hour

Plan B: Straight \$9.50 per hour

Suppose that the job takes n hours. For what values of n is Plan B better for the mason?

5.3

COMPOUND INEQUALITIES

OBJECTIVES

After finishing Section 5.3, you should be able to:

i **Solve and graph conjunctions of inequalities.**

ii **Solve and graph disjunctions of inequalities.**

We now consider compound inequalities, that is, sentences formed by two or more inequalities, joined with the word *and* or the word *or*. We solve such sentences; that is, we find the set of all solutions.

i Conjunctions of Sentences and Intersections of Sets

When two or more sentences are joined by the word *and* to make a compound sentence, the new sentence is called a *conjunction* of the sentences. Here are some examples:

The moon is red *and* the night is cold;

$x + y = 5$ *and* $x - y = 2$;

$-2 < x$ *and* $x < 1$.

For a conjunction to be true, all of the individual sentences must be true. Let's look at the solution sets and graphs of a conjunction and its individual sentences:

$$-2 < x \quad \text{and} \quad x < 1.$$

$\{x \mid -2 < x\}$

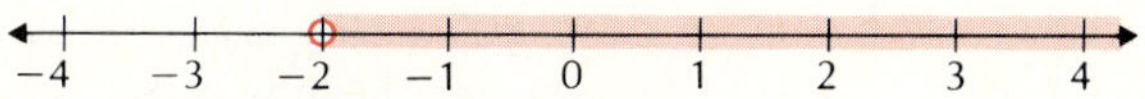

$\{x \mid x < 1\}$

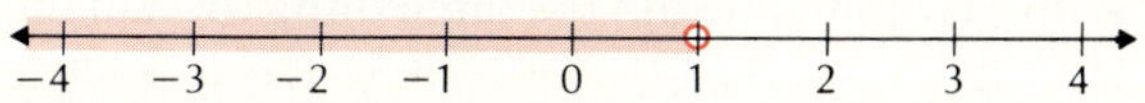

In order for a number to be a solution of the conjunction, it must be in *both* solution sets. The solution set of the conjunction is the part common to the individual solution sets.

$\{x \mid -2 < x \text{ and } x < 1\}$

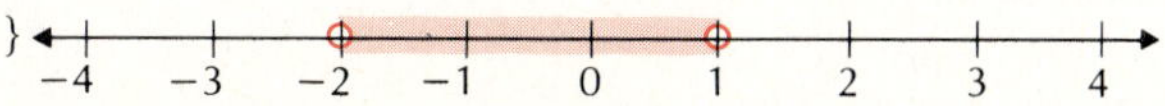

The common part of two or more sets is called their *intersection*. For two sets A and B, we can name the intersection $A \cap B$. If sets have no common members, we say that their intersection is empty, or the *empty set*, which can be named $\emptyset$.

The solution set of $-2 < x$ and $x < 1$ is the intersection

$$\{x \mid -2 < x\} \cap \{x \mid x < 1\}.$$

The conjunction "$-2 < x$ *and* $x < 1$" can be abbreviated as

$$-2 < x < 1.$$

EXAMPLE 1 Graph: $-3 \leqslant x < 4$.

The inequality is an abbreviation for the following conjunction:

$$-3 \leqslant x \quad \text{and} \quad x < 4.$$

The word *and* corresponds to set *intersection*. The solution set is thus the intersection of the solution sets:

$$\{x \mid -3 \leqslant x\} \cap \{x \mid x < 4\}.$$

The graph is the intersection of the individual graphs.

$\{x \mid -3 \leq x\}$ 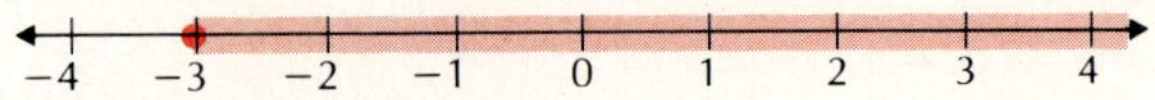

$\{x \mid x < 4\}$ 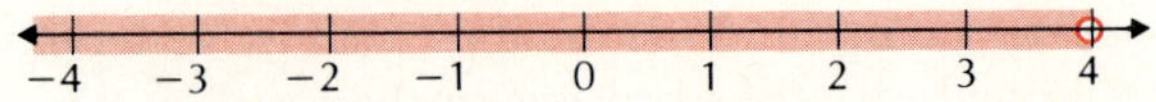

$\{x \mid -3 \leq x < 4\}$ 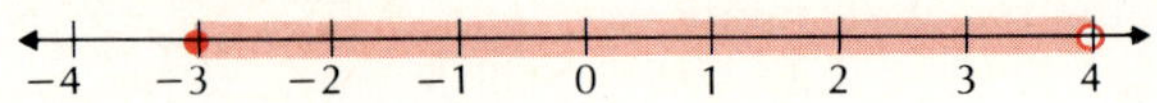

DO EXERCISES 1 AND 2.

EXAMPLE 2 Solve: $-3 < 2x + 5 < 7$.

Method 1: We write the conjunction with the word *and*:

$$-3 < 2x + 5 \quad and \quad 2x + 5 < 7.$$

Now we solve the individual inequalities separately.

$$\begin{array}{rcr} -3 < 2x + 5 & \text{and} & 2x + 5 < 7 \\ -3 + (-5) < 2x + 5 + (-5) & \text{and} & 2x + 5 + (-5) < 7 + (-5) \\ -8 < 2x & \text{and} & 2x < 2 \\ -4 < x & \text{and} & x < 1 \end{array}$$

We now abbreviate the answer:

$$-4 < x < 1.$$

Method 2: With Method 1, we did the same thing to each inequality. We can shorten the writing as follows:

$$-3 < 2x + 5 < 7$$
$$-3 + (-5) < 2x + 5 + (-5) < 7 + (-5) \qquad \text{Adding } -5$$
$$-8 < 2x < 2$$
$$-4 < x < 1 \qquad \text{Multiplying by } \tfrac{1}{2}$$

The solution set is $\{x \mid -4 < x < 1\}$.

DO EXERCISE 3.

ii Disjunctions of Sentences and Unions of Sets

When two or more sentences are joined by the word *or* to make a compound sentence, the new sentence is called a *disjunction* of the sentences. Here are some examples.

It is raining *or* the wind is blowing;

y is an odd number *or* y is a prime number;

$x < -3$ *or* $x > 3$.

In order for a disjunction to be true, at least one of the individual sentences must be true. Let's look at the solution sets and graphs of a disjunction and its individual sentences:

$$x < -3 \quad or \quad x > 3.$$

$\{x \mid x < -3\}$ 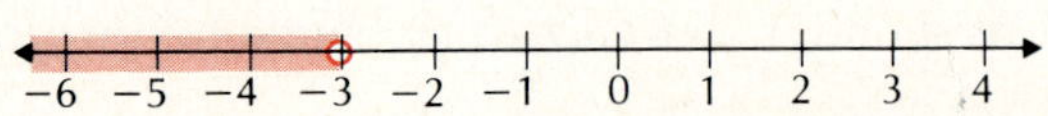

$\{x \mid x > 3\}$ 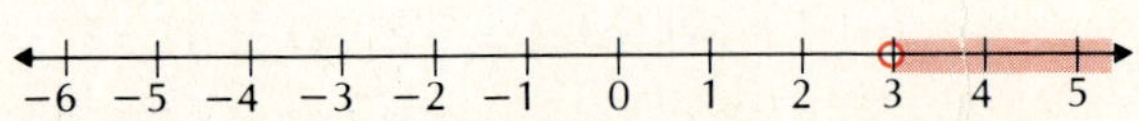

Graph.

1. $-1 < x < 4$

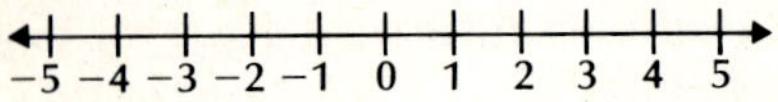

2. $-2 \leqslant y < 5$

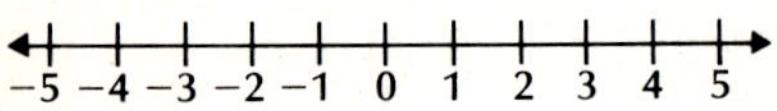

Solve using both methods as in Example 2.

3. $-2 < 3x + 4 < 7$

If a number is in either of the solution sets (or both), it is in the solution set of the disjunction. The solution set of the disjunction is the set we get by joining the other two sets.

$\{x \mid x < -3\} \cup \{x \mid x > 3\}$ 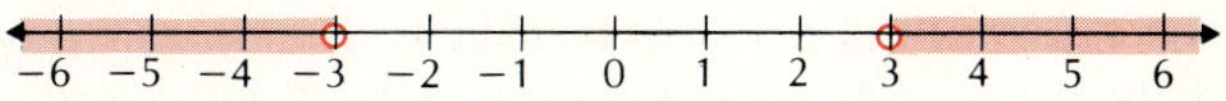

The set obtained by joining two or more sets is called their *union*. For two sets A and B, we can name the union $A \cup B$.

The solution set of $x < -3$ *or* $x > 3$ is the union

$$\{x|x < -3\} \cup \{x|x > 3\}, \quad \text{or} \quad \{x|x < -3 \text{ or } x > 3\}.$$

It is tempting to abbreviate a disjunction without writing the word *or* but there is no simple way to do that.

EXAMPLE 3 Graph: $x \leq 2$ or $x \geq 5$.

The graph consists of the union of the individual graphs.

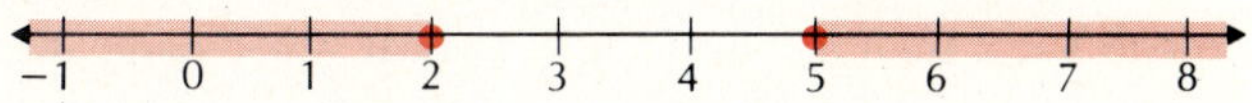

DO EXERCISES 4 AND 5.

EXAMPLE 4 Solve: $-2x - 5 < -2$ or $x - 3 < 2$.

We solve the individual inequalities separately, but we keep writing the word *or*.

$$-2x - 5 + 5 < -2 + 5 \quad \text{or} \quad x - 3 + 3 < 2 + 3$$
$$-2x < 3 \quad \text{or} \quad x < 5$$
$$x > -\tfrac{3}{2} \quad \text{or} \quad x < 5$$

The solution set is $\{x|x > -\frac{3}{2}\} \cup \{x|x < 5\}$, or $\{x|x > -\frac{3}{2}$ or $x < 5\}$ i.e., the set of all real numbers.

DO EXERCISES 6 AND 7.

Graph.

4. $x \leq -2$ or $x > 4$

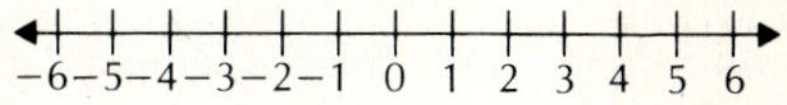

5. $x < -4$ or $x \geq 6$

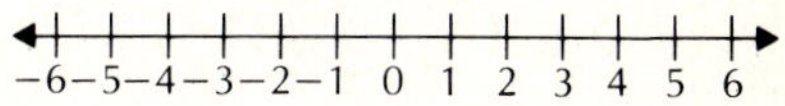

Solve.

6. $x - 4 < -3$ or $x - 3 \geq 3$

7. $-2x + 4 \leq -3$ or $x + 5 < 3$

EXERCISE SET 5.3

i Graph.

1. $1 < x < 6$
2. $0 \leq y \leq 3$
3. $-7 \leq y \leq -3$
4. $-9 \leq x < -5$
5. $-4 \leq -x < 3$
6. $-x > 3$ or $x < -3$
7. $6 > -x \geq -2$
8. $x < -4$ or $x > 2$
9. $-2 \leq x < 5$
10. $3 > x \geq 0$
11. $x > 5$ or $x \leq 1$
12. $x \leq -2$ or $x > 2$

Solve.

13. $-2 < x + 2 < 8$
14. $-1 < x + 1 \leq 6$
15. $1 < 2y + 5 \leq 9$
16. $3 \leq 5x + 3 \leq 8$
17. $-10 \leq 3x - 5 \leq -1$
18. $-18 \leq -2x - 7 < 0$
19. $2 < x + 3 \leq 9$
20. $-6 \leq x + 1 < 9$
21. $x + 7 \leq -2$ or $x + 7 > 2$
22. $x + 9 < -4$ or $x + 9 \geq 4$

ii Graph.

23. $x < -1$ or $x > 2$
24. $x < -2$ or $x > 0$
25. $x \quad -3$ or $x > 1$
26. $x \leq -1$ or $x > 3$
27. $x < -8$ or $x > -2$
28. $t \geq -10$ or $t \geq -5$

Solve.

29. $x + 7 < -2$ or $x + 7 > 2$
30. $x + 9 < -4$ or $x + 9 > 4$
31. $2x - 8 \leq -3$ or $x - 8 \geq 3$
32. $x - 7 \leq -2$ or $3x - 7 \geq 2$
33. $3x - 9 < -5$ or $x - 9 > 6$
34. $4x - 4 < -8$ or $x - 4 > 12$

35. $-6 \leqslant 2x - 3 < 6$

36. $7 > -4x + 5$ or $10 \leqslant -4x + 5$

37. $4 > -3m - 7 \geqslant 2$

38. $2x - 1 < 6$ or $2x - 1 \geqslant -4$

Problem-solving practice

39. The formula for changing a Celsius temperature to Fahrenheit is $F = \frac{9}{5}C + 32$. If temperatures range between $-5°C$ and $45°C$, what is their corresponding range in Fahrenheit temperatures?

40. Find all the numbers on a number line from which you can subtract 3, and still be at least a distance of 8 from the zero point on the number line.

Solve and graph.

41. $4a - 2 \leqslant a + 1 \leqslant 3a + 4$

42. $4m - 8 > 6m + 5$ or $5m - 8 < -2$

43. $x - 10 < 5x + 6 \leqslant x + 10$

44. $2[5(3 - y) - 2(y - 2)] > y + 4$

45. $-\frac{2}{15} \leqslant \frac{2}{3}x - \frac{2}{5} \leqslant \frac{2}{15}$

46. $2x - \frac{3}{4} < -\frac{1}{10}$ or $2x - \frac{3}{4} > \frac{1}{10}$

47. $3x < 4 - 5x < 5 + 3x$

48. $(x + 6)(x - 4) > (x + 1)(x - 3)$

Determine whether the following are true or false.

49. If $b > c$, then $b \nleq c$.

50. If $-b < -a$, then $a < b$.

51. If $c \neq a$, then $a < c$.

52. If $a < c$, and $c < b$, then $b \ngeq a$.

53. If $a < c$ and $b < c$, then $a < b$.

54. If $-a < c$ and $-c > b$, then $a < b$.

Solve.

55. $[4x - 2 < 8$ or $3(x - 1) < -2]$ and $-2 \leqslant 5x \leqslant 10$

56. $-2 \leqslant 4m + 3 < 7$ and $[m - 5 \geqslant 4$ or $3 - m > 12]$

57. $\frac{1}{4}x > \frac{1}{5} - \frac{1}{3}x$

58. $-\frac{2}{3}x + 4 \leqslant \frac{3}{4}$

59. $-3x^2 + 12 < -3$

60. $-3x + 2 < 15 < 4x - 3$

61. $-4 - 5x < -3x^2 - 2$

Solve.

62. $x + 6.922 \leqslant -1.998$ or $x + 7.0123 > 2.212$

63. $-5.888 \leqslant x + 1.003 < 9.354$

64. $-\frac{1}{3} < x - \frac{1}{6} \leqslant 6$

65. $0.9x > 1.8 - 0.1x$

66. $3x - \frac{1}{8} \leqslant \frac{3}{8} + 2x$

67. $2x + 3 \leqslant x - 6$ or $3x - 2 \leqslant 4x + 5$

68. $x + 4 < 2x - 6 \leqslant x + 12$

69. $2x - 3 < \frac{13}{4}x + 10 - 1.25x$

70. $7 - 3x \geqslant 5.7x + 8 - 8.7x$

71. $x + \frac{1}{10} \leqslant -\frac{1}{10}$ or $x + \frac{1}{10} \geqslant \frac{1}{10}$

OBJECTIVES

After finishing Section 5.4, you should be able to:

- **i** **Simplify expressions containing absolute-value symbols.**
- **ii** **Find the distance between two points on a coordinate line.**
- **iii** **Solve and graph equations and inequalities containing absolute-value symbols.**

5.4

ABSOLUTE VALUE

i The Absolute-Value Function

Intuitively we can think of the absolute value of a number as its distance from zero on a number line. The definition of the absolute-value function is as follows.

The absolute-value function $f(x) = |x|$ is defined as follows:

a) $|x| = x$ **for all nonnegative numbers** x

b) $|x| = -x$ **for all negative numbers** x

In other words, the absolute value of a nonnegative number is the number itself. The absolute value of a negative number is the additive inverse of that number. For the absolute-value function, there are no negative function values.

We can construct a graph of the absolute-value function by finding some values, plotting points, and connecting those points as follows.

x	-5	-3	0	2	5	7
$\lvert x\rvert$	5	3	0	2	5	7

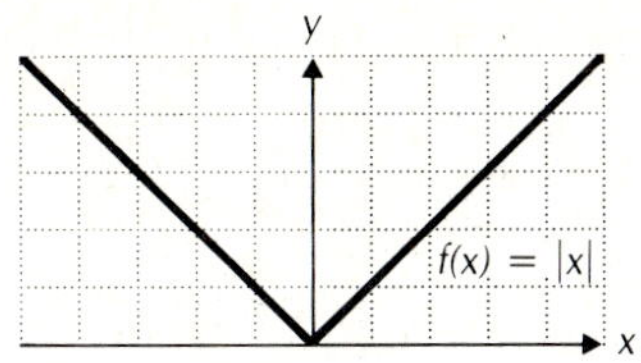

Note that the graph consists of portions of two straight lines,

$$y = x \quad \text{and} \quad y = -x.$$

Properties of Absolute Value. Some simple properties of the absolute-value function allow us to simplify algebraic expressions. Those properties follow from the definition, and are as follows.

PROPERTIES OF ABSOLUTE VALUE

a) **For any real numbers a and b, $|ab| = |a| \cdot |b|$.** **(The absolute value of a product is the product of the absolute values.)**

b) $\left|\frac{a}{b}\right| = \frac{|a|}{|b|}$, **provided that $b \neq 0$.** **(The absolute value of a quotient is the quotient of the absolute values.)**

EXAMPLES Simplify, leaving as little as possible inside the absolute-value signs.

1. $|5x| = |5| \cdot |x| = 5|x|$
2. $|-3y| = |-3| \cdot |y| = 3|y|$
3. $|7x^2| = |7| \cdot |x^2| = 7|x^2|$
   $= 7x^2$ Since x^2 is never negative for any number x
4. $\left|\frac{6x}{-3x^2}\right| = \left|\frac{2}{-x}\right|$
   $= \frac{|2|}{|-x|}$ Taking the quotient of the absolute values
   $= \frac{2}{|x|}$

DO EXERCISES 1–5.

Simplify, leaving as little as possible inside absolute-value signs.

1. $|7x|$

2. $|x^8|$

3. $|5a^2b|$

4. $\left|\frac{7a}{b^2}\right|$

5. $|-9x|$

ii Distance on a Line

The number on a line corresponding to a point of the line is called the *coordinate* of the point. Another useful property of absolute value concerns the distance between two points on a line.

The distance between any two points on a line having coordinates a and b is $|a - b|$ (the absolute value of the difference of the coordinates).

We should note that the distance is also $|b - a|$, because $a - b$ and $b - a$ are additive inverses of each other; hence have the same absolute value.

EXAMPLE 5 Find the distance between points having coordinates 10 and 3.

The distance is $|10 - 3|$, or $|7|$, which is 7.

Find the distance between the points with the specified coordinates.

6. $-6, -35$

7. 19, 14

8. $-3, 17$

9. Solve $|x| = 6$. Then graph using a number line.

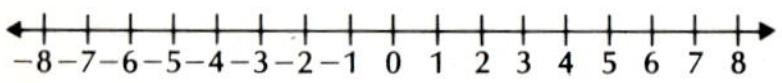

10. Solve $|x| < 5$. Then graph.

11. Solve $|y| \geqslant 8$. Then graph.

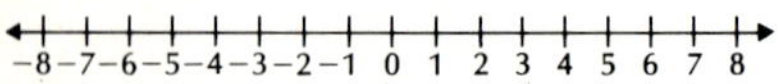

EXAMPLE 6 Find the distance between points having coordinates -8 and 92.

The distance is $|-8 - 92|$, or $|-100|$, which is 100.

DO EXERCISES 6–8.

iii Equations and Inequalities

EXAMPLE 7 Solve $|x| = 4$. Then graph, using a number line.

Note that $|x| = |x - 0|$, so that $|x - 0|$ is the distance from x to 0. The solutions of the equation are those numbers x whose distance from 0 is 4. Those numbers are 4 and -4. The solution set is $\{4, -4\}$. The graph consists of just two points, as shown.

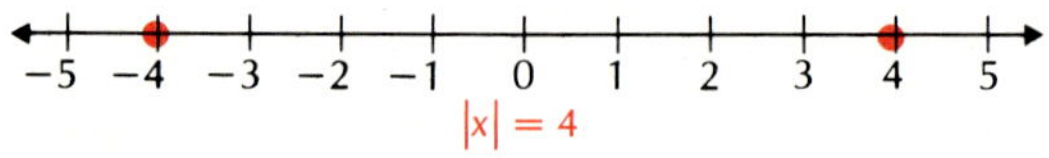

DO EXERCISE 9.

EXAMPLE 8 Solve $|x| < 4$. Then graph.

Again, note that $|x| = |x - 0|$, so $|x - 0|$ is the distance from x to 0. Thus the solutions of $|x - 0| < 4$ are those numbers x whose distance from 0 is less than 4. The solutions are those numbers x such that $-4 < x < 4$. The solution set is $\{x \mid -4 < x < 4\}$. The graph is as follows.

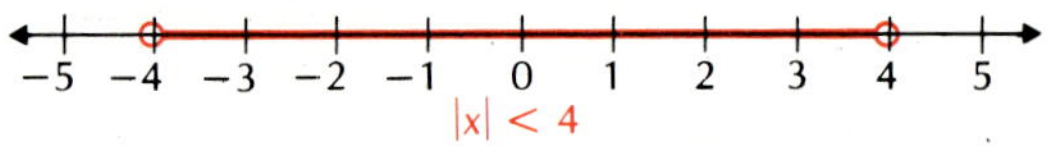

DO EXERCISE 10.

EXAMPLE 9 Solve $|x| \geqslant 4$. Then graph.

Since $|x| = |x - 0|$, the solutions of $|x - 0| \geqslant 4$ are those numbers x whose distance from 0 is greater than or equal to 4, in other words, those numbers x such that $x \leqslant -4$ *or* $x \geqslant 4$. The solution set is $\{x \mid x \leqslant -4 \text{ or } x \geqslant 4\}$. The graph is as follows.

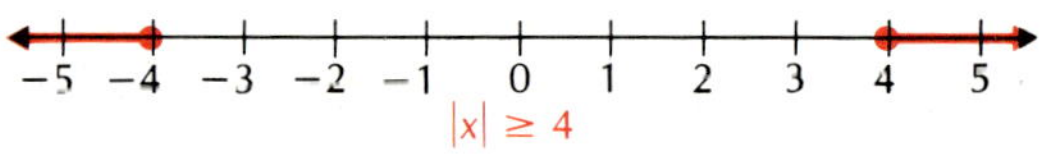

DO EXERCISE 11.

Examples 7–9 illustrate three cases of solving equations and inequalities with absolute value. The expression inside the absolute-value signs can be something besides a single variable. The following is a general principle for solving.

> **For any positive number b, and any expression X,**
>
> **a) The solutions of $|X| = b$ are those numbers that satisfy $X = -b$ or $X = b$.**
>
> **b) The solutions of $|X| < b$ are those numbers that satisfy $-b < X < b$.**
>
> **c) The solutions of $|X| > b$ are those numbers that satisfy $X < -b$ or $X > b$.**

a) As an example to illustrate part (a), the solutions of $|5x - 1| = 8$ are those numbers x for which

$$5x - 1 = 8 \quad \text{or} \quad 5x - 1 = -8.$$

b) As an example to illustrate part (b), the solutions of $|6x + 7| < 5$ are those numbers x for which

$$-5 < 6x + 7 < 5.$$

c) As an example to illustrate part (c), the solutions of $|2x - 9| > 4$ are those numbers x for which

$$2x - 9 < -4 \quad \text{or} \quad 2x - 9 > 4.$$

EXAMPLE 10 Solve $|5x - 4| = 11$. Then graph.

We use part (a). In this case, b is 11 and X is $5x - 4$:

$$|X| = b$$

$$|5x - 4| = 11 \qquad \text{Replacing } X \text{ by } 5x - 4 \text{ and } b \text{ by } 11$$

$$5x - 4 = 11 \quad \text{or} \quad 5x - 4 = -11 \qquad \text{Part (a)}$$

$$5x = 15 \quad \text{or} \quad 5x = -7 \qquad \text{Adding 4}$$

$$x = 3 \quad \text{or} \quad x = -\tfrac{7}{5} \qquad \text{Multiplying by } \tfrac{1}{5}$$

The solution set is $\{3, -\frac{7}{5}\}$. The graph is as follows.

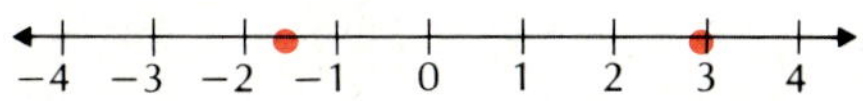

$|5x - 4| = 11$

DO EXERCISE 12.

EXAMPLE 11 Solve $|3x - 2| < 4$. Then graph.

We use part (b). In this case, b is 4 and X is $3x - 2$:

$$|X| < b$$

$$|3x - 2| < 4 \qquad \text{Replacing } X \text{ by } 3x - 2 \text{ and } b \text{ by } 4$$

$$-4 < 3x - 2 < 4 \qquad \text{Part (b)}$$

$$-2 < 3x < 6 \qquad \text{Adding 2}$$

$$-\tfrac{2}{3} < x < 2 \qquad \text{Multiplying by } \tfrac{1}{3}$$

The graph is as follows.

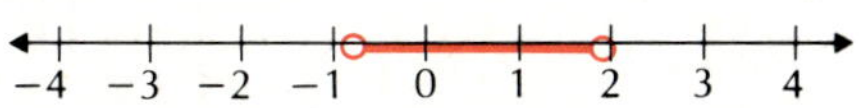

$|3x - 2| < 4$

EXAMPLE 12 Solve $|4x + 2| \geqslant 6$. Then graph.

We use part (c). In this case, b is 6 and X is $4x + 2$:

$$|X| \geqslant b$$

$$|4x + 2| \geqslant 6 \qquad \text{Replacing } X \text{ by } 4x + 2 \text{ and } b \text{ by } 6$$

$$4x + 2 \leqslant -6 \quad \text{or} \quad 4x + 2 \geqslant 6 \qquad \text{Part (c)}$$

$$4x \leqslant -8 \quad \text{or} \quad 4x \geqslant 4 \qquad \text{Adding } -2$$

$$x \leqslant -2 \quad \text{or} \quad x \geqslant 1 \qquad \text{Multiplying by } \tfrac{1}{4}$$

The graph is as follows.

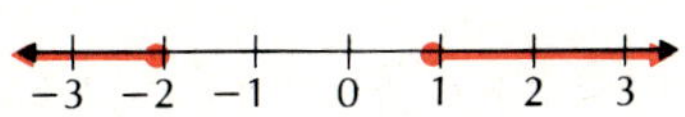

$|4x + 2| \geqq 6$

DO EXERCISES 13 AND 14.

Solve.

12. $|3x + 4| = 9$

Solve.

13. $|2x - 3| < 7$

14. $|2x - 4| > 7$

EXERCISE SET 5.4

i Simplify, leaving as little as possible inside absolute-value signs.

1. $|3x|$ **2.** $|17x|$ **3.** $|9x^2|$ **4.** $|6x^2|$ **5.** $|-4x^2|$ **6.** $|-10x^2|$

7. $|-8y|$ **8.** $|-13y|$ **9.** $\left|\frac{-4}{x}\right|$ **10.** $\left|\frac{y}{7}\right|$ **11.** $\left|\frac{x^2}{y}\right|$ **12.** $\left|\frac{x^4}{y}\right|$

13. $\left|\frac{-8x^2}{2x}\right|$ **14.** $\left|\frac{9y}{3y^2}\right|$

ii Find the distance between the points having the given coordinates.

15. $-8, -42$ **16.** $-9, -36$ **17.** $26, 15$ **18.** $54, 18$ **19.** $-9, 24$ **20.** $-18, -37$

21. $-5, 0$ **22.** $0, 23$

iii Solve. Then graph.

23. $|x| = 3$ **24.** $|x| = 5$ **25.** $|x| < 3$ **26.** $|x| \leqslant 5$ **27.** $|x| \geqslant 2$

28. $|y| > 8$ **29.** $|t| \geqslant 5.5$ **30.** $|m| > 0$ **31.** $|x - 3| = 12$ **32.** $|3x - 2| = 6$

33. $|2x - 3| \leqslant 4$ **34.** $|5x + 2| \leqslant 3$ **35.** $|2y - 7| > 10$ **36.** $|3y - 4| > 8$ **37.** $|4x - 9| \geqslant 14$

38. $|9y - 2| \geqslant 17$

Problem-solving practice

39. Jane needs to maintain an average of at least 3.0 in order to stay on the tennis team. Her average the first year in school was 3.3. Her average the second year was 2.9. For what averages during her third year will Jane be able to maintain the necessary cumulative average? Assume that she took the same course load each year.

40. Tickets to a concert sell for \$5.00. The concert promoter considers the concert a failure if there is more than a 10% loss and a success if there is at least a 10% profit. If his costs for the concert are \$20,000, for what numbers of tickets sold would the concert be considered a success? A failure?

○

Simplify, leaving as little as possible inside the absolute-value signs.

41. $|3x^4|$ **42.** $|-5y^6|$ **43.** $|3x^3|$ **44.** $|-5y^3|$

Solve.

45. $|m + 5| + 9 \leqslant 16$ **46.** $|t - 7| + 3 \geqslant 4$ **47.** $|g + 7| + 13 = 4$

48. $2|2x - 7| + 11 = 25$ **49.** $\left|\frac{2x - 1}{3}\right| \leqslant 1$ **50.** $\left|\frac{3x - 2}{5}\right| \leqslant 2$

51. $\left|\frac{2x - 1}{0.0059}\right| \leqslant 1$ **52.** $\left|\frac{3x - 2}{5}\right| \geqslant 1$ **53.** $\left|\frac{0.005x - 0.004}{0.0059}\right| \leqslant 0.0043$

54. $3\left|\frac{0.0064x - 0.0009}{0.0023}\right| \geqslant 0.0089$ **55.** $|3x - 4| > -2$ **56.** $|x - 6| \leqslant -8$

57. $|\frac{5}{9} + 3x| < \frac{1}{6}$ **58.** $1 - |\frac{1}{4}x + 8| > \frac{3}{4}$ **59.** $|x + 5| > x$

60. $2 \leqslant |x - 1| \leqslant 5$ **61.** $|7x - 2| = x + 4$ **62.** $|x - 1| - 2 = |2x - 5|$

63. $|x + 1| \leqslant |x - 3|$

OBJECTIVES

After finishing Section 5.5, you should be able to:

i **Determine whether an ordered pair of numbers is a solution of an inequality in two variables.**

ii **Graph linear inequalities.**

iii **Graph systems of linear inequalities.**

5.5 INEQUALITIES IN TWO VARIABLES

We now consider inequalities in two variables and their graphs. As we shall see, systems of such inequalities are useful in problem solving.

i Solutions

The solutions of equations in two variables are ordered pairs of numbers. The same is true for inequalities in two variables.

EXAMPLE 1 Determine whether $(3, -2)$ is a solution of $5x - 4y > 13$.

We replace x by 3 and y by -2, using alphabetical order:

$5x - 4y$	> 13
$5(3) - 4(-2)$	13
$15 - (-8)$	
23	

Since $23 > 13$ is true, $(3, -2)$ is a solution.

DO EXERCISE 1.

1. Determine whether $(1, -4)$ is a solution of the inequality $4x - 5y < 12$.

ii Graphs

A *linear inequality* is an inequality that is equivalent to

$$ax + by < c \quad \text{or} \quad ax + by > c,$$

that is, there is a linear polynomial on one side and c on the other. The inequality signs can, of course, be $\leqslant$ or $\geqslant$. To graph linear inequalities, we use what we already know about graphing linear equations.

EXAMPLE 2 Graph: $y < x$.

For comparison, we first graph the line $y = x$. We draw it dashed. Every solution of $y = x$ is an ordered pair having the same first and second coordinates.

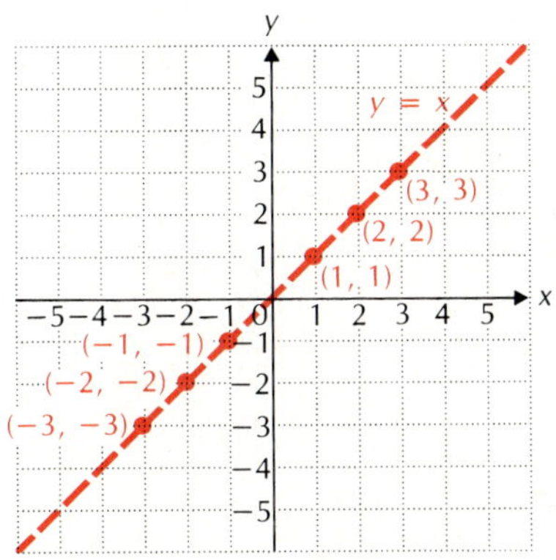

Now look at a vertical line and some ordered pairs on it. The one shown goes through $(2, 2)$.

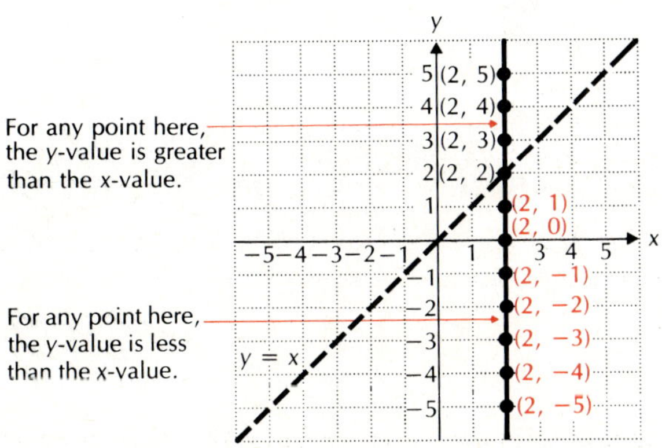

2. Graph $y > x$.

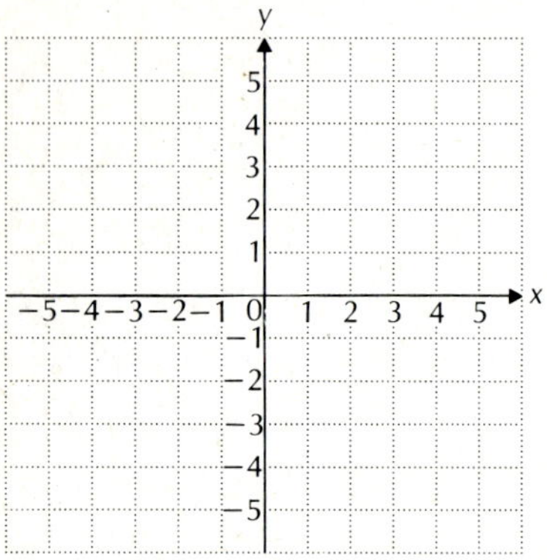

For all points below $y = x$, the second coordinate is less than the first, that is, $y < x$. For points above the line, $y > x$. The same thing happens for any vertical line. Thus all points below the line $y = x$ are solutions. We shade the half-plane below $y = x$. This is the graph of the inequality $y < x$. The points on the line $y = x$ are not included. That is why we drew the line dashed.

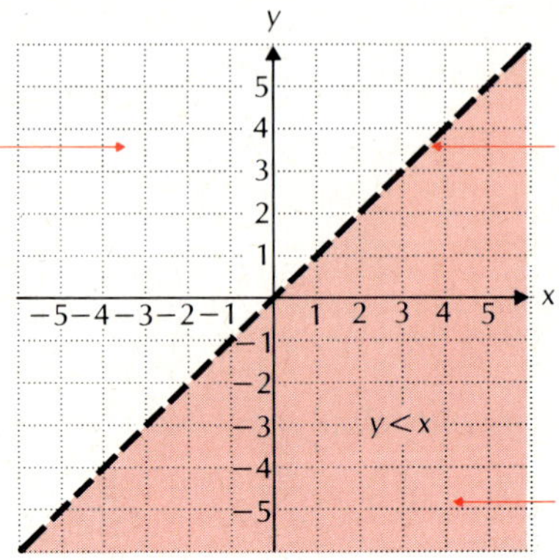

DO EXERCISE 2.

To graph a linear inequality, we first replace the inequality sign by an equals sign. Then we graph the resulting line. If the inequality symbol is $<$ or $>$, we draw the line dashed. If the inequality symbol is $\leqslant$ or $\geqslant$, we draw the line solid. The graph consists of a half-plane as in Example 2 and, if the line is solid, the line as well.

EXAMPLE 3 Graph: $6x - 2y < 12$.

a) We first graph the line $6x - 2y = 12$. Since the inequality symbol is $<$, we draw it dashed. The line is most easily graphed using intercepts.

b) We determine which half-plane to shade by trying some point off the line. The point (0, 0) is easy to check unless the line goes through the origin. *Any* point not on the line can be used as a check.

$6x - 2y$	< 12
$6 \cdot 0 - 2 \cdot 0$	12
0	

Since $0 < 12$, the point is in the graph. We shade that half-plane. The graph consists of the shaded half-plane, but not the dashed line.

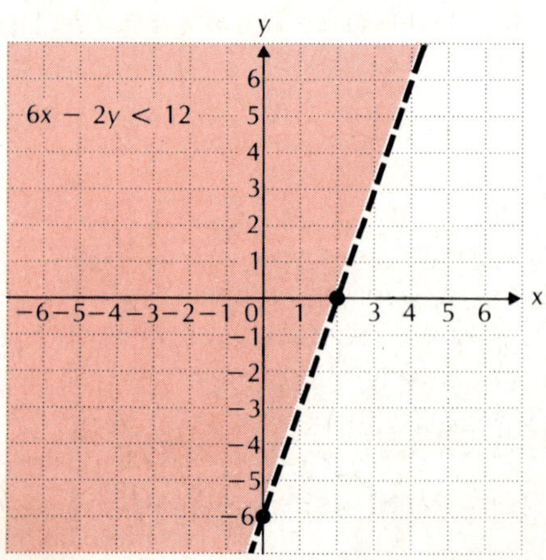

EXAMPLE 4 Graph: $8x + 3y \geqslant 24$.

a) We graph the line $8x + 3y = 24$. Since the inequality symbol is $\geqslant$, we draw the line solid.

b) The point (0, 0) is not in the graph, because $8 \cdot 0 + 3 \cdot 0 = 0$, which is less than 24. We shade the other half-plane. The graph consists of the shaded half-plane and the solid line as well.

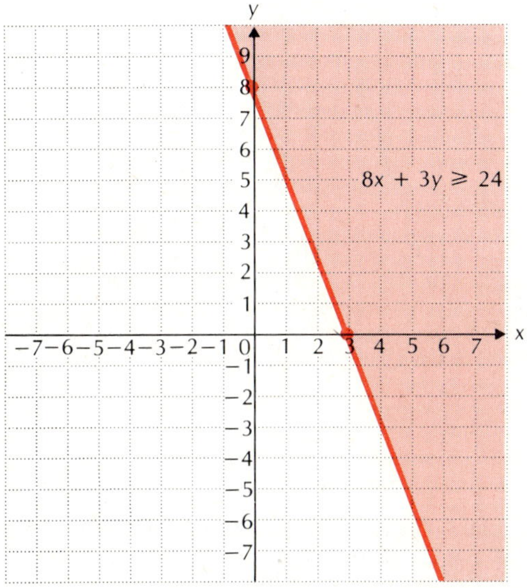

The graph of any inequality in two variables contains a *half-plane*, that is, a set of all points on one side of a line, as in the previous example. The *edge* of the half-plane, which is the line that divides the plane, may or may not be a part of the graph. In Example 4, the edge is a part of the graph, but in Example 3 the edge of the half-plane is not part of the graph.

To graph a linear inequality, we first graph the edge, which is a line. Then we determine which half-plane is in the graph. We can do that by choosing some point in one of the half-planes and determining whether it satisfies the inequality. The origin (0, 0) is often an easy point to use in such checking.

TO GRAPH A LINEAR INEQUALITY

a) **Graph the line that is the edge. Draw it dashed.**

b) **Choose some point (a, b) off the line and see whether it satisfies the inequality. If it does, the half-plane containing (a, b) is in the graph. (The origin (0, 0) is generally the easiest point to use.)**

c) **Shade the proper half-plane. If the edge is part of the graph, make the line solid.**

DO EXERCISES 3 AND 4.

Systems of Inequalities

A system of inequalities generally has a very large solution set. The most useful thing to do in problem situations involving such inequalities is to graph the solution set. To graph a system of equations, we graph the individual equations and then find the intersection of the individual graphs. We do the same thing for a system of inequalities, that is, we graph them and look for the intersection. It may help to keep in mind that a system of equations or inequalities is a conjunction of sentences. It would be logically quite proper to join the individual sentences with the word *and.*

Graph.

3. $6x - 3y < 18$

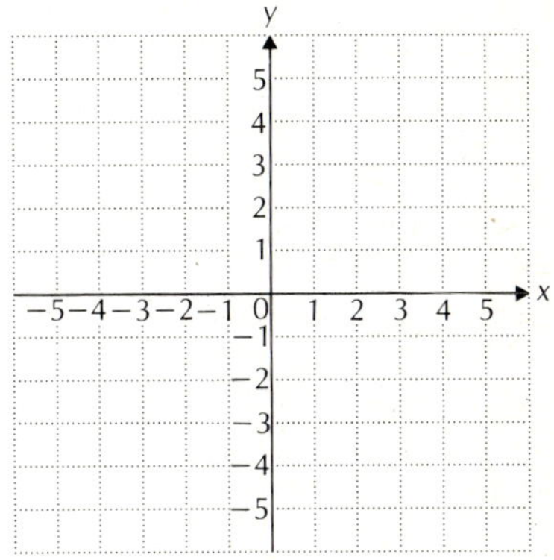

4. $4x + 3y \geqslant 12$

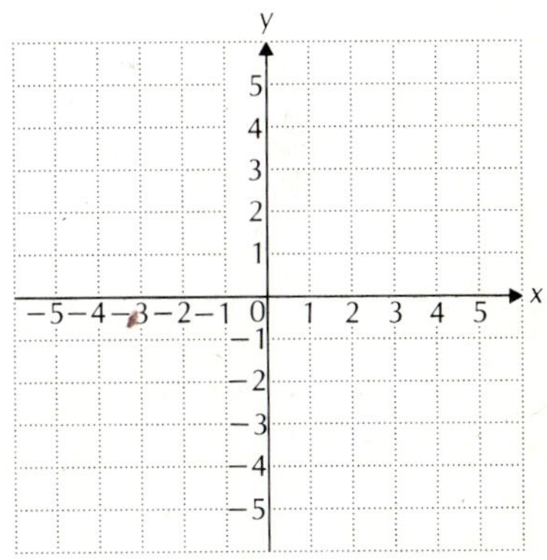

Graph.

5. $x + y \geqslant 1$
$y - x \geqslant 2$

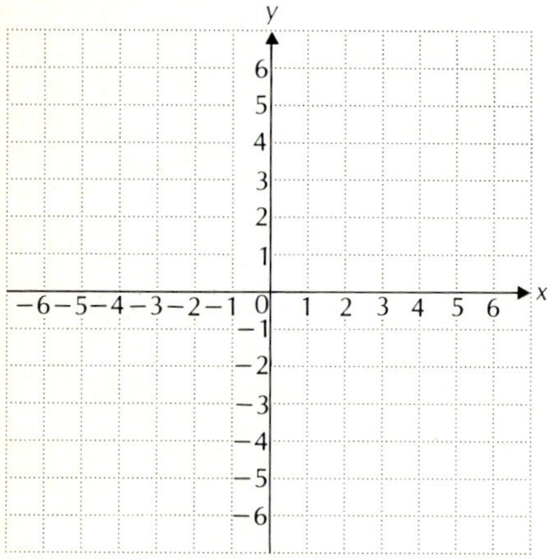

6. $5x + 6y \leqslant 30$
$0 \leqslant y \leqslant 3$
$0 \leqslant x \leqslant 4$

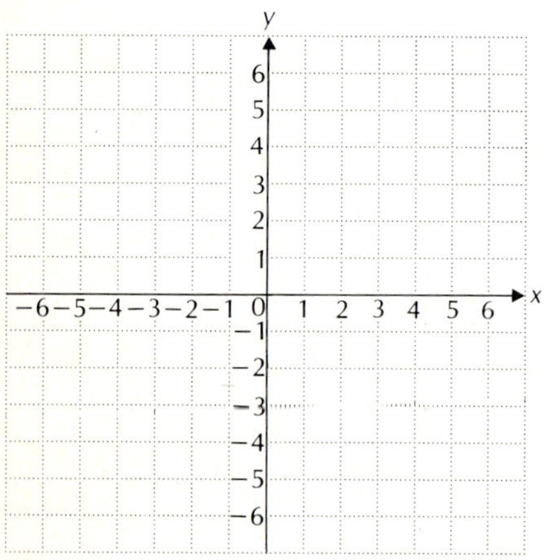

EXAMPLE 5 Graph the system of inequalities:

$$6x - 2y < 12,$$
$$y - 3 < 0,$$
$$x + y \geqslant 0.$$

We graph the individual inequalities. Note that the first one has already been done, in Example 3. The individual graphs are as follows.

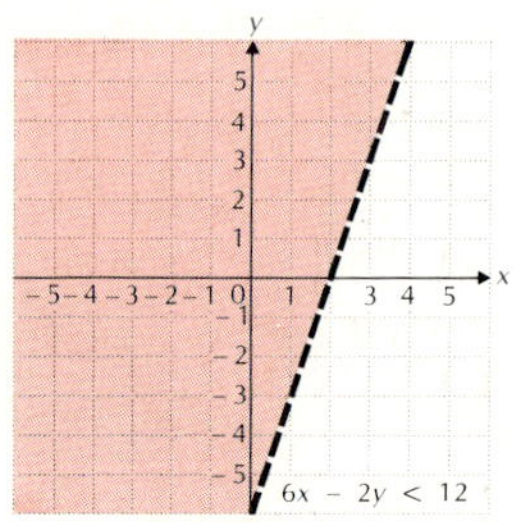

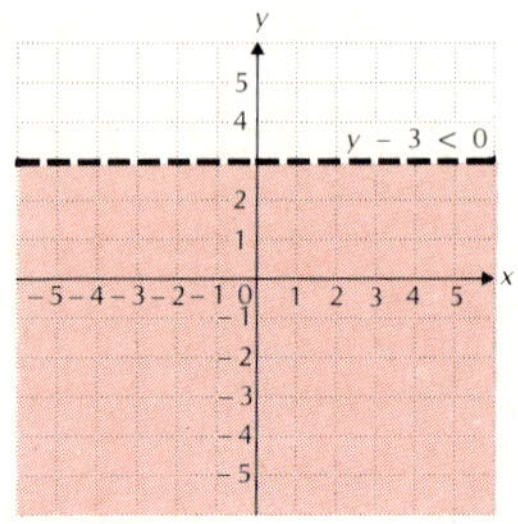

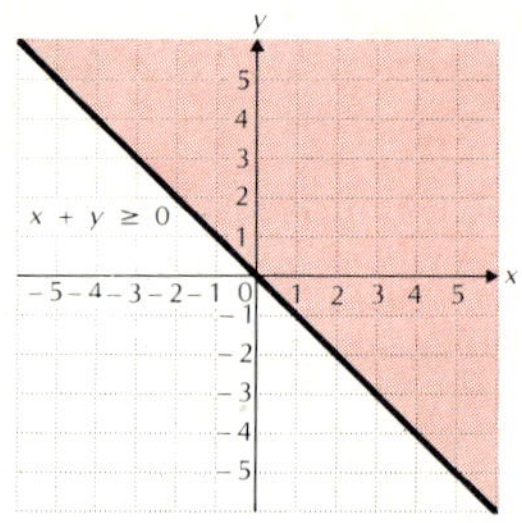

We now find the intersection of the three graphs.

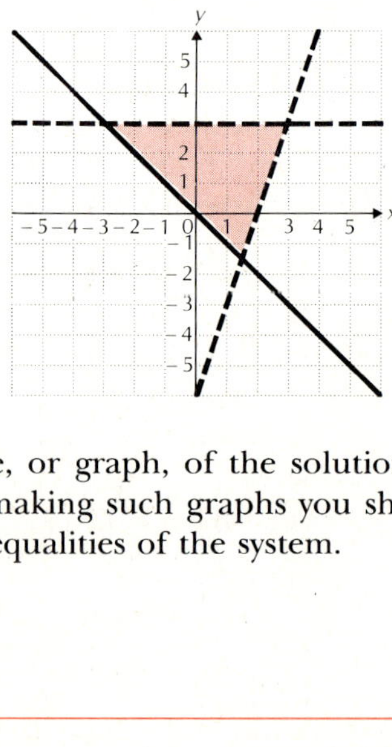

We now have a picture, or graph, of the solution set of the conjunction, or system, of inequalities. In making such graphs you should generally use the same set of axes for all of the inequalities of the system.

DO EXERCISES 5 AND 6.

EXERCISE SET 5.5

i Determine whether the ordered pair is a solution of the inequality.

1. $(-4, 2)$; $2x + y < -5$ **2.** $(3, -6)$; $4x + 2y \geqslant 0$ **3.** $(8, 14)$; $2y - 3x > 5$ **4.** $(7, 20)$; $3x - y > -1$

ii Graph these inequalities, using graph paper.

5. $y > 2x$ **6.** $y < 3x$ **7.** $y < x + 1$ **8.** $y \leqslant x - 3$ **9.** $y > x - 2$

10. $y \geqslant x + 4$ **11.** $x + y < 4$ **12.** $x - y \geqslant 3$ **13.** $3x + 4y \leqslant 12$ **14.** $2x + 3y < 6$

15. $2y - 3x > 6$ **16.** $2y - x \leqslant 4$ **17.** $3x - 2 \leqslant 5x + y$ **18.** $2x - 2y \geqslant 8 + 2y$ **19.** $x < -4$
20. $y \geqslant 5$ **21.** $x > -2$ **22.** $-4 < y < -1$ **23.** $-2 < y < 3$ **24.** $-3 \leqslant x \leqslant 3$
25. $-4 \leqslant x \leqslant 4$

iii Graph these systems of inequalities.

26. $y < x$, $y > -x + 3$
27. $y > x$, $y < -x + 1$
28. $y \geqslant x$, $y < -x + 4$
29. $y \geqslant x$, $y < -x + 2$
30. $y \geqslant -2$, $x > 1$
31. $y \leqslant -2$, $x > 2$
32. $x < 3$, $y \geqslant -3x + 2$
33. $x > -2$, $y \leqslant -2x + 3$
34. $y \geqslant -2$, $y \geqslant x + 3$
35. $y \leqslant 4$, $y \geqslant -x + 2$
36. $x + y \leqslant 1$, $x - y \leqslant 2$
37. $x + y < 3$, $x - y < 4$
38. $y - 2x > 1$, $y - 2x < 3$
39. $y + 3x > 0$, $y + 3x < 2$
40. $2y - x \leqslant 2$, $y - 3x \geqslant -1$
41. $y \leqslant 2x + 1$, $y \geqslant -2x + 1$, $x \leqslant 2$
42. $x - y \leqslant 2$, $x + 2y \geqslant 8$, $y \leqslant 4$
43. $x + 2y \leqslant 12$, $2x + y \leqslant 12$, $x \geqslant 0$, $y \geqslant 0$
44. $4y - 3x \geqslant -12$, $4y + 3x \geqslant -36$, $y \leqslant 0$, $x \leqslant 0$
45. $8x + 5y \leqslant 40$, $x + 2y \leqslant 8$, $x \geqslant 0$, $y \geqslant 0$
46. $3x + 4y \geqslant 12$, $5x + 6y \leqslant 30$, $1 \leqslant x \leqslant 3$

Problem-solving practice

47. One side of a square is 5 less than a side of an equilateral triangle. If the perimeter of the square is the same as the perimeter of the triangle, what is the length of a side of the square?

48. A bus traveled 630 miles in 12 hours. If it traveled at the same average speed, how far would it have been able to go in 15 hours?

Graph.

49. $y \geqslant |x|$ **50.** $y > |x| + 3$ **51.** $|x + y| \leqslant 1$ **52.** $|x| + |y| \leqslant 1$
53. $|x| > |y|$ **54.** $y > x^2 - 3$ **55.** $y \leqslant \frac{1}{4}x^2 - 1$

56. Solve $2x^2 + x < 6$ by graphing.

5.6

PROBLEM SOLVING USING LINEAR PROGRAMMING

OBJECTIVES

After finishing Section 5.6, you should be able to:

i **Find maximum and minimum values of functions of two variables on a domain whose graph is a convex polygonal region.**

ii **Solve certain maximum and minimum problems using linear programming.**

There are many problems in real life in which we want to find a greatest value (a maximum) or a least value (a minimum). For example, if you are in business you would like to know how to make the *most* profit. Or, you might like to know how to make your expenses the *least* possible. Some such problems can be solved using systems of inequalities.

Functions of Two Variables

For a function of two variables, the inputs are ordered pairs of numbers. An example is the area of a rectangle, a function $A(l, w)$:

$$A(l, w) = l \cdot w.$$

The numbers l and w of the input are the length and width of the rectangle. To find the function value, we multiply l and w.

As another example, consider the profit function,

$$P(x, y) = 5x + 20y,$$

for a company that makes two kinds of products. It makes a profit of \$5 per item from selling x items of one product and a profit of \$20 per item from selling y items of the other product. The total profit is given by $5x + 20y$. To find the function value, we multiply x by 5 and y by 20, and add.

In the problems we are about to consider, there will be a function of two variables and we will be looking for a maximum function value or a minimum function value. The domain of a function of two variables is, of course, a set of ordered pairs, so we may be able to graph the domain on a plane.

i Linear Programming

The domain of a function of two variables sometimes consists of a convex polygon* and its interior, as shown below.

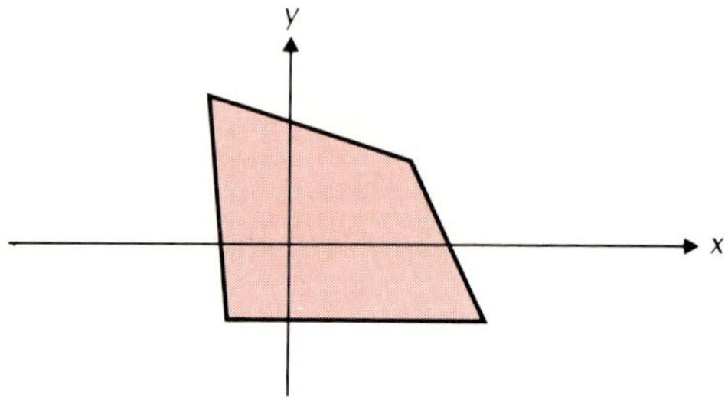

If it does, then the function has both a maximum value and a minimum value, and those values occur at the vertices of the polygon. (That fact is a theorem that was proved about the time of World War II.)

EXAMPLE 1 Find the maximum and minimum values of the function $F(x, y) = 5x - 2y$ and the values of x and y for which they occur, subject to

$$5x + 6y \leqslant 30$$

$$0 \leqslant y \leqslant 3$$

$$0 \leqslant x \leqslant 4.$$

The three inequalities form a system, called the *constraints*. We graph the system, thus illustrating the domain of the function F, and determine the vertices.

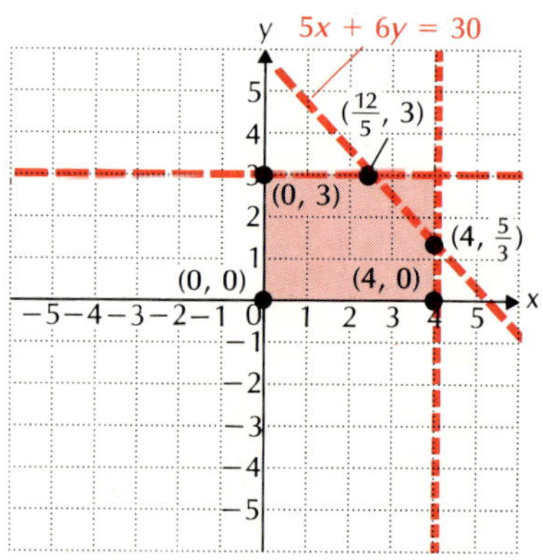

The vertices are $(0, 0)$, $(0, 3)$, $(4, 0)$, $(4, \frac{5}{3})$, $(\frac{12}{5}, 3)$.

We now find the function values at these points.

$$F(0, 0) = 5 \cdot 0 - 2 \cdot 0 = 0$$

$$F(0, 3) = 5 \cdot 0 - 2 \cdot 3 = -6$$

$$F(4, 0) = 5 \cdot 4 - 2 \cdot 0 = 20$$

$$F(4, \tfrac{5}{3}) = 5 \cdot 4 - 2 \cdot \tfrac{5}{3} = \tfrac{50}{3} = 16\tfrac{2}{3}$$

$$F(\tfrac{12}{5}, 3) = 5 \cdot \tfrac{12}{5} - 2 \cdot 3 = 6$$

* A *convex polygon* is one that has no indentations.

The maximum of these function values is also the maximum of the function F in the described domain. That maximum is 20, and it occurs at $(4, 0)$. The minimum is -6 and it occurs at $(0, 3)$.

We will now use the fact that maximum or minimum values of a function of two variables on a domain whose graph is a convex polygonal region occur at the vertices. It will enable us to solve some problems. The branch of mathematics that uses this approach is called *linear programming.*

DO EXERCISE 1.

1. Find the maximum and the minimum function values of $F(x, y) = 2y - 4x$ on the domain described by the system of constraints

$$y \leqslant 2x + 1$$
$$y \geqslant -2x + 1$$
$$x \leqslant 2.$$

ii Problem Solving

In problem solving we often take advantage of some mathematical results that mathematicians have proved. That is the case here. The five-step procedure is still appropriate, but we now have some new methods to use in translating and in carrying out a mathematical procedure.

EXAMPLE 2 You are taking a test in which items of type A are worth 10 points and items of type B are worth 15 points. It takes you 3 minutes to answer each item of type A and 6 minutes to answer each item of type B. The total time allowed is 60 minutes and you are not allowed to answer more than 16 questions. Assuming that all of your answers are correct, how many items of each type should you answer to get the best score?

1. **Familiarize** Tabulating information will help us to see the picture.

Type	Number of points for each	Time required for each	Number answered
A	10	3 min	x
B	15	6 min	y
Total time: 60 minutes			
Total number of items: 16 or fewer			

Note that we have used x to represent the number of items of type A and y to represent the number of items of type B.

2. **Translate** In this case it will help to extend the table.

Type	Number of points for each	Time required for each	Number answered	Total time for type	Total points for type
A	10	3 min	x	$3x$	$10x$
B	15	6 min	y	$6y$	$15y$
Total			$x + y \leqslant 16$	$3x + 6y \leqslant 60$	$10x + 15y$
			↑ Because no more than 16 items can be answered	↑ Because the time cannot be more than 60 minutes	↑ This is the total score on the test

Suppose that the total score on the test is T. We have T as a function of two variables x and y:

$$T(x, y) = 10x + 15y.$$

Let us consider the domain of the function. We know these things (called *constraints*) about x and y:

$$x + y \leqslant 16,$$
$$3x + 6y \leqslant 60,$$
$$\left.\begin{array}{l} x \geqslant 0, \\ y \geqslant 0. \end{array}\right\}$$ Because the number of items answered cannot be negative.

This system of inequalities describes the domain of T.

3. **Carry out** The mathematical manipulation that we will do here consists of graphing the domain and then computing the function values at the vertices. The graph of the domain is as follows.

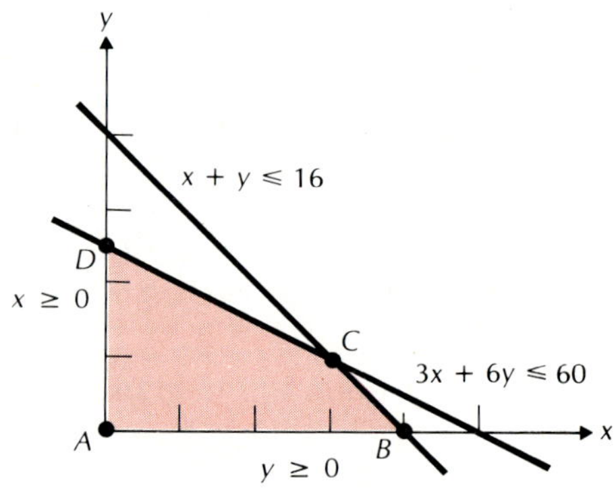

We need to find the coordinates of each vertex. Note that each vertex is the intersection of two lines. Thus we can find each one by solving a system of two linear equations. The coordinates of point A are obviously $(0, 0)$. Let us find the coordinates of point C. We solve the system

$$3x + 6y = 60,$$
$$x + y = 16$$

as follows:

$$\begin{array}{rl} 3x + 6y = & 60 \\ -3x - 3y = & -48 \\ \hline 3y = & 12 \\ y = & 4 \end{array}$$

Multiplying by -3; Adding

Then we find that $x = 12$. Thus the coordinates of vertex C are $(12, 4)$.

Continuing to find the coordinates of the vertices and computing the test score for each ordered pair, we obtain the following:

Vertices (x, y)	Score $T = 10x + 15y$
$A\ (0, 0)$	0
$B\ (16, 0)$	160
$C\ (12, 4)$	180
$D\ (0, 10)$	150

The greatest score in the table is 180, obtained when 12 items of type A are answered and 4 items of type B are answered.

4. **Check** We can go back to the original conditions of the problem and calculate the scores, using the ordered pairs in the table above. We can also check our work, the algebra and the arithmetic. In this case there is no further checking that we can do, without an undue amount of work.

5. **State** The answer is that you should answer 12 items of type A and 4 items of type B.

DO EXERCISE 2.

2. A snack bar sells hamburgers and hotdogs during football games. To stay in business it must sell at least 10 hamburgers but cannot cook more than 40. It must also sell at least 30 hotdogs but cannot cook more than 70. It cannot cook more than 90 sandwiches altogether. The profit is \$0.33 on a hamburger and \$0.21 on a hotdog. How many of each kind of sandwich should it sell in order to make the maximum profit?

EXERCISE SET 5.6

i Find the maximum and minimum values of the function and the values of x and y for which they occur.

1. $F(x, y) = 4x + 28y$, subject to
$5x + 3y \leqslant 34$,
$3x + 5y \leqslant 30$,
$x \geqslant 0$,
$y \geqslant 0$.

2. $G(x, y) = 14x + 16y$, subject to
$3x + 2y \leqslant 12$,
$7x + 5y \leqslant 29$,
$x \geqslant 0$,
$y \geqslant 0$.

3. $P(x, y) = 16x - 2y + 40$, subject to
$6x + 8y \leqslant 48$,
$0 \leqslant y \leqslant 4$,
$0 \leqslant x \leqslant 7$.

4. $Q(x, y) = 24x - 3y + 52$, subject to
$5x + 4y \geqslant 20$,
$0 \leqslant y \leqslant 4$,
$0 \leqslant x \leqslant 3$.

ii ***Problem-solving practice***

5. You are about to take a test that contains questions of type A worth 4 points and questions of type B worth 7 points. You must do at least 5 questions of type A but time restricts doing more than 10. You must do at least 3 questions of type B but time restricts doing more than 10. In total, you can do no more than 18 questions. How many of each type of question must you do to maximize your score? What is this maximum score?

6. You are about to take a test that contains questions of type A worth 10 points and questions of type B worth 25 points. You must do at least 3 questions of type A but time restricts doing more than 12. You must do at least 4 questions of type B but time restricts doing more than 15. In total you can do no more than 20 questions. How many of each type of question must you do to maximize your score? What is this maximum score?

7. A man is planning to invest up to \$22,000 in bank X or bank Y or both. He wants to invest at least \$2000 but no more than \$14,000 in bank X. Bank Y does not insure more than a \$15,000 investment so he will invest no more than that in bank Y. The interest in bank X is 6% and in bank Y it is $6\frac{1}{2}$% and this will be simple interest for one year. How much should he invest in each bank to maximize his income? What is the maximum income?

8. A woman is planning to invest up to \$40,000 in corporate or municipal bonds or both. The least she is allowed to invest in corporate bonds is \$6000 and she does not want to invest more than \$22,000 in corporate bonds. She also does not want to invest more than \$30,000 in municipal bonds. The interest on corporate bonds is 8% and on municipal bonds it is $7\frac{1}{2}$%. This is simple interest for one year. How much should she invest in each type of bond to maximize her income? What is the maximum income?

9. It takes a tailoring firm 2 hr of cutting and 4 hr of sewing to make a knit suit. To make a worsted suit it takes 4 hr of cutting and 2 hr of sewing. At most, 20 hr per day are available for cutting and, at most, 16 hr per day are available for sewing. The profit on a knit suit is \$34 and on a worsted suit is \$31. How many of each kind of suit should be made to maximize profit? What is the maximum profit?

10. A pipe tobacco company has 3000 lb of English tobacco, 2000 lb of Virginia tobacco, and 500 lb of Latakia tobacco. To make one batch of SMELLO tobacco it takes 12 lb of English tobacco and 4 lb of Latakia. To make one batch of ROPPO tobacco it takes 8 lb of English and 8 lb of Virginia tobacco. The profit is \$10.56 per batch for SMELLO and \$6.40 for ROPPO. How many batches of each kind of tobacco should be made to yield maximum profit? What is the maximum profit? (*Hint:* Organize the information in a table.)

11. An airline with two types of airplanes, P-1 and P-2, has contracted with a tour group to provide accommodations for a minimum of each of 2000 first-class, 1500 tourist, and 2400 economy-class passengers. Airplane P-1 costs \$12,000 per mile to operate and can accommodate 40 first-class, 40 tourist, and 120 economy-class passengers, whereas airplane P-2 costs \$10,000 per mile to operate and can accommodate 80 first-class, 30 tourist, and 40 economy-class passengers. How many of each type of airplane should be used to minimize the operating cost?

12. A new airplane P-3 becomes available, having an operating cost of \$15,000 per mile and accommodating 40 first-class, 80 tourist, and 80 economy-class passengers. If airplane P-1 of Exercise 11 were replaced by airplane P-3, how many of P-2 and P-3 would be needed to minimize the operating cost?

SUMMARY AND REVIEW: CHAPTER 5

The following contains a summary of what you should be able to do after completing this chapter. The review exercises are for practice. Answers are at the back of the book. If you miss an exercise, restudy the section indicated alongside the answer.

You should be able to:

Solve inequalities and graph the solution sets.

Find the solution set. Graph.

1. $x \leqslant -4$ **2.** $x + 5 > 6$ **3.** $a + 7 \leqslant -14$ **4.** $y - 5 \geqslant -12$

5. $4y > -15$ **6.** $-0.3y < 9$ **7.** $-6x - 5 < 13$ **8.** $4y + 3 < -6y - 9$

9. $-\frac{1}{2}x - \frac{1}{4} > \frac{1}{2} - \frac{1}{4}x$ **10.** $0.3y - 7 < 2.6y + 15$

Solve problems involving inequalities.

Solve.

11. You are taking a biology course, in which there will be 5 tests. You have scores of 91, 93, 86, and 88. You must make a total of 450 in order to get an A. What scores on the last test will give you an A?

12. You are going to invest \$30,000, part at 13% and part at 15%. What is the most that can be invested at 13% in order to make at least \$4300 interest per year?

Solve compound inequalities.

Solve.

13. $-4 < x + 3 \leqslant 5$ **14.** $-14 < -4x - 5 < 0$

15. $3x < -9$ or $-5x < -5$ **16.** $2x + 5 < -17$ or $-4x + 10 \leqslant 34$

17. $x + 5 < -6$ or $x + 5 > 6$ **18.** $2x + 7 < -5$ or $x + 7 > 15$

Simplify absolute-value expressions.

Simplify.

19. $\left|-\frac{3}{x}\right|$ **20.** $\left|\frac{2x}{y^2}\right|$ **21.** $\left|\frac{12y}{-3y^2}\right|$

Solve equations and inequalities involving absolute value.

Solve.

22. $|x| = 6$ **23.** $|x| < 0$ **24.** $|x| \geqslant 3.5$

25. $|x - 2| = 7$ **26.** $|2x + 5| < 12$ **27.** $|3x - 4| \leqslant 15$

Graph systems of inequalities.

Graph.

28. $y \geqslant -3,$ $x \geqslant 2$

29. $x + 3y > -1,$ $x + 3y < 4$

30. $x - 3y \leqslant 3,$ $x + 3y \geqslant 9$ $y \leqslant 6$

Solve.

31. $\left|\frac{4}{3} + x\right| \leqslant \frac{2}{5}$

32. $|2x + 5| \leqslant |x + 3|$

TEST: CHAPTER 5

Solve.

1. $x - 2 < 12$

2. $-0.6y < 30$

3. $-4y - 3 \geqslant 5$

4. $3a - 5 \leqslant -2a + 6$

5. $-5y - 1 > -9y + 3$

6. You can rent a car for either \$40 per day with unlimited mileage, or for \$30 per day with an extra charge of 15¢ a mile. For what numbers of miles traveled would the unlimited mileage plan save you money?

7. $-3 < x - 2 < 4$

8. $-11 \leqslant -5x - 2 < 0$

9. $-3x > 12$ or $4x > -10$

10. $x - 7 < -5$ or $x - 7 > -10$

11. $3x - 2 < 7$ or $x - 2 > 4$

Simplify.

12. $\left|\frac{7}{x}\right|$

13. $\left|\frac{-6x^2}{3x}\right|$

Solve.

14. $|x| = 9$

15. $|x| > 3$

16. $|4x - 1| < 4.5$

17. $|-5x - 3| \geqslant 10$

Graph.

18. $x + y > 3,$
$x - y > 5$

19. $2y - x \geqslant -7,$
$2y + 3x < 15,$
$y \leqslant 0,$
$x \leqslant 0$

How can you make up a number "trick" to amaze your friends? One way is to use rational expressions.

6 RATIONAL EXPRESSIONS, EQUATIONS, AND FUNCTIONS

A fractional expression is an expression that indicates division, as the fractional symbols in arithmetic do. In this chapter we consider algebraic fractional expressions, also called *rational expressions*. You will learn, or review, how to add them, multiply them, simplify them, and use them in equations or in describing functions. Then, of course, you will use the expressions, equations, and functions in problem solving. As you will see, there are problems for which you need to use fractional expression in the translation.

OBJECTIVES

After finishing Section 6.1, you should be able to:

- **i** **Find function values for rational functions.**
- **ii** **Multiply fractional expressions.**
- **iii** **Multiply a fractional expression by 1, using an expression like *A*/*A*.**
- **iv** **Simplify fractional expressions.**
- **v** **Multiply fractional expressions and simplify.**
- **vi** **Divide fractional expressions and simplify.**

1. For the rational function

$$f(x) = \frac{2x^2 - 5x + 1}{2x},$$

find each of the following.

a) $f(1)$

b) $f(-1)$

c) $f(3)$

d) $f(0)$

6.1

RATIONAL EXPRESSIONS: MULTIPLYING AND DIVIDING

An expression that is a quotient of two expressions is called a fractional expression, or a *rational expression.* The term *rational* comes from the word *ratio.* The following are rational expressions:

$$\frac{7}{8}, \quad \frac{a}{b}, \quad \frac{8}{y+5}, \quad \frac{x^2 + 7xy - 4}{x^3 - y^3}, \quad \frac{1 + \dfrac{3}{x}}{y - \dfrac{y}{3}}.$$

In most of the expressions that we will consider, the numerator and denominator will be polynomials.

Rational expressions indicate division. For example,

$$\frac{7}{8} \text{ means } 7 \div 8 \quad \text{and} \quad \frac{2x+y}{x-y} \text{ means } (2x+y) \div (x-y).$$

Certain substitutions are not sensible in rational expressions, since division by zero is not defined.

i Rational Functions

Like polynomials, rational expressions can be used to define or describe functions. A function that can be described by a rational expression is known as a *rational function.*

EXAMPLE 1 The rational function

$$v(t) = \frac{3t^2 + 2t + 5}{t}$$

gives the velocity of a car, in ft/sec, after t sec have gone by. Find the following function values (velocities): **(a)** $v(1)$; **(b)** $v(3)$; **(c)** $v(5)$.

a) $v(1) = \dfrac{3(1)^2 + 2(1) + 5}{1} = 10$ ft/sec

b) $v(3) = \dfrac{3(3)^2 + 2(3) + 5}{3} = \dfrac{38}{3}$ ft/sec

c) $v(5) = \dfrac{3(5)^2 + 2(5) + 5}{5} = 18$ ft/sec

DO EXERCISE 1.

ii Multiplying

Most of the calculations we do with rational expressions are very much like the calculations we do with fractional notation in arithmetic. By proceeding as in arithmetic, we will know that when we obtain one rational expression from another, the two expressions will be equivalent, that is, they will name the same number for all sensible replacements. The sensible replacements are, of course, those numbers that are in the domains of both functions.

For any rational expressions A/B and C/D,

$$\frac{A}{B}\cdot\frac{C}{D}=\frac{A\cdot C}{B\cdot D}.$$

(To multiply two rational expressions, multiply numerators and multiply denominators.)

EXAMPLE 2 Multiply.

$$\frac{x+3}{y-4}\cdot\frac{x^3}{y+5}=\frac{(x+3)x^3}{(y-4)(y+5)}$$ Multiplying numerators and multiplying denominators

DO EXERCISES 2 and 3.

iii Multiplying by 1.

Any number multiplied by 1 is that same number. Any expression with the same numerator and denominator names the number 1:

$$\frac{y+5}{y+5},\quad\frac{4x^2-5}{4x^2-5},\quad\frac{-1}{-1}$$ All name the number 1 for all sensible replacements.

We can multiply by 1 to get equivalent expressions. For example, let us multiply $(x+y)/5$ by 1:

$$\frac{x+y}{5}\cdot\frac{x-y}{x-y}=\frac{(x+y)(x-y)}{5(x-y)}$$ Multiplying by $\frac{x-y}{x-y}$, which is 1

We know that

$$\frac{x+y}{5}\quad\text{and}\quad\frac{(x+y)(x-y)}{5(x-y)}$$

are equivalent. That means they will name the same number for all replacements that do not make a denominator 0.

EXAMPLES Multiply to obtain equivalent expressions.

3. $\frac{x^2+3}{x-1}\cdot\frac{x+1}{x+1}=\frac{(x^2+3)(x+1)}{(x-1)(x+1)}$

4. $\frac{-1}{-1}\cdot\frac{x-4}{x-y}=\frac{-1\cdot(x-4)}{-1\cdot(x-y)}$

DO EXERCISES 4–6.

iv Simplifying

We can simplify rational expressions by reversing the procedure of multiplying by 1. We "remove" a factor of 1. We first factor the numerator and denominator, and then factor the rational expression, finding a factor that is equal to 1.

Multiply.

2. $\frac{x-2}{5}\cdot\frac{x+2}{x+4}$

3. $\frac{x+y}{x+3}\cdot\frac{x+y}{x-3}$

Multiply.

4. $\frac{3x+2y}{5x+4y}\cdot\frac{x}{x}$

5. $\frac{2x^2-y}{3x+4}\cdot\frac{3x+2}{3x+2}$

6. $\frac{-1}{-1}\cdot\frac{2a-5}{a-b}$

Simplify by removing factors of 1.

7. $\dfrac{7x^2}{x}$

8. $\dfrac{6a+9}{3}$

EXAMPLES Simplify by removing factors equal to 1.

5. $\dfrac{5x^2}{x} = \dfrac{5x \cdot x}{1 \cdot x}$ Factoring numerator and denominator

$= \dfrac{5x}{1} \cdot \dfrac{x}{x}$ Factoring the fractional expression

$= 5x \cdot 1$ $\dfrac{x}{x} = 1$ for all sensible replacements

$= 5x$ "Removing" a factor of 1

In this example we supplied a 1 in the denominator. This can always be done, if necessary.

6. $\dfrac{4a+8}{2} = \dfrac{2(2a+4)}{2 \cdot 1}$ Factoring numerator and denominator

$= \dfrac{2}{2} \cdot \dfrac{2a+4}{1}$ Factoring the fractional expression

$= \dfrac{2a+4}{1}$ "Removing" a factor of 1

$= 2a + 4$

A common error in Example 6 is to get $2a + 8$. That happens if you take an incorrect shortcut.

DO EXERCISES 7 AND 8.

Simplify by removing factors of 1.

9. $\dfrac{6x^2+4x}{4x^2+8x}$

10. $\dfrac{2y^2+6y+4}{y^2-1}$

7. $\dfrac{2x^2+4x}{6x^2+2x} = \dfrac{2x(x+2)}{2x(3x+1)}$ Factoring numerator and denominator

$= \dfrac{2x}{2x} \cdot \dfrac{x+2}{3x+1}$ Factoring the fractional expression

$= \dfrac{x+2}{3x+1}$ "Removing" a factor of 1

8. $\dfrac{x^2-1}{2x^2-x-1} = \dfrac{(x-1)(x+1)}{(2x+1)(x-1)}$ Factoring numerator and denominator

$= \dfrac{x-1}{x-1} \cdot \dfrac{x+1}{2x+1}$ Factoring the fractional expression

$= \dfrac{x+1}{2x+1}$ "Removing" a factor of 1

9. $\dfrac{9x^2+6xy-3y^2}{12x^2-12y^2} = \dfrac{3(x+y)(3x-y)}{12(x+y)(x-y)}$ Factoring numerator and denominator

$= \dfrac{3(x+y)}{3(x+y)} \cdot \dfrac{3x-y}{4(x-y)}$ Factoring the fractional expression

$= \dfrac{3x-y}{4(x-y)}$ "Removing" a factor of 1

DO EXERCISES 9 AND 10.

v Multiplying and Simplifying

After multiplying we ordinarily simplify, if possible. That is why we do not multiply out the numerator and denominator too soon. If we did that, we might need to factor them again anyway, in order to simplify.

EXAMPLES Multiply. Then simplify by removing factors of 1.

10. $$\frac{x+2}{x-3}\cdot\frac{x^2-4}{x^2+x-2}=\frac{(x+2)(x^2-4)}{(x-3)(x^2+x-2)}$$ Multiplying numerators and also denominators

$$=\frac{(x+2)(x-2)(x+2)}{(x-3)(x+2)(x-1)}$$ Factoring numerators and denominators

$$=\frac{x+2}{x+2}\cdot\frac{(x+2)(x-2)}{(x-3)(x-1)}$$ Factoring the fractional expression

$$=\frac{(x+2)(x-2)}{(x-3)(x-1)},$$ "Removing" a factor of 1

or $$\frac{x^2-4}{x^2-4x+3}$$ Multiplying out numerator and denominator

We may not always multiply out numerator and denominator.

11. $$\frac{a^3-b^3}{a^2-b^2}\cdot\frac{a^2+2ab+b^2}{a^2+ab+b^2}=\frac{(a^3-b^3)(a^2+2ab+b^2)}{(a^2-b^2)(a^2+ab+b^2)}$$

$$=\frac{(a-b)(a^2+ab+b^2)(a+b)(a+b)}{(a-b)(a+b)(a^2+ab+b^2)}$$

$$=\frac{(a-b)(a^2+ab+b^2)(a+b)}{(a-b)(a^2+ab+b^2)(a+b)}\cdot\frac{a+b}{1}$$

$$=a+b$$

DO EXERCISES 11 AND 12.

vi Dividing and Simplifying

Two expressions are reciprocals of each other if their product is 1. As in arithmetic, to find the reciprocal of a rational expression, we interchange numerator and denominator.

The reciprocal of $y-8$ is $\dfrac{1}{y-8}$.

The reciprocal of $\dfrac{x}{x^2+3}$ is $\dfrac{x^2+3}{x}$.

DO EXERCISES 13–15.

It should be noted that any replacement that makes a numerator zero will not be a sensible replacement in the expression for the reciprocal, because it will make the denominator zero.

Multiply and simplify.

11. $\dfrac{(x-y)^3}{x+y}\cdot\dfrac{3x+3y}{x^2-y^2}$

12. $\dfrac{a^3+b^3}{a^2-b^2}\cdot\dfrac{a^2-2ab+b^2}{a^2-ab+b^2}$

Find the reciprocal.

13. $\dfrac{x+3}{x-5}$

14. $x+7$

15. $\dfrac{1}{y^3-9}$

Divide. Simplify by removing a factor of 1 if possible. Then multiply out the numerator and denominator.

16. $\dfrac{x^2+7x+10}{2x-4} \div \dfrac{x^2-3x-10}{x-2}$

17. $\dfrac{a^2-b^2}{ab} \div \dfrac{a^2-2ab+b^2}{2a^2b^2}$

For any rational expressions A/B and C/D,

$$\frac{A}{B} \div \frac{C}{D} = \frac{A}{B} \cdot \frac{D}{C}.$$

(To divide two rational expressions, multiply by the reciprocal of the divisor, that is, invert and multiply.)

EXAMPLES Divide. Simplify by removing a factor of 1 if possible.

12. $\dfrac{x-2}{x+1} \div \dfrac{x+5}{x-3} = \dfrac{x-2}{x+1} \cdot \dfrac{x-3}{x+5}$ Multiplying by the reciprocal

$= \dfrac{(x-2)(x-3)}{(x+1)(x+5)}$ Multiplying numerators and denominators

or $\dfrac{x^2-5x+6}{x^2+6x+5}$

13. $\dfrac{a^2-1}{a-1} \div \dfrac{a^2-2a+1}{a+1} = \dfrac{a^2-1}{a-1} \cdot \dfrac{a+1}{a^2-2a+1}$ Multiplying by the reciprocal

$= \dfrac{(a^2-1)(a+1)}{(a-1)(a^2-2a+1)}$ Multiplying numerators and denominators

$= \dfrac{(a+1)(a-1)(a+1)}{(a-1)(a-1)(a-1)}$ Factoring numerator and denominator

$= \dfrac{a-1}{a-1} \cdot \dfrac{(a+1)(a+1)}{(a-1)(a-1)}$ Factoring the fractional expression

$= \dfrac{(a+1)(a+1)}{(a-1)(a-1)},$ "Removing" a factor of 1

or $\dfrac{a^2+2a+1}{a^2-2a+1}$ Multiplying out numerator and denominator

DO EXERCISES 16 AND 17.

EXERCISE SET 6.1

i For each rational function, find the function values indicated.

1. $v(t) = \dfrac{4t^2-5t+2}{t+3}$; $v(0)$, $v(3)$, $v(7)$

2. $s(x) = \dfrac{5x^2+4x-12}{6-x}$; $s(4)$, $s(-1)$, $s(3)$

3. $r(y) = \dfrac{3y^3-2y}{y-5}$; $r(0)$, $r(4)$, $r(5)$

4. $f(r) = \dfrac{\pi r^2+2\pi rh}{r-1}$; (use 3.14 for π) $f(2)$, $f(5)$, $f(1)$

ii, **iii** Multiply to obtain equivalent expressions. Do not simplify.

5. $\dfrac{3x}{3x} \cdot \dfrac{x+1}{x+3}$

6. $\dfrac{4-y^2}{6-y} \cdot \dfrac{-1}{-1}$

7. $\dfrac{t-3}{t+2} \cdot \dfrac{t+3}{t+3}$

8. $\dfrac{p-4}{p-5} \cdot \dfrac{p+5}{p+5}$

iv Simplify by removing a factor of 1.

9. $\dfrac{9y^2}{15y}$

10. $\dfrac{6x^3}{18x^2}$

11. $\dfrac{2a-6}{2}$

12. $\dfrac{3a-6}{3}$

13. $\dfrac{4y-12}{4y+12}$ **14.** $\dfrac{8x+16}{8x-16}$ **15.** $\dfrac{t^2-16}{t^2-8t+16}$ **16.** $\dfrac{p^2-25}{p^2+10p+25}$

v Multiply. Simplify by removing a factor of 1 if possible.

17. $\dfrac{x^2-16}{x^2}\cdot\dfrac{x^2-4x}{x^2-x-12}$ **18.** $\dfrac{y^2+10y+25}{y^2-9}\cdot\dfrac{y+3}{y+5}$ **19.** $\dfrac{y^2-16}{2y+6}\cdot\dfrac{y+3}{y-4}$

20. $\dfrac{m^2-n^2}{4m+4n}\cdot\dfrac{m+n}{m-n}$ **21.** $\dfrac{x^2-2x-35}{2x^3-3x^2}\cdot\dfrac{4x^3-9x}{7x-49}$ **22.** $\dfrac{y^2-10y+9}{y^2-1}\cdot\dfrac{y+4}{y^2-5y-36}$

23. $\dfrac{c^3+8}{c^2-4}\cdot\dfrac{c^2-4c+4}{c^2-2c+4}$ **24.** $\dfrac{x^3-27}{x^2-9}\cdot\dfrac{x^2-6x+9}{x^2+3x+9}$ **25.** $\dfrac{x^2-y^2}{x^3-y^3}\cdot\dfrac{x^2+xy+y^2}{x^2+2xy+y^2}$

26. $\dfrac{4x^2-9y^2}{8x^3-27y^3}\cdot\dfrac{4x^2+6xy+9y^2}{4x^2+12xy+9y^2}$

vi Divide. Simplify by removing a factor of 1 if possible.

27. $\dfrac{3y+15}{y}\div\dfrac{y+5}{y}$ **28.** $\dfrac{6x+12}{x}\div\dfrac{x+2}{x^3}$ **29.** $\dfrac{y^2-9}{y}\div\dfrac{y+3}{y+2}$

30. $\dfrac{x^2-4}{x}\div\dfrac{x-2}{x+4}$ **31.** $\dfrac{4a^2-1}{a^2-4}\div\dfrac{2a-1}{a-2}$ **32.** $\dfrac{25x^2-4}{x^2-9}\div\dfrac{5x-2}{x+3}$

33. $\dfrac{x^2-16}{x^2-10x+25}\div\dfrac{3x-12}{x^2-3x-10}$ **34.** $\dfrac{y^2-36}{y^2-8y+16}\div\dfrac{3y-18}{y^2-y-12}$ **35.** $\dfrac{y^3+3y}{y^2-9}\div\dfrac{y^2+5y-14}{y^2+4y-21}$

36. $\dfrac{a^3+4a}{a^2-16}\div\dfrac{a^2+8a+15}{a^2+a-20}$ **37.** $\dfrac{x^3-64}{x^3+64}\div\dfrac{x^2-16}{x^2-4x+16}$ **38.** $\dfrac{8y^3-27}{64y^3-1}\div\dfrac{4y^2-9}{16y^2+4y+1}$

Perform the indicated operations and simplify.

39. $\left[\dfrac{r^2-4s^2}{r+2s}\div(r+2s)\right]\cdot\dfrac{2s}{r-2s}$ **40.** $\left[\dfrac{d^2-d}{d^2-6d+8}\cdot\dfrac{d-2}{d^2+5d}\right]\div\dfrac{5d}{d^2-9d+20}$

Multiply or divide.

41. $\dfrac{834x}{y-427.2}\cdot\dfrac{26.3x}{y+427.2}$ **42.** $\dfrac{527}{x+93.87}\div\dfrac{x-93.87}{468}$

43. $\dfrac{0.0049t}{t+0.007}\cdot\dfrac{27{,}000t}{t-0.007}$ **44.** $\dfrac{y+924.6}{0.003}\div\dfrac{0.421}{y-924.6}$

Simplify.

45. $\dfrac{x(x+1)-2(x+3)}{(x+1)(x+2)(x+3)}$ **46.** $\dfrac{2x-5(x+2)-(x-2)}{x^2-4}$ **47.** $\dfrac{m^2-t^2}{m^2+t^2+m+t+2mt}$

48. $\dfrac{a^3-2a^2+2a-4}{a^3-2a^2-3a+6}$ **49.** $\dfrac{x^3+x^2-y^3-y^2}{x^2-2xy+y^2}$ **50.** $\dfrac{u^6+v^6+2u^3v^3}{u^3-v^3+u^2v-uv^2}$

51. $\dfrac{x^5-x^3+x^2-1-(x^3-1)(x+1)^2}{(x^2-1)^2}$

52. Here is a number "trick." You tell a friend to think of a number, then add 3, double the result, then subtract 2, and finally divide by 2. Your friend tells you the answer. You then subtract 2 from that result and announce the original number.

How does this work? It depends on equivalent rational expressions. Here is what you had your friend do.

Original number

Double Add 3 Subtract 2

$$\frac{2(x+3)-2}{2}$$

Divide by 2

This expression can be simplified to $x+2$, so the expressions represent the same number (have the same function value) for every x. You knew the simplified expression, so you simply subtracted 2 to get x, the original number.

Make up two or three other such number "tricks."

OBJECTIVES

After finishing Section 6.2, you should be able to:

- **i** **Find the LCM of several numbers by factoring.**
- **ii** **Add numbers (fractional notation) by first finding a common denominator.**
- **iii** **Find the LCM of several algebraic expressions by factoring.**

Find the LCM by factoring.

1. 18, 30

6.2

LEAST COMMON MULTIPLES

i To add rational expressions when denominators are different, we first find a common denominator. Let us review the procedure in arithmetic first. To do the addition

$$\frac{5}{42}+\frac{7}{12}$$

we first find a common denominator. We look for a number that is a multiple of both 42 and 12 (a common multiple). Usually we try to get the smallest such number, or the *least common multiple* (LCM). To find the LCM we factor both numbers completely (into primes).

$$42 = 2\cdot 3\cdot 7 \longleftarrow \text{Any multiple of 42 has these factors.}$$
$$12 = 2\cdot 2\cdot 3 \longleftarrow \text{Any multiple of 12 has these factors.}$$

The LCM is the number that has 2 as a factor twice, 3 as a factor once, and 7 as a factor once. The LCM is $2\cdot 2\cdot 3\cdot 7$, or 84.

To obtain the LCM we use each factor the greatest number of times it occurs in any one factorization.

EXAMPLE 1 Find the LCM of 18 and 24.

$$\left.\begin{aligned}18 &= 3\cdot 3\cdot 2\\ 24 &= 2\cdot 2\cdot 2\cdot 3\end{aligned}\right\}\quad \text{The LCM is } 3\cdot 3\cdot 2\cdot 2\cdot 2, \quad \text{or} \quad 72.$$

2. 12, 18, 24

DO EXERCISES 1 AND 2.

ii Now let us return to adding $\frac{5}{42}$ and $\frac{7}{12}$:

$$\frac{5}{42}+\frac{7}{12}=\frac{5}{2\cdot 3\cdot 7}+\frac{7}{2\cdot 2\cdot 3}. \qquad \text{Factoring denominators}$$

The LCM is $2\cdot 2\cdot 3\cdot 7$. To get this LCM in the first denominator we need a 2. In the second denominator we need a 7. We multiply by 1, as follows:

$$\frac{5}{2\cdot 3\cdot 7}\cdot\frac{2}{2}+\frac{7}{2\cdot 2\cdot 3}\cdot\frac{7}{7}=\frac{10}{2\cdot 2\cdot 3\cdot 7}+\frac{49}{2\cdot 2\cdot 3\cdot 7}$$
$$=\frac{59}{2\cdot 2\cdot 3\cdot 7}$$
$$=\frac{59}{84}$$

Multiplying the first fraction by $\frac{2}{2}$ gave us a denominator that is the LCM. Multiplying the second fraction by $\frac{7}{7}$ also gave us a denominator that is the LCM. When we have a common denominator, we add the numerators.

DO EXERCISES 3 AND 4.

iii LCMs of Algebraic Expressions

To find the LCM of several algebraic expressions we proceed as in arithmetic, beginning by factoring the expressions.

EXAMPLE 2 Find the LCM of $12xy^2$ and $15x^3y$.

We factor each expression completely.

$$12xy^2 = 2 \cdot 2 \cdot 3 \cdot x \cdot y \cdot y$$

$$15x^3y = 3 \cdot 5 \cdot x \cdot x \cdot x \cdot y$$

$$\text{LCM} = 2 \cdot 2 \cdot 3 \cdot 5 \cdot x \cdot x \cdot x \cdot y \cdot y = 60x^3y^2$$

Remember: To find the LCM, use each factor the greatest number of times it occurs in any one factorization.

EXAMPLE 3 Find the LCM of $x^2 + 2x + 1$, $5x^2 - 5x$, and $x^2 - 1$.

$$\left.\begin{aligned} x^2 + 2x + 1 &= (x+1)(x+1) \\ 5x^2 - 5x &= 5x(x-1) \\ x^2 - 1 &= (x+1)(x-1) \end{aligned}\right\} \quad \text{Factoring}$$

$$\text{LCM} = 5x(x+1)(x+1)(x-1)$$

EXAMPLE 4 Find the LCM of $x^2 - y^2$, $x^3 + y^3$, and $x^2 + 2xy + y^2$.

$$\begin{aligned} x^2 - y^2 &= (x-y)(x+y) \\ x^3 + y^3 &= (x+y)(x^2 - xy + y^2) \\ x^2 + 2xy + y^2 &= (x+y)(x+y) \\ \text{LCM} &= (x-y)(x+y)(x+y)(x^2 - xy + y^2) \end{aligned}$$

DO EXERCISES 5–7.

The additive inverse of an LCM is also an LCM. For example, if $(x+2)(x-3)$ is an LCM, then $-(x+2)(x-3)$ is also an LCM. We can name the latter

$$(x+2)(-1)(x-3), \quad \text{or} \quad (x+2)(3-x).$$

If, when we are finding LCMs, factors that are additive inverses occur, we do not use them both. For example, if $(a-b)$ occurs in one factorization and $(b-a)$ occurs in another, we do not use them both since they are additive inverses.

EXAMPLE 5 Find the LCM of $x^2 - y^2$ and $3y - 3x$.

$$x^2 - y^2 = (x+y)(x-y)$$

$$3y - 3x = 3(y-x)$$

We can use $(x-y)$ or $(y-x)$ but we do not use both.

$$\text{LCM} = 3(x+y)(x-y), \quad \text{or} \quad 3(x+y)(y-x)$$

DO EXERCISE 8.

Add, first finding the LCM of the denominators.

3. $\dfrac{5}{12} + \dfrac{11}{30}$

4. $\dfrac{7}{12} + \dfrac{13}{18} + \dfrac{1}{24}$

Find the LCM.

5. a^2b^2, $5a^3b$

6. $y^2 + 7y + 12$, $y^2 + 8y + 16$, $y + 4$

7. $x^2 - 9$, $x^3 - x^2 - 6x$, $2x^2$

Find the LCM.

8. $a^2 - b^2$, $2b - 2a$

EXERCISE SET 6.2

i Find the LCM by factoring.

1. 12, 18 **2.** 15, 20 **3.** 18, 48 **4.** 45, 54 **5.** 24, 36

6. 30, 75 **7.** 9, 15, 5 **8.** 27, 35, 63 **9.** 24, 36, 42 **10.** 24, 42, 60

ii Add, first finding the LCM of the denominators.

11. $\frac{5}{6}+\frac{4}{15}$ **12.** $\frac{5}{36}+\frac{5}{24}$ **13.** $\frac{7}{12}+\frac{11}{18}$ **14.** $\frac{11}{30}+\frac{19}{75}$ **15.** $\frac{3}{4}+\frac{2}{5}+\frac{1}{6}$ **16.** $\frac{5}{8}+\frac{7}{12}+\frac{11}{40}$

iii Find the LCM.

17. $8x^2, 12x^3$ **18.** $4y^2, 24y^3$ **19.** $12x^2y, 4xy$

20. $18r^2s, 12rs^3$ **21.** $15ab^2, 3ab, 10a^3b$ **22.** $6x^2y^2, 9x^3y, 15y^3$

23. $a+b, a-b$ **24.** $x-4, x+4$ **25.** $3(y-2), 6(2-y)$

26. $5(y-1), 10(1-y)$ **27.** $3y+9, y^2-9$ **28.** $a^2-b^2, ab+b^2$

29. $5y-15, y^2-6y+9$ **30.** $4x-16, x^2-8x+16$ **31.** $(a+1), (a-1)^2, a^2-1$

32. $(x-2), (x+2)^2, x^2-4$ **33.** $x^2-4, 2-x$ **34.** $y^2-9, 3-y$

35. $x^2+10x+25, x^2+2x-15$ **36.** $y^2+8x+16, y^2-3y-28$ **37.** $2r^2-5r-12, 3r^2-13r+4$

38. $3x^2-4x-4, 4x^2-5x-6$

Problem-solving practice

39. A kitchen floor is 3 ft longer than it is wide. The area of the floor is 180 ft^2. Find the dimensions of the kitchen.

40. On an 18-hole golf course there are par-three holes, par-four holes, and par-five holes. A golfer who shoots par on every hole has a total score of 70. If there are twice as many par-four holes as par-three holes, how many of each type of hole are there on the golf course?

○

Find the LCM.

41. $2x^2-5x-3, 2x^2-x-1, x^2-6x+9$ **42.** $3x^2+4x-4, 2x^2+7x+6, x^2-4x+4$

43. $x^8-x^4, x^5-x^2, x^5-x^3, x^5+x^2$

44. The LCM of two expressions is $8a^4b^7$. One of the expressions is $2a^3b^7$. List all the possibilities for the other expression.

OBJECTIVES

After finishing Section 6.3, you should be able to:

i Add and subtract fractional expressions having the same denominator.

ii Add and subtract fractional expressions whose denominators are additive inverses of each other.

iii Add and subtract fractional expressions having different denominators.

6.3 ADDITION AND SUBTRACTION

Addition and subtraction of rational expressions are done with the same methods that we used for fractional notation in arithmetic.

i When Denominators Are the Same

If the denominators of two or more rational expressions are the same, we keep that denominator and add the numerators. That procedure works for rational numbers. It also works for other real-number replacements in rational expressions, and it will always give us equivalent expressions.

When denominators are the same, we add or subtract the numerators and keep the same denominator.

EXAMPLE 1 Add: $\dfrac{3+x}{x}+\dfrac{4}{x}$.

$$\frac{3+x}{x}+\frac{4}{x}=\frac{7+x}{x}$$ ⟵ This expression does *not* simplify to 7.

Example 1 shows that

$$\frac{3+x}{x}+\frac{4}{x} \quad \text{and} \quad \frac{7+x}{x}$$

are equivalent expressions. They name the same number for all replacements except 0. As functions

$$\frac{3+x}{x}+\frac{4}{x} \quad \text{and} \quad \frac{7+x}{x}$$

are the same function.

EXAMPLE 2 Add: $\dfrac{4x^2-5xy}{x^2-y^2}+\dfrac{2xy-y^2}{x^2-y^2}$.

$$\frac{4x^2-5xy}{x^2-y^2}+\frac{2xy-y^2}{x^2-y^2}=\frac{4x^2-3xy-y^2}{x^2-y^2}$$ Adding numerators

$$=\frac{(4x+y)(x-y)}{(x+y)(x-y)}$$ Factoring numerator and denominator

$$=\frac{x-y}{x-y}\cdot\frac{4x+y}{x+y}$$ Factoring the fractional expression

$$=\frac{4x+y}{x+y}$$ "Removing" a factor of 1

DO EXERCISES 1 AND 2.

EXAMPLE 3 Subtract: $\dfrac{4x+5}{x+3}-\dfrac{x-2}{x+3}$.

$$\frac{4x+5}{x+3}-\frac{x-2}{x+3}=\frac{4x+5-(x-2)}{x+3}$$ Subtracting numerators

$$=\frac{4x+5-x+2}{x+3}$$

$$=\frac{3x+7}{x+3}$$

The parentheses in the numerator in Example 3 are important. If they are forgotten, you may incorrectly subtract only *part* of the numerator $x-2$.

DO EXERCISES 3 AND 4.

ii When Denominators Are Additive Inverses of Each Other

When one denominator is the additive inverse of the other, we multiply one expression by $-1/-1$. That gives us a common denominator.

Add.

1. $\dfrac{5+y}{y}+\dfrac{7}{y}$

2. $\dfrac{2x^2+5x-9}{x-5}+\dfrac{x^2-19x+4}{x-5}$

Subtract.

3. $\dfrac{a}{b+2}-\dfrac{b}{b+2}$

4. $\dfrac{4y+7}{x^2+y^2}-\dfrac{3y-5}{x^2+y^2}$

Add.

5. $\frac{b}{3b} + \frac{b^3}{-3b}$

6. $\frac{3x^2 + 4}{x - 5} + \frac{x^2 - 7}{5 - x}$

Subtract.

7. $\frac{3}{4y} - \frac{7x}{-4y}$

8. $\frac{4x^2}{2x - y} - \frac{7x^2}{y - 2x}$

EXAMPLE 4 Add: $\frac{a}{2a} + \frac{a^3}{-2a}$.

$$\frac{a}{2a} + \frac{a^3}{-2a} = \frac{a}{2a} + \frac{-1}{-1} \cdot \frac{a^3}{-2a} \quad \text{Multiplying by } \frac{-1}{-1}\text{, which is 1}$$

$$= \frac{a}{2a} + \frac{-a^3}{2a}$$

$$= \frac{a - a^3}{2a} \quad \text{Adding numerators}$$

$$= \frac{a(1 - a^2)}{2a} \quad \text{Factoring numerator and denominator}$$

$$= \frac{a}{a} \cdot \frac{1 - a^2}{2} \quad \text{Factoring the expression}$$

$$= \frac{1 - a^2}{2} \quad \text{"Removing" a factor of 1}$$

DO EXERCISES 5 AND 6.

EXAMPLE 5 Subtract: $\frac{5x}{x - 2y} - \frac{3y - 7}{2y - x}$.

$$\frac{5x}{x - 2y} - \frac{3y - 7}{2y - x} = \frac{5x}{x - 2y} - \frac{-1}{-1} \cdot \frac{3y - 7}{2y - x}$$

$$= \frac{5x}{x - 2y} - \frac{7 - 3y}{x - 2y}$$

$$= \frac{5x - (7 - 3y)}{x - 2y} \quad \text{Subtracting numerators}$$

$$= \frac{5x - 7 + 3y}{x - 2y}.$$

DO EXERCISES 7 AND 8.

iii When Denominators Are Different

When denominators are different, but not additive inverses of each other, we first find a common denominator, using the LCM, and then add or subtract numerators.

EXAMPLE 6 Add: $\frac{2a}{5} + \frac{3b}{2a}$.

First find the LCM of the denominators.

$$\begin{matrix} 5 \\ 2a \end{matrix} \qquad \text{The LCM is } 5 \cdot 2a, \text{ or } 10a.$$

Now we multiply each expression by 1. We choose whatever symbol for 1 will give us the LCM in each denominator. In this case we use $2a/2a$ and $5/5$:

$$\frac{2a}{5} \cdot \frac{2a}{2a} + \frac{3b}{2a} \cdot \frac{5}{5} = \frac{4a^2}{10a} + \frac{15b}{10a} = \frac{4a^2 + 15b}{10a}.$$

Multiplying by $2a/2a$ in the first term gave us a denominator of $10a$. Multiplying by $5/5$ in the second term also gave us a denominator of $10a$.

DO EXERCISE 9.

Add.

9. $\frac{3x}{7} + \frac{4y}{3x}$

EXAMPLE 7 Add: $\frac{3x^2 + 3xy}{x^2 - y^2} + \frac{2 - 3x}{x - y}$.

We first find the LCM of the denominators.

$$\left.\begin{aligned} x^2 - y^2 &= (x + y)(x - y) \\ x - y &= x - y \end{aligned}\right\} \quad \text{The LCM is } (x + y)(x - y).$$

We now multiply by 1 to get the LCM in the second expression. Then we add and simplify if possible.

$$\frac{3x^2 + 3xy}{(x + y)(x - y)} + \frac{2 - 3x}{x - y} \cdot \frac{x + y}{x + y} \quad \text{Multiplying by 1 to get the LCM}$$

$$= \frac{3x^2 + 3xy}{(x + y)(x - y)} + \frac{(2 - 3x)(x + y)}{(x - y)(x + y)}$$

$$= \frac{3x^2 + 3xy}{(x + y)(x - y)} + \frac{2x + 2y - 3x^2 - 3xy}{(x - y)(x + y)} \quad \text{Multiplying out the numerator}$$

$$= \frac{3x^2 + 3xy + 2x + 2y - 3x^2 - 3xy}{(x + y)(x - y)} \quad \text{Adding numerators}$$

$$= \frac{2x + 2y}{(x + y)(x - y)} = \frac{2(x + y)}{(x + y)(x - y)} \quad \text{Combining like terms and then factoring the numerator}$$

$$= \frac{x + y}{x + y} \cdot \frac{2}{x - y} = \frac{2}{x - y} \quad \text{Simplifying by "removing" a factor of 1}$$

DO EXERCISES 10 AND 11.

Add.

10. $\frac{2xy - 2x^2}{x^2 - y^2} + \frac{2x + 3}{x + y}$

11. $\frac{1}{3y} + \frac{4y}{y^2 - 1} + \frac{7}{y - 1}$

EXAMPLE 8 Subtract: $\frac{2y + 1}{y^2 - 7y + 6} - \frac{y + 3}{y^2 - 5y - 6}$.

$$\frac{2y + 1}{y^2 - 7y + 6} - \frac{y + 3}{y^2 - 5y - 6}$$

$$= \frac{2y + 1}{(y - 6)(y - 1)} - \frac{y + 3}{(y - 6)(y + 1)} \quad \text{The LCM is } (y - 6)(y - 1)(y + 1).$$

$$= \frac{2y + 1}{(y - 6)(y - 1)} \cdot \frac{y + 1}{y + 1} - \frac{y + 3}{(y - 6)(y + 1)} \cdot \frac{y - 1}{y - 1}$$

$$= \frac{(2y + 1)(y + 1) - (y + 3)(y - 1)}{(y - 6)(y - 1)(y + 1)}$$

$$= \frac{2y^2 + 3y + 1 - (y^2 + 2y - 3)}{(y - 6)(y - 1)(y + 1)}$$

$$= \frac{2y^2 + 3y + 1 - y^2 - 2y + 3}{(y - 6)(y - 1)(y + 1)} = \frac{y^2 + y + 4}{(y - 6)(y - 1)(y + 1)}$$

Subtract.

12. $\dfrac{a}{a+3} - \dfrac{a-4}{a}$

13. $\dfrac{4y-5}{y^2-7y+12} - \dfrac{y+7}{y^2+2y-15}$

Do this calculation.

14. $\dfrac{8x}{x^2-1} + \dfrac{2}{1-x} - \dfrac{4}{x+1}$

We may not always multiply out a numerator or denominator. In fact, it will be helpful when we solve equations not to multiply out denominators.

DO EXERCISES 12 AND 13.

EXAMPLE 9 Calculate: $\dfrac{2x}{x^2-4} + \dfrac{5}{2-x} - \dfrac{1}{2+x}$.

$$\frac{2x}{x^2-4} + \frac{5}{2-x} - \frac{1}{2+x} = \frac{2x}{(x-2)(x+2)} + \frac{5}{2-x} - \frac{1}{2+x}$$

$$= \frac{2x}{(x-2)(x+2)} + \frac{-1}{-1}\cdot\frac{5}{(2-x)} - \frac{1}{x+2}$$

$$= \frac{2x}{(x-2)(x+2)} + \frac{-5}{x-2} - \frac{1}{x+2}$$

The LCM is $(x-2)(x+2)$.

$$= \frac{2x}{(x-2)(x+2)} + \frac{-5}{x-2}\cdot\frac{x+2}{x+2} - \frac{1}{x+2}\cdot\frac{x-2}{x-2}$$

$$= \frac{2x-5(x+2)-(x-2)}{(x-2)(x+2)}$$

$$= \frac{2x-5x-10-x+2}{(x-2)(x+2)}$$

$$= \frac{-4x-8}{(x-2)(x+2)} = \frac{-4(x+2)}{(x-2)(x+2)}$$

$$= \frac{-4}{x-2}\cdot\frac{x+2}{x+2} = \frac{-4}{x-2}$$

DO EXERCISE 14.

EXERCISE SET 6.3

i, **ii**, **iii** Perform the indicated operations. Simplify when possible. If a denominator has three or more factors (other than monomials), leave it factored.

1. $\dfrac{a-3b}{a+b} + \dfrac{a+5b}{a+b}$

2. $\dfrac{x-5y}{x+y} + \dfrac{x+7y}{x+y}$

3. $\dfrac{4y+3}{y-2} - \dfrac{y-2}{y-2}$

4. $\dfrac{3t+2}{t-4} - \dfrac{t-4}{t-4}$

5. $\dfrac{a^2}{a-b} + \dfrac{b^2}{b-a}$

6. $\dfrac{s^2}{r-s} + \dfrac{r^2}{s-r}$

7. $\dfrac{3}{x} - \dfrac{8}{-x}$

8. $\dfrac{2}{a} - \dfrac{5}{-a}$

9. $\dfrac{2x-10}{x^2-25} - \dfrac{5-x}{25-x^2}$

10. $\dfrac{y-9}{y^2-16} - \dfrac{7-y}{16-y^2}$

11. $\dfrac{y-2}{y+4} + \dfrac{y+3}{y-5}$

12. $\dfrac{x-2}{x+3} + \dfrac{x+2}{x-4}$

13. $\dfrac{4xy}{x^2-y^2} + \dfrac{x-y}{x+y}$

14. $\dfrac{5ab}{a^2-b^2} + \dfrac{a+b}{a-b}$

15. $\dfrac{9x+2}{3x^2-2x-8} + \dfrac{7}{3x^2+x-4}$

16. $\dfrac{3y+2}{2y^2-y-10} + \dfrac{8}{2y^2-7y+5}$

17. $\dfrac{4}{x+1} + \dfrac{x+2}{x^2-1} + \dfrac{3}{x-1}$

18. $\dfrac{-2}{y+2} + \dfrac{5}{y-2} + \dfrac{y+3}{y^2-4}$

19. $\dfrac{x-1}{3x+15} - \dfrac{x+3}{5x+25}$

20. $\dfrac{y-2}{4y+8} - \dfrac{y+6}{5y+10}$

21. $\dfrac{5ab}{a^2-b^2} - \dfrac{a-b}{a+b}$

22. $\dfrac{6xy}{x^2-y^2}-\dfrac{x+y}{x-y}$

23. $\dfrac{3y}{y^2-7y+10}-\dfrac{2y}{y^2-8y+15}$

24. $\dfrac{5x}{x^2-6x+8}-\dfrac{3x}{x^2-x-12}$

25. $\dfrac{y}{y^2-y-20}+\dfrac{2}{y+4}$

26. $\dfrac{6}{y^2+6y+9}+\dfrac{5}{y^2-9}$

27. $\dfrac{3y+2}{y^2+5y-24}+\dfrac{7}{y^2+4y-32}$

28. $\dfrac{3y+2}{y^2-7y+10}+\dfrac{2y}{y^2-8y+15}$

29. $\dfrac{3x-1}{x^2+2x-3}-\dfrac{x+4}{x^2-9}$

30. $\dfrac{3p-2}{p^2+2p-24}-\dfrac{p-3}{p^2-16}$

31. $\dfrac{1}{x+1}-\dfrac{x}{x-2}+\dfrac{x^2+2}{x^2-x-2}$

32. $\dfrac{2}{y+3}-\dfrac{y}{y-1}+\dfrac{y^2+2}{y^2+2y-3}$

33. $\dfrac{4x}{x^2-1}+\dfrac{3x}{1-x}-\dfrac{4}{x-1}$

34. $\dfrac{5y}{1-2y}-\dfrac{2y}{2y+1}+\dfrac{3}{4y^2-1}$

Problem-solving practice

35. There are 50 dimes in a roll of dimes, 40 nickels in a roll of nickels, and 40 quarters in a roll of quarters. Robert has a total of 12 rolls of coins with a total value of \$70.00. If he has three more rolls of nickels than dimes, how many of each roll of coins does he have?

36. Debbie wants to buy tapes to record her favorite music. She needs some 30-min tapes and some 60-min tapes. If she buys 12 tapes with a total recording time of 10 hr, how many of each length tape did she buy?

Perform the indicated operations and simplify.

37. $2x^{-2}+3x^{-2}y^{-2}-7xy^{-1}$

38. $5(x-3)^{-1}+4(x+3)^{-1}-2(x+3)^{-2}$

39. $4(y-1)(2y-5)^{-1}+5(2y+3)(5-2y)^{-1}+(y-4)(2y-5)^{-1}$

Simplify, using $A=x+y$ and $B=x-y$.

40. $\dfrac{A+B}{A-B}-\dfrac{A-B}{A+B}$

41. $\left(\dfrac{1}{A}+\dfrac{x}{B}\right)\div\left(\dfrac{1}{B}-\dfrac{x}{A}\right)$

6.4

COMPLEX FRACTIONAL EXPRESSIONS

OBJECTIVE

After finishing Section 6.4, you should be able to:

i **Simplify complex fractional expressions.**

i A *complex fractional expression* is one that has one or more fractional expressions somewhere in its numerator or denominator. The following are examples of complex fractional expressions:

$$\frac{x}{x-\frac{1}{3}},\quad \frac{2x-\frac{4x}{3y}}{\frac{5x^2+2x}{6y^2}},\quad \frac{\frac{1}{a}+\frac{1}{b}}{\frac{1}{a}-\frac{1}{b}},\quad \frac{\frac{5}{x}}{\frac{x}{y}}.$$

To simplify a complex fractional expression:

1. **Add or subtract, as necessary, to get a single fractional expression in the numerator.**
2. **Add or subtract, as necessary, to get a single fractional expression in the denominator.**
3. **Divide the numerator by the denominator.**
4. **If possible, simplify by removing a factor of 1.**

EXAMPLE 1 Simplify: $\dfrac{x+\frac{1}{5}}{x-\frac{1}{3}}$.

We have

$$\frac{x+\frac{1}{5}}{x-\frac{1}{3}} = \frac{x \cdot \frac{5}{5} + \frac{1}{5}}{x - \frac{1}{3}} = \frac{\frac{5x+1}{5}}{x-\frac{1}{3}}$$

Finding the LCM in the numerator, multiplying by 1, and adding

$$= \frac{\frac{5x+1}{5}}{x \cdot \frac{3}{3} - \frac{1}{3}} = \frac{\frac{5x+1}{5}}{\frac{3x-1}{3}}$$

Finding the LCM in the denominator, multiplying by 1, and subtracting

$$= \frac{5x+1}{5} \cdot \frac{3}{3x-1}$$

Multiplying by the reciprocal of the denominator

$$= \frac{15x+3}{15x-5}$$

Multiplying numerators and denominators

Simplifying by removing a factor of 1 is not possible in this case.

EXAMPLE 2 Simplify: $\dfrac{1+\frac{1}{x}}{1-\frac{1}{x^2}}$.

$$\frac{1+\frac{1}{x}}{1-\frac{1}{x^2}} = \frac{\frac{x}{x}+\frac{1}{x}}{\frac{x^2}{x^2}-\frac{1}{x^2}}$$

Finding the LCM and multiplying by 1

$$= \frac{\frac{x+1}{x}}{\frac{x^2-1}{x^2}}$$

Adding in the numerator and subtracting in the denominator

$$= \frac{x+1}{x} \cdot \frac{x^2}{x^2-1}$$

Multiplying by the reciprocal of the denominator

Then

$$\frac{x+1}{x} \cdot \frac{x^2}{x^2-1} = \frac{(x+1) \cdot x^2}{x(x^2-1)}$$

$$= \frac{x(x+1)}{x(x+1)} \cdot \frac{x}{x-1}$$

Factoring the fractional expression

$$= \frac{x}{x-1}$$

Removing a factor of 1

Example 2 shows that the expressions

$$\frac{1+\frac{1}{x}}{1-\frac{1}{x^2}} \quad \text{and} \quad \frac{x}{x-1}$$

are equivalent. They name the same number for all replacements except 0, 1, and -1. As functions, they are also the same except that the domain does not include 0, 1, or -1.

DO EXERCISES 1 AND 2.

EXAMPLE 3 Simplify: $\dfrac{\frac{1}{a}+\frac{1}{b}}{\frac{1}{a^3}+\frac{1}{b^3}}$.

$$\frac{\frac{1}{a}+\frac{1}{b}}{\frac{1}{a^3}+\frac{1}{b^3}} = \frac{\frac{1}{a}\cdot\frac{b}{b}+\frac{1}{b}\cdot\frac{a}{a}}{\frac{1}{a^3}\cdot\frac{b^3}{b^3}+\frac{1}{b^3}\cdot\frac{a^3}{a^3}} = \frac{\frac{b}{ab}+\frac{a}{ab}}{\frac{b^3}{a^3b^3}+\frac{a^3}{a^3b^3}}$$

$$= \frac{\frac{b+a}{ab}}{\frac{b^3+a^3}{a^3b^3}} \qquad \text{Adding in the numerator and denominator}$$

$$= \frac{b+a}{ab}\cdot\frac{a^3b^3}{b^3+a^3} \qquad \text{Multiplying by the reciprocal of the denominator}$$

$$= \frac{(b+a)a^3b^3}{ab(b^3+a^3)} = \frac{(b+a)\cdot ab\cdot a^2b^2}{ab(b+a)(b^2-ab+a^2)}$$

$$= \frac{(b+a)ab}{(b+a)ab}\cdot\frac{a^2b^2}{b^2-ab+a^2}$$

$$= \frac{a^2b^2}{b^2-ab+a^2}$$

DO EXERCISES 3 AND 4.

Simplify.

1. $\dfrac{y+\frac{1}{2}}{y-\frac{1}{7}}$

2. $\dfrac{1-\frac{1}{x}}{1-\frac{1}{x^2}}$

Simplify.

3. $\dfrac{\frac{1}{a}+\frac{1}{b}}{\frac{1}{a}-\frac{1}{b}}$

4. $\dfrac{\frac{1}{a}-\frac{1}{b}}{\frac{1}{a^3}-\frac{1}{b^3}}$

EXERCISE SET 6.4

i Simplify.

1. $\dfrac{\frac{1}{x}+4}{\frac{1}{x}-3}$

2. $\dfrac{\frac{1}{y}+7}{\frac{1}{y}-5}$

3. $\dfrac{x-\frac{1}{x}}{x+\frac{1}{x}}$

4. $\dfrac{y+\frac{1}{y}}{y-\frac{1}{y}}$

5. $\dfrac{\frac{3}{x}+\frac{4}{y}}{\frac{4}{x}-\frac{3}{y}}$

6. $\dfrac{\dfrac{2}{y}+\dfrac{5}{z}}{\dfrac{1}{y}-\dfrac{4}{z}}$

7. $\dfrac{\dfrac{x^2-y^2}{xy}}{\dfrac{x-y}{y}}$

8. $\dfrac{\dfrac{a^2-b^2}{ab}}{\dfrac{a-b}{b}}$

9. $\dfrac{a-\dfrac{3a}{b}}{b-\dfrac{b}{a}}$

10. $\dfrac{1-\dfrac{2}{3x}}{x-\dfrac{4}{9x}}$

11. $\dfrac{\dfrac{1}{a}+\dfrac{1}{b}}{\dfrac{a^2-b^2}{ab}}$

12. $\dfrac{\dfrac{1}{x}-\dfrac{1}{y}}{\dfrac{x^2-y^2}{xy}}$

13. $\dfrac{\dfrac{1}{x+h}-\dfrac{1}{x}}{h}$

14. $\dfrac{\dfrac{1}{a-h}-\dfrac{1}{a}}{h}$

15. $\dfrac{\dfrac{y^2-y-6}{y^2-5y-14}}{\dfrac{y^2+6y+5}{y^2-6y-7}}$

16. $\dfrac{\dfrac{x^2-x-12}{x^2-2x-15}}{\dfrac{x^2+8x+12}{x^2-5x-14}}$

Problem-solving practice

17. The length of one rectangle is 3 less than the length of a second rectangle. The width of the first rectangle is 4 less than the width of the second. If the perimeter of the second rectangle is 1 less than twice the perimeter of the first, what are their respective perimeters?

18. A waitress received \$3.50 in tips on Monday, \$2.75 in tips on Tuesday, and \$4.50 in tips on Wednesday. How much will she have to earn in tips on Thursday if her average tips for the four days is to be \$3.75?

Simplify.

19. $\dfrac{5x^{-1}-5y^{-1}+10x^{-1}y^{-1}}{6x^{-1}-6y^{-1}+12x^{-1}y^{-1}}$

20. $\dfrac{\dfrac{4a}{2a^2-a-1}-\dfrac{4}{a-1}}{\dfrac{1}{a-1}+\dfrac{6}{2a+1}}$

21. $2+\cfrac{2}{2+\cfrac{2}{2+\cfrac{2}{2+\cfrac{2}{x}}}}$

22. $\left[\dfrac{\dfrac{x+3}{x-3}+1}{\dfrac{x+3}{x-3}-1}\right]^4$

23. $(a^2-ab+b^2)^{-1}(a^2b^{-1}+b^2a^{-1})(a^{-2}-b^{-2})(a^{-2}+2a^{-1}b^{-1}+b^{-2})^{-1}$

Find the reciprocal and simplify.

24. $x^2-\dfrac{1}{x}$

25. $\dfrac{1-\dfrac{1}{a}}{a-1}$

26. $\dfrac{a^3+b^3}{a+b}$

27. $1+\cfrac{1}{1+\cfrac{1}{1+\cfrac{1}{1+\cfrac{1}{x}}}}$

28. For $f(x)=\dfrac{1}{1-x}$, find $f(f(x))$ and $f(f(f(x)))$.

Find and simplify $\dfrac{f(x+h)-f(x)}{h}$ for each rational function f.

29. $f(x)=\dfrac{3}{x^2}$

30. $f(x)=\dfrac{5}{x}$

31. $f(x)=\dfrac{1}{1-x}$

32. $f(x)=\dfrac{x}{1+x}$

6.5

SOLVING FRACTIONAL EQUATIONS

i A *fractional equation* is an equation that contains one or more fractional expressions. Here are some examples:

$$\frac{2}{3} - \frac{5}{6} = \frac{1}{x}, \qquad \frac{x-1}{x-5} = \frac{4}{x-5}.$$

To solve a fractional equation, we multiply on both sides by the LCM of all the denominators. This is called *clearing of fractions.*

EXAMPLE 1 Solve: $\frac{2}{3} - \frac{5}{6} = \frac{1}{x}$.

The LCM of all denominators is $6x$, or $2 \cdot 3 \cdot x$. We multiply on both sides of the equation by the LCM:

$$(2 \cdot 3 \cdot x) \cdot \left(\frac{2}{3} - \frac{5}{6}\right) = (2 \cdot 3 \cdot x) \cdot \frac{1}{x} \quad \text{Multiplying by the LCM}$$

$$2 \cdot 3 \cdot x \cdot \frac{2}{3} - 2 \cdot 3 \cdot x \cdot \frac{5}{6} = 2 \cdot 3 \cdot x \cdot \frac{1}{x} \quad \text{Multiplying to remove parentheses}$$

$$\frac{2 \cdot 3 \cdot x \cdot 2}{3} - \frac{2 \cdot 3 \cdot x \cdot 5}{6} = \frac{2 \cdot 3 \cdot x}{x}$$

$$4x - 5x = 6 \quad \text{Simplifying}$$

$$-x = 6$$

$$-1 \cdot x = 6$$

$$x = -6.$$

Check:

$\frac{2}{3} - \frac{5}{6}$	$=$	$\frac{1}{x}$
$\frac{2}{3} - \frac{5}{6}$		$\frac{1}{-6}$
$\frac{4}{6} - \frac{5}{6}$		$-\frac{1}{6}$
$-\frac{1}{6}$		

When clearing of fractions, as in Example 1, be sure to multiply *every* term in the equation by the LCM.

Note that when we *clear of fractions*, all the denominators disappear. Then we have an equation without fractional expressions, which we know how to solve. Note too that we have introduced a new use of the LCM here. Before, we used the LCM in adding or subtracting. Now we are working with equations. We clear of fractions by multiplying on both sides by the LCM.

DO EXERCISE 1.

OBJECTIVE

After finishing Section 6.5, you should be able to:

i Solve fractional equations.

Solve.

1. $\frac{2}{3} + \frac{5}{6} = \frac{1}{x}$

Solve.

2. $\dfrac{x-7}{x-9} = \dfrac{2}{x-9}$

3. $\dfrac{x^2}{x+3} = \dfrac{9}{x+3}$

EXAMPLE 2 Solve: $\dfrac{x+1}{2} - \dfrac{x-3}{3} = 3.$

The LCM of all the denominators is $2 \cdot 3$, or 6. We multiply on both sides of the equation by the LCM:

$$2 \cdot 3 \cdot \left(\frac{x+1}{2} - \frac{x-3}{3}\right) = 2 \cdot 3 \cdot 3 \qquad \text{Multiplying on both sides by the LCM}$$

$$2 \cdot 3 \cdot \frac{x+1}{2} - 2 \cdot 3 \cdot \frac{x-3}{3} = 2 \cdot 3 \cdot 3 \qquad \text{Multiplying to remove parentheses}$$

$$\frac{2 \cdot 3(x+1)}{2} - \frac{2 \cdot 3(x-3)}{3} = 18$$

$$3(x+1) - 2(x-3) = 18$$

$$\left.\begin{aligned} 3x + 3 - 2x + 6 &= 18 \\ x + 9 &= 18 \end{aligned}\right\} \qquad \text{Multiplying and collecting like terms}$$

$$x = 9.$$

Check:

$$\begin{array}{c|c} \dfrac{x+1}{2} - \dfrac{x-3}{3} & = \; 3 \\ \hline \dfrac{9+1}{2} - \dfrac{9-3}{3} & 3 \\ 5 - 2 & \\ 3 & \end{array}$$

In solving fractional equations, we often multiply on both sides by an expression containing a variable. When we do so, we may obtain an equation that is not equivalent to the first one. Thus it is extremely important to check possible solutions in the original equation. They may not check, even though we have made no error.

EXAMPLE 3 Solve: $\dfrac{x-1}{x-5} = \dfrac{4}{x-5}.$

The LCM of the denominators is $x - 5$. We multiply by $x - 5$:

$$(x-5) \cdot \frac{x-1}{x-5} = (x-5) \cdot \frac{4}{x-5}$$

$$x - 1 = 4$$

$$x = 5.$$

Check:

$$\begin{array}{c|c} \dfrac{x-1}{x-5} \; = & \dfrac{4}{x-5} \\ \hline \dfrac{5-1}{5-5} & \dfrac{4}{5-5} \\ \dfrac{4}{0} & \dfrac{4}{0} \end{array}$$

We know that 5 is not a solution of the original equation because it results in division by 0. In fact, the equation has no solution.

DO EXERCISES 2 AND 3.

In solving an equation, when we multiply by an expression containing a variable, we may get an equation having solutions that are not solutions of the original equation.

EXAMPLE 4 Solve: $\dfrac{x^2}{x-2} = \dfrac{4}{x-2}$.

The LCM of the denominators is $x - 2$. We multiply by $x - 2$:

$$(x-2) \cdot \frac{x^2}{x-2} = (x-2) \cdot \frac{4}{x-2}$$

$$x^2 = 4$$

$$x^2 - 4 = 0$$

$$(x+2)(x-2) = 0$$

$$x = -2 \quad \text{or} \quad x = 2 \qquad \text{Using the principle of zero products}$$

Check: For 2:

$$\begin{array}{c|c} \multicolumn{2}{c}{\dfrac{x^2}{x-2} = \dfrac{4}{x-2}} \\ \hline \dfrac{2^2}{2-2} & \dfrac{4}{2-2} \\ \dfrac{4}{0} & \dfrac{4}{0} \end{array}$$

For −2:

$$\begin{array}{c|c} \multicolumn{2}{c}{\dfrac{x^2}{x-2} = \dfrac{4}{x-2}} \\ \hline \dfrac{(-2)^2}{-2-2} & \dfrac{4}{-2-2} \\ \dfrac{4}{-4} & \dfrac{4}{-4} \end{array}$$

The number −2 is a solution, but 2 is not (it results in division by 0).

EXAMPLE 5 Solve: $x + \dfrac{6}{x} = 5$.

The LCM of the denominators is x. We multiply on both sides by x:

$$x\left(x + \frac{6}{x}\right) = 5 \cdot x \qquad \text{Multiplying on both sides by } x$$

$$x^2 + x \cdot \frac{6}{x} = 5x$$

$$x^2 + 6 = 5x \qquad \text{Simplifying}$$

$$x^2 - 5x + 6 = 0 \qquad \text{Getting 0 on one side}$$

$$(x-3)(x-2) = 0 \qquad \text{Factoring}$$

$$x = 3 \quad \text{or} \quad x = 2. \qquad \text{Using the principle of zero products}$$

Check: For 3:

$$\begin{array}{c|c} \multicolumn{2}{c}{x + \dfrac{6}{x} = 5} \\ \hline 3 + \dfrac{6}{3} & 5 \\ 3 + 2 & \\ 5 & \end{array}$$

For 2:

$$\begin{array}{c|c} \multicolumn{2}{c}{x + \dfrac{6}{x} = 5} \\ \hline 2 + \dfrac{6}{2} & 5 \\ 2 + 3 & \\ 5 & \end{array}$$

The solutions are 2 and 3.

DO EXERCISES 4 AND 5.

Solve.

4. $\dfrac{x^2}{x-3} = \dfrac{9}{x-3}$

5. $x - \dfrac{12}{x} = 1$

Solve.

6. $\dfrac{2}{x-1} = \dfrac{3}{x+2}$

7. $\dfrac{2}{x^2-9} + \dfrac{5}{x-3} = \dfrac{3}{x+3}$

EXAMPLE 6 Solve: $\dfrac{2}{x-1} = \dfrac{3}{x+1}$.

The LCM of the denominators is $(x-1)(x+1)$.

$$(x-1)(x+1)\cdot\frac{2}{(x-1)} = (x-1)(x+1)\cdot\frac{3}{x+1} \qquad \text{Multiplying}$$

$$\begin{aligned} 2(x+1) &= 3(x-1) \qquad \text{Simplifying} \\ 2x+2 &= 3x-3 \\ 5 &= x \end{aligned}$$

Check:

$$\begin{array}{c|c} \multicolumn{2}{c}{\dfrac{2}{x-1} = \dfrac{3}{x+1}} \\ \hline \dfrac{2}{5-1} & \dfrac{3}{5+1} \\ \dfrac{2}{4} & \dfrac{3}{6} \\ \dfrac{1}{2} & \dfrac{1}{2} \end{array}$$

The number 5 is the solution.

EXAMPLE 7 Solve: $\dfrac{2}{x+5} + \dfrac{1}{x-5} = \dfrac{16}{x^2-25}$.

The LCM is $(x+5)(x-5)$. We multiply by $(x+5)(x-5)$:

$$(x+5)(x-5)\cdot\left[\frac{2}{x+5} + \frac{1}{x-5}\right] = (x+5)(x-5)\cdot\frac{16}{x^2-25}$$

$$(x+5)(x-5)\cdot\frac{2}{x+5} + (x+5)(x-5)\cdot\frac{1}{x-5} = (x+5)(x-5)\cdot\frac{16}{x^2-25}$$

$$\begin{aligned} 2(x-5) + (x+5) &= 16 \\ 2x - 10 + x + 5 &= 16 \\ 3x - 5 &= 16 \\ 3x &= 21 \\ x &= 7 \end{aligned}$$

Check:

$$\begin{array}{c|c} \multicolumn{2}{c}{\dfrac{2}{x+5} + \dfrac{1}{x-5} = \dfrac{16}{x^2-25}} \\ \hline \dfrac{2}{7+5} + \dfrac{1}{7-5} & \dfrac{16}{7^2-25} \\ \dfrac{2}{12} + \dfrac{1}{2} & \dfrac{16}{49-25} \\ \dfrac{8}{12} & \dfrac{16}{24} \\ \dfrac{2}{3} & \dfrac{2}{3} \end{array}$$

The solution is 7.

DO EXERCISES 6 AND 7.

EXERCISE SET 6.5

i Solve.

1. $\frac{2}{5}+\frac{7}{8}=\frac{y}{20}$
2. $\frac{4}{5}+\frac{1}{3}=\frac{t}{9}$
3. $\frac{1}{3}-\frac{5}{6}=\frac{1}{x}$
4. $\frac{5}{8}-\frac{2}{5}=\frac{1}{y}$
5. $\frac{x}{3}-\frac{x}{4}=12$
6. $\frac{y}{5}-\frac{y}{3}=15$
7. $y+\frac{5}{y}=-6$
8. $x+\frac{4}{x}=-5$
9. $\frac{4}{z}+\frac{2}{z}=3$
10. $\frac{4}{3y}-\frac{3}{y}=\frac{10}{3}$
11. $\frac{x-3}{x+2}=\frac{1}{5}$
12. $\frac{y-5}{y+1}=\frac{3}{5}$
13. $\frac{3}{y+1}=\frac{2}{y-3}$
14. $\frac{4}{x-1}=\frac{3}{x+2}$
15. $\frac{y-1}{y-3}=\frac{2}{y-3}$
16. $\frac{x-2}{x-4}=\frac{2}{x-4}$
17. $\frac{x+1}{x}=\frac{3}{2}$
18. $\frac{y+2}{y}=\frac{5}{3}$
19. $\frac{2}{x}-\frac{3}{x}+\frac{4}{x}=5$
20. $\frac{4}{y}-\frac{6}{y}+\frac{8}{y}=8$
21. $\frac{1}{2}-\frac{4}{9x}=\frac{4}{9}-\frac{1}{6x}$
22. $-\frac{1}{3}-\frac{5}{4y}=\frac{3}{4}-\frac{1}{6y}$
23. $\frac{60}{x}-\frac{60}{x-5}=\frac{2}{x}$
24. $\frac{50}{y}-\frac{50}{y-2}=\frac{4}{y}$
25. $\frac{7}{5x-2}=\frac{5}{4x}$
26. $\frac{5}{y+4}=\frac{3}{y-2}$
27. $\frac{x}{x-2}+\frac{x}{x^2-4}=\frac{x+3}{x+2}$
28. $\frac{3}{y-2}+\frac{2y}{4-y^2}=\frac{5}{y+2}$
29. $\frac{a}{2a-6}-\frac{3}{a^2-6a+9}=\frac{a-2}{3a-9}$
30. $\frac{2}{x+4}+\frac{2x-1}{x^2+2x-8}=\frac{1}{x-2}$
31. $\frac{2x+3}{x-1}=\frac{10}{x^2-1}+\frac{2x-3}{x+1}$
32. $\frac{y}{y+1}+\frac{3y+5}{y^2+4y+3}=\frac{2}{y+3}$

Problem-solving practice

33. There are 70 questions on a test. The questions are either multiple choice, true–false, or fill-ins. There are twice as many true–false as fill-in and 5 fewer multiple choice than true–false. How many of each type of question are there on the test?

34. Find two consecutive even numbers whose product is 288.

○

Solve.

35. $\left(\frac{1}{1+x}+\frac{x}{1-x}\right)\div\left(\frac{x}{1+x}-\frac{1}{1-x}\right)=-1$
36. $\frac{x+3}{x+2}-\frac{x+4}{x+3}=\frac{x+5}{x+4}-\frac{x+6}{x+5}$
37. $\frac{2.315}{y}-\frac{12.6}{17.4}=\frac{6.71}{7}+0.763$
38. $\frac{6.034}{x}-43.17=\frac{0.793}{x}+18.15$
39. $\frac{x^3+8}{x+2}=x^2-2x+4$
40. $\frac{(x-3)^2}{x-3}=x-3$

Equations that are true for all sensible replacements of the variables are called *identities*. Determine which equations are identities.

41. $\frac{x^2+6x-16}{x-2}=x+8$
42. $\frac{x^3+8}{x^2-4}=\frac{x^2-2x+4}{x-2}$
43. $(x+3)^2=x^2+9$
44. $(y-5)^2=y^2-25$
45. $(x+3)^2=x^2+6x+9$
46. $(y-5)^2=y^2-10y+25$

6.6

SOLVING PROBLEMS

Now that we have considered equations containing fractional expressions, we can use that skill in solving certain kinds of problems that we could not have handled before. The problem-solving steps and strategy are the same.

i Problems Using Fractional Notation

EXAMPLE 1 If a certain number is added to 5 times the reciprocal of 2 more than that number, the result is 4. Find the number.

OBJECTIVES

After finishing Section 6.6, you should be able to:

i **Solve problems using fractional equations.**

ii **Solve certain problems involving amounts of work.**

iii **Solve certain problems involving motion.**

1. When 12 is divided by 5 more than a certain number, the result is 3 more than 2 less than the number. Find the number.

This problem is stated in such a way that we can translate directly to an equation.

2. **Translate**

A certain number added to five times the reciprocal of 2 more than the number is 4

$$y \quad + \quad 5 \cdot \frac{1}{y+2} = 4$$

3. **Carry out** We solve as follows:

$$y + 5 \cdot \frac{1}{y+2} = 4$$

$$(y+2) \cdot y + (y+2) \cdot 5 \cdot \frac{1}{y+2} = (y+2) \cdot 4 \qquad \text{Multiplying by the LCM}$$

$$y^2 + 2y + 5 = 4y + 8 \qquad \text{Simplifying}$$

$$y^2 - 2y - 3 = 0 \qquad \text{Collecting like terms all on one side}$$

$$(y-3)(y+1) = 0 \qquad \text{Factoring}$$

$$y - 3 = 0 \quad \text{or} \quad y + 1 = 0 \qquad \text{Principle of zero products}$$

$$y = 3 \quad \text{or} \quad y = -1$$

4. **Check** The possible solutions are 3 and -1. We check 3 in the conditions of the problem.

Number:	3
2 more than the number:	5
Reciprocal of 2 more than the number:	$\frac{1}{5}$
5 times the reciprocal of 2 more than the number:	$5 \cdot \frac{1}{5} = 1$
Sum of number and 5 times the reciprocal of 2 more than the number:	$3 + 1 = 4$

The number 3 checks. So does the number -1, but we leave that check to you.

5. **State** The answer is that there are two numbers satisfying the conditions of the problem. They are 3 and -1.

DO EXERCISE 1.

ii Problems Involving Work

EXAMPLE 2 Tom can mow a lawn in 4 hours. Penny can mow the same lawn in 5 hours. How long would it take them working together, with 2 mowers, to mow the lawn?

1. **Familiarize** We list the facts:

Tom: Can mow in 4 hours;

Penny: Can mow in 5 hours.

We want to know how long it will take them working together. Let's let t be that number of hours.

We might try a guess. Suppose we guess 9 hours (we get that by adding 4 and 5). It's a good guess, but we are way off. Either one of them can do it alone in less than 9 hours.

Let's try making some kind of table. First of all, we know that Tom should be able to mow $\frac{1}{4}$ of the lawn in 1 hour, since he can do it all in 4 hours.

Number of hours	1	2	3	4	5	
Tom	$\frac{1}{4}$	$\frac{2}{4}$	$\frac{3}{4}$	$\frac{4}{4}$		All in 4 hours
Penny	$\frac{1}{5}$	$\frac{2}{5}$	$\frac{3}{5}$	$\frac{4}{5}$	$\frac{5}{5}$	All in 5 hours

$\frac{1}{4}$ in 1 hour

$\frac{1}{5}$ in 1 hour

In two hours, Tom can do $\frac{2}{4}$ of the lawn. In 4 hours, he does it all, or $\frac{4}{4}$ of the lawn. Similarly, Penny can do $\frac{1}{5}$ of the lawn in 1 hour, $\frac{2}{5}$ in two hours, and so on. She can do $\frac{5}{5}$, or all of it, in 5 hours.

Since Tom can do $\frac{1}{4}$ of the work in 1 hour and Penny can do $\frac{1}{5}$ of it in 1 hour, then working together they should do

$$\frac{1}{4} + \frac{1}{5}$$

of the lawn in 1 hour. We are supposing that it takes them t hours working together, so they should do $1/t$ of the lawn in 1 hour.

2. **Translate** They can do $1/t$ of the lawn in 1 hour. They can also do $\frac{1}{4} + \frac{1}{5}$ of it in one hour. That gives us an equation:

$$\frac{1}{4} + \frac{1}{5} = \frac{1}{t}.$$

3. **Carry out** We solve the equation:

$$20t\left(\frac{1}{4} + \frac{1}{5}\right) = 20t \cdot \frac{1}{t}$$

$$\frac{20t}{4} + \frac{20t}{5} = \frac{20t}{t}$$

$$5t + 4t = 20$$

Multiplying on both sides by the LCM of the denominators and clearing of fractions

$$9t = 20$$

$$t = \frac{20}{9}$$

4. **Check** The possible solution is $\frac{20}{9}$, or $2\frac{2}{9}$ hours. If Tom works $\frac{20}{9}$ hours, he will do $\frac{1}{4} \cdot \frac{20}{9}$ of the work, or $\frac{5}{9}$ of it. If Penny works $\frac{20}{9}$ hours, she will do $\frac{1}{5} \cdot \frac{20}{9}$ of the work, or $\frac{4}{9}$ of it. Altogether, they will do $\frac{5}{9} + \frac{4}{9}$ of the work, or all of it.

5. **State** The answer is that it will take $\frac{20}{9}$, or $2\frac{2}{9}$ hours for Tom and Penny to mow the lawn when they work together.

DO EXERCISE 2.

EXAMPLE 3 At a factory, smokestack A pollutes the air twice as fast as smokestack B. When the stacks operate together, they yield a certain amount of pollution in

2. Fred does a certain typing job in 6 hours. Fran can do the same job in 4 hours. How long would it take them to do the same amount of typing working together?

15 hours. Find the time it would take each to yield that same amount of pollution operating individually.

1. **Familiarize** Smokestack A takes half as much time as smokestack B to pollute the same amount. Let's let t represent the time that it will take smokestack A. Then $2t$ represents the time it takes smokestack B. We list what we know.

Time for smokestack A:	t hours;
Time for smokestack B:	twice that long, or $2t$ hours;
Time together:	15 hours

Thus in one hour,

A does $\frac{1}{t}$ of the total;

B does $\frac{1}{2t}$ of the total;

Together they do $\frac{1}{15}$ of the total.

2. **Translate** In 1 hour stacks A and B will yield $\frac{1}{15}$ of the total. They will also yield

$$\frac{1}{t}+\frac{1}{2t}$$

of the total. We now have an equation:

$$\frac{1}{t}+\frac{1}{2t}=\frac{1}{15}.$$

3. **Carry out** We solve the equation:

$$30t\left(\frac{1}{t}+\frac{1}{2t}\right)=30t\cdot\frac{1}{15}$$

$$\frac{30t}{t}+\frac{30t}{2t}=\frac{30t}{15}$$

$$30+15=2t$$

Multiplying on both sides by the LCM of the denominators and clearing of fractions

$$45=2t.$$

$$t=\frac{45}{2},\quad\text{or}\quad 22\frac{1}{2}.$$

4. **Check** The solution of the equation is $\frac{45}{2}$, the time in hours for smokestack A. If it takes smokestack B twice as long, then its time will be 45. Let's try these numbers. In one hour, A will do $\frac{1}{45/2}$ of the total and B will do $\frac{1}{45}$ of the total. Together, they will do

$$\frac{1}{45/2} + \frac{1}{45}$$

of the total. Adding, we get

$$\frac{2}{45} + \frac{1}{45} = \frac{3}{45} = \frac{1}{15}.$$

The answers check.

5. **State** The answer is that it would take smokestack A $\frac{45}{2}$, or $22\frac{1}{2}$ hours and smokestack B 45 hours.

DO EXERCISE 3.

3. Two pipes carry water to the same tank. Pipe A, working alone, can fill the tank three times as fast as pipe B. Together the pipes can fill the tank in 24 hours. Find the time each would take to fill the tank alone.

iii Problems Involving Motion

When a problem deals with speed, distance, and time, we can expect to use the definition of speed, or something equivalent to it.

DEFINITION OF SPEED

$$\text{Speed} = \frac{\text{distance}}{\text{time}}, \quad \text{or} \quad r = \frac{d}{t}$$

The equation

$$r = \frac{d}{t}$$

can be solved for either d or t. That gives us two other equations we can use:

$$d = rt \quad \text{or} \quad t = \frac{d}{r}.$$

You can remember $d = rt$, as many do, or you can remember the definition of speed. In any case, you should be able to obtain any of the three equations as needed in a problem situation.

EXAMPLE 4 A train leaves Sioux City traveling east at 30 mph. Two hours later another train leaves Sioux City in the same direction on a parallel track at 45 mph. How far from Sioux City will the faster train catch the slower one?

4. A train leaves Gilville traveling east at 35 mph. One hour later a faster train leaves Gilville, also traveling east on a parallel track at 40 mph. How far from Gilville will the faster train catch the slower one?

1. **Familiarize** We make a drawing and label it.

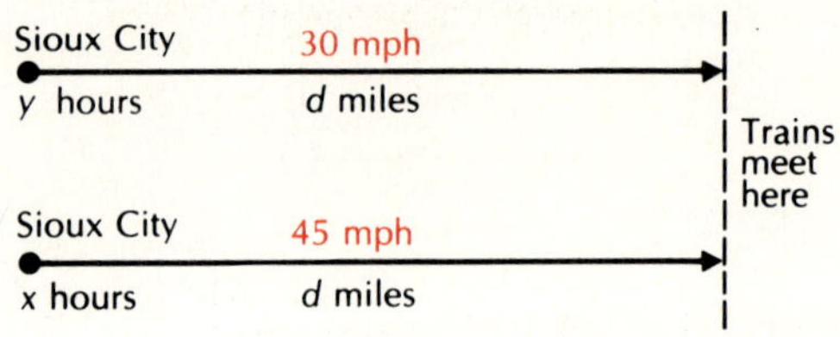

We have used x and y for the times and d for the distance. From the figure we see that the distance is the same for both trains. We can organize the information in a table.

	Distance	Speed	Time
Slow train	d	30	y (2 hours more than the fast train)
Fast train	d	45	x

2. **Translate** The table suggests that we use the relation $d = rt$. If we do, we get two equations:

$$d = 30y,$$
$$d = 45x.$$

Then we have

$$45x = 30y.$$

From the table, we also see that $y = x + 2$. Thus we have a system of equations:

$$45x = 30y,$$
$$y = x + 2.$$

3. **Carry out** We solve the system of equations. When we do we find that $x = 4$ and $y = 6$.

4. **Check** Since we have used x and y for the times, we know that the time for the slower train is 6 hours and the time for the faster train is 4 hours. Let's compute the distance in two cases:

Slow train: $d = rt = 30 \cdot 6 = 180$ miles;

Fast train: $d = rt = 45 \cdot 4 = 180$ miles.

We have an answer.

5. **State** The trains meet at a distance of 180 miles from Sioux City.

DO EXERCISE 4.

EXAMPLE 5 An airplane flies 1062 miles with the wind. In the same amount of time, it can fly 738 miles against the wind. The speed of the plane in still air is 200 mph. Find the speed of the wind.

1. **Familiarize** We know that the speed of the plane is 200, but we don't know the speed of the wind. Let's call it w. Now we make a drawing and label it.

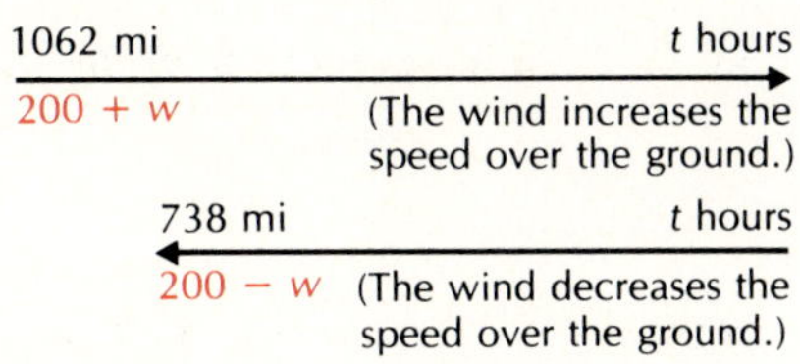

We can also keep the data in a table.

	Distance	Speed	Time
With the wind	1062	$200 + w$	t
Against the wind	738	$200 - w$	t

2. **Translate** The table suggests that we use

$$t = \frac{d}{r}.$$

Thus we get two equations:

$$t = \frac{1062}{200 + w} \quad \text{and} \quad t = \frac{738}{200 - w}.$$

3. **Carry out** We solve:

$$\frac{1062}{200 + w} = \frac{738}{200 - w} \qquad \text{Using substitution}$$

$$(200 + w)(200 - w)\left(\frac{1062}{200 + w}\right) = (200 + w)(200 - w)\left(\frac{738}{200 - w}\right) \qquad \text{Multiplying by the LCM to clear of fractions}$$

$$(200 - w)1062 = (200 + w)738$$

$$212{,}400 - 1062w = 147{,}600 + 738w$$

$$64{,}800 = 1800w$$

$$w = 36.$$

5. A boat travels 246 miles downstream in the same time that it takes to travel 180 miles upstream. The speed of the current in the stream is 5.5 mph. Find the speed of the boat in still water.

4. **Check** If it checks, the speed of the wind is 36 mph. Let's calculate:

With the wind: Speed of the plane is $200 + 36 = 236$ mph;
Against the wind: Speed of the plane is $200 - 36 = 164$ mph.

Now we calculate time, using

$$t = \frac{d}{r}.$$

236 mph: $t = \dfrac{1062}{236} = 4.5$ hours;

164 mph: $t = \dfrac{738}{164} = 4.5$ hours.

We have an answer.

5. **State** The answer is that the speed of the wind is 36 mph.

DO EXERCISE 5.

EXERCISE SET 6.6

Solve.

i

1. The reciprocal of 5 plus the reciprocal of 7 is the reciprocal of what number?

2. The reciprocal of 3 plus the reciprocal of 6 is the reciprocal of what number?

3. The sum of a number and 6 times its reciprocal is -5. Find the number.

4. The sum of a number and 21 times its reciprocal is -10. Find the number.

ii

5. Sam, an experienced shipping clerk, can fill a certain order in 5 hours. Willy, a new clerk, needs 9 hours to do the same job. Working together, how long will it take them to fill the order?

6. Paul can paint a room in 4 hours. Sally can paint the same room in 3 hours. Working together, how long will it take them to paint the room?

7. A swimming pool can be filled in 12 hours if water enters through a pipe alone, or in 30 hours if water enters through a hose alone. If water is entering through both the pipe and the hose, how long will it take to fill the pool?

8. A tank can be filled in 18 hours by pipe A alone and in 22 hours by pipe B alone. How long will it take to fill the tank if both pipes are working?

9. Bill can clear a lot in 5.5 hours. His partner can do the same job in 7.5 hours. How long will it take them to clear the lot working together?

10. One printing press can print an order of booklets in 4.5 hours. Another press can do the same job in 5.5 hours. How long will it take if both presses are used?

11. A can paint the neighbor's house 4 times as fast as B. The year they worked together it took them 8 days. How long would it take each to paint the house alone?

12. A can deliver papers 3 times as fast as B. If they work together, it takes them 1 hour. How long would it take each to deliver the papers alone?

iii

13. The speed of a stream is 3 mph. A boat travels 4 mi upstream in the same time it takes to travel 10 mi downstream. What is the speed of the boat in still water?

14. The speed of a stream is 4 mph. A boat travels 6 mi upstream in the same time it takes to travel 12 mi downstream. What is the speed of the boat in still water?

15. A train leaves a station and travels north at 75 km/h. Two hours later a second train leaves on a parallel track traveling north at 125 km/h. How far from the station will the second train overtake the first train?

16. A private airplane leaves an airport and flies due east at 180 km/h. Two hours later a jet leaves the same airport and flies due east at 900 km/h. How far from the airport will the jet overtake the private plane?

17. The speed of train A is 12 mph slower than the speed of train B. Train A travels 230 mi in the same time it takes train B to travel 290 mi. Find the speed of each train.

18. The speed of a passenger train is 14 mph faster than the speed of a freight train. The passenger train travels 400 mi in the same time it takes the freight train to travel 330 mi. Find the speed of each train.

19. Suzie has a boat that can move at a speed of 15 km/h in still water. She rides 140 km downstream in a river in the same time it takes to ride 35 km upstream. What is the speed of the river?

20. A paddleboat can move at a speed of 2 km/h in still water. The boat is paddled 4 km downstream in a river in the same time it takes to go 1 km upstream. What is the speed of the river?

21. At what time after 4:00 will the minute hand and the hour hand of a clock first be in the same position?

22. At what time after 10:30 will the hands of a clock first be perpendicular?

23. A boat travels 96 km downstream in 4 hr. It travels 28 km upstream in 7 hr. Find the speed of the boat and the speed of the stream.

24. An airplane carries enough fuel for 6 hours of flight time, and its speed in still air is 240 mph. It leaves an airport against a wind of 40 mph and returns to the same airport with a wind of 40 mph. How far can it fly under those conditions without refueling?

25. A motor boat travels 3 times as fast as the current. A trip up the river and back takes 10 hours, and the total distance of the trip is 100 km. Find the speed of the current.

26. An employee drives to work at 50 mph and arrives one minute late. The employee drives to work at 60 mph and arrives 5 minutes early. How far does the employee live from work?

Average speed is defined as *total distance divided by total time.*

27. A driver went 200 km. For the first half of the trip the driver traveled at a speed of 40 km/h. For the second half of the trip the driver traveled at a speed of 60 km/h. What was the average speed for the entire trip? (It is *not* 50 km/h.)

28. For half of a trip a driver travels at 40 mph. What speed would the driver have to travel for the last half of the trip so that the average speed for the entire trip would be 45 mph?

29. Three trucks A, B, and C, working together, can move a load of sand in t hours. When working alone, it takes A, 1 extra hour to move the sand; B, 6 extra hours; and C, t extra hours. Find t.

30. An employee drives to work at 45 mph and arrives one minute early. The employee drives to work at 40 mph and arrives one minute late. How far does the employee live from work?

6.7

DIVISION OF POLYNOMIALS

Remember that a rational expression indicates division. In some cases, it is useful to actually carry out that division. We now consider division of polynomials.

i Divisor a Monomial

To divide a polynomial by a monomial, we can first write a fractional expression, as in the following example.

EXAMPLE 1 Divide $12x^3 + 8x^2 + x + 4$ by $4x$.

$$\frac{12x^3 + 8x^2 + x + 4}{4x} \quad \text{Writing a fractional expression}$$

$$= \frac{12x^3}{4x} + \frac{8x^2}{4x} + \frac{x}{4x} + \frac{4}{4x} \quad \text{Doing the reverse of adding}$$

$$= 3x^2 + 2x + \frac{1}{4} + \frac{1}{x} \quad \text{Doing the four indicated divisions}$$

DO EXERCISE 1.

EXAMPLE 2 Divide: $(8x^4 - 3x^3 + 5x^2) \div x^2$.

$$\frac{8x^4 - 3x^3 + 5x^2}{x^2} = \frac{8x^4}{x^2} - \frac{3x^3}{x^2} + \frac{5x^2}{x^2}$$

$$= 8x^2 - 3x + 5$$

OBJECTIVES

After finishing Section 6.7, you should be able to:

i **Divide a polynomial by a monomial.**

ii **Divide a polynomial by a divisor that is not a monomial, and if there is a remainder, express the result in two ways.**

Divide.

1. $\dfrac{x^3 + 16x^2 + 6x}{2x}$

Divide.

2. $(15y^5 - 6y^4 + 18y^3) \div 3y^2$

3. $(x^4 + 10x^3 + 16x^2) \div 2x^2$

Divide and check.

4. $x - 2\,\overline{)\,x^2 + 3x - 10}$

It is not necessary to write out all of the steps shown in these examples. You should try to write only the answer.

To divide a polynomial by a monomial, we can divide each term by the monomial.

DO EXERCISES 2 AND 3.

ii Divisor Not a Monomial

When the divisor is not a monomial, we use a procedure very much like long division in arithmetic.

EXAMPLE 3 Divide $x^2 + 5x + 8$ by $x + 3$.

$$\begin{array}{r l}
x & \leftarrow \text{Divide the first term of the dividend by the first term of the divisor: } x^2/x = x.\\
x+3\,\overline{)\,x^2 + 5x + 8} & \\
x^2 + 3x & \leftarrow \text{Multiply } x \text{ above by divisor.}\\
\hline
2x & \leftarrow \text{Subtract.}
\end{array}$$

The subtraction we have done is $(x^2 + 5x) - (x^2 + 3x)$. Remember: To subtract, add the inverse (change the sign of every term, then add).

We now "bring down" the other terms of the dividend—in this case, 8:

$$\begin{array}{r l}
x + 2 & \leftarrow \text{Divide the first term by the first term: } 2x/x = 2.\\
x+3\,\overline{)\,x^2 + 5x + 8} & \\
x^2 + 3x & \\
\hline
2x + 8 & \leftarrow \text{The 8 has been "brought down."}\\
2x + 6 & \leftarrow \text{Multiply 2 by the divisor.}\\
\hline
2 & \leftarrow \text{Subtract: } (2x + 8) - (2x + 6).
\end{array}$$

Answer: Quotient $x + 2$, remainder 2.

To check, we multiply the quotient by the divisor and add the remainder, to see if we get the dividend.

$$\underbrace{(x + 3)}_{\text{Divisor}} \cdot \underbrace{(x + 2)}_{\text{Quotient}} + \underbrace{2}_{\text{Remainder}} = (x^2 + 5x + 6) + 2 = \underbrace{x^2 + 5x + 8}_{\text{Dividend}}$$

The answer checks.

DO EXERCISE 4.

You should remember the following.

1. **Arrange polynomials in descending order.**
2. **If there are missing terms in the dividend, either write them with zero coefficients or leave space for them.**

EXAMPLE 4 Divide: $(125y^3 - 8) \div (5y - 2)$.

a)

$$
\begin{array}{r@{}l}
 & 25y^2 + 10y + 4 \\
5y - 2 \,\big) & 125y^3 + 0y^2 + 0y - 8 \\
 & 125y^3 - 50y^2 \\
\hline
 & \quad 50y^2 + 0y - 8 \\
 & \quad 50y^2 - 20y \\
\hline
 & \qquad 20y - 8 \\
 & \qquad 20y - 8 \\
\hline
 & \qquad\quad 0
\end{array}
$$

This subtraction is $125y^3 - (125y^3 - 50y^2)$. We get $50y^2$

b)

$$
\begin{array}{r@{}l}
 & 25y^2 + 10y + 4 \\
5y - 2 \,\big) & 125y^3 \qquad\qquad - 8 \\
 & 125y^3 - 50y^2 \\
\hline
 & \quad 50y^2 \qquad\quad - 8 \\
 & \quad 50y^2 - 20y \\
\hline
 & \qquad 20y - 8 \\
 & \qquad 20y - 8 \\
\hline
 & \qquad\quad 0
\end{array}
$$

This subtraction is $50y^2 - (50y^2 - 20y)$. We get $20y$.

DO EXERCISE 5.

EXAMPLE 5 Divide: $(x^4 - 9x^2 - 5) \div (x - 2)$.

$$
\begin{array}{r@{}l}
 & x^3 + 2x^2 - 5x - 10 \\
x - 2 \,\big) & x^4 \qquad - 9x^2 \qquad - 5 \\
 & x^4 - 2x^3 \\
\hline
 & \quad 2x^3 - 9x^2 \qquad - 5 \\
 & \quad 2x^3 - 4x^2 \\
\hline
 & \qquad - 5x^2 \qquad - 5 \\
 & \qquad - 5x^2 + 10x \\
\hline
 & \qquad\qquad - 10x - 5 \\
 & \qquad\qquad - 10x + 20 \\
\hline
 & \qquad\qquad\qquad - 25
\end{array}
$$

The first subtraction is $x^4 - (x^4 - 2x^3)$.

The second subtraction is $(2x^3 - 9x^2) - (2x^3 - 4x^2)$.

The answer is $x^3 + 2x^2 - 5x - 10$, with R $= -25$, or

$$x^3 + 2x^2 - 5x - 10 + \frac{-25}{x - 2}.$$

DO EXERCISES 6 AND 7.

When dividing, we may "come out even" (have a remainder of 0), or we may not. If not, how long should we keep working? We continue until the degree of the remainder is less than the degree of the divisor, as in the next example. The answer can be given by writing the quotient and remainder, or by using a fractional expression, as you saw in Example 5 and as you will see in the next example.

EXAMPLE 6 Divide: $(x^3 + 9x^2 - 5) \div (x^2 - 1)$.

$$
\begin{array}{r@{}l}
 & x + 9 \\
x^2 - 1 \,\big) & x^3 + 9x^2 + 0x - 5 \\
 & x^3 \qquad\quad - x \\
\hline
 & \quad 9x^2 + x - 5 \\
 & \quad 9x^2 \qquad - 9 \\
\hline
 & \qquad\qquad x + 4
\end{array}
$$

Divide and check.

5. $(9y^4 + 14y^2 - 8) \div (3y + 2)$

Divide and check.

6. $(y^3 - 11y^2 + 6) \div (y - 3)$

7. $(x^3 + 9x^2 - 5) \div (x - 1)$

Divide and check.

8. $(y^3 - 11y^2 + 6) \div (y^2 - 3)$

The degree of the remainder is less than the degree of the divisor, so we are finished. The answer is $x + 9$, with $R = x + 4$, or

$$x + 9 + \frac{x+4}{x^2 - 1}.$$

The fractional expression is the remainder over the divisor.

DO EXERCISE 8.

EXERCISE SET 6.7

i Divide.

1. $\dfrac{30x^8 - 15x^6 + 40x^4}{5x^4}$
2. $\dfrac{24y^6 + 18y^5 - 36y^2}{6y^2}$
3. $\dfrac{-14a^3 + 28a^2 - 21a}{7a}$
4. $\dfrac{-32x^4 - 24x^3 - 12x^2}{4x}$
5. $(9y^4 - 18y^3 + 27y^2) \div 9y$
6. $(24a^3 + 28a^2 - 20a) \div 2a$
7. $(36x^6 - 18x^4 - 12x^2) \div -6x$
8. $(18y^7 - 27y^4 - 3y^2) \div -3y^2$
9. $(a^2b - a^3b^3 - a^5b^5) \div a^2b$
10. $(x^3y^2 - x^3y^3 - x^4y^2) \div x^2y^2$
11. $(6p^2q^2 - 9p^2q + 12pq^2) \div -3pq$
12. $(16y^4z^2 - 8y^6z^4 + 12y^8z^3) \div 4y^4z$

ii Divide and check.

13. $(x^2 + 10x + 21) \div (x + 3)$
14. $(y^2 - 8y + 16) \div (y - 4)$
15. $(a^2 - 8a - 16) \div (a + 4)$
16. $(y^2 - 10y - 25) \div (y - 5)$
17. $(y^2 - 25) \div (y + 5)$
18. $(a^2 - 81) \div (a - 9)$
19. $(y^3 - 4y^2 + 3y - 6) \div (y - 2)$
20. $(x^3 - 5x^2 + 4x - 7) \div (x - 3)$
21. $(a^3 - a + 12) \div (a - 4)$
22. $(x^3 - x + 6) \div (x + 2)$
23. $(8x^3 + 27) \div (2x + 3)$
24. $(64y^3 - 8) \div (4y - 2)$
25. $(x^4 - x^2 - 42) \div (x^2 - 7)$
26. $(y^4 - y^2 - 54) \div (y^2 - 3)$
27. $(x^4 - x^2 - x + 2) \div (x - 1)$
28. $(y^4 - y^2 - y + 3) \div (y + 1)$
29. $(10y^3 + 6y^2 - 9y + 10) \div (5y - 2)$
30. $(6x^3 - 11x^2 + 11x - 2) \div (2x - 3)$
31. $(2x^4 - x^3 - 5x^2 + x - 6) \div (x^2 + 2)$
32. $(3x^4 + 2x^3 - 11x^2 - 2x + 5) \div (x^2 - 2)$

Problem-solving practice

33. Find three consecutive positive integers such that the product of the first and second integers is 26 less than the product of the second and third integers.

34. For a certain fractional expression, the value of the numerator is 4 less than twice the value of the denominator. The reciprocal of the expression has a value $\frac{5}{6}$ greater than that of the original expression. What is the value of the original expression?

○

Divide.

35. $(x^4 - x^3y + x^2y^2 + 2x^2y - 2xy^2 + 2y^3) \div (x^2 - xy + y^2)$
36. $(4a^3b + 5a^2b^2 + a^4 + 2ab^3) \div (a^2 + 2b^2 + 3ab)$
37. $(x^4 - y^4) \div (x - y)$
38. $(a^7 + b^7) \div (a + b)$

Solve.

39. Find k so that when $x^3 - kx^2 + 3x + 7k$ is divided by $x + 2$, the remainder will be 0.

40. When $x^2 - 3x + 2k$ is divided by $x + 2$, the remainder is 7. Find k.

OBJECTIVE

After finishing Section 6.8, you should be able to:

i **Use synthetic division to divide a polynomial by a binomial of the type $x - a$.**

6.8 SYNTHETIC DIVISION

i The division procedure used in the preceding section can be shortened, or streamlined, in certain cases. If the divisor is a linear polynomial with leading coefficient 1, which we can express as $x - a$, then the shorter procedure will work.

Compare the following. In **A** we perform a division. In **B** we do not write the variables.

A.

$$\begin{array}{r}
4x^2 + 5x + 11 \\
x - 2\,\overline{)\,4x^3 - 3x^2 + x + 7} \\
\underline{4x^3 - 8x^2} \qquad\qquad \\
5x^2 + x \qquad \\
\underline{5x^2 - 10x} \qquad \\
11x + 7 \\
\underline{11x - 22} \\
29
\end{array}$$

B.

$$\begin{array}{r}
4 + 5 + 11 \\
1 - 2\,\overline{)\,4 - 3 + 1 + 7} \\
\underline{4 - 8} \qquad\quad \\
5 + 1 \qquad \\
\underline{5 - 10} \qquad \\
11 + 7 \\
\underline{11 - 22} \\
29
\end{array}$$

In **B** there is still some duplication of writing. Also, since we can subtract by adding an inverse, we can use 2 instead of -2 and then add instead of subtracting.

C. *Synthetic division*

a) $\underline{2}\rfloor\ 4 - 3 + 1 + 7$ — Write the 2 of $x - 2$ and the coefficients of the dividend.

$$\begin{array}{l}
\underline{} \\
4
\end{array}$$

Bring down the first coefficient.

b)

$$\begin{array}{r|rrrr}
2 & 4 & -3 & +1 & +7 \\
 & & 8 & & \\ \hline
 & 4 & 5 & &
\end{array}$$

Multiply 4 by 2, to get 8. Add 8 and -3.

c)

$$\begin{array}{r|rrrr}
2 & 4 & -3 & +1 & +7 \\
 & & 8 & 10 & \\ \hline
 & 4 & 5 & 11 &
\end{array}$$

Multiply 5 by 2 to get 10. Add 10 and 1.

d)

$$\begin{array}{r|rrrr}
2 & 4 & -3 & +1 & +7 \\
 & & 8 & 10 & 22 \\ \hline
 & 4 & 5 & 11 & 29
\end{array}$$

Multiply 11 by 2 to get 22. Add 22 and 7.

Quotient Remainder

The last number, 29, is the remainder. The other numbers are the coefficients of the quotient, with that of the term of highest degree first, as follows.

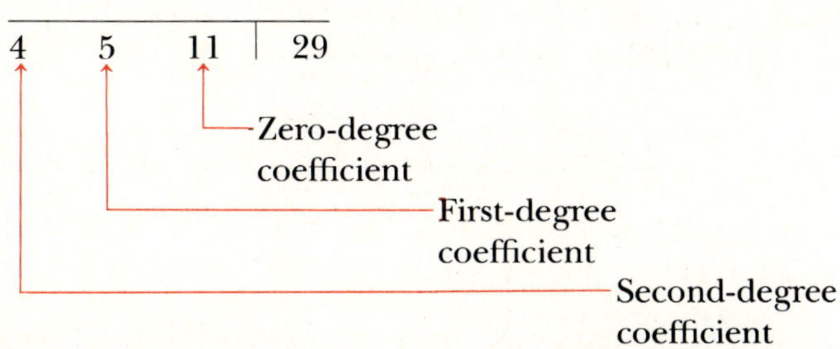

The quotient is $4x^2 + 5x + 11$. The remainder is 29.

It is important to remember that in order for this method to work, the divisor must be of the form $x - a$, that is, a variable minus a constant. The leading coefficient must be 1.

Use synthetic division to find the quotient and remainder.

1. $(2x^3 - 4x^2 + 8x - 8) \div (x - 3)$

EXAMPLE 1 Use synthetic division: $(x^3 + 6x^2 - x - 30) \div (x - 2)$.

$$\begin{array}{r|rrrr} 2 & 1 & 6 & -1 & -30 \\ & & 2 & 16 & 30 \\ \hline & 1 & 8 & 15 & \big| \; 0 \end{array}$$

The quotient is $x^2 + 8x + 15$. The remainder is 0.

DO EXERCISE 1.

EXAMPLES Use synthetic division to find the quotient and remainder.

2. $(2x^3 + 7x^2 - 5) \div (x + 3)$

There is no x-term, so we must write a 0 for the coefficient. Note that $x + 3 = x - (-3)$.

$$\begin{array}{r|rrrr} -3 & 2 & 7 & 0 & -5 \\ & & -6 & -3 & 9 \\ \hline & 2 & 1 & -3 & \big| \; 4 \end{array}$$

The quotient is $2x^2 + x - 3$. The remainder is 4.

Use synthetic division to find the quotient and remainder.

2. $(x^3 - 2x^2 + 5x - 4) \div (x + 2)$

3. $(x^3 + 4x^2 - x - 4) \div (x + 4)$

$$\begin{array}{r|rrrr} -4 & 1 & 4 & -1 & -4 \\ & & -4 & 0 & 4 \\ \hline & 1 & 0 & -1 & \big| \; 0 \end{array}$$

The quotient is $x^2 - 1$. The remainder is 0.

4. $(x^4 - 1) \div (x - 1)$

$$\begin{array}{r|rrrrr} 1 & 1 & 0 & 0 & 0 & -1 \\ & & 1 & 1 & 1 & 1 \\ \hline & 1 & 1 & 1 & 1 & \big| \; 0 \end{array}$$

3. $(y^3 + 1) \div (y + 1)$

The quotient is $x^3 + x^2 + x + 1$. The remainder is 0.

5. $(8x^5 - 6x^3 + x - 8) \div (x + 2)$

$$\begin{array}{r|rrrrrr} -2 & 8 & 0 & -6 & 0 & 1 & -8 \\ & & -16 & 32 & -52 & 104 & -210 \\ \hline & 8 & -16 & 26 & -52 & 105 & \big| \; -218 \end{array}$$

The quotient is $8x^4 - 16x^3 + 26x^2 - 52x + 105$. The remainder is -218.

DO EXERCISES 2 AND 3.

EXERCISE SET 6.8

i Use synthetic division to find the quotient and remainder.

1. $(x^3 - 2x^2 + 2x - 5) \div (x - 1)$

2. $(x^3 - 2x^2 + 2x - 5) \div (x + 1)$

3. $(a^2 + 11a - 19) \div (a + 4)$

4. $(a^2 + 11a - 19) \div (a - 4)$

5. $(x^3 - 7x^2 - 13x + 3) \div (x - 2)$

6. $(x^3 - 7x^2 - 13x + 3) \div (x + 2)$

7. $(3x^3 + 7x^2 - 4x + 3) \div (x + 3)$

8. $(3x^3 + 7x^2 - 4x + 3) \div (x - 3)$

9. $(y^3 - 3y + 10) \div (y - 2)$

10. $(x^3 - 2x^2 + 8) \div (x + 2)$

11. $(3x^4 - 25x^2 - 18) \div (x - 3)$

12. $(6y^4 + 15y^3 + 28y + 6) \div (y + 3)$

13. $(x^3 - 27) \div (x - 3)$

14. $(y^3 + 27) \div (y + 3)$

15. $(y^4 - 1) \div (y - 1)$

16. $(x^5 - 32) \div (x - 2)$

Problem-solving practice

17. The high temperatures during a five-day period in Alaska averaged 0°. The thermometer never rose above 10° during that time. If Monday's high was 5°, Tuesday's high 10°, and Wednesday's high −2°, what was the lowest possible high temperature that could have been registered on Thursday or Friday?

18. In a fractional expression, the value of the denominator is 2 more than 3 times the value of the numerator. If the value of both numerator and denominator is increased by 3, the value of the expression becomes $\frac{3}{7}$. Find the value of the original expression.

Divide, using synthetic division.

19. $(3.41x^4 - 24.25x^2 - 13.47) \div (x - 2.41)$

20. $(5.032x^5 + 11.414x^3 + 217.3) \div (x + 17.07)$

21. For the polynomial function

$$P(x) = 8x^5 - 3x^4 + 7x - 4:$$

a) By synthetic division, find the remainder when $P(x)$ is divided by $x - 2$.

b) Using nested evaluation, find $P(2)$.

c) Compare the answers and the algorithms. (An algorithm is a systematic procedure for doing a certain kind of computation.)

22. Devise a way to use synthetic division when the divisor is a linear polynomial $ax + b$, with leading coefficient different from 1.

6.9

FORMULAS AND PROBLEM SOLVING

OBJECTIVE

After finishing Section 6.9, you should be able to:

i Solve a formula containing a fractional expression for a specified letter, and use that skill in solving problems.

i To solve a formula, we use the same methods we would to solve any other equation. Many formulas contain fractional expressions, and to solve them we proceed as in solving any equations with fractional expressions, as follows.

TO SOLVE FRACTIONAL FORMULAS

1. **Multiply by the LCM to clear of fractions.**
2. **Multiply to remove parentheses, if necessary.**
3. **Get all terms with the unknown on one side of the equation.**
4. **Factor out the unknown.**
5. **Use the multiplication principle to get the unknown alone on one side.**

Remember: To solve a problem, part of the familiarization process may be to look up a formula and perhaps some definitions of the terms used in the problem. You may then have to solve a formula for a certain unknown.

EXAMPLE 1 A pitcher's earned run average is 3, and the pitcher has given up 12 earned runs. How many innings have been pitched?

1. **Familiarize** If we do not already know what "earned run average" means or what "earned runs" are, we must look up those terms. When we do, we find that an "earned run" is one that does not result from an error committed by someone on the defensive team, and that "earned run average" is defined by the formula

$$A = \frac{9R}{I},$$

where A is the earned run average (ERA), R is the number of earned runs, and I is the number of innings pitched.

1. The formula

$$\frac{PV}{T} = k$$

relates the pressure, volume, and temperature of a gas. Solve it for T. (*Hint:* Begin by clearing of fractions.)

In this problem, $R = 12$ and $A = 3$. We want to find I, the number of innings pitched.

2. **Translate** The translation in this case is the formula

$$A = \frac{9R}{I}.$$

3. **Carry out** We solve the formula for I:

$$A = \frac{9R}{I}$$

$$I \cdot A = I \cdot \frac{9R}{I} \qquad \text{Multiplying by the LCM to clear of fractions}$$

$$IA = 9R$$

$$IA \cdot \frac{1}{A} = 9R \cdot \frac{1}{A} \qquad \text{Using the multiplication principle to isolate } I$$

$$I = \frac{9R}{A}. \qquad \text{Simplifying}$$

We now substitute and calculate:

$$I = \frac{9 \cdot 12}{3}$$

$$I = 36.$$

4. **Check** The possible answer is 36. If there are 12 earned runs and the number of innings is 36, then the ERA will be

$$\frac{9 \cdot 12}{36}.$$

This simplifies to 3, so the answer checks.

5. **State** The pitcher has pitched 36 innings.

DO EXERCISE 1.

As Example 1 indicates, it is important to be able to solve a formula for a particular letter. In the following example, we illustrate that skill.

EXAMPLE 2 Solve the formula

$$I = \frac{pT}{M + pn}$$

for p.

We first clear of fractions by multiplying by the LCM, $M + pn$:

$$(M + pn)I = (M + pn)\,\frac{pT}{M + pn}$$

$$IM + Ipn = pT.$$

We now get all terms containing p alone on one side:

$$IM = pT - Ipn \qquad \text{Adding } -Ipn$$

$$IM = p(T - In) \qquad \text{Factoring out the unknown } p$$

$$\frac{IM}{T - In} = p \qquad \text{Multiplying by } \frac{1}{T - In}$$

We now have p alone on one side, and p does *not* appear on the other side, so we have solved the formula for p.

DO EXERCISE 2.

2. Solve $V = \dfrac{A}{n(T - t)}$ for T.

EXAMPLE 3 Two electrical resistors are connected in parallel. The resistance of one of them is 6 ohms and the resistance of the combination is 3.75 ohms. What is the resistance of the other resistor?

1. Familiarize In a book on electricity, radio, or physics we will find that resistors in parallel are diagrammed as shown here:

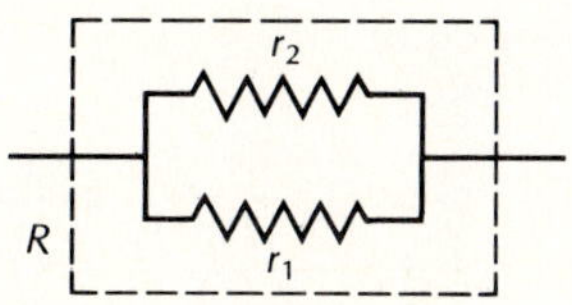

The rule, or law, for finding the resistance of a combination is given by the formula

$$\frac{1}{R} = \frac{1}{r_1} + \frac{1}{r_2},$$

where r_1 and r_2 are the resistances of the individual resistors and R is the resistance of the combination.

In this problem, we have the following:

$$R = 3.75 \text{ ohms};$$

$$r_1 = 6 \text{ ohms};$$

$$r_2 = ?$$

We could assign a letter to the unknown resistance, but we already have the letter r_2, so we will use it.

2. Translate The translation in this problem is the formula

$$\frac{1}{R} = \frac{1}{r_1} + \frac{1}{r_2}.$$

3. The formula

$$\frac{1}{p} + \frac{1}{q} = \frac{1}{f}$$

applies to a lens. The letters p and q represent the distances from an object to the lens and from the image to the lens, respectively. The letter f represents the focal length of the lens. Find the focal length of a lens that forms an image 15 cm from the lens when an object is placed 10 cm from the lens.

3. **Carry out** We solve the formula for the unknown resistance r_2. We multiply by the LCM, which is Rr_1r_2:

$$Rr_1r_2 \cdot \frac{1}{R} = Rr_1r_2 \cdot \left[\frac{1}{r_1} + \frac{1}{r_2}\right] \quad \text{Multiplying by the LCM}$$

$$Rr_1r_2 \cdot \frac{1}{R} = Rr_1r_2 \cdot \frac{1}{r_1} + Rr_1r_2 \cdot \frac{1}{r_2} \quad \text{Multiplying to remove parentheses}$$

$$\left.\begin{aligned} \frac{R}{R} \cdot r_1r_2 &= \frac{r_1}{r_1} \cdot Rr_2 + \frac{r_2}{r_2} \cdot Rr_1 \\ r_1r_2 &= Rr_2 + Rr_1. \end{aligned}\right\} \quad \text{Simplifying by removing factors of 1}$$

You might be tempted at this point to multiply by $1/r_1$ to get r_2 alone on the left, BUT note that there is an r_2 on the right. We must get all the terms involving r_2 on the *same side* of the equation:

$$r_1r_2 - Rr_2 = Rr_1 \quad \text{Adding } -Rr_2$$

$$r_2(r_1 - R) = Rr_1 \quad \text{Factoring out } r_2$$

$$r_2 = \frac{Rr_1}{r_1 - R}. \quad \text{Multiplying by } \frac{1}{r_1 - R} \text{ to get } r_2 \text{ alone}$$

We now substitute into the formula, as we have solved it, and compute:

$$r_2 = \frac{3.75 \cdot 6}{6 - 3.75}$$

$$r_2 = 10.$$

4. **Check** The possible answer is 10 ohms. We substitute into the original formula, as in checking an equation.

$\frac{1}{R}$	$= \frac{1}{r_1} + \frac{1}{r_2}$
$\frac{1}{3.75}$	$\frac{1}{6} + \frac{1}{10}$
$0.26666\ldots$	$\frac{5}{30} + \frac{3}{30}$
	$\frac{8}{30}$
	$0.266666\ldots$

The answer checks.

5. **State** The answer is that the resistor has a resistance of 10 ohms.

In the preceding example, you should note carefully that R, r_1, and r_2 are all different variables (or constants). We commonly use subscripts as in r_1 and r_2 to distinguish variables. It is important that you write the subscripts correctly and that you never write a capital letter in place of a lower-case letter.

DO EXERCISE 3.

EXERCISE SET 6.9

i Solve each formula for the specified letter.

1. $\frac{W_1}{W_2} = \frac{d_1}{d_2}$; d_1

2. $\frac{W_1}{W_2} = \frac{d_1}{d_2}$; W_2

3. $s = \frac{(v_1 + v_2)t}{2}$; t

4. $s = \frac{(v_1 + v_2)t}{2}$; v_1

5. $\frac{1}{R} = \frac{1}{r_1} + \frac{1}{r_2}$; r_1

6. $\frac{1}{R} = \frac{1}{r_1} + \frac{1}{r_2}$; R

7. $R = \frac{gs}{g + s}$; s

8. $R = \frac{gs}{g + s}$; g

9. $I = \frac{2V}{R + 2r}$; r

10. $I = \frac{2V}{R + 2r}$; R

11. $\frac{1}{p} + \frac{1}{q} = \frac{1}{f}$; f

12. $\frac{1}{p} + \frac{1}{q} = \frac{1}{f}$; p

13. $I = \frac{nE}{R + nr}$; r

14. $I = \frac{nE}{R + nr}$; n

15. $S = \frac{H}{m(t_1 - t_2)}$; H

16. $S = \frac{H}{m(t_1 - t_2)}$; t_1

17. $\frac{E}{e} = \frac{R + r}{r}$; e

18. $\frac{E}{e} = \frac{R + r}{r}$; r

19. $S = \frac{a - ar^n}{1 - r}$; a

20. $S = \frac{a}{1 - r}$; r

Problem-solving practice

21. A pitcher's earned run average is 2.4 and he has pitched 45 innings. How many earned runs were given up?

22. Two resistors are connected in parallel. Their resistances are, respectively, 8 ohms and 15 ohms. What is the resistance of the combination?

23. A resistor has a resistance of 50 ohms. What size resistor should be put with it, in parallel, in order to obtain a resistance of 5 ohms?

24. The area of a certain trapezoid is 25 cm^2. Its height is 5 cm and the length of one base is 4 cm. Find the length of the other base.

○

25. Pam can do a certain job in a hours working alone. Elaine, working alone, can do the same job in b hours. Working together it takes them t hours to do the job.

a) Find a formula for t.
b) Solve the formula for a.
c) Solve the formula for b.

26. The *harmonic mean* of two numbers a and b is a number M such that the reciprocal of M is the average of the reciprocals of a and b. Find a formula for the harmonic mean.

6.10

VARIATION AND PROBLEM SOLVING

OBJECTIVES

After finishing Section 6.10, you should be able to:

i **Given a description of direct variation, find an equation of variation.**

ii **Solve problems involving direct variation.**

iii **Given a description of inverse variation, write an equation of variation.**

iv **Solve problems involving inverse variation.**

We now see how to use certain functions in problem solving. Most of them involve fractional expressions in some way.

i Direct Variation

Let's say that a plumber earns \$18 per hour. In 1 hour \$18 is earned. In 2 hours \$36 is earned. In 3 hours \$54 is earned, and so on. This gives rise to a set of ordered pairs of numbers, all having the same ratio:

$$(1, 18), (2, 36), (3, 54), (4, 72), \quad \text{and so on.}$$

The ratio of earnings to time is $\frac{18}{1}$ in every case.

Whenever a situation gives rise to pairs of numbers in which the ratio is constant, we say that there is *direct variation*. Here the earnings *vary directly* as the time:

$$\frac{E}{t} = 18 \text{ (a constant)}, \quad \text{or} \quad E = 18t.$$

Note that the constant, 18, is positive. Note too that when one variable increases, so does the other. When one decreases, so does the other.

1. Find the variation constant and an equation of variation where y varies directly as x and $y = 8$ when $x = 20$.

2. Find the variation constant and an equation of variation where y varies directly as x and $y = 5.6$ when $x = 8$.

Whenever a situation gives rise to a linear function $f(x) = kx$, or $y = kx$, where k is a positive constant, we say that there is *direct variation*, or that y *varies directly as* x. The number k is called the *variation constant*.

EXAMPLE 1 Find the variation constant and an equation of variation where y varies directly as x, and where $y = 32$ and $x = 2$.

We know that (2, 32) is a solution of $y = kx$. Therefore,

$$32 = k \cdot 2 \qquad \text{Substituting}$$

$$\frac{32}{2} = k, \quad \text{or} \quad k = 16. \qquad \text{Solving for } k$$

The variation constant is 16. The equation of variation is $y = 16x$.

DO EXERCISES 1 AND 2.

Proportions. Suppose that y varies directly as x. Then we have $y = kx$. Suppose also that (x_1, y_1) and (x_2, y_2) are solutions of the equation. Then, substituting, we have

$$y_1 = kx_1 \quad \text{and} \quad y_2 = kx_2.$$

Dividing, we get

$$\frac{y_1}{y_2} = \frac{kx_1}{kx_2} = \frac{k}{k} \cdot \frac{x_1}{x_2} = \frac{x_1}{x_2}.$$

The equation

$$\frac{y_1}{y_2} = \frac{x_1}{x_2}$$

is called a *proportion*. Such equations are helpful in solving problems.

A helpful way to read the proportion above is

"y_1 is to y_2 as x_1 is to x_2."

ii Direct Variation and Problem Solving

There are many problem situations that translate to an equation of variation or a proportion, as in the following example.

EXAMPLE 2 The number of centimeters W of water produced from melting snow varies directly as S, the number of centimeters of snow. Meteorologists have found that 150 cm of snow will melt to 16.8 cm of water. To how many cm of water will 200 cm of snow melt?

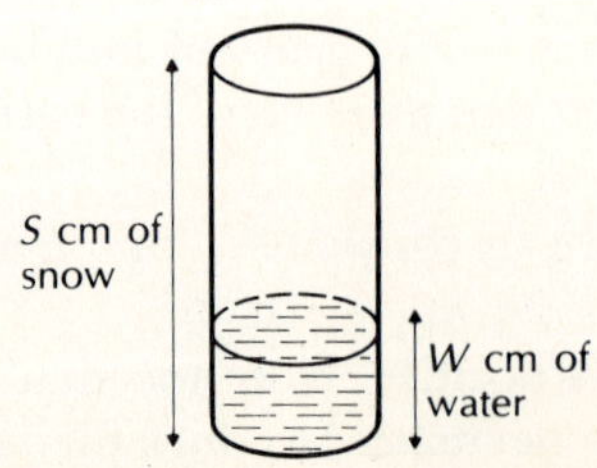

1. **Familiarize** Let's make a table, including the data from the problem and the data we need to find.

cm of snow $\xrightarrow{\text{melts to}}$	cm of water
150 or S_1	W_1 or 16.8
200 or S_2	W_2

We have used W_2 to represent the number of cm of water that will result when 200 cm of snow is melted.

We know from the statement of the problem that in this situation there is direct variation.

2. **Translate** We will translate twice, once with an equation of variation and once with a proportion.

Method 1. First find the variation constant using the data and then find an equation of variation.

$$W_1 = kS_1$$
$$16.8 = k \cdot 150 \qquad \text{Substituting}$$
$$\frac{16.8}{150} = k \qquad \text{Solving for } k$$
$$0.112 = k \qquad \text{This is the variation constant.}$$

The equation of variation is $W = 0.112S$. This is the translation.

Method 2. Write a proportion.

$$\frac{S_1}{S_2} = \frac{W_1}{W_2}$$

Amount of snow_1 is to amount of snow_2 as amount of water_1 is to amount of water_2. This is the translation.

3. **Carry out**
Method 1. The translation is $W = 0.112S$. We substitute and solve for W:

$$W = 0.112S$$
$$= 0.112(200) \qquad \text{Substituting}$$
$$= 22.4.$$

Method 2. The translation is

$$\frac{S_1}{S_2} = \frac{W_1}{W_2}.$$

We substitute and solve for W_2:

$$\frac{150}{200} = \frac{16.8}{W_2}$$
$$W_2 = \frac{200}{150} \cdot 16.8 \qquad \text{Solving}$$
$$= 22.4.$$

4. **Check** We already have a pretty good check, since we have done the translating and calculating two ways. Had we not done that, we would need to recheck all calculations, and perhaps do an estimate to see whether our answer is reasonable.

5. **State** The answer is that 200 cm of snow will melt into 22.4 cm of water.

DO EXERCISES 3 AND 4.

Solve two ways.

3. *Ohm's law* states that the voltage V in an electric circuit varies directly as the number of amperes I of electric current in the circuit. If the voltage is 10 volts when the current is 3 amperes, what is the voltage when the current is 15 amperes?

4. *An ecology problem.* The amount of garbage G produced in the United States varies directly as the number of people N who produce the garbage. It is known that 50 tons of garbage is produced by 200 people in 1 year. The population of San Francisco is 705,000. How much garbage is produced by San Francisco in 1 year?

iii Inverse Variation

To see what we mean by inverse variation, consider the following situation.

A bus is traveling a distance of 20 mi. At a speed of 20 mph, it will take 1 hour. At 40 mph it will take $\frac{1}{2}$ hour. At 60 mph it will take $\frac{1}{3}$ hour, and so on. This gives rise to a set of pairs of numbers, all having the same product:

$$(20, 1), (40, \tfrac{1}{2}), (60, \tfrac{1}{3}), (80, \tfrac{1}{4}), \quad \text{and so on.}$$

Whenever a situation gives rise to pairs of numbers whose product is constant, we say that there is *inverse variation.* Here the time *varies inversely* as the speed:

$$rt = 20 \text{ (a constant)}, \quad \text{or} \quad t = \frac{20}{r}.$$

Note that the constant, 20, is positive. Note too that when one variable increases, the other decreases.

Whenever a situation gives rise to a rational function $f(x) = k/x$, or $y = k/x$, where k is a positive constant, we say that there is *inverse variation*, or that y *varies inversely* as x. The number k is called the *variation constant.*

EXAMPLE 3 Find the variation constant and then an equation of variation where y varies inversely as x, and $y = 32$ when $x = 0.2$.

We know that $(0.2, 32)$ is a solution of

$$y = \frac{k}{x}.$$

Therefore,

$$32 = \frac{k}{0.2} \quad \text{Substituting}$$

$$(0.2)32 = k$$

$$6.4 = k. \quad \text{Solving}$$

The variation constant is 6.4. The equation of variation is

$$y = \frac{6.4}{x}.$$

Proportions. Suppose that y varies inversely as x. Then we have $y = k/x$. Suppose also that (x_1, y_1) and (x_2, y_2) are solutions of the equation. We substitute into $y = k/x$. Then we have

$$y_1 = \frac{k}{x_1} \quad \text{and} \quad y_2 = \frac{k}{x_2}.$$

Dividing, we get

$$\frac{y_1}{y_2} = \frac{\dfrac{k}{x_1}}{\dfrac{k}{x_2}} = \frac{k}{x_1} \cdot \frac{x_2}{k} = \frac{k}{k} \cdot \frac{x_2}{x_1} = \frac{x_2}{x_1}.$$

The equation

$$\frac{y_1}{y_2} = \frac{x_2}{x_1}$$

is called a *proportion*. It can be read

"y_1 is to y_2 as x_2 is to x_1."

Note that for direct variation the subscripts match, while for inverse variation the subscripts are interchanged.

DO EXERCISE 5.

5. Find the variation constant and an equation of variation where y varies inversely as x and $y = 0.012$ when $x = 50$.

iv Inverse Variation and Problem Solving

There are many problems that translate to an equation of inverse variation, as in the following example.

EXAMPLE 4 The time t required to do a certain job varies inversely as the number of people P who work on the job (assuming that all do the same amount of work). It takes 4 hours for 12 people to erect some football bleachers. How long would it take 3 people to do the same job?

This time we will not go through all five steps, but rather we will illustrate just the translation. We leave it to you to solve the problem.

2. Translate

Method 1. We first find the variation constant using the given data and then write an equation of variation,

$$t = \frac{k}{P},$$

because we know that t varies inversely as P. Thus

$$4 = \frac{k}{12} \qquad \text{Substituting}$$

$$48 = k. \qquad \text{Solving for } k\text{, the variation constant}$$

Then we have

$$t = \frac{48}{P}.$$

This is the translation.

Method 2. We write a proportion:

$$\frac{t_1}{t_2} = \frac{P_2}{P_1}.$$

t_1 is to t_2 as P_2 is to P_1. The proportion is the translation.

DO EXERCISE 6.

Solve two ways.

6. The time t required to drive a fixed distance varies inversely as the speed r. It takes 5 hours at 60 km/h to drive a fixed distance. How long would it take to drive that same distance at 40 km/h?

EXERCISE SET 6.10

i Find the variation constant and an equation of variation where y varies directly as x and the following are true.

1. $y = 24$ when $x = 3$ **2.** $y = 5$ when $x = 12$ **3.** $y = 3.6$ when $x = 1$ **4.** $y = 2$ when $x = 5$

5. $y = 15$ when $x = 3$ **6.** $y = 1$ when $x = 2$ **7.** $y = 30$ when $x = 8$ **8.** $y = 1$ when $x = \frac{1}{3}$

9. $y = 0.8$ when $x = 0.5$ **10.** $y = 0.6$ when $x = 0.4$

ii Solve.

11. The electric current I, in amperes, in a circuit varies directly as the voltage V. When 12 volts are applied, the current is 4 amperes. What is the current when 18 volts are applied?

12. Hooke's law states that the distance d that a spring is stretched by a hanging object varies directly as the weight w of the object. If the distance is 40 cm when the weight is 3 kg, what is the distance when the weight is 5 kg?

13. The number N of plastic straws produced by a machine varies directly as the amount of time t the machine is operating. If the machine produces 20,000 straws in 8 hours, how many straws can it produce in 50 hours?

14. The number N of aluminum cans used each year varies directly as the number of people using the cans. If 250 people use 60,000 cans in one year, how many cans are used each year in a city with a population of 850,000?

15. The amount of pollution A entering the atmosphere varies directly as the number of people N living in an area. If 60,000 people result in 42,600 tons of pollutants entering the atmosphere, how many tons enter the atmosphere in a city with a population of 750,000?

16. The weight M of an object on the moon varies directly as its weight E on earth. A person who weighs 95 kg on earth weighs 15.2 kg on the moon. How much would a 105-kg person weigh on the moon?

17. The weight M of an object on Mars varies directly as its weight E on earth. A person who weighs 95 kg on earth weighs 38 kg on Mars. How much would a 100-kg person weigh on Mars?

18. The number of kilograms of water W in a human body varies directly as the total weight. A person weighing 96 kg contains 64 kg of water. How many kilograms of water are in a person weighing 75 kg?

iii Find the variation constant and an equation of variation where y varies inversely as x, and the following are true.

19. $y = 6$ when $x = 10$ **20.** $y = 16$ when $x = 4$ **21.** $y = 4$ when $x = 3$ **22.** $y = 4$ when $x = 9$

23. $y = 12$ when $x = 3$ **24.** $y = 9$ when $x = 5$ **25.** $y = 27$ when $x = \frac{1}{3}$ **26.** $y = 81$ when $x = \frac{1}{9}$

iv Solve.

27. The current I in an electrical conductor varies inversely as the resistance R of the conductor. If the current is $\frac{1}{2}$ ampere when the resistance is 240 ohms, what is the current when the resistance is 540 ohms?

28. The time t required to empty a tank varies inversely as the rate r of pumping. If a pump can empty a tank in 45 min at the rate of 600 kL per min, how long will it take the pump to empty the same tank at the rate of 1000 kL per min?

29. The volume V of a gas varies inversely as the pressure P upon it. The volume of a gas is 200 cm^3 under a pressure of 32 kg/cm^2. What will be its volume under a pressure of 40 kg/cm^2?

30. The time T required to do a job varies inversely as the number of people P working. It takes 5 hours for 7 bricklayers to complete a certain job. How long will it take 10 bricklayers to complete the job?

31. The time t required to drive a fixed distance varies inversely as the speed r. It takes 5 hours at 80 km/h to drive a fixed distance. How long will it take to drive the fixed distance at 60 km/h?

32. The wavelength W of a radio wave varies inversely as its frequency F. A wave with a frequency of 1200 kilohertz per second has a length of 300 meters. What is the length of a wave with a frequency of 800 kilohertz per second?

A *unit price* is the price of one of something. It is also the ratio of price to the number of units:

$$\text{Unit price} = \frac{\text{Price}}{\text{Number of units}}.$$

33. A customer bought a 15-oz box of raisin bran for \$1.47. What is the unit price in cents per ounce? Round to the nearest hundredth.

34. A 14-oz can of baked beans costs 39¢. A 31-oz can of the same kind of beans costs 63¢. Which has the lower unit price? Find a way to use proportions to solve this problem.

35. To determine the number of deer in a forest, a conservationist catches 612 deer, tags them, and releases them. Later, 244 deer are caught, and it is found that 72 of them are tagged. Estimate how many deer are in the forest.

36. It is known that it takes 60 oz of grass seed to seed 3000 ft^2 of lawn. At this rate, how much would be needed for 5000 ft^2 of lawn?

37. Explain how you might use a proportion to estimate the number of words in a novel.

38. A loan of \$9000 earns simple interest of \$1665. Find an equation of variation that describes the simple interest I in terms of the loan principal P.

39. During the summer Steve sold 4 sweepers for every 3 sweepers sold by his sister, Stephanie. Together they sold 98 sweepers. How many did each sell?

40. A 96-fluid-ounce container of fabric softener costs \$2.99. A 128-fluid-ounce container of the same softener costs \$3.99. For which is the price per ounce less? Find a way to use proportions to solve this problem.

41. In each of the following equations, state whether y varies directly as x, inversely as x, or neither directly nor inversely as x.

a) $7xy = 14$
b) $x - 2y = 12$
c) $-2x + 3y = 0$
d) $x = \frac{3}{4}y$
e) $\frac{x}{y} = 2$

42.

a) If you graph an equation of direct variation $y = kx$, what will the graph look like?
b) Graph the equation of inverse variation $y = 2/x$. Describe the graph and compare it with a graph of direct variation.

SUMMARY AND REVIEW: CHAPTER 6

The following contains a summary of what you should be able to do after completing this chapter. The review exercises are for practice. Answers are at the back of the book. If you miss an exercise, restudy the section indicated alongside the answer.

You should be able to:

Simplify rational expressions by multiplying.

Simplify.

1. $\frac{y^2 - 64}{2y + 10} \cdot \frac{y + 5}{y + 8}$
2. $\frac{x^3 - 8}{x^2 - 25} \cdot \frac{x^2 + 10x + 25}{x^2 + 2x + 4}$

Simplify rational expressions by dividing.

Simplify.

3. $\frac{9a^2 - 1}{a^2 - 9} \div \frac{3a + 1}{a + 3}$
4. $\frac{x^3 - 64}{x^2 - 36} \div \frac{x^2 + 5x + 6}{x^2 - 3x - 18}$

Find the least common multiple of several polynomials.

Find the LCM.

5. $3y,\ 5y^2$
6. $6x^3,\ 16x^2$
7. $x^2 - 49,\ 3x + 21$
8. $x^2 + x - 20,\ x^2 + 3x - 10$

Add and subtract rational expressions, and simplify when possible.

Perform the indicated operation and simplify if possible.

9. $\frac{9xy}{x^2 - y^2} + \frac{x + y}{x - y}$
10. $\frac{2x^2}{x - y} + \frac{2y^2}{x + y}$
11. $\frac{3}{y + 4} - \frac{y}{y - 1} + \frac{y^2 + 3}{y^2 + 3y - 4}$

Simplify complex fractional expressions.

Simplify.

12. $\dfrac{\frac{1}{x} - x}{\frac{1}{x} + x}$
13. $x - \frac{3}{4x}$
14. $\dfrac{\frac{y^2 - 5y - 6}{y^2 - 7y - 18}}{\frac{y^2 + 3y + 2}{y^2 + 4y + 4}}$

Solve fractional equations.

Solve.

15. $\frac{6}{x} + \frac{4}{x} = 5$
16. $\frac{x}{7} + \frac{x}{4} = 1$
17. $\frac{5}{3x + 2} = \frac{3}{2x}$

Solve problems involving fractional equations.

Solve.

18. Dave can paint the outside of his house in 12 hours. Bill can paint the same house in 9 hours. How long would it take them to paint the house together?

19. A car gets on the turnpike traveling at a speed of 45 mph. Three hours later another car gets on at the same place traveling in the same direction at 55 mph. How far from the entrance to the turnpike will the second car overtake the first?

Divide polynomials.

Divide.

20. $(20r^2s^3 - 15r^2s^2 - 10r^3s^3) \div 5r^2s$

21. $(y^2 - 8y + 15) \div (y - 5)$

22. $(y^3 - 27) \div (y - 3)$

23. $(4x^3 + 3x^2 - 5x - 2) \div (x^2 + 1)$

Divide using synthetic division.

Divide.

24. $(x^3 + 3x^2 + 2x - 6) \div (x - 3)$

25. $(4x^3 + 6x^2 - 5) \div (x + 3)$

Solve formulas for specified variables.

Solve for the specified letter.

26. $R = \dfrac{gs}{g + s}$; g

27. $S = \dfrac{H}{m(t_1 - t_2)}$; m

Solve direct-variation problems.

Solve.

28. The score t on a test varies directly as the number of correct responses. If a student who gets 28 questions correct earns a score of 87, what is the score of a student who gets 25 questions correct?

29. The number of sodas sold at a restaurant varies directly as the number of hamburgers sold each day. If 69 hamburgers and 46 sodas are sold on Monday, how many sodas are sold on a day when the restaurant sells 87 hamburgers?

Simplify.

30. $\dfrac{\dfrac{1}{a - b} + \dfrac{1}{a + b}}{1 + \dfrac{b^2}{a^2 - b^2}}$

31. $\dfrac{2 - x}{x^2 - 2x + 1} - \dfrac{2x - 3}{x^2 - 3x + 2}$

32. $\dfrac{3x^7 - x^6 + 31x^4 + 21x + 5}{x + 2}$

TEST: CHAPTER 6

Simplify.

1. $\dfrac{4y^2 - 4}{3y + 9} \cdot \dfrac{y + 3}{y + 1}$

2. $\dfrac{x^3 + 27}{x^2 - 16} \div \dfrac{x^2 + 8x + 15}{x^2 + x - 20}$

Find the LCM.

3. $x^2 - 16,\ x^3 - 64$

4. $(x^2 + 8x - 33),\ (x^2 - 12x + 27)$

Perform the indicated operation and simplify.

5. $\dfrac{4ab}{a^2 - b^2} + \dfrac{a^2 + b^2}{a + b}$

6. $\dfrac{6}{x^3 - 64} - \dfrac{4}{x^2 - 16}$

7. $\dfrac{4}{y + 3} - \dfrac{y}{y - 2} + \dfrac{y^2 + 4}{y^2 + y - 6}$

Simplify.

8. $\dfrac{3 + \dfrac{3}{y}}{4 + \dfrac{4}{y}}$

9. $\dfrac{\dfrac{y^2 + 4y - 77}{y^2 - 10y + 25}}{\dfrac{y^2 - 5y - 14}{y^2 - 25}}$

Solve.

10. $\frac{1}{x} + \frac{3}{x} = 4$

11. $\frac{6}{5a+3} = \frac{4}{2a-5}$

12. Tom can mow the yard in 3.5 hours. Larry can mow the yard in 4.5 hours. How long will it take them to mow it together?

Divide.

13. $(16ab^3c - 10ab^2c^2 + 12a^2b^2c) \div (4a^2b)$

14. $(y^2 - 20y + 64) \div (y - 6)$

15. $(6x^4 + 3x^2 + 5x + 4) \div (x^2 + 2)$

16. Divide using synthetic division:

$$(x^3 + 5x^2 + 4x - 7) \div (x - 4).$$

17. Solve the formula for the specified letter:

$$A = \frac{h(b_1 + b_2)}{2}, \quad \text{for } b_1.$$

18. During January, the number of kilowatt hours of electricity used varies directly as the size of the house. If a 1300-ft^2 home uses 950 kWh of electricity, how much will a 2000-ft^2 home use? (Round your answer to the nearest kilowatt.)

CUMULATIVE REVIEW: CHAPTERS 1–6

Solve.

1. $4 + 3x = -2x - 5$
2. $6n - 4(2n - 3) = -8$
3. $-12 + 7b = -2(b - 4) - 7$
4. $-2.5x + 3.04 = 6.15 - 1.5x$
5. $\frac{1}{4}x - \frac{1}{3} = \frac{1}{2}(\frac{2}{3}x - \frac{1}{6})$
6. $8x - 1 = 16x^2$
7. $14 = -3x + 2x^2$
8. $x - 7 < 15$
9. $18 > 2 - 6x$
10. $-0.5y < 25$
11. $3a - 4 < 10 + 5a$
12. $\frac{1}{3}x - \frac{1}{5} > \frac{1}{5}x - \frac{1}{3}$
13. $-8 < x + 2 < 15$
14. $3x - 2 < -6$ or $x + 3 > 9$
15. $|x| > 6.4$
16. $|3x - 6| = 2$
17. $14 > |4x - 1|$
18. $4 \leqslant |3 - x| \leqslant 6$
19. $\frac{2}{n} - \frac{7}{n} = 3$
20. $\frac{6}{x-5} = \frac{2}{3x}$

Simplify.

21. $(-3x^{-5}y^2z^7)(4x^2y^{-1}z^4)^3$

22. $\frac{-15y^{a+b}y^{b-1}}{40y^{b+2}y^{a-4}}$

Graph.

23. $4x = -7y + 5$
24. $-y = 2x^2 - 3$
25. $y \geqslant -2$
26. $x - 4y \leqslant 3$

27. Find an equation of the line containing the points $(3, -2)$ and $(-1, -4)$.

28. Find an equation of the line with slope 4 and containing the point $(-2, 2)$.

Perform the indicated operations and simplify.

29. $(2x^2 - 3x + 1) + (6x - 3x^3 + 7x^2 - 4)$

30. $(5x^3y^2)(-3xy^2)$

31. $(3a + b - 2c) - (-4b + 3c - 2a)$

32. $(x - 4)^3$

33. $(5x^2 - 2x + 1)(3x^2 + x - 2)$

34. $(-5m^3n^2 - 3mn^3) + (-4m^2n^2 + 4m^3n^2) - (2mn^3 - 3m^2n^2)$

35. $\frac{y^2 - 36}{2y + 8} \cdot \frac{y + 4}{y + 6}$

36. $\frac{x^4 - 1}{x^2 - x - 2} \div \frac{x^2 + 1}{x - 2}$

37. $\frac{5ab}{a^2 - b^2} + \frac{a + b}{a - b}$

38. $\frac{2}{m+1} + \frac{3}{m-5} + \frac{m^2 - 1}{m^2 - 4m - 5}$

39. $y - \frac{2}{3y}$

40. $\dfrac{\frac{1}{x} - \frac{1}{y}}{x + y}$

41. $(9x^3 + 5x^2 + 2) \div (x + 2)$

Factor.

42. $4x^3 + 18x^2$

43. $x^2 + 8x - 84$

44. $16y^2 - 81$

45. $6x^2 + 11x - 10$

46. $64x^3 + 8$

47. $x^2 - 16x + 64$

48. $x^6 - x^2$

49. $0.027b^3 - 0.008c^3$

50. $20x^2 + 17x + 3$

51. $3x^2 - 5x - 28$

52. Solve $P = 2l + 2w$ for w.

Problem-solving practice

53. The sum of the squares of 3 consecutive even numbers is equal to 8 more than 3 times the square of the second number. Find the numbers.

54. A store purchases a case of shirts from a wholesaler for $135. There are 48 shirts in the case. The store wants to realize a profit by the time they sell half the shirts. What prices could they charge for each shirt in order to meet that goal?

55. How much pure alcohol must be added to 30 mL of a 40% alcohol solution to obtain a 50% alcohol solution?

56. One tractor can plow a field in 3 hr. Another tractor can plow the field in 1.5 hr. How long should it take them to plow the field together?

57. The cost c of an insurance policy varies directly as the age of the insured. If a 32-year-old person pays an annual premium of $152, what is the age of a person who pays $270?

7

EXPONENTIAL AND RADICAL SYMBOLISM; COMPLEX NUMBERS

Exponential and radical expressions and equations are used by surveyors to solve problems.

In this chapter we will be concerned with square roots, cube roots, and the like. To name such numbers, we use a radical symbol $\sqrt{\ }$, and we also use those symbols in algebraic expressions known as *radical expressions.* Curiously, certain radical expressions are equivalent to expressions with exponents. In this chapter we will study that connection. Of course, you will again have the opportunity to apply your knowledge of this kind of algebraic symbolism to the solving of problems.

OBJECTIVES

After finishing Section 7.1, you should be able to:

- **i** **Find principal square roots and their additive inverses.**
- **ii** **Find function values for square root functions.**
- **iii** **Simplify radical expressions with perfect-square radicands.**
- **iv** **Find cube roots, simplifying certain expressions.**
- **v** **Simplify expressions involving odd and even roots.**

Find the square roots of each number.

1. 9

2. 36

3. 121

Simplify.

4. $\sqrt{1}$

5. $\sqrt{36}$

6. $\sqrt{\frac{81}{100}}$

7. $\sqrt{0.0064}$

Find the following.

8. (a) $\sqrt{16}$; (b) $-\sqrt{16}$; (c) $\sqrt{-16}$

9. (a) $\sqrt{49}$; (b) $-\sqrt{49}$; (c) $\sqrt{-49}$

10. (a) $\sqrt{144}$; (b) $-\sqrt{144}$; (c) $\sqrt{-144}$

7.1

RADICAL EXPRESSIONS

In this section we consider roots, such as square roots and cube roots. We look at the symbolism that is used for them and the ways in which we can manipulate symbols to get equivalent expressions. All of that can be important in problem solving.

i Square Roots

A *square root* of a number a is a number c whose second power is a, that is, $c^2 = a$. For example,

5 is a square root of 25 because $5 \cdot 5 = 25$;

-5 is a square root of 25 because $(-5)(-5) = 25$;

-4 does not have a real-number square root because there is no real number such that $c^2 = -4$.

Later in this chapter we will see that there is a number system, other than the real-number system, in which negative numbers do have square roots.

Every positive real number has two real-number square roots. The number 0 has just one square root, 0 itself. Negative numbers do not have real-number square roots.

EXAMPLE 1 Find the two square roots of 64.

The square roots are 8 and -8, because $8^2 = 64$ and $(-8)^2 = 64$.

DO EXERCISES 1–3.

The *principal square root* of a nonnegative number is its nonnegative square root. The symbol $\sqrt{a}$ represents the principal square root of a. To name the negative square root of a, we write $-\sqrt{a}$.

EXAMPLES Simplify. Remember, $\sqrt{\ }$ indicates the principal square root.

2. $\sqrt{25} = 5$

3. $-\sqrt{64} = -8$

4. $\sqrt{\frac{25}{64}} = \frac{5}{8}$

5. $\sqrt{0.0049} = 0.07$

DO EXERCISES 4–10.

The symbol $\sqrt{\ }$ is called a *radical sign*. An expression written with a radical sign is called a *radical expression*. The expression written under the radical sign is called the *radicand*.

The following are radical expressions:

$$\sqrt{5}, \quad \sqrt{a}, \quad -\sqrt{5x}, \quad \sqrt{\frac{y^2 + 7}{\sqrt{x}}}.$$

ii Square Root Functions. For every nonnegative real number x, there is a principal square root $\sqrt{x}$. Thus there is a *square root function*,

$$f(x) = \sqrt{x}.$$

The domain is the set of all nonnegative real numbers, and the range is the set of all nonnegative real numbers.

There is another square root function,

$$f(x) = -\sqrt{x}.$$

The domain of this function is the set of all nonnegative real numbers and the range is the set of all nonpositive real numbers.

Years ago, the symbol $\sqrt{x}$ was used to represent *all* of the square roots of x. That usage has almost completely disappeared, because today we pay a lot more attention to functions.

By changing the radicand, we can write symbolism for still other square root functions. Examples are as follows:

$$f(x) = \sqrt{2x - 5}, \qquad g(x) = -\sqrt{3x^2 - 4x}.$$

These functions have different domains and ranges.

EXAMPLES For each function, determine whether the specified input is actually a member of the domain. If so, find the function value.

6. $f(x) = \sqrt{3x - 2}$; input 1

$f(1) = \sqrt{3 \cdot 1 - 2}$ Substituting

$= \sqrt{1} = 1$ Simplifying and taking the square root

7. $g(y) = \sqrt{5y^2 + 4}$; input -3

$g(-3) = \sqrt{5 \cdot (-3)^2 + 4}$ Substituting

$= \sqrt{49} = 7$ Simplifying and taking the square root

8. $f(x) = -\sqrt{6x + 4}$; input 2

$f(2) = -\sqrt{6(2) + 4}$ Substituting

$= -\sqrt{16} = -4$ Simplifying and taking the negative square root

9. $f(x) = \sqrt{5x - 12}$; input 1

$f(1) = \sqrt{5(1) - 12}$ Substituting

$= \sqrt{-7}$ Simplifying

Since negative numbers do not have real-number square roots, $f(1)$ does not exist. The number 1 is not in the domain of the function.

DO EXERCISES 11 AND 12.

Function Values, Calculators, and Tables. All but the simplest calculators give values for square roots, but they are, for the most part, approximations. For example, if you enter 5 in your calculator and then press the $\sqrt{\ }$ key, you will obtain something like

2.23606798,

depending on how your calculator rounds. The exact value is not given by any terminating decimal. The same will be true of the square root of any whole number, unless that square root is also a whole number (unless the number is a perfect square).

Determine whether the inputs are members of the domain of the function. If so, find the function values.

11. $f(x) = \sqrt{5x - 4}$

Inputs

a) 1

b) 0

c) 4

12. $g(y) = -\sqrt{2x + 16}$

Inputs

a) 0

b) 10

c) -6

d) -10

Simplify. Assume that letters can represent any real numbers.

13. $\sqrt{y^2}$

14. $\sqrt{(-24)^2}$

15. $\sqrt{(5y)^2}$

16. $\sqrt{16y^2}$

17. $\sqrt{(x+7)^2}$

18. $\sqrt{4(x-2)^2}$

19. $\sqrt{49(y+5)^2}$

20. $\sqrt{x^2-6x+9}$

Simplify. Assume that the variables and radicands do not represent negative numbers.

21. $\sqrt{z^2}$

22. $\sqrt{(3x+4)^2}$

23. $\sqrt{x^2-2x+1}$

Table 1 at the back of the book contains approximate values of square roots. That table can, of course, be used to find approximate function values for square-root functions. We shall have more to say about the table later.

iii Expressions $\sqrt{a^2}$

We now consider a special case, in which a radicand is a perfect square, an expression such as $\sqrt{a^2}$.

Suppose that $a = 5$; then we have $\sqrt{5^2}$, which is $\sqrt{25}$, or 5.

Suppose that $a = -5$; then we have $\sqrt{(-5)^2}$, which is $\sqrt{25}$, or 5.

The symbol $\sqrt{a^2}$ never represents a negative number, even though the input might be negative.

EXAMPLE 10 For the function $f(x) = \sqrt{x^2}$, find function values for the following inputs: (a) 6; (b) 0; (c) -2; (d) -10.

a) $f(6) = \sqrt{6^2} = \sqrt{36} = 6$
b) $f(0) = \sqrt{0^2} = \sqrt{0} = 0$
c) $f(-2) = \sqrt{(-2)^2} = \sqrt{4} = 2$
d) $f(-10) = \sqrt{(-10)^2} = \sqrt{100} = 10$

We have the following general rule, which we can use in simplifying radical expressions.

For any real number a,

$$\sqrt{a^2} = |a|.$$

(The principal square root of a^2 is the absolute value of a.)

When an expression contains variables in the radicand, we need to use absolute-value signs when simplifying unless we know (as we do in many applications) that the letters involved do not represent negative numbers.

EXAMPLES Simplify. Do not assume that the variables represent nonnegative numbers. (You may need absolute-value signs.)

11. $\sqrt{(x-1)^2} = |x-1|$
12. $\sqrt{x^2+8x+16} = \sqrt{(x+4)^2} = |x+4|$

DO EXERCISES 13–20.

EXAMPLES Simplify. Assume that the variables and radicands do not represent negative numbers.

13. $\sqrt{y^2} = y$
14. $\sqrt{(5x+2)^2} = 5x+2$
15. $\sqrt{x^2+2x+1} = \sqrt{(x+1)^2} = x+1$

DO EXERCISES 21–23.

iv Cube Roots

A number c is the *cube root* of a if its third power is a, that is, if $c^3 = a$. For example,

$$2 \text{ is the cube root of } 8 \text{ because } 2^3 = 2 \cdot 2 \cdot 2 = 8;$$

$$-4 \text{ is the cube root of } -64 \text{ because } (-4)^3 = (-4)(-4)(-4) = -64.$$

In the real-number system, every number has exactly one cube root. The cube root of a positive number is positive, and the cube root of a negative number is negative. In simplifying expressions involving cube roots, we need not use absolute-value signs.

EXAMPLE 16 For the function $f(y) = \sqrt[3]{y}$, find $f(8)$.

$$f(8) = \sqrt[3]{8} \quad \text{Substituting}$$
$$= 2 \quad \text{Taking the cube root}$$

EXAMPLE 17 For the function $g(x) = \sqrt[3]{x - 3}$, find $g(-24)$.

$$g(-24) = \sqrt[3]{-24 - 3} \quad \text{Substituting}$$
$$= \sqrt[3]{-27} = -3 \quad \text{Simplifying and taking the cube root}$$

EXAMPLE 18 Simplify: $\sqrt[3]{-8y^3}$.

$$\sqrt[3]{-8y^3} = -2y$$

(No absolute-value signs are needed.)

DO EXERCISES 24–27.

24. For the function $f(x) = \sqrt[3]{x - 2}$, find $f(66)$.

25. For the function $g(y) = \sqrt[3]{2y - 7}$, find $g(-10)$.

Simplify.

26. $\sqrt[3]{8(x + 2)^3}$

27. $\sqrt[3]{-27(y - 4)^3}$

v Odd and Even kth Roots

The fifth root of a number a is the number c for which $c^5 = a$. There are also 7th roots, 9th roots, and so on. Whenever the number k in $\sqrt[k]{}$ is an odd number, we say that we are taking an *odd* root. The number k is called the *index*. When the *index* is 2 we do not write it.

Every number has just one root for an odd k. If the number is positive, the root is positive. If the number is negative, the root is negative.

EXAMPLES Find each of the following.

19. $\sqrt[5]{32} = 2$

20. $\sqrt[5]{-32} = -2$

21. $-\sqrt[5]{32} = -2$

22. $-\sqrt[5]{-32} = -(-2) = 2$

23. $\sqrt[7]{x^7} = x$

24. $\sqrt[9]{(x - 1)^9} = x - 1$

Absolute-value signs are never needed when finding odd roots.

When the index k in $\sqrt[k]{}$ is an even number, we say that we are taking an *even* root. Every positive real number has two kth roots when k is even. One of those roots is positive and one is negative. Negative real numbers do not have kth roots when k is even. When finding even kth roots, absolute-value signs are sometimes necessary, as with square roots.

EXAMPLES Find the following. Assume that variables can represent any real numbers.

25. $\sqrt[4]{16} = 2$

26. $-\sqrt[4]{16} = -2$

27. $\sqrt[4]{-16}$ No real-number square roots

28. $\sqrt[4]{81x^4} = 3|x|$

29. $\sqrt[6]{(y+7)^6} = |y+7|$

DO EXERCISES 28–33.

> **For any real number a:**
>
> **a) $\sqrt[k]{a^k} = |a|$ when k is even. We use absolute value when k is even unless a is nonnegative.**
>
> **b) $\sqrt[k]{a^k} = a$ when k is odd. We do not use absolute value when k is odd.**

Find the following. Assume that letters can represent any real number.

28. $\sqrt[5]{-32x^5}$

29. $\sqrt[7]{(3x+2)^7}$

Find the following.

30. (a) $\sqrt[4]{81}$; (b) $-\sqrt[4]{81}$; (c) $\sqrt[4]{-81}$

31. $\sqrt[4]{16(x-2)^4}$

32. $\sqrt[6]{x^6}$

33. $\sqrt[8]{(x+3)^8}$

EXERCISE SET 7.1

i Find the square roots of each number.

1. 16 **2.** 225 **3.** 144 **4.** 9 **5.** 400 **6.** 81

Find the following square roots.

7. $-\sqrt{\dfrac{49}{36}}$ **8.** $-\sqrt{\dfrac{361}{9}}$ **9.** $\sqrt{196}$ **10.** $\sqrt{441}$ **11.** $-\sqrt{\dfrac{16}{81}}$

12. $-\sqrt{\dfrac{81}{144}}$ **13.** $\sqrt{0.09}$ **14.** $\sqrt{0.36}$ **15.** $-\sqrt{0.0049}$ **16.** $\sqrt{0.0144}$

Identify the radicand in each expression.

17. $5\sqrt{p^2+4}$ **18.** $-7\sqrt{y^2-8}$ **19.** $x^2y^2\sqrt{\dfrac{x}{y+4}}$ **20.** $a^2b^3\sqrt{\dfrac{a}{a^2-b}}$

ii For each function, determine whether the specified inputs are members of the domain. If so, find the function values.

21. $f(y) = \sqrt{5y-10}$; Inputs: 6, 2, 0, -3

22. $g(x) = \sqrt{x^2-25}$; Inputs: -5, 5, 0, 6, -6

23. $t(x) = -\sqrt{2x+1}$; Inputs: 4, -4, 0, 12

24. $p(z) = \sqrt{2z^2-20}$; Inputs: 0, 5, -5, 10, -10

25. $f(t) = \sqrt{t^2+1}$; Inputs: 5, -5, 0, 10, -10

26. $g(x) = -\sqrt{(x+1)^2}$; Inputs: 1, -1, 3, -3, 5, -5

iii Simplify. Do not assume that the variables represent nonnegative numbers.

27. $\sqrt{16x^2}$ **28.** $\sqrt{25t^2}$ **29.** $\sqrt{(-7c)^2}$ **30.** $\sqrt{(-6b)^2}$

31. $\sqrt{(a+1)^2}$ **32.** $\sqrt{(5-b)^2}$ **33.** $\sqrt{x^2-4x+4}$ **34.** $\sqrt{y^2+16y+64}$

35. $\sqrt{4x^2+28x+49}$ **36.** $\sqrt{9x^2-30x+25}$

Simplify. Assume that the variables and radicands do not represent negative numbers.

37. $\sqrt{16x^2}$ **38.** $\sqrt{25t^2}$ **39.** $\sqrt{(-6b)^2}$ **40.** $\sqrt{(-7c)^2}$

41. $\sqrt{(a+1)^2}$ **42.** $\sqrt{(5+b)^2}$ **43.** $\sqrt{4x^2+8x+4}$ **44.** $\sqrt{9x^2+36x+36}$

iv Simplify.

45. $\sqrt[3]{27}$ **46.** $-\sqrt[3]{64}$ **47.** $\sqrt[3]{-64x^3}$ **48.** $\sqrt[3]{-125y^3}$

49. $\sqrt[3]{-216}$ **50.** $-\sqrt[3]{-1000}$ **51.** $\sqrt[3]{0.343(x+1)^3}$ **52.** $\sqrt[3]{0.000008(y-2)^3}$

For each function, determine whether the specified inputs are members of the domain. If so, find the function values.

53. $f(x) = \sqrt[3]{x+1}$; Inputs: 7, 26, -9, -65

54. $g(x) = -\sqrt[3]{2x-1}$; Inputs: 0, -62, -13, 63

v Find each of the following. Do not assume that variables represent nonnegative numbers.

55. $\sqrt[4]{625}$ **56.** $-\sqrt[4]{256}$ **57.** $\sqrt[5]{-1}$ **58.** $-\sqrt[5]{-32}$ **59.** $\sqrt[5]{\frac{32}{243}}$

60. $\sqrt[5]{-\frac{1}{32}}$ **61.** $\sqrt[6]{x^6}$ **62.** $\sqrt[8]{y^8}$ **63.** $\sqrt[4]{(5a)^4}$ **64.** $\sqrt[4]{(7b)^4}$

65. $\sqrt[10]{(-6)^{10}}$ **66.** $\sqrt[12]{(-10)^{12}}$ **67.** $\sqrt[414]{(a+b)^{414}}$ **68.** $\sqrt[1976]{(2a+b)^{1976}}$ **69.** $\sqrt[7]{y^7}$

70. $\sqrt[3]{(-6)^3}$ **71.** $\sqrt[5]{(x-2)^5}$ **72.** $\sqrt[9]{(2xy)^9}$

73. (*An application: Parking*). A parking lot has attendants to park the cars. The number N of temporary stalls needed for waiting cars before attendants can get to them is given by the formula $N = 2.5\sqrt{A}$, where A is the number of arrivals in peak hours. Find the number of spaces needed for the given average number of arrivals in peak hours: (a) 25; (b) 36; (c) 49; (d) 64.

74. Graph the function $y = \sqrt{x}$. **75.** Graph the function $y = -\sqrt{x}$.

7.2

MULTIPLYING AND SIMPLIFYING WITH RADICAL NOTATION

i Multiplying

Notice that

$$\sqrt{4}\sqrt{25} = 2 \cdot 5 = 10 \quad \text{and} \quad \sqrt{4 \cdot 25} = \sqrt{100} = 10.$$

Likewise,

$$\sqrt[3]{27}\sqrt[3]{8} = 3 \cdot 2 = 6 \quad \text{and} \quad \sqrt[3]{27 \cdot 8} = \sqrt[3]{216} = 6.$$

These examples suggest the following.

For any nonnegative real numbers a and b and any index k,

$$\sqrt[k]{a} \cdot \sqrt[k]{b} = \sqrt[k]{a \cdot b}.$$

(To multiply, we multiply the radicands.)

EXAMPLES Multiply.

1. $\sqrt{3} \cdot \sqrt{5} = \sqrt{3 \cdot 5} = \sqrt{15}$
2. $\sqrt{x+2}\sqrt{x-2} = \sqrt{(x+2)(x-2)} = \sqrt{x^2-4}$
3. $\sqrt[3]{4}\sqrt[3]{5} = \sqrt[3]{4 \cdot 5} = \sqrt[3]{20}$
4. $\sqrt[4]{\frac{y}{5}}\sqrt[4]{\frac{7}{x}} = \sqrt[4]{\frac{y}{5} \cdot \frac{7}{x}} = \sqrt[4]{\frac{7y}{5x}}$

DO EXERCISES 1–4.

ii Simplifying by Factoring

Reversing the above statement, we have $\sqrt[k]{a \cdot b} = \sqrt[k]{a} \cdot \sqrt[k]{b}$. This shows a way to factor and thus simplify radical expressions. Consider $\sqrt{20}$. The number 20 has the factor 4, which is a perfect square. Therefore,

$$\sqrt{20} = \sqrt{4 \cdot 5} \quad \text{Factoring the radicand (4 is a perfect square)}$$
$$= \sqrt{4} \cdot \sqrt{5} \quad \text{Factoring into two radicals}$$
$$= 2\sqrt{5}. \quad \text{Taking the square root of 4}$$

OBJECTIVES

After finishing Section 7.2, you should be able to:

- **i** **Multiply with radical notation.**
- **ii** **Simplify radical expressions by factoring, assuming that all expressions represent nonnegative numbers.**
- **iii** **Multiply and simplify radical expressions.**
- **iv** **Identify rational and irrational numbers.**
- **v** **Approximate square roots using a table or calculator.**

Multiply.

1. $\sqrt{19}\sqrt{7}$

2. $\sqrt{x+2y}\sqrt{x-2y}$

3. $\sqrt[4]{403}\sqrt[4]{7}$

4. $\sqrt[3]{8x}\sqrt[3]{x^4+5}$

Simplify by factoring.

5. $\sqrt{32}$

6. $\sqrt[3]{80}$

Simplify by factoring. Assume that all expressions represent nonnegative numbers. Hence no absolute-value signs will be needed.

7. $\sqrt{300}$

8. $\sqrt{36y^2}$

9. $\sqrt{3x^2 + 12x + 12}$

10. $\sqrt{12ab^3c^2}$

11. $\sqrt[3]{16}$

12. $\sqrt[3]{81x^4y^8}$

13. $\sqrt[3]{(a+b)^4}$

Multiply and then simplify by factoring. Assume that all expressions under radicals represent nonnegative numbers.

14. $\sqrt{3}\sqrt{6}$

15. $\sqrt{18y}\sqrt{14y}$

16. $\sqrt[3]{3x^2y}\,\sqrt[3]{36x}$

17. $\sqrt{7a}\sqrt{21b}$

18. $\sqrt[3]{2(y+5)}\,\sqrt[3]{4(y+5)^4}$

To simplify a radical expression by factoring, look for factors of the radicand that are perfect *k*th powers (where *k* is the index). Then take the *k*th root of those factors.

EXAMPLES Simplify by factoring.

5. $\sqrt{50} = \sqrt{25 \cdot 2} = \sqrt{25} \cdot \sqrt{2} = 5\sqrt{2}$ — This factor is a perfect square.

6. $\sqrt[3]{32} = \sqrt[3]{8 \cdot 4} = \sqrt[3]{8} \cdot \sqrt[3]{4} = 2\sqrt[3]{4}$ — This factor is a perfect cube (third power).

DO EXERCISES 5 AND 6.

In many situations, expressions under radicals never represent negative numbers. In such cases, absolute-value notation is not necessary.

EXAMPLES Simplify by factoring. Assume that all expressions under radicals represent nonnegative numbers.

7. $\sqrt{5x^2} = \sqrt{x^2 \cdot 5}$ Factoring the radicand

$= \sqrt{x^2} \cdot \sqrt{5}$ Factoring into two radicals

$= x \cdot \sqrt{5}$ Taking the square root of x^2

8. $\sqrt{2x^2 - 4x + 2} = \sqrt{2(x-1)^2}$

$= \sqrt{(x-1)^2} \cdot \sqrt{2}$

$= (x-1) \cdot \sqrt{2}$

9. $\sqrt{216x^5y^3} = \sqrt{36 \cdot 6 \cdot x^4 \cdot x \cdot y^2 \cdot y}$

$= \sqrt{36 \cdot x^4 \cdot y^2 \cdot 6 \cdot x \cdot y}$ Factoring the radicand

$= \sqrt{36}\sqrt{x^4}\sqrt{y^2}\sqrt{6xy}$ Factoring into several radicals

$= 6x^2y\sqrt{6xy}$ Taking square roots

Note: Had we not seen that $216 = 36 \cdot 6$, where 36 is the largest square factor of 216, we could have found the prime factorization $2 \cdot 2 \cdot 2 \cdot 3 \cdot 3 \cdot 3$. Each pair of factors makes a square, so

$$\sqrt{2 \cdot 2 \cdot 2 \cdot 3 \cdot 3 \cdot 3} = \sqrt{2^2 \cdot 3^2 \cdot 2 \cdot 3} = 2 \cdot 3\sqrt{2 \cdot 3}.$$

DO EXERCISES 7–13.

iii Multiplying and Simplifying

Sometimes after we multiply we can then simplify by factoring.

EXAMPLES Multiply and then simplify by factoring. Assume that all expressions under radicals represent nonnegative numbers.

10. $\sqrt{15}\sqrt{6} = \sqrt{15 \cdot 6} = \sqrt{90} = \sqrt{9 \cdot 10} = 3\sqrt{10}$

11. $3\sqrt[3]{25} \cdot 2\sqrt[3]{5} = 6 \cdot \sqrt[3]{25 \cdot 5}$

$= 6 \cdot \sqrt[3]{125}$ Multiplying radicands

$= 6 \cdot 5$, or 30 Taking the cube root of 125

12. $\sqrt[3]{18y^3}\,\sqrt[3]{4x^2} = \sqrt[3]{18y^3 \cdot 4x^2}$

$= \sqrt[3]{72y^3x^2}$ Multiplying radicands

$= \sqrt[3]{8y^3 \cdot 9x^2}$ Factoring the radicand

$= \sqrt[3]{8y^3}\,\sqrt[3]{9x^2}$ Factoring into two radicals

$= 2y\sqrt[3]{9x^2}$ Taking the cube root

DO EXERCISES 14–18.

iv Rational and Irrational Numbers

The real numbers consist of the rational numbers and the irrational numbers. We can describe these kinds of numbers in several ways.

Rational numbers are:

1. **Those numbers that can be named with fractional notation a/b, where a and b are integers and $b \neq 0$. (Definition)**
2. **Those numbers for which decimal notation either ends or repeats.**

EXAMPLES

13. $\frac{5}{16} = 0.3125$ Ending (terminating) decimal

14. $\frac{8}{7} = 1.142857142857\ldots = 1.\overline{142857}$ Repeating decimal

15. $\frac{3}{11} = 0.2727\ldots = 0.\overline{27}$ Repeating decimal

We can also describe the irrational numbers in several ways.

Irrational numbers are:

1. **Those real numbers that are not rational.**
2. **Those real numbers that cannot be named with fractional notation a/b, where a and b are integers and $b \neq 0$.**
3. **Those real numbers for which decimal notation does not end and does not repeat.**

There are many irrational numbers. For example, $\sqrt{2}$ is irrational. We can find rational numbers a/b for which $(a/b) \cdot (a/b)$ is close to 2, but we cannot find such a number a/b for which $(a/b) \cdot (a/b)$ is *exactly* 2.

Unless a whole number is a perfect square, its square root is irrational. For example, $\sqrt{9}$ and $\sqrt{25}$ are rational, but all of the following are irrational:

$$\sqrt{3}, \qquad -\sqrt{14}, \qquad \sqrt{45}.$$

There are also many irrational numbers that we do not obtain by taking square roots. The number π is an example.* Decimal notation for π does not end and does not repeat.

EXAMPLES All of these are irrational.

16. $\pi = 3.1415926535\ldots$ Numeral does not repeat.

17. $1.101001000100001000001\ldots$ Numeral does not repeat.

18. $\sqrt{2} = 1.41421356\ldots$ Numeral does not repeat.

In Example 17 there is a pattern, but it is not a repeating pattern.

DO EXERCISES 19–25.

v Finding Square Roots

Not all square roots can be found directly from Table 1 or by using a calculator. In that event we proceed by factoring.

* 3.14 and $\frac{22}{7}$ are only rational approximations to π.

Which of the following are rational? Which are irrational?

19. $\frac{-4}{5}$

20. $\frac{59}{37}$

21. 7.42

22. 0.47474747...
(Numeral repeats)

23. 2.57340046631...
(Numeral does not repeat)

24. $\sqrt{7}$

25. $\sqrt{16}$

EXAMPLE 19 Use Table 1 to approximate $\sqrt{275}$ to the nearest tenth.

$$\begin{aligned}\sqrt{275} &= \sqrt{25 \cdot 11} && \text{Factoring the radicand (note that 25 is a perfect square)}\\ &= \sqrt{25}\cdot\sqrt{11} && \text{Factoring the expression}\\ &= 5\sqrt{11}\\ &\approx 5 \times 3.317 && \text{Using Table 1}\\ &\approx 16.6 && \text{Multiplying and rounding}\end{aligned}$$

DO EXERCISES 26 AND 27.

Use Table 1 to approximate to the nearest tenth.

26. $\sqrt{252}$

27. $\sqrt{405}$

EXAMPLE 20 Use a calculator to approximate $\sqrt{1{,}350{,}000{,}000}$.

If the number does not fit into a calculator, we factor the radicand, as follows:

$$\begin{aligned}\sqrt{1{,}350{,}000{,}000} &= \sqrt{1350 \cdot 10^6} && \text{Factoring the radicand (note that the exponent, 6, is even)}\\ &= \sqrt{1350}\cdot\sqrt{10^6} && \text{Factoring the expression}\\ &\approx 36.74 \cdot 10^3 && \text{Approximating } \sqrt{1350} \text{ with a calculator and finding } \sqrt{10^6}\\ &\approx 36{,}740. && \text{Multiplying}\end{aligned}$$

EXAMPLE 21 Use a calculator to approximate $\sqrt{0.000000005768}$.

If the number will not fit into a calculator, we factor:

$$\begin{aligned}\sqrt{0.000000005768} &= \sqrt{57.68 \times 10^{-10}} && \text{Factoring the radicand (note that the exponent, } -10\text{, is even)}\\ &= \sqrt{57.68}\times\sqrt{10^{-10}} && \text{Factoring the expression}\\ &\approx 7.595 \times 10^{-5} && \text{Approximating } \sqrt{57.68} \text{ with a calculator and finding } \sqrt{10^{-10}}\\ &\approx 0.00007595.\end{aligned}$$

DO EXERCISES 28 AND 29.

Use a calculator to approximate.

28. $\sqrt{473{,}100{,}000{,}000{,}000}$

29. $\sqrt{0.00000000000003142}$

EXERCISE SET 7.2

i Multiply.

1. $\sqrt{3}\sqrt{2}$ **2.** $\sqrt{5}\sqrt{7}$ **3.** $\sqrt[3]{2}\sqrt[3]{5}$ **4.** $\sqrt[3]{7}\sqrt[3]{2}$

5. $\sqrt[4]{8}\sqrt[4]{9}$ **6.** $\sqrt[4]{6}\sqrt[4]{3}$ **7.** $\sqrt{3a}\sqrt{10b}$ **8.** $\sqrt{2x}\sqrt{13y}$

9. $\sqrt[5]{9t^2}\sqrt[5]{2t}$ **10.** $\sqrt[5]{8y^3}\sqrt[5]{10y}$ **11.** $\sqrt{x-a}\sqrt{x+a}$ **12.** $\sqrt{y-b}\sqrt{y+b}$

13. $\sqrt[3]{0.3x}\sqrt[3]{0.2x}$ **14.** $\sqrt[3]{0.7y}\sqrt[3]{0.3y}$ **15.** $\sqrt[4]{x-1}\sqrt[4]{x^2+x+1}$ **16.** $\sqrt[5]{x-2}\sqrt[5]{(x-2)^2}$

17. $\sqrt{\frac{6}{x}}\sqrt{\frac{y}{5}}$ **18.** $\sqrt{\frac{7}{t}}\sqrt{\frac{s}{11}}$

ii Simplify by factoring. Assume that all variables and radicands represent nonnegative numbers.

19. $\sqrt{8}$ **20.** $\sqrt{18}$ **21.** $\sqrt{24}$ **22.** $\sqrt{20}$ **23.** $\sqrt{180x^4}$ **24.** $\sqrt{175y^6}$

25. $\sqrt[3]{54x^8}$ **26.** $\sqrt[3]{40y^3}$ **27.** $\sqrt[3]{80x^8}$ **28.** $\sqrt[3]{108m^5}$ **29.** $\sqrt[4]{32}$ **30.** $\sqrt[4]{80}$

31. $\sqrt[4]{162c^4d^6}$ **32.** $\sqrt[4]{243x^8y^{10}}$ **33.** $\sqrt[3]{(x+y)^4}$

iii Multiply and simplify by factoring. Assume that all variables and radicands represent nonnegative numbers.

34. $\sqrt{3}\sqrt{6}$ **35.** $\sqrt{5}\sqrt{10}$ **36.** $\sqrt{15}\sqrt{12}$ **37.** $\sqrt{2}\sqrt{32}$

38. $\sqrt{6}\sqrt{8}$ **39.** $\sqrt{18}\sqrt{14}$ **40.** $\sqrt[3]{3}\sqrt[3]{18}$ **41.** $\sqrt{45}\sqrt{60}$

42. $\sqrt{5b^3}\sqrt{10c^4}$
43. $\sqrt{2x^3y}\sqrt{12xy}$
44. $\sqrt[3]{y^4}\sqrt[3]{16y^5}$
45. $\sqrt[3]{5^2t^4}\sqrt[3]{5^4t^6}$
46. $\sqrt[3]{(b+3)^4}\sqrt[3]{(b+3)^2}$
47. $\sqrt[3]{(x+y)^3}\sqrt[3]{(x+y)^5}$
48. $\sqrt{12a^3b}\sqrt{8a^4b^2}$

iv Which of the following are irrational numbers? Which are rational?

49. $-\frac{5}{6}$
50. 8.93
51. $8.2\overline{323}$
52. $\sqrt{5}$
53. $\sqrt{36}$
54. $-\sqrt{15}$
55. 2.101001 . . . (Numeral does not repeat.)
56. 3.1415926 . . . (Numeral does not repeat.)

v Use Table 1 to approximate to the nearest tenth.

57. $\sqrt{252}$
58. $\sqrt{140}$
59. $\sqrt{189}$
60. $\sqrt{490}$
61. $\sqrt{350}$
62. $\sqrt{320}$
63. $\sqrt{891}$
64. $\sqrt{376}$
65. $\sqrt{240}$
66. $\sqrt{369}$

Use a calculator to approximate each of the following.

67. $\sqrt{24{,}500{,}000{,}000}$
68. $\sqrt{16{,}500{,}000{,}000}$
69. $\sqrt{468{,}200{,}000{,}000}$
70. $\sqrt{99{,}400{,}000{,}000}$
71. $\sqrt{175{,}420{,}000{,}000}$
72. $\sqrt{0.000004631}$
73. $\sqrt{0.0000000395}$
74. $\sqrt{0.0000001543}$
75. $\sqrt{0.0000005001}$
76. $\sqrt{0.000010101}$

Problem-solving practice

77. During a one-hour television show there were 12 commercials. Some of the commercials were 30 seconds long and the others were 60 seconds long. If the number of 30-second commercials was six less than the total number of minutes of commercial time during the show, how many 60-second commercials were used during the hour?

78. The length of a rectangle is one-fourth of the width. The area of the rectangle is twice the perimeter. Find the dimensions of the rectangle.

---

79. (*An application: Speed of a skidding car*). After an accident, police can estimate the speed at which a car was traveling by measuring its skid marks. The formula

$$r = 2\sqrt{5L}$$

can be used, where r is the speed in mph and L is the length of the skid marks in feet. Estimate the speed of a car that left skid marks (a) 20 ft long; (b) 70 ft long; (c) 90 ft long.

80. (*An application: Wind chill temperature*). In cold weather we feel colder if there is wind than if there is not. *Wind chill temperature* is the temperature at which, without wind, we would feel as cold in an actual situation with wind. Here is a formula for finding wind chill temperature:

$$T_w = 33 - \frac{(10.45 + 10\sqrt{v} - v)(33 - T)}{22},$$

where T is the actual temperature given in degrees Celsius and v is the wind speed in m/sec. Find the wind chill temperature for the given actual temperatures and wind speeds.

a) $T = 7°\text{C}$, $v = 8$ m/sec
b) $T = 0°\text{C}$, $v = 12$ m/sec
c) $T = -5°\text{C}$, $v = 14$ m/sec
d) $T = -23°\text{C}$, $v = 15$ m/sec

7.3

DIVIDING AND SIMPLIFYING WITH RADICAL NOTATION

In this section we will manipulate radical expressions using division rather than multiplication.

OBJECTIVES

After finishing Section 7.3, you should be able to:

i **Simplify radical expressions having a quotient for a radicand.**

ii **Divide and simplify radical expressions.**

iii **Calculate combinations of roots and powers.**

i Roots of Quotients

Notice that

$$\sqrt[3]{\frac{27}{8}} = \frac{3}{2} \quad \text{and} \quad \frac{\sqrt[3]{27}}{\sqrt[3]{8}} = \frac{3}{2}.$$

This example suggests that we can take the root of a quotient by taking the root of the numerator and denominator separately. This is true:

Rule A: For any nonnegative number a and any positive number b, and any index k,

$$\sqrt[k]{\frac{a}{b}} = \frac{\sqrt[k]{a}}{\sqrt[k]{b}}.$$

(We can take the kth root of the numerator and denominator separately.)

EXAMPLES Simplify by taking roots of the numerator and denominator. Assume that all expressions under radicals represent positive numbers.

1. $\sqrt[3]{\frac{27}{125}} = \frac{\sqrt[3]{27}}{\sqrt[3]{125}} = \frac{3}{5}$ Taking the cube root of the numerator and denominator

2. $\sqrt{\frac{25}{y^2}} = \frac{\sqrt{25}}{\sqrt{y^2}} = \frac{5}{y}$ Taking the square root of the numerator and denominator

3. $\sqrt{\frac{16x^3}{y^4}} = \frac{\sqrt{16x^3}}{\sqrt{y^4}} = \frac{\sqrt{16x^2 \cdot x}}{\sqrt{y^4}} = \frac{4x\sqrt{x}}{y^2}$

4. $\sqrt[3]{\frac{27y^5}{343x^3}} = \frac{\sqrt[3]{27y^5}}{\sqrt[3]{343x^3}} = \frac{\sqrt[3]{27y^3 \cdot y^2}}{\sqrt[3]{343x^3}} = \frac{\sqrt[3]{27y^3} \cdot \sqrt[3]{y^2}}{\sqrt[3]{343x^3}} = \frac{3y\sqrt[3]{y^2}}{7x}$

We are assuming that no expression represents zero or a negative number. Thus we need not be concerned about zero denominators or absolute-value signs.

DO EXERCISES 1–3.

Simplify by taking roots of numerator and denominator. Assume that all expressions under radicals represent positive numbers.

1. $\sqrt{\frac{25}{36}}$

2. $\sqrt{\frac{x^2}{100}}$

3. $\sqrt[3]{\frac{54x^5}{125}}$

ii Dividing Radical Expressions

Reversing Rule A shows a way to divide radical expressions as follows.

Rule B:

$$\frac{\sqrt[k]{a}}{\sqrt[k]{b}} = \sqrt[k]{\frac{a}{b}}$$

To divide, we divide the radicands. After doing so we can sometimes simplify by taking roots.

EXAMPLES Divide. Then simplify by taking roots if possible. Assume that all expressions under radicals represent positive numbers.

5. $\frac{\sqrt{80}}{\sqrt{5}} = \sqrt{\frac{80}{5}} = \sqrt{16} = 4$

6. $\frac{3\sqrt{2}}{5\sqrt{3}} = \frac{3}{5} \cdot \frac{\sqrt{2}}{\sqrt{3}} = \frac{3}{5} \cdot \sqrt{\frac{2}{3}}$

7. $\frac{5\sqrt[3]{32}}{\sqrt[3]{2}} = 5\sqrt[3]{\frac{32}{2}} = 5\sqrt[3]{16} = 5\sqrt[3]{8 \cdot 2} = 5\sqrt[3]{8}\sqrt[3]{2} = 5 \cdot 2\sqrt[3]{2} = 10\sqrt[3]{2}$

8. $\dfrac{\sqrt{72xy}}{2\sqrt{2}} = \dfrac{1}{2}\dfrac{\sqrt{72xy}}{\sqrt{2}} = \dfrac{1}{2}\sqrt{\dfrac{72xy}{2}} = \dfrac{1}{2}\sqrt{36xy} = \dfrac{1}{2}\sqrt{36}\sqrt{xy}$

$= \dfrac{1}{2}\cdot 6\sqrt{xy} = 3\sqrt{xy}$

9. $\dfrac{\sqrt[4]{33a^5b^3}}{\sqrt[4]{2b^{-1}}} = \sqrt[4]{\dfrac{33a^5b^3}{2b^{-1}}} = \sqrt[4]{\dfrac{33}{2}a^5b^4} = \sqrt[4]{a^4b^4}\sqrt[4]{\dfrac{33}{2}a} = ab\sqrt[4]{\dfrac{33}{2}a}$

DO EXERCISES 4–8.

iii Powers and Roots Combined

Consider the following:

$$\sqrt[3]{8^2} = \sqrt[3]{64} = 4,$$
$$(\sqrt[3]{8})^2 = (2)^2 = 4.$$

This suggests another important property of radical expressions.

Rule C: For any nonnegative number a, any index k, and any natural number m,

$$\sqrt[k]{a^m} = (\sqrt[k]{a})^m.$$

(We can raise to a power and then take a root, or we can take a root and then raise to a power.)

In some cases one way of calculating is easier than the other.

EXAMPLES Calculate as shown. Then use Rule C to calculate in another way. Assume that all expressions under radicals are positive.

10. a) $\sqrt[3]{27^2} = \sqrt[3]{729} = 9$ Finding 27^2 and then taking the cube root

b) $(\sqrt[3]{27})^2 = (3)^2 = 9$ Taking the cube root and then squaring

11. a) $\sqrt[3]{2^6} = \sqrt[3]{64} = 4$

b) $(\sqrt[3]{2})^6 = \sqrt[3]{2}\,\sqrt[3]{2}\,\sqrt[3]{2}\,\sqrt[3]{2}\,\sqrt[3]{2}\,\sqrt[3]{2} = 2\cdot 2 = 4$

12. a) $(\sqrt{5x})^3 = \sqrt{5x}\sqrt{5x}\sqrt{5x} = 5x\sqrt{5x}$

b) $\sqrt{(5x)^3} = \sqrt{5^3x^3} = \sqrt{5^2x^2}\,\sqrt{5x} = 5x\sqrt{5x}$

DO EXERCISES 9 AND 10.

Divide. Then simplify by taking roots if possible. Assume that all expressions under radicals represent positive numbers.

4. $\dfrac{\sqrt{75}}{\sqrt{3}}$

5. $\dfrac{14\sqrt{128xy}}{2\sqrt{2}}$

6. $\dfrac{\sqrt{50a^3}}{\sqrt{2a}}$

7. $\dfrac{4\sqrt[3]{250}}{7\sqrt[3]{2}}$

8. $\dfrac{\sqrt[3]{8a^3b}}{\sqrt[3]{27b^{-2}}}$

Calculate as shown. Then use Rule C to calculate another way.

9. $(\sqrt[3]{125})^2$

10. $(\sqrt{6y})^3$

EXERCISE SET 7.3

i Simplify by taking roots of the numerator and denominator. Assume that all expressions under radical signs represent positive numbers.

1. $\sqrt{\dfrac{16}{25}}$ **2.** $\sqrt{\dfrac{100}{81}}$ **3.** $\sqrt[3]{\dfrac{64}{27}}$ **4.** $\sqrt[3]{\dfrac{343}{512}}$ **5.** $\sqrt{\dfrac{49}{y^2}}$

6. $\sqrt{\dfrac{121}{x^2}}$ **7.** $\sqrt{\dfrac{25y^3}{x^4}}$ **8.** $\sqrt{\dfrac{36a^5}{b^6}}$ **9.** $\sqrt[3]{\dfrac{8x^5}{27y^3}}$ **10.** $\sqrt[3]{\dfrac{64x^7}{216y^6}}$

ii Divide. Then simplify by taking roots if possible. Assume that all expressions under radical signs represent positive numbers.

11. $\dfrac{\sqrt{21a}}{\sqrt{3a}}$ **12.** $\dfrac{\sqrt{28y}}{\sqrt{4y}}$ **13.** $\dfrac{\sqrt[3]{54}}{\sqrt[3]{2}}$ **14.** $\dfrac{\sqrt[3]{40}}{\sqrt[3]{5}}$ **15.** $\dfrac{\sqrt{40xy^3}}{\sqrt{8x}}$ **16.** $\dfrac{\sqrt{56ab^3}}{\sqrt{7a}}$

17. $\dfrac{\sqrt[3]{96a^4b^2}}{\sqrt[3]{12a^2b}}$ **18.** $\dfrac{\sqrt[3]{189x^5y^7}}{\sqrt[3]{7x^2y^2}}$ **19.** $\dfrac{\sqrt{72xy}}{2\sqrt{2}}$ **20.** $\dfrac{\sqrt{75ab}}{3\sqrt{3}}$ **21.** $\dfrac{\sqrt{x^3-y^3}}{\sqrt{x-y}}$ **22.** $\dfrac{\sqrt{r^3+s^3}}{\sqrt{r+s}}$

iii Calculate as shown. Then use Rule C to calculate in another way. Assume that all expressions under radicals represent positive numbers.

23. $\sqrt{(6a)^3}$ **24.** $\sqrt{(7y)^3}$ **25.** $(\sqrt[3]{16b^2})^2$ **26.** $(\sqrt[3]{25r^2})^2$

27. $\sqrt{(18a^2b)^3}$ **28.** $\sqrt{(12x^2y)^3}$ **29.** $(\sqrt[3]{(12c^2d)})^2$ **30.** $(\sqrt[3]{9x^2y})^2$

Problem-solving practice

31. Bob's salary increased 10% after his first year on the job. At the end of the second year he received another 10% raise. If his salary at the beginning of the third year was \$6.05 per hour, what did he earn per hour during his first year?

32. A fractional expression has a value of $\frac{3}{5}$. The value of the numerator is two less than three times a certain number, and the value of the denominator is one more than six times the same number. Find the values of the numerator and the denominator.

33. (*An application: Pendulums*). The *period* of a pendulum is the time it takes to complete one cycle, swinging to and fro. If a pendulum consists of a ball on a string, the period T is given by the following formula:

$$T = 2\pi\sqrt{\frac{L}{980}},$$

where T is in seconds and L is the length of the pendulum in centimeters. Find the period of a pendulum of the given lengths. Use 3.14 for π.

a) 65 cm **b)** 98 cm **c)** 120 cm

Divide and simplify.

34. $\dfrac{7\sqrt{a^2b}\sqrt{25xy}}{5\sqrt{a^{-4}b^{-1}}\sqrt{49x^{-1}y^{-3}}}$ **35.** $\dfrac{(\sqrt[3]{81mn^2})^2}{(\sqrt[3]{mn})^2}$ **36.** $\dfrac{\sqrt{44x^2y^9z}\sqrt{22y^9z^6}}{(\sqrt{11xy^8z^2})^2}$

OBJECTIVE

After finishing Section 7.4, you should be able to:

i **Add or subtract with radical notation and simplify.**

Add or subtract. Simplify by collecting like radical terms if possible.

1. $5\sqrt{2} + 8\sqrt{2}$

2. $7\sqrt[4]{5x} + 3\sqrt[4]{5x} - \sqrt{7}$

7.4

ADDITION AND SUBTRACTION INVOLVING RADICALS

i Any two real numbers can be added. For instance, the sum of 7 and $\sqrt{3}$ can be named

$$7 + \sqrt{3}.$$

We cannot simplify this name for the sum. However, when we have *like radicals* (radicals having the same index and radicand), we can use the distributive laws to simplify by collecting like radical terms.

EXAMPLES Add or subtract. Simplify by collecting like radical terms if possible.

1. $6\sqrt{7} + 4\sqrt{7} = (6+4)\sqrt{7}$ Using the distributive law (factoring out $\sqrt{7}$)
$= 10\sqrt{7}$

2. $8\sqrt[3]{2} - 7x\sqrt[3]{2} + 5\sqrt[3]{2} = (8 - 7x + 5)\sqrt[3]{2}$ Factoring out $\sqrt[3]{2}$
$= (13 - 7x)\sqrt[3]{2}$

3. $6\sqrt[5]{4x} + 4\sqrt[5]{4x} - \sqrt[3]{4x} = (6+4)\sqrt[5]{4x} - \sqrt[3]{4x}$
$= 10\sqrt[5]{4x} - \sqrt[3]{4x}$

DO EXERCISES 1 AND 2.

Sometimes we need to factor in order to have terms with like radicals.

EXAMPLES Add or subtract. Simplify by collecting like radical terms if possible.

4. $3\sqrt{8} - 5\sqrt{2} = 3\sqrt{4 \cdot 2} - 5\sqrt{2}$ Factoring 8

$= 3\sqrt{4} \cdot \sqrt{2} - 5\sqrt{2}$ Factoring $\sqrt{4 \cdot 2}$ into two radicals

$= 3 \cdot 2\sqrt{2} - 5\sqrt{2}$ Taking the square root of 4

$= 6\sqrt{2} - 5\sqrt{2}$

$= (6 - 5)\sqrt{2}$

$= \sqrt{2}$ Collecting like radical terms

5. $5\sqrt{2} - 4\sqrt{3}$ No simplification possible

6. $5\sqrt[3]{16y^4} + 7\sqrt[3]{2y} = 5\sqrt[3]{8y^3 \cdot 2y} + 7\sqrt[3]{2y}$

$= 5\sqrt[3]{8y^3} \cdot \sqrt[3]{2y} + 7\sqrt[3]{2y}$ Factoring the first radical

$= 5 \cdot 2y \cdot \sqrt[3]{2y} + 7\sqrt[3]{2y}$ Taking the cube root

$= 10y\sqrt[3]{2y} + 7\sqrt[3]{2y}$

$= (10y + 7)\sqrt[3]{2y}$ Collecting like radical terms

DO EXERCISES 3–5.

Add or subtract. Simplify by collecting like radical terms if possible.

3. $7\sqrt{45} - 2\sqrt{5}$

4. $3\sqrt[3]{y^5} + 4\sqrt[3]{y^2} + \sqrt[3]{8y^6}$

5. $\sqrt{25x - 25} - \sqrt{9x - 9}$

EXERCISE SET 7.4

i Add or subtract. Simplify by collecting like radical terms if possible, assuming that all expressions under radicals represent nonnegative numbers.

1. $6\sqrt{3} + 2\sqrt{3}$

2. $8\sqrt{5} + 9\sqrt{5}$

3. $9\sqrt[3]{5} - 6\sqrt[3]{5}$

4. $14\sqrt[5]{2} - 6\sqrt[5]{2}$

5. $4\sqrt[3]{y} + 9\sqrt[3]{y}$

6. $6\sqrt[4]{t} - 3\sqrt[4]{t}$

7. $8\sqrt{2} - 6\sqrt{2} + 5\sqrt{2}$

8. $2\sqrt{6} + 8\sqrt{6} - 3\sqrt{6}$

9. $4\sqrt[3]{3} - \sqrt{5} + 2\sqrt[3]{3} + \sqrt{5}$

10. $5\sqrt{7} - 8\sqrt[4]{11} + \sqrt{7} + 9\sqrt[4]{11}$

11. $8\sqrt{27} - 3\sqrt{3}$

12. $9\sqrt{50} - 4\sqrt{2}$

13. $8\sqrt{45} + 7\sqrt{20}$

14. $9\sqrt{12} + 16\sqrt{27}$

15. $18\sqrt{72} + 2\sqrt{98}$

16. $12\sqrt{45} - 8\sqrt{80}$

17. $3\sqrt[3]{16} + \sqrt[3]{54}$

18. $\sqrt[3]{27} - 5\sqrt[3]{8}$

19. $2\sqrt{128} - \sqrt{18} + 4\sqrt{32}$

20. $5\sqrt{50} - 2\sqrt{18} + 9\sqrt{32}$

21. $\sqrt{5a} + 2\sqrt{45a^3}$

22. $4\sqrt{3x^3} - \sqrt{12x}$

23. $\sqrt[3]{24x} - \sqrt[3]{3x^4}$

24. $\sqrt[3]{54x} - \sqrt[3]{2x^4}$

25. $\sqrt{8y - 8} + \sqrt{2y - 2}$

26. $\sqrt{12t + 12} + \sqrt{3t + 3}$

27. $\sqrt{x^3 - x^2} + \sqrt{9x - 9}$

28. $\sqrt{4x - 4} - \sqrt{x^3 - x^2}$

29. $6\sqrt{8} + 3\sqrt{18} - 4\sqrt{32}$

30. $4\sqrt{27} - 6\sqrt{75} + 6\sqrt{48}$

31. $5\sqrt[3]{32} - \sqrt[3]{108} + 2\sqrt[3]{256}$

32. $3\sqrt[3]{8x} - 4\sqrt[3]{27x} + 2\sqrt[3]{64x}$

33. $9\sqrt{45} - 10\sqrt{20} - 4\sqrt{80}$

34. $12\sqrt{44} + 4\sqrt{28} - 8\sqrt{99}$

35. $\sqrt{x^3 + x^2} + \sqrt{4x^3 + 4x^2} - \sqrt{9x^3 + 9x^2}$

36. $\sqrt{5x^2 + 4} - 5\sqrt{45x^2 + 36} + 3\sqrt{20x^2 + 16}$

Problem-solving practice

37. The perimeter of a hexagon with all six sides congruent is the same as the perimeter of a square. One side of the hexagon is three less than a side of the square. Find the perimeter of each polygon.

38. John is one-third as old as his father, Bill. Bill is three-fifths as old as his father, Sam. The sum of the three ages is 144. What are the ages of John, Bill, and Sam?

39. Find as many pairs of numbers, a and b, as you can for which

$$\sqrt{a} + \sqrt{b} = \sqrt{a + b}.$$

Find several pairs of numbers for which the above equation is false.

Add or subtract by collecting like terms.

40. $\sqrt{432} - \sqrt{6125} + \sqrt{845} - \sqrt{4800}$

41. $\sqrt{1250x^3y} - \sqrt{1800xy^3} - \sqrt{162x^3y^3}$

42. $\frac{1}{2}\sqrt{36a^5bc^4} - \frac{1}{2}\sqrt[3]{64a^4bc^6} + \frac{1}{6}\sqrt{144a^3bc^2}$

43. $7x\sqrt{(x + y)^3} - 5xy\sqrt{x + y} - 2y\sqrt{(x + y)^3}$

OBJECTIVE

After finishing Section 7.5, you should be able to:

i **Multiply expressions involving radicals, where some of the expressions contain more than one term.**

Multiply.

1. $\sqrt{2}(5\sqrt{3}+3\sqrt{7})$

2. $\sqrt[3]{a^2}(\sqrt[3]{3a}-\sqrt[3]{2})$

Multiply.

3. $(\sqrt{3}-5\sqrt{2})(2\sqrt{3}+\sqrt{2})$

4. $(\sqrt{a}+2\sqrt{3})(3\sqrt{b}-4\sqrt{3})$

Multiply.

5. $(\sqrt{2}+\sqrt{5})(\sqrt{2}-\sqrt{5})$

Multiply.

6. $(2\sqrt{5}-y)^2$

7. $(3\sqrt{6}+2)^2$

7.5

MORE ABOUT MULTIPLICATION WITH RADICALS

i We now consider multiplication of expressions in which some factors contain more than one term. To do that we use the procedures for multiplying polynomials, which are based on the distributive laws.

EXAMPLES Multiply.

1. $\sqrt{3}(x-\sqrt{5}) = \sqrt{3}\cdot x-\sqrt{3}\cdot\sqrt{5}$ Using the distributive law
$= x\sqrt{3}-\sqrt{15}$ Multiplying radicals

2. $\sqrt[3]{y}(\sqrt[3]{y^2}+\sqrt[3]{2}) = \sqrt[3]{y}\cdot\sqrt[3]{y^2}+\sqrt[3]{y}\cdot\sqrt[3]{2}$ Using the distributive law
$= \sqrt[3]{y^3}+\sqrt[3]{2y}$ Multiplying radicals
$= y+\sqrt[3]{2y}$ Simplifying $\sqrt[3]{y^3}$

DO EXERCISES 1 AND 2.

EXAMPLE 3 Multiply: $(4\sqrt{3}+\sqrt{2})(\sqrt{3}-5\sqrt{2})$.

F O I L

$$(4\sqrt{3}+\sqrt{2})(\sqrt{3}-5\sqrt{2}) = 4(\sqrt{3})^2-20\sqrt{3}\cdot\sqrt{2}+\sqrt{2}\cdot\sqrt{3}-5(\sqrt{2})^2$$
$$= 4\cdot 3-20\sqrt{6}+\sqrt{6}-5\cdot 2$$
$$= 12-20\sqrt{6}+\sqrt{6}-10 = 2-19\sqrt{6}$$

EXAMPLE 4 Multiply: $(\sqrt{a}+\sqrt{3})(\sqrt{b}+\sqrt{3})$. Assume that all expressions under radicals represent positive numbers.

$$(\sqrt{a}+\sqrt{3})(\sqrt{b}+\sqrt{3}) = \sqrt{a}\sqrt{b}+\sqrt{a}\sqrt{3}+\sqrt{3}\sqrt{b}+\sqrt{3}\sqrt{3}$$
$$= \sqrt{ab}+\sqrt{3a}+\sqrt{3b}+3$$

DO EXERCISES 3 AND 4.

EXAMPLE 5 Multiply: $(\sqrt{5}+\sqrt{7})(\sqrt{5}-\sqrt{7})$.

$$(\sqrt{5}+\sqrt{7})(\sqrt{5}-\sqrt{7}) = (\sqrt{5})^2-(\sqrt{7})^2$$ This is now a difference of two squares.
$$= 5-7$$
$$= -2$$

DO EXERCISE 5.

EXAMPLE 6 Multiply: $(\sqrt{3}+x)^2$.

$$(\sqrt{3}+x)^2 = (\sqrt{3})^2+2x\sqrt{3}+x^2$$ Squaring a binomial
$$= 3+2x\sqrt{3}+x^2$$

DO EXERCISES 6 AND 7.

EXERCISE SET 7.5

i Multiply.

1. $\sqrt{6}(2-3\sqrt{6})$
2. $\sqrt{3}(4+\sqrt{3})$
3. $\sqrt{2}(\sqrt{3}-\sqrt{5})$
4. $\sqrt{5}(\sqrt{5}-\sqrt{2})$
5. $\sqrt{3}(2\sqrt{5}-3\sqrt{4})$
6. $\sqrt{2}(3\sqrt{10}-2\sqrt{2})$

7. $\sqrt[3]{2}(\sqrt[3]{4} - 2\sqrt[3]{32})$
8. $\sqrt[3]{3}(\sqrt[3]{9} - 4\sqrt[3]{21})$
9. $\sqrt[3]{a}(\sqrt[3]{2a^2} + \sqrt[3]{16a^2})$
10. $\sqrt[3]{x}(\sqrt[3]{3x^2} - \sqrt[3]{81x^2})$
11. $(\sqrt{3} - \sqrt{2})(\sqrt{3} + \sqrt{2})$
12. $(\sqrt{5} + \sqrt{6})(\sqrt{5} - \sqrt{6})$
13. $(\sqrt{8} + 2\sqrt{5})(\sqrt{8} - 2\sqrt{5})$
14. $(\sqrt{18} + 3\sqrt{7})(\sqrt{18} - 3\sqrt{7})$

Multiply. Assume that all expressions under radicals represent positive numbers.

15. $(\sqrt{a} + \sqrt{b})(\sqrt{a} - \sqrt{b})$
16. $(\sqrt{x} - \sqrt{y})(\sqrt{x} + \sqrt{y})$
17. $(3 - \sqrt{5})(2 + \sqrt{5})$
18. $(2 + \sqrt{6})(4 - \sqrt{6})$
19. $(\sqrt{3} + 1)(2\sqrt{3} + 1)$
20. $(4\sqrt{3} + 5)(\sqrt{3} - 2)$
21. $(2\sqrt{7} - 4\sqrt{2})(3\sqrt{7} + 6\sqrt{2})$
22. $(4\sqrt{5} + 3\sqrt{3})(3\sqrt{5} - 4\sqrt{3})$
23. $(\sqrt{a} + \sqrt{2})(\sqrt{a} + \sqrt{3})$
24. $(2 - \sqrt{x})(1 - \sqrt{x})$
25. $(2\sqrt[3]{3} + \sqrt[3]{2})(\sqrt[3]{3} - 2\sqrt[3]{2})$
26. $(3\sqrt[4]{7} + \sqrt[4]{6})(2\sqrt[4]{9} - 3\sqrt[4]{6})$
27. $(2 + \sqrt{3})^2$
28. $(\sqrt{5} + 1)^2$

Problem-solving practice

29. The sum of the digits of a two-digit number is 10. If the digits are reversed, the new number is two less than three times the original number. Find the numbers.

30. Harry can fill his pool with a garden hose in 12 hours. If he also uses his neighbor's larger hose, it takes only 3 hours to fill the pool. How long would it take to fill the pool using only the neighbor's hose?

---

31. **a)** Calculate $(3 + \sqrt{2})^2$ by first finding $\sqrt{2}$, then adding 3, and then squaring.
b) Find $(3 + \sqrt{2})^2$ by expanding, to get $9 + 6\sqrt{2} + 2$, and then evaluating. Compare your answers.

Multiply and simplify.

32. $\sqrt{9 + 3\sqrt{5}}\sqrt{9 - 3\sqrt{5}}$
33. $(\sqrt{x + 2} - \sqrt{x - 2})^2$
34. $(\sqrt{3} + \sqrt{5} - \sqrt{6})^2$
35. $\sqrt[3]{y}(1 - \sqrt[3]{y})(1 + \sqrt[3]{y})$
36. $(\sqrt[3]{9} - 2)(\sqrt[3]{9} + 4)$
37. $[\sqrt{3 + \sqrt{2 + \sqrt{1}}}]^4$
38. $(\sqrt{y + 10} - \sqrt{10})(\sqrt{y + 10} + \sqrt{10})$
39. $(8\sqrt{a + 3} - a\sqrt{3})(2\sqrt{a + 3} + 5a\sqrt{3})$
40. $(8\sqrt{3} + 5\sqrt{2})(6\sqrt{7} - 2\sqrt{5})$
41. $(\sqrt{17} + \sqrt{2} - \sqrt{19})(\sqrt{17} + \sqrt{2} + \sqrt{19})$
42. Evaluate: $\sqrt{7 + 4\sqrt{3}} - \sqrt{7 - 4\sqrt{3}}$.

7.6

RATIONALIZING NUMERATORS OR DENOMINATORS

OBJECTIVES

After finishing Section 7.6, you should be able to:

i **Rationalize the denominator of a radical expression.**

ii **Rationalize the numerator of a radical expression.**

iii **Rationalize denominators or numerators having two terms.**

When a fractional expression contains a radical in its denominator, we often want to find an equivalent expression that does not have a radical in the denominator. When we do that, we say that we have *rationalized the denominator.* Similarly, if we find an equivalent expression without a radical sign in a numerator, we say that we have *rationalized the numerator.*

i Rationalizing Denominators

Let us consider several ways of finding $\sqrt{\dfrac{1}{2}}$.

1. $\sqrt{\dfrac{1}{2}} = \dfrac{\sqrt{1}}{\sqrt{2}} = \dfrac{1}{\sqrt{2}}$

We find an approximation to $\sqrt{2}$ to be 1.414. Then

$$\frac{1}{\sqrt{2}} \approx \frac{1}{1.414}.$$

By long division we can then find decimal notation for the answer. The division would be rather lengthy without a calculator.

Rationalize the denominator.

1. $\sqrt{\frac{2}{3}}$

2. $\sqrt{\frac{10}{3}}$

3. $\sqrt[3]{\frac{3}{6}}$

Rationalize the denominator. Assume that all expressions under radicals represent positive numbers.

4. $\sqrt{\frac{4a}{3b}}$

5. $\frac{\sqrt{4x^5}}{\sqrt{3y^3}}$

2. $\sqrt{\frac{1}{2}} = \sqrt{\frac{1}{2} \cdot \frac{2}{2}}$ Multiplying by $\frac{2}{2}$

$= \sqrt{\frac{2}{4}} = \frac{\sqrt{2}}{\sqrt{4}} = \frac{\sqrt{2}}{2}$

We find an approximation to $\sqrt{2}$ and then compute:

$$\frac{\sqrt{2}}{2} \approx \frac{1.414}{2} = 0.707.$$

This time the division is easy, even without a calculator.

3. $\sqrt{\frac{1}{2}} = \sqrt{0.5}$ Dividing

≈ 0.707 Using a calculator

Before the days of calculators, method 2 was almost always used. Today, with calculators it doesn't matter much which of the three methods we use for computing. It is still more or less standard, however, to write radical expressions with the denominator rationalized. The ability to rationalize a numerator or a denominator is a skill that is important in a number of situations.

EXAMPLE 1 Rationalize the denominator: $\sqrt[3]{\frac{7}{9}}$.

$$\sqrt[3]{\frac{7}{9}} = \sqrt[3]{\frac{7}{3 \cdot 3} \cdot \frac{3}{3}}$$ Multiplying by $\frac{3}{3}$ to make the denominator a perfect cube

$$= \sqrt[3]{\frac{21}{3 \cdot 3 \cdot 3}} = \frac{\sqrt[3]{21}}{\sqrt[3]{3^3}} = \frac{\sqrt[3]{21}}{3}$$

DO EXERCISES 1–3.

The idea in rationalizing a denominator is to multiply by 1 in order to make the denominator a perfect power. In Example 1 we multiplied by 1 under the radical. We can also multiply by 1 outside the radical, as in Example 2.

EXAMPLE 2 Rationalize the denominator: $\sqrt{\frac{2a}{5b}}$. Assume that all expressions under radicals represent positive numbers.

$\sqrt{\frac{2a}{5b}} = \frac{\sqrt{2a}}{\sqrt{5b}}$ Converting to a quotient of radicals

$= \frac{\sqrt{2a}}{\sqrt{5b}} \cdot \frac{\sqrt{5b}}{\sqrt{5b}}$ Multiplying by 1

$= \frac{\sqrt{10ab}}{\sqrt{25b^2}}$ The radicand in the denominator is a perfect square.

$= \frac{\sqrt{10ab}}{5b}$ Taking the square root of $25b^2$

DO EXERCISES 4 AND 5.

EXAMPLE 3 Rationalize the denominator: $\frac{\sqrt[3]{a}}{\sqrt[3]{9x}}$.

$$\frac{\sqrt[3]{a}}{\sqrt[3]{9x}} = \frac{\sqrt[3]{a}}{\sqrt[3]{9x}} \cdot \frac{\sqrt[3]{3x^2}}{\sqrt[3]{3x^2}} \qquad \text{Multiplying by 1}$$

To choose the symbol for 1, we look at the radicand $9x$. This is $3 \cdot 3 \cdot x$. To make it a cube we need another 3 and two more x's. Thus we multiply by $\sqrt[3]{3x^2}/\sqrt[3]{3x^2}$. We have

$$\frac{\sqrt[3]{3ax^2}}{\sqrt[3]{27x^3}},$$

which simplifies to

$$\frac{\sqrt[3]{3ax^2}}{3x}.$$

DO EXERCISES 6 AND 7.

Rationalize the denominator.

6. $\frac{\sqrt[3]{7}}{\sqrt[3]{2}}$

7. $\sqrt[3]{\frac{3x^5}{2y}}$

ii Rationalizing Numerators

Sometimes in advanced mathematics it is necessary to rationalize a numerator. We use a similar procedure.

EXAMPLE 4 Rationalize the numerator: $\frac{\sqrt{7}}{\sqrt{5}}$.

$$\frac{\sqrt{7}}{\sqrt{5}} = \frac{\sqrt{7}}{\sqrt{5}} \cdot \frac{\sqrt{7}}{\sqrt{7}} \qquad \text{Multiplying by 1}$$

$$= \frac{\sqrt{49}}{\sqrt{35}} \qquad \text{The radicand in the numerator is a perfect square.}$$

$$= \frac{7}{\sqrt{35}}$$

EXAMPLE 5 Rationalize the numerator: $\frac{\sqrt[3]{5}}{\sqrt[3]{3}}$.

$$\frac{\sqrt[3]{5}}{\sqrt[3]{3}} = \frac{\sqrt[3]{5}}{\sqrt[3]{3}} \cdot \frac{\sqrt[3]{5^2}}{\sqrt[3]{5^2}} \qquad \text{Multiplying by 1}$$

$$= \frac{\sqrt[3]{5^3}}{\sqrt[3]{3 \cdot 5^2}}$$

$$= \frac{\sqrt[3]{5^3}}{\sqrt[3]{75}}$$

$$= \frac{5}{\sqrt[3]{75}}$$

DO EXERCISES 8 AND 9.

Rationalize the numerator.

8. $\frac{\sqrt{11}}{\sqrt{6}}$

9. $\frac{\sqrt[3]{7}}{\sqrt[3]{4}}$

What symbol for 1 would you use to rationalize the denominator?

10. $\dfrac{\sqrt{5}+1}{\sqrt{3}-y}$

11. $\dfrac{1}{\sqrt{2}+\sqrt{3}}$

Rationalize the denominator.

12. $\dfrac{5}{1-\sqrt{2}}$

iii Rationalizing When There Are Two Terms

When the denominator to be rationalized has two terms, choose a symbol for 1 as illustrated in the following example.

EXAMPLE 6 Rationalize the denominator: $\dfrac{4+\sqrt{2}}{\sqrt{5}-\sqrt{2}}$.

$$\frac{4+\sqrt{2}}{\sqrt{5}-\sqrt{2}} = \frac{4+\sqrt{2}}{\sqrt{5}-\sqrt{2}} \cdot \frac{\sqrt{5}+\sqrt{2}}{\sqrt{5}+\sqrt{2}} \quad \text{Multiplying by 1}$$

$$= \frac{(4+\sqrt{2})(\sqrt{5}+\sqrt{2})}{(\sqrt{5}-\sqrt{2})(\sqrt{5}+\sqrt{2})} \quad \text{Multiplying numerators and denominators}$$

$$= \frac{4\sqrt{5}+4\sqrt{2}+\sqrt{2}\sqrt{5}+(\sqrt{2})^2}{(\sqrt{5})^2-(\sqrt{2})^2} \quad \text{Using FOIL}$$

$$= \frac{4\sqrt{5}+4\sqrt{2}+\sqrt{10}+2}{5-2} \quad \text{Squaring in the denominator}$$

$$= \frac{4\sqrt{5}+4\sqrt{2}+\sqrt{10}+2}{3}$$

Note that the denominator in this example was $\sqrt{5}-\sqrt{2}$. We chose a symbol for 1 that had $\sqrt{5}+\sqrt{2}$ in the numerator and denominator. If the denominator had been $\sqrt{5}+\sqrt{2}$ we would have chosen

$$\frac{\sqrt{5}-\sqrt{2}}{\sqrt{5}-\sqrt{2}}$$

for 1.

EXAMPLES What symbol for 1 would you use to rationalize the denominator?

	Expression	*Symbol for 1*
7.	$\dfrac{3}{x+\sqrt{7}}$	$\dfrac{x-\sqrt{7}}{x-\sqrt{7}}$
8.	$\dfrac{\sqrt{7}+4}{3-2\sqrt{5}}$	$\dfrac{3+2\sqrt{5}}{3+2\sqrt{5}}$

DO EXERCISES 10 AND 11.

EXAMPLE 9 Rationalize the denominator: $\dfrac{4}{\sqrt{3}+x}$.

We have

$$\frac{4}{\sqrt{3}+x} = \frac{4}{\sqrt{3}+x} \cdot \frac{\sqrt{3}-x}{\sqrt{3}-x}$$

$$= \frac{4(\sqrt{3}-x)}{(\sqrt{3}+x)(\sqrt{3}-x)} = \frac{4(\sqrt{3}-x)}{(\sqrt{3})^2-x^2} = \frac{4(\sqrt{3}-x)}{3-x^2}$$

DO EXERCISE 12.

We can also rationalize numerators. Of course we choose our symbol for 1 a little differently.

EXAMPLES What symbol for 1 would you use to rationalize the numerator?

	Expression	*Symbol for 1*
10.	$\dfrac{\sqrt{5}+\sqrt{2}}{1-\sqrt{3}}$	$\dfrac{\sqrt{5}-\sqrt{2}}{\sqrt{5}-\sqrt{2}}$
11.	$\dfrac{x-\sqrt{7}}{3}$	$\dfrac{x+\sqrt{7}}{x+\sqrt{7}}$
12.	$\dfrac{3-2\sqrt{5}}{\sqrt{7}+4}$	$\dfrac{3+2\sqrt{5}}{3+2\sqrt{5}}$

DO EXERCISE 13.

EXAMPLE 13 Rationalize the numerator: $\dfrac{4+\sqrt{2}}{\sqrt{5}-\sqrt{2}}$.

$$\frac{4+\sqrt{2}}{\sqrt{5}-\sqrt{2}} = \frac{4+\sqrt{2}}{\sqrt{5}-\sqrt{2}}\cdot\frac{4-\sqrt{2}}{4-\sqrt{2}}$$

$$= \frac{16-(\sqrt{2})^2}{4\sqrt{5}-\sqrt{5}\sqrt{2}-4\sqrt{2}+(\sqrt{2})^2}$$

$$= \frac{14}{4\sqrt{5}-\sqrt{10}-4\sqrt{2}+2}$$

DO EXERCISE 14.

What symbol for 1 would you use to rationalize the numerator?

13. $\dfrac{5-\sqrt{7}}{\sqrt{6}-3}$

Rationalize the numerator.

14. $\dfrac{3+\sqrt{5}}{\sqrt{2}-\sqrt{6}}$

EXERCISE SET 7.6

i Rationalize the denominator.

1. $\sqrt{\dfrac{6}{5}}$ **2.** $\sqrt{\dfrac{11}{6}}$ **3.** $\sqrt{\dfrac{10}{7}}$ **4.** $\sqrt{\dfrac{22}{3}}$ **5.** $\dfrac{6\sqrt{5}}{5\sqrt{3}}$

6. $\dfrac{2\sqrt{3}}{5\sqrt{2}}$ **7.** $\sqrt[3]{\dfrac{16}{9}}$ **8.** $\sqrt[3]{\dfrac{3}{9}}$ **9.** $\dfrac{\sqrt[3]{3a}}{\sqrt[3]{5c}}$ **10.** $\dfrac{\sqrt[3]{7x}}{\sqrt[3]{3y}}$

11. $\dfrac{\sqrt[3]{2y^4}}{\sqrt[3]{6x^4}}$ **12.** $\dfrac{\sqrt[3]{3a^4}}{\sqrt[3]{7b^2}}$ **13.** $\dfrac{1}{\sqrt[3]{xy}}$ **14.** $\dfrac{1}{\sqrt[3]{ab}}$

ii Rationalize the numerator.

15. $\dfrac{\sqrt{7}}{\sqrt{3x}}$ **16.** $\dfrac{\sqrt{6}}{\sqrt{5x}}$ **17.** $\sqrt{\dfrac{14}{21}}$ **18.** $\sqrt{\dfrac{12}{15}}$ **19.** $\dfrac{4\sqrt{13}}{3\sqrt{7}}$

20. $\dfrac{5\sqrt{21}}{2\sqrt{6}}$ **21.** $\dfrac{\sqrt[3]{7}}{\sqrt[3]{2}}$ **22.** $\dfrac{\sqrt[3]{5}}{\sqrt[3]{4}}$ **23.** $\sqrt{\dfrac{7x}{3y}}$ **24.** $\sqrt{\dfrac{6a}{2b}}$

25. $\dfrac{\sqrt[3]{5y^4}}{\sqrt[3]{6x^5}}$ **26.** $\dfrac{\sqrt[3]{3a^5}}{\sqrt[3]{7b^2}}$ **27.** $\dfrac{\sqrt{ab}}{3}$ **28.** $\dfrac{\sqrt{xy}}{5}$

iii Rationalize the denominator. Assume that all expressions under radicals represent positive numbers.

29. $\dfrac{5}{8-\sqrt{6}}$ **30.** $\dfrac{7}{9+\sqrt{10}}$ **31.** $\dfrac{-4\sqrt{7}}{\sqrt{5}-\sqrt{3}}$ **32.** $\dfrac{-3\sqrt{2}}{\sqrt{3}-\sqrt{5}}$ **33.** $\dfrac{\sqrt{5}-2\sqrt{6}}{\sqrt{3}-4\sqrt{5}}$

34. $\dfrac{\sqrt{6} - 3\sqrt{5}}{\sqrt{3} - 2\sqrt{7}}$ **35.** $\dfrac{\sqrt{x} - \sqrt{y}}{\sqrt{x} + \sqrt{y}}$ **36.** $\dfrac{\sqrt{a} + \sqrt{b}}{\sqrt{a} - \sqrt{b}}$ **37.** $\dfrac{5\sqrt{3} - 3\sqrt{2}}{3\sqrt{2} - 2\sqrt{3}}$ **38.** $\dfrac{7\sqrt{2} + 4\sqrt{3}}{4\sqrt{3} - 3\sqrt{2}}$

Rationalize the numerator. Assume that all expressions under radicals represent positive numbers.

39. $\dfrac{\sqrt{3} + 5}{8}$ **40.** $\dfrac{3 - \sqrt{2}}{5}$ **41.** $\dfrac{\sqrt{3} - 5}{\sqrt{2} + 5}$ **42.** $\dfrac{\sqrt{6} - 3}{\sqrt{3} + 7}$

43. $\dfrac{\sqrt{x} - \sqrt{y}}{\sqrt{x} + \sqrt{y}}$ **44.** $\dfrac{\sqrt{x} + \sqrt{y}}{\sqrt{x} - \sqrt{y}}$ **45.** $\dfrac{4\sqrt{6} - 5\sqrt{3}}{2\sqrt{3} + 7\sqrt{6}}$ **46.** $\dfrac{8\sqrt{2} + 5\sqrt{3}}{5\sqrt{3} - 7\sqrt{2}}$

Rationalize the denominator.

47. $\dfrac{\sqrt{5} + \sqrt{10} - \sqrt{6}}{\sqrt{50}}$ **48.** $\dfrac{3\sqrt{y} + 4\sqrt{yz}}{5\sqrt{y} - 2\sqrt{z}}$ **49.** $\dfrac{b + \sqrt{b}}{1 + b + \sqrt{b}}$ **50.** $\dfrac{12}{3 - \sqrt{3 - y}}$ **51.** $\dfrac{36a^2b}{\sqrt[3]{6a^2b}}$

Rationalize the numerator.

52. $\dfrac{\sqrt{y + 18} - \sqrt{y}}{18}$ **53.** $\dfrac{\sqrt{x + 6} - 5}{\sqrt{x + 6} + 5}$

Simplify. (*Hint:* Rationalize the denominator.)

54. $\sqrt{a^2 - 3} - \dfrac{a^2}{\sqrt{a^2 - 3}}$ **55.** $5\sqrt{\dfrac{x}{y}} + 4\sqrt{\dfrac{y}{x}} - \dfrac{3}{\sqrt{xy}}$

56. $\dfrac{\dfrac{1}{\sqrt{w}} - \sqrt{w}}{\dfrac{\sqrt{w} + 1}{\sqrt{w}}}$ **57.** $\dfrac{1}{4 + \sqrt{3}} + \dfrac{1}{\sqrt{3}} + \dfrac{1}{\sqrt{3} - 4}$

OBJECTIVES

After finishing Section 7.7, you should be able to:

i Write expressions with or without fractional exponents.

ii Write expressions without negative exponents.

iii Use the laws of exponents with fractional exponents.

iv Use fractional exponents to simplify radical expressions.

Rewrite without fractional exponents.

1. $y^{1/4}$
2. $(3a)^{1/2}$
3. $16^{1/4}$
4. $(125)^{1/3}$
5. $(a^3b^2c)^{1/5}$

7.7 RATIONAL NUMBERS AS EXPONENTS

i Fractional Exponents

Expressions like $a^{1/2}$, $5^{-1/4}$, and $(2y)^{4/5}$ have not yet been defined. We shall define such expressions in such a way that the usual properties of exponents hold.

Consider $a^{1/2} \cdot a^{1/2}$. If we still want to multiply by adding exponents, it must follow that $a^{1/2} \cdot a^{1/2} = a^{1/2 + 1/2}$ or a^1. Thus we should define $a^{1/2}$ to be a square root of a. Similarly, $a^{1/3} \cdot a^{1/3} \cdot a^{1/3} = a^{1/3 + 1/3 + 1/3}$ or a^1, so $a^{1/3}$ should be defined to mean $\sqrt[3]{a}$.

For any nonnegative number a and any index n, $a^{1/n}$ means $\sqrt[n]{a}$ (the nonnegative nth root of a).

Whenever we use fractional exponents we assume that the bases are nonnegative.

EXAMPLES Rewrite without fractional exponents.

1. $x^{1/2} = \sqrt{x}$
2. $27^{1/3} = \sqrt[3]{27}$, or 3
3. $(abc)^{1/5} = \sqrt[5]{abc}$

DO EXERCISES 1–5.

EXAMPLES Rewrite with fractional exponents.

4. $\sqrt[5]{7xy} = (7xy)^{1/5}$

5. $\sqrt[7]{\frac{x^3y}{9}} = \left(\frac{x^3y}{9}\right)^{1/7}$

DO EXERCISES 6–8.

How should we define $a^{2/3}$? If the usual properties of exponents are to hold, we have $a^{2/3} = (a^{1/3})^2$, or $(\sqrt[3]{a})^2$, or $\sqrt[3]{a^2}$. We make our definition accordingly.

For any natural numbers m and n ($n \neq 1$) and any nonnegative number a,

$$a^{m/n} \quad \textbf{means} \quad \sqrt[n]{a^m}, \quad \textbf{or } (\sqrt[n]{a})^m.$$

EXAMPLES Rewrite without fractional exponents.

6. $(27)^{2/3} = \sqrt[3]{27^2}$, or $(\sqrt[3]{27})^2$
$= 3^2$, or 9

7. $4^{3/2} = \sqrt[2]{4^3}$, or $(\sqrt[2]{4})^3$
$= 2^3$, or 8

DO EXERCISES 9–11.

EXAMPLES Rewrite with fractional exponents.

8. $\sqrt[3]{9^4} = 9^{4/3}$

9. $(\sqrt[4]{7xy})^5 = (7xy)^{5/4}$

DO EXERCISES 12 AND 13.

ii Negative Rational Exponents

Negative fractional exponents have a meaning similar to that of negative integer exponents. Changing the sign of an exponent amounts to finding a reciprocal.

For any rational number m/n and any positive real number a, $a^{-m/n}$ means $\frac{1}{a^{m/n}}$ ($a^{m/n}$ and $a^{-m/n}$ are reciprocals).

EXAMPLES Rewrite with positive exponents.

10. $4^{-1/2} = \frac{1}{4^{1/2}}$ $\quad$ $4^{-1/2}$ is the reciprocal of $4^{1/2}$.

Since $4^{1/2} = \sqrt{4} = 2$, the answer simplifies to $\frac{1}{2}$.

11. $(5xy)^{-4/5} = \frac{1}{(5xy)^{4/5}}$ $\quad$ $(5xy)^{-4/5}$ is the reciprocal of $(5xy)^{4/5}$.

DO EXERCISES 14 AND 15.

Rewrite with fractional exponents.

6. $\sqrt[3]{19}$

7. $\sqrt{abc}$

8. $\sqrt[5]{\frac{x^2y}{16}}$

Rewrite without fractional exponents.

9. $x^{3/2}$

10. $8^{2/3}$

11. $4^{5/2}$

Rewrite with fractional exponents.

12. $(\sqrt[3]{7abc})^4$

13. $\sqrt[5]{6^7}$

Rewrite with positive exponents.

14. $16^{-1/4}$

15. $(3xy)^{-7/8}$

Use the laws of exponents to simplify.

16. $7^{1/3} \cdot 7^{3/5}$

17. $\dfrac{5^{7/6}}{5^{5/6}}$

18. $(9^{3/5})^{2/3}$

Use fractional exponents to simplify.

19. $\sqrt[4]{a^2}$

20. $\sqrt[4]{x^4}$

21. $\sqrt[6]{8}$

Use fractional exponents to simplify.

22. $\sqrt[5]{a^5 b^{10}}$

23. $\sqrt[4]{x^4 y^{12}}$

24. $\sqrt[12]{x^3 y^6}$

iii Laws of Exponents

The same laws hold for rational-number exponents as for integer exponents. We list them for review.

For any real number a and any rational exponents m and n:

1. $a^m \cdot a^n = a^{m+n}$ **In multiplying, we can add exponents if the bases are the same.**
2. $\dfrac{a^m}{a^n} = a^{m-n}$ **In dividing, we can subtract exponents if the bases are the same.**
3. $(a^m)^n = a^{m \cdot n}$ **To raise a power to a power, we can multiply the exponents.**

EXAMPLES Use the laws of exponents to simplify.

12. $3^{1/5} \cdot 3^{3/5} = 3^{1/5+3/5} = 3^{4/5}$ Adding exponents

13. $5^{2/3} \cdot 5^{1/4} = 5^{2/3+1/4} = 5^{8/12+3/12} = 5^{11/12}$ Adding exponents

14. $\dfrac{7^{1/4}}{7^{1/2}} = 7^{1/4-1/2} = 7^{1/4-2/4} = 7^{-1/4}$ Subtracting exponents

15. $(7.2^{2/3})^{3/4} = 7.2^{2/3 \cdot 3/4} = 7.2^{6/12}$ Multiplying exponents

$= 7.2^{1/2}$

DO EXERCISES 16–18.

iv Simplifying Radical Expressions

Fractional exponents can be used to simplify some radical expressions. The procedure is as follows.

1. **Convert radical expressions to exponential expressions.**
2. **Use arithmetic and the laws of exponents to simplify.**
3. **Convert back to radical notation when appropriate.**

Important: **This works only when all expressions under radicals are nonnegative, since fractional exponents are not defined otherwise. No absolute-value signs will be needed.**

EXAMPLE 16 Use fractional exponents to simplify.

$\sqrt[6]{x^3} = x^{3/6}$ Converting to an exponential expression

$= x^{1/2}$ Using arithmetic to simplify the exponent

$= \sqrt{x}$ Converting back to radical notation

EXAMPLE 17 Use fractional exponents to simplify.

$\sqrt[8]{a^2 b^4} = (a^2 b^4)^{1/8}$ Converting to exponential notation

$= a^{2/8} \cdot b^{4/8}$ Using the third law of exponents

$= a^{1/4} \cdot b^{2/4}$ Using arithmetic to simplify the exponents

$= (ab^2)^{1/4}$ Using the third law of exponents (in reverse)

$= \sqrt[4]{ab^2}$ Converting back to radical notation

DO EXERCISES 19–24.

We can use properties of fractional exponents to write a single radical expression for a product or quotient.

EXAMPLES Use fractional exponents to write a single radical expression.

18. $\sqrt[3]{5} \cdot \sqrt{2} = 5^{1/3} \cdot 2^{1/2}$ Converting to exponential notation

$= 5^{2/6} \cdot 2^{3/6}$ Rewriting so that exponents have a common denominator

$= (5^2 \cdot 2^3)^{1/6}$ Using the third law of exponents

$= \sqrt[6]{5^2 \cdot 2^3}$ Converting back to radical notation

$= \sqrt[6]{200}$ Multiplying under the radical

19. $\dfrac{\sqrt[4]{(x+y)^3}}{\sqrt{x+y}} = \dfrac{(x+y)^{3/4}}{(x+y)^{1/2}}$ Converting to exponential notation

$= (x+y)^{3/4-1/2}$ Using the second law of exponents

$= (x+y)^{1/4}$ Subtracting exponents

$= \sqrt[4]{x+y}$ Converting back to radical notation

DO EXERCISES 25 AND 26.

EXAMPLE 20 Write a single radical expression.

$a^{1/2}b^{-1/2}c^{5/6} = a^{3/6}b^{-3/6}c^{5/6}$ Rewriting exponents with a common denominator

$= (a^3b^{-3}c^5)^{1/6}$ Using the third law of exponents

$= \sqrt[6]{a^3b^{-3}c^5}$ Converting to radical notation

DO EXERCISES 27 AND 28.

We have now seen several different methods of simplifying radical expressions. We list them.

SOME WAYS TO SIMPLIFY RADICAL EXPRESSIONS

1. ***Simplifying by factoring.*** **We factor the radicand, looking for factors that are perfect powers.**

   Example: $\sqrt[3]{16} = \sqrt[3]{8}\sqrt[3]{2}$
   $= 2\sqrt[3]{2}$

2. ***Rationalizing denominators.*** **Radical expressions are usually considered simpler if there are no radicals in the denominator.**

   Example: $\dfrac{1}{\sqrt{2}} = \dfrac{\sqrt{2}}{2}$

3. ***Collecting like radical terms.***

   Example: $\sqrt{8} + 3\sqrt{2} = \sqrt{4} \cdot \sqrt{2} + 3\sqrt{2}$
   $= 2\sqrt{2} + 3\sqrt{2}$
   $= 5\sqrt{2}$

4. ***Using fractional exponents to simplify.*** **We convert to exponential notation and then use arithmetic and the laws of exponents to simplify the exponents. Then we convert back to radical notation.** ***Caution:*** **This works only when there are only nonnegative expressions under radical signs.**

Use fractional exponents to write a single radical expression.

25. $\sqrt[4]{7} \cdot \sqrt{3}$

26. $\dfrac{\sqrt[4]{(a-b)^5}}{(a-b)}$

Write a single radical expression.

27. $x^{2/3}y^{1/2}z^{5/6}$

28. $\dfrac{a^{1/2}b^{3/8}}{a^{1/4}b^{1/8}}$

EXERCISE SET 7.7

i Rewrite without fractional exponents.

1. $x^{1/4}$ **2.** $y^{1/5}$ **3.** $(8)^{1/3}$ **4.** $(16)^{1/2}$ **5.** $(a^2b^2)^{1/5}$

6. $(x^3y^3)^{1/4}$ **7.** $a^{2/3}$ **8.** $b^{3/2}$ **9.** $16^{3/4}$ **10.** $4^{7/2}$

Rewrite with fractional exponents.

11. $\sqrt[3]{20}$ **12.** $\sqrt[3]{19}$ **13.** $\sqrt{17}$ **14.** $\sqrt{6}$ **15.** $\sqrt[4]{cd}$ **16.** $\sqrt[5]{xy}$

17. $\sqrt[5]{xy^2z}$ **18.** $\sqrt[7]{x^3y^2z^2}$ **19.** $(\sqrt{3mn})^3$ **20.** $(\sqrt[3]{7xy})^4$ **21.** $(\sqrt[7]{8x^2y})^5$ **22.** $(\sqrt[6]{2a^5b})^7$

ii Rewrite with positive exponents.

23. $x^{-1/3}$ **24.** $y^{-1/4}$ **25.** $(2rs)^{-3/4}$ **26.** $(5xy)^{-5/6}$

27. $\left(\dfrac{1}{10}\right)^{-2/3}$ **28.** $\left(\dfrac{1}{8}\right)^{-3/4}$ **29.** $\dfrac{1}{x^{-2/3}}$ **30.** $\dfrac{1}{x^{-5/6}}$

iii Use the properties of exponents to simplify.

31. $5^{3/4}\cdot 5^{1/8}$ **32.** $11^{2/3}\cdot 11^{1/2}$ **33.** $\dfrac{7^{5/8}}{7^{3/8}}$ **34.** $\dfrac{9^{9/11}}{9^{7/11}}$

35. $\dfrac{8.3^{3/4}}{8.3^{2/5}}$ **36.** $\dfrac{3.9^{3/5}}{3.9^{1/4}}$ **37.** $(10^{3/5})^{2/5}$ **38.** $(5^{5/4})^{3/7}$

iv Use fractional exponents to simplify.

39. $\sqrt[6]{a^4}$ **40.** $\sqrt[6]{y^2}$ **41.** $\sqrt[3]{8y^6}$ **42.** $\sqrt{x^4y^6}$ **43.** $\sqrt[4]{32}$ **44.** $\sqrt[8]{81}$

45. $\sqrt[6]{4x^2}$ **46.** $\sqrt[4]{16x^4y^2}$ **47.** $\sqrt[5]{32c^{10}d^{15}}$ **48.** $\sqrt[4]{16x^{12}y^{16}}$ **49.** $\sqrt[6]{\dfrac{m^{12}n^{24}}{64}}$ **50.** $\sqrt[5]{\dfrac{x^{15}y^{20}}{32}}$

51. $\sqrt[8]{r^4s^2}$ **52.** $\sqrt[12]{64t^6s^6}$ **53.** $\sqrt[3]{27a^3b^9}$ **54.** $\sqrt[4]{81x^8y^8}$

Use fractional exponents to write a single radical expression.

55. $\sqrt[3]{7}\cdot\sqrt{2}$ **56.** $\sqrt[3]{7}\cdot\sqrt[4]{5}$ **57.** $\sqrt{x\sqrt[3]{2x}}$ **58.** $\sqrt[3]{y\sqrt[5]{3y}}$ **59.** $\sqrt{x\sqrt[3]{x-2}}$ **60.** $\sqrt[4]{3x\sqrt{y+4}}$

61. $\dfrac{\sqrt[3]{(a+b)^2}}{\sqrt{(a+b)}}$ **62.** $\dfrac{\sqrt[3]{(x+y)^2}}{\sqrt[4]{(x+y)^3}}$ **63.** $a^{2/3}\cdot b^{3/4}$ **64.** $x^{1/3}\cdot y^{1/4}\cdot z^{1/6}$ **65.** $\dfrac{s^{7/12}\cdot t^{5/6}}{s^{1/3}\cdot t^{-1/6}}$ **66.** $\dfrac{x^{8/15}\cdot y^{4/5}}{x^{1/3}\cdot y^{-1/5}}$

Use fractional exponents to write a single radical expression and simplify.

67. $\sqrt[5]{yx^2\sqrt{xy}}$ **68.** $\sqrt{x^5\sqrt[3]{x^4}}$ **69.** $\dfrac{\sqrt{(a+b)^3}\sqrt[3]{(a+b)^2}}{\sqrt[4]{a+b}}$ **70.** $\sqrt[4]{\sqrt[3]{8x^3y^6}}$

Simplify.

71. $(-\sqrt[4]{7}\sqrt[3]{w})^{12}$ **72.** $\sqrt[12]{p^2+2pq+q^2}$ **73.** $\dfrac{1}{\sqrt[3]{3}-\sqrt[3]{2}}$

74. $[\sqrt[10]{\sqrt[5]{x^{15}}}]^5[\sqrt[5]{\sqrt[10]{x^{15}}}]^5$ **75.** $\sqrt[p]{x^{5p}y^{7p+1}z^{p+3}}$

76. (*An application: Road pavement messages*). In a psychological study it was determined that the proper length L of the letters of a word printed on pavement is given by

$$L = \frac{(0.00252)d^{2.27}}{h},$$

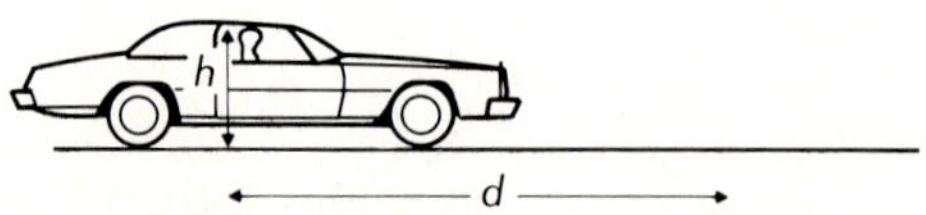

where d is the distance of a car from the lettering and h is the height of the eye above the surface of the road. All units are in meters. This formula says that if a person is h meters above the surface of the road and is to be able to recognize a message d meters away, that message will be the most recognizable if the length of the letters is L. Find L, given d and h.

a) $h = 1$ m, $d = 60$ m **b)** $h = 0.9906$ m, $d = 75$ m

c) $h = 2.4$ m, $d = 80$ m **d)** $h = 1.1$ m, $d = 100$ m

7.8

SOLVING RADICAL EQUATIONS

i A radical equation is an equation in which variables occur in one or more radicands. Here are some examples:

$$\sqrt[3]{2x} + 1 = 5, \qquad \sqrt{x} + \sqrt{4x - 2} = 7.$$

To solve such equations we need a new principle for equations. Suppose that the equation $a = b$ is true. When we square both sides we get another true equation, $a^2 = b^2$. Generalizing, we have the principle of powers.

THE PRINCIPLE OF POWERS

If an equation $a = b$ is true, then $a^n = b^n$ is true for any natural number n.

If an equation $a^n = b^n$ is true, it *may not* be true that $a = b$. For example, $3^2 = (-3)^2$ is true, but $3 = -3$ is not true.

EXAMPLE 1 Solve: $\sqrt{x} - 3 = 4$.

$$\sqrt{x} - 3 = 4$$
$$\sqrt{x} = 7 \quad \text{Adding to isolate the radical}$$
$$(\sqrt{x})^2 = 7^2 \quad \text{Using the principle of powers}$$
$$x = 49$$

Check:

$\sqrt{x} - 3 = 4$	
$\sqrt{49} - 3$	4
$7 - 3$	
4	

The solution is 49.

The principle of powers does not always produce equivalent equations.

EXAMPLE 2 Solve: $\sqrt{x} = -3$.

We might observe at the outset that this equation has no solution because the principal square root of a number is never negative. Let us continue as above, for comparison.

$$x = (-3)^2, \text{ or } 9 \quad \text{Principle of powers (squaring)}$$

Check:

$\sqrt{x} = -3$	
$\sqrt{9}$	-3
3	

The number 9 does not check. Hence the equation has no solution.

In solving radical equations, possible solutions found using the principle of powers *must* be checked!

DO EXERCISES 1 AND 2.

OBJECTIVES

After finishing Section 7.8, you should be able to:

i Solve radical equations with one radical term.

ii Solve radical equations with two radical terms.

iii Use the Pythagorean property of right triangles to find lengths of sides of triangles and to solve problems.

Solve.

1. $\sqrt{x} - 7 = 3$

2. $\sqrt{x} = -2$

Solve.

3. $x - 1 = \sqrt{x + 5}$

4. $\sqrt[4]{x - 1} - 2 = 0$

To solve an equation with a radical term, we first isolate the radical term on one side of the equation. Then we use the principle of powers.

EXAMPLE 3 Solve: $x = \sqrt{x + 7} + 5$.

$$
\begin{aligned}
x &= \sqrt{x + 7} + 5 \\
x - 5 &= \sqrt{x + 7} && \text{Adding } -5 \text{ to isolate the radical term} \\
(x - 5)^2 &= (\sqrt{x + 7})^2 && \text{Principle of powers: squaring both sides} \\
x^2 - 10x + 25 &= x + 7 \\
x^2 - 11x + 18 &= 0 \\
(x - 9)(x - 2) &= 0 && \text{Factoring} \\
x = 9 \quad \text{or} \quad x &= 2 && \text{Using the principle of zero products}
\end{aligned}
$$

The possible solutions are 9 and 2. Let us check.

For 9:

$x = \sqrt{x + 7} + 5$	
9	$\sqrt{9 + 7} + 5$
	9

For 2:

$x = \sqrt{x + 7} + 5$	
2	$\sqrt{2 + 7} + 5$
	8

Since 9 checks but 2 does not, the solution is 9.

EXAMPLE 4 Solve: $\sqrt[3]{2x + 1} + 5 = 0$.

$$
\begin{aligned}
\sqrt[3]{2x + 1} + 5 &= 0 \\
\sqrt[3]{2x + 1} &= -5 && \text{Adding } -5\text{; this isolates the radical term} \\
2x + 1 &= (-5)^3, \quad \text{or } -125 && \text{Principle of powers: raising to the third power} \\
2x &= -126 && \text{Adding } -1 \\
x &= -63
\end{aligned}
$$

Check:

$\sqrt[3]{2x + 1} + 5 = 0$	
$\sqrt[3]{2 \cdot (-63) + 1} + 5$	0
$\sqrt[3]{-125} + 5$	
$-5 + 5$	
0	

The solution is -63.

DO EXERCISES 3 AND 4.

ii Equations With Two Radical Terms

A general strategy for solving equations with two radical terms is as follows.

1. **Isolate one of the radical terms.**
2. **Use the principle of powers.**
3. **If a radical remains, perform steps 1 and 2 again.**
4. **Check possible solutions.**

EXAMPLE 5 Solve: $\sqrt{x-3} + \sqrt{x+5} = 4$.

$$\sqrt{x-3} + \sqrt{x+5} = 4$$

$$\sqrt{x-3} = 4 - \sqrt{x+5}$$ Adding $-\sqrt{x+5}$; this isolates one of the radical terms

$$(\sqrt{x-3})^2 = (4 - \sqrt{x+5})^2$$ Principle of powers: squaring both sides

In the above step 2, we are squaring a binomial. We square 4, then find twice the product of 4 and $\sqrt{x+5}$, and then the square of $\sqrt{x+5}$.

$$x - 3 = 16 - 8\sqrt{x+5} + (x+5)$$

$$-24 = -8\sqrt{x+5}$$ Isolating the remaining radical term

$$3 = \sqrt{x+5}$$

$$3^2 = x + 5$$ Squaring

$$4 = x$$

The number 4 checks, so it is the solution.

EXAMPLE 6 Solve: $\sqrt{2x-5} = 1 + \sqrt{x-3}$.

$$\sqrt{2x-5} = 1 + \sqrt{x-3}$$

$$(\sqrt{2x-5})^2 = (1 + \sqrt{x-3})^2$$ One radical is already isolated; we square both sides.

$$2x - 5 = 1 + 2\sqrt{x-3} + (x-3)$$

$$x - 3 = 2\sqrt{x-3}$$ Isolating the remaining radical term

$$(x-3)^2 = (2\sqrt{x-3})^2$$ Squaring both sides

$$x^2 - 6x + 9 = 4(x-3)$$

$$x^2 - 6x + 9 = 4x - 12$$

$$x^2 - 10x + 21 = 0$$

$$(x-7)(x-3) = 0$$ Factoring

$$x = 7 \quad \text{or} \quad x = 3$$ Using the principle of zero products

The numbers 7 and 3 check, so they are the solutions.

DO EXERCISES 5 AND 6.

Solve.

5. $\sqrt{x} - \sqrt{x-5} = 1$

6. $\sqrt{3x+1} = 1 + \sqrt{x+4}$

iii The Pythagorean Property of Right Triangles

A famous property of right triangles, relating the lengths of the sides, is named for Pythagoras, a mathematician of ancient Greece. His theorem is very useful in problem solving. As we shall see, the use of the property in problem solving involves the use of exponential and radical symbolism.

In a right triangle, the longest side is called the *hypotenuse,* and the other two sides are called the *legs*.

THE PYTHAGOREAN THEOREM

In a right triangle, if c is the length of the hypotenuse and a and b are the lengths of the legs, then

$$a^2 + b^2 = c^2.$$

If we know the lengths of any two sides of a right triangle, we can find the length of the other side.

7. Find the length of the hypotenuse of this right triangle. Approximate your answer to three decimal places.

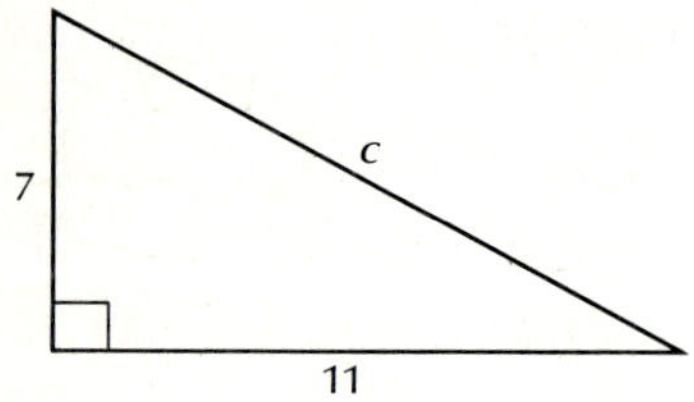

EXAMPLE 7 Find the length of the hypotenuse of the right triangle shown here. Approximate your answer to three decimal places.

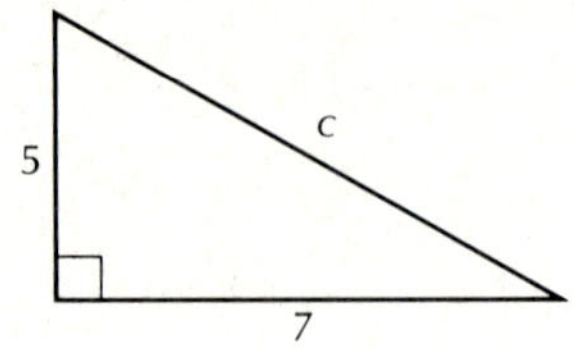

$a^2 + b^2 = c^2$	The Pythagorean equation
$7^2 + 5^2 = c^2$	Substituting
$49 + 25 = c^2$	
$74 = c^2$	
$\sqrt{74} = c$	Taking the principal square root
$8.602 \approx c$	Using Table 1 or a calculator

EXAMPLE 8 Find the length of the leg of the right triangle shown here. Approximate your answer to two decimal places.

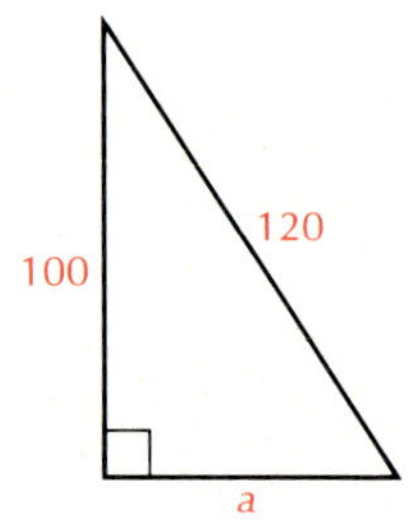

8. Find the length of the leg of this right triangle. Approximate your answer to two decimal places.

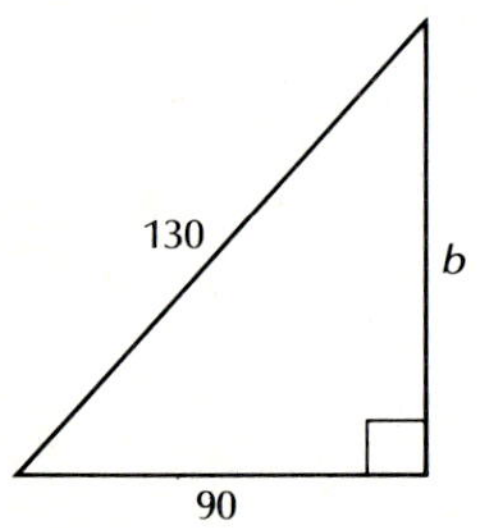

We use the Pythagorean equation

$$a^2 + b^2 = c^2.$$

We solve the equation for a:

$$a^2 = c^2 - b^2$$
$$a = \sqrt{c^2 - b^2}.$$

Now we substitute:

$$a = \sqrt{120^2 - 100^2}$$
$$= \sqrt{14{,}400 - 10{,}000}$$
$$= \sqrt{4400}, \quad \text{or } 20\sqrt{11}.$$

Using Table 1 or a calculator, we find that $a \approx 66.33$.

DO EXERCISES 7 AND 8.

Problem Solving

Right triangles occur in many problems. When that happens, we should always think of trying to use the Pythagorean theorem.

EXAMPLE 9 How much wire will be needed to make a guy wire for a telephone pole if the wire is to be fastened 20 ft above the ground and is to be anchored in the ground 9 ft from the base of the pole?

1. **Familiarize** We draw a picture and label it. In the picture we see a right triangle, so we should see if the Pythagorean theorem can be applied. We have called the length of the wire d.

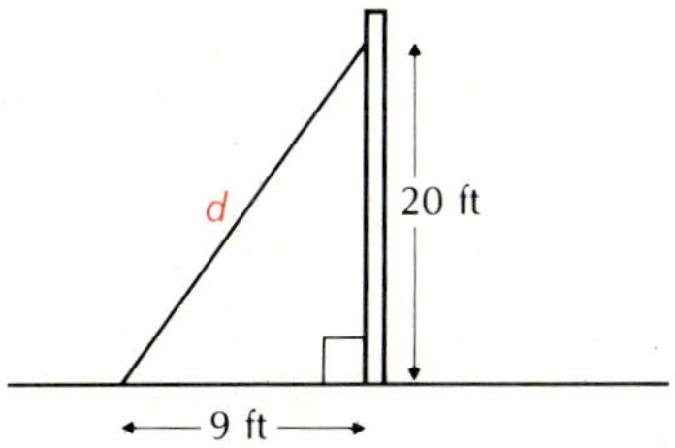

2. **Translate** From the right triangle in our drawing, we have

$$d^2 = 9^2 + 20^2. \quad \text{Substituting in the Pythagorean equation}$$

3. **Carry out** We solve the equation for d:

$$\begin{aligned} d &= \sqrt{9^2 + 20^2} && \text{Taking the principal square root on both sides} \\ &= \sqrt{481} \\ &\approx 21.93. && \text{Using Table 1 or a calculator} \end{aligned}$$

4. **Check** We square the lengths 9 and 20, obtaining 81 + 400, or 481. We then square the length 21.93:

$$21.93 \times 21.93 = 480.92.$$

This is very close to 481, so the answer checks.

5. **State** The answer is that the guy wire will be 21.93 ft long, plus an amount that will be needed for fastening it to the pole and the ground.

DO EXERCISE 9.

9. Hometown is 13 miles east of Georgetown. Northwood is 18 miles north of Georgetown. How far is it from Hometown to Northwood?

EXERCISE SET 7.8

i Solve.

1. $\sqrt{2x-3} = 1$
2. $\sqrt{x+3} = 6$
3. $\sqrt{y+1} - 5 = 8$
4. $\sqrt{x-2} - 7 = -4$
5. $\sqrt[3]{x+5} = 2$
6. $\sqrt[3]{x-2} = 3$
7. $\sqrt[4]{y-3} = 2$
8. $\sqrt[4]{x+3} = 3$
9. $\sqrt{3y+1} = 9$
10. $\sqrt{2y+1} = 13$
11. $3\sqrt{x} = 6$
12. $8\sqrt{y} = 2$
13. $\sqrt[3]{x} = -3$
14. $\sqrt[3]{y} = -4$
15. $\sqrt{y+3} - 20 = 0$
16. $\sqrt{x+4} - 11 = 0$
17. $\sqrt{x+2} = -4$
18. $\sqrt{y-3} = -2$
19. $8 = \dfrac{1}{\sqrt{x}}$
20. $3 = \dfrac{1}{\sqrt{y}}$
21. $\sqrt[3]{6x+9} + 8 = 5$
22. $\sqrt[3]{3y+6} + 2 = 3$

ii Solve.

23. $\sqrt{3y+1} = \sqrt{2y+6}$
24. $\sqrt{5x-3} = \sqrt{2x+3}$
25. $2\sqrt{1-x} = \sqrt{5}$
26. $2\sqrt{2y-3} = \sqrt{4y}$
27. $2\sqrt{t-1} = \sqrt{3t-1}$
28. $\sqrt{y+10} = 3\sqrt{2y+3}$
29. $\sqrt{y-5} + \sqrt{y} = 5$
30. $\sqrt{x-9} + \sqrt{x} = 1$
31. $3 + \sqrt{z-6} = \sqrt{z+9}$

32. $\sqrt{4x-3} = 2 + \sqrt{2x-5}$ **33.** $\sqrt{20-x+8} = \sqrt{9-x}+11$ **34.** $4 + \sqrt{10-x} = 6 + \sqrt{4-x}$

35. $\sqrt{x+2} + \sqrt{3x+4} = 2$ **36.** $\sqrt{6x+7} - \sqrt{3x+3} = 1$ **37.** $\sqrt{4y+1} - \sqrt{y-2} = 3$

38. $\sqrt{y+15} - \sqrt{2y+7} = 1$ **39.** $\sqrt{3x-5} + \sqrt{2x+3} + 1 = 0$ **40.** $\sqrt{2m-3} = \sqrt{m+7} - 2$

iii In each exercise, the lengths of two sides of a right triangle are given. Find the length of the other side (c is always the length of the hypotenuse). Use Table 1 or a calculator to find an approximation.

41. $a = 3, \quad b = 4$ **42.** $b = 8, \quad c = 10$ **43.** $a = 16, \quad c = 20$ **44.** $a = 5, \quad b = 12$ **45.** $a = 6, \quad b = 8$

46. $b = 20, \quad c = 25$ **47.** $a = 3, \quad b = 7$ **48.** $b = 11, \quad c = 12$ **49.** $a = 1, \quad b = 1$ **50.** $a = 4, \quad b = 7$

51. $a = 2, \quad b = 2\sqrt{2}$ **52.** $a = 3, \quad b = 3\sqrt{6}$

Problem solving

53. Triangle ABC has sides that are of lengths 25 ft, 25 ft, and 30 ft. Triangle PQR has sides that are of lengths 25 ft, 25 ft, and 40 ft. Which triangle has the greater area and by how much?

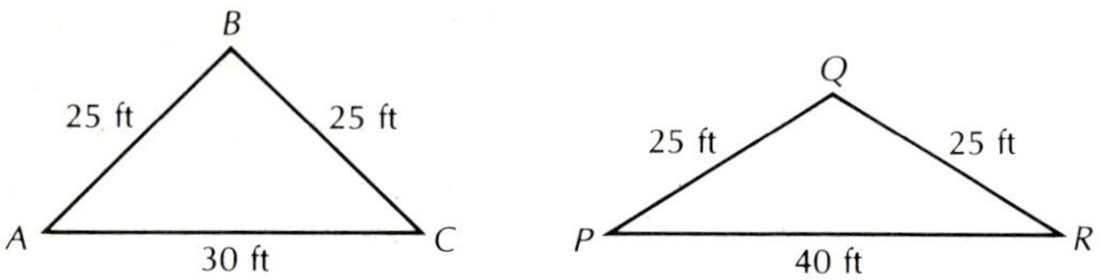

54. During the summer heat a 2-mile bridge expands 2 ft in length. Assuming the bulge occurs straight up from the middle, how high is the bulge?

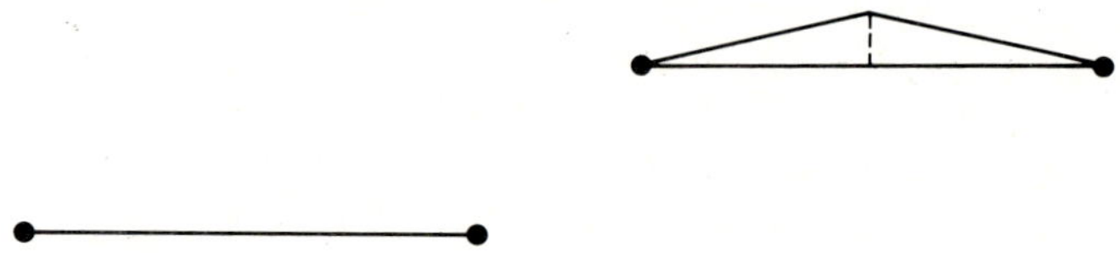

55. Two sides of a triangle have length s, and the height is h. Find a formula for the length of the third side.

56. Each side of a regular octagon has length s. Find the distance d between the parallel sides of the octagon.

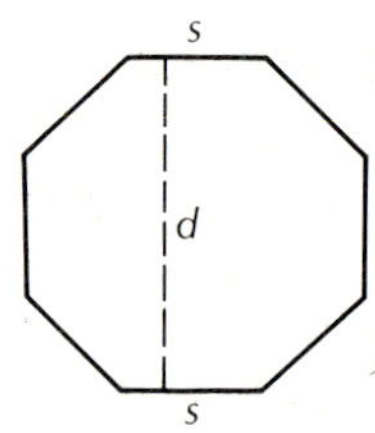

(*An application: Sighting to the horizon*). The formula $V = 1.2\sqrt{h}$ can be used to approximate the distance V, in miles, that a person can see to the horizon from a height h, in feet. Use a calculator for Exercises 57 and 58.

57. How far can you see to the horizon through an airplane window at a height of 30,000 ft?

58. How far can a sailor see to the horizon from the top of a 72-ft mast?

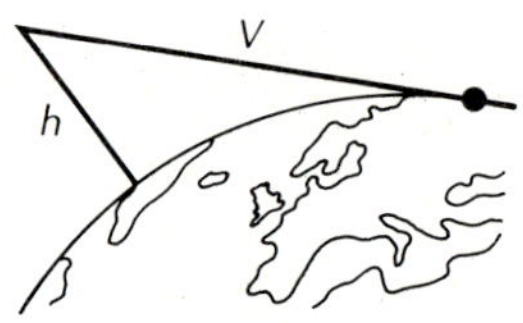

Solve.

59. $\dfrac{x + \sqrt{x+1}}{x - \sqrt{x+1}} = \dfrac{5}{11}$ **60.** $\sqrt[3]{\dfrac{z}{4}} - 10 = 2$ **61.** $\sqrt[4]{z^2+17} = 3$

62. $\sqrt{\sqrt{y} + 49} = 7$ **63.** $\sqrt[3]{x^2+x+15} - 3 = 0$ **64.** $x^2 - 5x - \sqrt{x^2-5x-2} = 4$

65. $\sqrt{8-b} = b\sqrt{8-b}$ **66.** $\sqrt{x-2} - \sqrt{x+2} + 2 = 0$ **67.** $6\sqrt{y} + 6y^{-1/2} = 37$

68. $\sqrt{a^2+30a} = a + \sqrt{5a}$ **69.** $\sqrt{\sqrt{x}+4} = \sqrt{x} - 2$ **70.** $\dfrac{x-1}{\sqrt{x^2+3x+6}} = \dfrac{1}{4}$

71. $\sqrt{x+1} - \dfrac{2}{\sqrt{x+1}} = 1$ **72.** $\sqrt{11x + \sqrt{6+x}} = 6$ **73.** $\sqrt{2x-2} + \sqrt{7x+4} = \sqrt{13x+10}$

(*The Pythagorean formula in three dimensions*). The length d of a diagonal of a rectangular box is given by $d = \sqrt{a^2+b^2+c^2}$, where a, b, and c are the lengths of the sides.

74. 🖩 Find the length of a diagonal of a box whose sides have lengths 2, 4, and 5.

75. 🖩 Will a baton 57 cm long fit inside an attache case that measures 12.5 cm by 45 cm by 31.5 cm?

7.9

THE COMPLEX NUMBERS

OBJECTIVES

After finishing Section 7.9, you should be able to:

- **i** **Express imaginary numbers as *bi*, where *b* is a real number, and complex numbers as *a* + *bi*, where *a* and *b* are real numbers.**
- **ii** **Add and subtract complex numbers.**
- **iii** **Multiply complex numbers.**
- **iv** **Find conjugates of complex numbers and divide complex numbers.**
- **v** **Determine whether a complex number is a solution of an equation.**

i Imaginary and Complex Numbers

Negative numbers do not have square roots in the system of real numbers. *Imaginary** numbers were invented so that negative numbers would have square roots. We begin by making up a number that is a square root of -1. We call this new number i. Thus $i^2 = -1$, or $i = \sqrt{-1}$. All other imaginary numbers can be expressed as a product of i and a real number.

EXAMPLES Express in terms of i.

1. $\sqrt{-5} = \sqrt{-1 \cdot 5} = \sqrt{-1} \cdot \sqrt{5} = i\sqrt{5}$ or $\sqrt{5}i$

2. $-\sqrt{-49} = -\sqrt{-1 \cdot 49} = -\sqrt{-1} \cdot \sqrt{49} = -i\sqrt{49} = -7i$

The procedure in Example 2 works, even though the general rule for factoring radicals does not hold for negative radicands.

DO EXERCISES 1–4.

An *imaginary number* is a number that can be named *bi*, where *b* is some real number and $b \neq 0$.

To form the system of *complex numbers* we take the imaginary numbers and the real numbers, and all possible sums of real and imaginary numbers. These are complex numbers:

$$3 - 4i, \qquad -\sqrt{46} + 2i, \qquad 14, \qquad i\sqrt{2}.$$

A *complex number* is any number that can be named *a* + *bi*, where *a* and *b* are any real numbers. (Note that *a* or *b* or both can be 0.)

ii Addition and Subtraction

The complex numbers follow the commutative and associative laws of addition. Thus we can add and subtract them as we do binomials in real numbers.

EXAMPLES Add or subtract.

3. $(8 + 6i) + (3 + 2i) = (8 + 3) + (6 + 2)i$ Collecting the real parts and the imaginary parts

$\qquad = 11 + 8i$

4. $(3 + 2i) - (5 - 2i) = (3 - 5) + [2 - (-2)]i$

$\qquad = -2 + 4i$

DO EXERCISES 5–8.

* Don't let the name *imaginary* fool you. The imaginary numbers are very important in such fields as engineering and the physical sciences.

Express in terms of i.

1. $\sqrt{-7}$

2. $\sqrt{-25}$

3. $-\sqrt{-11}$

4. $-\sqrt{-36}$

Add or subtract.

5. $(7 + 4i) + (8 - 7i)$

6. $(-5 - 6i) + (-7 + 12i)$

7. $(8 + 3i) - (5 + 8i)$

8. $(5 - 4i) - (-7 + 3i)$

Multiply.

9. $-3i(4 - 3i)$

10. $5i(-5 + 7i)$

11. $(1 + 3i)(1 + 5i)$

12. $(3 - 2i)(1 + 4i)$

Simplify.

13. i^5
[*Hint:* $i^5 = (i^2)(i^2)i$.]

14. $(2i)^3$

15. $(3 + 2i)^2$

16. $(2 - 5i)^2$

Find the conjugate.

17. $6 + 3i$

18. $-9 - 5i$

19. $\pi - \frac{1}{4}i$

iii Multiplying

The complex numbers also obey the commutative and associative laws of multiplication and the distributive laws. Thus we multiply them in much the same way as we do binomials in real numbers. We must remember, however, that $i^2 = -1$.

EXAMPLES Multiply.

5. $(-4i)(3 - 5i) = (-4i) \cdot 3 + (-4i)(-5i)$ Using a distributive law
$= -12i + 20i^2$
$= -12i - 20$ $i^2 = -1$
$= -20 - 12i$

6. $(1 + 2i)(1 + 3i) = 1 + 3i + 2i + 6i^2$ Multiplying each term of one number by every term of the other FOIL
$= 1 + 3i + 2i - 6$ $i^2 = -1$
$= -5 + 5i$ Collecting like terms

DO EXERCISES 9–12.

EXAMPLES Simplify.

7. $i^3 = (i^2)i = -1 \cdot i$
$= -i$

8. $i^4 = (i^2)(i^2) = (-1) \cdot (-1)$
$= 1$

9. $(3 - 2i)^2 = 3^2 - 2 \cdot (3)(2i) + (2i)^2$
$= 9 - 12i + 4i^2$
$= 9 - 12i - 4$
$= 5 - 12i$

DO EXERCISES 13–16.

We must remember that the property $\sqrt{a}\sqrt{b} = \sqrt{ab}$ does not hold for imaginary numbers. All square roots of negative numbers must be expressed in terms of i before multiplying. For example,

$$\sqrt{-2} \cdot \sqrt{-5} = i\sqrt{2} \cdot i\sqrt{5} = i^2\sqrt{10} = -\sqrt{10} \quad \text{is correct.}$$

But

$$\sqrt{-2} \cdot \sqrt{-5} = \sqrt{(-2)(-5)} = \sqrt{10} \quad \text{is wrong!}$$

iv Conjugates and Division

Conjugates of complex numbers are defined as follows.

The *conjugate* of a complex number $a + bi$ is $a - bi$, and the conjugate of $a - bi$ is $a + bi$.

EXAMPLES Find the conjugate.

10. $5 + 7i$ The conjugate is $5 - 7i$.
11. $14 - 3i$ The conjugate is $14 + 3i$.
12. $-3 - 9i$ The conjugate is $-3 + 9i$.
13. $4i$ The conjugate is $-4i$.

DO EXERCISES 17–19.

When we multiply a number by its conjugate, we get a real number.

EXAMPLES Multiply.

14. $(5 + 7i)(5 - 7i) = 5^2 - (7i)^2$
$= 25 - 49i^2$
$= 25 - (-49)$
$= 74$

15. $(2 - 3i)(2 + 3i) = 2^2 - (3i)^2$
$= 4 - 9i^2$
$= 4 - (-9)$
$= 13$

DO EXERCISES 20 AND 21.

We use conjugates in dividing complex numbers.

EXAMPLE 16 Divide: $\frac{-5 + 9i}{1 - 2i}$.

$$\frac{-5 + 9i}{1 - 2i} \cdot \frac{1 + 2i}{1 + 2i} = \frac{(-5 + 9i)(1 + 2i)}{(1 - 2i)(1 + 2i)} \qquad \text{Multiplying by 1}$$

$$= \frac{-23 - i}{5}$$

$$= -\frac{23}{5} - \frac{1}{5}i$$

Note the similarity between this example and rationalizing denominators. The symbol for the number 1 was chosen with reference to the divisor, $1 - 2i$. The symbol for 1 was formed using the conjugate of the divisor for both numerator and denominator.

EXAMPLE 17 What symbol for 1 would you use to divide?

Division to be done	*Symbol for 1*
$\frac{3 + 5i}{4 + 3i}$	$\frac{4 - 3i}{4 - 3i}$

DO EXERCISES 22 AND 23.

V Solutions of Equations

The equation $x^2 + 1 = 0$ has no real-number solution, but it has two nonreal solutions.

EXAMPLE 18 Determine whether i is a solution of the equation $x^2 + 1 = 0$.

We substitute i for x in the equation:

$x^2 + 1 = 0$	
$i^2 + 1$	0
$-1 + 1$	
0	

The number i is a solution.

DO EXERCISE 24.

Multiply.

20. $(7 - 2i)(7 + 2i)$

21. $(-3 - i)(-3 + i)$

Divide.

22. $\frac{6 + 2i}{1 - 3i}$

23. $\frac{2 + 3i}{-1 + 4i}$

24. Determine whether $-i$ is a solution of $x^2 + 1 = 0$.

25. Determine whether $1 - i$ is a solution of $x^2 - 2x + 2 = 0$.

Any equation consisting of a polynomial in one variable on one side and 0 on the other has complex-number solutions (some may be real). It is not always easy to find the solutions, but they always exist.

EXAMPLE 19 Determine whether $1 + i$ is a solution of $x^2 - 2x + 2 = 0$.

We substitute $1 + i$ for x in the equation:

$$\begin{array}{r|l} \multicolumn{2}{c}{x^2 - 2x + 2 = 0} \\ \hline (1 + i)^2 - 2(1 + i) + 2 & 0 \\ 1 + 2i + i^2 - 2 - 2i + 2 & \\ 1 + 2i - 1 - 2 - 2i + 2 & \\ (1 - 1 - 2 + 2) + (2 - 2)i & \\ 0 + 0i & \\ 0 & \end{array}$$

Thus, $1 + i$ is a solution.

EXAMPLE 20 Determine whether $2i$ is a solution of $x^2 + 3x - 4 = 0$.

$$\begin{array}{r|l} \multicolumn{2}{c}{x^2 + 3x - 4 = 0} \\ \hline (2i)^2 + 3(2i) - 4 & 0 \\ 4i^2 + 6i - 4 & \\ -4 + 6i - 4 & \\ -8 + 6i & \end{array}$$

Thus, $2i$ is not a solution.

DO EXERCISE 25.

EXERCISE SET 7.9

i Express in terms of i.

1. $\sqrt{-15}$ **2.** $\sqrt{-17}$ **3.** $\sqrt{-16}$ **4.** $\sqrt{-25}$ **5.** $-\sqrt{-12}$ **6.** $-\sqrt{-20}$

ii Add or subtract and simplify.

7. $(3 + 2i) + (5 - i)$ **8.** $(-2 + 3i) + (7 + 8i)$ **9.** $(4 - 3i) + (5 - 2i)$ **10.** $(-2 - 5i) + (1 - 3i)$

11. $(9 - i) + (-2 + 5i)$ **12.** $(6 + 4i) + (2 - 3i)$ **13.** $(3 - i) - (5 + 2i)$ **14.** $(-2 + 8i) - (7 + 3i)$

15. $(4 - 2i) - (5 - 3i)$ **16.** $(-2 - 3i) - (1 - 5i)$ **17.** $(9 + 5i) - (-2 - i)$ **18.** $(6 - 3i) - (2 + 4i)$

iii Multiply and simplify.

19. $(3 + 2i)(1 + i)$ **20.** $(4 + 3i)(2 + 5i)$ **21.** $(2 + 3i)(6 - 2i)$ **22.** $(5 + 6i)(2 - i)$

23. $(6 - 5i)(3 + 4i)$ **24.** $(5 - 6i)(2 + 5i)$ **25.** $(7 - 2i)(2 - 6i)$ **26.** $(-4 + 5i)(3 - 4i)$

27. i^6 **28.** $(-i)^4$ **29.** $(5i)^3$ **30.** $(-3i)^5$

31. $(3 - 2i)^2$ **32.** $(5 - 2i)^2$ **33.** $(2 + 3i)^2$ **34.** $(4 + 2i)^2$

35. $(-2 + 3i)^2$ **36.** $(-5 - 2i)^2$

iv Divide.

37. $\dfrac{3 + 2i}{2 + i}$ **38.** $\dfrac{4 + 5i}{5 - i}$ **39.** $\dfrac{5 - 2i}{2 + 5i}$ **40.** $\dfrac{3 - 2i}{4 + 3i}$ **41.** $\dfrac{8 - 3i}{7i}$ **42.** $\dfrac{3 + 8i}{5i}$

v Determine whether the complex number is a solution of the equation.

43. $1 + 2i$; $x^2 - 2x + 5 = 0$ **44.** $1 - 2i$; $x^2 - 2x + 5 = 0$ **45.** $1 - i$; $x^2 + 2x + 2 = 0$ **46.** $2 + i$; $x^2 - 4x - 5 = 0$

Problem-solving practice

47. The campus bookstore bought 40 dozen legal pads for a total cost of \$192.00. If the bookstore plans to sell them at 70¢ each, how many must they sell in order to realize a 50% profit?

48. A pitcher on a baseball team has pitched a total of 60 innings so far this season, giving up an average of 2.5 runs per 9-inning game. If he gives up 3 runs in 7 innings during his next outing, what is he now averaging in runs per 9-inning game?

---

49. A function g is given by $g(x) = \dfrac{x^4 - x^2}{x - 1}$. Find $g(2i)$; $g(i + 1)$; $g(2i - 1)$.

50. Evaluate $\dfrac{1}{w - w^2}$ when $w = \dfrac{1 - i}{10}$; when $w = \dfrac{1}{1 + i}$.

Express in terms of i.

51. $\frac{1}{8}(-24 - \sqrt{-1024})$

52. $12\sqrt{-\frac{1}{32}}$

53. $7\sqrt{-64} - 9\sqrt{-256}$

Simplify.

54. $\dfrac{i^5 + i^6 + i^7 + i^8}{(1 - i)^4}$

55. $(1 - i)^3(1 + i)^3$

56. $\dfrac{5 - \sqrt{5}i}{\sqrt{5}i}$

57. $\dfrac{6}{1 + \dfrac{3}{i}}$

58. $(\frac{1}{2} - \frac{1}{3}i)^2 - (\frac{1}{2} + \frac{1}{3}i)^2$

59. $\dfrac{i - i^{38}}{1 + i}$

60. $\dfrac{2}{1 - 3i} - \dfrac{3}{4 + 2i}$

61. $\dfrac{2\sqrt{6} + 4\sqrt{5}i}{2\sqrt{6} - 4\sqrt{5}i}$

62. $\dfrac{1 - 4i}{4i(1 + 4i)^{-1}}$

SUMMARY AND REVIEW: CHAPTER 7

The following contains a summary of what you should be able to do after completing this chapter. The review exercises are for practice. Answers are at the back of the book. If you miss an exercise, restudy the section indicated alongside the answer.

You should be able to:

Simplify radical expressions.

Simplify. Assume that expressions in radicals represent *any* real number.

1. $\sqrt{(-48)^2}$

2. $\sqrt{25x^2}$

3. $\sqrt[3]{\dfrac{-27}{64x^3}}$

4. $\sqrt[5]{-32}$

5. $\sqrt[4]{(-5x)^4}$

Multiply and simplify with radical notation.

Multiply.

6. $\sqrt{15x}\sqrt{30x}$

7. $\sqrt[3]{m^4n}\sqrt[3]{m^2n^8}$

8. $\sqrt[3]{9a^3b^4}\sqrt[3]{6ab^2}$

Approximate square roots.

Approximate.

9. $\sqrt{245}$

10. $\sqrt{640}$

11. $\sqrt{63{,}000{,}000{,}000}$

12. $\sqrt{0.000003033}$

Divide and simplify with radical notation.

Divide and simplify. Assume that expressions in radicands are positive for the remainder of this review.

13. $\dfrac{\sqrt[3]{54}}{\sqrt[3]{2}}$

14. $\sqrt{\dfrac{20x^4}{y^6}}$

15. $\sqrt{\dfrac{72a^7}{40a^4}}$

Add and subtract with radical notation.

Add and subtract.

16. $3\sqrt{242} - \sqrt{32} + \sqrt{98}$

17. $\sqrt[3]{128} - \sqrt[3]{54}$

18. $2\sqrt[3]{4x^4} - \sqrt[3]{32x}$

Multiply radical expressions whose factors contain more than one term.

Multiply.

19. $(3 + 5\sqrt{2})(3 - 5\sqrt{2})$

20. $(4\sqrt{6} - 3)^2$

21. $(3 + \sqrt{x})(2 + \sqrt{x})$

Rationalize the denominator in fractional expressions containing radicals.

Rationalize the denominator.

22. $\dfrac{\sqrt{7x}}{\sqrt{12x}}$ **23.** $\dfrac{8}{6+\sqrt{5}}$ **24.** $\dfrac{\sqrt{2}-5}{\sqrt{5}-2}$

Simplify radical expressions using fractional exponents.

Simplify.

25. $\sqrt[8]{x^5y^4z^3}$ **26.** $(\sqrt[3]{25})^2$ **27.** $\sqrt[10]{36}$ **28.** $\sqrt[4]{81x^{16}y^{20}}$

Use fractional exponents to write a single radical expression.

Write as a single radical expression.

29. $x^{1/4}y^{1/2}$ **30.** $\sqrt[3]{x}\sqrt{x}$ **31.** $\dfrac{\sqrt[4]{(a-b)^3}}{\sqrt[5]{(a-b)^2}}$

Solve radical equations.

Solve.

32. $\sqrt{4x-5}=3$ **33.** $\sqrt{5x+1}-\sqrt{x}=5$

Add and subtract complex numbers, simplifying answers.

Add or subtract.

34. $(2+2i)+(4-i)$ **35.** $(3-4i)-(5+7i)$ **36.** $(-2+6i)-(9-9i)$

Multiply and divide complex numbers, simplifying answers.

Multiply or divide.

37. $(4-4i)(2+3i)$ **38.** $(-6+7i)(-3-5i)$ **39.** $\dfrac{3+4i}{2-3i}$ **40.** $\dfrac{-3+7i}{5-4i}$

---

41. Multiply.

$$(\sqrt{x}-2\sqrt{y})(2\sqrt{x}+\sqrt{y})$$

42. Rationalize.

$$\frac{\sqrt{3xy}}{\sqrt{2x}-\sqrt{3y}}$$

43. Write fractional exponent notation with no denominator.

$$(4\sqrt{x})(3\sqrt{xy^{-1}})$$

TEST: CHAPTER 7

Simplify. Assume that expressions in radicands represent any real number.

1. $-\sqrt{144x^2}$ **2.** $\sqrt[3]{-216}$ **3.** $\sqrt{24x}\sqrt{48x}$ **4.** $\sqrt[3]{x^3y^5}\sqrt[3]{x^2y^2}$ **5.** $\dfrac{\sqrt{21x^7}}{\sqrt{28x^2}}$ **6.** $\dfrac{\sqrt[3]{810}}{\sqrt[3]{15}}$

Add or subtract. Assume that expressions in radicals are positive for the remainder of this test.

7. $\sqrt{56}+\sqrt{126}$ **8.** $\sqrt[3]{120x}-\sqrt[3]{15x}$

Multiply.

9. $(4+3\sqrt{2})(4-3\sqrt{2})$ **10.** $(\sqrt{4x}+\sqrt{9y})(\sqrt{4x}-\sqrt{9y})$

Rationalize the denominator.

11. $\dfrac{\sqrt{3}}{\sqrt{11}}$ **12.** $\dfrac{\sqrt[3]{3a^4}}{\sqrt[3]{6b^4}}$

Simplify.

13. $\sqrt[3]{6x^4y^2}$ **14.** $\sqrt[6]{49a^9b^9}$ **15.** $\sqrt[6]{25m^{18}n^{24}}$

Solve.

16. $\sqrt{a+5} = a + 3$

17. $\sqrt{x-2} + 1 = \sqrt{2x-6}$

Add or subtract.

18. $(5 + 6i) + (6 - 5i)$

19. $(10 - 3i) - (-4 + 5i)$

Multiply or divide.

20. $(3 - 7i)(2 + 3i)$

21. $\dfrac{(-5 + 4i)}{(11 + 2i)}$

Solve.

22. $(\sqrt{x-1} - 3)(\sqrt{x-1} + 3) = 14$

How should you cut out corners to make the biggest possible box? Quadratic functions can be used to solve such problems.

8

QUADRATIC EQUATIONS AND FUNCTIONS

When you translate a problem situation to mathematical language, you may obtain a function or equation containing a polynomial of second degree. Such equations and functions are called *quadratic*. In this chapter you will consider quadratic equations and learn to solve them. Then you will have the opportunity to solve problems whose translation contains quadratic equations or functions.

OBJECTIVES

After finishing Section 8.1, you should be able to:

i **Solve equations of the type $ax^2 + c = 0$.**

ii **Solve equations of the type $ax^2 + bx = 0$.**

iii **Solve equations of the type $ax^2 + bx + c = 0$.**

iv **Translate problems to quadratic equations and solve.**

Solve.

1. $5x^2 = 15$

2. $7x^2 = 0$

8.1

QUADRATIC EQUATIONS

Polynomial equations $P(x) = 0$ are called *quadratic* if they are of second degree. The standard form of a quadratic equation is

$$ax^2 + bx + c = 0,$$

where the coefficients a, b, and c are real numbers. The coefficients b and c might be 0, but a cannot be. If it were, the polynomial would not be quadratic.

i Solving Equations $ax^2 + c = 0$

We first consider equations in which the first-degree term is missing, that is, $ax^2 + c = 0$. Such equations can be solved by taking square roots on both sides. To see how that works, we shall proceed by factoring:

$$ax^2 = -c$$

$$x^2 = -\frac{c}{a}.$$

Here we have x^2 on one side and some number on the other. To simplify writing, we will write k for the constant. Now we solve by factoring:

$$x^2 = k$$

$$x^2 - k = 0$$

$$(x + \sqrt{k})(x - \sqrt{k}) = 0 \qquad \text{Factoring}$$

$$x + \sqrt{k} = 0 \quad \text{or} \quad x - \sqrt{k} = 0 \qquad \text{Using the principle of zero products}$$

$$x = -\sqrt{k} \quad \text{or} \quad x = \sqrt{k}.$$

There are two solutions. They are the principal square root $\sqrt{k}$ and its additive inverse $-\sqrt{k}$.

EXAMPLE 1 Solve: $3x^2 = 6$.

$$3x^2 = 6$$

$$x^2 = 2 \qquad \text{Multiplying by } \tfrac{1}{3}$$

$$x = \sqrt{2} \quad \text{or} \quad x = -\sqrt{2} \qquad \text{Taking square roots}$$

These numbers check, so the solutions are $\sqrt{2}$ and $-\sqrt{2}$.

DO EXERCISES 1 AND 2.

In a situation like Example 1, we sometimes use the symbol $\pm\sqrt{k}$ to represent both of the square roots of k. Such a shorthand use of symbolism is handy, but we must not use it in connection with square-root *functions.*

Sometimes we rationalize denominators to simplify answers.

EXAMPLE 2 Solve: $-5x^2 + 2 = 0$.

$$-5x^2 + 2 = 0$$

$$x^2 = \tfrac{2}{5} \qquad \text{Adding } -2 \text{ and multiplying by } -\tfrac{1}{5}$$

$$x = \sqrt{\tfrac{2}{5}} \quad \text{or} \quad x = -\sqrt{\tfrac{2}{5}} \qquad \text{Taking square roots}$$

$$x = \sqrt{\tfrac{2}{5} \cdot \tfrac{5}{5}} \quad \text{or} \quad x = -\sqrt{\tfrac{2}{5} \cdot \tfrac{5}{5}} \qquad \text{Rationalizing the denominators}$$

$$x = \frac{\sqrt{10}}{5} \quad \text{or} \quad x = -\frac{\sqrt{10}}{5}$$

The checks can be done together.

Check:

$$\begin{array}{r|l} -5x^2 + 2 = 0 & \\ \hline -5\left(\pm\frac{\sqrt{10}}{5}\right)^2 + 2 & 0 \\ -5\left(\frac{10}{25}\right) + 2 & \\ -2 + 2 & \\ 0 & \end{array}$$

The solutions are $\frac{\sqrt{10}}{5}$ and $-\frac{\sqrt{10}}{5}$, or $\pm\frac{\sqrt{10}}{5}$.

DO EXERCISE 3.

Sometimes we get solutions that are not real numbers.

EXAMPLE 3 Solve: $4x^2 + 9 = 0$.

$$4x^2 + 9 = 0$$

$$x^2 = -\tfrac{9}{4} \quad \text{Adding } -9 \text{ and multiplying by } \tfrac{1}{4}$$

$$x = \sqrt{-\tfrac{9}{4}} \quad \text{or} \quad x = -\sqrt{-\tfrac{9}{4}} \quad \text{Taking square roots}$$

$$x = \tfrac{3}{2}i \quad \text{or} \quad x = -\tfrac{3}{2}i \quad \text{Simplifying}$$

The numbers $\frac{3}{2}i$ and $-\frac{3}{2}i$ check, so they are solutions.

DO EXERCISE 4.

ii Solving Equations $ax^2 + bx = 0$

When c is 0 but b is not, we can factor and use the principle of zero products.

EXAMPLE 4 Solve: $3x^2 + 5x = 0$.

$$3x^2 + 5x = 0$$

$$x(3x + 5) = 0 \quad \text{Factoring}$$

$$x = 0 \quad \text{or} \quad 3x + 5 = 0 \quad \text{Principle of zero products}$$

$$x = 0 \quad \text{or} \quad 3x = -5$$

$$x = 0 \quad \text{or} \quad x = -\tfrac{5}{3}$$

Check: For 0:

$$\begin{array}{r|l} 3x^2 + 5x = 0 & \\ \hline 3 \cdot (0)^2 + 5(0) & 0 \\ 0 + 0 & \\ 0 & \end{array}$$

For $-\frac{5}{3}$:

$$\begin{array}{r|l} 3x^2 + 5x = 0 & \\ \hline 3(-\frac{5}{3})^2 + 5(-\frac{5}{3}) & 0 \\ 3(\frac{25}{9}) + (-\frac{25}{3}) & \\ \frac{25}{3} + (-\frac{25}{3}) & \\ 0 & \end{array}$$

The solutions are 0 and $-\frac{5}{3}$.

A quadratic equation of the above type will always have 0 as one solution and a nonzero number as the other solution.

DO EXERCISE 5.

Solve.

3. $-3x^2 + 8 = 0$

Solve.

4. $2x^2 + 1 = 0$

Solve.

5. $4x^2 - 3x = 0$

Solve.

6. $5x^2 - 8x + 3 = 0$

iii Solving Equations $ax^2 + bx + c = 0$

When none of the constants a, b, or c is 0, we can try factoring.

EXAMPLE 5 Solve: $3x^2 + x - 2 = 0$.

$$\begin{aligned} 3x^2 + x - 2 &= 0 \\ (3x - 2)(x + 1) &= 0 \qquad \text{Factoring} \\ 3x - 2 = 0 \quad &\text{or} \quad x + 1 = 0 \\ 3x = 2 \quad &\text{or} \quad x = -1 \\ x = \tfrac{2}{3} \quad &\text{or} \quad x = -1 \end{aligned}$$

Check: For -1:

$$\begin{array}{c|c} 3x^2 + x - 2 = 0 & \\ \hline 3(-1)^2 + (-1) - 2 & 0 \\ 3 - 1 - 2 & \\ 0 & \end{array}$$

For $\frac{2}{3}$:

$$\begin{array}{c|c} 3x^2 + x - 2 = 0 & \\ \hline 3(\frac{2}{3})^2 + (\frac{2}{3}) - 2 & 0 \\ 3 \cdot \frac{4}{9} + \frac{2}{3} - 2 & \\ \frac{4}{3} + \frac{2}{3} - 2 & \\ \frac{6}{3} - 2 & \\ 0 & \end{array}$$

The solutions are -1 and $\frac{2}{3}$.

DO EXERCISE 6.

Sometimes it helps to find the standard form before factoring.

EXAMPLE 6 Solve: $(x - 1)(x + 1) = 5(x - 1)$.

$$\begin{aligned} (x - 1)(x + 1) &= 5(x - 1) \\ x^2 - 1 &= 5x - 5 \qquad \text{Multiplying} \\ x^2 - 5x + 4 &= 0 \qquad \text{Finding standard form} \\ (x - 4)(x - 1) &= 0 \qquad \text{Factoring} \\ x = 4 \quad \text{or} \quad x &= 1 \qquad \text{Principle of zero products} \end{aligned}$$

Solve.

7. $6(y - 3) = (y - 3)(y - 2)$

The solutions are 4 and 1 (the check has been left for the reader).

DO EXERCISE 7.

iv Solving Problems

It often happens that when we translate a problem to mathematical language we get a quadratic equation. Of course, we use the five-step problem-solving process as usual.

EXAMPLE 7 A rectangular lawn measures 60 ft by 80 ft. Part of the lawn is torn up to install a sidewalk of uniform width around it. The area of the new lawn is one-sixth the area of the old. How wide is the sidewalk?

1. **Familiarize** We make a drawing and label it with both known and unknown information. We have called the width of the sidewalk x.

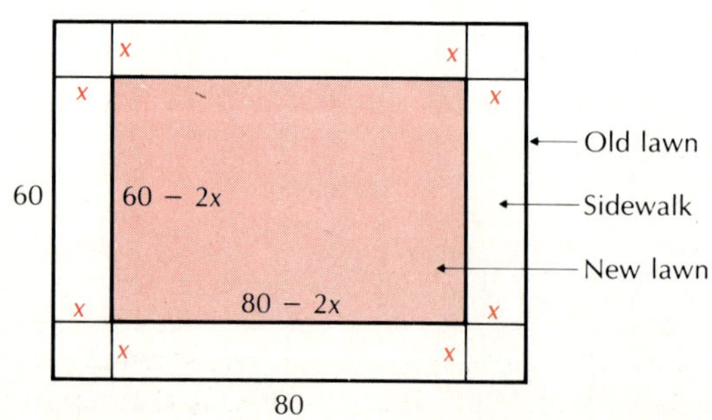

We will need a formula for the area of a rectangle. It is

$$A = l \cdot w \quad \text{(length} \times \text{width)}.$$

2. **Translate** We know that the area of the original lawn is $60 \cdot 80$:

$$\text{Area of the new lawn is } (60 - 2x)(80 - 2x).$$

Since the area of the new lawn is $\frac{1}{6}$ of the old, we have

$$(60 - 2x)(80 - 2x) = \tfrac{1}{6} \cdot 60 \cdot 80.$$

This is the translation.

3. **Carry out** We solve the equation:

$4800 - 120x - 160x + 4x^2 = 800$	Multiplying binomials on the left
$4x^2 - 280x + 4000 = 0$	Collecting like terms
$x^2 - 70x + 1000 = 0$	Multiplying on both sides by $\frac{1}{4}$
$(x - 20)(x - 50) = 0$	Factoring
$x = 20 \quad \text{or} \quad x = 50.$	Principle of zero products

4. **Check** We now check in the original problem. We see that 50 is not a solution because when $x = 50$, $60 - 2x = -40$, and the width of the sidewalk cannot be negative.

Let's see whether 20 checks. If the sidewalk is 20 ft wide, the lawn itself will have length $80 - 2 \cdot 20$, or 40 ft. The width will be $60 - 2 \cdot 20$, or 20 ft. The new area is thus $20 \cdot 40$, or 800 ft^2. The old area was $60 \cdot 80$, or 4800 ft^2. The new area of 800 ft^2 is $\frac{1}{6}$ of this, so the number 20 checks.

5. **State** The answer is that the sidewalk is 20 ft wide.

DO EXERCISE 8.

8. An open box is to be made from a 10-ft-by-20-ft rectangular piece of cardboard by cutting a square from each corner. The area of the bottom of the box is to be 96 sq ft. What is the length of the sides of the squares that are cut from the corners?

EXAMPLE 8 A wire is stretched from the ground to the top of a building, as shown. The wire is 20 ft long. The height of the building is 4 ft greater than the distance d from the building. Find the distance d and the height of the building.

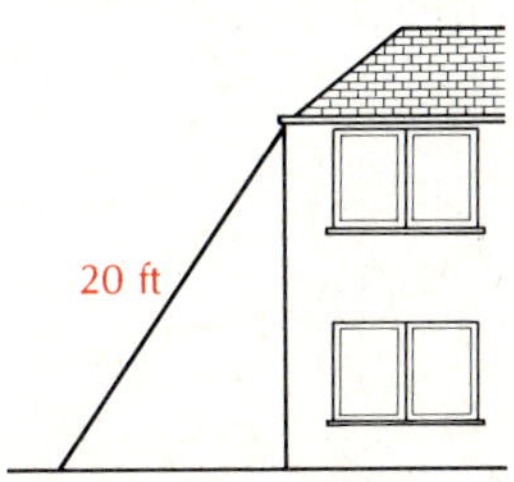

1. **Familiarize** We make a drawing and label it. We do not know the distance d or the height of the building, but the height must be $d + 4$.

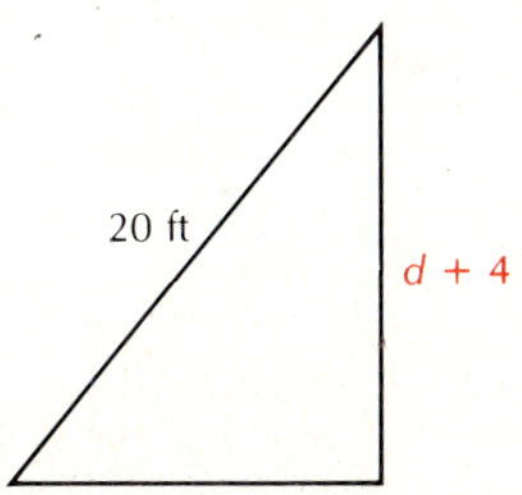

As we look at the drawing we see that a right triangle is formed. That should bring to mind the Pythagorean theorem. It says that in any right triangle, the

9. Three towns A, B, and C are situated as shown. The roads at A form a right angle. The distance from A to B is two miles less than the distance from A to C. The distance from B to C is 10 miles. Find the distance from A to B and the distance from A to C.

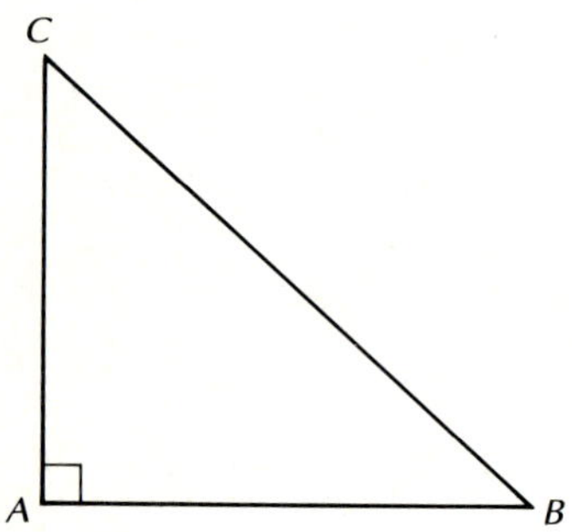

square of the longest side c (the hypotenuse) is the sum of the squares of the other two sides a and b (the legs):

$$c^2 = a^2 + b^2.$$

2. **Translate** In this problem the longest side has a length of 20. One of the legs has length d and the other has length $d + 4$. Substituting into the Pythagorean equation, we have

$$20^2 = d^2 + (d+4)^2.$$

3. **Carry out** We solve the equation:

$$\begin{aligned} 400 &= d^2 + d^2 + 8d + 16 && \text{Squaring} \\ 2d^2 + 8d - 384 &= 0 && \text{Finding standard form} \\ d^2 + 4d - 192 &= 0 && \text{Multiplying by } \tfrac{1}{2} \\ (d+16)(d-12) &= 0 && \text{Factoring} \\ d + 16 = 0 \quad &\text{or} \quad d - 12 = 0 && \text{Principle of zero products} \\ d = -16 \quad &\text{or} \quad d = 12. \end{aligned}$$

4. **Check** We know that -16 is not an answer because distance cannot be negative. We try 12. If $d = 12$, then the height is $d + 4$, or 16:

$$\begin{aligned} 12^2 + 16^2 &= 144 + 256 \\ &= 400. \end{aligned}$$

Is $400 = 20^2$? It is, so the answer checks.

5. **State** The height of the building is 16 ft and the distance d is 12 ft.

DO EXERCISE 9.

EXERCISE SET 8.1

i Solve.

1. $4x^2 = 20$
2. $3x^2 = 21$
3. $10x^2 = 0$
4. $9x^2 = 0$
5. $2x^2 - 3 = 0$
6. $3x^2 - 7 = 0$
7. $-3x^2 + 5 = 0$
8. $-2x^2 + 1 = 0$
9. $25x^2 + 4 = 0$
10. $9x^2 + 16 = 0$
11. $3x^2 + 1 = 0$
12. $5x^2 + 1 = 0$
13. $x^2 + 5 = 0$
14. $x^2 + 6 = 0$
15. $2x^2 + 14 = 0$
16. $3x^2 + 15 = 0$
17. $\frac{4}{9}x^2 - 1 = 0$
18. $\frac{16}{25}x^2 - 1 = 0$

ii Solve.

19. $x^2 - 5x = 0$
20. $x^2 - 6x = 0$
21. $5x^2 + 10x = 0$
22. $3x^2 + 12x = 0$
23. $3x^2 - 2x = 0$
24. $7x^2 - 3x = 0$
25. $14x^2 + 9x = 0$
26. $19x^2 + 8x = 0$
27. $11x^2 - 55x = 0$

iii Solve.

28. $x^2 - 6x + 5 = 0$
29. $x^2 - 7x + 6 = 0$
30. $x^2 - 4x - 5 = 0$
31. $x^2 - 6x - 7 = 0$
32. $x^2 + 8x + 15 = 0$
33. $x^2 + 9x + 14 = 0$
34. $6x^2 - x - 2 = 0$
35. $2x^2 + 13x + 15 = 0$
36. $9t^2 + 15t + 4 = 0$
37. $3y^2 + 10y - 8 = 0$
38. $6x^2 + 4x = 10$
39. $3x^2 + 7x = 20$
40. $2x(4x - 5) = 3$
41. $t(2t + 9) = -7$
42. $(p - 3)(p - 4) = 42$
43. $16(t - 1) = t(t + 8)$
44. $4x(x - 2) - 5x(x - 1) = 2$
45. $14(x - 4) - (x + 2) = (x + 2)(x - 4)$

iv ***Problem solving***

46. A picture frame measures 14 cm by 20 cm and 160 cm^2 of picture shows. Find the width of the frame.

47. A picture frame measures 12 cm by 20 cm and 84 cm^2 of picture shows. Find the width of the frame.

48. The width of a rectangle is 4 m less than the length. The area is 12 m^2. Find the length and width.

49. The width of a rectangle is 5 m less than the length. The area is 24 m^2. Find the length and width.

50. The length of a rectangle is twice the width. The area is 288 m^2. Find the length and width.

51. The length of a rectangle is twice the width. The area is 338 km^2. Find the length and width.

52. The hypotenuse of a right triangle is 5 m long. One leg is 1 m shorter than the other. Find the lengths of the legs.

53. The hypotenuse of a right triangle is 13 m long. One leg is 7 m longer than the other. Find the lengths of the legs.

54. The hypotenuse of a right triangle is 26 m long. The length of one leg is 14 m longer than the other. Find the lengths of the legs.

55. The hypotenuse of a right triangle is 25 km long. The length of one leg is 17 km shorter than the other. Find the lengths of the legs.

Solve.

56. $25.55x^2 - 1635.2 = 0$

57. $(3x^2 - 7x - 20)(2x - 5) = 0$

58. $(x)(2x^2 + 9x - 56)(3x + 10) = 0$

59. $(x - \frac{1}{3})(x - \frac{1}{3}) + (x - \frac{1}{3})(x + \frac{2}{9}) = 0$

Solve for x.

60. $ax^2 - b = 0$

61. $ax^2 - bx = 0$

62. Boats A and B leave the same point at the same time at right angles, but B travels 7 km/h slower than A. After 4 hr they are 68 km apart. Find the speed of each boat.

63. Find three consecutive integers such that the square of the first plus the product of the other two is 67.

64. An equilateral triangle is inscribed in a circle whose circumference is 6π. Find the area of the triangle.

8.2

THE QUADRATIC FORMULA

OBJECTIVES

After finishing Section 8.2, you should be able to:

i **Solve quadratic equations using the quadratic formula.**

ii **Find approximate solutions using a table or a calculator.**

i Some quadratic equations are difficult to solve by factoring. In fact some of them are impossible to solve by factoring, if we look only for integer coefficients in the factors. However, the following formula can be used to find the solutions of any quadratic equation.

THE QUADRATIC FORMULA

The solutions of a quadratic equation $ax^2 + bx + c = 0$ are given by

$$x = \frac{-b \pm \sqrt{b^2 - 4ac}}{2a}.$$

The quadratic formula can be obtained in several different ways. The following derivation is one of the shorter ways. It involves completing the square. We solve the equation $ax^2 + bx + c = 0$:

$$ax^2 + bx + c = 0$$

$$4a^2x^2 + 4abx + 4ac = 0 \qquad \text{Multiplying on both sides by } 4a \text{, to make the first term a square}$$

$$4a^2x^2 + 4abx + (b^2 - b^2) + 4ac = 0 \qquad \text{Adding } b^2 - b^2 \text{ on the left, to complete the square}$$

$$4a^2x^2 + 4abx + b^2 - (b^2 - 4ac) = 0 \qquad \text{Rearranging}$$

$$4a^2x^2 + 4abx + b^2 = b^2 - 4ac \qquad \text{Adding } b^2 - 4ac \text{ on both sides}$$

$$(2ax + b)^2 = b^2 - 4ac. \qquad \text{Factoring the trinomial square on the left}$$

Solve using the quadratic formula.

1. $3x^2 + 2x = 7$

We now take square roots:

$$2ax + b = \sqrt{b^2 - 4ac} \quad \text{or} \quad 2ax + b = -\sqrt{b^2 - 4ac}.$$

Solving for x, we get

$$x = \frac{-b + \sqrt{b^2 - 4ac}}{2a} \quad \text{or} \quad x = \frac{-b - \sqrt{b^2 - 4ac}}{2a}.$$

These are the solutions. We abbreviate as follows:

$$x = \frac{-b \pm \sqrt{b^2 - 4ac}}{2a}.$$

EXAMPLE 1 Solve: $5x^2 - 8x = 3$.

First find standard form and determine a, b, and c:

$$5x^2 - 8x - 3 = 0; \qquad a = 5, \quad b = -8, \quad c = -3.$$

Then substitute into the quadratic formula:

$$x = \frac{-(-8) \pm \sqrt{(-8)^2 - 4 \cdot 5 \cdot (-3)}}{2 \cdot 5}$$

$$x = \frac{8 \pm \sqrt{64 + 60}}{10} = \frac{8 \pm \sqrt{124}}{10}$$

$$x = \frac{8 \pm \sqrt{4 \cdot 31}}{10} = \frac{8 \pm 2\sqrt{31}}{10}$$

$$x = \frac{2(4 \pm \sqrt{31})}{2 \cdot 5} = \frac{4 \pm \sqrt{31}}{5}$$

The checking of possible solutions such as these is cumbersome. When we use the quadratic formula, we will never obtain any numbers that are not solutions of the original equation unless we have made a mistake. Therefore, it is usually better to check our work, rather than to check by substituting into the original equation.

The solutions are

$$\frac{4 + \sqrt{31}}{5} \quad \text{and} \quad \frac{4 - \sqrt{31}}{5}.$$

DO EXERCISE 1.

Some quadratic equations have solutions that are not real numbers.

EXAMPLE 2 Solve: $x^2 + x + 1 = 0$.

$$a = 1, \quad b = 1, \quad c = 1$$

$$x = \frac{-b \pm \sqrt{b^2 - 4ac}}{2a}$$

$$x = \frac{-1 \pm \sqrt{1^2 - 4 \cdot 1 \cdot 1}}{2 \cdot 1} = \frac{-1 \pm \sqrt{1 - 4}}{2} = \frac{-1 \pm \sqrt{-3}}{2}$$

$$x = \frac{-1 \pm i\sqrt{3}}{2}$$

The solutions are

$$\frac{-1+i\sqrt{3}}{2} \quad \text{and} \quad \frac{-1-i\sqrt{3}}{2}.$$

DO EXERCISE 2.

Solve.

2. $x^2 - x + 2 = 0$

EXAMPLE 3 Solve: $2 + \frac{7}{x} = \frac{4}{x^2}$.

First find standard form.

$$2x^2 + 7x = 4 \qquad \text{Multiplying by } x^2\text{, the LCM of the denominators}$$

$$2x^2 + 7x - 4 = 0 \qquad \text{Adding } -4$$

$$a = 2, \quad b = 7, \quad c = -4$$

$$x = \frac{-7 \pm \sqrt{7^2 - 4 \cdot 2 \cdot (-4)}}{2 \cdot 2}$$

$$x = \frac{-7 \pm \sqrt{49 + 32}}{4}$$

$$x = \frac{-7 \pm \sqrt{81}}{4} = \frac{-7 \pm 9}{4}$$

$$x = \frac{-7+9}{4} = \frac{1}{2} \quad \text{or} \quad x = \frac{-7-9}{4} = -4$$

In this case, since we started with a fractional equation, we *do* need to check.* At least we need to show that neither of the numbers makes a denominator 0. Since neither of them does, the solutions are $\frac{1}{2}$ and -4.

Solve.

3. $3 - \frac{5}{x} = \frac{2}{x^2}$

The solutions of a quadratic equation can *always* be found using the quadratic formula. They are not always easy to find by factoring. A general strategy for solving quadratic equations is as follows.

To solve a quadratic equation:

a) Try factoring.

b) If factoring seems difficult, use the quadratic formula; it *always works*!

DO EXERCISE 3.

Approximating Solutions

Many solutions of quadratic equations are irrational numbers, because they involve square roots. In such cases we may wish to find rational-number approximations to the solutions, using a table or a calculator.

EXAMPLE 4 Approximate, to the nearest tenth, the solutions of the equation in Example 1.

* The quadratic formula always gives correct results when we start with the standard form. In such cases we need check only to detect errors. In this example, however, we cleared of fractions before obtaining standard form, and this step could possibly introduce numbers that do not check in the original equation.

Approximate to the nearest tenth.

4. $\dfrac{8 \pm \sqrt{47}}{7}$

5. Approximate the solutions to Margin Exercise 1. Round to the nearest tenth.

Using a calculator, we find that $\sqrt{31} \approx 5.568$. Thus,

$$\frac{4+\sqrt{31}}{5} \approx \frac{4+5.568}{5}$$

$$\approx \frac{9.568}{5} \quad \text{Adding}$$

$$\approx 1.913 \quad \text{Dividing}$$

$$\approx 1.9 \quad \text{Rounding to the nearest tenth}$$

$$\frac{4-\sqrt{31}}{5} \approx \frac{4-5.568}{5}$$

$$\approx \frac{-1.568}{5} \quad \text{Subtracting}$$

$$\approx -0.313 \quad \text{Dividing}$$

$$\approx -0.3 \quad \text{Rounding to the nearest tenth}$$

DO EXERCISES 4 AND 5.

EXERCISE SET 8.2

i Solve.

1. $x^2 + 6x + 4 = 0$
2. $x^2 - 6x - 4 = 0$
3. $x^2 + 4x - 5 = 0$
4. $x^2 - 2x - 15 = 0$
5. $y^2 + 7y = 30$
6. $y^2 - 7y = 30$
7. $2t^2 - 3t - 2 = 0$
8. $5m^2 + 3m - 2 = 0$
9. $3p^2 = -8p - 5$
10. $3u^2 = 18u - 6$
11. $x^2 - x + 1 = 0$
12. $x^2 + x + 2 = 0$
13. $x^2 + 13 = 4x$
14. $x^2 + 13 = 6x$
15. $z^2 + 5 = 0$
16. $t^2 + 3 = 0$
17. $r^2 + 3r = 8$
18. $h^2 + 4 = 6h$
19. $1 + \dfrac{2}{x} + \dfrac{5}{x^2} = 0$
20. $1 + \dfrac{5}{x^2} = \dfrac{2}{x}$
21. $x^2 - 2x + 5 = 0$
22. $x^2 - 4x + 5 = 0$
23. $5 = 2x^2$
24. $2 = 3x^2$
25. $x(x - 2) = -3x$
26. $x(x - 3) = -4x$
27. $2x + 1 = -5x^2$
28. $x + 2 = -3x^2$
29. $(2t - 3)^2 + 17t = 15$
30. $2y^2 - (y + 2)(y - 3) = 12$
31. $(x - 2)^2 + (x + 1)^2 = 0$
32. $(x + 3)^2 + (x - 1)^2 = 0$
33. $x + \dfrac{1}{x} = \dfrac{13}{6}$
34. $\dfrac{3}{x} + \dfrac{x}{3} = \dfrac{5}{2}$

ii Approximate solutions to the nearest tenth.

35. $x^2 + 4x - 7 = 0$
36. $x^2 + 6x + 4 = 0$
37. $x^2 - 6x + 4 = 0$
38. $x^2 - 4x + 1 = 0$
39. $2x^2 - 3x - 7 = 0$
40. $3x^2 - 3x - 2 = 0$

Problem-solving practice

41. A store has one type of coffee worth \$1.50 per lb and another type worth \$2.50 per lb. They want to make a blend of coffee worth \$1.90 per lb. How much of each kind of coffee should be used to make 50 lb of this blend?

42. Sam determines that it will take 210 yd of fence to enclose a rectangular field. If the length is 15 yd more than twice the width, what is the area of the field?

Solve.

43. $2.2x^2 + 0.5x - 1 = 0$
44. $5.33x^2 - 8.23x - 3.24 = 0$
45. $t^2 + 0.2t - 0.3 = 0$
46. $p^2 + 0.3p - 0.2 = 0$
47. $x^2 - 0.75x - 0.5 = 0$
48. $z^2 + 0.84z - 0.4 = 0$
49. $x^2 + x - \sqrt{2} = 0$
50. $x^2 - x - \sqrt{3} = 0$
51. $x^2 + \sqrt{5}\,x - \sqrt{3} = 0$
52. $\sqrt{2}\,x^2 + 5x + \sqrt{2} = 0$
53. $x^2 + 3x + i = 0$
54. $ix^2 - 2x + 1 = 0$

55. Solve for x in terms of y: $3x^2 + xy + 4y^2 = 9$.

Solve.

56. $\frac{5}{x} + \frac{x}{4} = \frac{11}{7}$
57. $2x^2 - x - \sqrt{5} = 0$
58. $\sqrt{3}\,x^2 + 6x + \sqrt{3} = 0$
59. $ix^2 - x - 1 = 0$

60. $(1 + \sqrt{3})x^2 - (3 + 2\sqrt{3})x + 3 = 0$

61. One solution of $kx^2 + 3x - k = 0$ is -2. Find the other.

8.3

FRACTIONAL EQUATIONS AND PROBLEM SOLVING

OBJECTIVES

After finishing Section 8.3, you should be able to:

i **Solve fractional equations that are quadratic after clearing of fractions.**

ii **Solve problems that translate to this kind of equation.**

It often happens that when we clear of fractions in a fractional equation, we get a quadratic equation. When we do, we proceed as in solving any quadratic equation. It will be important to check possible solutions in these cases, however.

i Fractional Equations That Are Quadratic

The following example illustrates how a fractional equation can actually be quadratic.

EXAMPLE 1 Solve: $\dfrac{14}{x+2} - \dfrac{1}{x-4} = 1.$

We multiply by the LCM of the denominators, which is $(x + 2)(x - 4)$:

$$(x+2)(x-4)\cdot\left[\frac{14}{x+2} - \frac{1}{x-4}\right] = (x+2)(x-4)\cdot 1$$

$$(x+2)(x-4)\cdot\frac{14}{x+2} - (x+2)(x-4)\cdot\frac{1}{x-4}$$

$$= (x+2)(x-4)\cdot 1 \qquad \text{Using a distributive law}$$

$$14(x-4) - (x+2) = (x+2)(x-4) \qquad \text{Simplifying by removing factors of 1}$$

$$14x - 56 - x - 2 = x^2 - 2x - 8 \qquad \text{Multiplying}$$

$$13x - 58 = x^2 - 2x - 8 \qquad \text{Collecting like terms}$$

$$0 = x^2 - 15x + 50 \qquad \text{Writing standard form}$$

$$0 = (x-5)(x-10) \qquad \text{Factoring}$$

$$x = 5 \quad \text{or} \quad x = 10. \qquad \text{Using the principle of zero products}$$

Since we started with a fractional equation, we might have introduced some numbers that are not solutions of the original equation when we cleared of fractions. This time, we must check.

Check:

For 5:

$$\frac{14}{x+2} - \frac{1}{x-4} = 1$$

$\frac{14}{5+2} - \frac{1}{5-4}$	1
$\frac{14}{7} - 1$	
$2 - 1$	
1	

For 10:

$$\frac{14}{x+2} - \frac{1}{x-4} = 1$$

$\frac{14}{10+2} - \frac{1}{10-4}$	1
$\frac{14}{12} - \frac{1}{6}$	
$\frac{14}{12} - \frac{2}{12}$	
1	

The solutions are 5 and 10.

If for a fractional equation, clearing of fractions produces a quadratic equation, we solve that resulting equation as we would solve any quadratic equation. However, it is necessary to check possible solutions by substituting into the original equation.

DO EXERCISE 1.

Solve.

1. $\frac{24}{r-2} + \frac{24}{r+2} = 5$

ii Solving Problems

Some problems translate to fractional equations that are actually quadratic, like the equation in Example 1. The following examples illustrate.

EXAMPLE 2 Gene and Sally work together to type a short story, and it takes them 4 hr. It would take Sally 6 hr more than Gene to do the job alone. How long would each need to do the job if they worked alone?

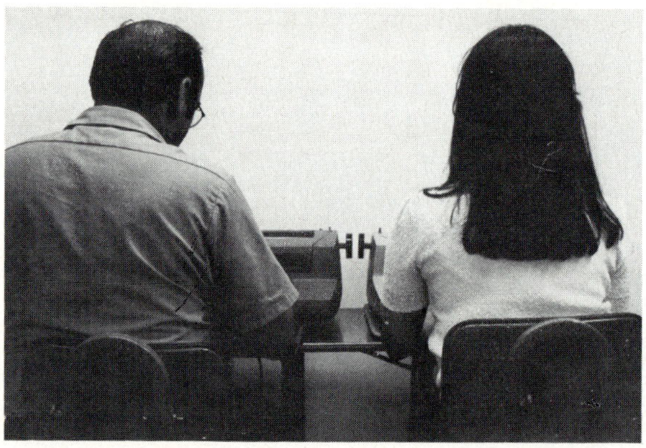

1. **Familiarize** Let's make a guess. Suppose that it takes Gene 5 hr. Then it would take Sally 6 more hr, or 11 hr. If this is a correct answer, then Gene will do $\frac{4}{5}$ of the job in 4 hr and Sally will do $\frac{4}{11}$ of the job in 4 hr:

$$\begin{aligned}\tfrac{4}{5} + \tfrac{4}{11} &= \tfrac{44}{55} + \tfrac{20}{55} \\ &= \tfrac{64}{55}.\end{aligned}$$

We do not have an answer. If we did, the sum would be $\frac{55}{55}$ of the job in 4 hr.

Let us suppose that Gene takes x hr. Then Sally will take 6 more hr, or $(x + 6)$ hr. It takes them 4 hr working together, so they will do $\frac{1}{4}$ of the job in

1 hr. Gene will do $1/x$ of the job in 1 hr and Sally will do $1/(x + 6)$ of the job in 1 hr.

2. **Translate** We have an equation:

$$\frac{1}{x} + \frac{1}{x+6} = \frac{1}{4}.$$

3. **Carry out** We solve the equation:

$$4x(x+6)\cdot\left[\frac{1}{x} + \frac{1}{x+6}\right] = 4x(x+6)\cdot\frac{1}{4} \quad \text{Multiplying by the LCM}$$

$$4x(x+6)\cdot\frac{1}{x} + 4x(x+6)\cdot\frac{1}{x+6} = 4x(x+6)\cdot\frac{1}{4}$$

$$4(x+6) + 4x = x(x+6)$$

$$4x + 24 + 4x = x^2 + 6x$$

$$8x + 24 = x^2 + 6x$$

$$0 = x^2 - 2x - 24 \quad \text{Standard form}$$

$$0 = (x-6)(x+4) \quad \text{Factoring}$$

$$x = 6 \quad \text{or} \quad x = -4 \quad \text{Principle of zero products}$$

4. **Check** Since negative time has no meaning in this problem, -4 is not a solution. Let's see if 6 checks. This is the time for Gene to do the job alone. Then Sally would take 12 hr alone. Thus Gene would do $\frac{4}{6}$ of the job in 4 hr and Sally would do $\frac{4}{12}$ of it. Thus in 4 hr they would do $\frac{4}{6} + \frac{4}{12}$, or $\frac{8}{12} + \frac{4}{12}$ of the job. This is all of it, so the numbers check.

5. **State** The answer is that it would take Gene 6 hr alone and Sally 12 hr alone.

DO EXERCISE 2.

2. It takes an apprentice mason 9 hr longer to build a wall than it takes a master mason. If the two work together, they can build the wall in 20 hr. How long would it take the master mason, working alone, to build the wall?

EXAMPLE 3 A car travels 300 mi at a certain speed. If the speed had been 10 mph faster, the trip would have been made in 1 hr less time. Find the speed.

1. **Familiarize** We make a drawing, labeling it with the known and unknown information. We can also organize the information in a table.

Distance	Speed	Time
300	r	t
300	$r + 10$	$t - 1$

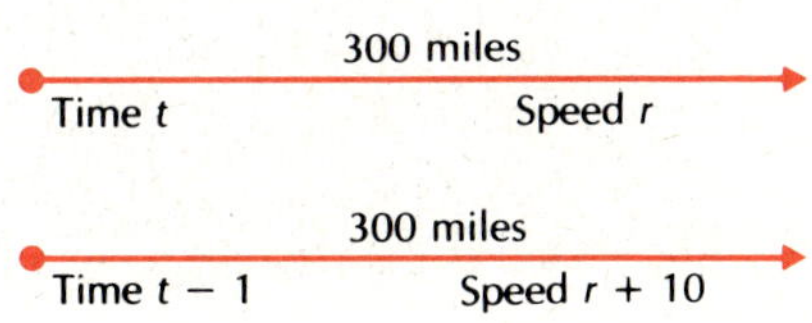

The definition of speed, $r = d/t$, relates the three quantities.

2. **Translate** From the first line of the table, we obtain

$$r = \frac{300}{t}.$$

From the second line we get

$$r + 10 = \frac{300}{t-1}.$$

3. Two ships make the same voyage, a distance of 3000 nautical miles. The faster ship travels 10 knots faster than the slower one (a *knot* is 1 nautical mile per hour). The faster ship makes the voyage in 50 hours less time than the slower one. Find the speeds of the two ships.

3. **Carry out** We have a system of equations. We substitute for r from the first equation into the second and solve the resulting equation:

$$\frac{300}{t} + 10 = \frac{300}{t-1}$$

$$t(t-1)\left[\frac{300}{t} + 10\right] = t(t-1) \cdot \frac{300}{t-1} \qquad \text{Multiplying by the LCM}$$

$$t(t-1) \cdot \frac{300}{t} + t(t-1) \cdot 10 = t(t-1) \cdot \frac{300}{t-1}$$

$$300(t-1) + 10(t^2 - t) = 300t$$

$$10t^2 - 10t - 300 = 0 \qquad \text{Standard form}$$

$$t^2 - t - 30 = 0 \qquad \text{Multiplying by } \tfrac{1}{10}$$

$$(t-6)(t+5) = 0 \qquad \text{Factoring}$$

$$t = 6 \quad \text{or} \quad t = -5 \qquad \text{Principle of zero products}$$

4. **Check** Since negative time has no meaning in this problem, we try 6 hr. Remembering that $r = d/t$, we get $r = 300/6 = 50$ mph.

To check, we take the speed 10 mph faster, which is 60 mph, and see how long the trip would have taken at that speed:

$$t = \frac{d}{r} = \frac{300}{60} = 5 \text{ hr.}$$

This is 1 hr less than the trip actually took, so we have an answer to the problem.

5. **State** The answer is that the speed was 50 mph.

DO EXERCISE 3.

EXERCISE SET 8.3

i Solve.

1. $\frac{1}{x} = \frac{x-2}{24}$

2. $\frac{x+3}{14} = \frac{2}{x}$

3. $\frac{1}{2x-1} - \frac{1}{2x+1} = \frac{1}{4}$

4. $\frac{1}{4-x} - \frac{1}{2+x} = \frac{1}{4}$

5. $\frac{50}{x} - \frac{50}{x-5} = -\frac{1}{2}$

6. $3x + \frac{10}{x-1} = \frac{16}{x-1}$

7. $\frac{x+2}{x} = \frac{x-1}{2}$

8. $\frac{x}{3} - \frac{6}{x} = 1$

9. $x - 6 = \frac{1}{x+6}$

10. $x + 7 = \frac{1}{x-7}$

11. $\frac{2}{x} = \frac{x+3}{5}$

12. $\frac{x+3}{x} = \frac{x-4}{3}$

13. $\frac{3}{3x+1} + \frac{6x}{11x-1} = 1$

14. $x - 8 = \frac{1}{x+8}$

15. $\frac{40}{x} - \frac{20}{x-3} = \frac{8}{7}$

16. $\frac{x+2}{x-1} = \frac{x-2}{x+3}$

ii Solve.

17. During the first part of a trip a canoeist travels 80 mi at a certain speed. The canoeist travels 35 mi on the second part of the trip at a speed 5 mph slower. The total time for the trip is 3 hr. What was the speed on each part of the trip?

18. During the first part of a trip a car travels 120 mi at a certain speed. It travels 100 mi on the second part of the trip at a speed 10 mph slower. The total time for the trip is 4 hr. What was the speed on each part of the trip?

19. A car travels 280 mi at a certain speed. If the speed had been 5 mph faster, the trip would have been made in 1 hr less time. Find the speed.

20. A car travels 200 mi at a certain speed. If the speed had been 10 mph faster, the trip would have been made in 1 hr less time. Find the speed.

21. Airplane A travels 2800 km at a certain speed. Plane B travels 2000 km at a speed 50 km/h faster than plane A in 3 hr less time. Find the speed of each plane.

22. Airplane A travels 600 mi at a certain speed. Plane B travels 1000 mi at a speed 50 mph faster, but takes 1 hr longer. Find the speed of each plane.

23. Two pipes are connected to the same tank. When working together they can fill the tank in 2 hr. The larger pipe, working alone, can fill the tank in 3 hr less time than the smaller one. How long would the smaller one take, working alone, to fill the tank?

24. Two hoses are connected to a swimming pool. When working together they can fill the pool in 4 hr. The larger hose, working alone, can fill the pool in 6 hr less time than the smaller one. How long would the smaller one take, working alone, to fill the pool?

25. A boat travels 1 km upstream and 1 km back. The time for the roundtrip is 1 hr. The speed of the stream is 2 km/h. What is the speed of the boat in still water?

26. A boat travels 10 mi upstream and 10 mi back. The time required for the roundtrip is 3 hr. The speed of the stream is 5 mph. What is the speed of the boat in still water?

---

27. At a factory, smokestack A pollutes the air 2.13 times as fast as smokestack B. Together they yield a certain amount of pollution in 16.3 hr. How long would it take each to pollute this much alone?

28. At Pizza Perfect, Ron Italio can make 100 large pizza crusts in 1.2 hr less time than Chester Mosserelli. Together they can do the job in 1.8 hr. How long does it take each to do the job alone?

Solve.

29. $\frac{12}{x^2-9} = 1 + \frac{3}{x-3}$

30. $\frac{x^2}{x-2} - \frac{x+4}{2} + \frac{2-4x}{x-2} + 1 = 0$

31. $\frac{4}{2x+i} - \frac{1}{x-i} = \frac{2}{x+i}$

32. Find a when the reciprocal of $a - 1$ is $a + 1$.

33. A discount store bought a quantity of beach towels for $250 and sold all but 15 at a profit of $3.50 per towel. With the total amount received the manager could buy 4 more than twice as many as were bought before. Find the cost per towel.

8.4

SOLUTIONS OF QUADRATIC EQUATIONS

OBJECTIVES

After finishing Section 8.4, you should be able to:

i Determine the nature of the solutions of a quadratic equation with real-number coefficients, without solving it.

ii Write a quadratic equation having two numbers specified as solutions.

We now consider quadratic equations with real-number coefficients. We shall see what the quadratic formula can tell us about the solutions, without actually solving the equation.

i The Discriminant

From the quadratic formula we know that the solutions x_1 and x_2 of a quadratic equation are given by

$$x_1 = \frac{-b + \sqrt{b^2 - 4ac}}{2a} \quad \text{and} \quad x_2 = \frac{-b - \sqrt{b^2 - 4ac}}{2a}.$$

The expression $b^2 - 4ac$ shows the nature of the solutions. This expression is called the *discriminant.* If it is 0, then it doesn't matter whether we choose the plus or minus sign in the formula; hence there is just one solution. If the discriminant is positive, there will be two real solutions. If it is negative, we will be taking the square root of a negative number; hence there will be two nonreal solutions, and they will be complex conjugates. We summarize.

Discriminant $b^2 - 4ac$	Nature of solutions
0	Only one solution; it is a real number.
Positive	Two different real-number solutions.
Negative	Two different nonreal-number solutions (complex conjugates).

Determine the nature of the solutions without solving.

1. $x^2 + 5x - 3 = 0$

2. $9x^2 - 6x + 1 = 0$

3. $3x^2 - 2x + 1 = 0$

EXAMPLE 1 Determine the nature of the solutions of $9x^2 - 12x + 4 = 0$.

We have

$$a = 9, \quad b = -12, \quad \text{and} \quad c = 4.$$

We compute the discriminant:

$$\begin{aligned} b^2 - 4ac &= (-12)^2 - 4 \cdot 9 \cdot 4 \\ &= 144 - 144 = 0. \end{aligned}$$

There is just one solution and it is a real number.

EXAMPLE 2 Determine the nature of the solutions of $x^2 + 5x + 8 = 0$.

We have

$$a = 1, \quad b = 5, \quad \text{and} \quad c = 8.$$

We compute the discriminant:

$$\begin{aligned} b^2 - 4ac &= 5^2 - 4 \cdot 1 \cdot 8 \\ &= 25 - 32 = -7. \end{aligned}$$

Since the discriminant is negative, there are two nonreal solutions.

EXAMPLE 3 Determine the nature of the solutions of $x^2 + 5x + 6 = 0$.

$$a = 1, \quad b = 5, \quad \text{and} \quad c = 6$$
$$b^2 - 4ac = 5^2 - 4 \cdot 1 \cdot 6 = 1$$

Since the discriminant is positive, there are two solutions and they are real numbers.

DO EXERCISES 1–3.

ii Writing Equations from Solutions

We know by the principle of zero products that $(x - 2)(x + 3) = 0$ has solutions 2 and -3. If we know the solutions of an equation, we can write the equation.

EXAMPLE 4 Write a quadratic equation whose solutions are $2i$ and $-2i$.

$$\begin{aligned} x = 2i \quad &\text{or} \quad x = -2i \\ x - 2i = 0 \quad &\text{or} \quad x + 2i = 0 && \text{Getting the 0's on one side} \\ (x - 2i)(x + 2i) &= 0 && \text{Principle of zero products (multiplying)} \\ x^2 + 2ix - 2ix - (2i)^2 = x^2 - (2i)^2 &= 0 && \text{Using FOIL} \\ x^2 + 4 &= 0 \end{aligned}$$

EXAMPLE 5 Find a quadratic equation whose solutions are 3 and $-\frac{2}{5}$.

$$\begin{aligned} x = 3 \quad &\text{or} \quad x = -\tfrac{2}{5} \\ x - 3 = 0 \quad &\text{or} \quad x + \tfrac{2}{5} = 0 && \text{Getting the 0's on one side} \\ (x - 3)(x + \tfrac{2}{5}) &= 0 && \text{Principle of zero products (multiplying)} \\ x^2 + \tfrac{2}{5}x - 3x - 3 \cdot \tfrac{2}{5} &= 0 && \text{Using FOIL} \\ x^2 - \tfrac{13}{5}x - \tfrac{6}{5} &= 0, \quad \text{or} \\ 5x^2 - 13x - 6 &= 0 && \text{Collecting like terms and multiplying by 5} \end{aligned}$$

EXAMPLE 6 Write a quadratic equation whose solutions are $\sqrt{3}$ and $-2\sqrt{3}$.

$$x = \sqrt{3} \quad \text{or} \quad x = -2\sqrt{3}$$
$$x - \sqrt{3} = 0 \quad \text{or} \quad x + 2\sqrt{3} = 0 \qquad \text{Getting the 0's on one side}$$
$$(x - \sqrt{3})(x + 2\sqrt{3}) = 0 \qquad \text{Principle of zero products}$$
$$x^2 + 2\sqrt{3}x - \sqrt{3}x - 2(\sqrt{3})^2 = 0 \qquad \text{Using FOIL}$$
$$x^2 + \sqrt{3}x - 6 = 0 \qquad \text{Collecting like terms}$$

DO EXERCISES 4–6.

Find a quadratic equation having the following solutions.

4. -4 and $\frac{5}{3}$

5. $5i$ and $-5i$

6. $-2\sqrt{2}$ and $\sqrt{2}$

EXERCISE SET 8.4

i Determine the nature of the solutions of each equation.

1. $x^2 - 6x + 9 = 0$
2. $x^2 + 10x + 25 = 0$
3. $x^2 + 7 = 0$
4. $x^2 + 2 = 0$
5. $x^2 - 2 = 0$
6. $x^2 - 5 = 0$
7. $4x^2 - 12x + 9 = 0$
8. $4x^2 + 8x - 5 = 0$
9. $x^2 - 2x + 4 = 0$
10. $x^2 + 3x + 4 = 0$
11. $9t^2 - 3t = 0$
12. $4m^2 + 7m = 0$
13. $y^2 = \frac{1}{2}y + \frac{3}{5}$
14. $y^2 + \frac{9}{4} = 4y$
15. $4x^2 - 4\sqrt{3}x + 3 = 0$
16. $6y^2 - 2\sqrt{3}y - 1 = 0$

ii Write a quadratic equation having the given numbers as solutions.

17. $-11,\ 9$
18. $-4,\ 4$
19. 7, only solution
[*Hint:* It must be a double solution, that is, $(x - 7)(x - 7) = 0$.]
20. -5, only solution
21. $-\frac{2}{5},\ \frac{6}{5}$
22. $-\frac{1}{4},\ -\frac{1}{2}$
23. $\frac{c}{2},\ \frac{d}{2}$
24. $\frac{k}{3},\ \frac{m}{4}$
25. $\sqrt{2},\ 3\sqrt{2}$
26. $-\sqrt{3},\ 2\sqrt{3}$

Problem-solving practice

27. Sara boiled six quarts of a 40% saline solution as part of a science experiment. After one hour the percentage of salt in the water was 60%. How much water evaporated during the boiling process?

28. If 4 is subtracted from the value of the numerator of a fractional expression, the value of the expression becomes 2/3 of the original value. Find the original values of the numerator and denominator.

29. Prove each of the following.
a) The sum of the solutions of $ax^2 + bx + c = 0$ is $-b/a$.
b) The product of the solutions of $ax^2 + bx + c = 0$ is c/a.

For each equation under the given condition, (a) find k, and (b) find the other solution.

30. $kx^2 - 17x + 33 = 0$; one solution is 3
31. $kx^2 - 2x + k = 0$; one solution is -3
32. $x^2 - kx + 2 = 0$; one solution is $1 + i$
33. $x^2 - (6 + 3i)x + k = 0$; one solution is 3

33. Find k for which $kx^2 - 4x + (2k - 1) = 0$ and the product of the solutions is 3.

35. Find a quadratic equation for which the sum of the solutions is $\sqrt{3}$ and the product is 8.

36. Write a quadratic equation having the given numbers as solutions:
a) $\frac{2 + \sqrt{3}}{2},\ \frac{2 - \sqrt{3}}{2}$
b) $\frac{g}{h},\ -\frac{h}{g}$
c) $2 - 5i,\ \ 2 + 5i$

37. Find h and k, where $3x^2 - hx + 4k = 0$ and the sum of the solutions is -12 and the product of the solutions is 20.

38. One solution of the equation $p(q - r)y^2 + q(r - p)y + r(p - q) = 0$ is 2. Find the other.

39. The sum of the squares of the solutions of $x^2 + 2kx - 5 = 0$ is 26. Find the absolute value of k.

40. The solutions of $x^2 + px + q = 0$ are p and q, where $p \neq 0$ and $q \neq 0$. Find the sum of the solutions.

41. When the solutions of each of the equations $x^2 + kx + 8 = 0$ and $x^2 - kx + 8 = 0$ are listed in increasing order, each solution of the second equation is 6 more than the corresponding solution of the first equation. Find k.

OBJECTIVES

After finishing Section 8.5, you should be able to:

i **Solve a formula for a specified letter.**

ii **Solve problems in which the solving of a formula is needed.**

8.5

FORMULAS AND PROBLEM SOLVING

We can now apply our knowledge and skill with quadratic equations to solving formulas and solving problems.

i Solving Formulas

Recall that to solve a formula for a certain letter, we use the principles for solving equations to get that letter alone on one side. When square roots appear, we can usually eliminate the radical signs by squaring both sides.

EXAMPLE 1 (*A pendulum formula*). Solve $T = 2\pi\sqrt{\frac{l}{g}}$ for l.

$$T = 2\pi\sqrt{\frac{l}{g}}$$

$$T^2 = \left(2\pi\sqrt{\frac{l}{g}}\right)^2 \quad \text{Principle of powers (squaring)}$$

$$T^2 = 2^2\pi^2\frac{l}{g}$$

$$gT^2 = 4\pi^2 l \quad \text{Clearing of fractions}$$

$$\frac{gT^2}{4\pi^2} = l \quad \text{Multiplying by } \frac{1}{4\pi^2}$$

We now have l alone on one side and l does not appear on the other side, so the formula is solved for l.

In most formulas the letters represent nonnegative numbers, so we do not need to use absolute-value signs when taking square roots.

DO EXERCISE 1.

1. Solve $A = \sqrt{\frac{w_1}{w_2}}$ for w_2.

EXAMPLE 2 (*A motion formula*). Solve $s = gt + 16t^2$ for t.

This time we use the quadratic formula to get t alone on one side of the equation:

$$16t^2 + gt - s = 0 \quad \text{Writing standard form}$$

$$a = 16, \quad b = g, \quad \text{and} \quad c = -s$$

$$t = \frac{-g \pm \sqrt{g^2 - 4 \cdot 16 \cdot (-s)}}{2 \cdot 16} \quad \text{Using the quadratic formula}$$

Since taking the negative square root would result in a negative answer, we take the positive one:

$$t = \frac{-g + \sqrt{g^2 + 64s}}{32}.$$

DO EXERCISES 2 AND 3.

2. Solve $V = \pi r^2 h$ for r.

3. Solve $Ls^2 + Rs = -\frac{1}{C}$ for s.

4. Solve $\dfrac{b}{\sqrt{a^2 - b^2}} = p$ for b.

The following list of steps should help you when solving formulas. Remember that when solving formulas you do the same things you would do to solve any equation.

TO SOLVE A FORMULA FOR A LETTER, SAY b:

1. **Clear of fractions (if needed).**
2. **Use the principle of powers until radicals are eliminated. (*Note:* You may sometimes do step 2 before step 1.)**
3. **Collect all terms with b^2 in them. Also collect all terms with b in them.**
4. **If b^2 does not appear, finish by using the addition and multiplication principles.**
5. **If b^2 appears but b does not, solve for b^2 and take the square root.**
6. **If b^2 and b both appear, write standard form and factor or use the quadratic formula.**

EXAMPLE 3 Solve the following formula for a: $q = \dfrac{a}{\sqrt{a^2 + b^2}}$.

In this case, we could either clear of fractions first or use the principle of powers first. Let us clear of fractions. We then have

$$q\sqrt{a^2 + b^2} = a.$$

Now we square both sides and then continue:

$$\begin{aligned}
(q\sqrt{a^2 + b^2})^2 &= a^2 && \text{Squaring} \\
q^2(a^2 + b^2) &= a^2 \\
q^2a^2 + q^2b^2 &= a^2 \\
q^2b^2 &= a^2 - q^2a^2 && \text{Getting all } a^2 \text{ terms together} \\
q^2b^2 &= a^2(1 - q^2) && \text{Factoring out } a^2 \\
\frac{q^2b^2}{1 - q^2} &= a^2 && \text{Multiplying by } \frac{1}{1 - q^2} \\
\sqrt{\frac{q^2b^2}{1 - q^2}} &= a && \text{Taking the square root}
\end{aligned}$$

Since we have a alone on one side and a does not appear on the other side, we have solved the formula for a.

DO EXERCISE 4.

ii Solving Problems

In problem solving we often need to use a formula, and it is not uncommon for us to have to solve a formula for a certain letter. Let's look at an example.

A function that gives the distance, in meters, that an object will fall in t seconds is as follows:

$$s(t) = 4.9t^2 + v_0t,$$

where v_0 is the initial speed, or velocity, and is hence a constant. This function gives us a formula that we can solve for t if necessary.

EXAMPLE 4 An object is dropped from the top of the Gateway Arch in St. Louis, which is 195 meters high. How long will it take to reach the ground?

5. The lateral area of a cone is 157 cm^2, and the slant height is 10 cm. What is the radius of the base?
[*Hint:* A formula for the lateral area of a cone is $A = \pi rs$, where r is the radius of the base and s is the slant height. Use 3.14 for π.]

1. **Familiarize** We make a sketch, and label it.

Part of the familiarization consists of finding the proper formula to use. If it had not been given to us in the above discussion, we would need to look it up somewhere, probably in an encyclopedia or physics book. The formula is $s = 4.9t^2 + v_0t$.

2. **Translate** We have the formula $s = 4.9t^2 + v_0t$ and we know that $v_0 = 0$. Substituting, we have the translation

$$s = 4.9t^2 + 0 \cdot t, \quad \text{or} \quad 4.9t^2.$$

3. **Carry out** We solve the formula for t:

$$s = 4.9t^2$$

$$\frac{s}{4.9} = t^2 \qquad \text{Multiplying by } \frac{1}{4.9}$$

$$\sqrt{\frac{s}{4.9}} = t \qquad \text{Taking the principal square root}$$

Next, we substitute our value of s and calculate:

$$t = \sqrt{\frac{195}{4.9}} \approx 6.31.$$

4. **Check** We go back to the original formula, or function, and substitute 6.31:

$$\begin{aligned} s = 4.9t^2 + v_0t &= 4.9t^2 + 0 \cdot t \\ &= 4.9(6.31)^2 \\ &= 195.1. \end{aligned}$$

Our answer is of course approximate, but it checks.

5. **State** The answer is that it will take approximately 6.31 seconds for the object to reach the ground.

DO EXERCISE 5.

EXERCISE SET 8.5

i Solve each formula for the indicated letter. Assume that all variables represent nonnegative numbers.

1. $A = 6s^2$, for s
(Area of a cube)

2. $A = 4\pi r^2$, for r
(Area of a sphere)

3. $F = \dfrac{Gm_1m_2}{r^2}$, for r
(Law of gravity)

4. $N = \dfrac{kQ_1Q_2}{s^2}$, for s
(Number of phone calls between two cities)

5. $E = mc^2$, for c
(Energy–mass relationship)

6. $A = \pi r^2$, for r
(Area of a circle)

7. $a^2 + b^2 = c^2$, for b
(Pythagorean formula in two dimensions)

8. $a^2 + b^2 + c^2 = d^2$, for c
(Pythagorean formula in three dimensions)

9. $N = \dfrac{k^2 - 3k}{2}$, for k
(Number of diagonals of a polygon)

10. $s = v_0t + \dfrac{gt^2}{2}$, for t
(A motion formula)

11. $A = 2\pi r^2 + 2\pi rh$, for r
(Area of a cylinder)

12. $A = \pi r^2 + \pi rs$, for r
(Area of a cone)

13. $T = 2\pi\sqrt{\dfrac{l}{g}}$, for g
(A pendulum formula)

14. $W = \sqrt{\dfrac{l}{LC}}$, for L
(An electricity formula)

15. $P_1 - P_2 = \dfrac{32LV}{gD^2}$, for D

16. $N + p = \dfrac{6.2A^2}{pR^2}$, for R

17. $m = \dfrac{m_0}{\sqrt{1 - \dfrac{v^2}{c^2}}}$, for v
(A relativity formula)

18. Solve the formula given in Exercise 17 for c.

ii Solve.

19. a) An object is dropped 75 m from an airplane. How long does it take to reach the ground?
b) An object is thrown downward 75 m from the plane at an initial velocity of 30 m/sec. How long does it take to reach the ground?
c) How far will an object fall in 2 sec, thrown downward at an initial velocity of 30 m/sec?

20. a) An object is dropped 500 m from an airplane. How long does it take to reach the ground?
b) An object is thrown downward 500 m from the plane at an initial velocity of 30 m/sec. How long does it take to reach the ground?
c) How far will an object fall in 5 sec, thrown downward at an initial velocity of 30 m/sec?

21. An amount of money P is invested at interest rate r. In t years it will grow to the amount A given by $A = P(1 + r)^t$, where interest is compounded annually. For the following situations, find the interest rate for interest compounded annually.

a) \$2560 grows to \$3610 in 2 years
b) \$1000 grows to \$1210 in 2 years
c) \$8000 grows to \$9856.80 in 2 years (Use a calculator.)
d) \$1000 grows to \$1271.26 in 2 years (Use a calculator.)

22. A ladder 10 ft long leans against a wall. The bottom of the ladder is 6 ft from the wall. How much would the lower end of the ladder have to be pulled away so that the top end would be pulled down the same amount?

23. A ladder 13 ft long leans against a wall. The bottom of the ladder is 5 ft from the wall. How much would the lower end of the ladder have to be pulled away so that the top end would be pulled down the same amount?

24. The area of a triangle is 18 cm^2. The base is 3 cm longer than the height. Find the height.

25. A baseball diamond is a square 90 ft on a side. How far is it directly from second base to home?

26. Trains A and B leave the same city at right angles at the same time. Train B travels 5 mph faster than train A. After 2 hr they are 50 mi apart. Find the speed of each train.

8.6

EQUATIONS REDUCIBLE TO QUADRATIC

OBJECTIVE

After finishing Section 8.6, you should be able to:

i **Solve equations that are reducible to quadratic.**

i Certain equations that are not really quadratic can be thought of in such a way that they can be solved as quadratic. For example, consider this fourth-degree equation.

Solve.

1. $x^4 - 10x^2 + 9 = 0$

$$\begin{array}{cccc} x^4 & -\ 9x^2 & +\ 8 & = 0 \\ \downarrow & \downarrow & \downarrow & \downarrow \\ (x^2)^2 & -\ 9(x^2) & +\ 8 & = 0 \\ \downarrow & \downarrow & \downarrow & \downarrow \\ u^2 & -\ 9u & +\ 8 & = 0 \end{array}$$

Thinking of x^4 as $(x^2)^2$

To make this clearer, write u instead of x^2.

The equation $u^2 - 9u + 8 = 0$ can be solved by factoring or the quadratic formula. After that, we can find x by remembering that $x^2 = u$. Equations that can be solved like this are said to be *reducible to quadratic.*

EXAMPLE 1 Solve: $x^4 - 9x^2 + 8 = 0$.

Let $u = x^2$. Then we solve the equation found by substituting u for x^2:

$$u^2 - 9u + 8 = 0$$

$$(u-8)(u-1) = 0 \quad \text{Factoring}$$

$$u - 8 = 0 \quad \text{or} \quad u - 1 = 0 \quad \text{Principle of zero products}$$

$$u = 8 \quad \text{or} \quad u = 1$$

Now we substitute x^2 for u and solve these equations:

$$x^2 = 8 \quad \text{or} \quad x^2 = 1$$

$$x = \pm\sqrt{8} \quad \text{or} \quad x = \pm 1$$

$$x = \pm 2\sqrt{2} \quad \text{or} \quad x = \pm 1$$

To check first note that when $x = 2\sqrt{2}$, $x^2 = 8$ and $x^4 = 64$. Also, when $x = -2\sqrt{2}$, $x^2 = 8$ and $x^4 = 64$. Similarly, when $x = 1$, $x^2 = 1$, and $x^4 = 1$, and when $x = -1$, $x^2 = 1$, and $x^4 = 1$. Thus instead of making four checks we need make only two.

Check: For $\pm 2\sqrt{2}$:

$x^4 - 9x^2 + 8 = 0$	
$(\pm 2\sqrt{2})^4 - 9(\pm 2\sqrt{2})^2 + 8$	0
$64 - 9 \cdot 8 + 8$	
0	

For ± 1:

$x^4 - 9x^2 + 8 = 0$	
$(\pm 1)^4 - 9(\pm 1)^2 + 8$	0
$1 - 9 + 8$	
0	

The solutions are 1, -1, $2\sqrt{2}$, and $-2\sqrt{2}$.

DO EXERCISE 1.

A check by substituting is necessary when solving equations reducible to quadratic. Reducing the equation to quadratic can sometimes introduce numbers that are not solutions of the original equation.

EXAMPLE 2 Solve: $x - 3\sqrt{x} - 4 = 0$.

Let $u = \sqrt{x}$. Then we solve the equation found by substituting u for $\sqrt{x}$ (and, of course, u^2 for x):

$$u^2 - 3u - 4 = 0$$

$$(u-4)(u+1) = 0$$

$$u = 4 \quad \text{or} \quad u = -1.$$

Now we substitute $\sqrt{x}$ for u and solve these equations:

$$\sqrt{x} = 4 \quad \text{or} \quad \sqrt{x} = -1.$$

Squaring the first equation we get $x = 16$. The second equation has no real solution since principal square roots are never negative.

The number 16 checks, so it is the solution.

DO EXERCISE 2.

EXAMPLE 3 Solve: $(x^2 - 1)^2 - (x^2 - 1) - 2 = 0$.

Let $u = x^2 - 1$. Then we solve the equation found by substituting u for $x^2 - 1$:

$$u^2 - u - 2 = 0$$
$$(u - 2)(u + 1) = 0$$
$$u = 2 \quad \text{or} \quad u = -1.$$

Now we substitute $x^2 - 1$ for u and solve these equations:

$$x^2 - 1 = 2 \quad \text{or} \quad x^2 - 1 = -1$$
$$x^2 = 3 \quad \text{or} \quad x^2 = 0$$
$$x = \pm\sqrt{3} \quad \text{or} \quad x = 0.$$

The numbers $\sqrt{3}$, $-\sqrt{3}$, and 0 check. They are the solutions.

DO EXERCISE 3.

EXAMPLE 4 Solve: $y^{-2} - y^{-1} - 2 = 0$.

Let $u = y^{-1}$. Then we solve the equation found by substituting u for y^{-1} and u^2 for y^{-2}:

$$u^2 - u - 2 = 0$$
$$(u - 2)(u + 1) = 0$$
$$u = 2 \quad \text{or} \quad u = -1.$$

Now we substitute y^{-1} or $1/y$ for u and solve these equations:

$$\frac{1}{y} = 2 \quad \text{or} \quad \frac{1}{y} = -1.$$

Solving, we get

$$y = \frac{1}{2} \quad \text{or} \quad y = \frac{1}{(-1)} = -1.$$

The numbers $\frac{1}{2}$ and -1 both check. They are the solutions.

DO EXERCISE 4.

2. Solve $x + 3\sqrt{x} - 10 = 0$. Be sure to check.

Solve.

3. $(x^2 - x)^2 - 14(x^2 - x) + 24 = 0$

Solve.

4. $x^{-2} + x^{-1} - 6 = 0$

EXERCISE SET 8.6

i Solve.

1. $x - 10\sqrt{x} + 9 = 0$
2. $2x - 9\sqrt{x} + 4 = 0$
3. $x^4 - 10x^2 + 25 = 0$
4. $x^4 - 3x^2 + 2 = 0$
5. $t^{2/3} + t^{1/3} - 6 = 0$
6. $w^{2/3} - 2w^{1/3} - 8 = 0$
7. $z^{1/2} - z^{1/4} - 2 = 0$
8. $m^{1/3} - m^{1/6} - 6 = 0$
9. $(x^2 - 6x)^2 - 2(x^2 - 6x) - 35 = 0$
10. $(1 + \sqrt{x})^2 + (1 + \sqrt{x}) - 6 = 0$
11. $(y^2 - 5y)^2 - 2(y^2 - 5y) - 24 = 0$
12. $(2t^2 + t)^2 - 4(2t^2 + t) + 3 = 0$
13. $w^4 - 4w^2 - 2 = 0$
14. $t^4 - 5t^2 + 5 = 0$
15. $x^{-2} - x^{-1} - 6 = 0$
16. $4x^{-2} - x^{-1} - 5 = 0$
17. $2x^{-2} + x^{-1} - 1 = 0$
18. $m^{-2} + 9m^{-1} - 10 = 0$

Problem-solving practice

19. A department store having a sale reduced all their appliances by 20% and all their hardware by 25%. After the discount, the price of a drill was the same as the price of a toaster. If the original price of the drill was \$1.50 more than the toaster, what does each appliance cost during the sale?

20. The average of two positive integers is 171. If one of the numbers is the square root of the other, find the numbers.

Solve. Check possible solutions by substituting into the original equation.

21. $6.75x - 35\sqrt{x} - 5.36 = 0$

22. $\pi x^4 - \pi^2 x^2 - \sqrt{99.3} = 0$

Solve.

23. $\left(\frac{y^2-1}{y}\right)^2 - 4\left(\frac{y^2-1}{y}\right) - 12 = 0$

24. $\left(\sqrt{\frac{x}{x-3}}\right)^2 - 24 = 10\sqrt{\frac{x}{x-3}}$

25. $\left(\frac{x^2-1}{x}\right)^2 - \left(\frac{x^2-1}{x}\right) - 2 = 0$

26. $\left(\frac{x^2-2}{x}\right)^2 - 7\left(\frac{x^2-2}{x}\right) - 18 = 0$

27. $\left(\frac{x^2+1}{x}\right)^2 - 8\left(\frac{x^2+1}{x}\right) + 15 = 0$

28. $\frac{x}{x-1} - 6\sqrt{\frac{x}{x-1}} - 40 = 0$

29. $\left(\frac{x+1}{x-1}\right)^2 + \left(\frac{x+1}{x-1}\right) - 2 = 0$

30. $5\left(\frac{x+2}{x-2}\right)^2 - 3\left(\frac{x+2}{x-2}\right) - 2 = 0$

31. $9x^{3/2} - 8 = x^3$

32. $\sqrt[3]{2x+3} = \sqrt[6]{2x+3}$

33. $\sqrt{x-3} - \sqrt[4]{x-3} = 2$

34. $a^3 - 26a^{3/2} - 27 = 0$

35. $\frac{2x+1}{x} = 3 + 7\sqrt{\frac{2x+1}{x}}$

OBJECTIVES

After finishing Section 8.7, you should be able to:

- **i** **Given a situation of variation, write an equation of variation or a proportion.**
- **ii** **Solve problems involving variation.**

8.7

OTHER KINDS OF VARIATION

We have already studied direct variation and inverse variation. We now consider variation in which the equations involved are of second degree or greater. In some cases, we also consider equations with variables in the denominators.

Consider the equation for the area of a circle, where A and r are variables and π is a constant:

$$A = \pi r^2.$$

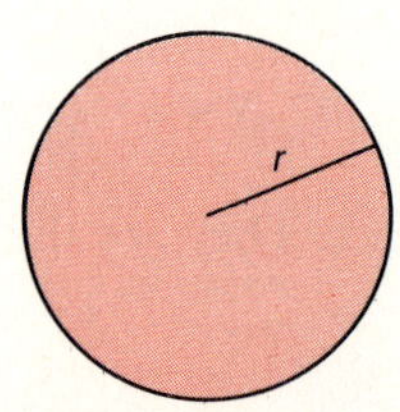

We say that the area varies directly as the square of the radius. When (r_1, A_1) and (r_2, A_2) are solutions of the equation, we have $A_1 = \pi r_1^2$ and $A_2 = \pi r_2^2$. Then dividing we get

$$\frac{A_1}{A_2} = \frac{\pi r_1^2}{\pi r_2^2} = \frac{r_1^2}{r_2^2}.$$

A proportion like

$$\frac{A_1}{A_2} = \frac{r_1^2}{r_2^2}$$

can be helpful in solving problems.

y varies directly as the square of x if there is some positive constant k such that $y = kx^2$.

EXAMPLE 1 Find an equation of variation where y varies directly as the square of x, and $y = 12$ when $x = 2$.

We write an equation of variation and find k:

$$y = kx^2, \qquad \text{so } 12 = k \cdot 2^2 \quad \text{and} \quad 3 = k.$$

Thus $y = 3x^2$.

DO EXERCISE 1.

1. Find an equation of variation where y varies directly as the square of x and $y = 175$ when $x = 5$.

From the law of gravity, we know that the weight W of an object varies inversely as the square of its distance d from the center of the earth:

$$W = \frac{k}{d^2}.$$

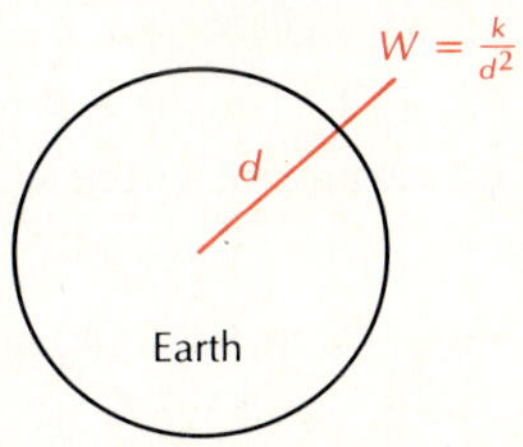

When (d_1, W_1) and (d_2, W_2) are solutions of the equation we have $W_1 = k/d_1^2$ and $W_2 = k/d_2^2$. Then

$$\frac{W_1}{W_2} = \frac{\frac{k}{d_1^2}}{\frac{k}{d_2^2}} = \frac{k}{d_1^2} \cdot \frac{d_2^2}{k} = \frac{d_2^2}{d_1^2}.$$

A proportion like

$$\frac{W_1}{W_2} = \frac{d_2^2}{d_1^2}$$

can be helpful in solving problems.

2. Find an equation of variation where y varies inversely as the square of x and $y = \frac{1}{4}$ when $x = 6$.

y varies inversely as the square of x if there is some positive constant k such that $y = k/x^2$.

EXAMPLE 2 Find an equation of variation where W varies inversely as the square of d and $W = 3$ when $d = 5$.

$$W = \frac{k}{d^2}, \qquad \text{so } 3 = \frac{k}{5^2} \quad \text{and} \quad k = 75$$

Thus

$$W = \frac{75}{d^2}.$$

DO EXERCISE 2.

3. Find an equation of variation where y varies jointly as x and z and $y = 65$ when $x = 10$ and $z = 13$.

4. Find an equation of variation where y varies jointly as x and the square of z and inversely as w and $y = 80$ when $x = 4$, $z = 10$, and $w = 25$.

Consider the equation for the area A of a triangle with height h and base b:

$$A = \tfrac{1}{2}bh.$$

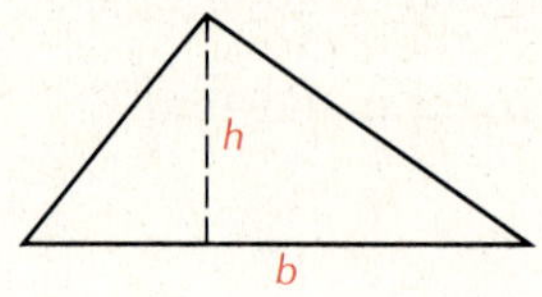

We say that the area varies *jointly* as the height and the base. From two solutions of the equation we get the proportion

$$\frac{A_1}{A_2} = \frac{\frac{1}{2}b_1h_1}{\frac{1}{2}b_2h_2} = \frac{b_1h_1}{b_2h_2}.$$

y varies jointly as x and z if there is some positive constant k such that $y = kxz$.

EXAMPLE 3 Find an equation of variation where y varies jointly as x and z, and $y = 42$ when $x = 2$ and $z = 3$.

$$y = kxz, \quad \text{so} \quad 42 = k \cdot 2 \cdot 3 \quad \text{and} \quad k = 7$$

Thus $y = 7xz$.

DO EXERCISE 3.

The equation

$$y = k \cdot \frac{xz^2}{w}$$

asserts that y varies jointly as x and the square of z, and inversely as w.

EXAMPLE 4 Find an equation of variation where y varies jointly as x and z and inversely as the square of w, and $y = 105$ when $x = 3$, $z = 20$, and $w = 2$.

$$y = k \cdot \frac{xz}{w^2}, \quad \text{so} \quad 105 = k \cdot \frac{3 \cdot 20}{2^2} \quad \text{and} \quad k = 7.$$

Thus

$$y = 7 \cdot \frac{xz}{w^2}.$$

DO EXERCISE 4.

ii Problem Solving

There are many problem situations that translate to mathematical language in the form of equations of variation. The following is an example.

EXAMPLE 5 The volume of wood V in a tree trunk varies jointly as the height h and the square of the girth g (girth is distance around). If the volume of a redwood tree is 216 m^3 when the height is 30 m and the girth is 1.5 m, what is the height of a tree whose volume is 960 m^3 and girth is 2 m?

1. **Familiarize** We'll make a table, including the data from the problem and the data we need to find.

Volume of wood	Height of tree	Girth of tree
216 m^3, or V_1	30 m, or h_1	1.5 m, or g_1
960 m^3, or V_2	h_2	2 m, or g_2

We have of course used h_2 to represent the height of the tree in question, a quantity that we do not know and wish to determine.

We know from the statement of the problem that in this situation the volume varies jointly as the height and the square of the girth.

2. **Translate** We will do this twice, once with an equation of variation and once with a proportion.

Method 1. First we find k using the first set of data. Then we solve for h using the second set of data:

$$V = khg^2$$
$$216 = k \cdot 30 \cdot 1.5^2$$
$$3.2 = k. \quad \text{This is the variation constant.}$$

The equation of variation is $V = 3.2hg^2$.

Method 2. We first write a proportion:

$$\frac{V_1}{V_2} = \frac{h_1 \cdot g_1^2}{h_2 \cdot g_2^2}.$$

This is the translation.

3. **Carry out**

Method 1. The translation is $V = 3.2hg^2$. We substitute and solve for h:

$$960 = 3.2 \cdot h \cdot 2^2$$
$$75 = h.$$

5. The distance s that an object falls when dropped from some point above the ground varies directly as the square of the time t it falls. If the object falls 19.6 m in 2 sec, how far will the object fall in 10 sec?

6. At a fixed temperature, the resistance R of a wire varies directly as the length l and inversely as the square of its diameter d. If the resistance is 0.1 ohm when the diameter is 1 mm and the length is 50 cm, what is the resistance when the length is 2000 cm and the diameter is 2 mm?

Method 2. The translation is

$$\frac{V_1}{V_2} = \frac{h_1 \cdot g_1^2}{h_2 \cdot g_2^2}.$$

We substitute and solve for h_2:

$$\frac{216}{960} = \frac{30 \cdot 1.5^2}{h_2 \cdot 2^2}$$

$$h_2 = \frac{960 \cdot 67.5}{4 \cdot 216}$$

$$h_2 = 75.$$

4. **Check** We already have a pretty good check, since we have done the translating and calculating in two different ways. Had we not done that, we would need to recheck all calculations and perhaps do an estimate to see whether our answer is reasonable.

5. **State** The answer is that the height of the tree is 75 meters.

DO EXERCISES 5 AND 6.

EXERCISE SET 8.7

i Find an equation of variation in which:

1. y varies directly as the square of x and $y = 0.15$ when $x = 0.1$.

2. y varies directly as the square of x and $y = 6$ when $x = 3$.

3. y varies inversely as the square of x and $y = 0.15$ when $x = 0.1$.

4. y varies inversely as the square of x and $y = 6$ when $x = 3$.

5. y varies jointly as x and z and $y = 56$ when $x = 7$ and $z = 8$.

6. y varies directly as x and inversely as z and $y = 4$ when $x = 12$ and $z = 15$.

7. y varies jointly as x and the square of z and $y = 105$ when $x = 14$ and $z = 5$.

8. y varies jointly as x and z and inversely as w and $y = \frac{3}{2}$ when $x = 2$, $z = 3$, and $w = 4$.

9. y varies jointly as x and z and inversely as the product of w and p, and $y = \frac{3}{28}$ when $x = 3$, $z = 10$, $w = 7$, and $p = 8$.

10. y varies jointly as x and z and inversely as the square of w, and $y = \frac{12}{5}$ when $x = 16$, $z = 3$, and $w = 5$.

ii ***Problem solving***

11. (*Stopping distance of a car*). The stopping distance d of a car after the brakes have been applied varies directly as the square of the speed r. If a car traveling 60 mph can stop in 200 ft, how many feet will it take the same car to stop when it is traveling 80 mph?

12. (*Area of a cube*). The area of a cube varies directly as the square of the length of a side. If a cube has an area 168.54 in^2 when the length of a side is 5.3 in., what will the area be when the length of a side is 10.2 in.?

13. (*Weight of an astronaut*). The weight W of an object varies inversely as the square of the distance d from the center of the earth. At sea level (6400 km from the center of the earth) an astronaut weighs 100 kg. Find his weight when he is 200 km above the surface of the earth and the spacecraft is not in motion.

14. (*Intensity of light*). The intensity I of light from a light bulb varies inversely as the square of the distance d from the bulb. Suppose I is 90 W/m^2 when the distance is 5 m. Find the intensity at a distance of 10 m.

15. (*Earned run average*). A pitcher's earned run average A varies directly as the number R of earned runs allowed and inversely as the number I of innings pitched. In a recent year Tom Seaver had an earned run average of 2.92. He gave up 85 earned runs in 262 innings. How many earned runs would he have given up had he pitched 300 innings with the same average? Round to the nearest whole number.

16. (*Volume of a gas*). The volume V of a given mass of a gas varies directly as the temperature T and inversely as the pressure P. If $V = 231$ cm^3 when $T = 42°$ and $P = 20$ kg/cm^2, what is the volume when $T = 30°$ and $P = 15$ kg/cm^2?

17. Suppose y varies directly as x and x is tripled. What is the effect on y?

18. Suppose y varies inversely as x and x is tripied. What is the effect on y?

19. Suppose y varies inversely as the square of x and x is multiplied by n. What is the effect on y?

20. Suppose y varies directly as the square of x and x is multiplied by n. What is the effect on y?

21. The area of a circle varies directly as the square of the length of a diameter. What is the variation constant?

22. A peanut butter jar in the shape of a right circular cylinder is 4 in. high and 3 in. in diameter and sells for \$1.20. Assuming the same ratio of volume of peanut butter to cost, how much should a jar 6 in. high and 6 in. in diameter cost?

23. Suppose y varies inversely as the cube of x and x is multiplied by 0.5. What is the effect on y?

24. (*An application: The gravity model*). It has been determined that the average number of telephone calls in a day N, between two cities, is directly proportional to the populations P_1 and P_2 of the cities and inversely proportional to the square of the distance between the cities. That is, $N = kP_1P_2/d^2$. This model is called the "gravity model" because the equation of variation resembles the equation that applies to Newton's law of gravity.

a) The population of Indianapolis is 744,624 and the population of Cincinnati is 452,524, and the distance between the cities is 174 km. The average number of daily phone calls between the two cities is 11,153. Find the value k and write the equation of variation.

b) The average number of daily phone calls between Indianapolis and New York is 4270 and the population of New York is 7,895,563. Find the distance between Indianapolis and New York.

8.8

QUADRATIC FUNCTIONS

A *quadratic function* is a function that can be described by a quadratic polynomial, as follows:

$$f(x) = ax^2 + bx + c.$$

The constants b and c may be 0, but of course a cannot be 0, or else the polynomial will not be quadratic. In this section we will look at the graphs of quadratic functions.

OBJECTIVES

After finishing Section 8.8, you should be able to:

- **i** **Graph functions like $y = ax^2$ or $f(x) = ax^2$.**
- **ii** **Graph functions like $f(x) = a(x - h)^2$.**
- **iii** **Graph functions like $f(x) = a(x - h)^2 + k$ and determine the vertex, line of symmetry, and the maximum or minimum function values.**
- **iv** **Use translations to graph functions or to determine properties of the function without graphing.**

i Graphs of $f(x) = ax^2$

We first consider graphs of quadratic functions in which b and c are both 0. In fact, let's start with a very simple one, a function for which $a = 1$.

EXAMPLE 1 Graph: $f(x) = x^2$.

We choose some numbers for x, some positive and some negative, and for each number compute $f(x)$.

x	0	1	2	3	−1	−2	−3
$f(x)$	0	1	4	9	1	4	9

1. Graph $y = \frac{1}{2}x^2$.

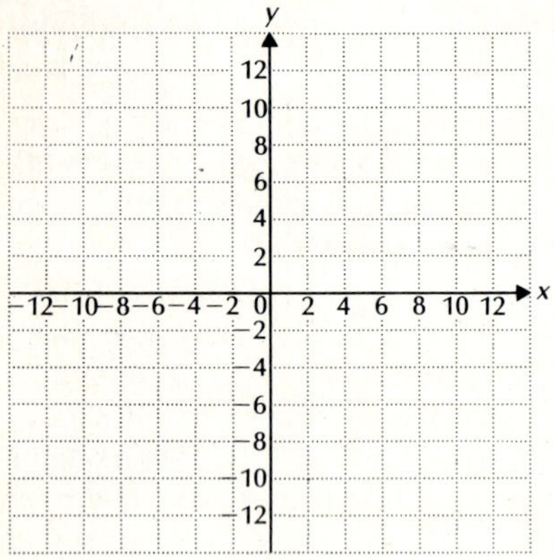

We plot these ordered pairs and connect the points with a smooth curve.

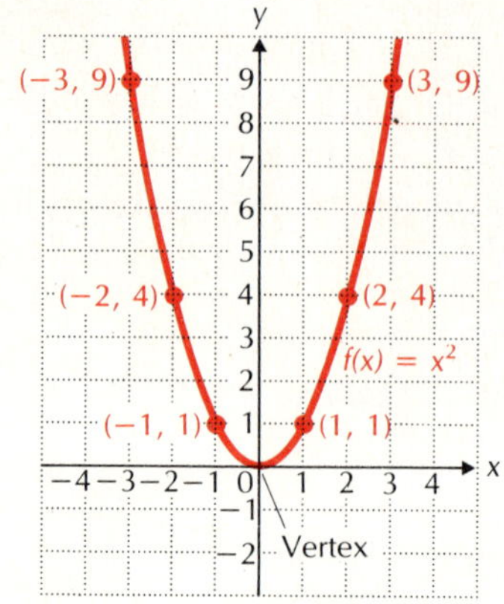

All quadratic functions have graphs similar to the one in Example 1. Such curves are called *parabolas*. They are smooth, cup-shaped curves, with a line of symmetry. In the graph above, the vertical axis is the *line of symmetry*. If the paper were folded on this line, the two halves of the curve would match. The point (0, 0) is known as the *vertex* of the parabola.

Some common errors in drawing graphs of quadratic functions are shown in the figure below.

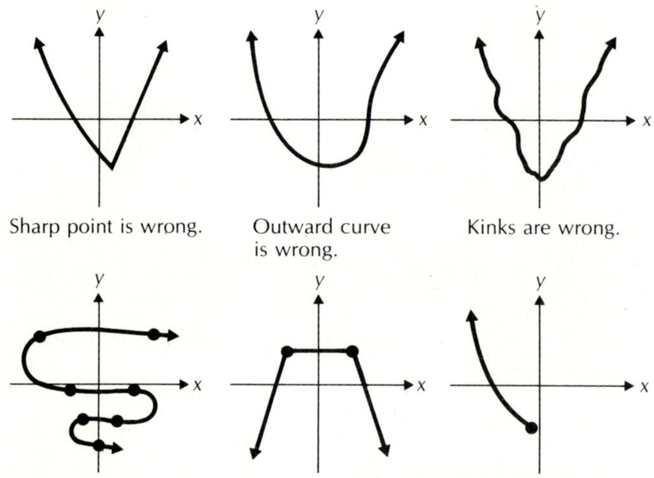

Sharp point is wrong. Outward curve is wrong. Kinks are wrong.

S-shaped curve is wrong. Flat nose is wrong. Half a cup-shaped curve is wrong.

DO EXERCISE 1.

The graph of $f(x) = \frac{1}{2}x^2$ is a flatter parabola than the graph of $y = x^2$. The graph of $f(x) = 2x^2$ is narrower, as shown here, but the vertex and line of symmetry have not changed.

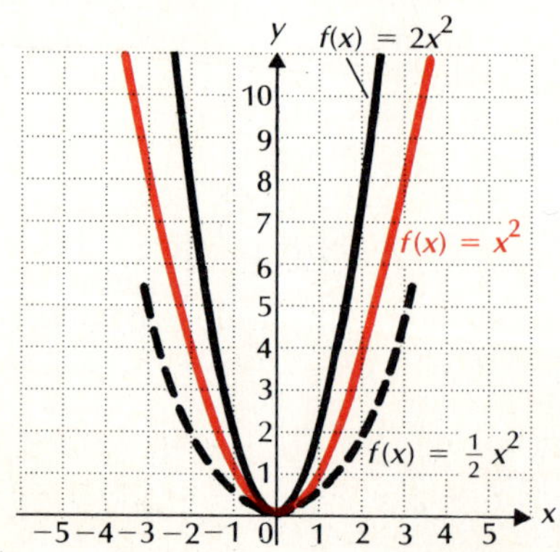

The graph of $y = ax^2$ is a parabola with the vertical axis as its line of symmetry and its vertex at the origin.

If $|a|$ is greater than 1, the parabola is narrower than $y = x^2$.

If $|a|$ is between 0 and 1, the parabola is flatter than $y = x^2$.

If a is positive, the parabola opens upward; if a is negative, the parabola opens downward.

DO EXERCISES 2 AND 3.

ii Graphs of $f(x) = a(x - h)^2$

Why not now consider graphs of $f(x) = ax^2 + bx + c$, where b and c are not both 0? In effect, we shall do that, but in a disguised form. It turns out to be convenient to consider functions $f(x) = a(x - h)^2$, that is, where we start with ax^2 but then replace x by $x - h$, where h is some constant.

EXAMPLE 2 Graph: $f(x) = 2(x - 3)^2$.

We choose some values of x and compute $f(x)$.

x	0	1	2	3	4	5	6
$f(x)$, or $2(x-3)^2$	18	8	2	0	2	8	18

Now we plot the points and draw the curve.

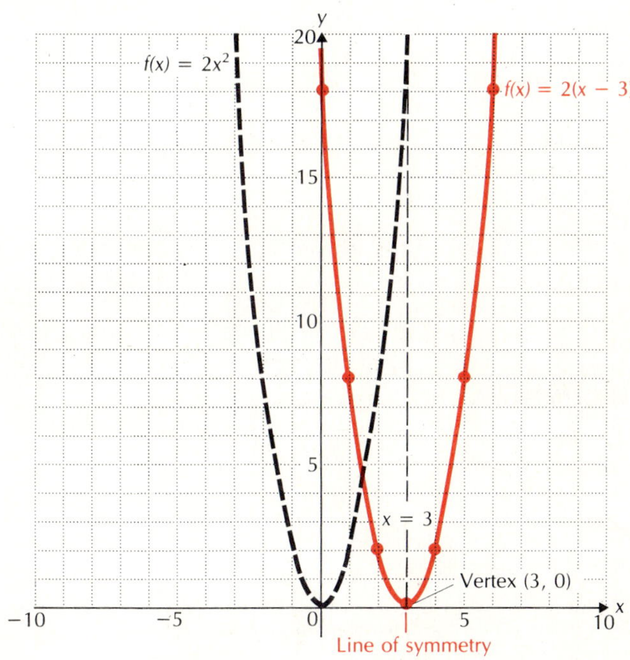

The curve in Example 2 looks just like the graph of $f(x) = 2x^2$, except that it is moved three units to the right. The line of symmetry is now the line $x = 3$ and the vertex is the point $(3, 0)$.

2. Graph $y = 3x^2$.

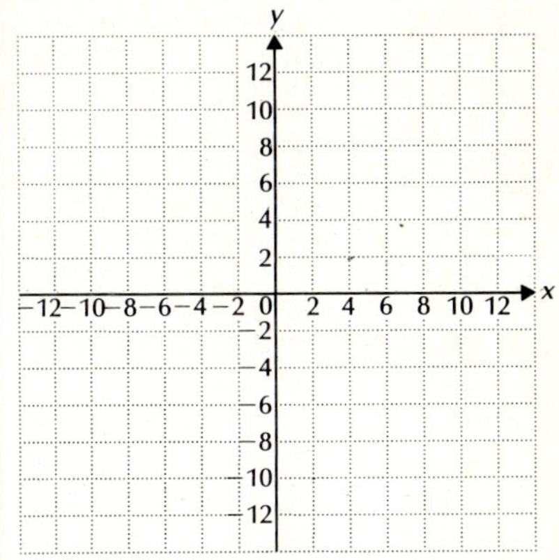

3. Graph $y = -2x^2$.

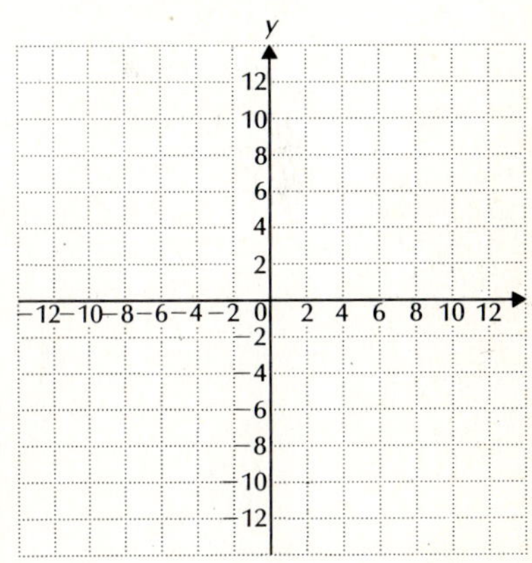

4\. a) Graph $y = \frac{1}{2}(x - 4)^2$. Label the vertex and draw the line of symmetry.

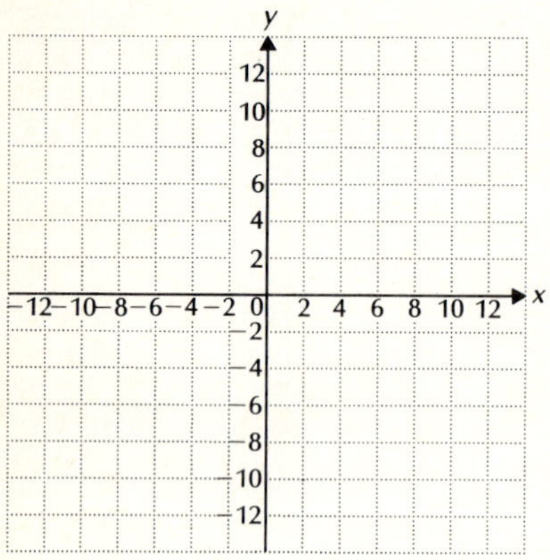

b) Graph $y = -\frac{1}{2}(x - 4)^2$.

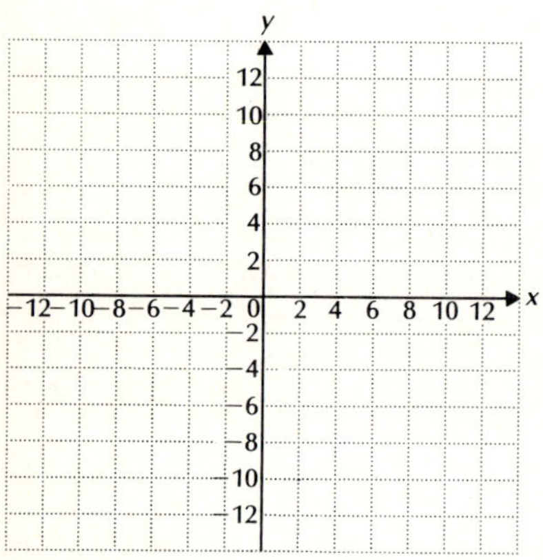

The graph of $y = a(x - h)^2$ looks just like the graph of $y = ax^2$, except that it is moved to the right or left. If h is positive, it is moved to the right. If h is negative, it is moved to the left. The vertex is now $(h, 0)$ and the line of symmetry is $x = h$.

EXAMPLE 3 Graph: $f(x) = 2(x + 3)^2$.

We know that the graph looks like that of $f(x) = 2x^2$, but moved right or left. Think of $x + 3$ as $x - (-3)$. The number we are *subtracting* is negative, so we move *left*. We can draw $f(x) = 2x^2$ lightly, then move it three units to the left. See the following graph. We should compute a few values as a check.

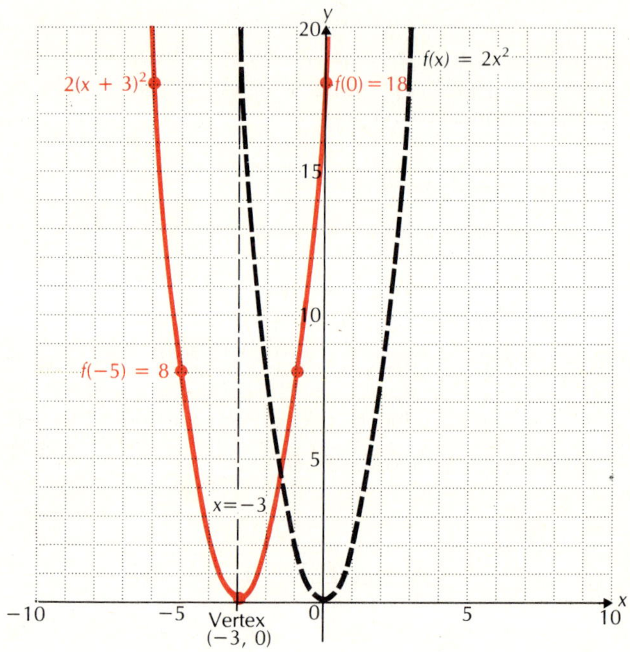

EXAMPLE 4 Graph: $y = -2(x + 2)^2$.

We know that the graph looks like that of $y = 2x^2$, but moved to the left two units, and it will also open downward, because of the negative coefficient, -2. See the graph below.

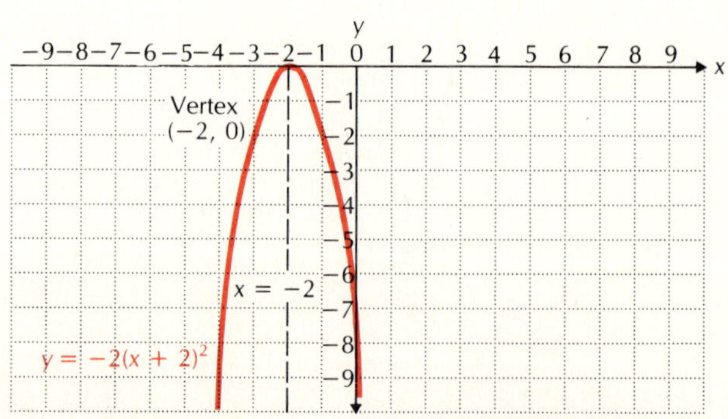

DO EXERCISE 4.

iii Graphs of $f(x) = a(x - h)^2 + k$

Given a graph of $f(x) = a(x - h)^2$, what happens to it if we add a constant k? Suppose we add 2. This increases each function value $f(x)$ by 2, so the curve is moved up. If we add a negative number, the curve is moved down. The vertex of the parabola will be at the point (h, k) and the line of symmetry will be $x = h$.

Note that if a parabola opens upward $(a > 0)$, the function value, or y-value, at the vertex is a least, or *minimum* value. That is, it is less than the y-value at any other point. If the parabola opens downward $(a < 0)$, the function value at the vertex will be a greatest, or *maximum* value.

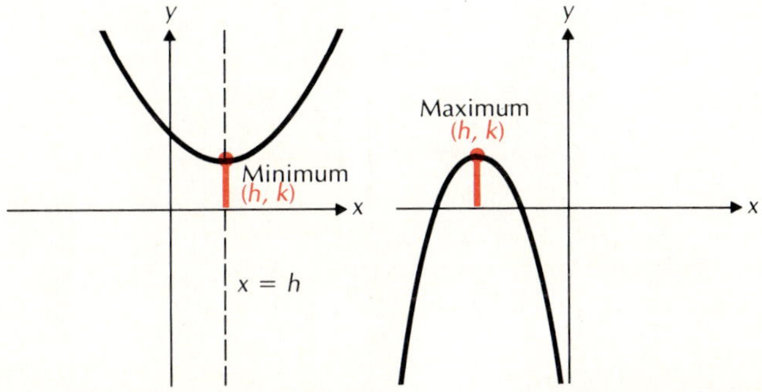

The graph of $f(x) = a(x - h)^2 + k$ looks just like the graph of $f(x) = a(x - h)^2$, except that it is moved up or down. If k is positive, the curve is moved up k units. If k is negative, the curve is moved down. The vertex is at (h, k) and the line of symmetry is $x = h$. If $a > 0$, then k is the minimum function value. If $a < 0$, then k is the maximum function value.

EXAMPLE 5 Graph $f(x) = 2(x + 3)^2 - 5$ and find the minimum function value.

We know that the graph looks like that of $f(x) = 2(x + 3)^2$ (this is shown in Example 3) but moved down five units. The vertex is now $(-3, -5)$, and the minimum function value is -5.

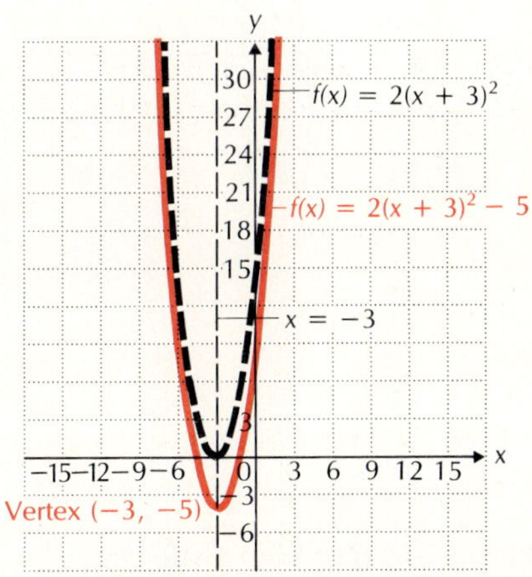

EXAMPLE 6 Graph $f(x) = \frac{1}{2}(x - 3)^2 + 5$ and find the minimum function value.

We know that the graph looks just like that of $f(x) = \frac{1}{2}x^2$, but moved to the right three units and up five units. The vertex is (3, 5) and the line of symmetry is

5. **a)** Graph $y = \frac{1}{2}(x + 2)^2 - 4$.

b) Label the vertex and draw the line of symmetry.

c) Find the minimum function value.

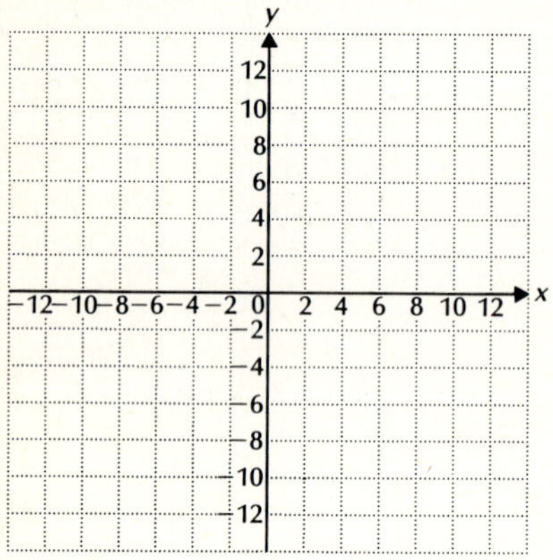

the line $x = 3$. We draw $f(x) = \frac{1}{2}x^2$ and then move the curve over and up. We plot a few points as a check. The minimum function value is 5.

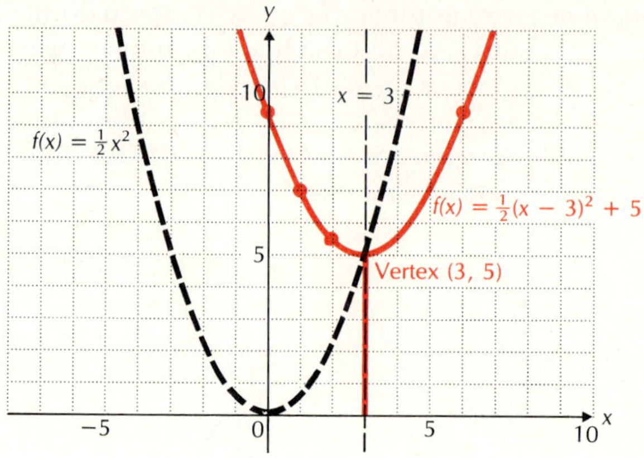

x	0	1	2	6
$f(x)$	$9\frac{1}{2}$	7	$5\frac{1}{2}$	$9\frac{1}{2}$

EXAMPLE 7 Graph $y = -2(x + 3)^2 + 5$ and find the maximum function value.

We know that the graph looks like that of $f(x) = 2x^2$, but moved to the left three units and up five units, except that it now opens downward. The vertex is $(-3, 5)$. This time there is a *greatest* (maximum) function value, 5.

6. **a)** Graph $f(x) = -2(x - 5)^2 + 3$.

b) Label the vertex and draw the line of symmetry.

c) Find the maximum function value.

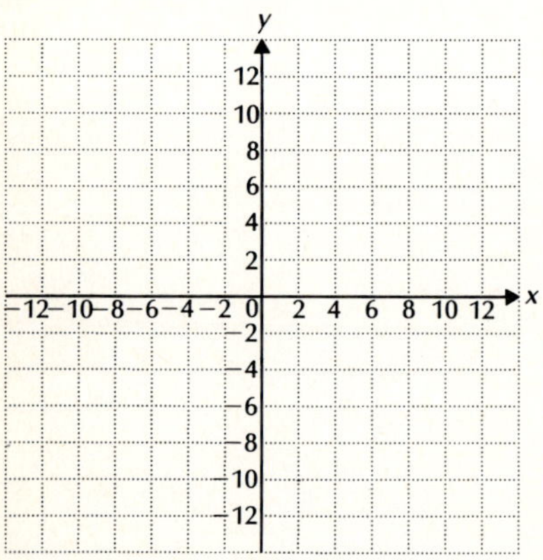

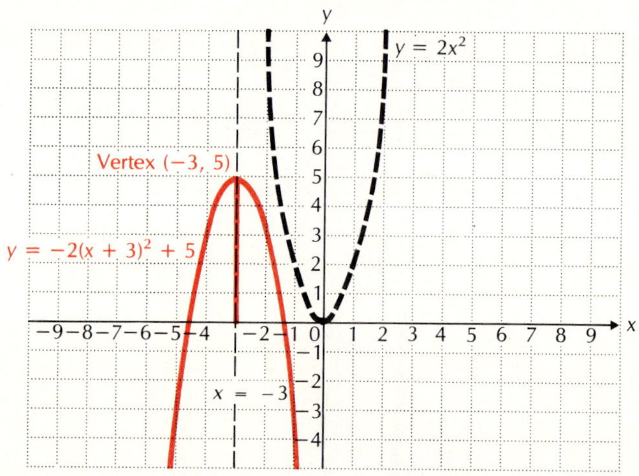

DO EXERCISES 5 AND 6.

iv Translations

In this section we have seen a comparison of the graphs of

$$f(x) = ax^2 \quad \text{and} \quad f(x) = a(x - h)^2.$$

The graphs are alike, except that the second one has been moved horizontally. We have also seen a comparison of

$$f(x) = a(x - h)^2 \quad \text{and} \quad f(x) = a(x - h)^2 + k.$$

Again, the graphs look alike, but in this case the second one has been moved vertically.

A motion of a graph like those we are considering, in which we do not change the size or shape of a graph and in which there is no rotation, is called a *translation.* What we have seen with parabolas holds for any other function. It also holds for other graphs, whether or not they are graphs of functions. We summarize as follows. Although the properties are stated for functions, they hold in general.

For a function $y = f(x)$, if we replace x by $x - h$, where h is a constant, the graph will be moved horizontally. If h is positive, the graph will be moved in the positive x-direction. If h is negative, the graph will be moved in the negative x-direction.

EXAMPLE 8 The following figure is a graph of a function $y = g(x)$. Draw graphs for (a) $y = g(x - 4)$; (b) $y = g(x + 3)$.

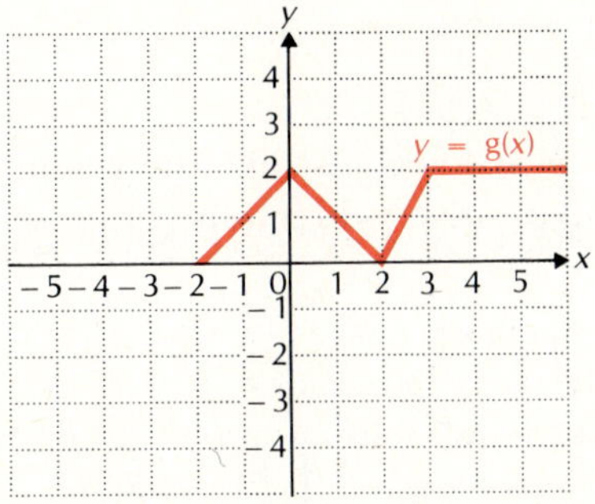

a) The graph of $y = g(x - 4)$ looks just like the graph of $y = g(x)$, but it will be moved four units in the positive x-direction (since we are subtracting a positive constant from x, in $x - 4$).

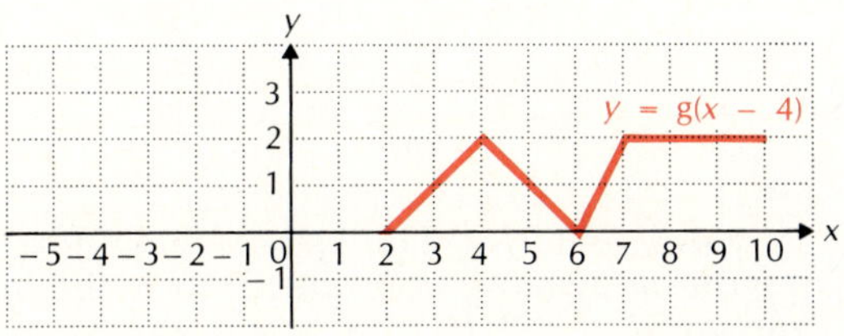

b) We can think of the graph of $y = g(x + 3)$ as the graph of $y = g(x - (-3))$. Since we are subtracting a negative constant from x, the graph will be moved in the negative x-direction.

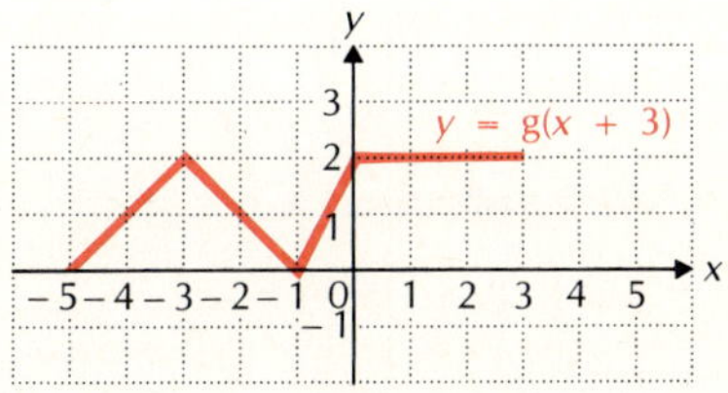

DO EXERCISE 7.

7. This is the graph of a function $y = f(x)$.

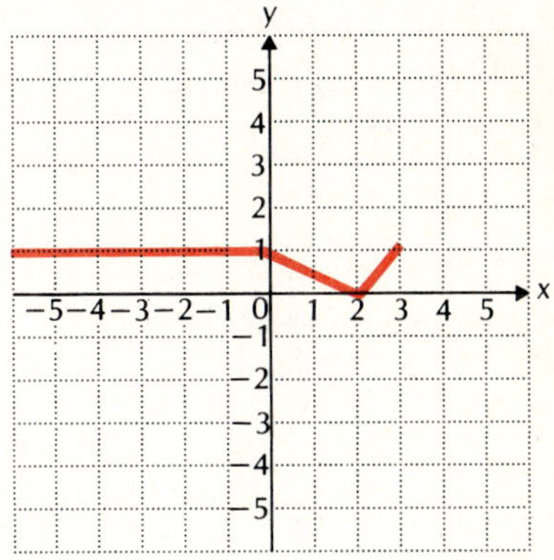

Draw the graph of each of the following.

a) $y = f(x + 2)$

b) $y = f(x - 3)$

8. Use the function of Exercise 7. Draw the graph of each of the following.

a) $y + 3 = f(x)$, or $y = f(x) - 3$

b) $y - 2 = f(x)$, or $y = f(x) + 2$

What we have said about replacing x by $x - h$ also holds for replacing y by $y - k$, or replacing $f(x)$ by $f(x) - k$, where k is some constant.

For a function $y = f(x)$, if we replace y by $y - k$, where k is a constant, the graph will be moved vertically. If k is positive, the graph will be moved in the positive y-direction. If k is negative, the graph will be moved in the negative y-direction.

For functions, we are most likely to write $y = f(x) + k$ than $y - k = f(x)$. The rule is easier to remember, however, if you write, or at least think of writing, $y - k = f(x)$. In equations that do not represent functions, it will often be *necessary* to write $y - k$ in place of y.

EXAMPLE 9 Use the function $y = g(x)$ in Example 8 to draw graphs for (a) $y - 2 = g(x)$, or $y = g(x) + 2$; (b) $y + 4 = g(x)$, or $y = g(x) - 4$.

a) The graph of $y - 2 = g(x)$ looks just like the graph of $y = g(x)$, but it will be moved two units in the positive y-direction (since we are subtracting a positive constant from y, in $y - 2$).

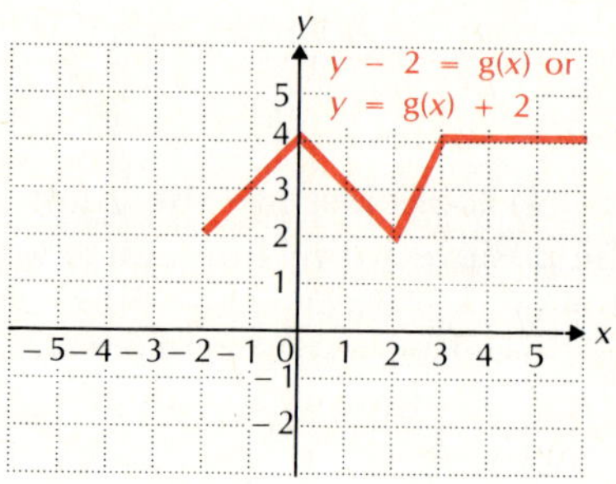

b) The graph of $y + 4 = g(x)$ can be thought of as the graph of $y - (-4) = g(x)$. Since we are subtracting a negative constant from y, the graph will be moved in the negative y-direction.

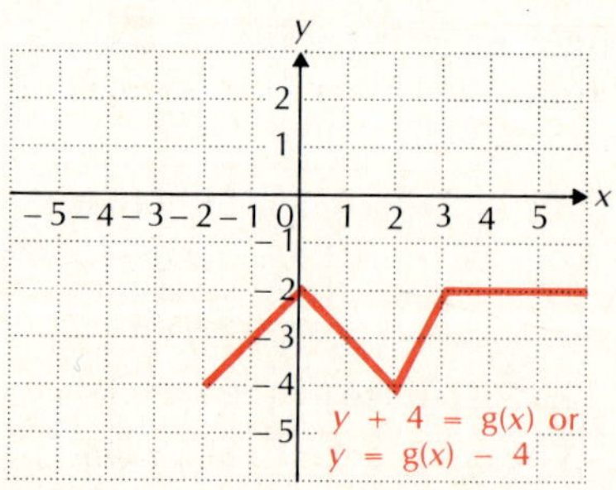

DO EXERCISE 8.

EXERCISE SET 8.8

i, **ii** For each of the following, graph the function, find the vertex, and find the line of symmetry.

1. $f(x) = x^2$ **2.** $f(x) = -x^2$ **3.** $f(x) = -4x^2$ **4.** $f(x) = 2x^2$

5. $f(x) = (x - 3)^2$ **6.** $f(x) = (x - 7)^2$ **7.** $f(x) = -(x + 4)^2$ **8.** $f(x) = -(x - 2)^2$

9. $f(x) = 2(x - 3)^2$ **10.** $f(x) = -4(x - 7)^2$ **11.** $f(x) = -2(x + 9)^2$ **12.** $f(x) = 2(x + 7)^2$

13. $f(x) = 3(x - 1)^2$ **14.** $f(x) = -4(x - 2)^2$ **15.** $f(x) = -3(x - \frac{1}{2})^2$ **16.** $f(x) = -2(x + \frac{1}{2})^2$

17. $f(x) = \frac{1}{2}(x + 1)^2$ **18.** $f(x) = \frac{1}{3}(x - 2)^2$

iii For each of the following, graph the function, find the vertex, find the line of symmetry, and find the maximum value or the minimum value.

19. $f(x) = (x - 3)^2 + 1$ **20.** $f(x) = (x + 2)^2 - 3$ **21.** $f(x) = (x + 1)^2 - 2$ **22.** $f(x) = (x - 1)^2 + 2$

23. $f(x) = 2(x - 1)^2 - 3$ **24.** $f(x) = 2(x + 1)^2 + 4$ **25.** $f(x) = -3(x + 4)^2 + 1$ **26.** $f(x) = -2(x - 5)^2 - 3$

Without graphing, find the vertex, find the line of symmetry, and find the maximum value or the minimum value.

27. $f(x) = 8(x - 9)^2 + 5$ **28.** $f(x) = 10(x + 5)^2 - 8$ **29.** $f(x) = 5(x + \frac{1}{4})^2 - 13$

30. $f(x) = 6(x - \frac{1}{4})^2 + 19$ **31.** $f(x) = -7(x - 10)^2 - 20$ **32.** $f(x) = -9(x + 12)^2 + 23$

33. $f(x) = \sqrt{2}\,(x + 4.58)^2 + 65\pi$ **34.** $f(x) = 4\pi(x - 38.2)^2 - \sqrt{34}$

iv Use the graph of the function $y = h(x)$ below for Exercises 35–40.

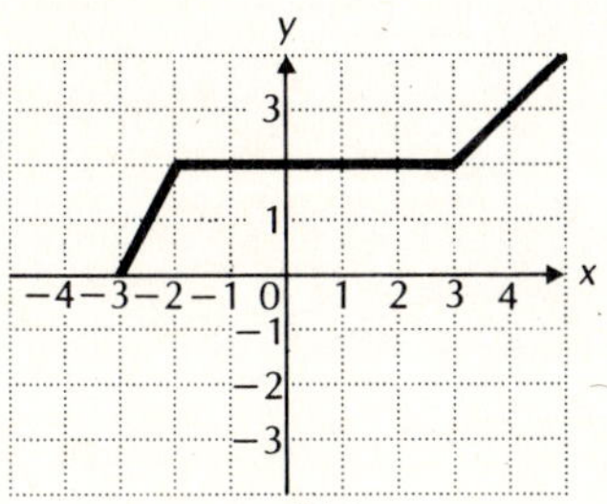

$y = h(x)$

Draw a graph of each of the following.

35. $y = h(x + 2)$ **36.** $y = h(x - 3)$ **37.** $y - 3 = h(x)$

38. $y = h(x) + 4$ **39.** $y + 2 = h(x - 4)$ **40.** $y = h(x - 2) + 3$

Problem-solving practice

41. A restaurant sells 12-oz, 20-oz, and 32-oz sodas for 45¢, 60¢, and 80¢, respectively. In one day they sold a total of 27 sodas for $17.80. If the restaurant sold 5 gal of soda that day, how much of each size drink did they sell?

42. Given four consecutive odd numbers, the product of the first and third is 56 less than the product of the second and fourth. Find the numbers.

○

For each of the following, write the equation of the parabola that is a translation of $f(x) = 2x^2$ or $f(x) = -2x^2$ and has a maximum or minimum value at the specific point.

43. Maximum: (0, 4) **44.** Minimum: (2, 0) **45.** Minimum: (6, 0)

46. Maximum: (0, 3) **47.** Maximum: (3, 8) **48.** Minimum: (−2, 3)

49. Minimum: (−3, 6) **50.** Maximum: (−4, −3) **51.** Minimum: (2, −3)

For each of the following, write an equation of the parabola.

52. The parabola has a minimum value at the same point as $f(x) = 3(x - 4)^2$, but for all x in the domain except 4 the function values are doubled.

53. The parabola is a translation of

$$f(x) = -\frac{1}{2}(x - 2)^2 + 4$$

and has a maximum value at the same point as $f(x) = -2(x - 1)^2 - 6$.

OBJECTIVES

After finishing Section 8.9, you should be able to:

i **Given a quadratic function $f(x) = ax^2 + bx + c$, complete the square to put it into the form**

$$f(x) = a(x - h)^2 + k,$$

and then graph the function.

ii **Find the x-intercepts of a quadratic function.**

1. For $f(x) = x^2 - 4x + 7$:
   a) find an equation of the type $f(x) = a(x - h)^2 + k$;
   b) find the vertex and line of symmetry;
   c) graph the function.

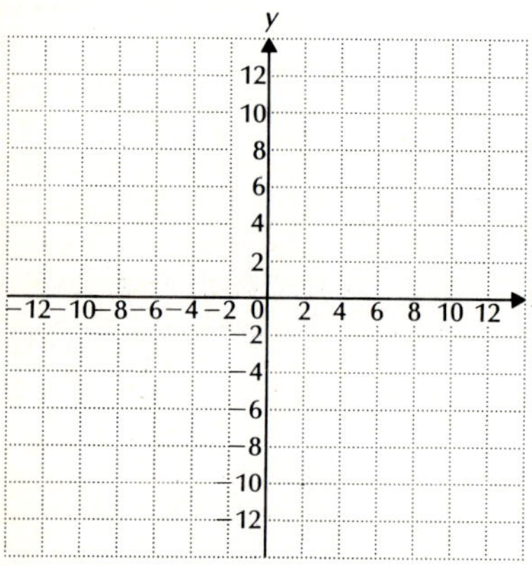

8.9 MORE ABOUT GRAPHING QUADRATIC FUNCTIONS

Now that we know how to graph $f(x) = a(x - h)^2 + k$, we are in a position to graph any quadratic function

$$f(x) = ax^2 + bx + c.$$

All we have to do is find a way to rewrite a quadratic polynomial as $a(x - h)^2 + k$.

i Completing the Square

Actually, we already know a way to rewrite a quadratic function as described: The method is that of *completing the square.*

EXAMPLE 1 Graph: $f(x) = x^2 - 6x + 4$.

$$f(x) = x^2 - 6x + 4$$
$$= (x^2 - 6x) + 4$$

We complete the square inside the parentheses. We take half the x-coefficient:

$$\tfrac{1}{2} \cdot (-6) = -3.$$

We square it:

$$(-3)^2 = 9.$$

Then we add $9 - 9$ inside the parentheses.

$$f(x) = (x^2 - 6x + 9 - 9) + 4$$
$$= (x^2 - 6x + 9) + (-9 + 4) \qquad \text{Rearranging terms}$$
$$= 1 \cdot (x - 3)^2 - 5$$

The vertex is $(3, -5)$. The line of symmetry is $x = 3$.

We plot a few points as a check, and draw the curve.

x	$f(x)$
0	4
1	−1
5	−1
6	4

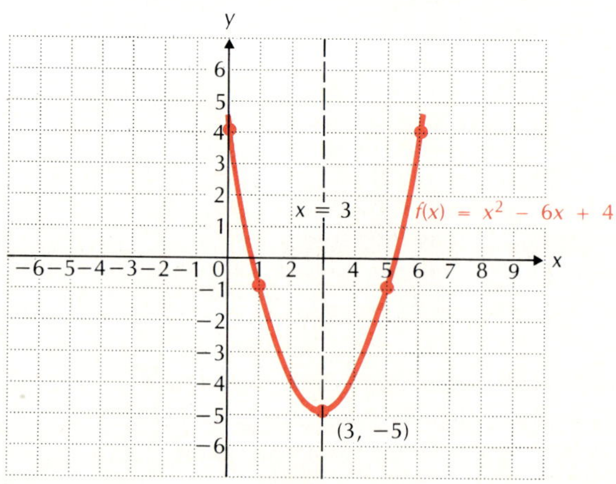

DO EXERCISE 1.

EXAMPLE 2 Graph: $f(x) = -2x^2 + 10x - 7$.

We first factor the expression $-2x^2 + 10x$ by "removing" -2 from the first two terms. This makes the coefficient of x^2 inside the parentheses 1. Then we proceed as before.

$$f(x) = -2x^2 + 10x - 7$$

and

$$f(x) = -2(x^2 - 5x) - 7.$$

We take half the x-coefficient and square it, to get $\frac{25}{4}$. Then we add $\frac{25}{4} - \frac{25}{4}$ inside the parentheses:

$$\begin{aligned} f(x) &= -2(x^2 - 5x + \tfrac{25}{4} - \tfrac{25}{4}) - 7 \\ &= -2(x^2 - 5x + \tfrac{25}{4}) + 2(\tfrac{25}{4}) - 7 \quad \text{Multiplying by } -2, \text{ using the distributive law, and rearranging terms} \\ &= -2(x - \tfrac{5}{2})^2 + \tfrac{11}{2} \end{aligned}$$

The vertex is $(\frac{5}{2}, \frac{11}{2})$. The line of symmetry is $x = \frac{5}{2}$. The coefficient -2 is negative, so the graph opens downward.

We plot a few points as a check and draw the curve.

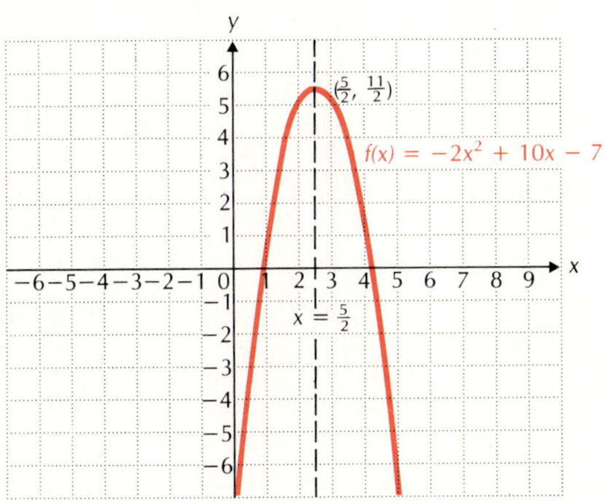

DO EXERCISE 2.

2. For $f(x) = -4x^2 + 12x - 5$:

a) find an equation of the type $f(x) = a(x - h)^2 + k$;

b) find the vertex and line of symmetry;

c) graph the function.

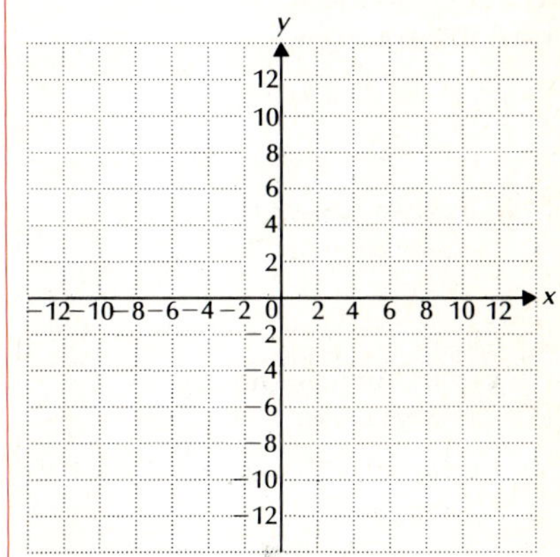

ii Intercepts

The points at which a graph crosses the x-axis are called its *x-intercepts*. These are, of course, the points at which $y = 0$.

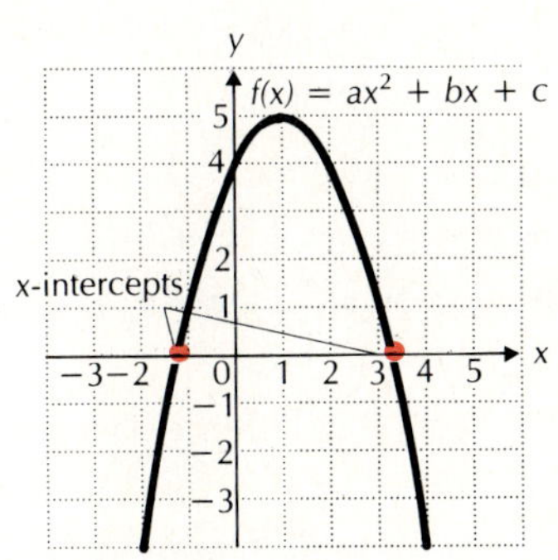

To find the x-intercepts of a quadratic function $f(x) = ax^2 + bx + c$ we solve the equation

$$0 = ax^2 + bx + c.$$

EXAMPLE 3 Find the x-intercepts of the graph of $f(x) = x^2 - 2x - 2$.

We solve the equation

$$0 = x^2 - 2x - 2.$$

Find the x-intercepts.

3. $f(x) = x^2 - 2x - 5$

Find the x-intercepts if they exist.

4. $f(x) = x^2 - 2x - 3$

5. $f(x) = x^2 + 8x + 16$

6. $f(x) = -2x^2 - 4x - 3$

The equation is difficult to factor, so we use the quadratic formula and get $x = 1 \pm \sqrt{3}$. Thus the x-intercepts are $(1 - \sqrt{3}, 0)$ and $(1 + \sqrt{3}, 0)$.

For plotting, we approximate, to get $(-0.7, 0)$ and $(2.7, 0)$. We sometimes refer to the x-coordinates as *intercepts*. It is useful to have these points when graphing a function.

DO EXERCISE 3.

The discriminant, $b^2 - 4ac$, tells us how many real-number solutions the equation $0 = ax^2 + bx + c$ has, so it also indicates how many intercepts there are. Compare.

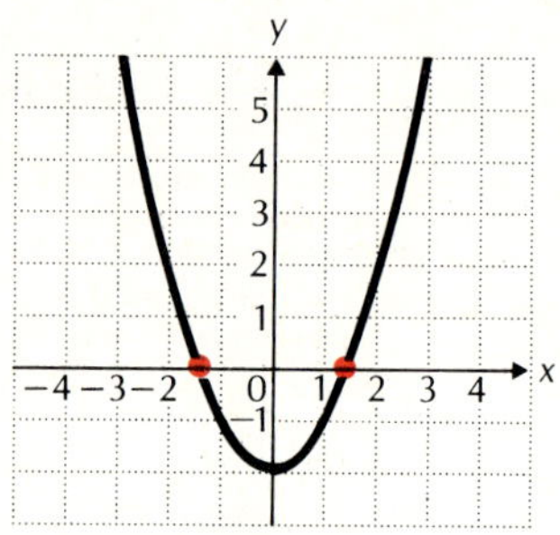

$y = ax^2 + bx + c$
$b^2 - 4ac > 0$
Two real solutions
Two x-intercepts

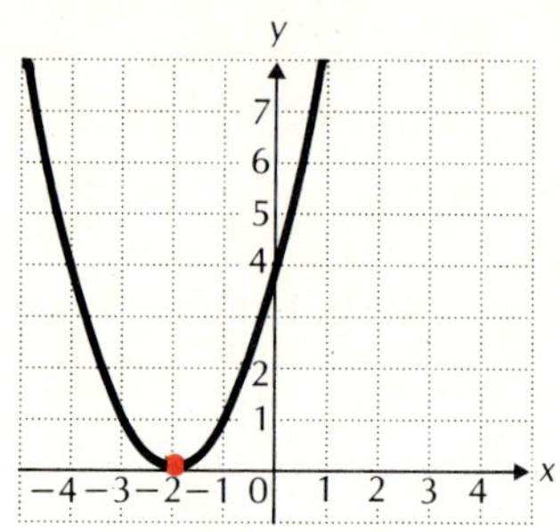

$y = ax^2 + bx + c$
$b^2 - 4ac = 0$
One real solution
One x-intercept

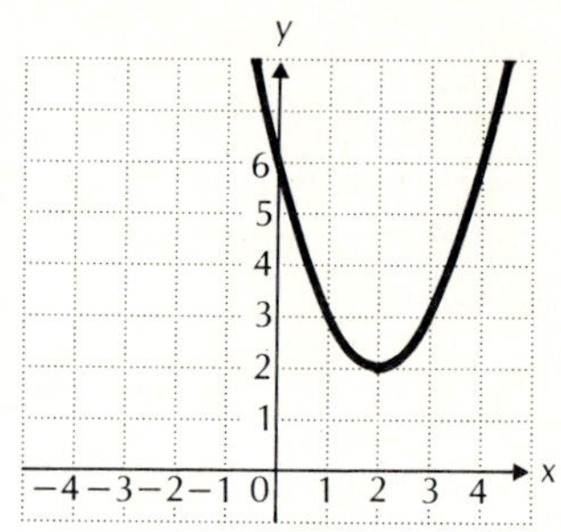

$y = ax^2 + bx + c$
$b^2 - 4ac < 0$
No real solutions
No x-intercepts

DO EXERCISES 4–6.

EXERCISE SET 8.9

i Complete the square. Find the vertex, the line of symmetry, and the maximum or minimum function value.

1. $f(x) = x^2 - 2x - 3$
2. $f(x) = x^2 + 2x - 5$
3. $f(x) = -x^2 + 4x + 6$
4. $f(x) = -x^2 - 4x + 3$
5. $f(x) = x^2 + 3x - 10$
6. $f(x) = x^2 + 5x + 4$
7. $f(x) = x^2 - 9x$
8. $f(x) = x^2 + x$
9. $f(x) = 3x^2 - 24x + 50$
10. $f(x) = 4x^2 + 8x - 3$
11. $f(x) = \frac{3}{4}x^2 + 9x$
12. $f(x) = -2x^2 + 2x + 1$

ii Find the x-intercepts.

13. $f(x) = x^2 - 4x + 1$
14. $f(x) = x^2 + 6x + 10$
15. $f(x) = -x^2 + 2x + 3$
16. $f(x) = x^2 - 2x - 5$
17. $f(x) = x^2 - 3x - 4$
18. $f(x) = x^2 - 8x + 5$
19. $f(x) = -x^2 + 3x + 4$
20. $f(x) = 2x^2 - 4x + 6$
21. $f(x) = 2x^2 + 4x - 1$
22. $f(x) = x^2 - x + 2$
23. $f(x) = x^2 - x + 1$
24. $f(x) = 4x^2 + 12x + 9$
25. $f(x) = -x^2 - 3x - 3$
26. $f(x) = -5x^2 + 6x - 5$

Problem-solving practice

27. The length of a diagonal of a rectangle is one more than the length of the rectangle and three more than twice the width of the rectangle. Find the dimensions of the rectangle.

28. Bill leaves the hotel parking lot going due north at 30 mph. Mary leaves the same lot going due east at 40 mph. How long will they have to travel before they are 240 mi apart?

○

Find the maximum or minimum value.

29. ▦ $f(x) = 2.31x^2 - 3.135x - 5.89$

30. ▦ $f(x) = -18.8x^2 + 7.92x + 6.18$

Find the x-intercepts.

31. ▦ $f(x) = 0.05x^2 - 4.735x + 100.23$

32. ▦ $f(x) = 1.13x^2 + 2.809x - 7.114$

33. ▦ $f(x) = 2.12x^2 + 3.21x + 9.73$

34. ▦ $f(x) = 0.13x^2 - 0.071x - 0.12$

35. Graph the function $f(x) = x^2 - x - 6$. Then use your graph to approximate the solutions to the following equations.

a) $x^2 - x - 6 = 2$ **b)** $x^2 - x - 6 = -3$

36. Graph the function

$$f(x) = \frac{x^2}{8} + \frac{x}{4} - \frac{3}{8}.$$

Use your graph to approximate solutions to the following equations.

a) $\frac{x^2}{8} + \frac{x}{4} - \frac{3}{8} = 0$ **b)** $\frac{x^2}{8} + \frac{x}{4} - \frac{3}{8} = 1$ **c)** $\frac{x^2}{8} + \frac{x}{4} - \frac{3}{8} = 2$

Find an equation of the type $f(x) = a(x - h)^2 + k$.

37. $f(x) = ax^2 + bx + c$ **38.** $f(x) = 3x^2 + mx + m^2$

Graph.

39. $f(x) = |x^2 - 1|$ **40.** $f(x) = |3 - 2x - x^2|$

Graph these quadratic inequalities.

41. $y < x^2 - 4x - 1$ **42.** $y \geqslant x^2 + 3x - 4$ **43.** $y \leqslant x^2 + 5x + 6$ **44.** $y < -x^2 - 2x + 3$

45. $y > 3x^2 + 6x + 2$ **46.** $y > 2x^2 + 4x - 2$ **47.** $y \geqslant 4x^2 + 8x + 3$ **48.** $y < 2x^2 - 4x + 2$

8.10

PROBLEM SOLVING AND QUADRATIC FUNCTIONS

OBJECTIVES

After finishing Section 8.10, you should be able to:

i **Solve maximum and minimum problems.**

ii **Solve problems by fitting a quadratic function to three data points.**

There are many ways in which quadratic functions are used in problem solving. In this section we will consider two of those methods.

i Maximum and Minimum Problems

For a quadratic function, the value $f(x)$ at the vertex will either be greater than any other $f(x)$ or less than all other $f(x)$. If the graph opens upward, $f(x)$ will be a minimum. If the graph opens downward, $f(x)$ will be a maximum. In certain problems we want to find a maximum or a minimum. If a problem situation can be translated to a quadratic function, we can solve by finding $f(x)$ at the vertex.

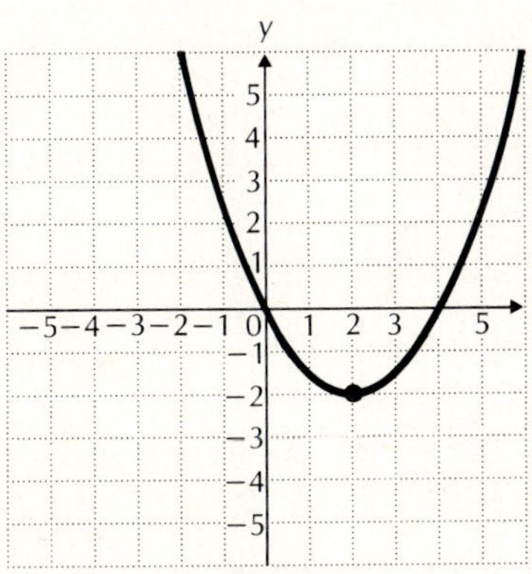

$f(x)$ at the vertex
a minimum

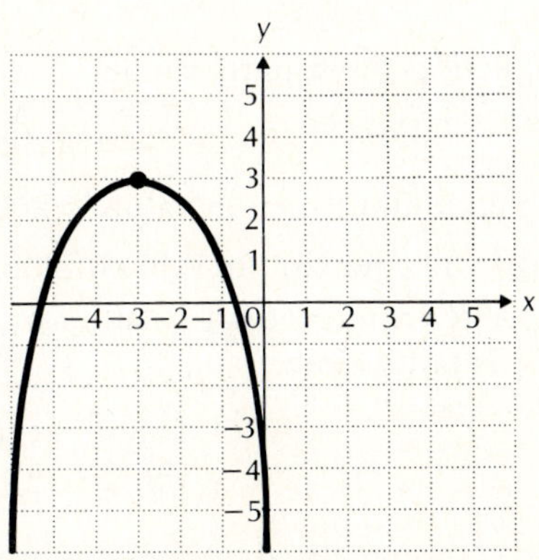

$f(x)$ at the vertex
a maximum

EXAMPLE 1 What are the dimensions of the largest rectangular pen that can be enclosed with 64 m of fence?

1. What is the maximum product of two numbers whose sum is 30?

1. Familiarize We make a drawing and label it.

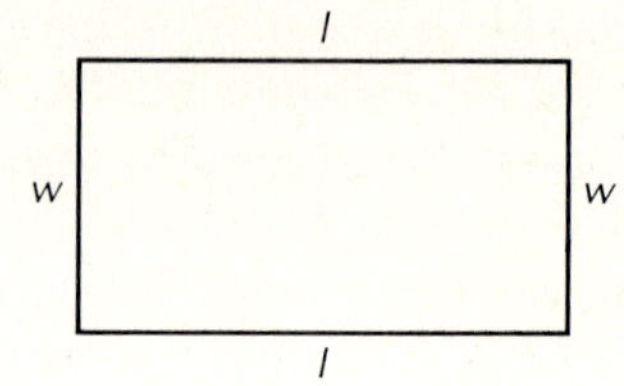

Perimeter: $2w + 2l = 64$ m

Area: $A = l \cdot w$

We may need to look up definitions or formulas in step 1. In addition, we may be able to get a better idea of the problem by looking at some possible dimensions for a rectangular pen that can be enclosed with 64 m of fence.

l	w	A
10	22	220
12	20	240
14	18	252
16	16	256
17	15	255
16.4	15.6	255.84

2. What are the dimensions of the largest rectangular pen that can be enclosed with 100 m of fence?

2. Translate We have a system of equations

$$2w + 2l = 64,$$
$$A = l \cdot w.$$

3. Carry out We will use substitution. Solving the first equation for l, we get $l = 32 - w$. Substituting for l in the second equation we get a quadratic function, $A(w)$, or just A:

$$A = (32 - w)w$$
$$A = -w^2 + 32w.$$

Completing the square, we get

$$A = -(w - 16)^2 + 256.$$

The maximum function value is 256.

4. Check The maximum value of 256 occurs when $w = 16$. We can check a function value for x less than 16 and for x greater than 16. That is not a 100% check, but it is fairly close:

$$A(15) = -(15)^2 + 32 \cdot 15 = 255,$$
$$A(17) = -(17)^2 + 32 \cdot 17 = 255.$$

Since 256 is greater than these numbers, it looks as though we have a maximum.

5. State The answer is that the dimensions of the largest pen that can be enclosed are 16 m by 16 m (the pen will be square).

DO EXERCISES 1 AND 2.

ii Fitting Quadratic Functions to Data

In many problem situations, we can make the translation by finding a quadratic function that describes how the variables behave. If we know that a quadratic function is appropriate in a certain situation, then we can proceed to find that function. If we know three inputs and their outputs, we can do so. Each such ordered pair is called a *data point.*

The following example provides a good illustration. (See Example 11 in Section 2.5.)

EXAMPLE 2 The instruction booklet for a video cassette recorder (VCR) includes a table relating the counter reading to the time the tape has run.

Time tape has run (hr)	Counter reading
0	000
1	300
2	500
3	675
4	800

Find a quadratic function that fits the data. Then find the counter reading after the tape has run for $1\frac{1}{2}$ hr.

We look for a function

$$N = at^2 + bt + c,$$

where N is the number of revolutions shown on the counter and t is the time in hours. We substitute some values of t and N:

$$0 = a(0)^2 + b(0) + c,$$
$$300 = a(1)^2 + b(1) + c,$$
$$500 = a(2)^2 + b(2) + c.$$

That gives us a system of three equations in the unknowns a, b, and c. When we solve that system, we find that

$$a = -50, \quad b = 350, \quad \text{and} \quad c = 0.$$

Therefore, the function we are looking for is

$$N = -50t^2 + 350t.$$

To find the counter reading after $1\frac{1}{2}$ hr, we substitute:

$$N = -50(1.5)^2 + 350(1.5)$$
$$= 412.5.$$

Thus after $1\frac{1}{2}$ hr the counter should read about 412.

DO EXERCISES 3 AND 4.

3. Find the quadratic function that fits the data points (1, 0), (−1, 4), and (2, 1).

4. The following table shows the accident records in a city. It has values that a quadratic function will fit.

Age of driver	Number of accidents (in a year)
20	400
40	150
60	400

a) Assuming that a quadratic function will describe the situation, find the number of accidents as a function of age.

b) Use the function to calculate the number of accidents in which a typical 16-year-old is involved.

EXERCISE SET 8.10

i Solve.

1. A rancher is fencing off a rectangular field with a fixed perimeter of 76 ft. What dimensions will yield the maximum area? What is the maximum area?

2. A carpenter is building a rectangular room with a fixed perimeter of 68 ft. What dimensions will yield the maximum area? What is the maximum area?

3. What is the maximum product of two numbers whose sum is 22? What numbers yield this product?

4. What is the maximum product of two numbers whose sum is 45? What numbers yield this product?

5. What is the minimum product of two numbers whose difference is 4? What are the numbers?

6. What is the minimum product of two numbers whose difference is 6? What are the numbers?

7. What is the minimum product of two numbers whose difference is 5? What are the numbers?

8. What is the minimum product of two numbers whose difference is 7? What are the numbers?

ii Find the quadratic function that fits each set of data points.

9. $(1, 4)$, $(-1, -2)$, $(2, 13)$

10. $(1, 4)$, $(-1, 6)$, $(-2, 16)$

11. $(1, 5)$, $(2, 9)$, $(3, 7)$

12. $(1, -4)$, $(2, -6)$, $(3, -6)$

13. (*Predicting earnings*). A business earns \$38 in the first week, \$66 in the second week, and \$86 in the third week. The manager graphs the points (1, 38), (2, 66), and (3, 86) and uses a quadratic function to describe the situation.

a) Find a quadratic function that fits the data.
b) Using the function, predict the earnings for the fourth week.

14. (*Predicting earnings*). A business earns \$1000 in its first month, \$2000 in the second month, and \$8000 in the third month. The manager plots the points (1, 1000), (2, 2000), and (3, 8000) and uses a quadratic function to describe the situation.

a) Find a quadratic function that fits the data.
b) Using the function, predict the earnings for the fourth month.

15. a) Find a quadratic function that fits the following data.

Travel speed in km/h	Number of daytime accidents (for every 200 million km)
60	100
80	130
100	200

b) Use the function to calculate the number of daytime accidents that occur at 50 km/h.

16. a) Find a quadratic function that fits the following data.

Travel speed in km/h	Number of nighttime accidents (for every 200 million km)
60	250
80	250
100	400

b) Use the function to calculate the number of nighttime accidents that occur at 50 km/h.

Find the maximum or minimum value for each function.

17. $f(x) = 2.31x^2 - 3.105x - 5.98$

18. $f(x) = -17.7x^2 + 6.29x + 6.08$

Find the quadratic function that fits each set of data points.

19. $(20.34, -5.86)$, $(34.67, -6.02)$, $(28.55, -8.46)$

20. $(0.789, 245)$, $(0.988, 350)$, $(1.233, 404)$

21. The sum of the base and the height of a triangle is 38 cm. Find the dimensions for which the area is a maximum, and find the maximum area.

22. The perimeter of a rectangle is 44 ft. Find the least possible length of a diagonal.

23. A horticulturist has 180 ft of fencing with which to form a rectangular garden. A greenhouse will provide one side of the garden, and the fencing will be used for the other three sides. What is the area of the largest region that can be enclosed?

24. (*Maximizing yield*). An orange grower finds that she gets an average yield of 40 bu per tree when she plants 20 trees on an acre of ground. Each time she adds a tree to an acre the yield per tree decreases by 1 bu, due to congestion. How many trees per acre should she plant for maximum yield?

25. (*Maximizing revenue*). When a theater owner charges \$2 for admission he averages 100 people attending. For each 10¢ increase in admission price the average number attending decreases by 1. What should he charge in order to make the most money?

26. A farmer wants to build a rectangular fence near a river, and will use 120 ft of fencing. What is the area of the largest region that can be enclosed? (The side next to the river is not fenced.)

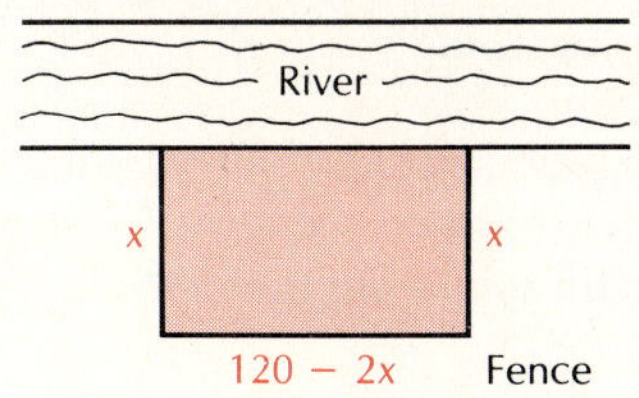

SUMMARY AND REVIEW: CHAPTER 8

The following contains a summary of what you should be able to do after completing this chapter. The review exercises are for practice. Answers are at the back of the book. If you miss an exercise, restudy the section indicated alongside the answer.

You should be able to:

Solve quadratic equations.

Solve.

1. $2x^2 - 7 = 0$ **2.** $14x^2 + 5x = 0$ **3.** $x^2 - 12x + 27 = 0$ **4.** $(x - 2)(x + 3) = 50$

Solve problems using quadratic equations.

Solve.

5. The hypotenuse of a right triangle is 41 m long. One leg is 31 m longer than the other. Find the lengths of the legs.

Use the quadratic formula to solve quadratic equations.

Solve.

6. $x^2 + 4x + 2 = 0$ **7.** $-3m^2 - 2m - 3 = 0$

8. $2x^2 = 7$ **9.** $(x - 3)^2 + (x - 1)^2 = 0$

Solve fractional equations that are quadratic.

Solve.

10. $x + 5 = \frac{1}{x - 5}$ **11.** $15 + \frac{6}{x - 2} = \frac{8}{x + 2}$ **12.** $\frac{x}{4} - \frac{4}{x} = 2$

Solve problems using fractional equations that are quadratic.

Solve.

13. An airplane travels 500 mi at a certain speed. A larger plane travels 800 mi at a speed 100 mi faster, and takes an hour less to reach its destination. Find the speed of each plane.

Determine the nature of the solutions of quadratic equations.

Determine the nature of the solutions.

14. $x^2 - 8 = 0$ **15.** $x^2 - 8x - 4 = 0$ **16.** $3x^2 + 5x + 5 = 0$ **17.** $y^2 = \frac{3}{5}y - \frac{1}{2}$

Solve problems using formulas.

Solve.

18. The area of a right circular cone is given by $A = \frac{1}{3}\pi r^2 h$. Solve for r.

19. The area of a right circular cone with height 8 is 45. Find the radius of the cone.

Solve equations reducible to quadratic.

Solve.

20. $x^4 - 12x^2 + 36 = 0$ **21.** $3x^{-2} - x^{-1} - 2 = 0$ **22.** $(x^2 - 4)^2 - (x^2 - 4) - 6 = 0$

Solve problems using quadratic variation.

Solve.

23. (*Area of an equilateral triangle*). The area of an equilateral triangle varies directly as the square of the length of a side. If an equilateral triangle has an area of 18.3 in^2 when the length of a side is 6.5 in., what will be the area when the length of a side is 7.4 in.?

Graph quadratic functions.

Graph.

24. $f(x) = 2(x + 2)^2$

25. $g(x) = -(x - 3)^2 + 2$

Solve maximum and minimum problems.

Solve.

26. The perimeter of a rectangle is 87. What dimensions will yield the maximum area of the rectangle?

27. Find the roots, if any exist, of the equation

$$x^2 + 2(k + 3)x + 12k = 0.$$

28. Find the equation whose roots are $0, \frac{b}{a}, \pm c$.

TEST: CHAPTER 8

Solve.

1. $x^2 + 8x - 33 = 0$

2. $(x - 3)(x + 5) = 84$

3. $x^2 - 3x + 5 = 0$

4. $(x + 4)^2 + (x - 5)^2 = 0$

5. $3x^2 - 7x + 14 = 0$

6. $x + 3 = \frac{2}{x - 3}$

7. $\frac{x}{2} - \frac{6}{x} = 4$

8. $x^2 = \frac{1}{2}x - \frac{5}{6}$

9. $x^4 - 8x^2 + 15 = 0$

10. $(x^2 - 1)^2 - 4(x^2 - 1) - 5 = 0$

11. The diagonal of a rectangle has a length of $\sqrt{34}$. The length of the rectangle is 2 more than the width. Find the dimensions of the rectangle.

12. A ladder 25 ft long leans against a wall. The bottom of the ladder is 7 ft from the wall. How much would the lower end of the ladder have to be pulled away so that the top end would be pulled down the same amount?

Graph.

13. $f(x) = -(x - 3)^2$

14. $g(x) = 2(x - 4)^2 - 2$

15. What is the minimum product of two numbers whose difference is 8? What are the numbers?

Problems concerning compound interest can be solved using exponential functions.

9 EXPONENTIAL AND LOGARITHM FUNCTIONS

The functions that we consider in this chapter are interesting not only from a purely intellectual point of view, but also for their applications to many fields. We will look at such applications as compound interest and business growth, but there are many others.

The basis of the theory concerns exponents. We define some functions having variable exponents (*exponential functions*); the rest follows from those functions and their properties.

OBJECTIVES

After finishing Section 9.1, you should be able to:

i **Graph exponential functions.**

ii **Graph inverses of exponential functions.**

iii **Find formulas for inverses of certain functions.**

9.1

EXPONENTIAL FUNCTIONS

i In Chapter 7 we gave meaning to expressions with rational-number exponents such as

$$5^{1/4}, \quad 3^{-3/4}, \quad 7^{2.34}, \quad 8^{1.73}.$$

For example, $5^{1.73}$, or $5^{173/100}$, means to raise 5 to the 173rd power and then take the 100th root. We now give meaning to expressions with irrational exponents, such as

$$5^{\sqrt{3}}, \quad 7^{\pi}, \quad 9^{-\sqrt{2}}.$$

Consider $5^{\sqrt{3}}$. Let us think of rational numbers r close to $\sqrt{3}$ and look at 5^r. As r gets closer to $\sqrt{3}$, 5^r gets closer to some real number.

r closes in on $\sqrt{3}$.	5^r closes in on some real number p.
r	5^r
$1 < \sqrt{3} < 2$	$5^1 < p < 5^2$
$1.7 < \sqrt{3} < 1.8$	$5^{1.7} < p < 5^{1.8}$
$1.73 < \sqrt{3} < 1.74$	$5^{1.73} < p < 5^{1.74}$
$1.732 < \sqrt{3} < 1.733$	$5^{1.732} < p < 5^{1.733}$

As r closes in on $\sqrt{3}$, 5^r closes in on some real number p. We define $5^{\sqrt{3}}$ to be the number p.

Any positive irrational exponent can be defined in a similar way. Negative irrational exponents are then defined in the same way as negative integer exponents. Then the expression a^x has meaning for any real number x. The usual laws of exponents still hold, but we will not prove that here. We now define exponential functions.

The function $f(x) = a^x$, where a is a positive constant, is called the *exponential function*, base a.

EXAMPLE 1 Graph the exponential function $y = 2^x$.

We compute some function values, thinking of y as $f(x)$:

$$f(1) = 2^1 = 2;$$

$$f(-2) = 2^{-2} = \frac{1}{2^2} = \frac{1}{4};$$

$$f(3) = 2^3 = 8.$$

We can keep a record of values in a table.

x	0	1	2	-1	-2	-3
y, or $f(x)$	1	2	4	$\frac{1}{2}$	$\frac{1}{4}$	$\frac{1}{8}$

Next, we plot these points and connect them with a smooth curve.

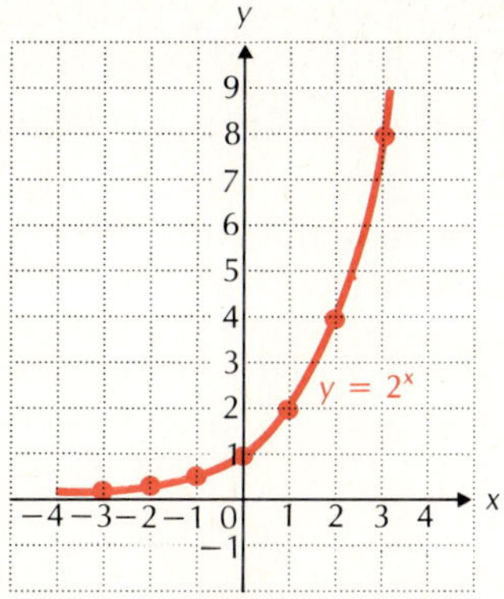

Note that as x increases, the function values increase indefinitely. As x decreases, the function values decrease, getting very close to 0. The x-axis is an *asymptote*, meaning that the curve comes very close to it.

DO EXERCISE 1.

EXAMPLE 2 Graph the exponential function $y = (\frac{1}{2})^x$.

We compute some function values, thinking of y as $f(x)$, and keep the results in a table:

$$f(2) = (\tfrac{1}{2})^2 = \tfrac{1}{4};$$
$$f(3) = (\tfrac{1}{2})^3 = \tfrac{1}{8};$$
$$f(0) = (\tfrac{1}{2})^0 = 1;$$
$$f(-3) = (\tfrac{1}{2})^{-3} = \frac{1}{(\frac{1}{2})^3} = \frac{1}{\frac{1}{8}} = 8.$$

x	-3	-2	-1	0	1	2	3
y, or $f(x)$	8	4	2	1	$\frac{1}{2}$	$\frac{1}{4}$	$\frac{1}{8}$

We plot these points and draw the curve. Note that this graph is a reflection of the one in Example 1.

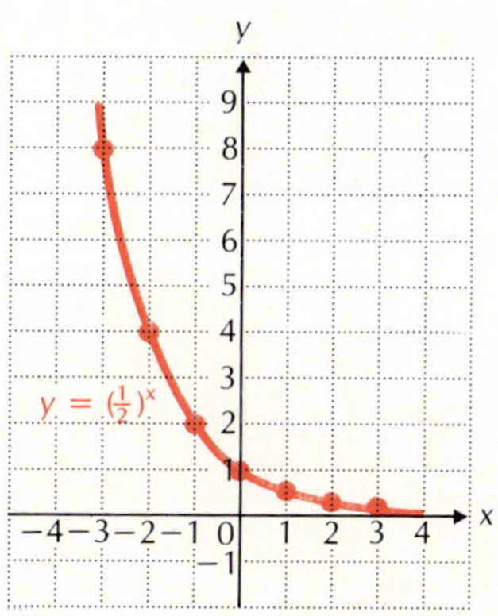

DO EXERCISE 2.

1. Graph $y = 3^x$.

a) Complete this table of solutions.

x	0	1	2	-1	-2	-3
y						

b) Plot the points from the table and connect them with a smooth curve.

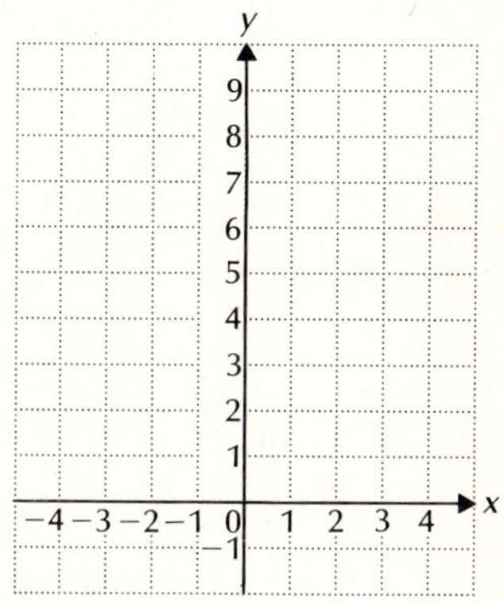

2. Graph $y = \left(\frac{1}{3}\right)^x$.

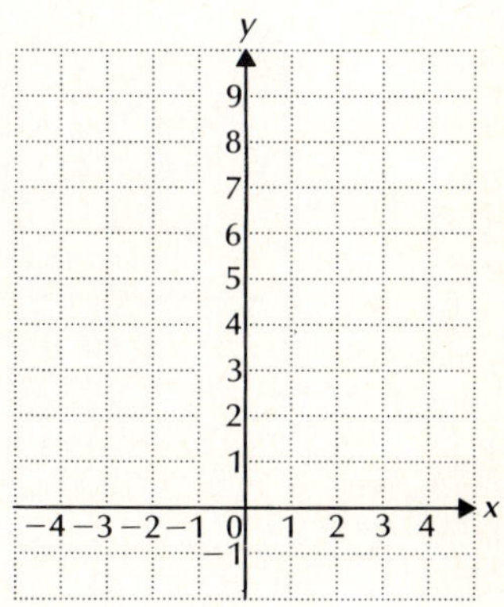

3. Graph $y = 2^{x+2}$.

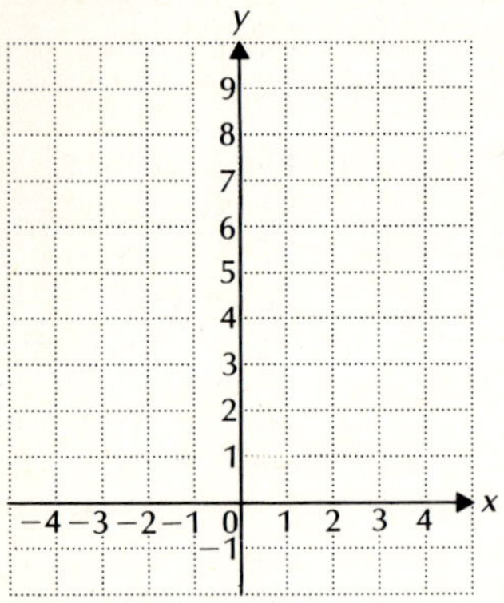

The preceding examples illustrate exponential functions of various bases. If $a = 1$, the graph is the line $y = 1$. For other positive values of a, the graphs are more interesting.

A When $a > 1$, the function $f(x) = a^x$ increases from left to right. See the color curves below. The greater the value of a, the steeper the curve.

B When $a < 1$, the function $f(x) = a^x$ decreases from left to right. See the black curve below. As a approaches 1, the curve becomes less steep. All such functions go through the point (0, 1).

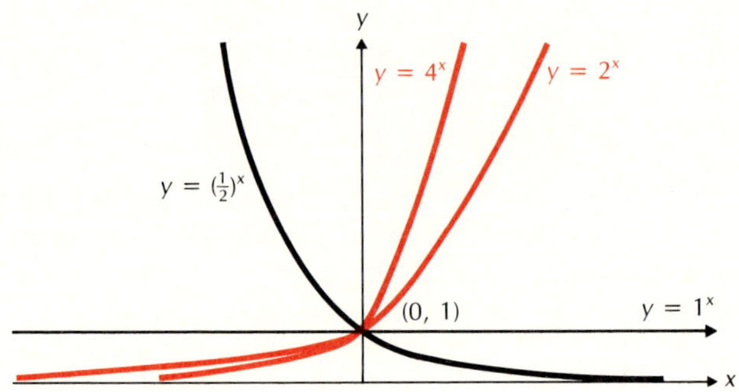

EXAMPLE 3 Graph: $y = 2^{x-2}$.

We know what the graph of $y = 2^x$ looks like. Here we have replaced x by $x - 2$. Hence we have a translation in the x-direction, 2 units to the right.

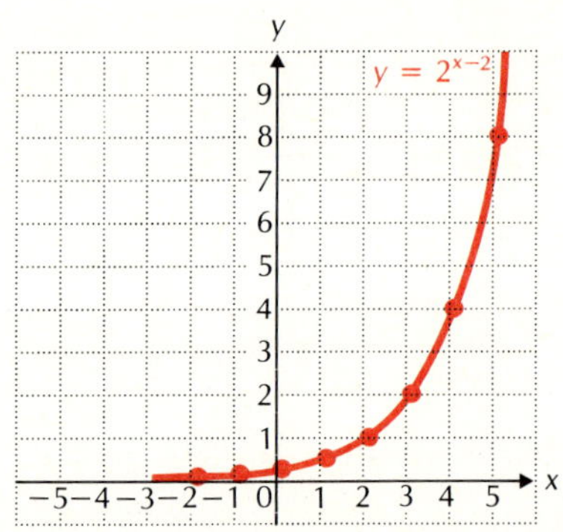

DO EXERCISE 3.

ii Inverses of Functions

If in $y = a^x$ we interchange x and y, we get another function, called the *inverse* of the exponential function.

EXAMPLE 4 Graph: $x = 2^y$.

To construct a table of values, we can use the one we constructed in Example 1, interchanging x and y.

x	1	2	4	$\frac{1}{2}$	$\frac{1}{4}$	$\frac{1}{8}$
y, or $f(x)$	0	1	2	-1	-2	-3

We plot these points and connect them with a smooth curve.

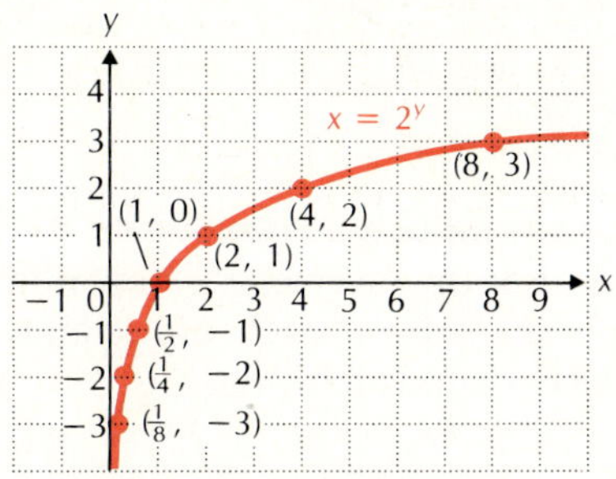

4. Graph $x = 3^y$.

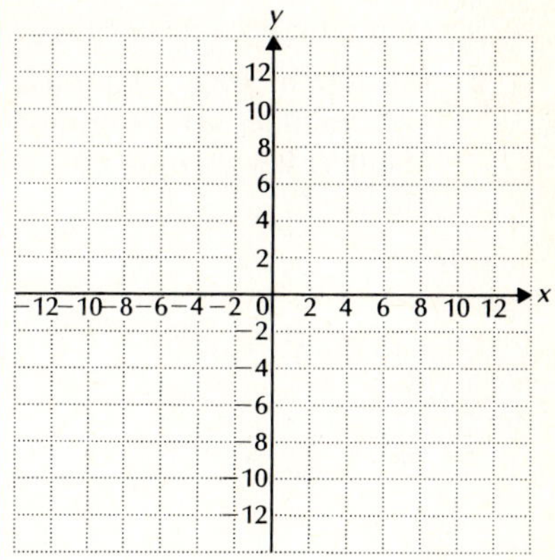

Note that this curve looks just like the graph of $y = 2^x$, except that it is reflected across the line $y = x$, as shown.

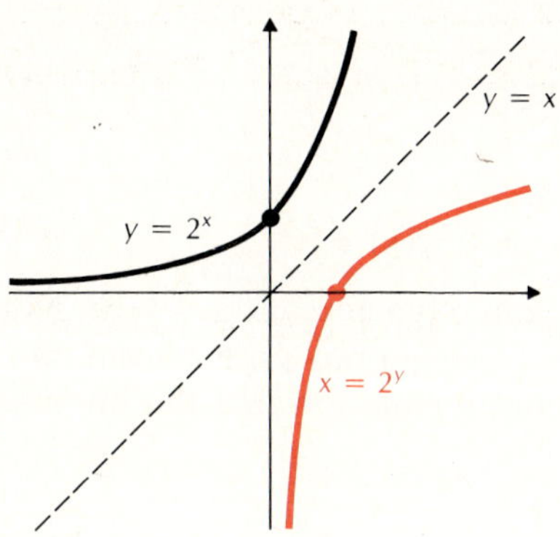

DO EXERCISE 4.

iii Formulas for Inverses

5. Let $f(x) = x - 5$. Find a formula for $f^{-1}(x)$.

When we interchange variables in any equation with two variables, we obtain an equation whose graph is a reflection of the original across the line $y = x$. If what we started with was an equation of a function f, the inverse may or may not also be a function. If the inverse is a function, it can be named f^{-1} (read "f-inverse"). Note that the -1 in f^{-1} is *not* an exponent.

EXAMPLE 5 Given $f(x) = x + 1$, find a formula for $f^{-1}(x)$.

a) Let us think of this as $y = x + 1$.

b) To find the inverse, we interchange x and y: $x = y + 1$.

c) Now we solve for y: $y = x - 1$.

d) Thus $f^{-1}(x) = x - 1$.

Note in Example 5 that any input x has an output $x + 1$ (this function adds 1 to each number of the domain). For the inverse, f^{-1}, any input x has an output $x - 1$ (this inverse function subtracts 1 from each member of its domain). Thus the function and its inverse do opposite things.

DO EXERCISE 5.

6. Let $q(x) = \dfrac{x+2}{3}$.

a) Find a formula for $q^{-1}(x)$.

b) Describe in words what each function does and determine whether they do opposite things.

EXAMPLE 6 Let $g(x) = 2x + 3$. Find a formula for $g^{-1}(x)$.

a) Let us think of this as $y = 2x + 3$.

b) To find the inverse, we interchange x and y: $x = 2y + 3$.

c) Now we solve for y: $y = (x - 3)/2$.

d) Thus $g^{-1}(x) = (x - 3)/2$.

Note in Example 6 that any input x has an output $2x + 3$ (this function doubles each input and then adds 3). For the inverse, g^{-1}, any input has an output $(x - 3)/2$ (this inverse function subtracts 3 from each input and then divides by 2). Thus the function and its inverse do opposite things.

DO EXERCISE 6.

To find a formula for the inverse of a function, if the inverse is also a function:

1. **Write y instead of $f(x)$.**
2. **Interchange x and y.**
3. **Solve for y.**
4. **Write $f^{-1}(x)$ instead of y.**

EXAMPLE 7 Find the inverse of $f(x) = x^2$.

To find the inverse we interchange x and y: $x = y^2$. Now we solve for y: $y = \pm\sqrt{x}$. Note that for each positive x we get two y's. For example, if x is 4, y can be either 2 or -2, as shown in the graph. Note too that the inverse fails the vertical-line test.

7. Find the inverse of $g(x) = x^4$. Is the inverse a function?

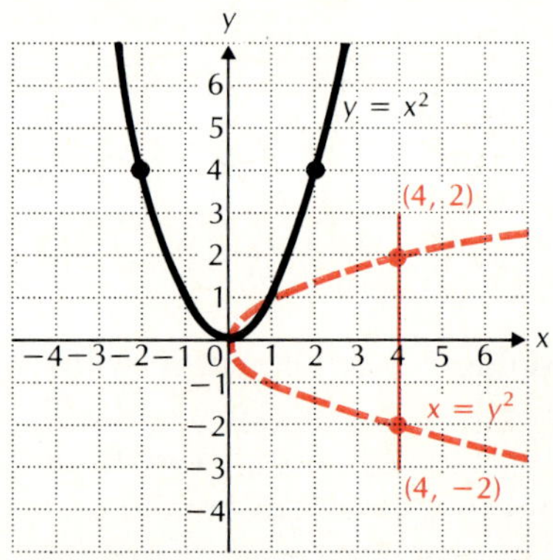

If we restrict the domain of $f(x) = x^2$ to nonnegative numbers, then its inverse is a function, $f^{-1}(x) = \sqrt{x}$.

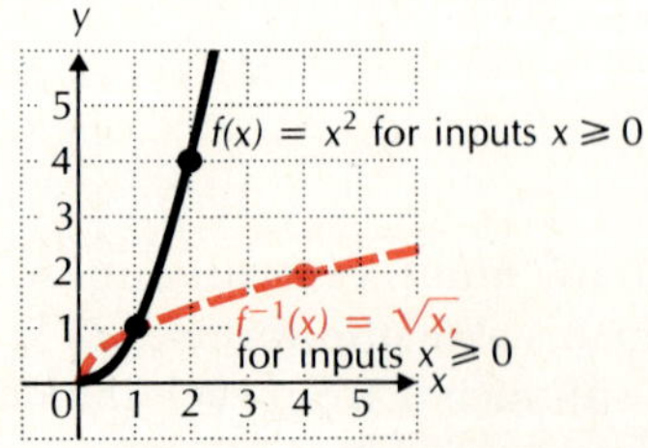

DO EXERCISE 7.

Let us consider inverses of functions in terms of function machines. Suppose that the function f programmed into a machine has an inverse that is also a function. Suppose then that the function machine has a reverse switch. When the switch is thrown, the machine is then programmed to do the inverse mapping f^{-1}. Inputs then enter at the opposite end, and the entire process is reversed.

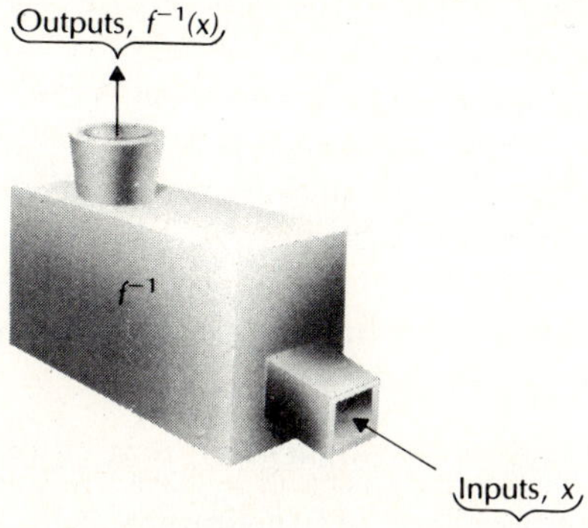

EXERCISE SET 9.1

i Graph.

1. $y = 2^x$
2. $y = 3^x$
3. $y = 5^x$
4. $y = 6^x$
5. $y = 2^{x+1}$
6. $y = 2^{x-1}$
7. $y = (\frac{1}{2})^x$
8. $y = (\frac{1}{3})^x$
9. $y = (\frac{1}{4})^x$
10. $y = (\frac{1}{5})^x$

ii Graph.

11. $x = 4^y$
12. $x = 5^y$
13. $x = (\frac{1}{4})^y$
14. $x = (\frac{1}{2})^y$

iii Find a formula for the inverse of each function.

15. $f(x) = x - 1$
16. $f(x) = x - 2$
17. $f(x) = x + 4$
18. $f(x) = x + 3$
19. $f(x) = x + 8$
20. $f(x) = x + 7$
21. $f(x) = 2x + 5$
22. $f(x) = 3x + 2$

Find a formula for the inverse of each function. Describe in words what each function does and determine whether the function and its inverse do opposite things.

23. $g(x) = \dfrac{2x+1}{3}$
24. $f(x) = \dfrac{3x+2}{5}$
25. $g(x) = 2x^2 + 3$
26. $f(x) = 3x^2 - 2$

Problem-solving practice

27. A 176-ft cable is to be divided into three pieces such that the second piece is 5 ft shorter than twice the length of the first piece and the length of the third piece is 19 less than twice the length of the second piece. How long is each piece of cable?

28. The telephone company charges \$0.40 for the first minute and \$0.25 for every other minute of a long distance call placed before 5 P.M. The rates after 5 P.M. drop to \$0.30 for the first minute and \$0.20 for every other minute of the call. A certain call placed before 5 P.M. cost \$5.65. The same length call placed to the same location after 5 P.M. cost \$4.50. How long did the call last?

○

Graph both equations using the same set of axes.

29. $y = (\frac{1}{4})^x, \quad x = (\frac{1}{4})^y$
30. $y = 3^{-(x-1)}, \quad x = 3^{-(y-1)}$

Graph.

31. $y = 2^{2x-1}$
32. $y = 3^x + 3^{-x}$
33. $y = 2^{-(x-1)^2}$
34. $y = |2^{x^2} - 1|$
35. $y = |2^x - 2|$

36. ▦ Approximate each of the following to six decimal places.

a) 2^3 b) $2^{3.1}$ c) $2^{3.14}$ d) $2^{3.141}$ e) $2^{3.1415}$ f) $2^{3.14159}$

▦ Determine which of the two numbers is larger.

37. 5^π or π^5
38. $\sqrt{8^3}$ or $8^{\sqrt{3}}$

OBJECTIVES

After finishing Section 9.2, you should be able to:

- **i** **Graph logarithm functions.**
- **ii** **Convert from exponential equations to logarithmic equations.**
- **iii** **Convert from logarithmic equations to exponential equations.**
- **iv** **Solve certain logarithmic equations.**
- **v** **Simplify expressions $a^{\log_a x}$ and $\log_a a^x$.**

9.2

LOGARITHM FUNCTIONS

In this section we consider a kind of function called a *logarithm function*, or *logarithmic function*. Such functions have many applications to problem solving.

i Graphs of Logarithm Functions

For any exponential function $y = a^x$, the inverse is called a *logarithm function.* The graph of the inverse can, of course, be obtained by reflecting the graph of $y = a^x$ across the line $y = x$.

The inverse of $y = a^x$ is $x = a^y$, but we use a special name and symbolism for the latter, as follows:

$$y = \log_a x \qquad \text{“logarithm, base } a\text{, of } x\text{.”}$$

To be consistent with our former use of function notation, we should probably write $y = \log_a (x)$, but unless the parentheses are actually needed for clarity they are ordinarily omitted. Almost always, we use a number a that is greater than 1 for a logarithm base.

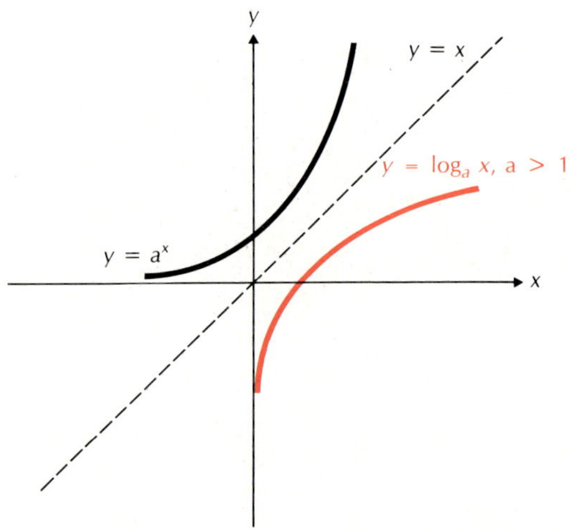

We define $y = \log_a x$ as $x = a^y$, where $x > 0$ and a is a positive constant other than 1.

It is helpful in dealing with logarithm functions to remember that the logarithm of a number is an *exponent.* It is the exponent y in a^y. You might also think to yourself, "The logarithm, base a, of a number x is the power to which a must be raised in order to get x."

EXAMPLE 1 Graph: $y = \log_5 x$.

The equation $y = \log_5 x$ is equivalent to $5^y = x$. The graph is a reflection of $y = 5^x$ across the line $y = x$. We make a table of values for $y = 5^x$ and then interchange x and y.

For $y = 5^x$:

x	0	1	2	-1	-2
y	1	5	25	$\frac{1}{5}$	$\frac{1}{25}$

For $y = \log_5 x$ (or $5^y = x$):

x	1	5	25	$\frac{1}{5}$	$\frac{1}{25}$
y	0	1	2	-1	-2

We plot the latter set of ordered pairs and connect them with a smooth curve. The graph of $y = 5^x$ has been shown only for reference.

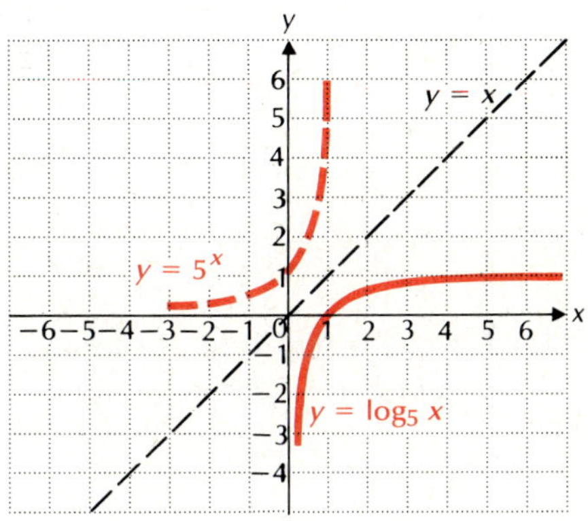

DO EXERCISE 1.

ii From Exponential to Logarithmic Equations

We use the definition of logarithm functions to convert from exponential to logarithmic equations.

EXAMPLES Convert to logarithmic equations.

2. $8 = 2^x \rightarrow x = \log_2 8$ The exponent is the logarithm. The base remains the base.

3. $y^{-1} = 4 \rightarrow -1 = \log_y 4$

4. $a^b = c \rightarrow b = \log_a c$

DO EXERCISES 2–5.

iii From Logarithmic to Exponential Equations

We also use the definition to convert from logarithmic to exponential equations.

EXAMPLES Convert to exponential equations.

5. $y = \log_3 5 \rightarrow 3^y = 5$ The logarithm is the exponent. The base does not change.

6. $-2 = \log_a 7 \rightarrow a^{-2} = 7$

7. $a = \log_b d \rightarrow b^a = d$

DO EXERCISES 6–9.

1. Graph $y = \log_3 x$.

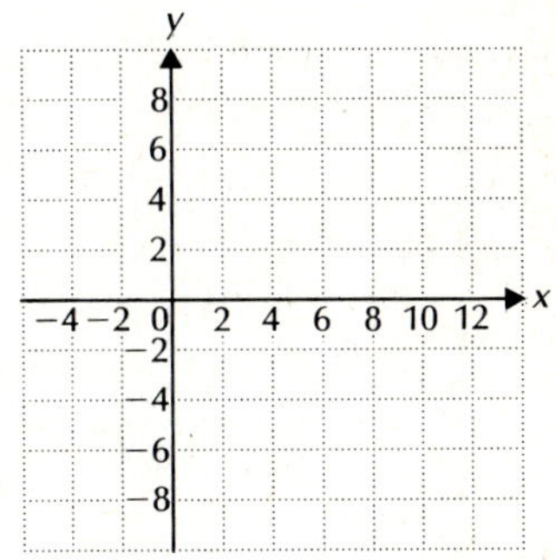

Convert to logarithmic equations.

2. $6^0 = 1$

3. $10^{-3} = 0.001$

4. $16^{0.25} = 2$

5. $m^T = P$

Convert to exponential equations.

6. $\log_2 32 = 5$

7. $\log_{10} 1000 = 3$

8. $\log_a Q = 7$

9. $\log_t M = x$

Solve.

10. $\log_{10} x = 4$

11. $\log_x 81 = 4$

Find the following.

12. $\log_{10} 10{,}000$

13. $\log_{10} 0.0001$

14. $\log_7 1$

iv Solving Logarithmic Equations

Certain equations involving logarithms can be solved by first converting to exponential equations.

EXAMPLE 8 Solve: $\log_2 x = -3$.

$$\log_2 x = -3$$

$$2^{-3} = x \quad \text{Converting to an exponential equation}$$

$$\tfrac{1}{8} = x \quad \text{Computing } 2^{-3}$$

This checks, so the solution is $\frac{1}{8}$.

EXAMPLE 9 Solve: $\log_x 16 = 2$.

$$\log_x 16 = 2$$

$$x^2 = 16 \quad \text{Converting to an exponential equation}$$

$$x = 4 \quad \text{or} \quad x = -4 \quad \text{Solving}$$

Check. $\log_4 16 = 2$ because $4^2 = 16$. Thus 4 is a solution. Since all logarithm bases must be positive, $\log_{-4} 16$ is not defined. Therefore, -4 is not a solution.

DO EXERCISES 10–11.

Solving an equation like $\log_b a = x$ amounts to finding the logarithm, base b, of the number a. You have done this before in graphing logarithm functions. To think of finding logarithms as solving equations may help in some cases.

EXAMPLE 10 Find $\log_{10} 1000$.

Let $\log_{10} 1000 = x$. Then,

$$10^x = 1000 \quad \text{Converting to an exponential equation}$$

$$x = 3.$$

Therefore, $\log_{10} 1000 = 3$.

EXAMPLE 11 Find $\log_{10} 0.01$.

Let $\log_{10} 0.01 = x$. Then,

$$10^x = 0.01 \quad \text{or} \quad 10^x = \tfrac{1}{100}, \quad \text{or} \quad 10^x = 10^{-2}$$

$$x = -2.$$

Therefore, $\log_{10} 0.01 = -2$.

EXAMPLE 12 Find $\log_5 1$.

Let $\log_5 1 = x$. Then,

$$5^x = 1 \quad \text{Converting to an exponential equation}$$

$$x = 0.$$

Therefore, $\log_5 1 = 0$.

DO EXERCISES 12–14.

v Simplifying $a^{\log_a x}$ and $\log_a a^x$

Let us recall that a function and its inverse (if the inverse is also a function) do opposite things. For a function f, suppose that we find a function value at x, that is, we find $f(x)$. Then we take that result and find the inverse function value for it. That is, we find $f^{-1}(f(x))$. We will then be back where we started, at the number x. Therefore, we have the following.

For any function f whose inverse f^{-1} is also a function:

$$f^{-1}(f(x)) = x$$

and

$$f(f^{-1}(x)) = x,$$

where x is any number for which the function values are defined.

The expressions

$$a^{\log_a x} \quad \text{and} \quad \log_a a^x$$

are expressions like those in the box above. Therefore, they both simplify to x.

For any number a suitable as a logarithm base:

$$a^{\log_a x} = x \quad \text{for any positive number } x,$$

and

$$\log_a a^x = x \quad \text{for any number } x.$$

You should remember this fact, because it will help in simplifying expressions and in solving equations.

EXAMPLES Simplify.

13. $2^{\log_2 5} = 5$

14. $10^{\log_{10} t} = t$

15. $\log_{10} 10^{5.6} = 5.6$

DO EXERCISES 15–22.

Simplify.

15. $5^{\log_5 3}$

16. $8^{\log_8 17}$

17. $e^{\log_e 4.2}$

18. $10^{\log_{10}(3x^2-4)}$

19. $\log_2 2^6$

20. $\log_{10} 10^{3.2}$

21. $\log_e e^{12}$

22. $\log_e e^{(5x-y^2)}$

EXERCISE SET 9.2

i Graph.

1. $y = \log_2 x$ **2.** $y = \log_{10} x$ **3.** $y = \log_6 x$ **4.** $y = \log_3 x$

ii Convert to logarithmic equations.

5. $10^3 = 1000$ **6.** $10^2 = 100$ **7.** $5^{-3} = \frac{1}{125}$ **8.** $4^{-5} = \frac{1}{1024}$

9. $8^{1/3} = 2$ **10.** $16^{1/4} = 2$ **11.** $10^{0.3010} = 2$ **12.** $10^{0.4771} = 3$

iii Convert to exponential equations.

13. $t = \log_3 8$ **14.** $h = \log_7 10$ **15.** $\log_5 25 = 2$ **16.** $\log_6 6 = 1$

17. $\log_{10} 0.1 = -1$ **18.** $\log_{10} 0.01 = -2$ **19.** $\log_{10} 7 = 0.845$ **20.** $\log_{10} 3 = 0.4771$

iv Solve.

21. $\log_3 x = 2$ **22.** $\log_4 x = 3$ **23.** $\log_x 16 = 2$ **24.** $\log_x 64 = 3$

25. $\log_2 x = -1$ **26.** $\log_3 x = -2$ **27.** $\log_8 x = \frac{1}{3}$ **28.** $\log_{32} x = \frac{1}{5}$

Find the following.

29. $\log_{10} 100$
30. $\log_{10} 100{,}000$
31. $\log_{10} 0.1$
32. $\log_{10} 0.001$
33. $\log_{10} 1$
34. $\log_{10} 10$
35. $\log_5 625$
36. $\log_2 64$
37. $\log_5 \frac{1}{25}$
38. $\log_2 \frac{1}{16}$
39. $\log_3 1$
40. $\log_4 4$

v Simplify.

41. $3^{\log_3 4}$
42. $7^{\log_7 10}$
43. $\log_t t^9$
44. $\log_p p^a$
45. $e^{\log_e (5x^2 + 7y^3)}$
46. $\log_e e^{(5x^4 + z^5)}$

Problem-solving practice

47. A right angle is divided into three different angles such that the second angle is 3 times the first and the third angle is 15 less than the second. Find the three angles.

48. A student's scores on four tests are 91, 93, 82, and 75. What must the score be on the fifth test in order that the student average 85?

49. Graph both equations using the same set of axes.

$$y = (\tfrac{3}{2})^x, \quad y = \log_{3/2} x$$

Graph.

50. $y = \log_2 (x - 1)$
51. $y = \log_3|x + 1|$

Solve.

52. $|\log_3 x| = 3$
53. $\log_{125} x = \frac{2}{3}$
54. $\log_\pi \pi^4 = x$
55. $\log_{\sqrt{5}} x = -3$
56. $\log_b b^{2x^2} = x$
57. $\log_4 (3x - 2) = 2$
58. $\log_8 (2x + 1) = -1$
59. $\log_x \sqrt[5]{36} = \frac{1}{10}$
60. $\log_{10} (x^2 + 21x) = 2$

Simplify.

61. $\log_{27} 9$
62. $\log_{1/4} \frac{1}{64}$
63. $\log_3 3^{x^2}$
64. $\log_{81} 3 \cdot \log_3 81$
65. $\log_{10} (\log_4 (\log_3 81))$
66. $8^{\log_8 (3x - 8)}$
67. $\log_2 (\log_2 (\log_4 256))$
68. $\log_{\sqrt{3}} \frac{1}{81}$
69. $\log_{1/5} 25$

OBJECTIVES

After finishing Section 9.3, you should be able to:

i Express the logarithm of a product as a sum of logarithms, and conversely.

ii Express the logarithm of a power as a product.

iii Express the logarithm of a quotient as a difference of logarithms, and conversely.

iv Convert from logarithms of products, quotients, and powers to expressions in terms of individual logarithms, and conversely.

9.3

PROPERTIES OF LOGARITHM FUNCTIONS

Logarithm functions have some interesting properties, properties that enable us to do some manipulations that we cannot do with functions in general.

i Logarithms of Products

One of the interesting properties special to logarithm functions is the following.

PROPERTY 1

For any positive numbers M and N,

$$\log_a M \cdot N = \log_a M + \log_a N.$$

(The logarithm of a product is the sum of the logarithms of the factors. The number a can be any logarithm base.)

EXAMPLE 1 Express as a sum of logarithms: $\log_2 (4 \cdot 16)$.

$$\log_2 (4 \cdot 16) = \log_2 4 + \log_2 16 \qquad \text{By Property 1}$$

EXAMPLE 2 Express as a single logarithm: $\log_{10} 0.01 + \log_{10} 1000$.

$$\log_{10} 0.01 + \log_{10} 1000 = \log_{10} (0.01 \times 1000) \quad \text{By Property 1}$$
$$= \log_{10} 10$$

DO EXERCISES 1–4.

A proof of Property 1 (optional). Let $\log_a M = x$ and $\log_a N = y$. Converting to exponential equations, we have $a^x = M$ and $a^y = N$.

Now we multiply the latter two equations, to obtain

$$M \cdot N = a^x \cdot a^y = a^{x+y}.$$

Converting back to a logarithmic equation, we get

$$\log_a M \cdot N = x + y.$$

Remembering what x and y represent, we get

$$\log_a M \cdot N = \log_a M + \log_a N,$$

which was to be shown.

ii Logarithms of Powers

The second basic property is as follows.

PROPERTY 2

For any positive number M and any real number p,

$$\log_a M^p = p \cdot \log_a M.$$

(The logarithm of a power of M is the exponent times the logarithm of M. The number a can be any logarithm base.)

EXAMPLES Express as a product.

3. $\log_a 9^{-5} = -5 \log_a 9$ By Property 2

4. $\log_a \sqrt[4]{5} = \log_a 5^{1/4}$ Writing exponential notation
$= \frac{1}{4} \log_a 5$ By Property 2

DO EXERCISES 5 AND 6.

A proof of Property 2 (optional). Let $x = \log_a M$. Then convert to an exponential equation, to get $a^x = M$. Raising both sides to the pth power, we obtain

$$(a^x)^p = M^p, \quad \text{or} \quad a^{xp} = M^p.$$

Converting back to a logarithmic equation, we get

$$\log_a M^p = xp.$$

But $x = \log_a M$, so

$$\log_a M^p = (\log_a M)p,$$

or

$$p \cdot \log_a M,$$

which was to be shown.

Express as a sum of logarithms.

1. $\log_5 (25 \cdot 5)$

2. $\log_b PQ$

Express as a single logarithm.

3. $\log_3 7 + \log_3 5$

4. $\log_a C + \log_a A + \log_a B + \log_a I + \log_a N$

Express as a product.

5. $\log_7 4^5$

6. $\log_a \sqrt{5}$

Express as a difference of logarithms.

7. $\log_b \frac{P}{Q}$

Express as a single logarithm.

8. $\log_2 x - \log_2 25$

Express in terms of logarithms of x, y, and z.

9. $\log_a \sqrt{\frac{z^3}{xy}}$

iii Logarithms of Quotients

Here is the third basic property.

PROPERTY 3

For any positive numbers M and N,

$$\log_a \frac{M}{N} = \log_a M - \log_a N.$$

(The logarithm of a quotient is the logarithm of the dividend minus the logarithm of the divisor. The number a can be any logarithm base.)

EXAMPLE 5 Express as a difference of logarithms: $\log_t \frac{6}{U}$.

$$\log_t \frac{6}{U} = \log_t 6 - \log_t U \qquad \text{By Property 3}$$

EXAMPLE 6 Express as a single logarithm: $\log_b 17 - \log_b 27$.

$$\log_b 17 - \log_b 27 = \log_b \frac{17}{27} \qquad \text{By Property 3}$$

EXAMPLE 7 Express as a single logarithm: $\log_{10} 10{,}000 - \log_{10} 100$.

$$\log_{10} 10{,}000 - \log_{10} 100 = \log_{10} \frac{10{,}000}{100} = \log_{10} 100$$

DO EXERCISES 7 AND 8.

iv Using the Properties Together

EXAMPLES Express in terms of logarithms of x, y, and z.

8.
$$\begin{aligned} \log_a \frac{x^2y^3}{z^4} &= \log_a (x^2y^3) - \log_a z^4 && \text{Using Property 3} \\ &= \log_a x^2 + \log_a y^3 - \log_a z^4 && \text{Using Property 1} \\ &= 2 \log_a x + 3 \log_a y - 4 \log_a z && \text{Using Property 2} \end{aligned}$$

9.
$$\begin{aligned} \log_a \sqrt[4]{\frac{xy}{z^3}} &= \log_a \left(\frac{xy}{z^3}\right)^{1/4} && \text{Writing exponential notation} \\ &= \frac{1}{4} \cdot \log_a \frac{xy}{z^3} && \text{Using Property 2} \\ &= \tfrac{1}{4}(\log_a xy - \log_a z^3) && \text{Using Property 3} \\ &= \tfrac{1}{4}(\log_a x + \log_a y - 3 \log_a z) && \text{Using Properties 1 and 2} \end{aligned}$$

DO EXERCISE 9.

EXAMPLES Express as a single logarithm.

10.
$$\begin{aligned} \tfrac{1}{2} \log_a x - 7 \log_a y + \log_a z &= \log_a \sqrt{x} - \log_a y^7 + \log_a z && \text{Using Property 2} \\ &= \log_a \frac{\sqrt{x}}{y^7} + \log_a z && \text{Using Property 3} \\ &= \log_a \frac{z\sqrt{x}}{y^7} && \text{Using Property 1} \end{aligned}$$

11. $\log_a \dfrac{b}{\sqrt{x}} + \log_a \sqrt{bx} = \log_a b - \log_a \sqrt{x} + \log_a \sqrt{bx}$ Using Property 3

$= \log_a b - \frac{1}{2}\log_a x + \frac{1}{2}\log_a (bx)$ Using Property 2

$= \log_a b - \frac{1}{2}\log_a x + \frac{1}{2}(\log_a b + \log_a x)$ Using Property 1

$= \log_a b - \frac{1}{2}\log_a x + \frac{1}{2}\log_a b + \frac{1}{2}\log_a x$

$= \frac{3}{2}\log_a b$ Collecting like terms

$= \log_a b^{3/2}$ Using Property 2

Example 11 could also be done as follows:

$$\log_a \frac{b}{\sqrt{x}} + \log_a \sqrt{bx} = \log_a \frac{b}{\sqrt{x}}\sqrt{bx} \quad \text{Using Property 1}$$
$$= \log_a b\sqrt{b}$$
$$= \log_a b^{3/2}$$

DO EXERCISE 10.

The following properties are important, and they do not depend on what the logarithm base happens to be.

For any base a:

$$\log_a a = 1 \quad \text{and} \quad \log_a 1 = 0.$$

Express as a single logarithm.

10. $5\log_a x - \log_a y + \dfrac{1}{4}\log_a z$

EXERCISE SET 9.3

i Express as a sum of logarithms.

1. $\log_2 (32 \cdot 8)$ **2.** $\log_3 (27 \cdot 81)$ **3.** $\log_4 (64 \cdot 16)$

4. $\log_5 (25 \cdot 125)$ **5.** $\log_c Bx$ **6.** $\log_t 5Y$

Express as a single logarithm.

7. $\log_a 6 + \log_a 70$ **8.** $\log_b 65 + \log_b 2$ **9.** $\log_c K + \log_c y$ **10.** $\log_t H + \log_t M$

ii Express as a product.

11. $\log_a x^3$ **12.** $\log_b t^5$ **13.** $\log_c y^6$ **14.** $\log_{10} y^7$ **15.** $\log_b C^{-3}$ **16.** $\log_c M^{-5}$

iii Express as a difference of logarithms.

17. $\log_a \dfrac{67}{5}$ **18.** $\log_t \dfrac{T}{7}$ **19.** $\log_b \dfrac{3}{4}$ **20.** $\log_a \dfrac{y}{x}$

iv Express as a single logarithm.

21. $\log_a 15 - \log_a 7$ **22.** $\log_b 42 - \log_b 7$

Express in terms of logarithms of x, y, and z.

23. $\log_a x^2y^3z$ **24.** $\log_a 5xy^4z^3$ **25.** $\log_b \dfrac{xy^2}{z^3}$ **26.** $\log_c \sqrt[3]{\dfrac{x^4}{y^3z^2}}$

Express as a single logarithm and simplify if possible.

27. $\frac{2}{3}\log_a x - \frac{1}{2}\log_a y$ **28.** $\frac{1}{2}\log_a x + 3\log_a y - 2\log_a x$ **29.** $\log_a 2x + 3(\log_a x - \log_a y)$

30. $\log_a x^2 - 2\log_a \sqrt{x}$ **31.** $\log_a \dfrac{a}{\sqrt{x}} - \log_a \sqrt{ax}$ **32.** $\log_a (x^2 - 4) - \log_a (x - 2)$

Given $\log_{10} 2 = 0.301$, $\log_{10} 3 = 0.477$, and $\log_{10} 10 = 1$, find the following.

33. $\log_{10} 4$ **34.** $\log_{10} 5$ **35.** $\log_{10} 50$ **36.** $\log_{10} 12$ **37.** $\log_{10} 60$ **38.** $\log_{10} \frac{1}{3}$

39. $\log_{10} \sqrt{\frac{2}{3}}$ **40.** $\log_{10} \sqrt[5]{12}$ **41.** $\log_{10} 90$ **42.** $\log_{10} \frac{9}{8}$ **43.** $\log_{10} \frac{1}{4}$ **44.** $\log_{10} \frac{9}{10}$

Problem-solving practice

45. A board 5 m long is divided into two pieces such that one piece is $\frac{3}{8}$ as long as the other. Find the length of each piece.

46. A rectangular swimming pool is 45 ft long and has a uniform depth of 10 ft. If the pool holds 15,750 ft^3 of water, how wide is the pool?

Express as a single logarithm and simplify if possible.

47. $\log_a (x^8 - y^8) - \log_a (x^2 + y^2)$

48. $\log_a (x + y) + \log_a (x^2 - xy + y^2)$

Express as a sum or difference of logarithms.

49. $\log_a \sqrt{1 - s^2}$

50. $\log_a \dfrac{c - d}{\sqrt{c^2 - d^2}}$

51. If $\log_a x = 2$, $\log_a y = 3$, and $\log_a z = 4$, what is $\log_a \dfrac{\sqrt[3]{x^2 z}}{\sqrt[3]{y^2 z^{-2}}}$?

Solve.

52. $\log_\pi \pi^{2x+3} = 4$ **53.** $3^{\log_3 (8x-4)} = 5$ **54.** $4^{2 \log_4 x} = 7$

55. $8^{2 \log_8 x + \log_8 x} = 27$ **56.** $(x + 3) \cdot \log_a a^x = x$ **57.** $\log_a 5x = \log_a 5 + \log_a x$

58. If $\log_a x = 2$, what is $\log_a \left(\dfrac{1}{x}\right)$?

59. If $\log_a x = 2$, what is $\log_{1/a} x$?

OBJECTIVES

After finishing Section 9.4, you should be able to:

i Find logarithms and antilogarithms, base 10, on a calculator.

ii Find logarithms and antilogarithms, base *e*, on a calculator.

iii Find logarithms and antilogarithms, base 10, using a table.

iv Find logarithms and antilogarithms, base *e*, using a base-10 table.

Using a calculator, find each of the following.

1. log 8,021,000

2. log 0.000895

9.4 LOGARITHM FUNCTION VALUES; TABLES AND CALCULATORS

i Common Logarithms on a Calculator

Logarithms to the base 10 are called *common logarithms.* They were invented many years ago for the purpose of simplifying computations. Today they are no longer used for that purpose, but are used instead to solve certain equations and certain kinds of problems.

The abbreviation **log** is generally used with common logarithms. Thus the symbol

$$\log 23$$

means $\log_{10} 23$. On scientific calculators the key for the common logarithm function is marked [log]. To find the common logarithm of a number, you enter that number and then press the [log] key.

EXAMPLE 1 Find log 475,000.

We enter 475,000 and then press the [log] key. We find that $\log 475{,}000 \approx 5.6767$ (we have rounded the answer).

EXAMPLE 2 Find log 0.00372.

We enter 0.00372 and then press the [log] key. We find that $\log 0.00372 \approx -2.4295$ (we have again rounded the answer).

DO EXERCISES 1 AND 2.

Antilogarithms. The inverse of a logarithm function is, of course, an exponential function. The inverse of finding a logarithm is also called finding an *antilogarithm.* To find an antilogarithm, we *exponentiate*:

$$\text{antilog}_{10}\, x = 10^x.$$

Generally, there is no key on a calculator marked "antilog." It is up to you to know that to find the inverse, or antilogarithm, you must use the $\boxed{10^x}$ key, if there is one. If there is no such key, then you must raise 10 to the x power using an exponential key.

EXAMPLE 3 Find the antilogarithm, base 10, of 3.2546.

a) (*Using the* 10^x *key*). We enter 3.2546 and then press the $\boxed{10^x}$ key. We find that

$$\text{antilog}\, 3.2546 = 10^{3.2546} \approx 1797.21.$$

b) (*Using an exponential key*). Enter 3.2546 and also 10. Then, pressing keys in the order appropriate for your particular calculator (you must read the instructions), you will find that

$$\text{antilog}\, 3.2546 = 10^{3.2546} \approx 1797.21.$$

DO EXERCISES 3 AND 4.

Using a calculator, find each of the following.

3. $\text{antilog}_{10}\, 5.1053$

4. $\text{antilog}_{10}\, 0.001832$

ii Natural Logarithms on a Calculator

A logarithm base called e occurs in a great many applications. The number e may seem like a strange one to use as a logarithm base, because it is an irrational number. Therefore, its decimal representation does not terminate or repeat:

$$e \approx 2.71828\ldots.$$

Logarithms to the base e are called *natural logarithms.*

The abbreviation **ln** is generally used with natural logarithms. Thus the symbol

$$\ln 23$$

denotes the logarithm, base e, of 23.

On scientific calculators the key for the natural logarithm function is marked $\boxed{\ln}$.

EXAMPLE 4 Find ln 3568.

We enter 3568 and then press the $\boxed{\ln}$ key. We find that $\ln 3568 \approx 8.1798$ (we have rounded the answer).

EXAMPLE 5 Find ln 0.0007659.

We enter 0.0007659 and then press the $\boxed{\ln}$ key. We find that $\ln 0.0007659 \approx -7.1745$ (again, we have rounded the answer).

DO EXERCISES 5 AND 6.

Using a calculator, find each of the following.

5. ln 65,030

6. ln 0.001042

Antilogarithms. To find the antilogarithm, base e, we use the $\boxed{e^x}$ key, if there is one.

EXAMPLE 6 Find the antilogarithm, base e, of 2.4837.

a) (*Using the* e^x *key*). We enter 2.4837 and then press the $\boxed{e^x}$ key. We find that

$$\text{antilog}_e\, 2.4837 = e^{2.4837} \approx 11.9855.$$

Using a calculator, find each of the following.

7. antilog_e 5.1485

8. antilog_e 0.05051

b) (*Using an exponential key*). Enter 2.4837 and also an approximate value of e, say, 2.71828. Then, pressing keys in the order appropriate for your calculator (you must read the instructions), you will find that

$$\text{antilog}_e\ 2.4837 = e^{2.4837} \approx 11.9855.$$

DO EXERCISES 7 AND 8.

iii Tables of Common Logarithms (Optional)

Table 2 at the back of the book is a table of common (base 10) logarithms, for numbers from 1 to 10. Part of that table is shown here.

x	0	1	2	3	4	5	6	7	8	9
5.0	0.6990	0.6998	0.7007	0.7016	0.7024	0.7033	0.7042	0.7050	0.7059	0.7067
5.1	0.7076	0.7084	0.7093	0.7101	0.7110	0.7118	0.7126	0.7135	0.7143	0.7152
5.2	0.7160	0.7168	0.7177	0.7185	0.7193	0.7202	0.7210	0.7218	0.7226	0.7235
5.3	0.7243	0.7251	0.7259	0.7267	0.7275	0.7284	0.7292	0.7300	0.7308	0.7316
5.4	0.7324	0.7332	0.7340	0.7348	0.7356	0.7364	0.7372	0.7380	0.7388	0.7396

Using Table 2, find each of the following.

9. log 4.38

10. log 9.22

The values in the table are the same as those found on a calculator, except that the table values have been rounded to four decimal places.

At this point a question will surely arise: "Since calculators are rather easily available, why bother learning to use a table?" The answer is that, if all you are interested in is obtaining values to solve problems when a calculator is available, then there is little reason to fuss with a table. The educational value in learning to use a table lies in essentially two realms:

1. So that you could use a table if necessary, when a calculator is inoperative or unavailable.
2. So that you have some facility in working with tables in general. There are many functions, including functions that you might devise for yourself, through experiment in business or science, whose values are not stored in a calculator. For such functions, tables are important, and your ability to use tables is therefore also important. The following material on the use of the tables is optional.

Now let us see how to use the table of common logarithms.

Using Table 2, find each of the following.

11. antilog 0.8882

12. antilog 0.4757

EXAMPLE 7 Find log 5.24.

At the left of the table, we find the row headed 5.2. Then we move across to the column headed (at the top of the table) 4. We find log 5.24 (the color entry in the table):

$$\log 5.24 \approx 0.7193.$$

DO EXERCISES 9 AND 10.

EXAMPLE 8 Find antilog 0.7193.

We reverse the procedure used in Example 7. We first look for 0.7193 in the body of the table (the color entry). We find the heading 5.2 at the left and the heading 4 at the top:

$$\text{antilog}\ 0.7193 = 10^{0.7193} \approx 5.24.$$

DO EXERCISES 11 AND 12.

Numbers Not Between 1 and 10. Table 2 contains logarithms of numbers from 1 to 10 only. How do we find logarithms of other numbers? The answer is that we write scientific notation for the number and then use the properties of the logarithm function.

First, we should note that for powers of ten, the common logarithms are integers. For example:

$\log 10 = 1,$ because $10 = 10^1$ (the log is the exponent);
$\log 100 = 2,$ because $100 = 10^2$;
$\log 1000 = 3,$ because $1000 = 10^3$;
$\log 0.1 = -1,$ because $0.1 = 10^{-1}$;
$\log 0.001 = -3,$ because $0.001 = 10^{-3}$.

EXAMPLE 9 Find log 5430.

We first write scientific notation for the number:

$$5430 = 5.43 \times 10^3.$$

By Property 1,

$$\begin{aligned}\log(5.43 \times 10^3) &= \log 5.43 + \log 10^3\\ &= \log 5.43 + 3 && \text{The log of } 10^3 \text{ is 3.}\\ &\approx 0.7348 + 3 && \text{Finding log 5.43 from the table}\\ &\approx 3.7348 && \text{Adding}\end{aligned}$$

EXAMPLE 10 Find log 0.000507.

We first write scientific notation for the number:

$$0.000507 = 5.07 \times 10^{-4}.$$

By Property 1,

$$\begin{aligned}\log(5.07 \times 10^{-4}) &= \log 5.07 + \log 10^{-4}\\ &= \log 5.07 + (-4) && \text{The log of } 10^{-4} \text{ is } -4.\\ &\approx 0.7050 + (-4) && \text{Finding log 5.07 from the table}\\ &\approx -3.2950 && \text{Doing the addition}\end{aligned}$$

DO EXERCISES 13 AND 14.

Antilogarithms. To find antilogarithms, we reverse the procedure of Example 9 or Example 10.

EXAMPLE 11 Find antilog 4.7308.

By the definition of antilog, we have

$$\begin{aligned}\text{antilog } 4.7308 &= 10^{4.7308}\\ &= 10^{4+0.7308} && \text{Separating the exponent into an integer and a number between 0 and 1}\\ &= 10^{0.7308} \times 10^4 && \text{Using a property of exponents } (a^x \cdot a^y = a^{x+y})\end{aligned}$$

Now, using the table, we find $10^{0.7308}$. It is 5.38. Therefore, antilog 4.7308 ≈ 5.38×10^4, or 53,800.

Using Table 2, find each of the following.

13. log 7,640

14. log 0.00105

Using Table 2, find each of the following.

15. antilog 5.9232

16. antilog (-2.3261)

EXAMPLE 12 Find antilog (-1.2874).

By the definition of antilog, we have

$$\text{antilog}\,(-1.2874) = 10^{-1.2874}.$$

We need to separate -1.2874 into two parts, one of which is an integer and one of which is a number between 0 and 1. We do that by adding and subtracting 2 (or any integer greater than 1):

$$2 - 1.2874 - 2.$$

Doing the subtraction $2 - 1.2874$, we get

$$0.7126 - 2.$$

Now we have

$$\begin{aligned}\text{antilog}\,(-1.2874) &= \text{antilog}\,(0.7126 - 2)\\ &= 10^{(0.7126-2)}\\ &= 10^{0.7126} \times 10^{-2} \qquad \text{Using a property of exponents}\end{aligned}$$

Using the table, we find $10^{0.7126}$. It is 5.16. Therefore,

$$\begin{aligned}\text{antilog}\,(-1.2874) &= \text{antilog}\,(0.7126 - 2)\\ &\approx 5.16 \times 10^{-2}, \quad \text{or } 0.0516.\end{aligned}$$

DO EXERCISES 15 AND 16.

iv Natural Logarithms and Tables (Optional)

Using Table 2, find each of the following.

17. ln 86,200

18. ln 0.0514

There are tables of natural logarithms, but they are not nearly as easy to use as tables of common logarithms. The reason for the difficulty lies in the fact that natural logarithms of powers of 10 are not integers. To find the natural logarithm of a number from a table, it is usually easier to find the base-10 logarithm from a table and then apply a conversion formula, as follows.

TO CONVERT FROM COMMON LOGARITHMS TO NATURAL LOGARITHMS

For a positive number *M*,

$$\ln M = \frac{\log M}{\log e} \approx 2.3026 \log M.$$

EXAMPLE 13 Find ln 5480.

We first find log 5480, or 5.48×10^3:

$$\log 5480 \approx 3.7388.$$

We then multiply by 2.3026, and obtain 8.6090. Therefore,

$$\ln 5480 \approx 8.6090.$$

DO EXERCISES 17 AND 18.

Antilogarithms. The natural antilogarithm of a number x is, of course, e^x. If we take the common logarithm we obtain $\log e^x$. If we then take the base-10 antilogarithm, we have

$$10^{\log e^x}, \text{ which is also } e^x.$$

Also, by Property 2,

$$10^{\log e^x} = 10^{x \log e}.$$

All of this tells us that to find the natural antilogarithm e^x of a number x, we can multiply x by $\log e$ and then take the base-10 antilogarithm.

TO FIND THE BASE-e ANTILOGARITHM

To find e^x (the antilogarithm of x), multiply x by $\log e$ (which is about 0.4343) and then find the base-10 antilogarithm.

EXAMPLE 14 Find $\text{antilog}_e\ 8.5867$.

We first multiply by $\log e$, which is about 0.4343:

$$8.5867 \times 0.4343 \approx 3.7292.$$

Now we find the base-10 antilog of 3.7292. It is

$$5.36 \times 10^3, \quad \text{or} \quad 5360.$$

Thus

$$\text{antilog}_e\ 8.5867 = e^{8.5867} \approx 5360.$$

DO EXERCISES 19 AND 20.

Using Table 2, find each of the following.

19. $\text{antilog}_e\ 10.6827$

20. $\text{antilog}_e\ 0.5596$

EXERCISE SET 9.4

i Use a calculator to find the following common logarithms and antilogarithms.

1. log 6.34 **2.** log 5.02 **3.** log 45 **4.** log 74
5. log 62.4 **6.** log 11.4 **7.** log 437 **8.** log 295
9. log 13,400 **10.** log 93,100 **11.** log 0.57 **12.** log 0.69
13. log 0.052 **14.** log 0.387 **15.** log 0.00621 **16.** log 0.00483
17. antilog 3 **18.** antilog 5 **19.** antilog 2.7 **20.** antilog (−4.8)
21. antilog 0.4771 **22.** antilog (−0.6532) **23.** antilog (−0.5465) **24.** antilog (−0.3404)
25. antilog (−2.9523) **26.** antilog (−4.8982)

ii Find the following logarithms and antilogarithms, base e, using a calculator.

27. ln 62 **28.** ln 30 **29.** ln 4365 **30.** ln 901.2
31. ln 0.0062 **32.** ln 0.00073 **33.** $\text{antilog}_e\ 3.6052$ **34.** $\text{antilog}_e\ 4.9312$
35. $\text{antilog}_e\ (-6.0751)$ **36.** $\text{antilog}_e\ (-2.3001)$

iii Using the table on page 413, find the following common logarithms and antilogarithms.

37. log 6.23 **38.** log 4.85 **39.** log 9.87 **40.** log 1.42
41. log 3.41 **42.** log 7.92 **43.** log 2.74 **44.** log 5.33
45. antilog 0.6998 **46.** antilog 0.8727 **47.** antilog 0.4518 **48.** antilog 0.3075
49. antilog 0.7825 **50.** antilog 0.4281 **51.** antilog 0.2430 **52.** antilog 0.9763

Using the table on page 413, and the properties of logarithms, find the following common logarithms and antilogarithms.

53. log 48.3 **54.** log 69.4 **55.** log 567 **56.** log 935

57. log 41,200 **58.** log 58,700 **59.** log 0.00642 **60.** log 0.000333

61. antilog 1.9969 **62.** antilog 3.7789 **63.** antilog 2.4502 **64.** antilog 4.0531

65. antilog (−0.4473) **66.** antilog (−0.0846) **67.** antilog (−2.1952) **68.** antilog (−3.6556)

iv Using the table on page 413, and the properties of logarithms, find the following natural logarithms and antilogarithms.

69. ln 3460 **70.** ln 9030 **71.** ln 0.0351 **72.** ln 0.00468

73. antilog_e 7.4012 **74.** antilog_e 6.3058 **75.** antilog_e (−2.0325) **76.** antilog_e (−1.3783)

Solve for x.

77. $\log 374x = 4.2931$ **78.** $\log 95x^2 = 3.0177$ **79.** $\log 692 + \log x = \log 3450$

80. $\dfrac{4.31}{\ln x} = \dfrac{28}{3.01}$ **81.** $\sqrt{x\sqrt[3]{x\sqrt[4]{x\sqrt[5]{x}}}} = 146$

OBJECTIVES

After finishing Section 9.5, you should be able to:

i **Use interpolation and Table 2 to find logarithms when four digits occur.**

ii **Use interpolation and Table 2 to find antilogarithms when four digits occur.**

9.5 INTERPOLATION (OPTIONAL)

What we do in this section applies not only to tables of logarithms, but to tables in general. By a process called *interpolation,* we will find values between those given in a table.

No calculator or computer will ever hold all tables. What we do here is important for using tables in general. There are many functions, including those that you might devise for yourself in business or science, whose values are not stored in a calculator or computer.

Linear Interpolation

Let us consider how a table of values for any function is made. We select members of the domain x_1, x_2, x_3, and so on. Then we compute or somehow determine the corresponding function values $f(x_1)$, $f(x_2)$, $f(x_3)$, and so on. Then we tabulate the results. We might also graph the results.

x	x_1	x_2	x_3	x_4	...
$f(x)$	$f(x_1)$	$f(x_2)$	$f(x_3)$	$f(x_4)$	...

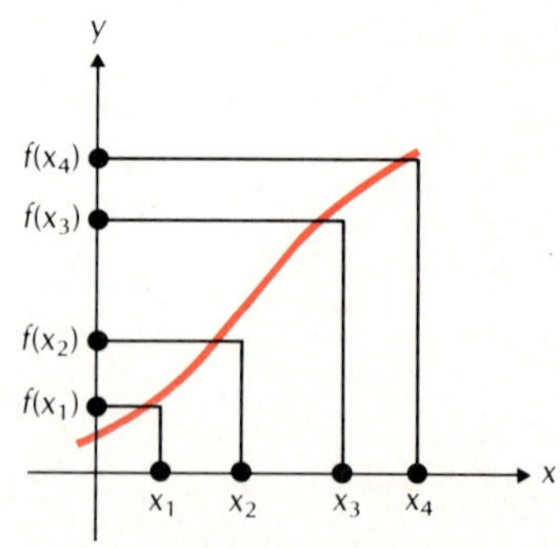

Suppose we want to find the function value $f(x)$ for an x not in the table. If x is halfway between x_1 and x_2, then we can take the number halfway between $f(x_1)$ and $f(x_2)$ as an approximation to $f(x)$. If x is one-fifth of the way between x_2 and

x_3, we take the number that is one-fifth of the way between $f(x_2)$ and $f(x_3)$ as an approximation to $f(x)$. What we do is divide the length from x_2 to x_3 in a certain ratio, and then divide the length from $f(x_2)$ to $f(x_3)$ in the same ratio. This is *linear interpolation.*

We can show this geometrically. The length from x_1 to x_2 is divided in a certain ratio by x. The length from $f(x_1)$ to $f(x_2)$ is divided in the same ratio by y. The number y approximates $f(x)$ with the noted error.

Note the slanted line in the figure. The approximation y comes from this line. This explains the use of the term *linear interpolation.* Let us apply linear interpolation to Table 2 of common logarithms.

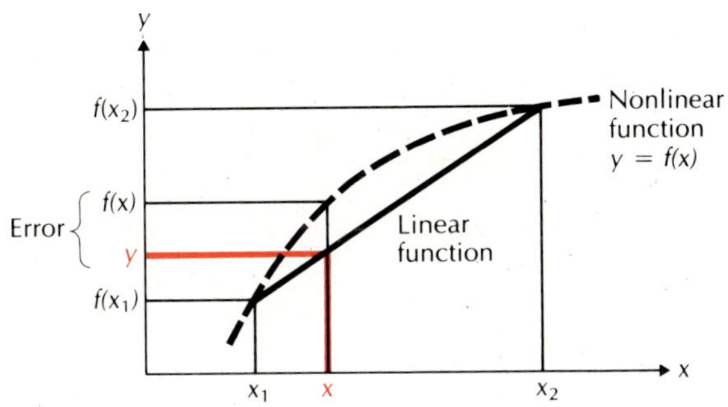

1. Find log 4562.

i Logarithms

EXAMPLE 1 Find log 34870.

a) We write scientific notation for the number:

$$3.487 \times 10^4.$$

Then

$$\log (3.487 \times 10^4) = \log 3.487 + 4.$$

b) We find log 3.487. From Table 2, we have the following:

$$\log 3.480 \approx 0.5416$$

0.007

0.010 $\qquad \log 3.487 \approx 0.54??$ $\qquad$ The tabular difference is 0.0012.

$$\log 3.490 \approx 0.5428$$

The tabular difference (the difference between consecutive values in the table) is 0.0012. Now 3.487 is $\frac{7}{10}$ of the way from 3.480 to 3.490. So we take 0.7 of 0.0012, which is 0.00084, and round it to 0.0008. We add this to 0.5416. Thus $\log 3.487 \approx 0.5424$.

c) We add the integer 4 from step (a):

$$\log (3.487 \times 10^4) \approx 4.5424.$$

DO EXERCISE 1.

EXAMPLE 2 Find log 0.009543.

a) We write scientific notation for the number:

$$9.543 \times 10^{-3}.$$

2. Find log 0.02387.

Then

$$\log (9.543 \times 10^{-3}) = \log 9.543 - 3.$$

b) We find log 9.543. From Table 2 we have the following:

$$\left.\begin{array}{ll} & \log 9.540 \approx 0.9795 \\ 0.010 \quad 0.003 & \\ & \log 9.543 \approx 0.97?? \\ & \log 9.550 \approx 0.9800 \end{array}\right\}$$

The tabular difference is 0.0005.

Now 9.543 is $\frac{3}{10}$ of the way from 9.540 to 9.550, so we take 0.3 of 0.0005, which is 0.00015, and round it to 0.0002. We add this to 0.9795. The result is 0.9797. Thus $\log 9.543 \approx 0.9797$.

c) We add the integer -3 from step (a):

$$\log (9.543 \times 10^{-3}) \approx -2.0203.$$

DO EXERCISE 2.

ii Antilogarithms. We interpolate when finding antilogarithms, using the table in reverse.

EXAMPLE 3 Find antilog 4.9164.

3. Find antilog 3.7749.

a) We are setting out to find $10^{4.9164}$, which is

$$10^{0.9164} \times 10^{4}.$$

Separating the exponent into an integer and a number between 0 and 1

b) We use Table 2 to find $10^{0.9164}$:

$$\left.\begin{array}{ll} & \text{antilog } 0.9159 \approx 8.240 \\ 0.0006 \quad 0.0005 & \\ & \text{antilog } 0.9164 \approx 8.24? \\ & \text{antilog } 0.9165 \approx 8.250 \end{array}\right\}$$

The difference is 0.010.

The difference between 0.9159 and 0.9165 is 0.0006. Thus 0.9164 is $\frac{0.0005}{0.0006}$, or $\frac{5}{6}$, of the way between 0.9159 and 0.9165. Then antilog 0.9164 is $\frac{5}{6}$ of the way between 8.240 and 8.250, so we take $\frac{5}{6}(0.010)$, which is 0.00833. . . , and round it to 0.008. Thus $10^{0.9164} \approx 8.248$.

We now have

$$\begin{aligned} \text{antilog } 4.9164 &= 10^{0.9164} \times 10^{4} \\ &\approx 8.248 \times 10^{4}, \quad \text{or} \quad 82{,}480. \end{aligned}$$

DO EXERCISE 3.

EXAMPLE 4 Find antilog (-2.5878).

a) We are setting out to find $10^{-2.5878}$. We need to separate -2.5878 into two parts, one of which is an integer and one of which is a number between 0 and

1. We can do that by adding and subtracting 3 (or any integer greater than 2):

$$3 - 2.5878 - 3 = 0.4122 - 3. \quad \text{Doing the first subtraction}$$

Now we have

$$\begin{aligned}\text{antilog}\,(-2.5878) &= \text{antilog}\,(0.4122 - 3)\\ &= 10^{(0.4122-3)}\\ &= 10^{0.4122} \times 10^{-3}.\end{aligned}$$

b) Next we use Table 2:

$$\left.\begin{aligned}\text{antilog}\ 0.4116 &\approx 2.580\\ \text{antilog}\ 0.4122 &\approx 2.58?\\ \text{antilog}\ 0.4133 &\approx 2.590\end{aligned}\right\}\ \text{The difference is } 0.010.$$

0.0017 0.0006

The difference between 0.4116 and 0.4133 is 0.0017. Thus 0.4122 is $\frac{0.0006}{0.0017}$, or $\frac{6}{17}$, of the way between 0.4116 and 0.4133. Then antilog 0.4122 is $\frac{6}{17}$ of the way between 2.580 and 2.590, so we take $\frac{6}{17}(0.010)$, which is 0.0035 to four places. We round it to 0.004. Thus $10^{0.4122} \approx 2.584$.

We now have

$$\begin{aligned}\text{antilog}\,(-2.5878) &= 10^{0.4122} \times 10^{-3}\\ &\approx 2.584 \times 10^{-3}, \quad \text{or} \quad 0.002584.\end{aligned}$$

DO EXERCISE 4.

4. Find antilog (6.4557 − 10).

EXERCISE SET 9.5

i Find the following logarithms using interpolation.

1. log 41.63 **2.** log 472.1 **3.** log 2.944 **4.** log 21.76 **5.** log 650.2 **6.** log 37.37

7. log 0.1425 **8.** log 0.09045 **9.** log 0.004257 **10.** log 4518 **11.** log 0.1776 **12.** log 0.08356

13. log 600.6 **14.** log 800.1

ii Find the following antilogarithms using interpolation.

15. antilog 1.6350 **16.** antilog 2.3512 **17.** antilog 0.6478 **18.** antilog 1.1624

19. antilog 0.0342 **20.** antilog 4.8453 **21.** antilog (−0.1436) **22.** antilog (−1.0341)

23. antilog (−2.5872) **24.** antilog (−0.2722) **25.** antilog (−1.7990) **26.** antilog (−2.1370)

Problem-solving practice

27. The perimeter of a right triangle is 120 cm. The longer leg is 50 cm less than 3 times the shorter leg. The hypotenuse is 20 cm longer than the shorter leg. What is the area of the triangle?

28. Mary and Bob both received a raise after one year on the job. Mary received an 8% raise and Bob a 7% raise. Before the raise Mary was earning $500 a year less than Bob, but after the raise their salaries are the same. What were their respective salaries before the raise?

○

Find.

29. log (log 3) **30.** log (log 5) **31.** log (log 7)

OBJECTIVES

After finishing Section 9.6, you should be able to:

- **i** **Solve exponential equations.**
- **ii** **Solve logarithmic equations.**
- **iii** **Solve certain applied problems involving exponential and logarithmic equations.**

1. Solve $4^{2x-3} = 64$.

9.6

EXPONENTIAL AND LOGARITHMIC EQUATIONS

i Exponential Equations

Equations with variables in exponents, such as $3^{2x-1} = 4$, are called *exponential equations.* We solve such equations by taking a logarithm on both sides and then using Property 2.

EXAMPLE 1 Solve: $2^{3x-5} = 16$.

Method 1.

$$\log 2^{3x-5} = \log 16 \qquad \text{Taking the log on both sides}$$

$$(3x-5)\log 2 = \log 16 \qquad \text{Using Property 2}$$

$$3x - 5 = \frac{\log 16}{\log 2}$$

$$x = \frac{\dfrac{\log 16}{\log 2} + 5}{3} \qquad \text{Solving for } x$$

$$x \approx \frac{\dfrac{1.2041}{0.3010} + 5}{3} \qquad \text{Finding logs}$$

$$x \approx 3.0001 \qquad \text{Calculating}$$

The answer is approximate because the logarithms are approximate.

Method 2. Note that $16 = 2^4$. Then we have

$$2^{3x-5} = 2^4.$$

Since the base is the same, 2, on both sides, the exponents must be the same. Thus

$$3x - 5 = 4.$$

We solve this equation to get

$$x = 3.$$

This answer is exact.

DO EXERCISE 1.

ii Logarithmic Equations

Equations containing logarithmic expressions are called *logarithmic equations.* To solve them, we obtain a single logarithmic expression on one side and then take the antilogarithm, or exponentiate.

EXAMPLE 2 Solve: $\log x + \log (x-3) = 1$.

$$\log x(x-3) = 1 \qquad \text{Using Property 1 to obtain a single logarithm}$$

$$x(x-3) = 10^1 \qquad \text{Taking the antilog on both sides}$$

$$x^2 - 3x - 10 = 0$$

$$(x+2)(x-5) = 0 \qquad \text{Factoring}$$

$$x = -2 \quad \text{or} \quad x = 5 \qquad \text{Principle of zero products}$$

Check:

$\log x + \log (x - 3) = 1$	
$\log (-2) + \log (-2 - 3)$	1

$\log x + \log (x - 3) = 1$	
$\log 5 + \log (5 - 3)$	1
$\log 5 + \log 2$	
$\log 10$	
1	

The number -2 is not a solution because negative numbers do not have logarithms. The solution is 5.

DO EXERCISE 2.

2. Solve $\log x + \log (x + 3) = 1$.

iii Problem Solving

There are many problem situations that can be translated to exponential or logarithmic equations. We now consider two of them.

EXAMPLE 3 (*Compound interest*). The amount of money A that a principal P will be worth after t years at interest rate r, compounded annually, is given by the formula

$$A = P(1 + r)^t.$$

Suppose that \$4000 principal is invested at 12% interest compounded annually and is eventually worth \$10,706. How long was it invested?

1. **Familiarize** We have the formula $A = P(1 + r)^t$. If we did not have it, we would have to look it up.
   Let's list the known and unknown quantities:

   A = Total worth = \$10,706;
   P = Principal = \$4000;
   r = Interest rate = 12%;
   t = time, in years.

2. **Translate** The formula is the translation in this case.

3. **Carry out** Since it is time that we want to find, we solve the formula for t:

$$A = P(1 + r)^t$$

$$\log A = \log P(1 + r)^t \quad \text{Taking the logarithm on both sides}$$

$$\log A = \log P + \log (1 + r)^t \quad \text{Using Property 1}$$

$$\log A = \log P + t \log (1 + r) \quad \text{Using Property 2}$$

$$\log A - \log P = t \log (1 + r) \quad \text{Subtracting } \log P$$

$$\frac{\log A - \log P}{\log (1 + r)} = t. \quad \text{Dividing by } \log (1 + r)$$

3. Suppose \$5000 principal is invested at 8% interest compounded and yields \$8569. How many years was it invested?

We now substitute into the formula as we have it solved for t:

$$\frac{\log 10{,}706 - \log 4000}{\log(1 + 0.12)} = t \qquad \text{Substituting}$$

$$\frac{4.0296 - 3.6021}{0.0492} \approx t \qquad \text{Finding the logarithms}$$

$$8.69 \approx t. \qquad \text{Calculating}$$

4. **Check** We have a possible answer of 8.69 years. We can check by substituting into the original formula:

$$\begin{aligned} A &= P(1 + r)^t \\ &= 4000(1.12)^{8.69} \\ &= 10{,}709. \end{aligned}$$

This is pretty close to \$10,706, so we have a check.

5. **State** The answer is that the money was invested for about 8.69 years.

DO EXERCISE 3.

EXAMPLE 4 (*Population growth*). The exponential equation

$$P = P_0 e^{kt},$$

under certain conditions, allows us to calculate a population P, the number of people (or other living organism) at time t, where P_0 is the population when $t = 0$ and k is a positive constant that depends on the situation. k is called the exponential growth rate.

The population of the United States was 208 million in 1970. It was 225 million in 1980. Use these data to predict what the population will be in the year 2000.

1. **Familiarize** Let's list the things that we know and things that we want to find:

   $P =$ the number of people (population) of the U.S. in the year 2000;
   1970 population was 208 million;
   1980 population was 225 million;
   $t =$ the time in years;
   $k =$ some constant, depending on the situation.

   It looks as if we will have to determine the constant k, in a manner similar to what we do for equations of variation.

   We must first decide when to start the time clock running. That is, in what year will t be 0? When we decide that, we must then calculate in order to find what t is in the year 2000.

2. **Translate** The formula is the translation.

3. **Carry out** Let us start the clock in 1970. Then in 1980, t will be 10 and in 2000, t will be 30. We'll solve for k and then substitute to find k:

$$P = P_0 e^{kt}$$

$$\ln P = \ln P_0 e^{kt} \qquad \text{Taking the natural logarithm on both sides}$$

$$\ln P = \ln P_0 + \ln e^{kt} \qquad \text{Using Property 1}$$

$$\ln P = \ln P_0 + kt \qquad \text{Simplifying } (\ln e^{kt} = kt)$$

$$\ln P - \ln P_0 = kt \qquad \text{Subtracting } \ln P_0$$

$$\frac{\ln P - \ln P_0}{t} = k. \qquad \text{Dividing by } t$$

In 1980, we have $t = 10$ and $P = 225$ million. Of course, we know that $P_0 = 208$ million, the population in 1970. We substitute and calculate:

$$\frac{\ln 225 - \ln 208}{10} = k \qquad \text{We are omitting the "millions."}$$

$$\frac{5.4161 - 5.3375}{10} \approx k \qquad \text{Finding the logarithms}$$

$$0.008 \approx k. \qquad \text{Calculating}$$

The population equation in this situation is now

$$P = P_0 e^{0.008t} = 208e^{0.008t}.$$

In the year 2000, t will be 30, so for the population in 2000 we have

$$P = 208e^{0.008(30)} = 208e^{0.24}.$$

We find that $e^{0.24} \approx 1.2712$. Thus

$$P \approx 208 \times 1.2712 \approx 264.$$

4. **Check** In this case, the best we can do is recheck the calculations.

5. **State** The answer is that, according to the population equation, the population of the United States should be about 264 million in the year 2000.

DO EXERCISE 4.

4. In a laboratory experiment, there are about 1500 bacteria in a Petri dish at 8:00 A.M. At 1:00 P.M. of the same day, there are about 2500 bacteria. According to the population equation, how many bacteria will there be at 8:00 P.M.?

EXERCISE SET 9.6

i Solve.

1. $2^x = 8$
2. $2^x = 32$
3. $2^x = 10$
4. $2^x = 33$
5. $5^{4x-7} = 125$
6. $4^{3x+5} = 16$
7. $3^{x^2+4x} = \frac{1}{27}$
8. $3^{5x} \cdot 9^{x^2} = 27$
9. $4^x = 7$
10. $8^x = 10$
11. $2^x = 3^{x-1}$
12. $3^{x+2} = 5^{x-1}$
13. $(2.8)^x = 41$
14. $(3.4)^x = 80$
15. $(1.7)^x = 20$
16. $(4.5)^y = 125$

ii Solve.

17. $\log x + \log (x - 9) = 1$
18. $\log x + \log (x + 9) = 1$
19. $\log x - \log (x + 3) = -1$
20. $\log (x + 9) - \log x = 1$
21. $\log_4 (x + 3) + \log_4 (x - 3) = 2$
22. $\log_5 (x + 4) + \log_5 (x - 4) = 2$
23. $\log \sqrt[3]{x} = \sqrt{\log x}$
24. $\log \sqrt[4]{x} = \sqrt{\log x}$
25. $\log_5 \sqrt{x^2 + 1} = 1$
26. $\log \sqrt[3]{x^2} + \log \sqrt[3]{x^4} = \log 2^{-3}$

iii ***Problem solving***

27. (*Doubling time*). How many years will it take an investment of $1000 to double itself when interest is compounded annually at 12%?

28. (*Tripling time*). How many years will it take an investment of $1000 to triple itself when interest is compounded annually at 11%?

29. (*Population growth*). The approximate population of Dallas in 1960 was 680,000. In 1969 it was 815,000. Find k in the growth formula and estimate the population in 1990.

30. (*Population growth*). The approximate population of Kansas City in 1960 was 475,000. In 1970 it was 507,000. Find k in the growth formula and estimate the population in 2000.

31. (*Business*). In 1970 the cost of a double-dip ice cream cone was 52¢. In 1978 it was 66¢. Assuming that an exponential function like that for population growth fits this situation:

a) Find the cost of a cone in 1996.
b) When will the cost of a cone be twice that of 1978?

32. (*Business*). It is estimated that there were 714,000 accountants employed in 1972 and 935,000 employed in 1985. Assuming that an exponential function like that for population growth fits the situation:

a) Find the number of accountants needed in 1990.
b) When will the number of accountants needed be twice that of 1972?

33. (*Psychology*). Bornstein and Bornstein found in a study that the average walking speed v of a person living in a city of population p, in thousands, is given by

$$v(p) = 0.37 \ln p + 0.05,$$

where v is in feet per second.

a) The population of Seattle is 531,000. What is the average walking speed of a person living in Seattle? Find $v(531)$.
b) The population of New York is 7,900,000. What is the average walking speed of a person living in New York?

34. (*Psychology*). Students in college botany took a final exam. They took equivalent forms of the exam in monthly intervals thereafter. The average score, $S(t)$, in percent after t months was found to be given by

$$S(t) = 68 - 20 \ln (t + 1), \qquad t \geqslant 0.$$

a) What was the average score when they initially took the test, $t = 0$?
b) What was the average score after 4 months?
c) What was the average score after 24 months?
d) What percentage of the initial score did they retain after 2 years (24 months)?

35. (*Business*). The cost of a candy bar in 1962 was \$0.05, and was increasing at an exponential growth rate $k = 9.7\%$. What will the cost of a candy bar be in 1987? in 1997?

36. (*Business*). Peter Minuit, of the Dutch West India Company, purchased Manhattan Island from the Indians in 1626 for \$24 worth of merchandise. Assuming an exponential rate of inflation $k = 8\%$, how much would Manhattan be worth in 1990?

Solve.

37. $8^x = 16^{3x+9}$
38. $27^x = 81^{2x-3}$
39. $\log_2 x + \log_2 (x - 2) = 3$
40. $\log_4 (x + 6) - \log_4 x = 2$
41. $\log_6 (\log_2 x) = 0$
42. $\log_x (\log_3 27) = 3$
43. $\log_5 \sqrt{x^2 - 9} = 1$
44. $x \log \frac{1}{8} = \log 8$
45. $\log (\log x) = 5$
46. $2^{x^2+4x} = \frac{1}{8}$
47. $3^{\log_3 (8x-4)} = 5$
48. $\log x^2 = (\log x)^2$
49. $10^{\log 14} = 3x - 19$
50. $\log_5 |x| = 4$
51. $\log x^{\log x} = 25$
52. $\log \sqrt{2x} = \sqrt{\log 2x}$
53. $\log_a a^{x^2+4x} = 21$
54. $(\log_a x)^{-1} = \log_a x$
55. $x^{\log_{10} x} = \dfrac{x^{-4}}{1000}$
56. $3^{2x} - 8 \cdot 3^x + 15 = 0$
57. $(81^{x-2})(27^{x+1}) = 9^{2x-3}$
58. $25^{x+2} = 3120 + 25^x$
59. $3^{2x} - 3^{2x-1} = 18$

60. Given that $2^y = 16^{x-3}$ and $3^{y+2} = 27^x$, find the value of $x + y$.

61. If $x = (\log_{125} 5)^{\log_5 125}$, what is the value of $\log_3 x$?

62. Suppose an amount P_0 is invested in a savings account at interest rate k, compounded continuously. The balance P, after t years, is given by

$$P = P_0 e^{kt}.$$

Suppose \$1000 is invested at 14%.

a) What is the balance after 1 year? 2 years? 10 years?
b) After what amount of time will the \$1000 double itself?
c) Suppose P_0 dollars are invested at interest rate k and that amount grows to $2P_0$ dollars in time T. Find a formula relating T and k.

SUMMARY AND REVIEW: CHAPTER 9

The following contains a summary of what you should be able to do after completing this chapter. The review exercises are for practice. Answers are at the back of the book. If you miss an exercise, restudy the section indicated alongside the answer.

You should be able to:

Graph an exponential function.

Graph.

1. $y = 4^x$
2. $y = 3^{x+1}$

Find the inverse of a function.

Find the inverse.

3. $f(x) = x + 2$
4. $f(x) = 3x - 1$
5. $f(x) = x + 5$
6. $f(x) = 2x + 3$

Convert from an exponential equation to a logarithmic equation.

Convert to logarithmic equations.

7. $10^4 = 10{,}000$ **8.** $6^3 = 216$ **9.** $25^{1/2} = 5$ **10.** $81^{1/4} = 3$

Convert from a logarithmic equation to an exponential equation.

Convert to exponential equations.

11. $\log_4 16 = x$ **12.** $\log_8 8 = 1$ **13.** $\log_{10} 2 = 0.3010$ **14.** $\log_{10} 0.0001 = -4$

Use the properties of logarithms.

Express as single logarithms.

15. $\log_a 8 + \log_a 15$ **16.** $\log_x 72 - \log_x 12$

Express in terms of logarithms of x, y, and z.

17. $\log_a x^4y^2z^3$ **18.** $\log_a 7xyz^5$ **19.** $\log_a \frac{xy}{z^2}$ **20.** $\log_a \frac{7x^3}{y^2z}$

Find common logarithms and antilogarithms.

Find the following.

21. log 7.05 **22.** log 3.37 **23.** log 45.3 **24.** log 99.4

25. log 0.0047 **26.** log 0.0564 **27.** log 4564 **28.** log 0.6064

29. antilog 0.5955 **30.** antilog 0.9248 **31.** antilog 2.6739 **32.** antilog (−1.2636)

33. antilog 0.6351 **34.** antilog (−2.3424)

Find natural logarithms and antilogarithms.

Find the following.

35. ln 67 **36.** ln 405 **37.** ln 0.0345 **38.** ln 0.246

39. $\text{antilog}_e\ 0.6752$ **40.** $\text{antilog}_e\ 0.2135$ **41.** $\text{antilog}_e\ 2.5555$ **42.** $\text{antilog}_e\ (-3.5001)$

Solve exponential and logarithmic equations.

Solve.

43. $2^x = 24$ **44.** $4^{x+2} = 64$

45. $\log x + \log (x - 15) = 2$ **46.** $\log_3 (x - 4) + \log_3 (x + 4) = 3$

Solve.

47. $5^{x+y} = 100,$
$2^{2x-y} = 10$

48. $(0.2)^x < \frac{5}{3}$

49. A single cell of bacteria divides into two every 15 min. If the same rate of division is maintained for 8 hr, how many cells will be produced from a single cell?

TEST: CHAPTER 9

Graph.

1. $y = 3^x$ **2.** $y = 2^{x+3}$

Find the inverse.

3. $f(x) = x - 4$ **4.** $f(x) = 4x - 3$

Convert to logarithmic equations.

5. $10^{-2} = 0.01$ **6.** $256^{1/2} = 16$

Convert to exponential equations.

7. $m = \log_7 49$ **8.** $\log_{10} 100{,}000 = 5$

Express as single logarithms.

9. $\log_a 1.2 + \log_a 0.603$

10. $(\log_b 62 - \log_b 495) + \log_b 0.101$

Express in terms of logarithms of x, y, and z.

11. $\log_a x^{1/2}y^{1/4}z$

12. $\log_b 50xy^2z^7$

Find the following logarithms and antilogarithms.

13. log 9.64

14. log 145

15. log 6240

16. log 0.0562

17. log 34.67

18. log 0.9528

19. ln 256

20. ln 0.567

21. antilog 2.5705

22. antilog −3.3530

23. antilog_e 0.5454

24. antilog_e 2.0555

Solve.

25. $10^{x+2} = 45$

26. $\log(x+2) - \log(x^2-4) = 2$

27. The population of Tempe, Arizona was 25 thousand in 1960. In 1969, it was 52 thousand. Use these data to predict what the population will be in the year 1990.

CUMULATIVE REVIEW: CHAPTERS 1–9

Solve.

1. $3n - 7 = 6 - 4n$

2. $-(2x-5) = 4x - (3+x)$

3. $y + 8 < -6 + 5y$

4. $3 - 4a > -(8a+1)$

5. $7 \geqslant |9 - 3x| \geqslant 1$

6. $-3x = 4x^2 + x^3$

7. $\frac{x}{7} + \frac{x}{3} = 2$

8. $\sqrt{3x-1} = 5$

9. $\sqrt{3-5x} = \sqrt{5-3x}$

10. $t^2 - 5 = 0$

11. $5x^2 = 2$

12. $(x+2)^2 + (x-1)^2 = 0$

13. $x^2 + 6x + 3 = 0$

14. $\frac{x-3}{x} = \frac{x-4}{4}$

15. $x^4 - 8x^2 + 7 = 0$

16. $\log_2(x-3) + \log_2(x+3) = 4$

17. Solve $a^2 + b^2 = c^2$ for a.

18. Solve $A = 2r^2 + 2rh$ for h.

Simplify.

19. $(a+b)(a^2 - ab + b^2)$

20. $(-3x^2y + 4xy^2) - (6xy^2 + 2x^2y) - (9xy - 5x^2y)$

21. $\dfrac{x-y}{\frac{1}{x} - \frac{1}{y}}$

22. $\sqrt{81x^4}$

23. $\sqrt[3]{-125}$

24. $\sqrt[3]{4m^3n^4}\,\sqrt[3]{16mn^5}$

25. $\sqrt{\dfrac{32a^5}{12a^2}}$

26. $2\sqrt{56} + \sqrt{48} - 5\sqrt{96}$

27. $(2\sqrt{3} + 2)^2$

28. $(5 + 5i) + (6 - i)$

29. $(-3 + 4i) - (8 - 7i)$

30. $(2 - 9i)(3 - i)$

31. $\dfrac{5 - 2i}{-4 + 8i}$

Graph.

32. $y = (x+2)^2$

33. $x > y + 1$

34. $y^2 = x^2 - 8x + 16$

35. $f(x) = -(x+3)^2$

36. $f(x) = 2(x-5)^2$

37. $f(x) = \frac{1}{4}(x-2)^2$

38. $f(x) = -3(x+1)^2 - 2$

39. $f(x) = 4(x - \frac{1}{2})^2 - 10$

40. $f(x) = x^2 - 4x + 1$

41. $y = 3^x$

Find the inverse.

42. $f(x) = 2x - 5$

43. $f(x) = -x + 4$

Convert to logarithmic equations.

44. $10^{-5} = 0.00001$

45. $169^{1/2} = 13$

Convert to exponential equations.

46. $\log_3 27 = x$

47. $\log_{10} 3 = 0.4771$

Find the logarithms and antilogarithms.

48. log 90.5

49. antilog 0.7007

50. log 0.0509

51. ln 391

52. antilog_e 0.6780

53. ln 0.00586

Problem solving

54. The length of a rectangle is 5 cm more than the width. The area is 266 cm^2. Find the length and width.

55. Airplane A travels 500 mi at a certain speed. Airplane B travels 900 mi at a speed 50 mph faster but takes 1 hr longer. Find the speed of each plane.

56. Marie has 110 ft of fence with which to enclose a garden. What dimensions will yield the maximum garden area? What is the maximum area?

57. A 150-ft cable is divided into 3 pieces. The first piece is 3 ft longer than the third piece. The second piece is 12 ft less than the sum of the lengths of the other two pieces. What is the length of the smallest piece of cable?

10

EQUATIONS OF SECOND DEGREE AND THEIR GRAPHS

Problems concerning the orbit of a satellite can be solved using equations of second degree.

This chapter is largely devoted to a discussion of graphing. We graph equations of second degree, that is, equations in two variables in which one or both variables are raised to the second degree (at most). It is a curious fact that the graphs of such equations are almost always the kinds of figures you could obtain by cutting a cone with a plane (conic sections).

These equations and graphs have a great many interesting applications. We will take a look at a few of them here and solve some problems.

OBJECTIVE

After finishing Section 10.1, you should be able to:

i **Use the distance formula to find the distance between two points whose coordinates are known.**

Find the distance between each pair of points.

1. (7, 12) and (7, −2)

2. (6, 2) and (−5, 2)

10.1

THE DISTANCE FORMULA

i For our work with graphs, we need a formula for finding the distance between two points when we know their coordinates.

To derive the distance formula, we begin by considering two points on a vertical or horizontal line.

If points are on a vertical line, they have the same first coordinate. In the figure on the left, we can find the distance between the points by subtracting their second coordinates:

$$6-(-2)=8. \qquad \text{The distance is 8.}$$

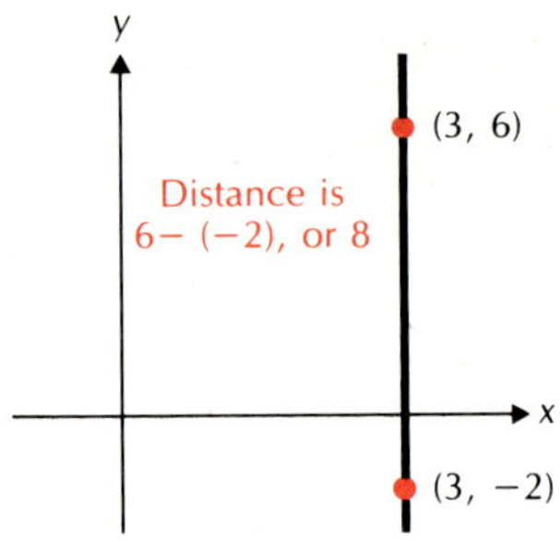

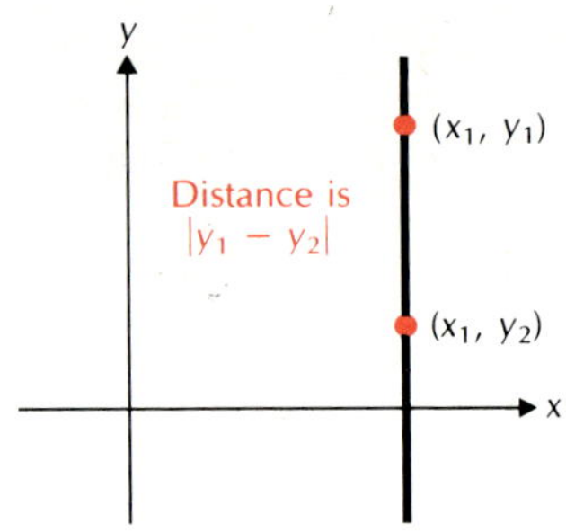

If we subtract the opposite way, we get

$$-2-6=-8. \qquad \text{The distance is } |-8|\text{, or 8.}$$

If we use absolute value, as in the figure on the right, we need not worry about the order. We can subtract either way.

If points are on a horizontal line, we subtract their first coordinates and take the absolute value.

EXAMPLES Find the distance between these points.

1. (−5, 13) and (−5, 2)

We subtract second coordinates, since the first coordinates are the same. The distance is

$$|13-2|, \quad \text{or} \quad 11.$$

2. (7, −3) and (−5, −3)

Since the second coordinates are the same, we subtract first coordinates, to get

$$|-5-7|=|-12|, \quad \text{or} \quad 12.$$

DO EXERCISES 1 AND 2.

Next we consider two points that are not on either a vertical or horizontal line, such as (x_1, y_1) and (x_2, y_2) in the figure below. By drawing horizontal and vertical

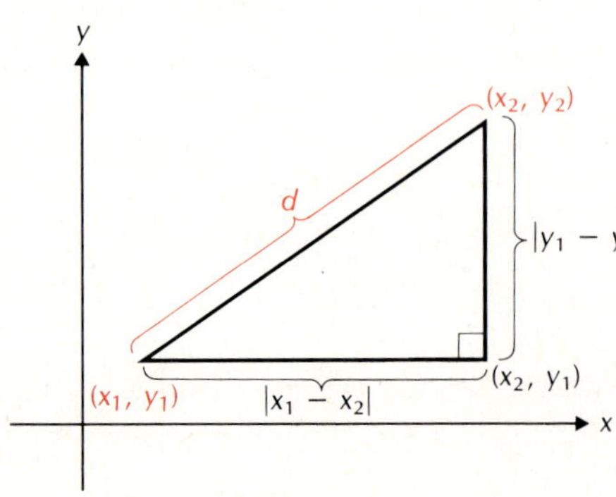

lines through these points, we form a right triangle. The vertex of the right angle of this triangle has coordinates (x_2, y_1). The legs have lengths $|x_1 - x_2|$ and $|y_1 - y_2|$. The distance we want is the length of the hypotenuse. By the Pythagorean Theorem,

$$d^2 = |x_1 - x_2|^2 + |y_1 - y_2|^2.$$

Since squares of numbers are never negative, we don't really need the absolute-value signs. Thus we have

$$d^2 = (x_1 - x_2)^2 + (y_1 - y_2)^2.$$

Taking the principal square root, we get the distance formula.

THE DISTANCE FORMULA

The distance between any two points (x_1, y_1) and (x_2, y_2) is given by

$$d = \sqrt{(x_1 - x_2)^2 + (y_1 - y_2)^2}.$$

This formula holds even when the two points are on a vertical or horizontal line.

EXAMPLE 3 Find the distance between $(4, -3)$ and $(-5, 4)$.

We substitute into the distance formula:

$$d = \sqrt{[4 - (-5)]^2 + (-3 - 4)^2} \quad \text{Substituting}$$
$$d = \sqrt{9^2 + (-7)^2}$$
$$d = \sqrt{130}$$
$$d \approx 11.402.$$

DO EXERCISES 3 and 4.

Find the distance between each pair of points. Where appropriate, use Table 1 or a calculator to approximate.

3. $(2, 6)$ and $(-4, -2)$

4. $(-2, 1)$ and $(4, 2)$

EXERCISE SET 10.1

Find the distance between each pair of points. Where appropriate, use Table 1 or a calculator to find an approximation.

1. $(-3, -2)$ and $(1, 1)$ **2.** $(5, 9)$ and $(-1, 6)$ **3.** $(0, -7)$ and $(3, -4)$

4. $(2, 2)$ and $(-2, -2)$ **5.** $(9, 5)$ and $(6, 1)$ **6.** $(1, 10)$ and $(7, 2)$

7. $(5, 6)$ and $(5, -2)$ **8.** $(5, 6)$ and $(0, 6)$ **9.** $(a, -3)$ and $(2a, 5)$

10. $(5, 2k)$ and $(-3, k)$ **11.** $(0, 0)$ and (a, b) **12.** $(\sqrt{2}, \sqrt{3})$ and $(0, 0)$

13. $(\sqrt{a}, \sqrt{b})$ and $(-\sqrt{a}, \sqrt{b})$ **14.** $(c - d, c + d)$ and $(c + d, d - c)$

Problem-solving practice

15. A grocery store purchased 96 watermelons at \$10.50 per dozen. If the store plans to sell each melon for \$2.50, how many will they have to sell in order to make a \$116 profit?

16. A runner completes a 5-mi race in 44 min. He ran the first 3 mi at an average speed that was 25% faster than his average speed for the last 2 mi. What was his average speed over the last 2 mi?

The converse of the Pythagorean theorem is true. That is, if the sides of a triangle have lengths a, b, and c, and $a^2 + b^2 = c^2$, then the triangle is a right triangle. Determine whether the points are vertices of a right triangle.

17. $(9, 6)$, $(-1, 2)$, and $(1, -3)$

18. $(-8, -5)$, $(6, 1)$, and $(-4, 5)$

19. Explain why the distance formula holds even if two points are on a vertical or horizontal line.

20. Find the distance between these points.

a) $(-1, 3k)$ and $(6, 2k)$

b) $(0, \sqrt{7})$ and $(\sqrt{6}, 0)$

c) $(6m, -7n)$ and $(-2m, n)$

d) $(\sqrt{d}, -\sqrt{3c})$ and $(\sqrt{d}, \sqrt{3c})$

e) $(\frac{5}{7}, \frac{1}{14})$ and $(\frac{1}{7}, \frac{11}{14})$

f) $(-3\sqrt{3}, 1 - \sqrt{6})$ and $(\sqrt{3}, 1 + \sqrt{6})$

g) (calculator) $(5.989, 2.001)$ and $(3.712, -7.784)$

21. Prove: If the endpoints of a segment are (x_1, y_1) and (x_2, y_2), then the coordinates of the midpoint are

$$\left(\frac{x_1 + x_2}{2}, \frac{y_1 + y_2}{2}\right).$$

22. Find the midpoints of the segments having the following endpoints.

a) $(-12, -7)$ and $(-3, 13)$

b) $(-a, b)$ and (a, b)

c) $(2 - \sqrt{3}, 5\sqrt{2})$ and $(2 + \sqrt{3}, 3\sqrt{2})$

d) $(\frac{1}{3}, -\frac{7}{11})$ and $(-\frac{2}{3}, \frac{3}{4})$

23. Determine whether the points $(6, 3)$, $(-4, -1)$, and $(-2, -6)$ are vertices of a right triangle.

24. Find the point on the y-axis that is equidistant from $(2, 10)$ and $(6, 2)$.

OBJECTIVES

After finishing Section 10.2, you should be able to:

i **Given an equation of a circle in standard form, find its center and radius and graph it.**

ii **Given the center and radius of a circle, write an equation of the circle.**

iii **Given an equation of a circle not in standard form, complete the square to find standard form.**

iv **Given the standard equation of an ellipse, graph the ellipse.**

10.2

CONIC SECTIONS: CIRCLES AND ELLIPSES

In this section and the following one, we will study certain geometric figures that can be formed by cross-sections of cones.

i Circles

A circle is defined as the set of all points in a plane that are at a fixed distance from a point in that plane. Circles are called *conic sections* because they can be formed by cutting a cone by a plane, as shown.

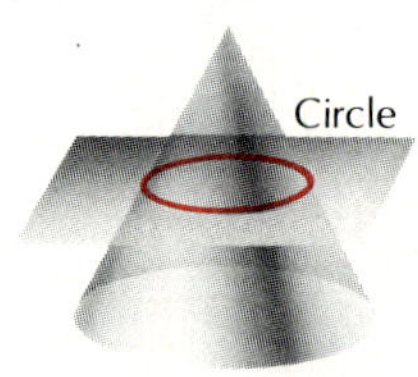

Let us find an equation for a circle. We call the center (a, b) and we suppose that the radius has length r. Suppose (x, y) is any point on the circle. By the distance formula, we have

$$\sqrt{(x - a)^2 + (y - b)^2} = r.$$

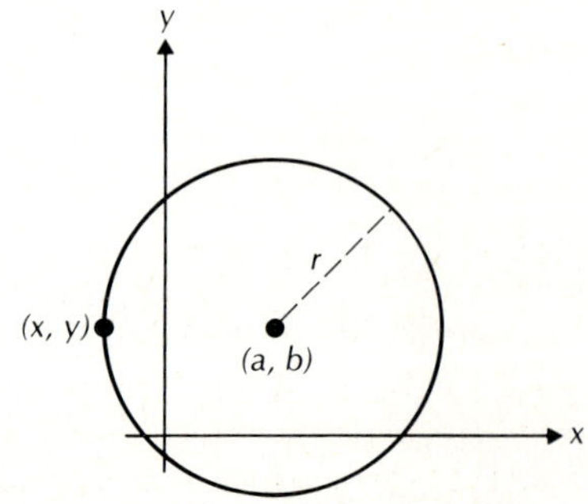

Squaring both sides gives an equation of the circle in standard form.

EQUATION OF A CIRCLE

A circle with center (a, b) and radius r has an equation

$$(x - a)^2 + (y - b)^2 = r^2. \qquad \text{(Standard form)}$$

EXAMPLE 1 Find the center and radius and graph this circle:

$$(x + 2)^2 + (y - 3)^2 = 16.$$

We rewrite to get the standard form given above:

$$[x - (-2)]^2 + (y - 3)^2 = 4^2.$$

Thus the center is $(-2, 3)$ and the radius is 4. The graph is as shown. It can be drawn by locating the center and then using a compass.

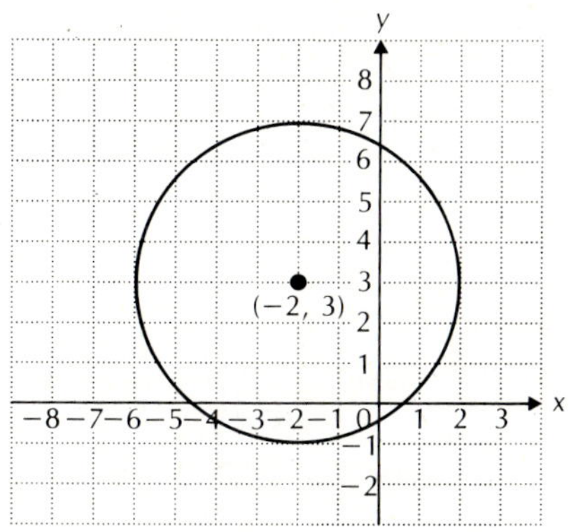

DO EXERCISE 1.

ii Writing Equations

EXAMPLE 2 Write an equation of a circle with center $(9, -5)$ and radius $\sqrt{2}$.

We use the standard form $(x - a)^2 + (y - b)^2 = r^2$:

$$(x - 9)^2 + [y - (-5)]^2 = (\sqrt{2})^2 \qquad \text{Substituting}$$

$$(x - 9)^2 + (y + 5)^2 = 2. \qquad \text{Simplifying}$$

DO EXERCISE 2.

If a circle is centered at the origin, then $(a, b) = (0, 0)$.

A circle centered at the origin with radius r has an equation

$$x^2 + y^2 = r^2.$$

DO EXERCISE 3.

iii Standard Form by Completing the Square

By completing the square, we can show that certain equations have graphs that are circles.

EXAMPLE 3 Show that this is an equation of a circle. Find the center and radius.

$$x^2 + y^2 + 6x - 4y + 9 = 0$$

1. Find the center and radius of the circle $(x - 5)^2 + (y + \frac{1}{2})^2 = 9$. Graph the circle.

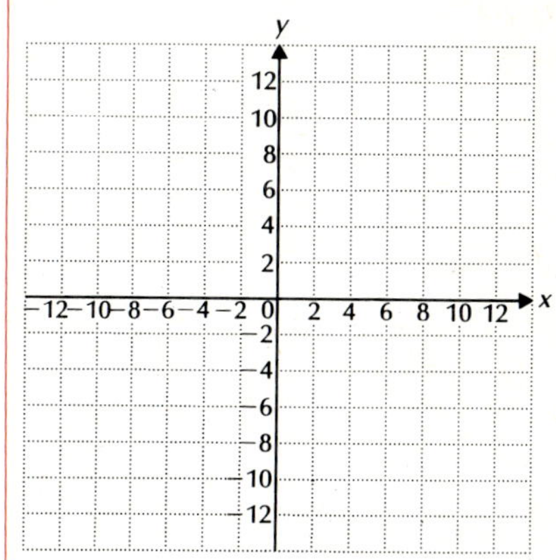

2. Find an equation of a circle with center $(-3, 1)$ and radius 6.

3. Find the center and radius of the circle $x^2 + y^2 = 64$.

4. Show that this is an equation of a circle. Find the center and radius.

$x^2 + y^2 + 8x - 10y - 8 = 0$

We first regroup the terms and then complete the square twice, once with $x^2 + 6x$ and once with $y^2 - 4y$.

$$(x^2 + 6x) + (y^2 - 4y) + 9 = 0$$
$$(x^2 + 6x + 9 - 9) + (y^2 - 4y + 4 - 4) + 9 = 0$$
$$(x^2 + 6x + 9) + (y^2 - 4y + 4) - 9 - 4 + 9 = 0$$
$$(x + 3)^2 + (y - 2)^2 = 4$$
$$[x - (-3)]^2 + (y - 2)^2 = 2^2$$

This shows that the graph of the equation is a circle. The center is $(-3, 2)$ and the radius is 2.

DO EXERCISE 4.

iv Ellipses

When a cone is cut at an angle, as shown, the conic section formed is an *ellipse.*

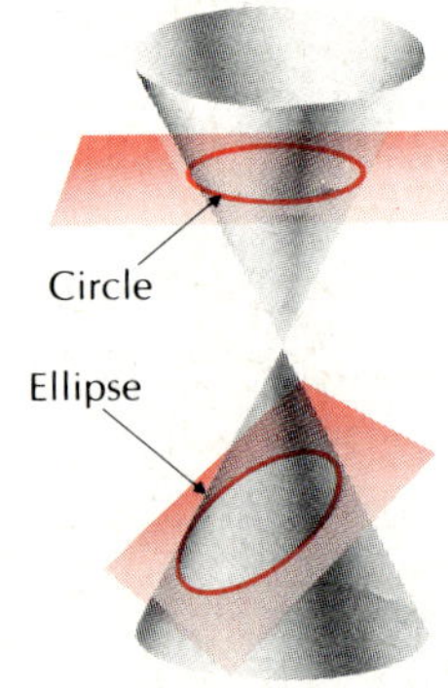

You can draw an ellipse by sticking two tacks in a piece of cardboard. Then tie a string to the tacks, place a pencil as shown, and draw.

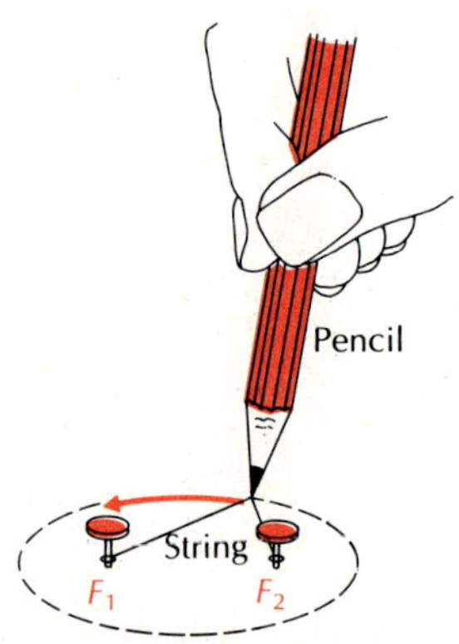

The formal mathematical definition is related to this method of drawing the ellipse. The ellipse is defined as the set of all points in a plane such that the *sum* of the distances from two fixed points F_1 and F_2 (called *foci*) is constant.

Ellipses have equations as follows.

EQUATION OF AN ELLIPSE

An ellipse with its center at the origin has an equation

$$\frac{x^2}{a^2} + \frac{y^2}{b^2} = 1, \quad a, b > 0. \qquad \text{(Standard form)}$$

In graphing ellipses, it helps to first find the intercepts. If we replace x by 0, we can find the y-intercepts:

$$\frac{0^2}{a^2} + \frac{y^2}{b^2} = 1$$

$$\frac{y^2}{b^2} = 1$$

$$y^2 = b^2$$

$$y = \pm b.$$

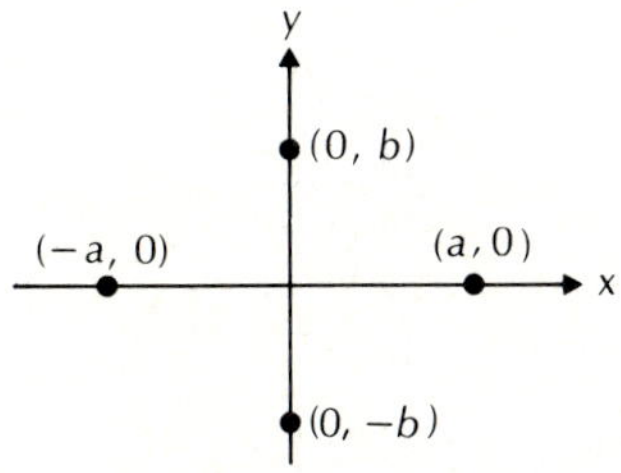

Thus the y-intercepts are $(0, b)$ and $(0, -b)$. Similarly, the x-intercepts are $(a, 0)$ and $(-a, 0)$. Plotting these points and filling in an oval-shaped curve, we get a graph of the ellipse. If a more precise graph is desired, we can plot more points.

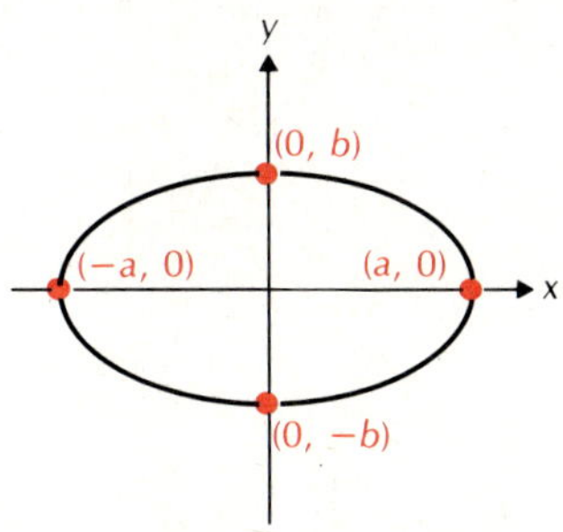

EXAMPLE 4 Graph the ellipse

$$\frac{x^2}{4} + \frac{y^2}{9} = 1.$$

Note that

$$\frac{x^2}{4} + \frac{y^2}{9} = \frac{x^2}{2^2} + \frac{y^2}{3^2}.$$

Thus the x-intercepts are $(2, 0)$ and $(-2, 0)$, and the y-intercepts are $(0, 3)$ and $(0, -3)$.

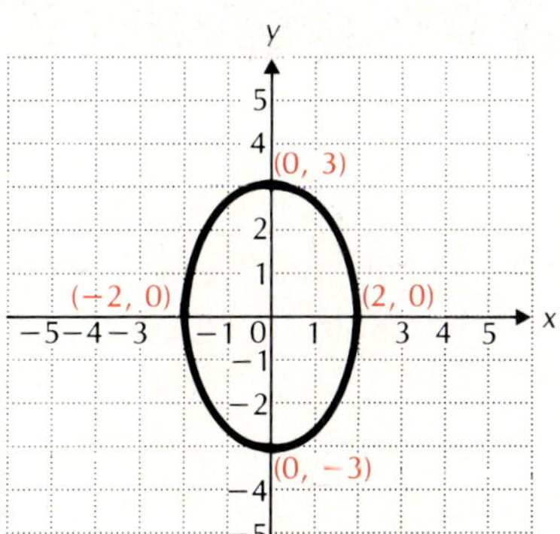

DO EXERCISES 5 AND 6.

5. Graph the ellipse

$$\frac{x^2}{9} + \frac{y^2}{4} = 1.$$

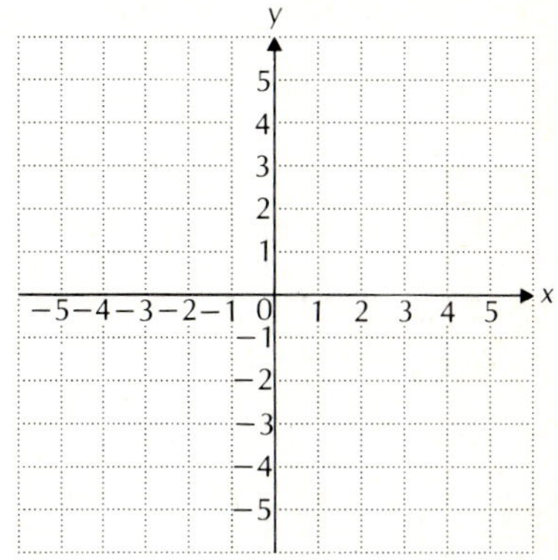

6. Graph the ellipse

$$\frac{x^2}{9} + \frac{y^2}{25} = 1.$$

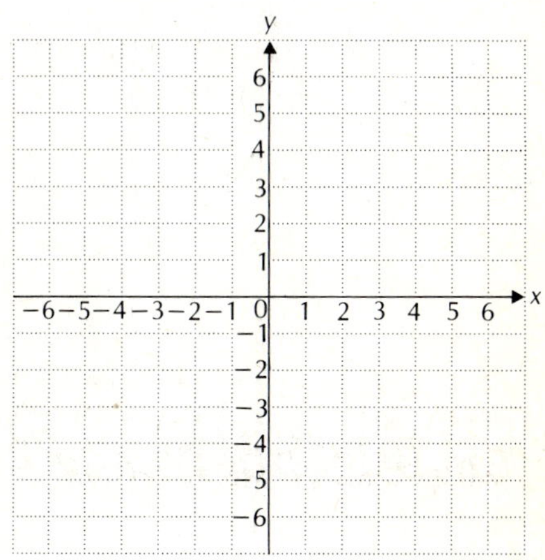

EXERCISE SET 10.2

i, **iii** Find the center and radius of each circle. Then graph the circle.

1. $(x+1)^2+(y+3)^2=4$
2. $(x-2)^2+(y+3)^2=1$
3. $(x-8)^2+(y+3)^2=40$
4. $(x+5)^2+(y-1)^2=75$
5. $x^2+y^2=2$
6. $x^2+y^2=3$
7. $(x-5)^2+y^2=\frac{1}{4}$
8. $x^2+(y-1)^2=\frac{1}{25}$
9. $x^2+y^2+8x-6y-15=0$
10. $x^2+y^2+6x-4y-15=0$
11. $x^2+y^2-8x+2y+13=0$
12. $x^2+y^2+6x+4y+12=0$
13. $x^2+y^2-4x=0$
14. $x^2+y^2+10y-75=0$

ii Find an equation of the circle having the given center and radius.

15. Center $(0, 0)$, radius 7
16. Center $(0, 0)$, radius 4
17. Center $(-2, 7)$, radius $\sqrt{5}$
18. Center $(5, 6)$, radius $2\sqrt{3}$

iv Graph each ellipse.

19. $\frac{x^2}{4}+\frac{y^2}{1}=1$
20. $\frac{x^2}{1}+\frac{y^2}{4}=1$
21. $16x^2+9y^2=144$
22. $9x^2+16y^2=144$
23. $2x^2+3y^2=6$
24. $5x^2+7y^2=35$
25. $4x^2+9y^2=1$
26. $25x^2+16y^2=1$

Problem-solving practice

27. A rectangle 6 in. long and 10 in. wide is bordered by a strip of uniform width. If the perimeter of the larger rectangle is twice that of the smaller rectangle, what is the width of the border?

28. One airplane flies 60 mph faster than another. The faster plane takes 4 hr and the slower plane 4 hr and 24 min to fly a certain distance. What is the distance?

---

29. Find an equation of a circle satisfying these conditions.

a) Center $(0, 0)$, passing through $\left(\frac{1}{4}, \frac{\sqrt{31}}{4}\right)$
b) Center $(-4, 1)$, passing through $(2, -5)$
c) Center $(-3, -2)$ and tangent to the y-axis
d) The endpoints of a diameter are $(7, 3)$ and $(-1, -3)$.

The equation (standard form) of an ellipse centered at the origin is

$$\frac{x^2}{a^2}+\frac{y^2}{b^2}=1.$$

The vertices are $(a, 0)$, $(-a, 0)$, $(0, b)$, and $(0, -b)$. The equation (standard form) of an ellipse with center (h, k) is

$$\frac{(x-h)^2}{a^2}+\frac{(y-k)^2}{b^2}=1.$$

The vertices are $(a+h, k)$, $(-a+h, k)$, $(h, b+k)$, and $(h, -b+k)$.

30. For each ellipse write standard form, and find the center and vertices. Then graph the ellipse.

a) $4x^2+9y^2=36$
b) $4(x-1)^2+9(y+2)^2=36$
c) $16x^2+y^2+96x-8y+144=0$
d) $4x^2+25y^2-8x+50y-71=0$

31. Find the equation of the ellipse with the following vertices.

a) $(-8, 0)$, $(8, 0)$, $(0, -2)$, $(0, 2)$
b) $(0, 2)$, $(5, -1)$, $(10, 2)$, $(5, 5)$

32. Find an equation of the ellipse that contains the point $(1, 2\sqrt{3})$ and two of whose vertices are $(2, 0)$ and $(-2, 0)$.

33. Describe the graph of $\frac{x^2}{a^2}+\frac{y^2}{b^2}=1$ when $a^2=b^2$.

Determine whether each of the following points lies on the *unit circle* $x^2 + y^2 = 1$.

34. $(0, -1)$ **35.** $\left(\frac{\sqrt{3}}{2}, -\frac{1}{2}\right)$ **36.** $(\sqrt{2} + \sqrt{3}, 0)$ **37.** $\left(\frac{\pi}{4}, \frac{4}{\pi}\right)$ **38.** $\left(\frac{\sqrt{2}}{2}, \frac{\sqrt{2}}{2}\right)$

39. Prove that $\angle ABC$ is a right angle. Assume that point B is on the circle whose radius is a and whose center is at the origin. (*Hint:* Use slopes and an equation of the circle.)

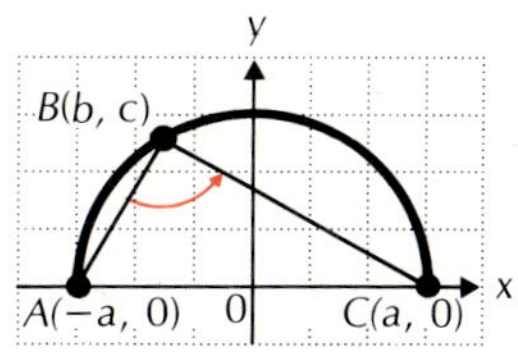

10.3

CONIC SECTIONS: PARABOLAS AND HYPERBOLAS

OBJECTIVES

After finishing Section 10.3, you should be able to:

i **Graph parabolas.**

ii **Given an equation of a hyperbola in standard form, graph it.**

iii **Given an equation of a hyperbola $xy = c$, graph it.**

In this section, we continue our study of figures that can be formed by taking cross-sections of cones.

i Parabolas

When a cone is cut as shown in the middle below, the conic section formed is a *parabola.* If both parts of the cone are cut by a plane, as shown on the right, the conic section is called a *hyperbola.* We do not show it here, but graphs of quadratic functions (see Chapter 8) are actually parabolas. General equations of parabolas are quadratic.

Many antennas have a parabolic cross-section.

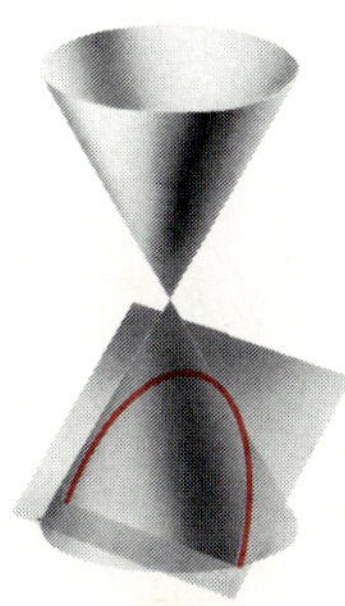

Parabola

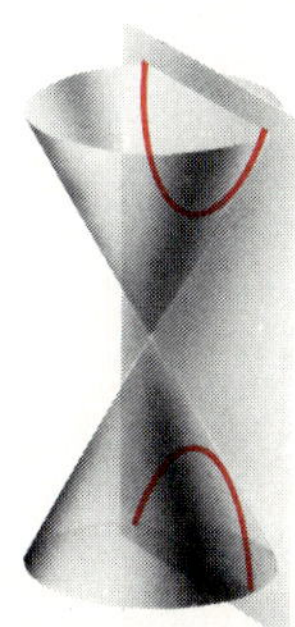

Hyperbola

PARABOLAS

Parabolas have equations as follows.

$y = ax^2 + bx + c$ **(Line of symmetry parallel to y-axis)**

$x = ay^2 + by + c$ **(Line of symmetry parallel to x-axis)**

EXAMPLE 1 Graph $y = x^2 - 4x + 8$. Then graph $x = y^2 - 4y + 8$.

a) We complete the square:

$$y = (x^2 - 4x + 4) + 4 = (x - 2)^2 + 4.$$

The vertex is at $(2, 4)$ and the curve opens upward, as shown in Figure 1.

Parabolas have many applications in electricity, mechanics, and optics. The cross section of a radar reflector is a parabola. Cables that support bridges are shaped like parabolas. (Free hanging cables have a different shape, called a "catenary.")

Graph.

1. $y = x^2 + 4x + 7$

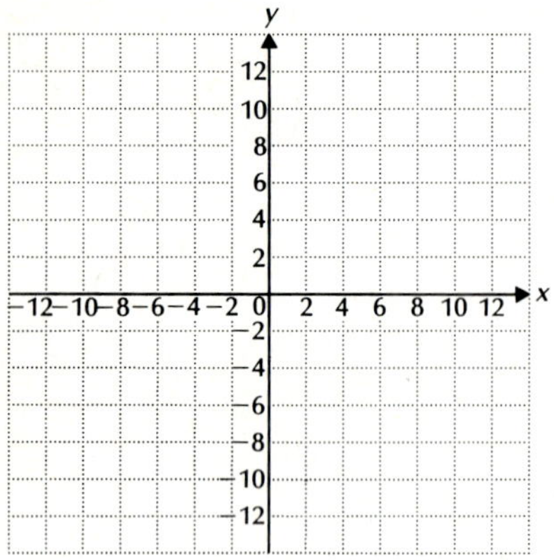

2. $x = y^2 + 4y + 7$

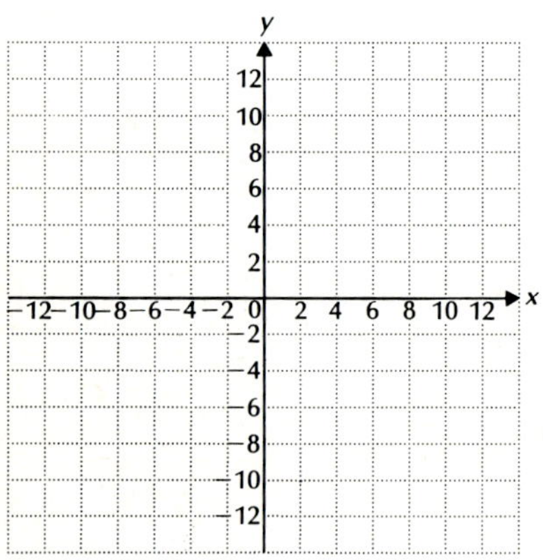

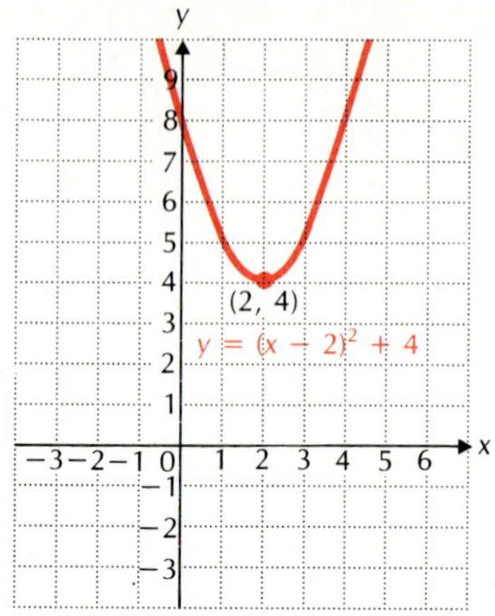

FIGURE 1

b) To graph $x = y^2 - 4y + 8$, we complete the square on y:

$$x = (y - 2)^2 + 4.$$

Since x and y are merely interchanged, this graph looks just like the first one, but is turned horizontally. The vertex is (4, 2). It helps to plot a few points. (See Figure 2.)

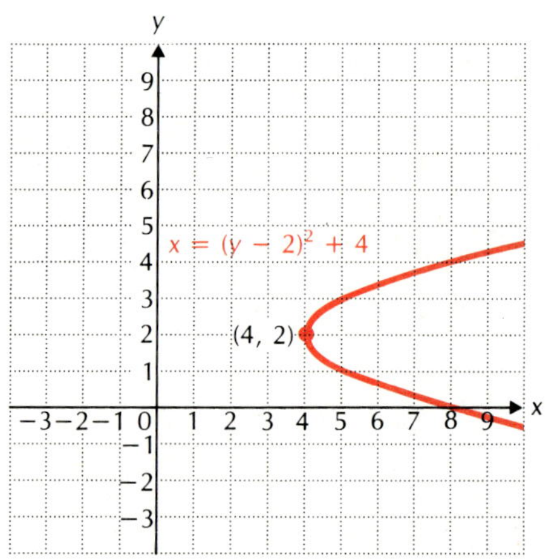

FIGURE 2

DO EXERCISES 1 AND 2.

ii Hyperbolas (Standard Form)

A hyperbola looks somewhat like a pair of parabolas, but the actual shapes are different. A hyperbola has two vertices, and the line through the vertices is known as an *axis.* The point halfway between the vertices is called the *center.* We consider only hyperbolas centered at the origin.

HYPERBOLAS

Hyperbolas with their centers at the origin have equations as follows.

$$\frac{x^2}{a^2} - \frac{y^2}{b^2} = 1 \qquad \textbf{(Axis horizontal)}$$

$$\frac{y^2}{b^2} - \frac{x^2}{a^2} = 1 \qquad \textbf{(Axis vertical)}$$

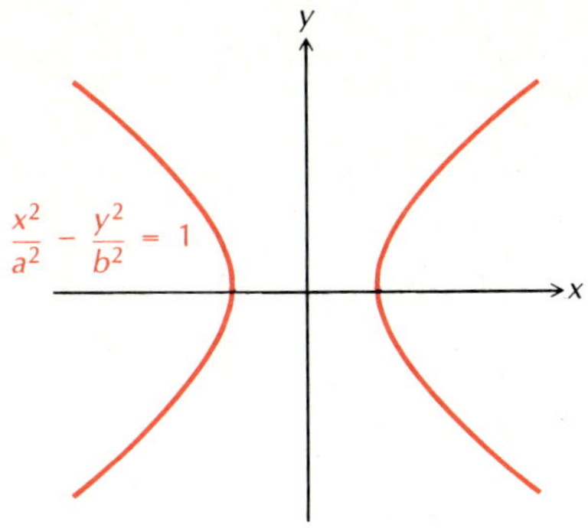

Axis horizontal

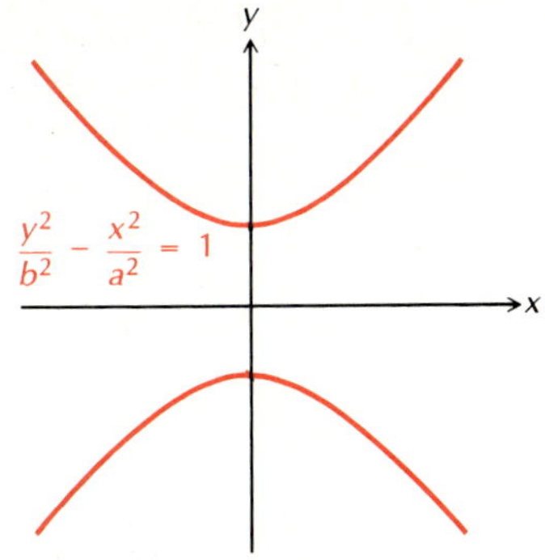

Axis vertical

To graph a hyperbola, it helps to begin by graphing the lines called *asymptotes*.

ASYMPTOTES OF A HYPERBOLA

For hyperbolas with equations as given above, the asymptotes are the lines

$$y = \frac{b}{a}x \quad \text{and} \quad y = -\frac{b}{a}x.$$

As a hyperbola gets farther away from the origin, it gets closer and closer to its asymptotes.

The next thing to do is to find the vertices. Then it is easy to sketch the curve.

EXAMPLE 2 Graph: $\dfrac{x^2}{4} - \dfrac{y^2}{9} = 1.$

a) Note that

$$\frac{x^2}{4} - \frac{y^2}{9} = \frac{x^2}{2^2} - \frac{y^2}{3^2},$$

so $a = 2$ and $b = 3$. The asymptotes are thus

$$y = \frac{3}{2}x \quad \text{and} \quad y = -\frac{3}{2}x.$$

We sketch them.

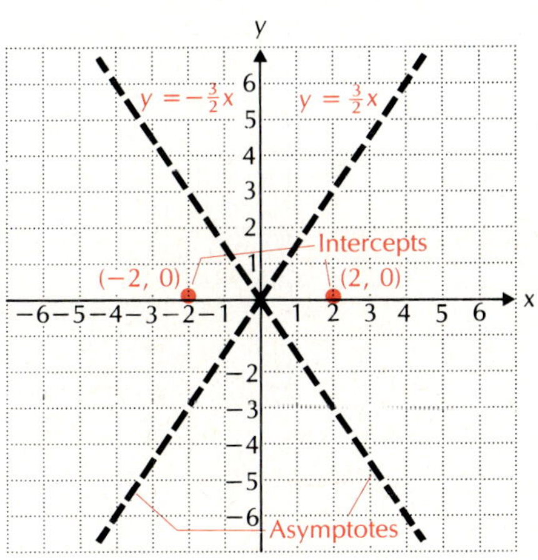

3. Graph

$$\frac{x^2}{16} - \frac{y^2}{25} = 1.$$

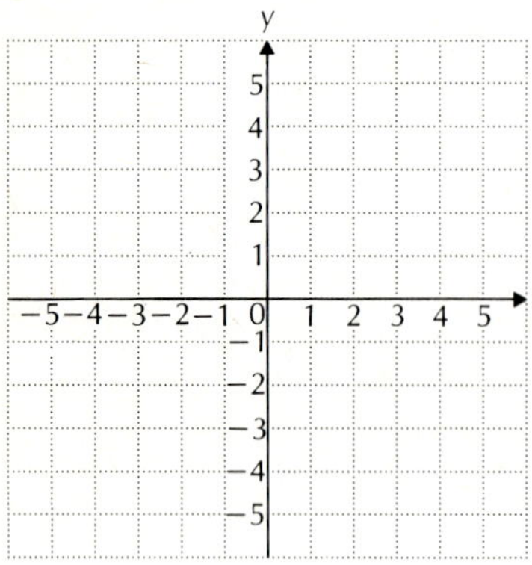

b) Next we find the intercepts. Let $y = 0$ (simply cover up the term containing y), and we see that $x^2/2^2 = 1$, so $x = \pm 2$. The intercepts are $(2, 0)$ and $(-2, 0)$. There are intercepts on only one axis. If we cover up the terms containing x, we see that $y^2/9 = -1$, and this equation has no real-number solutions.

c) We plot the intercepts and sketch the graph. Then through each intercept we draw a smooth curve that approaches the asymptotes closely.

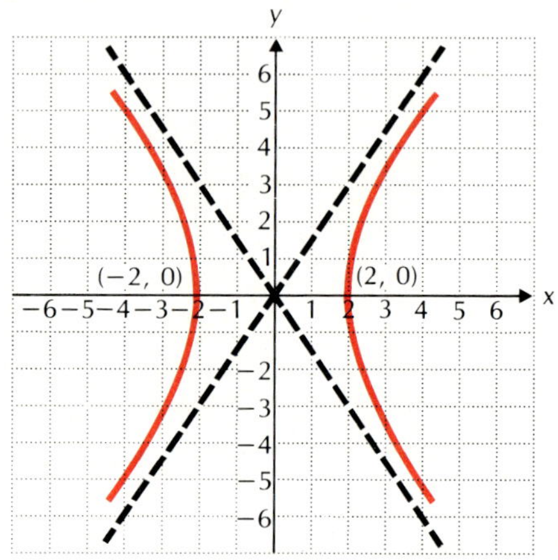

DO EXERCISE 3.

Graph: $$\frac{y^2}{36} - \frac{x^2}{4} = 1.$$

a) Note that

$$\frac{y^2}{36} - \frac{x^2}{4} = \frac{y^2}{6^2} - \frac{x^2}{2^2} = 1.$$

The intercept distance is the number in color in the term without the minus sign. There is a y in that term, so the intercepts are on the y-axis. The slopes of the asymptotes are $\frac{6}{2}$ and $-\frac{6}{2}$.

b) Using the numbers 6 and 2, we can quickly sketch a rectangle to use as a guide. To find the corners, we go 6 units (up or down) in the y-direction and 2 units (right or left) in the x-direction. The asymptotes go through the corners.

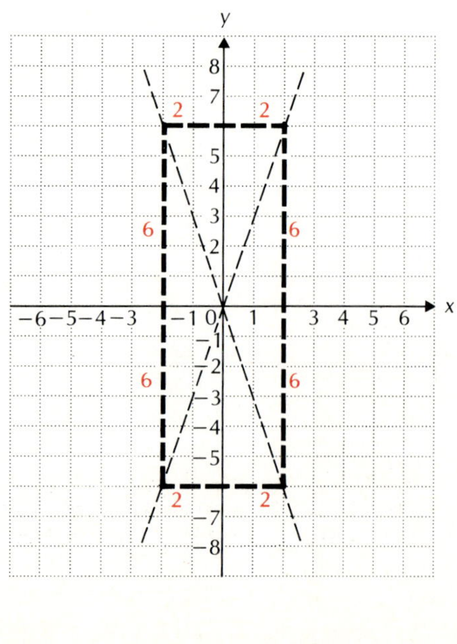

c) Now we draw curves through the intercepts toward the asymptotes as shown.

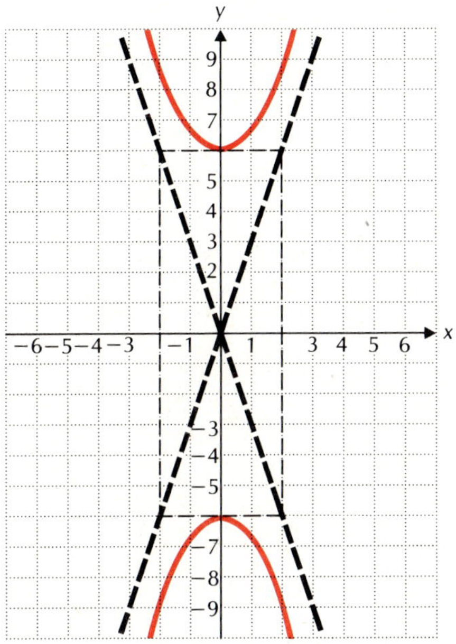

DO EXERCISE 4.

4. Graph.

a) $\dfrac{y^2}{9} - \dfrac{x^2}{49} = 1$

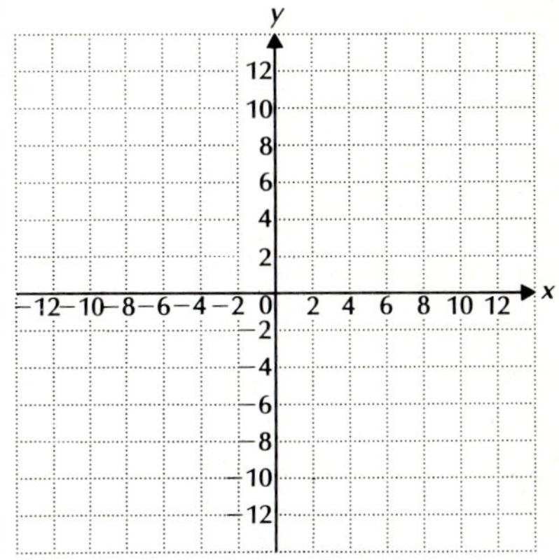

b) $\dfrac{x^2}{49} - \dfrac{y^2}{9} = 1$

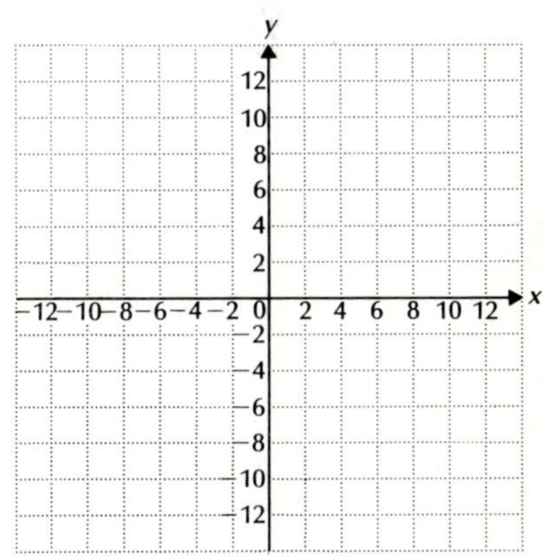

iii Hyperbolas (Nonstandard Form)

The equations we have already seen for hyperbolas are the standard ones, but there are other hyperbolas. We consider some of them.

Hyperbolas having the x- and y-axes as asymptotes have equations as follows:

$xy = c$, where c is a nonzero constant.

EXAMPLE 4 Graph: $xy = -8$.

First solve for y:

$$y = -\frac{8}{x}.$$

Next find some solutions, keeping the results in a table. (Note that we cannot use 0 for x.)

x	2	−2	4	−4	1	−1
y	−4	4	−2	2	−8	8

Now plot the points. Connect those points in the second quadrant with a smooth curve. Similarly, connect those in the fourth quadrant. Remember that the x-axis and the y-axis are asymptotes.

Graph.

5. $xy = -12$

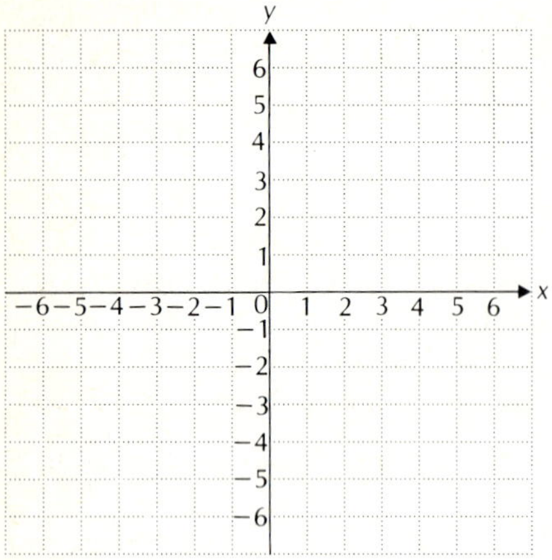

6. $xy = 10$

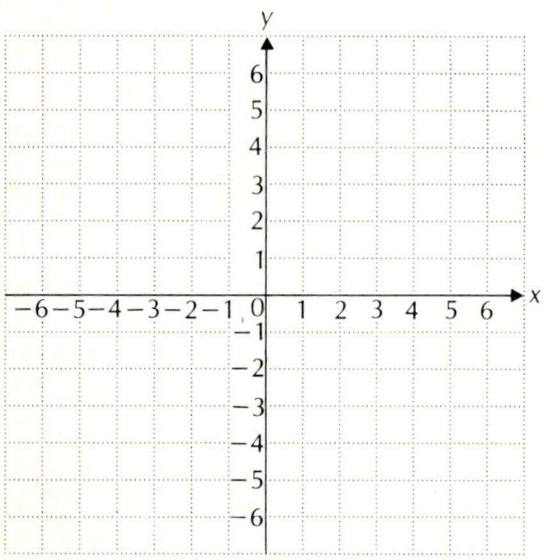

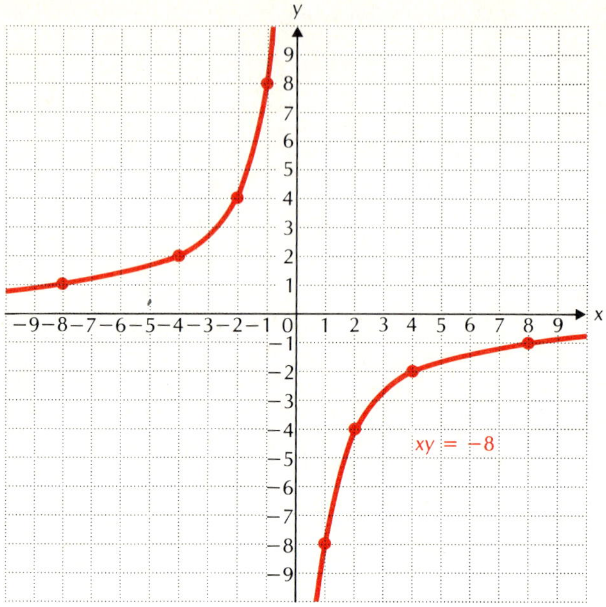

In Example 4, the two parts, or *branches,* of the hyperbola are in the second and fourth quadrants. If the constant c is positive, the two branches will be in the first and third quadrants, as in the graph shown below.

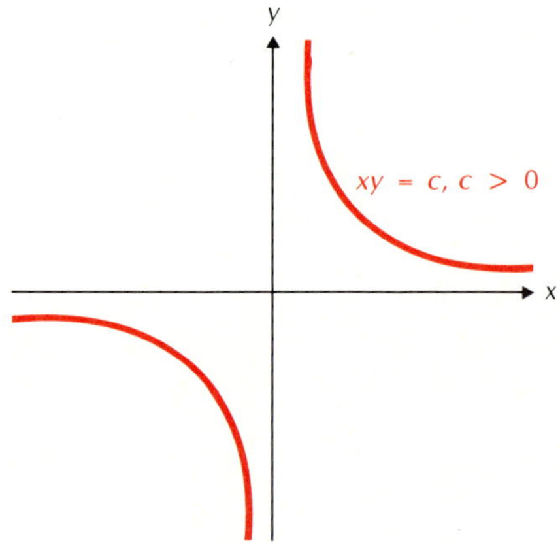

DO EXERCISES 5 AND 6.

EXERCISE SET 10.3

Graph each of the following.

i

1. $y = x^2$

2. $x = y^2$

3. $x = y^2 + 4y + 1$

4. $y = x^2 - 2x + 3$

5. $y = -x^2 + 4x - 5$

6. $x = 4 - 3y - y^2$

7. $x = -3y^2 - 6y - 1$

8. $y = -5 - 8x - 2x^2$

ii

9. $\dfrac{x^2}{16} - \dfrac{y^2}{16} = 1$

10. $\dfrac{y^2}{9} - \dfrac{x^2}{9} = 1$

11. $\dfrac{y^2}{16} - \dfrac{x^2}{9} = 1$

12. $\dfrac{x^2}{9} - \dfrac{y^2}{4} = 1$

13. $\dfrac{x^2}{25} - \dfrac{y^2}{36} = 1$

14. $\dfrac{y^2}{9} - \dfrac{x^2}{25} = 1$

iii

15. $xy = 6$

16. $xy = -4$

17. $xy = -8$

18. $xy = 3$

Problem-solving practice

19. Julie dilutes 7 oz of ammonia with 4 oz of water. How much water will she need to add in order for her mixture to be 25% ammonia?

20. The sum of two numbers is 33 and the difference of their squares is 264. Find the numbers.

21. (*An application: Horsepower of an engine*). The horsepower of a certain kind of engine is given by the formula

$$H = \frac{D^2N}{2.5},$$

where N is the number of cylinders and D is the diameter of each piston. Graph this equation, assuming that $N = 6$ (a 6-cylinder engine). Let D run from 2.5 to 8.

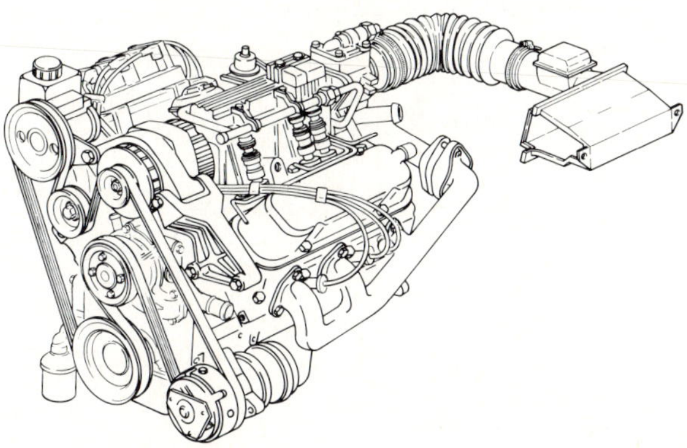

22. Find an equation of a parabola satisfying the given conditions.

a) Line of symmetry parallel to the y-axis and passing through the points $(0, 3)$, $(-1, 6)$, $(2, 9)$.

b) Line of symmetry parallel to the x-axis, vertex $(2, 1)$, and passing through the point $(18, -1)$.

23. Classify the graph of each of the following equations as a circle, an ellipse, a parabola, or a hyperbola.

a) $x^2 + y^2 - 10x + 8y - 40 = 0$

b) $y + 1 = 2x^2$

c) $9x^2 - 4y^2 - 36x + 24y - 36 = 0$

d) $1 - 3y = 2y^2 - x$

e) $4x^2 + 25y^2 - 8x - 100y + 4 = 0$

f) $\dfrac{x^2}{7} + \dfrac{y^2}{7} = 1$

10.4

SYSTEMS OF EQUATIONS

The systems of equations that we have studied so far have all been linear. We now consider some other kinds of systems.

OBJECTIVES

After finishing Section 10.4, you should be able to:

i Solve systems of one first-degree and one second-degree equation.

ii Solve problems involving such systems of equations.

i Algebraic Solutions

We consider systems of one first-degree and one second-degree equation. For example, the graphs may be a circle and a line. If so, there are three possibilities. (See the figure on the following page.)

Solve.

1. $x^2 + y^2 = 25$
   $y - x = -1$

2. $y = x^2 - 2x - 1$
   $y = x + 3$

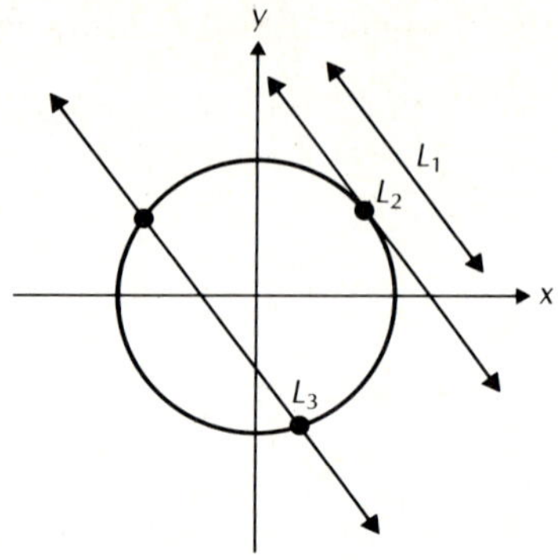

For L_1 there is no point of intersection, hence no solution in real numbers. For L_2 there is one point of intersection, hence one real-number solution. For L_3 there are two points of intersection, hence two real-number solutions.

These systems can be solved graphically by finding the points of intersection. In solving algebraically we use the substitution method.

EXAMPLE 1 Solve the following system:

$$x^2 + y^2 = 25, \quad (1)$$
$$3x - 4y = 0. \quad (2)$$

First solve the linear equation (2) for x:

$$x = \tfrac{4}{3}y. \quad (3)$$

Then substitute $\frac{4}{3}y$ for x in equation (1) and solve for y:

$$\left(\tfrac{4}{3}y\right)^2 + y^2 = 25$$
$$\tfrac{16}{9}y^2 + y^2 = 25$$
$$\tfrac{25}{9}y^2 = 25$$
$$y^2 = 9$$
$$y = \pm 3.$$

Now substitute these numbers into equation (3) and solve for x:

$$x = \tfrac{4}{3}(3), \text{ or } 4; \qquad x = \tfrac{4}{3}(-3), \text{ or } -4.$$

The pairs $(4, 3)$ and $(-4, -3)$ check, so they are solutions.

DO EXERCISES 1 AND 2.

EXAMPLE 2 Solve the following system:

$$y + 3 = 2x, \quad (1)$$
$$x^2 + 2xy = -1. \quad (2)$$

First solve the linear equation (1) for y:

$$y = 2x - 3. \quad (3)$$

Then substitute $2x - 3$ for y in equation (2) and solve for x:

$$x^2 + 2x(2x - 3) = -1$$
$$5x^2 - 6x + 1 = 0$$
$$(5x - 1)(x - 1) = 0 \quad \text{Factoring}$$
$$5x - 1 = 0 \text{ or } x - 1 = 0 \quad \text{Using the principle of zero products}$$
$$x = \tfrac{1}{5} \text{ or } x = 1.$$

Now substitute these numbers into equation (3) and solve for y:

$$y = 2(\tfrac{1}{5}) - 3, \quad \text{or } -\tfrac{13}{5},$$
$$y = 2(1) - 3, \quad \text{or } -1.$$

The pairs $(\frac{1}{5}, -\frac{13}{5})$ and $(1, -1)$ check, so they are solutions.

DO EXERCISE 3.

ii Problem Solving

We now consider solving problems in which the translation is to a system of equations, one quadratic and one linear.

EXAMPLE 3 The sum of two numbers is 14 and the sum of their squares is 106. What are the numbers?

1. **Familiarize** We should read the problem carefully. We then realize that we are looking for two numbers satisfying two conditions.

2. **Translate** Let's use x and y for the numbers. We translate to a system of equations:

The sum of two numbers is 14.
$$x + y = 14$$

The sum of the squares is 106.
$$x^2 + y^2 = 106$$

3. **Carry out** We solve the system:

$$x + y = 14,$$
$$x^2 + y^2 = 106$$

as follows:

$$(14 - y)^2 + y^2 = 106 \quad \text{Solving the first equation for } x \text{ and substituting into the second}$$
$$14^2 - 28y + y^2 + y^2 = 106 \quad \text{Squaring}$$
$$2y^2 - 28y + 90 = 0 \quad \text{Standard form}$$
$$y^2 - 14y + 45 = 0 \quad \text{Multiplying by } \tfrac{1}{2}$$
$$(y - 5)(y - 9) = 0 \quad \text{Factoring}$$
$$y = 5 \quad \text{or} \quad y = 9. \quad \text{Principle of zero products}$$

If $y = 5$, then $x = 9$; if $y = 9$, then $x = 5$. There are two solutions to the system of equations: (5, 9) and (9, 5).

4. **Check** If the numbers are 5 and 9, the sum is 14. The sum of the squares is $5^2 + 9^2 = 25 + 81 = 106$. The numbers check. The pair (9, 5) does not give us another solution.

5. **State** The answer is that the numbers are 5 and 9.

DO EXERCISE 4.

EXAMPLE 4 The perimeter of a rectangular field is 204 m and the area is 2565 m^2. Find the dimensions of the field.

Solve.

3. $y + 3x = 1$
$x^2 - 2xy = 5$

4. The difference of two numbers is 4 and the difference of their squares is 72. What are the numbers?

5. The perimeter of a rectangular field is 34 m and the length of a diagonal is 13 m. Find the dimensions of the field.

1. **Familiarize** We draw a picture of the field, labeling the drawing.

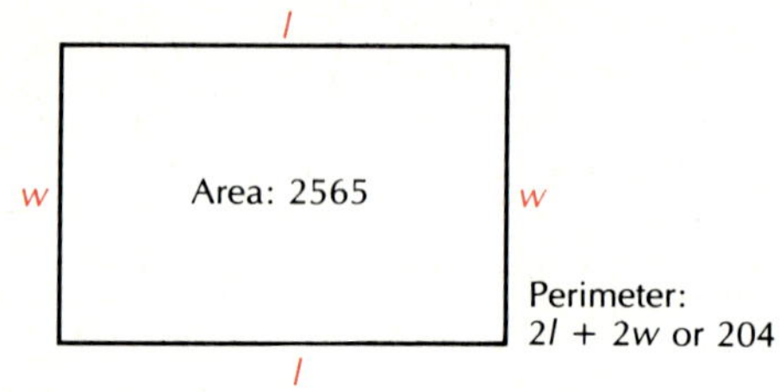

2. **Translate** We have used l and w for the length and width. We then have

Perimeter: $2w + 2l = 204;$

Area: $l \cdot w = 2565.$

3. **Carry out** We solve the system

$$2w + 2l = 204,$$
$$lw = 2565$$

as follows:

$$2w + 2\left(\frac{2565}{w}\right) = 204 \qquad \text{Solving the second equation for } l \text{ and substituting into the first}$$

$$2w^2 + 2 \times 2565 = 204w \qquad \text{Multiplying by } w$$

$$2w^2 - 204w + 2 \times 2565 = 0 \qquad \text{Standard form}$$

$$w^2 - 102w + 2565 = 0 \qquad \text{Multiplying by } \tfrac{1}{2}$$

$$w = \frac{102 \pm \sqrt{102^2 - 4 \cdot (1) \cdot 2565}}{2} \qquad \text{Quadratic formula}$$

$$w = \frac{102 \pm 12}{2}$$

$$w = 57 \quad \text{or} \quad w = 45.$$

If $w = 57$, then $l = 45$; if $w = 45$, then $l = 57$. Since we want w less than l, we have $(57, 45)$.

4. **Check** If $l = 57$ and $w = 45$, then the perimeter is $57 + 57 + 45 + 45$, or 204. The area is $57 \cdot 45 = 2565$. The numbers check.

5. **State** The answer is that the length is 57 m and the width is 45 m.

DO EXERCISE 5.

EXERCISE SET 10.4

i Solve.

1. $x^2 + y^2 = 25,$ $y - x = 1$

2. $x^2 + y^2 = 100,$ $y - x = 2$

3. $4x^2 + 9y^2 = 36,$ $3y + 2x = 6$

4. $9x^2 + 4y^2 = 36,$ $3x + 2y = 6$

5. $y^2 = x + 3,$ $2y = x + 4$

6. $y = x^2,$ $3x = y + 2$

7. $x^2 - xy + 3y^2 = 27,$ $x - y = 2$

8. $x + y = -6,$ $xy = -7$

9. $x^2 + 4y^2 = 25,$
$x + 2y = 7$

10. $y^2 - x^2 = 16,$
$2x - y = 1$

11. $x^2 - xy + 3y^2 = 5,$
$x - y = 2$

12. $2y^2 + xy + x^2 = 7,$
$x - 2y = 5$

13. $3x + y = 7,$
$4x^2 + 5y = 24$

14. $2y^2 + xy = 5,$
$4y + x = 7$

ii Solve.

15. The sum of two numbers is 12 and the sum of their squares is 90. What are the numbers?

16. The sum of two numbers is 15 and the difference of their squares is also 15. What are the numbers?

17. A rectangle has a perimeter of 28 cm and the length of a diagonal is 10 cm. What are its dimensions?

18. A rectangle has a perimeter of 6 m and the length of a diagonal is $\sqrt{5}$ m. What are its dimensions?

19. A rectangle has an area of 20 in^2 and a perimeter of 18 in. Find its dimensions.

20. A rectangle has an area of 2 yd^2 and a perimeter of 6 yd. Find its dimensions.

21. Find an equation of the circle that passes through the points $(4, 6)$, $(-6, 2)$, and $(1, -3)$.

22. Find two numbers whose product is 2 and the sum of whose reciprocals is $\frac{33}{8}$.

23. Find an equation of a circle that passes through $(-2, 3)$ and $(-4, 1)$ and whose center is on the line $5x + 8y = -2$.

24. The sum of two numbers is 1, and their product is 1. Find the sum of their cubes. There is a method to solve this problem that is easier than solving a system of one first-degree equation and one second-degree equation. Can you discover it?

10.5

SYSTEMS OF SECOND-DEGREE EQUATIONS

OBJECTIVES

After finishing Section 10.5, you should be able to:

- **i** **Solve systems of two second-degree equations.**
- **ii** **Solve problems involving such systems.**

i Algebraic Solutions

Two second-degree equations can have common solutions in various ways. If the graphs happen to be a circle and a hyperbola, for example, there are five possibilities, as shown below.

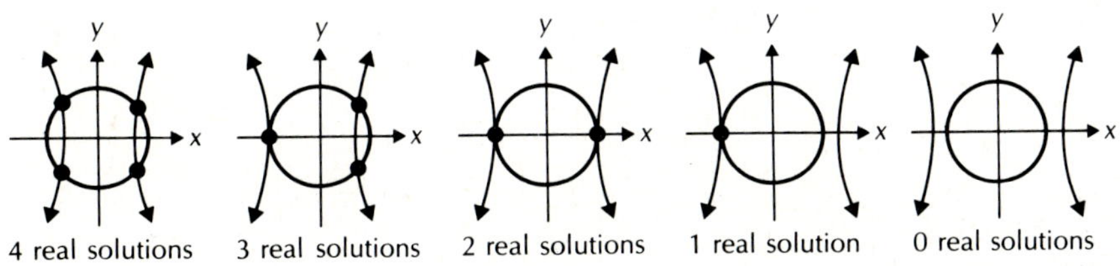

To solve systems of two second-degree equations, we can use either the substitution method or the addition method.

EXAMPLE 1 Solve the following system:

$$2x^2 + 5y^2 = 22, \quad (1)$$
$$3x^2 - y^2 = -1. \quad (2)$$

Here we use the addition method:

$$2x^2 + 5y^2 = 22$$
$$15x^2 - 5y^2 = -5 \quad \text{Multiplying by 5 on both sides of (2)}$$
$$17x^2 = 17 \quad \text{Adding}$$
$$x^2 = 1$$
$$x = \pm 1.$$

Solve.

1. $2y^2 - 3x^2 = 6$
$5y^2 + 2x^2 = 53$

If $x = 1$, $x^2 = 1$, and if $x = -1$, $x^2 = 1$, so substituting 1 or -1 for x in equation (2) we have

$$3 \cdot 1 - y^2 = -1$$
$$y^2 = 4$$
$$y = \pm 2.$$

Thus if $x = 1$, $y = 2$ or $y = -2$, and if $x = -1$, $y = 2$ or $y = -2$. The possible solutions are $(1, 2)$, $(1, -2)$, $(-1, 2)$, and $(-1, -2)$.

Check. Since $(2)^2 = 4$, $(-2)^2 = 4$, $(1)^2 = 1$, and $(-1)^2 = 1$, we can check all four pairs at one time.

$2x^2 + 5y^2 = 22$	
$2(\pm 1)^2 + 5(\pm 2)^2$	22
$2 + 20$	
22	

$3x^2 - y^2 = -1$	
$3(\pm 1)^2 - (\pm 2)^2$	-1
$3 - 4$	
-1	

The solutions are $(1, 2)$, $(1, -2)$, $(-1, 2)$, and $(-1, -2)$.

DO EXERCISE 1.

EXAMPLE 2 Solve this system:

$$x^2 + 4y^2 = 20, \qquad (1)$$
$$xy = 4. \qquad (2)$$

Here we use the substitution method. First we solve equation (2) for y:

$$y = \frac{4}{x}.$$

Then we substitute $4/x$ for y in equation (1) and solve for x:

$$x^2 + 4\left(\frac{4}{x}\right)^2 = 20$$
$$x^2 + \frac{64}{x^2} = 20$$
$$x^4 + 64 = 20x^2 \qquad \text{Multiplying by } x^2$$
$$x^4 - 20x^2 + 64 = 0$$
$$u^2 - 20u + 64 = 0 \qquad \text{Letting } u = x^2$$
$$(u - 16)(u - 4) = 0 \qquad \text{Factoring}$$
$$u = 16 \quad \text{or} \quad u = 4 \qquad \text{Principle of zero products}$$

Then $x = 4$ or $x = -4$ or $x = 2$ or $x = -2$. Since $y = 4/x$, if $x = 4$, $y = 1$; if $x = -4$, $y = -1$; if $x = 2$, $y = 2$; and if $x = -2$, $y = -2$. The ordered pairs $(4, 1)$, $(-4, -1)$, $(2, 2)$, $(-2, -2)$ check. They are the solutions.

Solve.

2. $x^2 + xy + y^2 = 19$
$xy = 6$

DO EXERCISE 2.

ii Problem Solving

EXAMPLE 3 The area of a rectangle is 300 yd^2 and the length of a diagonal is 25 yd. Find the dimensions.

1. **Familiarize** We draw a picture and label it. We note that there is a right triangle in the figure.

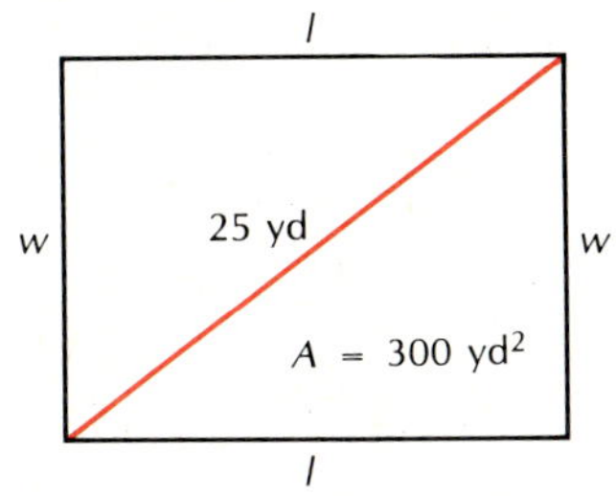

2. **Translate** We translate to a system of equations.

From the Pythagorean theorem: $l^2 + w^2 = 25^2$,

Area: $l \cdot w = 300$.

3. **Carry out** We solve the system

$$l^2 + w^2 = 625,$$
$$lw = 300.$$

We get (20, 15) and (−20, −15).

4. **Check** Lengths of sides cannot be negative, so we need check only (20, 15). In the right triangle, $15^2 + 20^2 = 225 + 400 = 625$, which is 25^2. The area is $15 \cdot 20 = 300$, so we have a solution.

5. **State** The answer is that the length is 20 yd and the width is 15 yd.

DO EXERCISE 3.

3. The area of a rectangle is 2 in^2 and the length of a diagonal is $\sqrt{5}$ in. Find the dimensions of the rectangle.

EXERCISE SET 10.5

i Solve.

1. $x^2 + y^2 = 25,\ y^2 = x + 5$
2. $y = x^2,\ x = y^2$
3. $x^2 + y^2 = 9,\ x^2 - y^2 = 9$
4. $y^2 - 4x^2 = 4,\ 4x^2 + y^2 = 4$
5. $x^2 + y^2 = 25,\ xy = 12$
6. $x^2 - y^2 = 16,\ x + y^2 = 4$
7. $x^2 + y^2 = 4,\ 16x^2 + 9y^2 = 144$
8. $x^2 + y^2 = 25,\ 25x^2 + 16y^2 = 400$
9. $x^2 + y^2 = 16,\ y^2 - 2x^2 = 10$
10. $x^2 + y^2 = 14,\ x^2 - y^2 = 4$
11. $x^2 + y^2 = 5,\ xy = 2$
12. $x^2 + y^2 = 20,\ xy = 8$
13. $x^2 + y^2 = 13,\ xy = 6$
14. $x^2 + 4y^2 = 20,\ xy = 4$
15. $x^2 + y^2 + 6y + 5 = 0,\ x^2 + y^2 - 2x - 8 = 0$
16. $2xy + 3y^2 = 7,\ 3xy - 2y^2 = 4$

ii Solve.

17. The product of two numbers is 156. The sum of their squares is 313. Find the numbers.

18. The product of two numbers is 60. The sum of their squares is 136. Find the numbers.

19. The area of a rectangle is $\sqrt{3}$ m^2 and the length of a diagonal is 2 m. Find the dimensions.

20. The area of a rectangle is $\sqrt{2}$ m^2 and the length of a diagonal is $\sqrt{3}$ m. Find the dimensions.

21. A garden contains two square peanut beds. Find the length of each bed if the sum of their areas is 832 ft^2, and the difference of their areas is 320 ft^2.

22. A certain amount of money saved for 1 yr at a certain interest rate yielded \$7.50. If the principal had been \$25 more and the interest rate 1% less, the interest would have been the same. Find the principal and the rate.

23. The square of a certain number exceeds twice the square of another number by $\frac{1}{8}$. The sum of their squares is $\frac{5}{16}$. Find the numbers.

24. Find an equation of a circle that passes through $(4, -20)$, $(10, -2)$, and $(-4, -4)$.

25. Four squares with sides 5 in. long are cut from the corners of a rectangular metal sheet that has an area of 340 in^2. The edges are bent up to form an open box with a volume of 350 in^3. Find the dimensions of the box.

OBJECTIVES

After finishing Section 10.6, you should be able to:

i **Solve quadratic and other polynomial inequalities.**

ii **Solve rational inequalities.**

10.6

QUADRATIC AND RATIONAL INEQUALITIES

i Quadratic and Certain Other Polynomial Inequalities

Inequalities such as the following are called *quadratic inequalities:*

$$x^2 + 3x - 10 < 0, \qquad 5x^2 - 3x + 2 \geqslant 0.$$

In each case we have a polynomial of degree 2 on the left. One way to solve a quadratic inequality is to graph. Consider the quadratic function $f(x) = x^2 + 3x - 10$ and its graph. Its graph opens upward since the leading coefficient is positive. Function values will be positive to the left and right of the intercepts as shown. We find the intercepts by setting the polynomial equal to 0 and solving:

$$x^2 + 3x - 10 = 0$$

$$(x + 5)(x - 2) = 0$$

$$x + 5 = 0 \quad \text{or} \quad x - 2 = 0$$

$$x = -5 \quad \text{or} \quad x = 2.$$

We sketch the graph using methods we learned in Sections 8.8 and 8.9.

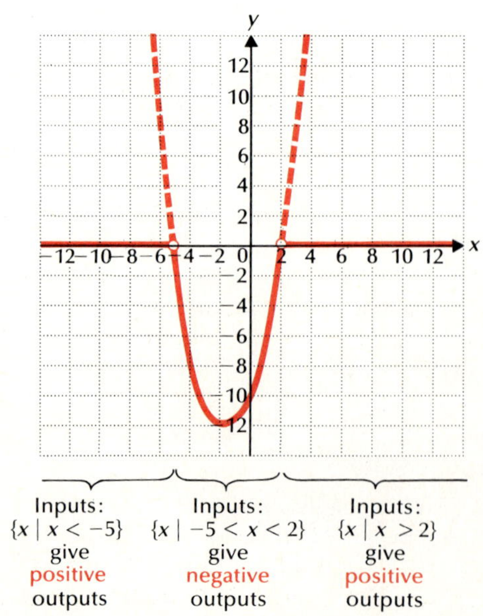

EXAMPLE 1 Solve $x^2 + 3x - 10 < 0$.

Looking at the graph or at least visualizing it tells us that the function is negative for those inputs x such that x is between -5 and 2. That is, the solution set of $x^2 + 3x - 10 < 0$ is

$$\{x \mid -5 < x < 2\}.$$

EXAMPLE 2 Solve $x^2 + 3x - 10 \geqslant 0$.

Looking at the graph or at least visualizing it tells us that the function is nonnegative for those inputs x that are less than or equal to -5 or greater than or equal to 2. The function is 0 at the x-intercepts. The solution set of $x^2 + 3x - 10 \geqslant 0$ is

$$\{x \mid x \leqslant -5 \text{ or } x \geqslant 2\}.$$

DO EXERCISES 1 AND 2.

Let us now consider another method of solving inequalities that works for any polynomial that is factored into a product of first-degree polynomials.

EXAMPLE 3 Solve $x^2 + 3x - 10 < 0$.

We factor the inequality, obtaining $(x + 5)(x - 2) < 0$. The solutions of $(x + 5)(x - 2) = 0$ are -5 and 2. They are not solutions of the inequality, but they divide the real-number line in a natural way, pictured as follows. The product $(x + 5)(x - 2)$ is positive or negative for values other than -5 and 2, depending on the signs of the factors $x + 5$ and $x - 2$. We can determine this efficiency with a diagram as follows.

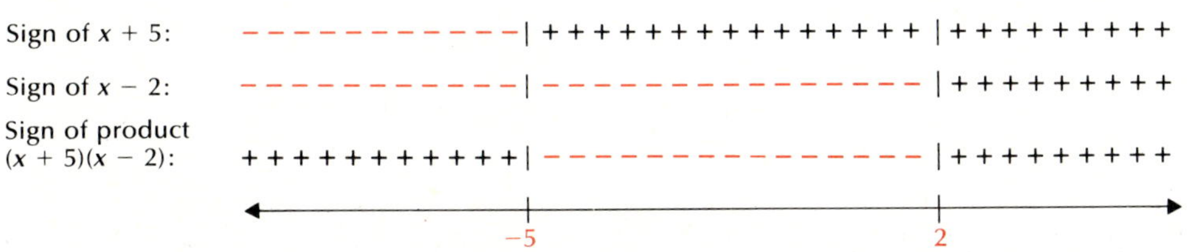

To set up the diagram, solve $x + 5 > 0$. We get $x > -5$. Thus $x + 5$ is positive for all numbers to the right of -5. We indicate that with the $+$ signs. Accordingly, $x + 5 < 0$ for all numbers to the left of -5. We indicate that with the $-$ signs.

Similarly, we solve $x - 2 > 0$, and get $x > 2$. Thus $x - 2$ is positive for all numbers to the right of 2. Accordingly, $x - 2 < 0$ for all numbers to the left of 2. We indicate this with the $+$ and $-$ signs.

Finally, we determine the signs of the product by using the rules for multiplication. In order for the product $(x + 5)(x - 2)$ to be negative, one factor must be positive and the other negative. In the table this situation occurs only when $-5 < x < 2$. The solution set of the inequality is $\{x \mid -5 < x < 2\}$.

Note that setting up a diagram in this manner also tells us the solutions of $(x + 5)(x - 2) > 0$. That solution set is $\{x \mid x < -5 \text{ or } x > 2\}$.

DO EXFRCISES 3 AND 4.

EXAMPLE 4 Solve $7x(x + 3)(x - 2) \geqslant 0$.

The solutions of $7x(x + 3)(x - 2) = 0$ are -3, 0, and 2. They divide the real-number line in a natural way, pictured as follows. The product $7x(x + 3)(x - 2)$ is

Solve by graphing.

1. $x^2 + 2x - 3 > 0$

2. $x^2 + 2x - 3 \leqslant 0$

Solve. Use the method of Example 3.

3. $x^2 + 3x > 4$

4. $x^2 + 3x \leqslant 4$

Solve.

5. $4x(x+1)(x-1) < 0$

positive or negative depending on the signs of the factors $7x$, $x+3$, and $x-2$. We determine this efficiently using a diagram.

Sign of $7x$:	− − − − − − − −	− − − − − − − − − − −	+ + + + + + +	+ + + + + + + + + +
Sign of $x + 3$:	− − − − − − − −	+ + + + + + + + + + +	+ + + + + + +	+ + + + + + + + + +
Sign of $x - 2$:	− − − − − − − −	− − − − − − − − − − −	− − − − − − −	+ + + + + + + + + +
Sign of product $7x(x + 3)(x - 2)$:	− − − − − − − −	+ + + + + + + + + + +	− − − − − − −	+ + + + + + + + + +

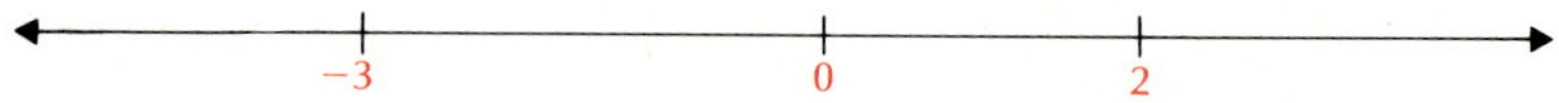

The product of three numbers is positive when either all three are positive or two are negative and one is positive. We see from the diagram that numbers satisfying $-3 < x < 0$ or $2 < x$ are solutions. The intercepts are also solutions. The solution set of the inequality is

$$\{x \mid -3 \leqslant x \leqslant 0 \text{ or } 2 \leqslant x\}.$$

DO EXERCISE 5.

ii Rational Inequalities

We can also use the preceding method when an inequality involves fractional expressions that are quotients of polynomials. We call these *rational inequalities.*

EXAMPLE 5 Solve $\dfrac{x-3}{x+4} \geqslant 2.$

We add -2 to get 0 on one side:

$$\frac{x-3}{x+4} - 2 \geqslant 0.$$

Now we obtain a single fractional expression:

$$\begin{aligned}\frac{x-3}{x+4} - 2 \cdot \frac{x+4}{x+4} &= \frac{x-3}{x+4} - \frac{2(x+4)}{x+4} \\ &= \frac{x-3-2x-8}{x+4} \\ &= \frac{-x-11}{x+4} \geqslant 0.\end{aligned}$$

Let us consider the equality portion:

$$\frac{-x-11}{x+4} = 0.$$

This has the solution -11. Thus -11 is in the solution set. Next, consider the inequality portion:

$$\frac{-x-11}{x+4} > 0.$$

The solution of $x + 4 = 0$ is -4. This cannot be a solution to the inequality because it makes the denominator 0, but -4 and -11 divide the real-number line in a natural way, pictured as follows. The quotient is positive or negative depending on the signs of $-x - 11$ and $x + 4$. We determine this efficiently using a diagram as follows.

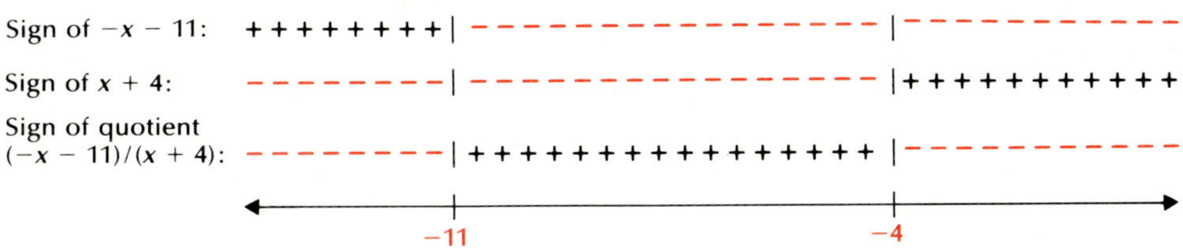

We see from the diagram that the solution set of the inequality

$$(-x - 11)/(x + 4) > 0$$

is $\{x \mid -11 < x < -4\}$. Thus the solution of the original inequality is

$$\{x \mid -11 \leqslant x < -4\}.$$

DO EXERCISES 6 AND 7.

Solve.

6. $\dfrac{x+1}{x-2} \geqslant 3$

7. $\dfrac{x}{x-5} < 2$

EXERCISE SET 10.6

i Solve.

1. $(x-5)(x+3) > 0$

2. $(x-4)(x+1) > 0$

3. $(x+1)(x-2) \leqslant 0$

4. $(x-5)(x+3) \leqslant 0$

5. $x^2 - x - 2 < 0$

6. $x^2 + x - 2 < 0$

7. $9 - x^2 \leqslant 0$

8. $4 - x^2 \geqslant 0$

9. $x^2 - 2x + 1 \geqslant 0$

10. $x^2 + 6x + 9 < 0$

11. $x^2 + 8 < 6x$

12. $x^2 - 12 > 4x$

13. $3x(x+2)(x-2) < 0$

14. $5x(x+1)(x-1) > 0$

15. $(x+3)(x-2)(x+1) > 0$

16. $(x-1)(x+2)(x-4) < 0$

17. $(x+3)(x+2)(x-1) < 0$

18. $(x-2)(x-3)(x+1) < 0$

ii Solve.

19. $\dfrac{1}{x-4} < 0$

20. $\dfrac{1}{x+5} > 0$

21. $\dfrac{x+1}{x-3} > 0$

22. $\dfrac{x-2}{x+5} < 0$

23. $\dfrac{3x+2}{x-3} \leqslant 0$

24. $\dfrac{5-2x}{4x+3} \leqslant 0$

25. $\dfrac{x-1}{x-2} > 3$

26. $\dfrac{x+1}{2x-3} < 1$

27. $\dfrac{(x-2)(x+1)}{x-5} < 0$

28. $\dfrac{(x+4)(x-1)}{x+3} > 0$

29. $\dfrac{x}{x-2} \geqslant 0$

30. $\dfrac{x+3}{x} \leqslant 0$

31. $\dfrac{x-5}{x} < 1$

32. $\dfrac{x}{x-1} > 2$

33. $\dfrac{x-1}{(x-3)(x+4)} < 0$

34. $\dfrac{x+2}{(x-2)(x+7)} > 0$

35. $2 < \dfrac{1}{x}$

36. $\dfrac{1}{x} \leqslant 3$

Solve.

37. $x^2 - 2x \leqslant 2$

38. $x^2 + 2x > 4$

SUMMARY AND REVIEW: CHAPTER 10

The following contains a summary of what you should be able to do after completing this chapter. The review exercises are for practice. Answers are at the back of the book. If you miss an exercise, restudy the section indicated alongside the answer.

You should be able to:

Use the distance formula to find the distance between two points.

Find the distance between the given points.

1. $(2, 6)$ and $(6, 6)$
2. $(-1, 1)$ and $(-5, 4)$
3. $(4, 7)$ and $(-3, -2)$
4. $(2, 3a)$ and $(-1, a)$

Find the center and radius of a circle, given an equation of a circle.

Find the center and radius.

5. $(x + 2)^2 + (y - 3)^2 = 2$
6. $(x - 5)^2 + y^2 = 49$
7. $x^2 + y^2 + 6x - 2y + 1 = 0$
8. $x^2 + y^2 + 8x - 6y - 10 = 0$

Graph a circle and an ellipse.

Graph.

9. $4x^2 + 4y^2 = 100$
10. $9x^2 + 2y^2 = 18$

Graph a hyperbola and a parabola.

Graph.

11. $y = x^2 - 2x + 5$
12. $\dfrac{x^2}{64} - \dfrac{y^2}{9} = 1$

Solve a system of equations involving one linear and one quadratic equation.

Solve.

13. $x^2 - y^2 = 33,$
    $x + y = 11$
14. $x^2 - 2x + 2y^2 = 8,$
    $2x + y = 6$

Solve a problem by solving a system of equations, one linear and one quadratic.

Solve.

15. A rectangle has a perimeter of 38 m and an area of 84 m^2. What are its dimensions?
16. Find two positive integers whose sum is 12 and the sum of whose reciprocals is $\frac{3}{8}$.

Solve a system of quadratic equations.

Solve.

17. $x^2 + y^2 = 100,$
    $2x^2 - 3y^2 = -120$
18. $x^2 + 2y^2 = 12,$
    $xy = 4$

Solve quadratic and rational inequalities.

Solve.

19. $x^2 < 6x + 7$
20. $\dfrac{x - 5}{x + 3} < 0$

○

21. Solve.

$$3x^2 + 4y^2 = 8,$$
$$x^2 - y^2 = 5$$

22. Find the points whose distance from $(8, 0)$ and from $(-8, 0)$ is 10.

TEST: CHAPTER 10

Find the distance between the given points.

1. $(4, -1)$ and $(-5, 8)$
2. $(3, -a)$ and $(-3, a)$

Find the center and radius of the circle.

3. $(x + 2)^2 + (y - 3)^2 = 64$

4. $x^2 + y^2 + 4x - 6y + 4 = 0$

Graph.

5. $y = x^2 + 3x + 5$

6. $x^2 + y^2 + 2x + 6y + 6 = 0$

7. $\dfrac{x^2}{36} - \dfrac{y^2}{25} = 1$

8. $8x^2 + 4y^2 = 32$

Solve.

9. $x^2 + y^2 = 157,$
$x - y = 5$

10. $3x^2 + 2y^2 = 8,$
$-4x^2 + 3y^2 = 2$

11. In a fractional expression, the sum of the values of the numerator and denominator is 23. The product of their values is 120. Find the values of the numerator and denominator.

12. A rectangle with diagonal of length $5\sqrt{5}$ has an area of 22. Find the dimensions of the rectangle.

13. Solve: $(x + 2)(x - 1)(x - 2) > 0$.

14. Solve: $x^2 + 2x > 5$.

11

SEQUENCES, SERIES, AND PROBABILITY

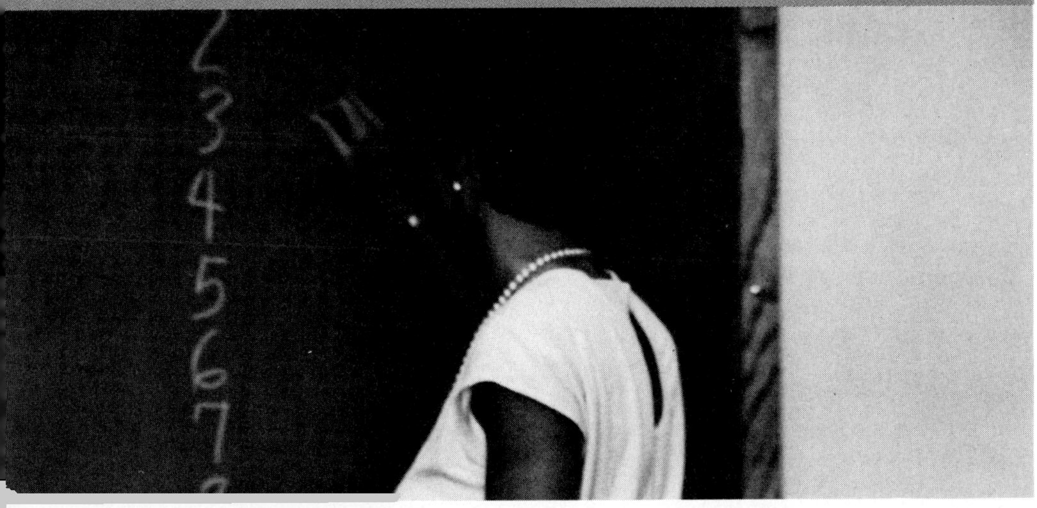

What is the sum of the first 375 natural numbers? The algebra of series can be used to solve such a problem without doing the addition.

By a *sequence* of numbers, we mean a list, or set of numbers, in order. That list might be short or it might be long. It can even be unending. If we add the numbers in such a set, we say that we have a *series*. The applications of sequences and series are far-reaching, so in this chapter we present a brief introduction to the subject. Even though we do not delve deeply into the subject, we will be able to use the new knowledge to solve some problems.

In this chapter, we also present a brief introduction to the idea of probability. We define probability and then do some calculations that should prove interesting.

OBJECTIVES

After finishing Section 11.1, you should be able to:

i Given a formula for the nth term (general term) of a sequence, find any term in the sequence, given a value for n.

ii Given a sequence, look for a pattern and try to guess a rule or formula for the general term.

iii Find sums of sequences.

iv Convert between sigma ($\sum$) notation and other notation for the sum of a sequence.

1. A sequence is given by

$$a(n) = 2n - 1.$$

a) Find the first 3 terms.

b) Find the 34th term.

2. A sequence is given by

$$a_n = \frac{(-1)^n}{n-1}, \quad n \geqslant 2.$$

a) Find the first 4 terms.

b) Find the 48th term.

11.1

SEQUENCES AND SERIES

In this section we discuss sets of numbers, considered in some order, and their sums.

i Sequences

When the elements of a set are considered in order, we say that we have a *sequence*. We specify which element is first, which is second, and so on. That amounts to assigning elements of the set to certain natural numbers. Here is an example:

red, white, blue, purple

is a sequence in which red is assigned to the number 1, white is assigned to the number 2, and so on. A sequence is thus a *function*, where the domain is some set of natural numbers.

A *sequence* is a function having for its domain some set of natural numbers.

As another example, consider the sequence given by*

$$a(n) = 2^n, \quad \text{or} \quad a_n = 2^n.$$

Some of the function values (also known as *terms* of the sequence) are as follows:

$$a_1 = 2^1 = 2, \qquad a_3 = 2^3 = 8,$$
$$a_2 = 2^2 = 4, \qquad a_6 = 2^6 = 64.$$

The first term of the sequence is a_1, the fifth term is a_5, and the nth term or *general* term is a_n. This sequence can also be denoted in the following ways:

a) $2, 4, 8, \ldots;$

b) $2, 4, 8, \ldots, 2^n.$

EXAMPLE 1 Find the first 4 terms and the 57th term of the sequence whose general term is given by $a_n = (-1)^n/(n+1)$.

$$a_1 = \frac{(-1)^1}{1+1} = -\frac{1}{2},$$

$$a_2 = \frac{(-1)^2}{2+1} = \frac{1}{3},$$

$$a_3 = \frac{(-1)^3}{3+1} = -\frac{1}{4},$$

$$a_4 = \frac{(-1)^4}{4+1} = \frac{1}{5},$$

$$a_{57} = \frac{(-1)^{57}}{57+1} = -\frac{1}{58}.$$

DO EXERCISES 1 AND 2.

* The notation a_n means the same as $a(n)$ but is more commonly used with sequences.

ii When a sequence is described merely by naming the first few terms, we do not know for sure what the general term is, but the reader is expected to make a guess by looking for a pattern.

EXAMPLES For each sequence, make a guess at the general term.

2. 1, 4, 9, 16, 25, . . .

These are squares of numbers, so the general term may be n^2.

3. $\sqrt{1}, \sqrt{2}, \sqrt{3}, \sqrt{4}, \ldots$

These are square roots of numbers, so the general term may be $\sqrt{n}$.

4. $-1, 2, -4, 8, -16, \ldots$

These are powers of 2 with alternating signs, so the general term may be $(-1)^n[2^{n-1}]$.

5. 2, 4, 8, . . .

If we see the pattern of powers of 2, we will see 16 as the next term and guess 2^n for the general term. If we see that we can get the second term by adding 2, the third term by adding 4, and the next term by adding 6, and so on, we will see 14 as the next term. A general term for the sequence is then $n^2 - n + 2$.

DO EXERCISES 3–7.

iii Sums, or Series

With any sequence there is associated another sequence, obtained by taking sums of terms of the given sequence. The general term of this sequence of "partial sums" is denoted S_n. Consider the sequence

$$3, 5, 7, 9, \ldots, 2n + 1.$$

We construct some terms of the sequence of partial sums:

$S_1 = 3,$ This is the first term of the given sequence.

$S_2 = 3 + 5 = 8,$ This is the sum of the first two terms.

$S_3 = 3 + 5 + 7 = 15,$ The sum of the first three terms.

$S_4 = 3 + 5 + 7 + 9 = 24.$ The sum of the first four terms.

A sum like any of the above is called a *series.*

EXAMPLE 6 For the sequence $-2, 4, -6, 8, -10, 12, -14$, (a) find S_3; (b) find S_5.

a) $S_3 = -2 + 4 + (-6) = -4$

b) $S_5 = -2 + 4 + (-6) + 8 + (-10) = -6$

DO EXERCISE 8.

iv Sigma Notation

The Greek letter $\sum$ (sigma) can be used to simplify notation when a series has a formula for the general term.

The sum of the first four terms of the sequence $3, 5, 7, 9, \ldots, 2k + 1$ can be named as follows:

$$\sum_{k=1}^{4} (2k + 1).$$

This is read "the sum as k goes from 1 to 4 of $(2k + 1)$."

For each sequence try to find a rule for finding the general term, or the nth term. Answers may vary.

3. 2, 4, 6, 8, 10, . . .

4. 1, 2, 3, 4, 5, 6, . . .

5. 1, 8, 27, 64, 125, . . .

6. $x, \frac{x^2}{2}, \frac{x^3}{3}, \frac{x^4}{4}, \frac{x^5}{5}, \ldots$

7. 1, 2, 4, 8, 16, 32, . . .

8. For the sequence

$1, 2, -1, -2, 3, -3, 4, -4,$

find each of the following.

a) S_4

b) S_7

EXAMPLES Rename the following sums without using sigma notation.

7. $\sum_{k=1}^{5} k^2 = 1^2 + 2^2 + 3^2 + 4^2 + 5^2$

8. $\sum_{k=1}^{4} (-1)^k(2k) = (-1)^1(2 \cdot 1) + (-1)^2(2 \cdot 2) + (-1)^3(2 \cdot 3) + (-1)^4(2 \cdot 4)$
$= -2 + 4 - 6 + 8$

EXAMPLES Write sigma notation for each sum.

9. $1 + 4 + 9 + 16 + 25$

This is a sum of squares, $1^2 + 2^2 + 3^2 + 4^2 + 5^2$, so the general term is k^2. Sigma notation is

$$\sum_{k=1}^{5} k^2.$$

10. $-1 + 3 - 5 + 7$

These are odd integers with alternating signs. Therefore, the general term is $(-1)^k(2k - 1)$ beginning with $k = 1$. Sigma notation is

$$\sum_{k=1}^{4} (-1)^k(2k - 1).$$

DO EXERCISES 9–15.

Name each series without using $\sum$.

9. $\sum_{k=1}^{3} 2 + \frac{1}{k}$

10. $\sum_{k=0}^{4} 5^k$

11. $\sum_{k=8}^{11} k^3$

Name each series with sigma notation.

12. $2 + 4 + 6 + 8 + 10$

13. $1 + 8 + 27 + 64$

14. $x + \frac{x^2}{2} + \frac{x^3}{3} + \frac{x^4}{4} + \frac{x^5}{5} + \frac{x^6}{6}$

15. $2 + 3 + 4 + 5$

EXERCISE SET 11.1

i In each of the following, the nth term of a sequence is given. In each case find the first 4 terms, the 10th term, a_{10}, and the 15th term, a_{15}.

1. $a_n = 3n + 1$ **2.** $a_n = 3n - 1$ **3.** $a_n = \frac{n}{n+1}$ **4.** $a_n = n^2 + 1$

5. $a_n = n^2 - 2n$ **6.** $a_n = \frac{n^2 - 1}{n^2 + 1}$ **7.** $a_n = n + \frac{1}{n}$ **8.** $a_n = \left(-\frac{1}{2}\right)^{n-1}$

ii For each sequence, find the general term, or nth term, a_n, or a rule for finding a_n. Answers may vary.

9. $1, 3, 5, 7, 9, \ldots$ **10.** $3, 9, 27, 81, 243, \ldots$ **11.** $\frac{2}{3}, \frac{3}{4}, \frac{4}{5}, \frac{5}{6}, \frac{6}{7}, \ldots$

12. $\sqrt{2}, \sqrt{4}, \sqrt{6}, \sqrt{8}, \sqrt{10}, \ldots$ **13.** $\sqrt{3}, 3, 3\sqrt{3}, 9, 9\sqrt{3}, \ldots$ **14.** $1 \cdot 2, 2 \cdot 3, 3 \cdot 4, 4 \cdot 5, \ldots$

15. $-1, -4, -7, -10, -13, \ldots$ **16.** $\log 1, \log 10, \log 100, \log 1000, \ldots$

iii For each sequence, find the associated sum.

17. $1, 2, 3, 4, 5, 6, 7, \ldots$. Find S_7. **18.** $1, -3, 5, -7, 9, -11, \ldots$. Find S_8.

19. $2, 4, 6, 8, \ldots$. Find S_5. **20.** $1, \frac{1}{4}, \frac{1}{9}, \frac{1}{16}, \frac{1}{25}, \ldots$. Find S_5.

iv Rename each series without using $\sum$.

21. $\sum_{k=1}^{5} \frac{1}{2k}$ **22.** $\sum_{k=1}^{6} \frac{1}{2k+1}$ **23.** $\sum_{k=0}^{5} 2^k$ **24.** $\sum_{k=4}^{7} \sqrt{2k-1}$ **25.** $\sum_{k=7}^{10} \log k$ **26.** $\sum_{k=0}^{4} \pi k$

Write sigma notation.

27. $\frac{1}{2} + \frac{2}{3} + \frac{3}{4} + \frac{4}{5} + \frac{5}{6} + \frac{6}{7}$ **28.** $3 + 6 + 9 + 12 + 15$ **29.** $-2 + 4 - 8 + 16 - 32 + 64$

30. $\frac{1}{1^2} + \frac{1}{2^2} + \frac{1}{3^2} + \frac{1}{4^2} + \frac{1}{5^2}$ **31.** $4 - 9 + 16 - 25 + \cdots (-1)^n n^2$ **32.** $9 - 16 + 25 + \cdots + (-1)^{n+1} n^2$

Problem-solving practice

33. A cable television company has 20,000 subscribers. To offset the cost of installing a new satellite dish, the company can either increase its number of subscribers by 15% or increase its monthly rate to $8.82. What is the present monthly rate for cable television subscribers?

34. The area of one square is 60 more than the perimeter of a second square. If a side of the first square is 7 more than a side of the second square, what is the area of the second square?

Find the first five terms of each sequence, and then find S_5.

35. $a_n = \frac{1}{2n} \log 1000^n$

36. $a_n = i^n,\ i = \sqrt{-1}$

37. $a_n = \ln(1 \cdot 2 \cdot 3 \cdots n)$

38. **a)** Find the first few terms of the sequence $a_n = n^2 - n + 41$.
b) What pattern do you observe?
c) Find the 41st term. Does the pattern you found in (b) still hold?

Find decimal notation, rounded to six decimal places, for the first six terms of each sequence.

39. 🖩 $a_n = \left(1 + \frac{1}{n}\right)^n$

40. 🖩 $a_n = \sqrt{n+1} - \sqrt{n}$

41. Find a formula for S_n, given that $a_n = \frac{1}{n} \cdot \frac{1}{n+1}$.

11.2

ARITHMETIC SEQUENCES AND SERIES

In this section, we concentrate on a particular kind of sequence and its associated series.

i Arithmetic Sequences

Consider this sequence:

$$2, 5, 8, 11, 14, 17, \ldots.$$

Note that adding 3 to any term produces the following term. Or, the difference between any term and the preceding one is 3. Sequences in which the difference between successive terms is constant are called *arithmetic sequences.*

EXAMPLES The following are arithmetic sequences. Identify the first term, a_1, and the common difference, d.

Sequence	First term, a_1	Common difference, d
1. 4, 9, 14, 19, 24, . . .	4	5
2. 34, 27, 20, 13, 6, −1, −8, . . .	34	−7
3. 2, $2\frac{1}{2}$, 3, $3\frac{1}{2}$, 4, $4\frac{1}{2}$, . . .	2	$\frac{1}{2}$

We now find a formula for the general, or nth, term of any arithmetic sequence. Let us denote the difference between successive terms (called the *common difference*) by d, and write out the first few terms:

$$a_1,$$
$$a_2 = a_1 + d,$$
$$a_3 = a_2 + d = (a_1 + d) + d = a_1 + 2d,$$
$$a_4 = a_3 + d = (a_1 + 2d) + d = a_1 + 3d.$$

DO EXERCISE 1.

OBJECTIVES

After finishing Section 11.2, you should be able to:

i **Use Formula 1 to find, for any arithmetic sequence,**
a) the *n*th term when *n* is given;
b) *n*, when an *n*th term is given.
c) Given two terms of a sequence, find the common difference and construct the sequence.

ii **Use Formulas 2 and 3 for arithmetic sequences to find the sum of any arithmetic sequence.**

iii **Solve problems using arithmetic sequences.**

1. In the following arithmetic sequence, identify the first term a_1 and the common difference d.

3.1, 3.9, 4.7, 5.5, 6.3

2. Find the 13th term of the sequence 2, 6, 10, 14, ...

Generalizing, we obtain the following.

1. The *n*th term of an arithmetic sequence is given by

$$a_n = a_1 + (n - 1)d,$$

where *d* is the common difference between successive terms.

EXAMPLE 4 Find the 14th term of the arithmetic sequence 4, 7, 10, 13, ...

First note that $a_1 = 4$, $d = 3$, and $n = 14$. Then using formula (1), we obtain

$$\begin{aligned} a_{14} &= 4 + (14 - 1) \cdot 3 \\ &= 4 + 39 \\ &= 43. \end{aligned}$$

DO EXERCISE 2.

3. In the sequence given in Exercise 2, what term would be 298? That is, what is n if $a_n = 298$?

EXAMPLE 5 In the sequence in Example 4, which term is 301? That is, what is n if $a_n = 301$?

We substitute into formula (1) and solve for n:

$$\begin{aligned} a_n &= a_1 + (n - 1)d \\ 301 &= 4 + (n - 1) \cdot 3 \\ 301 &= 4 + 3n - 3 \\ 300 &= 3n \\ 100 &= n. \end{aligned}$$

The 100th term is 301.

DO EXERCISE 3.

Given two terms and their places in an arithmetic sequence, we can construct the sequence.

4. The 7th term of an arithmetic sequence is 79 and the 13th term is 151. Find a_1 and d. Construct the sequence.

EXAMPLE 6 The third term of an arithmetic sequence is 8, and the sixteenth term is 47. Find a_1 and d and construct the sequence.

We know that $a_3 = 8$ and $a_{16} = 47$. Thus we would have to add d thirteen times to get from 8 to 47. That is,

$$13d = 47 - 8 = 39.$$

Solving, we obtain

$$d = 3.$$

Since $a_3 = 8$, we subtract d twice to get to a_1. Thus

$$a_1 = 2.$$

The sequence is 2, 5, 8, 11, ...

DO EXERCISE 4.

ii Sums, or Series

We now look for a formula for the sum of the first n terms of an arithmetic sequence. Let us denote the terms as follows:

$$a_1, \quad (a_1 + d), \quad (a_1 + 2d), \ldots, (a_n - 2d), \quad (a_n - d), \quad a_n.$$

Then the sum of the first n terms is given by

$$S_n = a_1 + (a_1 + d) + (a_1 + 2d) + \cdots + (a_n - 2d) + (a_n - d) + a_n.$$

Reversing the right side, we obtain

$$S_n = a_n + (a_n - d) + (a_n - 2d) + \cdots + (a_1 + 2d) + (a_1 + d) + a_1.$$

Adding these two equations, we obtain

$$2S_n = (a_1 + a_n) + (a_1 + a_n) + \cdots + (a_1 + a_n),$$

a total of n terms on the right. Thus $2S_n = n(a_1 + a_n)$. Dividing by 2 gives us the formula we seek.

2. The sum of the first n terms of an arithmetic sequence is given by

$$S_n = \frac{n}{2}(a_1 + a_n).$$

We now derive a variant of this formula, which we can use in case a_n is not known. We substitute the expression for a_n given in formula (1), $a_n = a_1 + (n - 1)d$, into formula (2):

$$S_n = \frac{n}{2}(a_1 + [a_1 + (n - 1)d]).$$

Thus we have the following.

3. The sum of the first n terms of an arithmetic sequence is given by

$$S_n = \frac{n}{2}[2a_1 + (n - 1)d].$$

EXAMPLE 7 Find the sum of the first 100 natural numbers.

The series is $1 + 2 + 3 + \cdots + 100$, so $a_1 = 1$, $a_n = 100$, and $n = 100$. We substitute into formula (2):

$$S_n = \frac{n}{2}(a_1 + a_n)$$

$$S_{100} = \frac{100}{2}(1 + 100) = 5050.$$

EXAMPLE 8 Find the sum of the first 14 terms of the arithmetic sequence 2, 5, 8, 11, 14, ...

We note that $a_1 = 2$, $d = 3$, and $n = 14$. We use formula (3):

$$S_n = \frac{n}{2}[2a_1 + (n - 1)d]$$

$$S_{14} = \frac{14}{2} \cdot [2 \cdot 2 + (14 - 1)3] = 7 \cdot [4 + 13 \cdot 3] = 7 \cdot 43$$

$$S_{14} = 301.$$

DO EXERCISES 5 AND 6.

5. Find the sum of the first 200 natural numbers.

6. Find the sum of the first 15 terms of the arithmetic sequence

1, 3, 5, 7, 9, ...

iii Problem Solving

For some problem situations, the translation may involve sequences or series. We look at some examples.

EXAMPLE 9 You take a job, starting with an hourly wage of \$4.25. You are promised a raise of 5¢ per hour every 2 months for 5 years. At the end of 5 years, what will be your hourly wage?

1. **Familiarize** One thing to do is write down your hourly wage for several two-month time periods.

   Beginning: 4.25,

   After two months: 4.30,

   After four months: 4.35,

   and so on.

   What appears is a *sequence of numbers*: 4.25, 4.30, 4.35, ... Is it an arithmetic sequence? Yes, because we add 0.05 each time in order to get the next term.

   We ask ourselves what we know about arithmetic sequences. We recall, or if necessary look up, the pertinent formula(s). There are two:

   $$a_n = a_1 + (n-1)d \quad \text{and} \quad S_n = \frac{n}{2}(a_1 + a_n).$$

   In this case, we are not looking for a *sum*, so it is probably the first formula that will give us our answer. We want to know the last term in a sequence. We will need to know a_1, n, and d. From our list above, we see that

   $$a_1 = 4.25 \quad \text{and} \quad d = 0.05.$$

   What is n? That is, how many terms are in the sequence? Each year there are 6 raises, since you get a raise every 2 months. There are 5 years, so the total number of raises will be 5×6, or 30. There will be 31 terms: the original wage and 30 increased rates.

2. **Translate** We want to find the 31st term of an arithmetic sequence, with $a_1 = 4.25$, $d = 0.05$, and $n = 31$. We substitute into the formula:

   $$a_{31} = 4.25 + (31 - 1) \times 0.05.$$

3. **Carry out** We calculate, to obtain 5.75.

4. **Check** We can check the calculations. We can also calculate in a slightly different way for another check. For example, at the end of a year, there will be 6 raises, for a total raise of 30¢. At the end of 5 years, the total raise will be 5×30, or 1.50. If we add that to the original wage of 4.25, we obtain 5.75. The answer checks.

5. **State** At the end of 5 years, your hourly wage will be \$5.75.

Example 9 is one in which the calculations or the translation could be done in a number of different ways. It is possible to solve the problem in several different, and very good ways. Here is a point to remember: There is often a variety of ways in which a problem can be solved. You should use the one that is best or easiest for you. In this chapter, however, we will concentrate on the use of sequences and series in problem solving.

EXAMPLE 10 A stack of telephone poles has 30 poles in the bottom row. There are 29 poles in the second row, 28 in the next row, and so on. How many poles are in the stack?

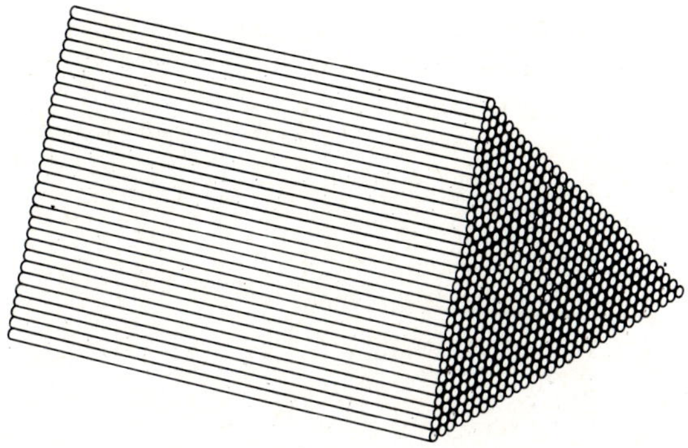

1. **Familiarize** A picture will help in this case. The following figure shows the ends of the poles and the way in which they stack. There are 30 poles on the bottom, and we see that there will be one fewer in each succeeding row. How many rows will there be?

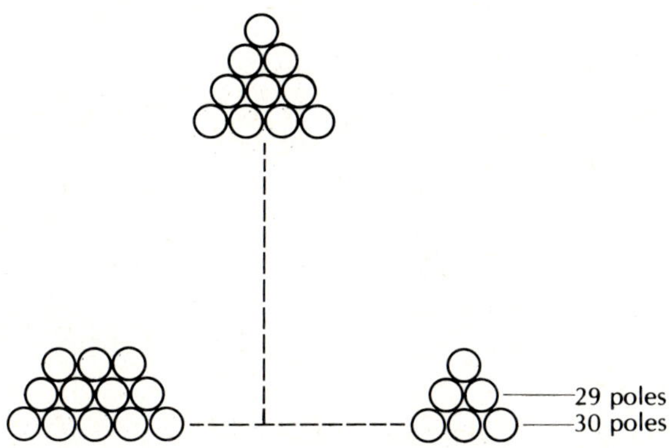

We go from 30 poles in a row, down to one pole in the top row, so there must be 30 rows.

We want the sum

$$30 + 29 + 28 + \cdots + 1.$$

Thus we want the sum of an arithmetic sequence. We recall, or look up, if necessary, the formula for the sum of an arithmetic sequence:

$$S_n = \frac{n}{2}(a_1 + a_n).$$

2. **Translate** We want to find the sum of an arithmetic sequence, where $a_1 = 30$ and, since there are 30 terms, $n = 30$. There is just one pole on top, so $a_{30} = 1$.

Substituting into the formula, we have

$$S_{30} = \frac{30}{2}(30 + 1).$$

7. A stack of iron rods has 112 rods on the bottom row, 111 rods on the next row, and so on. The top row has 48 rods. How many rods are in the stack?

3. **Carry out** We calculate, and find that $S_{30} = 465$.

4. **Check** In this case we can check the calculations by doing them again. A longer, harder way would be to do the entire addition, $30 + 29 + 28 + \cdots + 1$.

5. **State** The answer is that there are 465 poles in the stack.

DO EXERCISE 7.

EXERCISE SET 11.2

i For the arithmetic sequences in Exercises 1–6, find the first term and the common difference.

1. $2, 7, 12, 17, \ldots$

2. $1.06, 1.12, 1.18, 1.24, \ldots$

3. $7, 3, -1, -5, \ldots$

4. $-9, -6, -3, 0, \ldots$

5. $\frac{3}{2}, \frac{9}{4}, 3, \frac{15}{4}, \ldots$

6. $\frac{3}{5}, \frac{1}{10}, -\frac{2}{5}, \ldots$

7. Find the 12th term of the arithmetic sequence $2, 6, 10, \ldots$.

8. Find the 11th term of the arithmetic sequence $0.07, 0.12, 0.17, \ldots$.

9. Find the 17th term of the arithmetic sequence $7, 4, 1, \ldots$.

10. Find the 14th term of the arithmetic sequence $3, \frac{7}{3}, \frac{5}{3}, \ldots$.

11. In the sequence of Exercise 7, what term is 106?

12. In the sequence of Exercise 8, what term is 1.67?

13. In the sequence of Exercise 9, what term is -296?

14. In the sequence of Exercise 10, what term is -27?

15. Find a_{17} when $a_1 = 5$ and $d = 6$.

16. Find a_{20} when $a_1 = 14$ and $d = -3$.

17. Find a_1 when $d = 4$ and $a_8 = 33$.

18. Find d when $a_1 = 8$ and $a_{11} = 26$.

19. Find n when $a_1 = 5$, $d = -3$, and $a_n = -76$.

20. Find n when $a_1 = 25$, $d = -14$, and $a_n = -507$.

21. In an arithmetic sequence $a_{17} = -40$ and $a_{28} = -73$. Find a_1 and d. Write the first 5 terms of the sequence.

22. In an arithmetic sequence $a_{17} = \frac{25}{3}$ and $a_{32} = \frac{95}{6}$. Find a_1 and d. Write the first 5 terms of the sequence.

ii

23. Find the sum of the first 20 terms of the series $5 + 8 + 11 + 14 + \cdots$.

24. Find the sum of the first 14 terms of the series $11 + 7 + 3 + \cdots$.

25. Find the sum of the even numbers from 2 to 100, inclusive.

26. Find the sum of the odd numbers from 1 to 99, inclusive.

27. Find the sum of the multiples of 7, from 7 to 98 inclusive.

28. Find the sum of all multiples of 4 that are between 14 and 523.

29. If an arithmetic series has $a_1 = 2$, $d = 5$, and $n = 20$, what is S_n?

30. If an arithmetic series has $a_1 = 7$, $d = -3$, and $n = 32$, what is S_n?

iii

31. A gardener is making a triangular planting, with 35 plants in the front row, 31 in the second row, 27 in the third row, and so on. If the pattern is consistent, how many plants will there be in the last row?

32. A formation of a marching band has 14 marchers in the front row, 16 in the second row, 18 in the third row, and so on, for 25 rows. How many marchers are in the last row?

33. How many poles will be in a pile of telephone poles if there are 50 in the first layer, 49 in the second, and so on, until there is one in the last layer?

34. If 10¢ is saved on October 1, 20¢ on October 2, 30¢ on October 3, and so on, how much is saved during October? (October has 31 days.)

○

35. Find a formula for the sum of the first n odd natural numbers.

36. Find three numbers in an arithmetic sequence such that the sum of the first and third is 10 and the product of the first and second is 15.

37. Find the first term and the common difference for the arithmetic sequence where $a_2 = 4p - 3q$ and $a_4 = 10p + q$.

38. 🖩 Find the first 10 terms of the arithmetic sequence for which $a_1 = \$8760$ and $d = -\$798.23$.

39. 🖩 Find the sum of the first 10 terms of the sequence given in Exercise 38.

40. Prove that if p, m, and q form an arithmetic sequence, then

$$m = \frac{p + q}{2}.$$

11.3

GEOMETRIC SEQUENCES AND SERIES

There are many kinds of sequences and series. In this section we consider another type.

OBJECTIVES

After finishing Section 11.3, you should be able to:

i Use Formula 4 for geometric sequences to find a given term.

ii Find sums of geometric sequences.

iii Use geometric sequences to solve problems.

i Geometric Sequences

Consider this sequence:

$$3, 6, 12, 24, 48, 96, \ldots .$$

Note that multiplying any term by 2 produces the following term. That is, the ratio of any term and the preceding one is 2. Sequences in which the ratio of successive terms is constant (but different from 1) are called *geometric sequences*.

EXAMPLES Each of the following are geometric sequences. Identify the common ratio.

	Sequence	Common ratio
1.	$3, 6, 12, 24, \ldots$	2
2.	$3, -6, 12, -24, \ldots$	-2
3.	$1, \frac{1}{2}, \frac{1}{4}, \frac{1}{8}, \ldots$	$\frac{1}{2}$

We now find a formula for the general, or nth, term of any geometric sequence. Let us denote the ratio of successive terms (called the *common ratio*) by r, and write out the first few terms:

$$a_1,$$
$$a_2 = a_1 r,$$
$$a_3 = a_2 r = (a_1 r)r = a_1 r^2,$$
$$a_4 = a_3 r = (a_1 r^2)r = a_1 r^3.$$

Generalizing, we obtain the following.

4. The nth term of a geometric sequence is given by

$$a_n = a_1 r^{n-1},$$

where r is the common ratio of successive terms.

EXAMPLE 4 Find the 6th term of the geometric sequence 4, 20, 100, ...

First note that $a_1 = 4$, $n = 6$, and $r = 5$. We use formula (4):

$$a_n = a_1 r^{n-1}$$
$$a_6 = 4 \cdot 5^{6-1} = 4 \cdot 5^5$$
$$= 12{,}500.$$

1. Find the 8th term of the geometric sequence

2, 4, 8, 16, ...

EXAMPLE 5 Find the 11th term of the geometric sequence 64, −32, 16, −8, ...

Note that $a_1 = 64$, $n = 11$, and $r = -\frac{1}{2}$. We use formula (4):

$$a_n = a_1 r^{n-1}$$

$$a_{11} = 64 \cdot \left(-\frac{1}{2}\right)^{11-1}$$

$$= 64 \cdot \left(-\frac{1}{2}\right)^{10}$$

$$= 2^6 \cdot \frac{1}{2^{10}}$$

$$= \frac{1}{2^4} \quad \text{or} \quad \frac{1}{16}.$$

DO EXERCISE 1.

ii Sums, or Series

We now look for a formula for the sum of the first n terms of a geometric sequence. Let us denote the terms as follows:

$$a_1, \quad a_1 r, \quad a_1 r^2, \quad a_1 r^3, \ldots, \quad a_1 r^{n-1}.$$

Then the sum of the first n terms is given by

$$S_n = a_1 + a_1 r + a_1 r^2 + \cdots + a_1 r^{n-2} + a_1 r^{n-1}.$$

We next multiply on both sides of this equation by $-r$, obtaining

$$-rS_n = -a_1 r - a_1 r^2 - \cdots - a_1 r^{n-1} - a_1 r^n.$$

We now add the two equations, obtaining

$$S_n - rS_n = a_1 - a_1 r^n.$$

Solving for S_n gives us the formula we seek.

5. The sum of the first n terms of a geometric sequence is given by

$$S_n = \frac{a_1 - a_1 r^n}{1 - r}, \qquad r \neq 1.$$

We now derive a variant of this formula, which we can use when a_n is known. Using formula (4), we substitute a_n for $a_1 r^{n-1}$ into formula (5). We then have the following.

6. The sum of the first n terms of a geometric sequence is given by

$$S_n = \frac{a_1 - ra_n}{1 - r}, \qquad r \neq 1.$$

EXAMPLE 6 Find the sum of the first 6 terms of the geometric sequence 3, 6, 12, 24, ...

Note that $a_1 = 3$, $n = 6$, and $r = 2$. We use formula (5):

$$S_n = \frac{a_1 - a_1 r^n}{1 - r}$$

$$S_6 = \frac{3 - 3 \cdot 2^6}{1 - 2} = \frac{3 - 192}{-1}, \quad \text{or} \quad 189.$$

DO EXERCISES 2 AND 3.

2. Find the sum of the first 6 terms of the geometric sequence

$$3, 15, 75, 375, \ldots .$$

3. Find the sum of the first 10 terms of the geometric sequence

$$2, -1, \frac{1}{2}, -\frac{1}{4}, \ldots .$$

Recall the definition of a *sequence* as a function having for its domain some set of natural numbers. If that domain is an infinite (unending) set of natural numbers, then the sequence is called an *infinite sequence.* In the examples that follow, the domain will be the set of *all* natural numbers, even though that is not a requirement of the definition.

Let us look at two examples of infinite geometric sequences and their sequences of partial sums.

A. $2, 4, 8, 16, \ldots, 2^n, \ldots$

$S_1 = 2$

$S_2 = 6$

$S_3 = 14$

$S_4 = 30$

$S_5 = 62$

The terms of the sequence S_n get larger and larger, without bound. Now let us look at the next sequence.

B. $\frac{1}{2}, \frac{1}{4}, \frac{1}{8}, \frac{1}{16}, \ldots, \frac{1}{2^n}, \ldots$

$$S_1 = \frac{1}{2}$$

$$S_2 = \frac{1}{2} + \frac{1}{4} = \frac{3}{4}$$

$$S_3 = S_2 + \frac{1}{8} = \frac{3}{4} + \frac{1}{8} = \frac{7}{8}$$

$$S_4 = S_3 + \frac{1}{16} = \frac{15}{16}$$

$$S_5 = S_4 + \frac{1}{32} = \frac{31}{32}$$

It appears that we can write a formula for S_n as follows:

$$S_n = \frac{2^n - 1}{2^n}.$$

Not all infinite sequences have sums (for example, sequence A above). An infinite sum is also called an *infinite series.*

We now find a formula for the "sum" of an infinite geometric sequence by looking at values of S_n as n becomes very large. Recall that

$$S_n = \frac{a_1 - a_1 r^n}{1 - r} = \frac{a_1}{1 - r} - \frac{a_1 r^n}{1 - r}.$$

Find the sum.

4. $1 + \frac{1}{3} + \frac{1}{9} + \frac{1}{27} + \cdots$

5. $4 - 1 + \frac{1}{4} - \frac{1}{16} \cdots$

As n becomes large, r^n will also become large if $r > 1$. If r is negative and $|r| > 1$, then the values of r^n will alternate between positive and negative numbers having large absolute value. In either case, S_n will not approach a limit. If $r = 1$, the formula does not hold, and the sequence will not be geometric. However, if $|r| < 1$, then the values of r^n approach 0 as n becomes large, and we have the following.

7. An infinite geometric sequence has a sum if and only if $|r| < 1$. That sum is given by

$$S_\infty = \frac{a_1}{1 - r}.$$

EXAMPLE 7 Find the sum of the infinite geometric sequence $5, \frac{5}{2}, \frac{5}{4}, \frac{5}{8}, \ldots$, if it exists.

Note that $a_1 = 5$. The sum exists because $r = \frac{1}{2}$, which is less than 1. We use formula (7):

$$S_\infty = \frac{a_1}{1 - r}$$

$$= \frac{5}{1 - \frac{1}{2}} = 10.$$

DO EXERCISES 4 AND 5.

iii Problem Solving

For some problem-solving situations, the translation may involve geometric sequences or series.

EXAMPLE 8 A student loan is in the amount of \$600. Interest is to be 12% compounded annually, and the entire amount is to be paid after 11 years. How much is to be paid back?

1. **Familiarize** Suppose we let P represent any principal amount. At the end of a year the amount owed will be $P + 0.12P$, or $1.12P$. That amount will be the principal for the second year. The amount owed at the end of the second year will be $1.12(1.12P)$, or $(1.12^2)P$. Thus the principal at the beginnings of successive years is as follows:

   P, $1.12P$, 1.12^2P, 1.12^3P, and so on.

   We have a geometric sequence. The principal at the beginning of the 12th year will be the amount owed at the end of the 11th year.

2. **Translate** We have a geometric sequence with $a_1 = 600$, $r = 1.12$, and $n = 12$. The appropriate formula is

   $$a_n = a_1 r^{n-1}.$$

3. **Carry out** We substitute and calculate:

   $$a_{12} = 600(1.12)^{11}$$
   $$\approx 600(3.4785)$$
   $$= 2087.13, \quad \text{rounded to the nearest hundredth.}$$

4. **Check** We repeat the calculations. We might also estimate an answer to see whether the result is reasonable.

5. **State** The answer is that \$2087.13 is to be paid back at the end of 11 years.

EXAMPLE 9 A super-rebound ball rebounds to 60% of the height from which it has fallen or is dropped. The ball is dropped from a height of 20 ft and begins to bounce.

a) After bouncing and rebounding 12 times, how far in all has it traveled upward (the total distance bounced)?

b) Approximately how far will the ball have traveled upward (bounced) before it comes to rest?

1. Familiarize Let's do some calculations and look for a pattern.

First fall:	20 ft
First rebound:	0.6×20, or 12 ft
Second fall:	12 ft, or 0.6×20
Second rebound:	0.6×12, or $0.6(0.6 \times 20)$, which is 7.2 ft
Third fall:	7.2 ft, or $0.6(0.6 \times 20)$
Third rebound:	0.6×7.2, or $0.6(0.6(0.6 \times 20))$, which is 4.32 ft

The rebound distances form a geometric sequence:

$$0.6 \times 20, \quad 0.6^2 \times 20, \quad 0.6^3 \times 20, \quad 0.6^4 \times 20, \ldots,$$

or

$$12, \quad 0.6 \times 12, \quad 0.6^2 \times 12, \ldots .$$

2. Translate

a) The total rebound distance is the sum of a geometric sequence. The first term is 12 and the common ratio is 0.6. There will be 12 terms, so we can use formula (5):

$$S_n = \frac{a_1 - a_1 r^n}{1 - r}.$$

b) Theoretically, the ball will never stop bouncing. Actually, it will eventually stop. We can approximate the total distance bounced by considering an infinite number of bounces. After just a few bounces the numbers will be very small, so if we ignore them we will have a pretty fair approximation. We can use formula (7):

$$S_\infty = \frac{a_1}{1 - r}.$$

3. Carry out

a) We substitute into the formula and calculate:

$$S_{12} = \frac{12 - 12(0.6)^{12}}{1 - 0.6}$$

$$\approx 29.93. \qquad \text{Using a calculator}$$

b) We substitute and calculate:

$$S_\infty = \frac{12}{1 - 0.6}$$

$$= 30.$$

4. Check We can do the calculations again. For part (a), we could also recalculate using formula (6).

5. State The answers are as follows.

a) In 12 bounces the ball will travel a total distance upward of about 29.93 ft.

b) The ball will travel upward about 30 ft before it comes to rest.

DO EXERCISE 6.

6. A college student borrows $400 at 6% interest compounded annually. She pays off the loan at the end of 3 yr. What is the amount paid?

EXERCISE SET 11.3

i For each geometric sequence, find the common ratio.

1. $2, 4, 8, 16, \ldots$
2. $12, -4, \frac{4}{3}, -\frac{4}{9}, \ldots$
3. $1, -1, 1, -1, \ldots$
4. $-5, -0.5, -0.05, -0.005, \ldots$
5. $\frac{1}{2}, -\frac{1}{4}, \frac{1}{8}, -\frac{1}{16}, \ldots$
6. $\frac{2}{3}, -\frac{4}{3}, \frac{8}{3}, -\frac{16}{3}, \ldots$
7. $75, 15, 3, \frac{3}{5}, \ldots$
8. $6.275, 0.6275, 0.06275, \ldots$
9. $\frac{1}{x}, \frac{1}{x^2}, \frac{1}{x^3}, \ldots$
10. $5, \frac{5m}{2}, \frac{5m^2}{4}, \frac{5m^3}{8}$

For each geometric sequence, find the nth, or general, term.

11. $1, 3, 9, \ldots$
12. $25, 5, 1, \ldots$
13. $1, -1, 1, -1, \ldots$
14. $2, 4, 8, \ldots$
15. $\frac{1}{x}, \frac{1}{x^2}, \frac{1}{x^3}, \ldots$
16. $5, \frac{5m}{2}, \frac{5m^2}{4}, \ldots$

For each geometric sequence, find the indicated term.

17. $2, 4, 8, 16, \ldots$; the 6th term
18. $2, -10, 50, -250, \ldots$; the 9th term
19. $2, 2\sqrt{3}, 6, \ldots$; the 9th term
20. $1, -1, 1, -1, \ldots$; the 57th term
21. $\frac{8}{243}, \frac{8}{81}, \frac{8}{27}, \ldots$; the 10th term

22. Find the sum of the first 7 terms of the geometric series
$$6 + 12 + 24 + \cdots.$$

23. Find the sum of the first 6 terms of the geometric series
$$16 - 8 + 4 - \cdots.$$

24. Find the sum of the first 7 terms of the geometric series
$$\tfrac{1}{18} - \tfrac{1}{6} + \tfrac{1}{2} - \cdots.$$

25. Find the sum of the first 5 terms of the geometric series
$$6 + 0.6 + 0.06 + \cdots.$$

26. Find the sum of the first 8 terms of the series
$$1 + x + x^2 + x^3 + \cdots.$$

27. Find the sum of the first 10 terms of the series
$$1 + x^2 + x^4 + x^6 + \cdots.$$

ii Find these infinite sums. (The series are geometric.)

28. $4 + 2 + 1 + \cdots$
29. $7 + 3 + \frac{9}{7} + \cdots$
30. $25 + 20 + 16 + \cdots$
31. $12 + 9 + \frac{27}{4} + \cdots$
32. $100 - 10 + 1 - \frac{1}{10} + \cdots$
33. $-6 + 3 - \frac{3}{2} + \frac{3}{4} - \cdots$
34. $0.3 + 0.03 + 0.003 + \cdots$
35. $9 + 0.9 + 0.09 + 0.009 + \cdots$

iii ***Problem solving***

36. A ping-pong ball is dropped from a height of 16 ft and always rebounds $\frac{1}{4}$ of the distance fallen. How high does it rebound the 6th time?

37. Approximate the total of the rebound heights of the ball in Exercise 36.

38. Gaintown has a population of 100,000 now and the population is increasing by 3% each year. What will the population be in 15 years?

39. How long will it take for the population of Gaintown to double? (See Exercise 38.)

40. A student borrows \$1200. The loan is to be repaid in 13 years, at 12% interest compounded annually. How much will be repaid at the end of 13 years?

41. (*An annuity*). A person decides to save money in a savings account for retirement. At the beginning of each year, \$1000 is invested at 11%, compounded annually. How much will be in the retirement fund at the end of 40 years?

42. Find the sum of the first n terms of
$$1 + x + x^2 + \cdots.$$

43. A piece of paper is 0.01 in. thick. It is folded repeatedly in such a way that its thickness is doubled each time for 20 times. How thick is the result?

44. A superball dropped from the top of the Washington Monument (556 ft high) always rebounds $\frac{3}{4}$ of the distance fallen. How far (up and down) will the ball have traveled when it hits the ground for the 6th time?

45. Approximate the total distance that the ball of Exercise 44 will have traveled when it comes to rest.

Find fractional notation for the following infinite sums. (*Hint:* These are geometric series.)

46. 0.3333 . . .

47. 9.999999 . . .

48. 0.55555 . . .

49. 0.66666 . . .

50. 0.15151515 . . .

51. 0.12121212 . . .

52. The sides of a square are each 16 cm long. A second square is inscribed by joining the midpoints of the sides, successively. In the second square we repeat the process, inscribing a third square. If this process is continued indefinitely, what is the sum of all of the area of all the squares? (*Hint:* Use an infinite geometric series.)

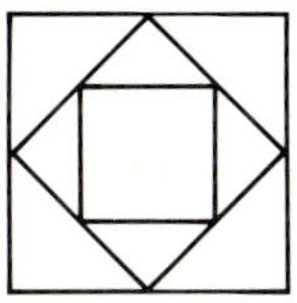

53. The infinite series

$$2+\frac{1}{2}+\frac{1}{2\cdot 3}+\frac{1}{2\cdot 3\cdot 4}+\frac{1}{2\cdot 3\cdot 4\cdot 5}+\frac{1}{2\cdot 3\cdot 4\cdot 5\cdot 6}+\cdots$$

is not geometric, but does have a sum. Find values of S_1, S_2, S_3, S_4, S_5, and S_6. Make a conjecture about the value of S.

11.4 PROBABILITY

OBJECTIVES

After finishing Section 11.4, you should be able to:

i **Compute the probability of a simple event.**

ii **Use the Fundamental Counting Principle to determine how many ways an event may occur.**

We say that when a coin is tossed the chances that it will fall heads are 1 out of 2, or that the *probability* that it will fall heads is $\frac{1}{2}$. Of course this does not mean that if a coin is tossed 10 times it will necessarily fall heads exactly 5 times. If the coin is tossed a great number of times, however, it will fall heads very nearly half of them.

i Experimental and Theoretical Probability

If we toss a coin a large number of times, say 1000, and count the number of heads, we can determine the probability. If there are 503 heads, we would calculate the probability to be

$$\frac{503}{1000} \quad \text{or} \quad 0.503.$$

If we plant 100 seeds and 81 of them sprout, we calculate the probability that a seed will sprout to be

$$\frac{81}{100} \quad \text{or} \quad 0.81.$$

These are examples of *experimental* determinations of probability.

If we consider a coin and reason that it is just as likely to fall heads as tails, we would calculate the probability to be $\frac{1}{2}$. This is a *theoretical* determination of probability. You might ask, "What is the *true* probability?" In fact there is none. Experimentally we can determine probabilities within certain limits. These may or may not agree with what we obtain theoretically.

We need some terminology before we continue. Suppose we perform an experiment such as flipping a coin, throwing a dart, drawing a card from a deck, or checking an item off an assembly line for quality. The results of an experiment are called the *outcomes.* The set of all possible outcomes is called the *sample space.* An *event* is a set of outcomes; that is, a subset of the sample space. For example, for the experiment "throwing a dart," suppose the board was A in the figure below. Then one event is "hitting red," which is a subset of the sample space "red, white, gray," assuming the dart hits the target somewhere.

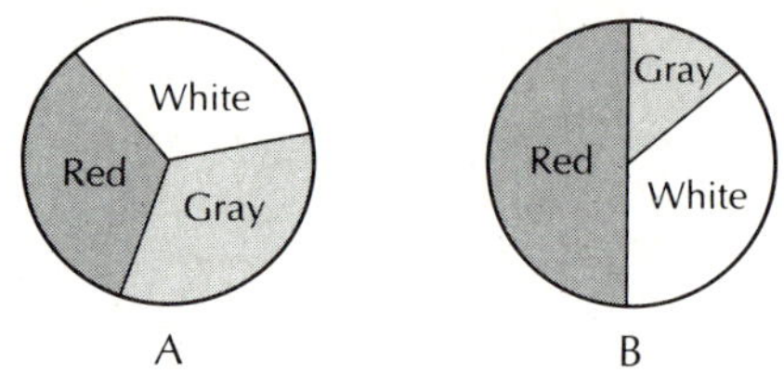

We denote the probability that an event E occurs as $P(E)$. For example, "getting a head" may be denoted by H. Then $P(\text{H})$ represents the probability of getting a head. When the outcomes of an experiment all have the same probability of occurring, we say that they are *equally likely.* To see the distinction between events that are equally likely and those that are not, consider the dartboards shown here. For dartboard A, the events, *hitting red, hitting white,* and *hitting gray* are equally likely, but for board B they are not. A sample space that can be expressed as a union of equally likely events allows us to calculate probabilities of other events.

PRINCIPLE *P*

If an event E can occur m ways out of n possible equally likely outcomes of a sample space S, the probability of that event is given by

$$P(E) = \frac{m}{n}.$$

A die (pl., dice) is a cube, with six faces, each containing a number of dots from 1 to 6, as shown.

EXAMPLE 1 What is the probability of rolling a 3 on a die?

On a fair die there are 6 equally likely outcomes, so the sample space contains the 6 items, getting a 1, 2, 3, 4, 5, or 6. There is just one way to get a 3. By Principle P, $P(3) = \frac{1}{6}$.

EXAMPLE 2 What is the probability of rolling an even number on a die?

The sample space is the same as that of Example 1. The *event* in this case is getting an even number. That can occur in 3 ways. Or, to say it another way, the event consists of the three elements of the sample space, 2, 4, and 6. By Principle *P*, $P(\text{even}) = \frac{3}{6}$, or $\frac{1}{2}$.

DO EXERCISE 1.

1. What is the probability of rolling a prime number on a die?

EXAMPLE 3 What is the probability of getting a total of 8 on a roll of a pair of dice? (Assume that the dice are different, say one red and one black.)

On each die there are 6 outcomes. The outcomes are paired, so there are $6 \cdot 6$, or 36 elements in the sample space, as shown.

Red die						
6	(1, 6)	(2, 6)	(3, 6)	(4, 6)	(5, 6)	(6, 6)
5	(1, 5)	(2, 5)	(3, 5)	(4, 5)	(5, 5)	(6, 5)
4	(1, 4)	(2, 4)	(3, 4)	(4, 4)	(5, 4)	(6, 4)
3	(1, 3)	(2, 3)	(3, 3)	(4, 3)	(5, 3)	(6, 3)
2	(1, 2)	(2, 2)	(3, 2)	(4, 2)	(5, 2)	(6, 2)
1	(1, 1)	(2, 1)	(3, 1)	(4, 1)	(5, 1)	(6, 1)
	1	2	3	4	5	6 Black die

The pairs that total 8 are shaded. Those elements taken together constitute the *event* of rolling an 8. There are 5 of them. By Principle *P*, $P(8) = \frac{5}{36}$.

DO EXERCISE 2.

2. What is the probability of getting a total of 7 on a roll of a pair of dice?

The following are some results that follow from Principle *P*.

If an event E cannot ever occur, then $P(E) = 0$.

If an event E is certain to occur (every trial is a success), then $P(E) = 1$.

The probability of any event is a number from 0 to 1:

$$0 \leq P(E) \leq 1.$$

ii The Fundamental Counting Principle

A basic principle that helps determine the number of ways in which a compound event may occur is as follows.

THE FUNDAMENTAL COUNTING PRINCIPLE

In a compound event in which the first event can occur in m ways and a second can occur in n ways, the number of ways in which the first and then the second can occur is $m \cdot n$.

This principle can be extended to events consisting of more than two separate events.

EXAMPLE 4 There are 11 names in a hat. You draw one name and then another. In how many different ways can that be done?

The first name can be drawn in 11 ways. After that, there are only 10 names left, so the second can be drawn in 10 ways. By the Fundamental Counting Principle, the total number of ways of drawing a first and a second name is $11 \cdot 10$, or 110.

EXAMPLE 5 There are 11 names in a hat. You draw one name and then put it back in the hat. Then you draw a second name. In how many ways can that be done?

The first name can be drawn in 11 ways. The second can also be drawn in 11 ways since the first name was replaced. By the Fundamental Counting Principle, the total number of ways is $11 \cdot 11$, or 121.

DO EXERCISE 3.

3. How many 3-digit numbers can be named using the digits 5, 6, 7:
   a) without repetition?
   b) with repetition?

EXAMPLE 6 Two people are to be selected at random from a group consisting of 6 men and 4 women. What is the probability of selecting two women?

The total number of ways of selecting two people is $10 \cdot 9$; or 90. Thus the sample space contains 90 elements. The number of ways of selecting two women is $4 \cdot 3$, or 12. By Principle P the probability is $\frac{12}{90}$, or $\frac{2}{15}$.

EXAMPLE 7 Suppose you draw 2 cards from a well-shuffled deck of 52 playing cards. What is the probability that you draw two spades?

There are 13 spades in the deck. The number of ways of drawing 2 spades from 13 spades is $13 \cdot 12$, or 156.

The number of ways of drawing 2 cards from the deck is $52 \cdot 51$, or 2652. The probability is $\frac{156}{2652}$, or $\frac{1}{17}$.

DO EXERCISE 4.

4. Suppose 2 people are selected at random from a group that consists of 8 men and 6 women. What is the probability that both of them are women?

The Origin and Use of Probability

A desire to calculate odds in games of chance gave rise to the theory of probability. Today the theory of probability and its closely related field, mathematical statistics, have many applications, most of them not related to games of chance. Opinion polls, with such uses as predicting elections, are a familiar example. Quality control, in which a prediction about the percentage of faulty items manufactured is made without testing them all, is an important application, among many, in business. Still other applications are in the areas of genetics, medicine, and the kinetic theory of gases.

EXERCISE SET 11.4

i Suppose we select, without looking, one marble from a bag containing 4 red marbles and 10 green marbles.

1. What is the probability of selecting a red marble?
2. What is the probability of selecting a green marble?
3. What is the probability of selecting a purple marble?
4. What is the probability of selecting a white marble?

Suppose 4 cards are drawn from a well-shuffled deck of 52 cards.

5. What is the probability that all 4 are spades?
6. What is the probability that all 4 are hearts?
7. If 4 marbles are drawn at random all at once from a bag containing 8 white marbles and 6 black marbles, what is the probability that all 4 are black?
8. From a group of 8 men and 7 women, a committee of 3 is chosen. What is the probability that all 3 are women?
9. What is the probability of getting a total of 6 on a roll of a pair of dice?
10. What is the probability of getting a total of 3 on a roll of a pair of dice?
11. What is the probability of getting snake eyes (a total of 2) on a roll of a pair of dice?
12. What is the probability of getting boxcars (a total of 12) on a roll of a pair of dice?

13. From a bag containing 5 nickels, 8 dimes, and 7 quarters, 5 coins are drawn at random. What is the probability that all 5 are dimes?

14. From a bag containing 6 nickels, 10 dimes, and 4 quarters, 6 coins are drawn at random. What is the probability that all 6 are nickels?

15. Suppose you draw 2 cards from a well-shuffled deck of 52 cards. What is the probability that you will get 2 aces?

16. Suppose you draw 2 cards from a well-shuffled deck of 52 cards. What is the probability that they will both the red?

(*Roulette.*) A roulette wheel contains slots numbered 00, 0, 1, 2, 3, . . . , 35, 36. Eighteen of the slots numbered 1 through 36 are colored red and eighteen are colored black. The 00 and 0 slots are uncolored. The wheel is spun, and a ball is rolled around the rim until it falls into a slot. What is the probability that the ball falls in:

17. a black slot?

18. a red slot?

19. a red or black slot?

20. the 00 slot?

21. the 0 slot?

22. either the 00 or 0 slot? (Here the house always wins.)

23. an odd-numbered slot?

11.5

THE BINOMIAL THEOREM

OBJECTIVES

After finishing Section 11.5, you should be able to:

i **Expand a power of a binomial $(a + b)^n$ using Pascal's triangle.**

ii **Expand a power of a binomial $(a + b)^n$ using preceding coefficient.**

iii **Evaluate and simplify notation involving factorials, expand a binomial using factorial notation, and find a specified term of a binomial expansion.**

i Binomial Expansions Using Pascal's Triangle

Consider the following expanded powers of $(a + b)^n$, where $a + b$ is any binomial and n is a natural number. Look for patterns.

$$(a+b)^0 = 1$$

$$(a+b)^1 = a + b$$

$$(a+b)^2 = a^2 + 2ab + b^2$$

$$(a+b)^3 = a^3 + 3a^2b + 3ab^2 + b^3$$

$$(a+b)^4 = a^4 + 4a^3b + 6a^2b^2 + 4ab^3 + b^4$$

$$(a+b)^5 = a^5 + 5a^4b + 10a^3b^2 + 10a^2b^3 + 5ab^4 + b^5$$

Each expansion is a polynomial. There are some patterns to be noted in the expansions.

1. In each term, the sum of the exponents is n.
2. The exponents of a start with n and decrease to 0. The last term has no factor of a. The first term has no factor of b. The exponents of b start in the second term with 1 and increase to n, or we can think of them starting in the first term with 0 and increasing to n.
3. There is one more term than the power n. That is, there are $n + 1$ terms in the expansion of $(a + b)^n$.
4. Now we consider the coefficients. The first and last coefficients are 1 and the coefficients have a symmetry to them. They start at 1 and increase through certain values about "half"-way and then decrease through these same values back to 1. Let us explore this further.

Suppose we wanted to find an expansion of $(a + b)^8$. If the patterns we have noticed were to continue, then we know that there are 9 terms in the expansion, which would be in the following form:

$$a^8 + c_1a^7b + c_2a^6b^2 + c_3a^5b^3 + c_4a^4b^4 + c_3a^3b^5 + c_2a^2b^6 + c_1ab^7 + b^8.$$

How can we determine these coefficients? There are actually three different ways that we will answer this question. Your instructor may direct you regarding which to

1. Find as many patterns as you can in Pascal's triangle. See if you can use the patterns to write three more rows of the triangle.

learn. The first method seems to be the easiest, but is not always. It involves writing down the coefficients in a triangular array as follows. We get what is known as *Pascal's Triangle:*

$(a+b)^0$:						1					
$(a+b)^1$:					1		1				
$(a+b)^2$:				1		2		1			
$(a+b)^3$:			1		3		3		1		
$(a+b)^4$:		1		4		6		4		1	
$(a+b)^5$:	1		5		10		10		5		1

There are many patterns in the triangle. Find as many as you can.

DO EXERCISE 1.

Perhaps you discovered a way to write the next row of numbers, given the numbers in the row above it. There are always 1's on the outsides. Each remaining number is found by adding the two numbers above. This is shown as follows:

1

1 1

1 2 1

1 3 3 1

1 4 6 4 1

1 5 10 10 5 1

1 6 15 20 15 6 1

We see that

the 1st number is 1;

the 2nd number is 1 + 5, or 6;

the 3rd number is 5 + 10, or 15;

the 4th number is 10 + 10, or 20;

the 5th number is 10 + 5, or 15; and

the 6th number is 5 + 1, or 6.

Thus the expansion of $(a+b)^6$ is

$$(a+b)^6 = a^6 + 6a^5b + 15a^4b^2 + 20a^3b^3 + 15a^2b^4 + 6ab^5 + b^6.$$

To find the expansion for $(a+b)^8$, we complete two more rows of Pascal's triangle:

1

1 1

1 2 1

1 3 3 1

1 4 6 4 1

1 5 10 10 5 1

1 6 15 20 15 6 1

1 7 21 35 35 21 7 1

1 8 28 56 70 56 28 8 1

Thus the expansion of $(a+b)^8$ is

$$(a+b)^8 = a^8 + 8a^7b + 28a^6b^2 + 56a^5b^3 + 70a^4b^4 + 56a^3b^5 + 28a^2b^6 + 8ab^7 + b^8.$$

We can generalize our results as follows:

THE BINOMIAL THEOREM

For any binomial $a+b$ and any natural number n,

$$(a+b)^n = c_0a^nb^0 + c_1a^{n-1}b + c_2a^{n-2}b^2 + \cdots + c_{n-1}a^1b^{n-1} + c_na^0b^n,$$

where the numbers $c_0, c_1, c_2, \ldots, c_n$ are from the $(n+1)$st row of Pascal's triangle, and $c_0 = c_n = 1$.

DO EXERCISE 2.

EXAMPLE 1 Expand $(u-v)^5$.

Note that $a = u$, $b = -v$, and $n = 5$. We use the 6th row of Pascal's triangle: 1 5 10 10 5 1. Then we have

$$\begin{aligned}(u-v)^5 &= (u)^5 + 5(u)^4(-v)^1 + 10(u)^3(-v)^2 + 10(u)^2(-v)^3 + 5(u)(-v)^4 + (-v)^5\\ &= u^5 - 5u^4v + 10u^3v^2 - 10u^2v^3 + 5uv^4 - v^5.\end{aligned}$$

Note that the signs of the terms alternate between + and −. When the power of b, or $-v$, is odd the sign is −.

DO EXERCISE 3.

EXAMPLE 2 Expand $\left(2t + \frac{3}{t}\right)^6$.

Note that $a = 2t$ and $b = 3/t$. We use the 7th row of Pascal's triangle: 1 6 15 20 15 6 1. Then we have

$$\begin{aligned}\left(2t+\frac{3}{t}\right)^6 &= (2t)^6 + 6(2t)^5\left(\frac{3}{t}\right)^1 + 15(2t)^4\left(\frac{3}{t}\right)^2 + 20(2t)^3\left(\frac{3}{t}\right)^3 + 15(2t)^2\left(\frac{3}{t}\right)^4\\ &\quad + 6(2t)^1\left(\frac{3}{t}\right)^5 + \left(\frac{3}{t}\right)^6\\ &= 64t^6 + 6(32t^5)\left(\frac{3}{t}\right) + 15(16t^4)\left(\frac{9}{t^2}\right) + 20(8t^3)\left(\frac{27}{t^3}\right) + 15(4t^2)\left(\frac{81}{t^4}\right)\\ &\quad + 6(2t)\left(\frac{243}{t^5}\right) + \frac{729}{t^6}\\ &= 64t^6 + 576t^4 + 2160t^2 + 4320 + 4860t^{-2} + 2916t^{-4} + 729t^{-6}.\end{aligned}$$

DO EXERCISE 4.

ii Binomial Expansion Using the Preceding Coefficient

The disadvantage in using Pascal's triangle is that one must compute all the preceding rows in the table to obtain the row needed for the expansion. The following two methods avoid this difficulty. To understand one of these methods, let us look at the expansion of $(a+b)^5$:

$$a^5 + 5a^4b + 10a^3b^2 + 10a^2b^3 + 5ab^4 + b^5.$$

Term: 1st 2nd 3rd 4th 5th 6th.

The method of finding a coefficient involves considering the preceding coefficient.

The *first-term coefficient* is 1.

The *second-term coefficient* is the preceding coefficient, 1, times the exponent of

---

2. Write one more row of Pascal's triangle and use it to expand $(a+b)^9$.

3. Expand $(x-y)^4$.

4. Expand $\left(2t+\frac{1}{t}\right)^7$.

a in the preceding term, 5, divided by 1, which is

$$2 - 1, \text{ the number of the term minus 1: } \frac{1 \cdot 5}{1} = 5.$$

The *third-term coefficient* is the preceding coefficient, 5, times the exponent of a in the preceding term, 4, divided by 2, which is

$$3 - 1, \text{ the number of the term minus 1: } \frac{5 \cdot 4}{2} = 10.$$

The *fourth-term coefficient* is the preceding coefficient, 10, times the exponent of a in the preceding term, 3, divided by 3, which is

$$4 - 1, \text{ the number of the term minus 1: } \frac{10 \cdot 3}{3} = 10.$$

The *fifth-term coefficient* is the preceding coefficient, 10, times the exponent of a in the preceding term, 2, divided by 4, which is

$$5 - 1, \text{ the number of the term minus 1: } \frac{10 \cdot 2}{4} = 5.$$

The *sixth-term coefficient* is the preceding coefficient, 5, times the exponent of a in the preceding term, 1, divided by 5, which is

$$6 - 1, \text{ the number of the term minus 1: } \frac{5 \cdot 1}{5} = 1.$$

In general,

The nth coefficient of the binomial expansion is:

1. 1 if $n = 1$,

2. $\frac{AB}{C}$, where A is the preceding coefficient, B is the exponent of a in the preceding term, and C is $n - 1$.

EXAMPLE 3 Expand $(m + n)^{10}$.

We find each term in sequence. There are 11 terms in the expansion. When we get to the 6th coefficient, we can use the symmetry to obtain the others.

The 1st term is m^{10}.

$$\text{The 2nd term is } \frac{1 \cdot 10}{1} m^9 n^1 = 10m^9 n.$$

$$\text{The 3rd term is } \frac{10 \cdot 9}{2} m^8 n^2 = 45m^8 n^2.$$

$$\text{The 4th term is } \frac{45 \cdot 8}{3} m^7 n^3 = 120m^7 n^3.$$

$$\text{The 5th term is } \frac{120 \cdot 7}{4} m^6 n^4 = 210m^6 n^4.$$

$$\text{The 6th term is } \frac{210 \cdot 6}{5} m^5 n^5 = 252m^5 n^5.$$

The rest of the coefficients are 210, 120, 45, 10, and 1. The complete expansion is

$$(m+n)^{10} = m^{10} + 10m^9n + 45m^8n^2 + 120m^7n^3 + 210m^6n^4 + 252m^5n^5 + 210m^4n^6 + 120m^3n^7 + 45m^2n^8 + 10mn^9 + n^{10}.$$

DO EXERCISE 5.

5. Expand $(x-y)^9$.

iii Binomial Expansion Using Factorial Notation

Suppose all we wanted was the 8th term in the expansion $(a+b)^{11}$. The preceding methods would yield this term, but the computations would be lengthy. The last method we use will allow us to find the rth term of an expansion without computing any others. This last method is also useful in more advanced courses such as finite mathematics and calculus.

To develop this method, we need some new notation. Products of successive natural numbers, such as $6 \cdot 5 \cdot 4 \cdot 3 \cdot 2 \cdot 1$ and $8 \cdot 7 \cdot 6 \cdot 5 \cdot 4 \cdot 3 \cdot 2 \cdot 1$, are used in this method. We adopt a special notation for them. For the product $6 \cdot 5 \cdot 4 \cdot 3 \cdot 2 \cdot 1$ we write 6!, read "6 factorial."

$n! = n(n-1)(n-2)(n-3) \cdots 3 \cdot 2 \cdot 1$

Start at n. End at 1.

6. Evaluate 8!.

Here are some examples.

$$7! = 7 \cdot 6 \cdot 5 \cdot 4 \cdot 3 \cdot 2 \cdot 1 = 5040$$
$$6! = 6 \cdot 5 \cdot 4 \cdot 3 \cdot 2 \cdot 1 = 720$$
$$5! = 5 \cdot 4 \cdot 3 \cdot 2 \cdot 1 = 120$$
$$4! = 4 \cdot 3 \cdot 2 \cdot 1 = 24$$
$$3! = 3 \cdot 2 \cdot 1 = 6$$
$$2! = 2 \cdot 1 = 2$$
$$1! = 1 = 1$$

DO EXERCISES 6 AND 7.

7. Evaluate 9!.

We also define 0! to be 1 so that the method we are developing for binomial expansion can be stated concisely.

Note that $6! = 6 \cdot 5!$ We can show this using the definition of factorial notation:

$$6! = 6 \cdot 5 \cdot 4 \cdot 3 \cdot 2 \cdot 1 = 6 \cdot (5 \cdot 4 \cdot 3 \cdot 2 \cdot 1) = 6 \cdot 5!$$

We can generalize this as follows:

For any natural number n, $n! = n(n-1)!$.

A binomial coefficient can be given by an expression like

$$\frac{8!}{5!3!}$$

We need to be able to simplify such expressions.

8. Simplify $\frac{9!}{5!4!}$.

9. Simplify $\frac{8!}{6!2!}$.

Simplify.

10. $\frac{7!}{6!1!}$

11. $\frac{4!}{4!0!}$

EXAMPLE 4 Simplify: $\frac{8!}{5!3!}$.

$$\frac{8!}{5!3!} = \frac{8 \cdot 7 \cdot 6 \cdot 5 \cdot 4 \cdot 3 \cdot 2 \cdot 1}{(5 \cdot 4 \cdot 3 \cdot 2 \cdot 1)(3 \cdot 2 \cdot 1)} = \frac{8 \cdot 7 \cdot 6}{3 \cdot 2 \cdot 1} = 8 \cdot 7 = 56$$

This simplifying can be done quicker as follows:

$$\frac{8!}{5!3!} = \frac{8 \cdot 7 \cdot 6 \cdot 5!}{5!(3 \cdot 2 \cdot 1)} = \frac{8 \cdot 7 \cdot 6}{3 \cdot 2 \cdot 1} = 8 \cdot 7 = 56$$

DO EXERCISE 8.

EXAMPLE 5 Simplify: $\frac{7!}{5!2!}$.

$$\frac{7!}{5!2!} = \frac{7 \cdot 6 \cdot 5!}{5!(2 \cdot 1)} = \frac{7 \cdot 6}{2 \cdot 1} = 7 \cdot 3 = 21$$

DO EXERCISE 9.

EXAMPLE 6 Simplify: $\frac{9!}{8!1!}$.

$$\frac{9!}{8!1!} = \frac{9 \cdot 8!}{8!(1)} = 9$$

EXAMPLE 7 Simplify: $\frac{6!}{6!0!}$.

$$\frac{6!}{6!0!} = \frac{6!}{6!(1)} = 1$$

DO EXERCISES 10 AND 11.

The notation for a binomial coefficient is given as follows in terms of factorials.

In the binomial expansion of $(a + b)^n$, the coefficient of the term $a^{n-r}b^r$ is

$$\frac{n!}{(n-r)!r!}.$$

The symbol $\binom{n}{r}$ is an abbreviation for this quotient of factorials. Thus,

$$\binom{n}{r} = \frac{n!}{(n-r)!r!}.$$

EXAMPLE 8 Evaluate: $\binom{9}{6}$.

$$\binom{9}{6} = \frac{9!}{3!6!} = \frac{9 \cdot 8 \cdot 7 \cdot 6!}{(3 \cdot 2 \cdot 1) \cdot 6!} = \frac{9 \cdot 8 \cdot 7}{3 \cdot 2 \cdot 1} = 3 \cdot 4 \cdot 7 = 84$$

EXAMPLE 9 Evaluate: $\binom{7}{1}$.

$$\binom{7}{1} = \frac{7!}{6!1!} = \frac{7 \cdot 6!}{6!(1)} = 7$$

The notation $\binom{n}{r}$ does not mean $n \div r$ or $\frac{n}{r}$.

DO EXERCISES 12–15.

Now we can restate the binomial theorem in terms of our new notation.

THE BINOMIAL THEOREM

For any binomial $a + b$ and any natural number n,

$$(a+b)^n = \binom{n}{0}a^n + \binom{n}{1}a^{n-1}b + \binom{n}{2}a^{n-2}b^2 + \cdots \binom{n}{n}b^n.$$

Expand $(2x+3)^4$.

Note that $a = 2x$, $b = 3$, and $n = 4$. Then using the binomial theorem and the new notation we have

$$(2x+3)^4 = \binom{4}{0}(2x)^4 + \binom{4}{1}(2x)^3(3)^1 + \binom{4}{2}(2x)^2(3)^2 + \binom{4}{3}(2x)^1(3)^3 + \binom{4}{4}(3)^4$$

$$= \frac{4!}{4!0!}(16x^4) + \frac{4!}{3!1!}(8x^3)(3) + \frac{4!}{2!2!}(4x^2)(9) + \frac{4!}{1!3!}(2x)(27) + \frac{4!}{0!4!}(81)$$

$$= 16x^4 + 96x^3 + 216x^2 + 216x + 81$$

DO EXERCISE 16.

EXAMPLE 11 Expand $(x^2 - 2y)^5$.

Note that $a = x^2$, $b = -2y$, and $n = 5$. Using the binomial theorem we have

$$(x^2-2y)^5 = \binom{5}{0}(x^2)^5 + \binom{5}{1}(x^2)^4(-2y)^1 + \binom{5}{2}(x^2)^3(-2y)^2$$

$$+ \binom{5}{3}(x^2)^2(-2y)^3 + \binom{5}{4}(x^2)^1(-2y)^4 + \binom{5}{5}(-2y)^5$$

$$= \frac{5!}{5!0!}x^{10} + \frac{5!}{4!1!}x^8(-2y) + \frac{5!}{3!2!}x^6(4y^2) + \frac{5!}{2!3!}x^4(-8y^3)$$

$$+ \frac{5!}{1!4!}x^2(16y^4) + \frac{5!}{0!5!}(-32y^5)$$

$$= x^{10} - 10x^8y + 40x^6y^2 - 80x^4y^3 + 80x^2y^4 - 32y^5$$

DO EXERCISE 17.

Suppose we wanted just a particular term of an expansion. The method we have developed will allow us to find such a term without computing all the rows of Pascal's triangle or all the preceding coefficients.

Evaluate.

12. $\binom{8}{5}$

13. $\binom{7}{6}$

14. $\binom{9}{1}$

15. $\binom{6}{0}$

16. Expand $(x + 5b)^5$.

17. Expand $(x - y^2)^4$.

18. Find the 4th term of $(3a + 4b)^7$.

19. Find the 9th term of $(3x - 2)^{10}$.

The $(r + 1)$st term of $(a + b)^n$ is

$$\binom{n}{r} a^{n-r} b^r.$$

EXAMPLE 12 Find the 5th term in the expansion of $(2x - 5y)^6$.

First, we note that $5 = 4 + 1$, $a = 2x$, $b = -5y$, and $n = 6$. Then the 5th term of the expansion is

$$\binom{6}{4}(2x)^{6-4}(-5y)^4, \quad \text{or} \quad \frac{6!}{2!4!}(2x)^2(-5y)^4, \quad \text{or} \quad 37{,}500x^2y^4.$$

EXAMPLE 13 Find the 8th term in the expansion of $(3x - 2)^{10}$.

First, we note that $8 = 7 + 1$, $a = 3x$, $b = -2$, and $n = 10$. Then the 8th term of the expansion is

$$\binom{10}{7}(3x)^{10-7}(-2)^7, \quad \text{or} \quad \frac{10!}{3!7!}(3x)^3(-128), \quad \text{or} \quad -414{,}720x^3.$$

DO EXERCISES 18 AND 19.

EXERCISE SET 11.5

i, ii, iii Expand.

1. $(m + n)^5$ **2.** $(a - b)^4$ **3.** $(x - y)^6$

4. $(p + q)^7$ **5.** $(x^2 - 3y)^5$ **6.** $(3c + d)^6$

7. $(3c - d)^6$ **8.** $(t^{-2} + 2)^6$ **9.** $(x - y)^3$

10. $(x - y)^5$ **11.** $\left(\frac{1}{x} + y\right)^7$ **12.** $(2s - 3t^2)^3$

13. $\left(a - \frac{2}{a}\right)^9$ **14.** $\left(2x + \frac{1}{x}\right)^9$ **15.** $(1 - 1)^n$

16. $(1 + 3)^n$ **17.** $(\sqrt{3} - t)^4$ **18.** $(\sqrt{5} + t)^6$

iv Evaluate.

19. $6!$ **20.** $4!$ **21.** $1!$ **22.** $0!$

23. $\binom{5}{0}$ **24.** $\binom{7}{1}$ **25.** $\binom{8}{4}$ **26.** $\binom{14}{2}$

Find the indicated term of the expansion.

27. 3rd, $(a + b)^6$ **28.** 6th, $(x + y)^7$ **29.** 12th, $(a - 2)^{14}$

30. 11th, $(x - 3)^{12}$ **31.** 5th, $(2x^3 - \sqrt{y})^8$ **32.** 4th, $\left(\frac{1}{b^2} + \frac{b}{3}\right)^7$

○

33. Find the middle term of the expansion of $(2u - 3v^2)^{10}$.

34. Find the two middle terms of the expansion of $(\sqrt{x} + \sqrt{3})^5$.

Expand.

35. $(x^{-2} + x^2)^4$

36. $\left(\frac{1}{\sqrt{x}} - \sqrt{x}\right)^6$

37. $(\sqrt{2} + 1)^6 - (\sqrt{2} - 1)^6$

38. $(1 - \sqrt{2})^4 + (1 + \sqrt{2})^4$

39. Evaluate $\binom{16}{11}$ and $\binom{16}{5}$.

40. Prove that $\binom{n}{r} = \binom{n}{n-r}$ for any whole numbers n and r.

SUMMARY AND REVIEW: CHAPTER 11

The following contains a summary of what you should be able to do after completing this chapter. The review exercises are for practice. Answers are at the back of the book. If you miss an exercise, restudy the section indicated alongside the answer.

You should be able to:

Find an indicated term of a sequence, given the general term.

Find the first 4 terms, the 8th term, a_8, and the 12th term, a_{12}.

1. $a_n = 4n - 3$

2. $a_n = \dfrac{n-1}{n^2+1}$

Find the general term of a sequence.

Find the general term.

3. $-2, -4, -6, -8, -10, \ldots$

4. $1, 4, 9, 16, 25, \ldots$

Rename a series without using sigma notation.

Rename.

5. $\sum_{k=1}^{5} (-2)^k$

6. $\sum_{k=2}^{7} (1-2k)$

Write sigma notation for a sum.

Write sigma notation.

7. $4 + 8 + 12 + 16 + 20$

8. $\dfrac{-1}{2} + \dfrac{1}{4} + \dfrac{-1}{8} + \dfrac{1}{16} + \dfrac{-1}{32}$

Find the first term, the *n*th term, or the common difference of an arithmetic sequence.

Find the first term, the *n*th term, or the common difference.

9. Find the 14th term of the arithmetic sequence $-6, 1, 8, \ldots$.

10. Find d when $a_1 = 11$ and $a_{10} = 35$.

11. Find a_1 and d when $a_{12} = 25$ and $a_{24} = 40$.

Find the sum of the first *n* terms of an arithmetic sequence.

Find the sum.

12. Find the sum of the first 17 terms of the series $-8 + -11 + -14 + \cdots$.

13. Find the sum of all the multiples of 6 that are between 10 and 321.

Solve problems involving arithmetic sequence or series.

Solve.

14. An auditorium has 31 seats in the first row, 33 seats in the second row, 35 seats in the third row, and so on, for 18 rows. How many seats are there in the 17th row?

15. In Exercise 14, how many seats are there in all 18 rows of the auditorium?

Find the first term, the *n*th term, or the common ratio of a geometric sequence.

Find the first term, *n*th term, or common ratio.

16. Find the 20th term of the geometric sequence $2, 2\sqrt{2}, 4, \ldots$.

17. Find the common ratio of the geometric sequence $2, \frac{4}{3}, \frac{8}{9}, \ldots$.

18. Find the *n*th term of the geometric sequence $-2, 2, -2, \ldots$.

19. Find the first term of the geometric sequence where $a_4 = 16$ and $a_8 = 81$.

Find the sum of the first *n* terms of a geometric series.

20. Find the sum of the first 8 terms of the geometric series $-8 + -2 + -\frac{1}{2} + \cdots$.

21. Find the sum of the first 12 terms of the geometric series $3x + -6x + 12x + \cdots$.

Find the sum of an infinite geometric series.

22. $6 + 3 + 1.5 + 0.75 + \cdots$

23. $0.04 + 0.008 + 0.0016 + \cdots$

Solve problems involving geometric sequence or series.

Solve.

24. A factory has to cut back its workers by 4% over each of the next 5 years. If they presently have 3000 employees, how many will they have when their program to cut back has been completed?

25. A secretary was hired at a starting salary of \$5.50 per hour. Her salary was to increase 6% per hour for each of the first 5 years she worked, and increase 8% per hour every year after that. What will her salary be after 8 years?

Determine the probability that an event will occur.

Find the probability.

26. What is the probability of getting a total of 9 on a pair of dice?

27. From a group of 30 people, what is the probability that 2 of them will have a birthday in January? (Assume that a year has 365 days.)

Expand a power of a binomial, $(a + b)^n$, and find a specified term of such an expression.

28. Find the 6th term of $(a + b)^7$.

29. Expand $(x - 2y)^4$.

---

30. Express $0.4\overline{23}$ as a fractional expression.

31. A coin is tossed three times. Find the probability that all three are heads:

a) if it is known that the first is heads;
b) if it is known that the first two are heads;
c) if it is known that two of them are heads.

TEST: CHAPTER 11

1. Find the first 5 terms, the 6th term, and the 16th term of a sequence with general term $a_n = 6n - 5$.

2. Find the general term of the sequence $\frac{1}{64}, \frac{1}{32}, \frac{1}{16}, \ldots$.

3. Rename $\sum_{k=1}^{5} (3 - 2^k)$ without using sigma notation.

4. Write sigma notation for the sum $1 + 8 + 27 + 64 + 125$.

5. Find the 12th term, a_{12}, of the arithmetic sequence $9, 4, -1, \ldots$.

6. Find the common difference, d, when $a_1 = 9$ and $a_7 = 11\frac{1}{4}$.

7. Find a_1 and d when $a_5 = 16$ and $a_{10} = -3$.

8. Find the sum of all the multiples of 12 from 24 to 240 inclusive.

9. A university discovered that its enrollment increased by 250 students every year. If there are 13,450 students presently enrolled, how many students were enrolled six years ago?

10. Find the 6th term of the geometric sequence $72, 18, 4\frac{1}{2}, \ldots$.

11. Find the common ratio of the geometric sequence $22\frac{1}{2}, 15, 10, \ldots$.

12. Find the nth term of the geometric sequence $3, -9, 27, \ldots$.

13. Find the sum of the first 9 terms of the geometric series $(1 + x) + (2 + 2x) + (4 + 4x) + \cdots$.

14. Find the sum of the geometric series $0.5 + 0.25 + 0.125 + \cdots$.

15. The value of a home has increased by 5% during each of the last 5 years. If the home is presently valued at \$70,000, what was its value 3 years ago?

16. What is the probability of getting a total of 11 on a pair of dice?

17. Find the 4th term in the expansion of $(a + x)^{12}$.

18. Expand $(x^2 - 3y)^5$.

CUMULATIVE REVIEW: CHAPTERS 1–11

Solve.

1. $5n - 8 = 9 - 9n$

2. $9 = -6x + x^2$

3. $|a - 2| = 3$

4. $2 \leqslant |4 - 5x| \leqslant 10$

5. $\frac{4}{b} + \frac{6}{b} = 1$

6. $y^3 - 20y^2 + 96y = 0$

7. $\log_4 (x^2 - 25) - \log_4 (x + 5) = 3$

8. $81 = 16x^2$

9. $5(x - 3) + 6 = -(x - 3)^2$

10. $4 = \sqrt{3x - 4}$

Simplify.

11. $(2m + n)^3$

12. $\sqrt[5]{-243}$

13. $2\sqrt{72} + \sqrt{180} + 4\sqrt{108}$

14. $(9 - 3i)(4 + 5i)$

15. $(3a^3b^4 - 9a^4b^3) - (6 + a^3b^4) + (7a^4b^3 - 3)$

Factor.

16. $4x^2 - 196$

17. $12x^2 + 19x + 4$

18. $8x - 64x^4$

19. $25x^2 - 9 + 16x$

Graph.

20. $y = 3(x - 4)^2$

21. $f(x) = x^2 - 8x + 3$

22. $f(x) = \frac{1}{2}(x - 1)^2 + 4$

23. $y + 5 < |x|$

24. $7x^2 + 14y^2 = 98$

25. $\frac{x^2}{25} - \frac{y^2}{81} = 1$

Find the logarithm or antilogarithm, as indicated.

26. log 56,300

27. antilog_e 2.0807

28. antilog 0.9217

29. ln 4360

Find the distance between the given points.

30. (4, 6) and (−3, 7)

31. (0, −4) and (3, 0)

32. (−2, 3) and (9, −5)

33. (−11, −1) and (1, −11)

34. Find the center and radius of the circle whose equation is $(x + 4)^2 + (y - 2)^2 = 5$.

35. Find the center, the vertices, and the asymptotes of the hyperbola

$$4x^2 - 25y^2 - 8x - 100y - 196 = 0.$$

Solve.

36. $x^2 - y^2 = 45,$
$x + y = 45$

37. $x^2 + y^2 = 117,$
$3y^2 - 2x^2 = -54$

38. Find the 14th term of the arithmetic sequence 3, 9, 15, ...

39. Find the 21st term of the arithmetic sequence 19, 12, 5, ...

40. Find the sum of the first 25 terms of the arithmetic series $-1 + 2 + 5 + \cdots$.

41. Find the general term of the geometric sequence 16, 4, $\frac{1}{4}$, ...

42. Find the 15th term of the geometric sequence 0.0025, 0.005, 0.01, ...

43. Find the sum of the first 9 terms of the geometric series $x + 1.5x + 2.25x + \cdots$.

Problem solving

44. A rectangle has an area of 255 cm^2 and a perimeter of 64 cm. What are its dimensions?

45. Find two positive integers whose sum is 20 and the sum of whose reciprocals is $\frac{5}{24}$.

46. A post-office worker earned \$5.50 per hour during her first year on the job, \$6.35 during the second year, and \$7.20 during the third year. If the pattern continued, what will her salary be during her 10th year at the post office?

47. On Mark's 9th birthday, his grandmother opened a savings account for him with \$100. The account draws 6% interest compounded annually. If Mark neither adds to nor withdraws any money from the bank, how much will be in the account on his 18th birthday?

48. Suppose you draw 2 cards from a well-shuffled deck of 52 cards. What is the probability that they will both be hearts?

TABLES

TABLE 1 POWERS, ROOTS, AND RECIPROCALS

n	n^2	n^3	$\sqrt{n}$	$\sqrt[3]{n}$	$\sqrt{10n}$	$\frac{1}{n}$	n	n^2	n^3	$\sqrt{n}$	$\sqrt[3]{n}$	$\sqrt{10n}$	$\frac{1}{n}$
1	1	1	1.000	1.000	3.162	1.0000	**51**	2,601	132,651	7.141	3.708	22.583	.0196
2	4	8	1.414	1.260	4.472	.5000	**52**	2,704	140,608	7.211	3.733	22.804	.0192
3	9	27	1.732	1.442	5.477	.3333	**53**	2,809	148,877	7.280	3.756	23.022	.0189
4	16	64	2.000	1.587	6.325	.2500	**54**	2,916	157,464	7.348	3.780	23.238	.0185
5	25	125	2.236	1.710	7.071	.2000	**55**	3,025	166,375	7.416	3.803	23.452	.0182
6	36	216	2.449	1.817	7.746	.1667	**56**	3,136	175,616	7.483	3.826	23.664	.0179
7	49	343	2.646	1.913	8.367	.1429	**57**	3,249	185,193	7.550	3.849	23.875	.0175
8	64	512	2.828	2.000	8.944	.1250	**58**	3,364	195,112	7.616	3.871	24.083	.0172
9	81	729	3.000	2.080	9.487	.1111	**59**	3,481	205,379	7.681	3.893	24.290	.0169
10	100	1,000	3.162	2.154	10.000	.1000	**60**	3,600	216,000	7.746	3.915	24.495	.0167
11	121	1,331	3.317	2.224	10.488	.0909	**61**	3,721	226,981	7.810	3.936	24.698	.0164
12	144	1,728	3.464	2.289	10.954	.0833	**62**	3,844	238,328	7.874	3.958	24.900	.0161
13	169	2,197	3.606	2.351	11.402	.0769	**63**	3,969	250,047	7.937	3.979	25.100	.0159
14	196	2,744	3.742	2.410	11.832	.0714	**64**	4,096	262,144	8.000	4.000	25.298	.0156
15	225	3,375	3.873	2.466	12.247	.0667	**65**	4,225	274,625	8.062	4.021	25.495	.0154
16	256	4,096	4.000	2.520	12.648	.0625	**66**	4,356	287,496	8.124	4.041	25.690	.0152
17	289	4,913	4.123	2.571	13.038	.0588	**67**	4,489	300,763	8.185	4.062	25.884	.0149
18	324	5,832	4.243	2.621	13.416	.0556	**68**	4,624	314,432	8.246	4.082	26.077	.0147
19	361	6,859	4.359	2.668	13.784	.0526	**69**	4,761	328,509	8.307	4.102	26.268	.0145
20	400	8,000	4.472	2.714	14.142	.0500	**70**	4,900	343,000	8.367	4.121	26.458	.0143
21	441	9,261	4.583	2.759	14.491	.0476	**71**	5,041	357,911	8.426	4.141	26.646	.0141
22	484	10,648	4.690	2.802	14.832	.0455	**72**	5,184	373,248	8.485	4.160	26.833	.0139
23	529	12,167	4.796	2.844	15.166	.0435	**73**	5,329	389,017	8.544	4.179	27.019	.0137
24	576	13,824	4.899	2.884	15.492	.0417	**74**	5,476	405,224	8.602	4.198	27.203	.0135
25	625	15,625	5.000	2.924	15.811	.0400	**75**	5,625	421,875	8.660	4.217	27.386	.0133
26	676	17,576	5.099	2.962	16.125	.0385	**76**	5,776	438,976	8.718	4.236	27.568	.0132
27	729	19,683	5.196	3.000	16.432	.0370	**77**	5,929	456,533	8.775	4.254	27.749	.0130
28	784	21,952	5.292	3.037	16.733	.0357	**78**	6,084	474,552	8.832	4.273	27.928	.0128
29	841	24,389	5.385	3.072	17.029	.0345	**79**	6,241	493,039	8.888	4.291	28.107	.0127
30	900	27,000	5.477	3.107	17.321	.0333	**80**	6,400	512,000	8.944	4.309	28.284	.0125

(continued)

TABLE 1 (continued)

n	n^2	n^3	$\sqrt{n}$	$\sqrt[3]{n}$	$\sqrt{10n}$	$\frac{1}{n}$	n	n^2	n^3	$\sqrt{n}$	$\sqrt[3]{n}$	$\sqrt{10n}$	$\frac{1}{n}$
31	961	29,791	5.568	3.141	17.607	.0323	**81**	6,561	531,441	9.000	4.327	28.460	.0123
32	1,024	32,768	5.657	3.175	17.889	.0312	**82**	6,724	551,368	9.055	4.344	28.636	.0122
33	1,089	35,937	5.745	3.208	18.166	.0303	**83**	6,889	571,787	9.110	4.362	28.810	.0120
34	1,156	39,304	5.831	3.240	18.439	.0294	**84**	7,056	592,704	9.165	4.380	28.983	.0119
35	1,225	42,875	5.916	3.271	18.708	.0286	**85**	7,225	614,125	9.220	4.397	29.155	.0118
36	1,296	46,656	6.000	3.302	18.974	.0278	**86**	7,396	636,056	9.274	4.414	29.326	.0116
37	1,369	50,653	6.083	3.332	19.235	.0270	**87**	7,569	658,503	9.327	4.431	29.496	.0115
38	1,444	54,872	6.164	3.362	19.494	.0263	**88**	7,744	681,472	9.381	4.448	29.665	.0114
39	1,521	59,319	6.245	3.391	19.748	.0256	**89**	7,921	704,969	9.434	4.465	29.833	.0112
40	1,600	64,000	6.325	3.420	20.000	.0250	**90**	8,100	729,000	9.487	4.481	30.000	.0111
41	1,681	68,921	6.403	3.448	20.248	.0244	**91**	8,281	753,571	9.539	4.498	30.166	.0110
42	1,764	74,088	6.481	3.476	20.494	.0238	**92**	8,464	778,688	9.592	4.514	30.332	.0109
43	1,849	79,507	6.557	3.503	20.736	.0233	**93**	8,649	804,357	9.644	4.531	30.496	.0108
44	1,936	85,184	6.633	3.530	20.976	.0227	**94**	8,836	830,584	9.695	4.547	30.659	.0106
45	2,025	91,125	6.708	3.557	21.213	.0222	**95**	9,025	857,375	9.747	4.563	30.822	.0105
46	2,116	97,336	6.782	3.583	21.448	.0217	**96**	9,216	884,736	9.798	4.579	30.984	.0104
47	2,209	103,823	6.856	3.609	21.679	.0213	**97**	9,409	912,673	9.849	4.595	31.145	.0103
48	2,304	110,592	6.928	3.634	21.909	.0208	**98**	9,604	941,192	9.899	4.610	31.305	.0102
49	2,401	117,649	7.000	3.659	22.136	.0204	**99**	9,801	970,299	9.950	4.626	31.464	.0101
50	2,500	125,000	7.071	3.684	22.361	.0200	**100**	10,000	1,000,000	10.000	4.642	31.623	.0100

TABLE 2 COMMON LOGARITHMS

x	0	1	2	3	4	5	6	7	8	9
1.0	.0000	.0043	.0086	.0128	.0170	.0212	.0253	.0294	.0334	.0374
1.1	.0414	.0453	.0492	.0531	.0569	.0607	.0645	.0682	.0719	.0755
1.2	.0792	.0828	.0864	.0899	.0934	.0969	.1004	.1038	.1072	.1106
1.3	.1139	.1173	.1206	.1239	.1271	.1303	.1335	.1367	.1399	.1430
1.4	.1461	.1492	.1523	.1553	.1584	.1614	.1644	.1673	.1703	.1732
1.5	.1761	.1790	.1818	.1847	.1875	.1903	.1931	.1959	.1987	.2014
1.6	.2041	.2068	.2095	.2122	.2148	.2175	.2201	.2227	.2253	.2279
1.7	.2304	.2330	.2355	.2380	.2405	.2430	.2455	.2480	.2504	.2529
1.8	.2553	.2577	.2601	.2625	.2648	.2672	.2695	.2718	.2742	.2765
1.9	.2788	.2810	.2833	.2856	.2878	.2900	.2923	.2945	.2967	.2989
2.0	.3010	.3032	.3054	.3075	.3096	.3118	.3139	.3160	.3181	.3201
2.1	.3222	.3243	.3263	.3284	.3304	.3324	.3345	.3365	.3385	.3404
2.2	.3424	.3444	.3464	.3483	.3502	.3522	.3541	.3560	.3579	.3598
2.3	.3617	.3636	.3655	.3674	.3692	.3711	.3729	.3747	.3766	.3784
2.4	.3802	.3820	.3838	.3856	.3874	.3892	.3909	.3927	.3945	.3962
2.5	.3979	.3997	.4014	.4031	.4048	.4065	.4082	.4099	.4116	.4133
2.6	.4150	.4166	.4183	.4200	.4216	.4232	.4249	.4265	.4281	.4298
2.7	.4314	.4330	.4346	.4362	.4378	.4393	.4409	.4425	.4440	.4456
2.8	.4472	.4487	.4502	.4518	.4533	.4548	.4564	.4579	.4594	.4609
2.9	.4624	.4639	.4654	.4669	.4683	.4698	.4713	.4728	.4742	.4757
3.0	.4771	.4786	.4800	.4814	.4829	.4843	.4857	.4871	.4886	.4900
3.1	.4914	.4928	.4942	.4955	.4969	.4983	.4997	.5011	.5024	.5038
3.2	.5051	.5065	.5079	.5092	.5105	.5119	.5132	.5145	.5159	.5172
3.3	.5185	.5198	.5211	.5224	.5237	.5250	.5263	.5276	.5289	.5307
3.4	.5315	.5328	.5340	.5353	.5366	.5378	.5391	.5403	.5416	.5428
3.5	.5441	.5453	.5465	.5478	.5490	.5502	.5514	.5527	.5539	.5551
3.6	.5563	.5575	.5587	.5599	.5611	.5623	.5635	.5647	.5658	.5670
3.7	.5682	.5694	.5705	.5717	.5729	.5740	.5752	.5763	.5775	.5786
3.8	.5798	.5809	.5821	.5832	.5843	.5855	.5866	.5877	.5888	.5899
3.9	.5911	.5922	.5933	.5944	.5955	.5966	.5977	.5988	.5999	.6010
4.0	.6021	.6031	.6042	.6053	.6064	.6075	.6085	.6096	.6107	.6117
4.1	.6128	.6138	.6149	.6160	.6170	.6180	.6191	.6201	.6212	.6222
4.2	.6232	.6243	.6253	.6263	.6274	.6284	.6294	.6304	.6314	.6325
4.3	.6335	.6345	.6355	.6365	.6375	.6385	.6395	.6405	.6415	.6425
4.4	.6435	.6444	.6454	.6464	.6474	.6484	.6493	.6503	.6513	.6522
4.5	.6532	.6542	.6551	.6561	.6571	.6580	.6590	.6599	.6609	.6618
4.6	.6628	.6637	.6646	.6656	.6665	.6675	.6684	.6693	.6702	.6712
4.7	.6721	.6730	.6739	.6749	.6758	.6767	.6776	.6785	.6794	.6803
4.8	.6812	.6821	.6830	.6839	.6848	.6857	.6866	.6875	.6884	.6893
4.9	.6902	.6911	.6920	.6928	.6937	.6946	.6955	.6964	.6972	.6981
5.0	.6990	.6998	.7007	.7016	.7024	.7033	.7042	.7050	.7059	.7067
5.1	.7076	.7084	.7093	.7101	.7110	.7118	.7126	.7135	.7143	.7152
5.2	.7160	.7168	.7177	.7185	.7193	.7202	.7210	.7218	.7226	.7235
5.3	.7243	.7251	.7259	.7267	.7275	.7284	.7292	.7300	.7308	.7316
5.4	.7324	.7332	.7340	.7348	.7356	.7364	.7372	.7380	.7388	.7396
x	0	1	2	3	4	5	6	7	8	9

(continued)

TABLE 2 (continued)

x	0	1	2	3	4	5	6	7	8	9
5.5	.7404	.7412	.7419	.7427	.7435	.7443	.7451	.7459	.7466	.7474
5.6	.7482	.7490	.7497	.7505	.7513	.7520	.7528	.7536	.7543	.7551
5.7	.7559	.7566	.7574	.7582	.7589	.7597	.7604	.7612	.7619	.7627
5.8	.7634	.7642	.7649	.7657	.7664	.7672	.7679	.7686	.7694	.7701
5.9	.7709	.7716	.7723	.7731	.7738	.7745	.7752	.7760	.7767	.7774
6.0	.7782	.7789	.7796	.7803	.7810	.7818	.7825	.7832	.7839	.7846
6.1	.7853	.7860	.7868	.7875	.7882	.7889	.7896	.7903	.7910	.7917
6.2	.7924	.7931	.7938	.7945	.7952	.7959	.7966	.7973	.7980	.7987
6.3	.7993	.8000	.8007	.8014	.8021	.8028	.8035	.8041	.8048	.8055
6.4	.8062	.8069	.8075	.8082	.8089	.8096	.8102	.8109	.8116	.8122
6.5	.8129	.8136	.8142	.8149	.8156	.8162	.8169	.8176	.8182	.8189
6.6	.8195	.8202	.8209	.8215	.8222	.8228	.8235	.8241	.8248	.8254
6.7	.8261	.8267	.8274	.8280	.8287	.8293	.8299	.8306	.8312	.8319
6.8	.8325	.8331	.8338	.8344	.8351	.8357	.8363	.8370	.8376	.8382
6.9	.8388	.8395	.8401	.8407	.8414	.8420	.8426	.8432	.8439	.8445
7.0	.8451	.8457	.8463	.8470	.8476	.8482	.8488	.8494	.8500	.8506
7.1	.8513	.8519	.8525	.8531	.8537	.8543	.8549	.8555	.8561	.8567
7.2	.8573	.8579	.8585	.8591	.8597	.8603	.8609	.8615	.8621	.8627
7.3	.8633	.8639	.8645	.8651	.8657	.8663	.8669	.8675	.8681	.8686
7.4	.8692	.8698	.8704	.8710	.8716	.8722	.8727	.8733	.8739	.8745
7.5	.8751	.8756	.8762	.8768	.8774	.8779	.8785	.8791	.8797	.8802
7.6	.8808	.8814	.8820	.8825	.8831	.8837	.8842	.8848	.8854	.8859
7.7	.8865	.8871	.8876	.8882	.8887	.8893	.8899	.8904	.8910	.8915
7.8	.8921	.8927	.8932	.8938	.8943	.8949	.8954	.8960	.8965	.8971
7.9	.8976	.8982	.8987	.8993	.8998	.9004	.9009	.9015	.9020	.9025
8.0	.9031	.9036	.9042	.9047	.9053	.9058	.9063	.9069	.9074	.9079
8.1	.9085	.9090	.9096	.9101	.9106	.9112	.9117	.9122	.9128	.9133
8.2	.9138	.9143	.9149	.9154	.9159	.9165	.9170	.9175	.9180	.9186
8.3	.9191	.9196	.9201	.9206	.9212	.9217	.9222	.9227	.9232	.9238
8.4	.9243	.9248	.9253	.9258	.9263	.9269	.9274	.9279	.9284	.9289
8.5	.9294	.9299	.9304	.9309	.9315	.9320	.9325	.9330	.9335	.9340
8.6	.9345	.9350	.9355	.9360	.9365	.9370	.9375	.9380	.9385	.9390
8.7	.9395	.9400	.9405	.9410	.9415	.9420	.9425	.9430	.9435	.9440
8.8	.9445	.9450	.9455	.9460	.9465	.9469	.9474	.9479	.9484	.9489
8.9	.9494	.9499	.9504	.9509	.9513	.9518	.9523	.9528	.9533	.9538
9.0	.9542	.9547	.9552	.9557	.9562	.9566	.9571	.9576	.9581	.9586
9.1	.9590	.9595	.9600	.9605	.9609	.9614	.9619	.9624	.9628	.9633
9.2	.9638	.9643	.9647	.9652	.9657	.9661	.9666	.9671	.9675	.9680
9.3	.9685	.9689	.9694	.9699	.9703	.9708	.9713	.9717	.9722	.9727
9.4	.9731	.9736	.9741	.9745	.9750	.9754	.9759	.9763	.9768	.9773
9.5	.9777	.9782	.9786	.9791	.9795	.9800	.9805	.9809	.9814	.9818
9.6	.9823	.9827	.9832	.9836	.9841	.9845	.9850	.9854	.9859	.9863
9.7	.9868	.9872	.9877	.9881	.9886	.9890	.9894	.9899	.9903	.9908
9.8	.9912	.9917	.9921	.9926	.9930	.9934	.9939	.9943	.9948	.9952
9.9	.9956	.9961	.9965	.9969	.9974	.9978	.9983	.9987	.9991	.9996
x	0	1	2	3	4	5	6	7	8	9

ANSWERS

CHAPTER 1

Margin Exercises, Section 1.1

1. 0.875 **2.** $0.\overline{63}$ **3.** $\frac{59}{10}$ **4.** $\frac{125}{1000}$ or $\frac{1}{8}$
5. 0.035, $\frac{35}{1000}$, or $\frac{7}{200}$ **6.** 0.384, or $\frac{384}{1000}$, or $\frac{48}{125}$
7. 0.375, 37.5% **8.** 0.85, 85% **9.** 0.35% **10.** 2.35
11. -23 **12.** 41 **13.** $\frac{3}{2}$ **14.** $\frac{-5}{4}$ **15.** -9 **16.** 23
17. $\frac{3}{2}$ **18.** $\frac{-9}{5}$ **19.** -18 **20.** -17 **21.** 11
22. -18.6 **23.** $\frac{-29}{5}$ **24.** $-\frac{1}{10}$ **25.** $\frac{-3}{2}$ **26.** 0 **27.** 0
28. 0 **29.** -5 **30.** -3 **31.** 3 **32.** 17 **33.** -16
34. -29.6 **35.** 1 **36.** $\frac{-9}{10}$ **37.** $\frac{3}{2}$ **38.** **(a)** -6;
(b) -40; **(c)** 6 **39.** 42.77 **40.** $\frac{21}{28}$ **41.** $\frac{-8}{-6}$ **42.** $\frac{9}{-2}$
43. $\frac{-14}{5}$ **44.** $\frac{17}{25}$ **45.** 2 **46.** -25 **47.** -3 **48.** 0.2
49. $\frac{-6}{7}$ **50.** $\frac{36}{7}$ **51.** $\frac{45}{25}$ **52.** $\frac{18}{42}$ **53.** $\frac{-56}{-64}$ **54.** $\frac{84}{-49}$
55. $\frac{4}{3}$ **56.** $\frac{9}{2}$ **57.** $\frac{-2}{3}$ **58.** $\frac{-3}{14}$ **59.** $\frac{1}{-2}$ **60.** $\frac{2}{-5}$
61. $\{-2, -1, 0, 1, 2, 3, 4, 5, 6\}$
62. $\{x \mid x$ is an odd integer between 4 and 12$\}$

Exercise Set 1.1, pp. 8–9

1. 0.375 **3.** $1.\overline{6}$ **5.** $\frac{27}{10}$ **7.** $\frac{29}{200}$ **9.** 80% **11.** 37.5%
13. 240% **15.** 3.75% **17.** 41.2% **19.** 0.1% **21.** 125%
23. 0.567, $\frac{567}{1000}$ **25.** 0.0685, $\frac{685}{10,000}$ **27.** 1.2, $\frac{12}{10}$
29. 0.005, $\frac{5}{1000}$ **31.** -7 **33.** 2.8 **35.** -10 **37.** 0
39. $\frac{4}{3}$ **41.** $-\frac{8}{7}$ **43.** $\frac{1}{15}$ **45.** $\frac{1}{4.4}$, or $0.22\overline{72}$ **47.** -28
49. -16 **51.** 5 **53.** 4 **55.** -7 **57.** -24 **59.** 1.2
61. -8.86 **63.** $\frac{1}{7}$ **65.** $-\frac{4}{3}$ **67.** $\frac{1}{10}$ **69.** -2 **71.** -12
73. 5 **75.** 15 **77.** -11.6 **79.** -29:25 **81.** $-\frac{7}{2}$
83. $-\frac{5}{12}$ **85.** -21 **87.** -8 **89.** 16 **91.** -126
93. 34.2 **95.** 26.46 **97.** 2 **99.** 60 **101.** $-\frac{12}{35}$
103. 1 **105.** $-\frac{8}{27}$ **107.** -2 **109.** -7 **111.** 7
113. $\frac{3}{10}$ **115.** $-\frac{6}{77}$ **117.** 25 **119.** -6 **121.** 5
123. $\frac{-126}{81}$ **125.** $-\frac{13}{5}$ **127.** $-\frac{3}{2}$
129. $\{x \mid x$ is a multiple of 4 between 34 and 49$\}$
131. $\{x \mid x$ is an integer between -4 and 3$\}$

Margin Exercises, Section 1.2

1. 5, 8, 0, 2 **2.** 3 **3.** -7 **4.** -1 **5.** 38 **6.** 140.3
7. $\frac{10}{8}$, or $\frac{5}{4}$ **8.** -49 **9.** $\frac{6}{28}$, or $\frac{3}{14}$ **10.** -3 **11.** $\frac{4}{3}$
12. $4 - 5y = 2$ **13.** $63x - 91 = 30$ **14.** 3 **15.** $\frac{-29}{5}$
16. $\frac{17}{2}$ **17.** -2 **18.** $\frac{-19}{8}$ **19.** No solution
20. All real numbers are solutions

Exercise Set 1.2, pp. 13–14

1. 18 **3.** 5 **5.** 24 **7.** 7 **9.** 8 **11.** 21 **13.** 2
15. 2 **17.** $\frac{18}{5}$ **19.** 0 **21.** $\frac{4}{5}$ **23.** No solution
25. $\frac{4}{3}$ **27.** $\frac{37}{5}$ **29.** 13 **31.** 2 **33.** 2 **35.** 7 **37.** 5
39. $-\frac{51}{31}$ **41.** 5 **43.** 2 **45.** 0.007 **47.** -4.1761905
49. No **51.** Yes

Margin Exercises, Section 1.3

1. Note: There is no single correct answer to the question. The following are some of the things you might do.
1) Read the problem carefully.
2) Make a list of the information given and the information asked for.
3) List the information in a table.
4) Draw a picture of a triangle and label it with the information about angle measures.
5) Look up a geometric fact about angles of triangles.
6) Make a guess at the answer.
2. $35 = \frac{3}{5}x$ **3.** $87 + \frac{1}{2}y = 115$ **4.** $\frac{175}{3}$ **5.** 56
6.

$35 \stackrel{?}{=} \frac{3}{5} \cdot \frac{175}{3}$	
35	$\frac{175}{5} \cdot \frac{3}{3}$
	35

Yes

7.

$87 + \frac{1}{2} \cdot 56 \stackrel{?}{=} 115$	
$87 + 28$	115
115	

Yes

8. 35 is $\frac{3}{5}$ of $\frac{175}{3}$
9. 87 more than half of 56 is 115.
10. You should familiarize yourself with the problem,

probably assigning a letter to an unknown number and listing information in a table.

A translation is as follows: If x is the second number, then the first number is $3x$ and the third is $2x$. The sum of the first, twice the second, and three times the third is $3x + 2x + 3 \cdot 2x = 165$.

Solving the equation, we obtain 15. Then $2x = 30$ and $3x = 45$. A possible answer is that the numbers are 45, 15, and 30.

Checking in the conditions of the problem you can see that the numbers check.

We state the answer. The first number is 45. The second number is 15, and the third number is 30.

Exercise Set 1.3, pp. 20–21

1. $x + (x+4) = 12$ **3.** $3x + x + (2x - 12) = 180$ **5.** $x + 2(x+2) + 3(x+4) = 70$ **7.** $4x + 4(x+2) = 100$ **9.** $x + (3x - 6) + [\frac{2}{3}(3x - 6) + 2] = 172$ **11.** $93 + 89 + 72 + 80 + 96 + x = 6(88)$ **13.** 50.3 **15.** 320 **17.** 63 **19.** 165, 330 **21.** \$100 **23.** 24.8% **25.** 96°, 32°, 52° **27.** 90 in^2

Margin Exercises, Section 1.4

1. (a) $\frac{14x - 21}{7x + 7}$; (b) $\frac{4xy - 6y}{2xy + 2y}$ **2.** $2y$ **3.** (a) $7z + 5y$; (b) $b \cdot a$ **4.** $5 + (2y + 7z)$ **5.** There are many answers. Examples: $(2x + z) + y$, $(z + 2x) + y$, $(y + 2x) + z$ **6.** $6xy + 2xz$ **7.** $4y(z + 3)$ **8.** $xy + 9yz + 7z$ **9.** $-2x - y + 3$ **10.** (a) $y - x$; (b) $5y - 2x$; (c) $5z - 3y$ **11.** $3xy - 2x$ **12.** $5x + 46$ **13.** 50.25 **14.** 81 **15.** `(X + Y)^2/(3 - X * Y)` **16.** `(A - B^2) * (A^2 - 2 * B^3 + 4/A^2)`

Exercise Set 1.4, pp. 28–30

1. $\frac{6x + 18}{30}$ **3.** $\frac{4xy - 6y}{6xy + 4y}$ **5.** $\frac{y}{x}$ **7.** $\frac{5}{-2x}$ **9.** $2x + 5y$ **11.** $-3y + 2x$ **13.** $\frac{y}{3} \cdot \frac{x}{2}$ **15.** $(2 \cdot 8)x$ **17.** $(x + 2y) + 5$ **19.** $3a + 3$ **21.** $4x - 4y$ **23.** $-10a - 15b$ **25.** $2ab - 2ac + 2ad$ **27.** $2\pi rh + 2\pi r$ **29.** $\frac{1}{2}ha + \frac{1}{2}hb$ **31.** $8(x + y)$ **33.** $9(p - 1)$ **35.** $7(x - 3)$ **37.** $x(y + 1)$ **39.** $2(x - y + z)$ **41.** $3(x + 2y - 1)$ **43.** $9a$ **45.** $-3b$ **47.** $15y$ **49.** $11a$ **51.** $-8t$ **53.** $10x$ **55.** $8x - 8y$ **57.** $2c + 10d$ **59.** $22x + 18$ **61.** $2x - 33y$ **63.** $-a - 5$ **65.** $m + 1$ **67.** $5d - 12$ **69.** $-7x + 14$ **71.** $-9x + 21$ **73.** $44a - 22$ **75.** -190 **77.** $-12y - 145$ **79.** 80 **81.** 425 **83.** `A^3 + B^3` **85.** `((A + B^2)/(A^2 + B))^3` **87.** `(A^3/B) + (A^2/B^3)` **89.** 24.8% **91.** $\frac{1}{4}$ **93.** $0.01583x + 0.00001y$ **95.** $4x + 12y$ **97.** $-31a$

Margin Exercises, Section 1.5

1. $\frac{1}{10^4}$ **2.** $\frac{1}{(-4)^3}$ **3.** 4^{-3} **4.** $(-5)^{-4}$ **5.** 8^4 **6.** y^5 **7.** $-18x^{11}$ **8.** $-75x^{-14}$ **9.** $-10x^{-12}y^2$ **10.** $60y$ **11.** 4^3 **12.** 5^6 **13.** 10^{-6} **14.** $-2y^{10}x^{-4}$, or $\frac{-2y^{10}}{x^4}$ **15.** $\frac{3}{2}a^3b^2$ **16.** 3^{42} **17.** 8^{-14} **18.** $8x^3y^3$ **19.** $4^2x^{-4}y^{14}$, or $16x^{-4}y^{14}$, or $\frac{16y^{14}}{x^4}$ **20.** $(-2)^5x^{20}y^{10}$, or $-32x^{20}y^{10}$ **21.** $10^3x^{-12}y^{21}z^{-6}$, or $1000x^{-12}y^{21}z^{-6}$, or $\frac{1000y^{21}}{x^{12}z^6}$ **22.** $3^{2a}x^{4a^2}y^{2ab+2a}$ **23.** x^9y^{12} **24.** $\frac{9}{4}x^4y^{-16}$, or $\frac{9x^4}{4y^{16}}$ **25.** $\frac{2^{2a}x^{-4a}y^{10a}}{5^{2a}}$, or $\left(\frac{2}{5}\right)^{2a}\frac{y^{10a}}{x^{4a}}$ **26.** Area $= 6a^2$, Volume $= a^3$; $6a^2 = a^3$

Exercise Set 1.5, pp. 36–37

1. $\frac{1}{6^3}$ **3.** $\frac{1}{9^5}$ **5.** $\frac{1}{-11}$ **7.** $\frac{1}{(5x)^3}$ **9.** $\frac{x^2}{y^3}$ **11.** $\frac{x^2}{y^2}$ **13.** x^3y^2 **15.** $\frac{1}{x^2y^5}$ **17.** $\frac{x^3}{y^5}$ **19.** $\frac{y^7z^4}{x^2}$ **21.** 3^{-4} **23.** $(-16)^{-2}$ **25.** $\frac{1}{6^{-4}}$ **27.** $\frac{6}{x^{-2}}$ **29.** $(5y)^{-3}$ **31.** $\frac{y^{-4}}{3}$ **33.** 5^9 **35.** 8^{-4} **37.** 8^{-6} **39.** b^{-3} **41.** a^3 **43.** $72x^5$ **45.** $-28m^5n^5$ **47.** $-14x^{-11}$ **49.** 5^{3a} **51.** $6x^{2a-1}y^3$ **53.** 6^5 **55.** 4^5 **57.** 10^{-9} **59.** 9^2 **61.** a^5 **63.** 1 **65.** $\frac{-4x^9}{3y^2}$ **67.** $\frac{3x^3}{2y^2}$ **69.** 10^a **71.** $3a^{-x-4}$ **73.** $-5x^{a+1}y^{-1}$ **75.** 4^6 **77.** 8^{-12} **79.** 6^{12} **81.** $3^3x^6y^6$ or $27x^6y^6$ **83.** $(-2)^{-2}x^{-6}y^8$ or $\frac{1}{4}x^{-6}y^8$ **85.** $(-6)^{-2}a^4b^{-6}c^{-2}$ or $\frac{1}{36}a^4b^{-6}c^{-2}$ **87.** $\frac{1}{4^9 \cdot 3^{12}}$ **89.** $\frac{8x^9y^3}{27}$ **91.** 7^{2ab} **93.** 12^{6b-2ab} **95.** $5^{2c}x^{2ac-2c}y^{2bc+2c}$ **97.** $2x^{a+2}y^{b-2}$ **99.** $x^2 + (x + 50)^2 = 15{,}700$ **101.** $25\frac{1}{4}$ **103.** 7 **105.** $2^{-2a-2b+ab}$ **107.** $\frac{64x^{32a}y^{-68b}}{81}$

Margin Exercises, Section 1.6

1. 4.6×10^{11} **2.** 9.3×10^7 **3.** 1.235×10^{-8} **4.** 1.7×10^{-24} **5.** 789,300,000,000 **6.** 0.0000567 **7.** 7.462×10^{-13} **8.** 5.6×10^{-15} **9.** 2×10^3 **10.** 5.5×10^2 **11.** 3×10^{-4} **12.** 4.1567×10^{-5} m^3 or 4.1567×10^4 (mm)^3

Exercise Set 1.6, pp. 41–42

1. 4.7×10^{10} **3.** 8.63×10^{17} **5.** 1.6×10^{-8} **7.** 7×10^{-11} **9.** 4.07×10^{11} **11.** 0.0004 **13.** 673,000,000 **15.** 0.0000000008923 **17.** 90,300,000,000 **19.** 9.66×10^{-5} **21.** 1.3338×10^{-11} **23.** 8.32×10^{10} **25.** 1.9565×10^3 **27.** 2.5×10^3 **29.** 5×10^{-4} **31.** 3×10^{11} **33.** 2×10^{10} **35.** 4.2×10^2 **37.** 1.1×10^{11}

39. 6.7×10 **41.** 1 **43.** $0.024\ \text{m}^3$ **45.** 5.84×10^8 miles
47. 8×10^{-90}; 7.1×10^{-90} **49.** 8 **51.** $c = \dfrac{2a-4}{a}$
53. $\frac{2}{27}$

Margin Exercises, Section 1.7

1. $b = \dfrac{2A}{h}$ **2.** $\frac{5}{3}P - 10 = c$ **3.** $m = \dfrac{H-2r}{3}$
4. $Q = \dfrac{T}{iy+1}$ **5.** 12 cm

Exercise Set 1.7, pp. 44–45

1. $l = \dfrac{A}{w}$ **3.** $I = \dfrac{W}{E}$ **5.** $m = \dfrac{F}{a}$ **7.** $P = \dfrac{I}{rt}$ **9.** $m = \dfrac{E}{c^2}$
11. $l = \dfrac{P-2w}{2}$ **13.** $a^2 = c^2 - b^2$ **15.** $r^2 = \dfrac{A}{\pi}$
17. $h = \frac{2}{11}W + 40$ **19.** $r^3 = \dfrac{3V}{4\pi}$ **21.** $h = \dfrac{2A}{a+b}$
23. $m = \dfrac{rF}{v^2}$ **25.** \$1571.43 **27.** 9 ft **29.** $a = \dfrac{2s - 2v_i t}{t^2}$
31. $V_1 = \dfrac{T_1 P_2 V_2}{P_1 T_2}$ **33.** \$75
35. Take the square root on both sides.

Summary and Review: Chapter 1, pp. 45–47

1. [1.1] -26 **2.** [1.1] 2.2 **3.** [1.1] -1.63
4. [1.1] -37.2 **5.** [1.1] -27.5 **6.** [1.1] $-\frac{7}{15}$
7. [1.1] -11.07 **8.** [1.1] -54 **9.** [1.1] $\frac{21}{50}$ **10.** [1.1] 8
11. [1.1] 0.6 **12.** [1.1] $-\frac{8}{5}$ **13.** [1.4] $-28a - 21b$
14. [1.4] $-6xy - 15xz$ **15.** [1.4] $3x$ **16.** [1.4] $31x - 6y$
17. [1.2] -18 **18.** [1.2] 14
19. [1.3] $3x + x + (5x - 9) = 180$
20. [1.3] $85 + 91 + 75 + 83 + x = 5(84)$ **21.** [1.3] 62.4
22. [1.3] 41, 123 **23.** [1.3] 22, 24, 26 **24.** [1.4] $3n - 1$
25. [1.4] $-59y + 21$ **26.** [1.4] $-11x - 26$
27. [1.4] $25c - 31$ **28.** [1.5] $x^0 = 1$ **29.** [1.5] $-35xy^{-10}$
30. [1.5] $-21x^{2a+3}y^{2b+4}$ **31.** [1.5] y^{11} **32.** [1.5] $\dfrac{5a^{16}}{7b^4}$
33. [1.5] $-4b^{x-5}$ **34.** [1.5] $\dfrac{b^{12}}{64a^6}$ **35.** [1.5] $\dfrac{64x^{21}}{27y^{15}}$
36. [1.5] $625^c x^{8ac} y^{4bc-4c}$ **37.** [1.6] 1.08×10^{-2}
38. [1.6] 2.6378×10^{-2} **39.** [1.6] 5×10^{-12}
40. [1.6] 4×10^1 **41.** [1.6] 4.5×10^{-17}
42. [1.6] 1.3×10^1 **43.** [1.6] 3.0144×10^9 miles
44. [1.7] $\pi = \dfrac{C}{2r}$ **45.** [1.7] $V_2 = \dfrac{P_1 V_1 T_2}{P_2 T_1}$
46. [1.7] $s = \sqrt{\dfrac{4A}{\sqrt{3}}}$ **47.** $\frac{36}{25}$ **48.** $x = \dfrac{zy}{y-z}$
49. $y = \dfrac{c-ab}{a-b}$

CHAPTER 2

Margin Exercises, Section 2.1

1–9. **10.** Both negative

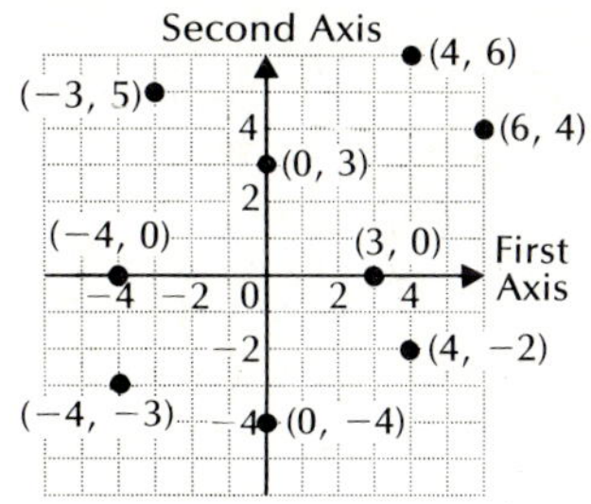

11. First positive, second negative **12.** No **13.** No
14. Yes
15.

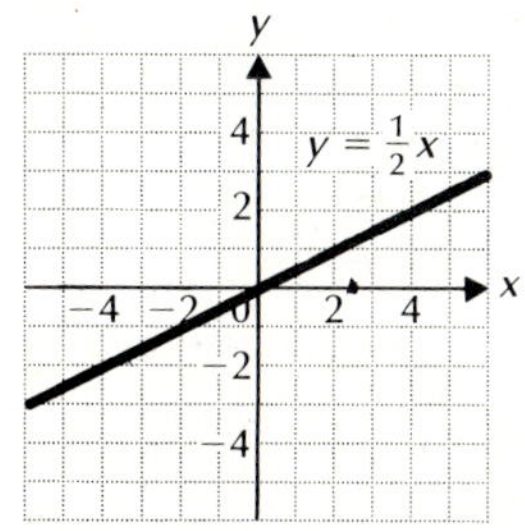

16.

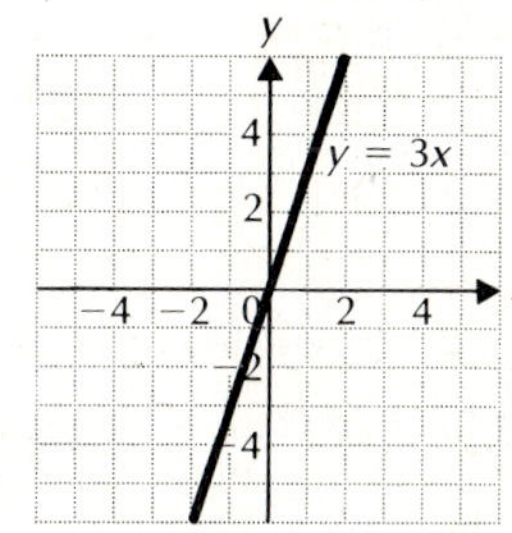

17.

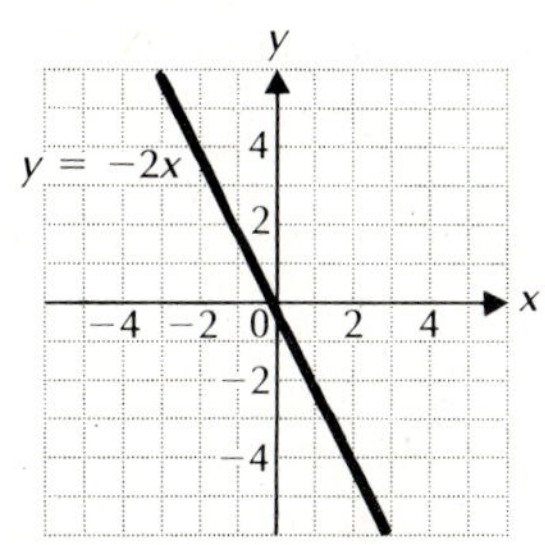

Exercise Set 2.1, p. 54

1.

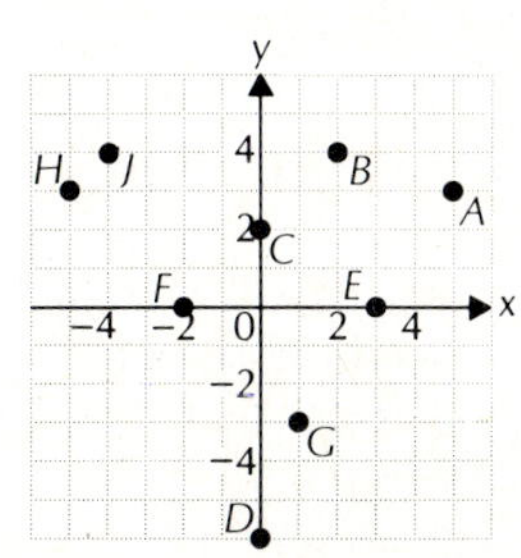

3.

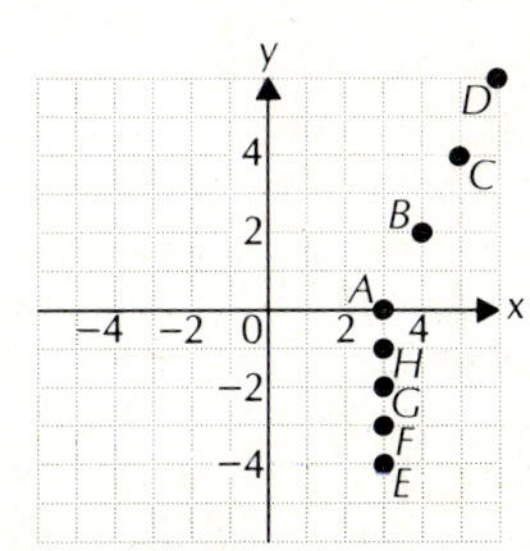

5. Triangle; area = 21 **7.** Yes **9.** No **11.** Yes **13.** Yes
15. Yes **17.** No **19.** Line through $(-1, 2)$ and $(1, -2)$
21. Line through $(2, 6)$ and $(-1, -3)$
23. Line through $(2, -3)$ and $(-2, 3)$
25. Line through $(5, 1)$ and $(-5, -1)$

27.

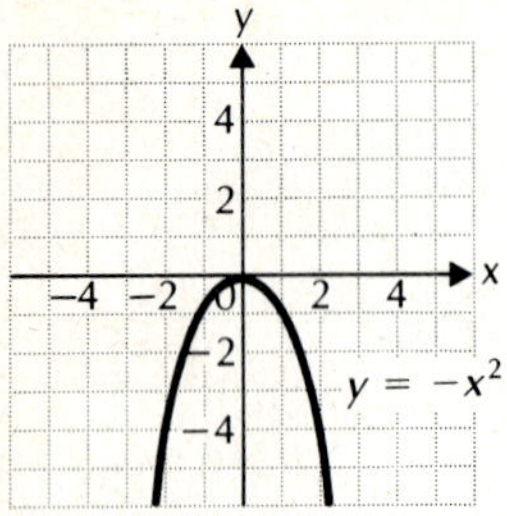

29.

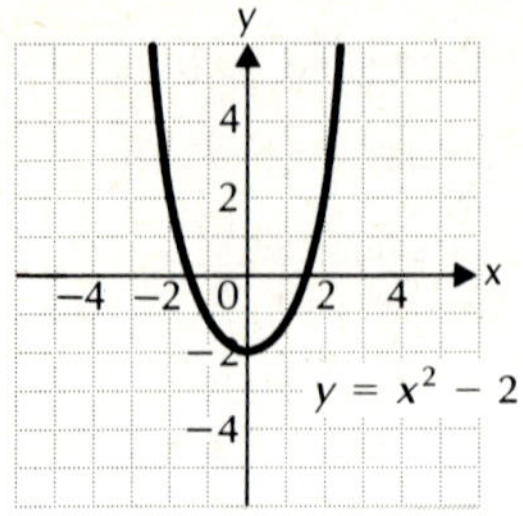

31. 26 ft **33.** $(-1, -2)$, $(13, 10)$, or $(-19, -2)$
35. (a); (d)
37. Positive: moves upward; negative: moves down
39.

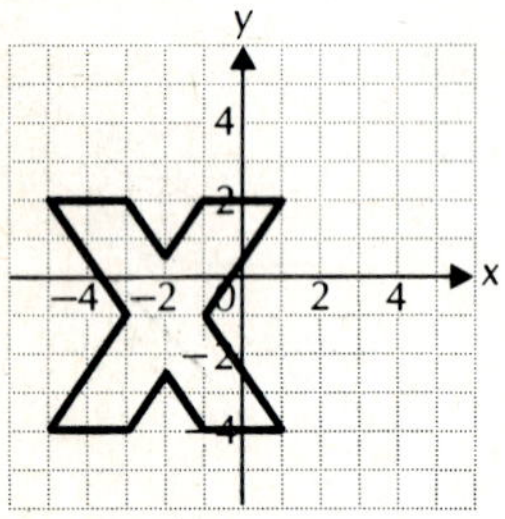

Margin Exercises, Section 2.2

1.

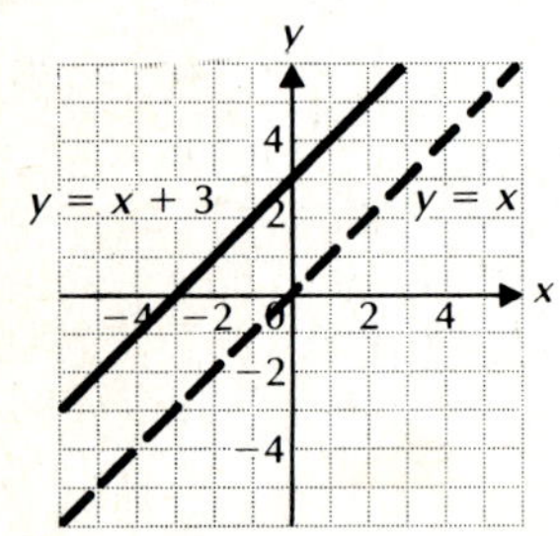

2.

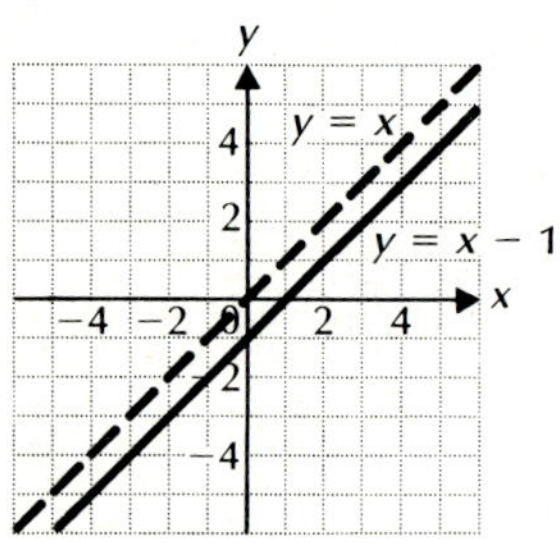

3.

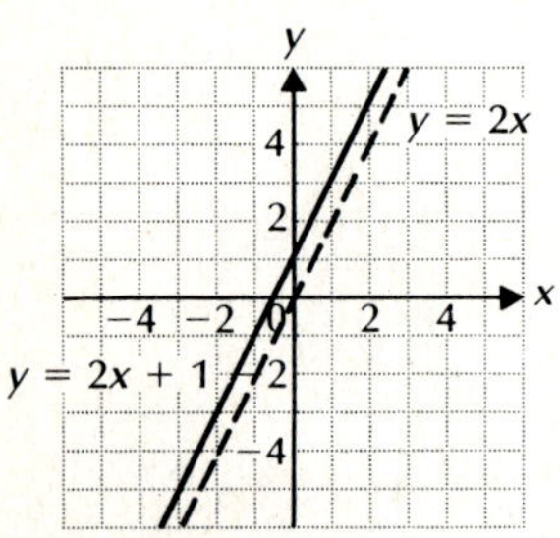

4.

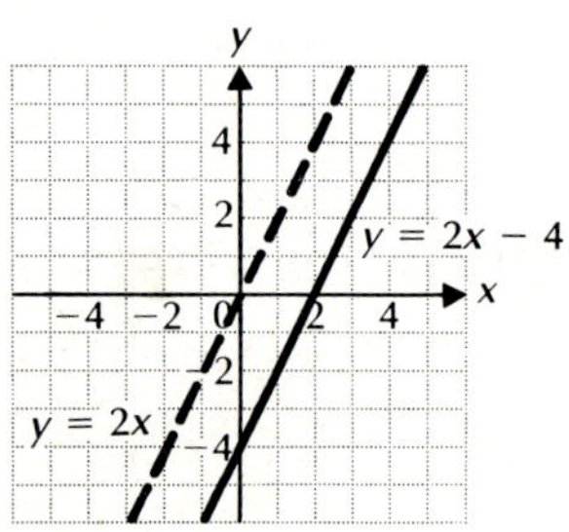

5. The point $(0, 7)$ **6.** The point $(0, -17)$
7. Slope is $\frac{3}{4}$; y-intercept is $(0, -2)$

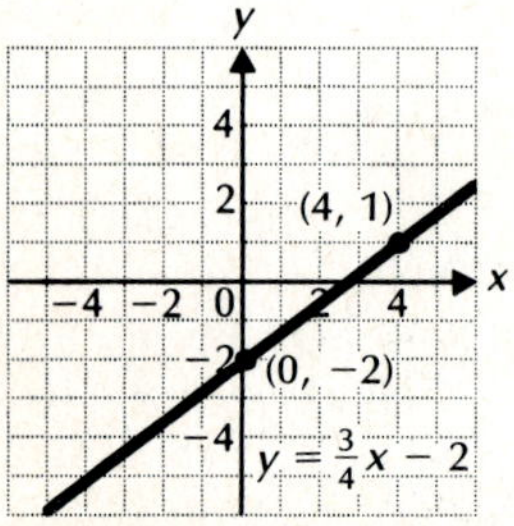

8. Slope is $\frac{3}{2}$; y-intercept is $(0, 1)$

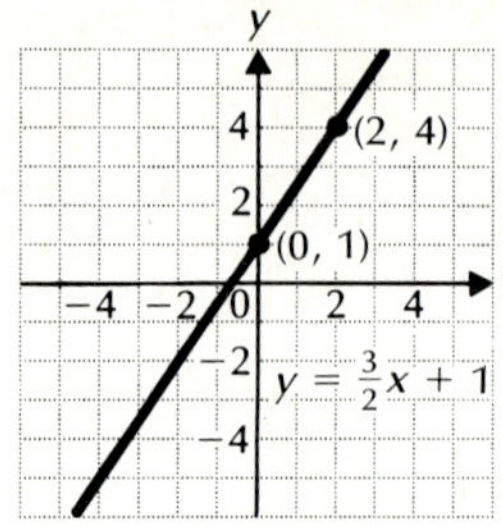

9. Slope is $-\frac{3}{5}$; y-intercept is $(0, 5)$

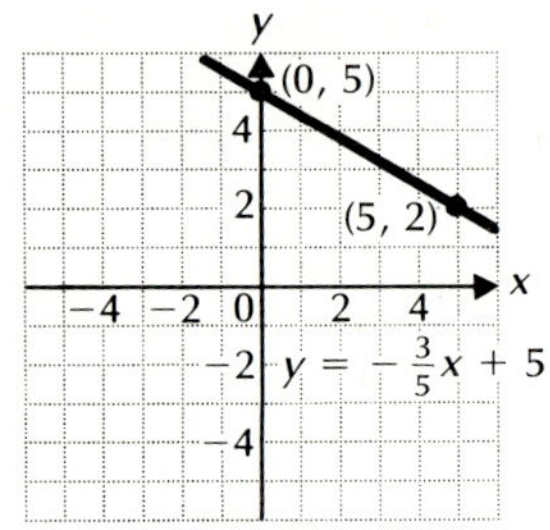

10. Slope is $-\frac{5}{3}$; y-intercept is $(0, -4)$

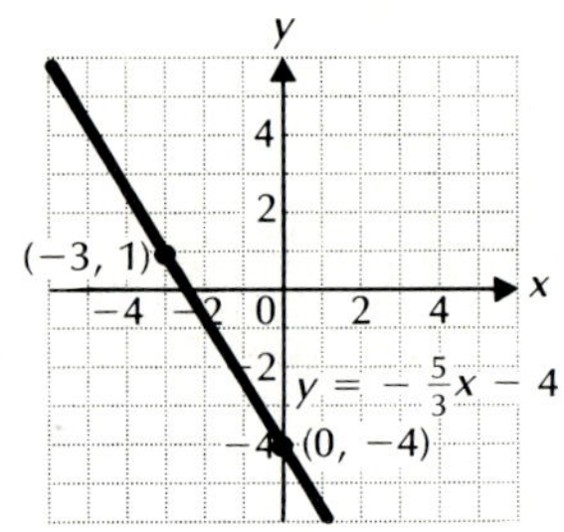

11.

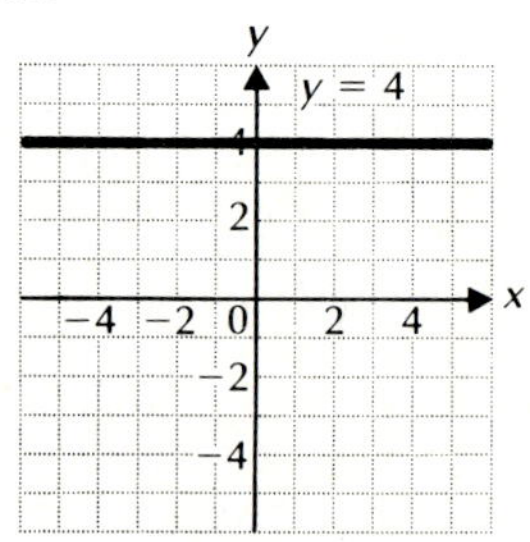

12.

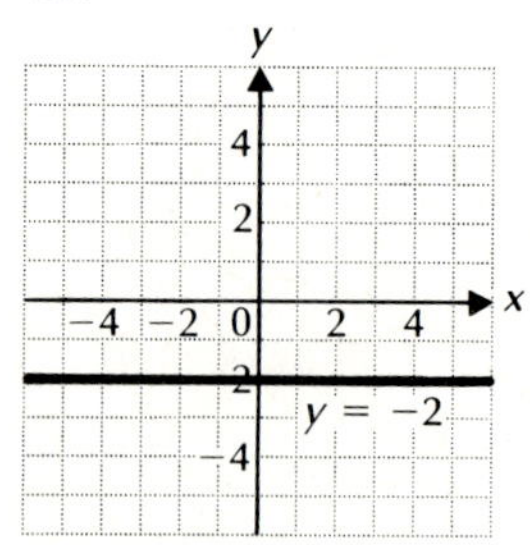

13.

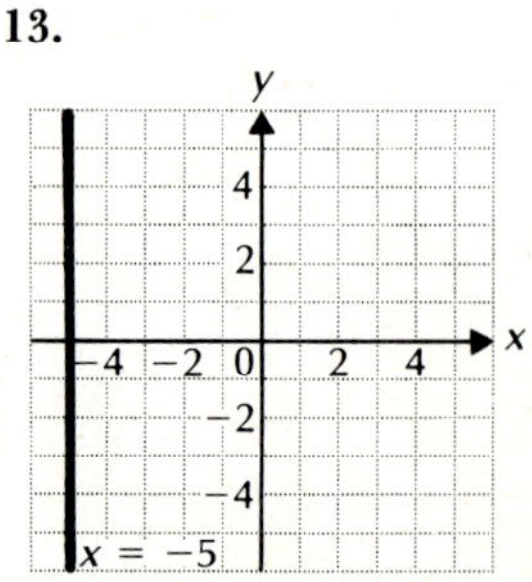

14. Linear, equivalent to $y = 2x + 3$ **15.** Not linear
16. Not linear

Exercise Set 2.2, pp. 60–61

1. Slope = 4, y-intercept = 5
3. Slope = -2, y-intercept = -6
5. Slope = $-\frac{6}{16}$, y-intercept = -0.2
7. Slope = $-\frac{3}{2}$, y-intercept = 3
9. Slope = 2, y-intercept = -8

11.

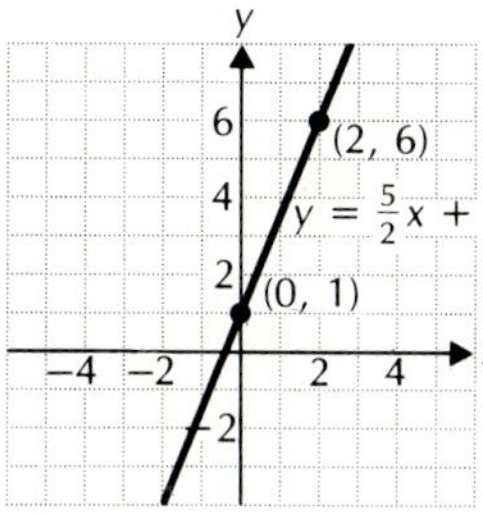

13.

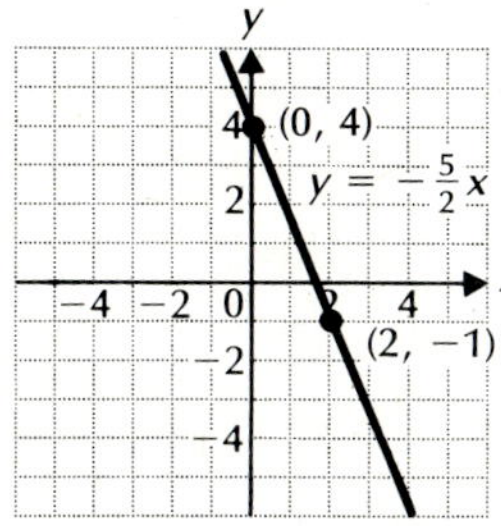

15.

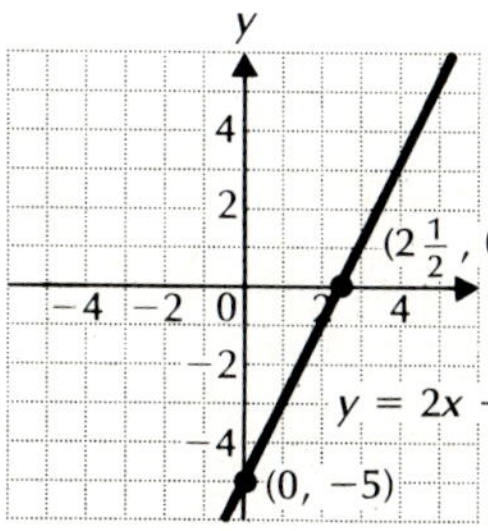

17.

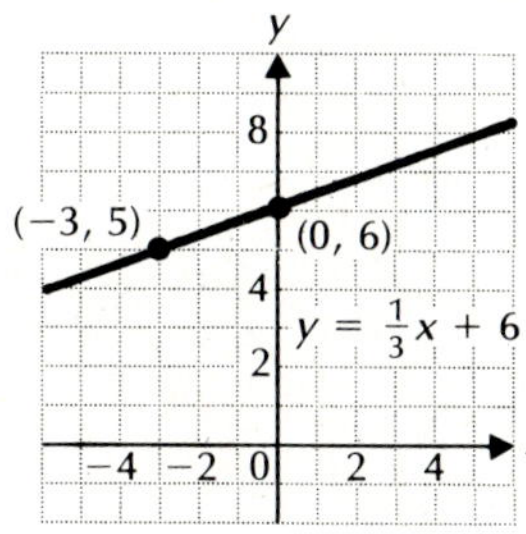

19.

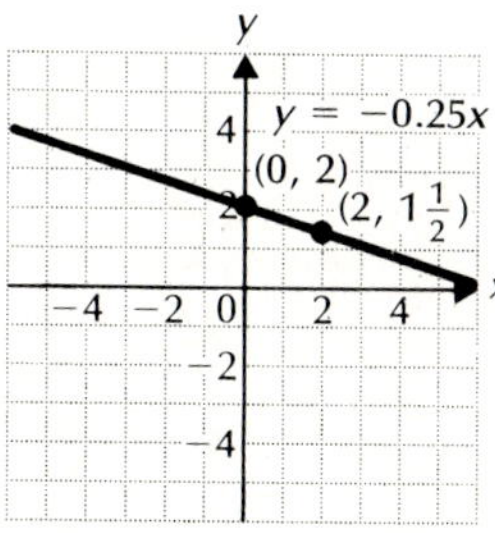

21.

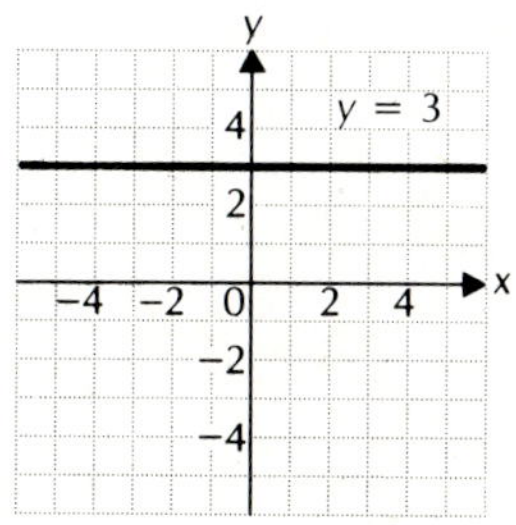

23.

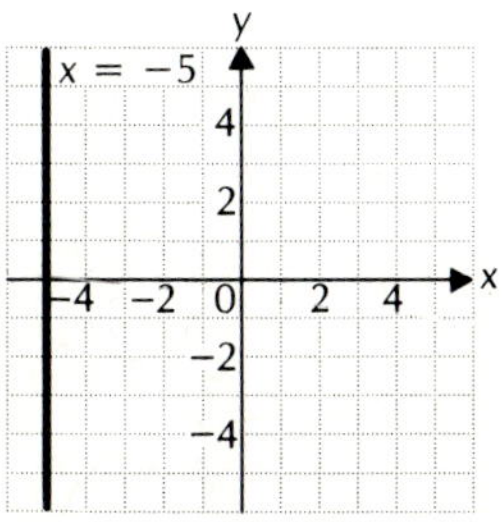

25.

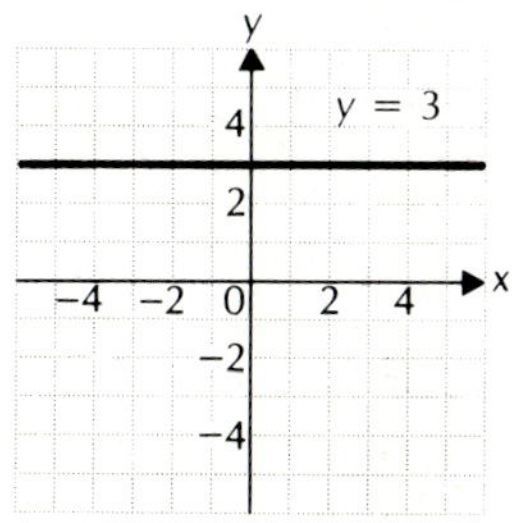

27. Yes **29.** Yes **31.** No **33.** No
35. No **37.** Slope = $-\dfrac{5}{a}$, y-intercept = $\dfrac{b}{a}$
39. Slope = $-\dfrac{a}{b}$, y-intercept = $\dfrac{c}{b}$ **41.** Yes **43.** Yes
45. Yes **47.** 34°, 68°, 78°

Margin Exercises, Section 2.3

1.

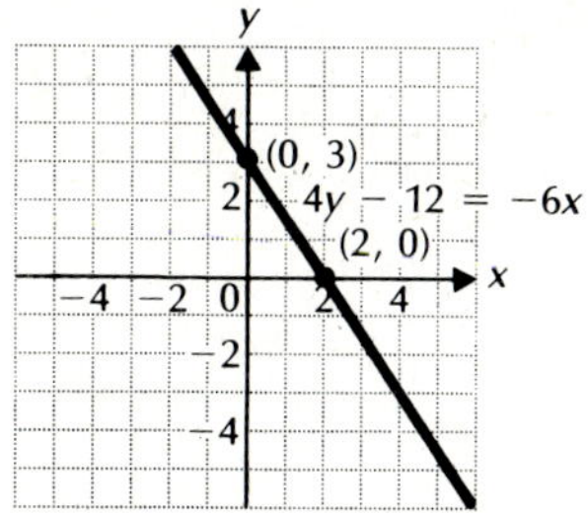

2.

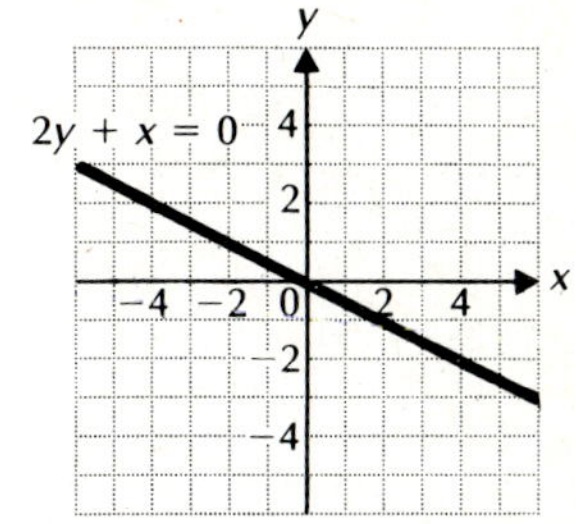

3.

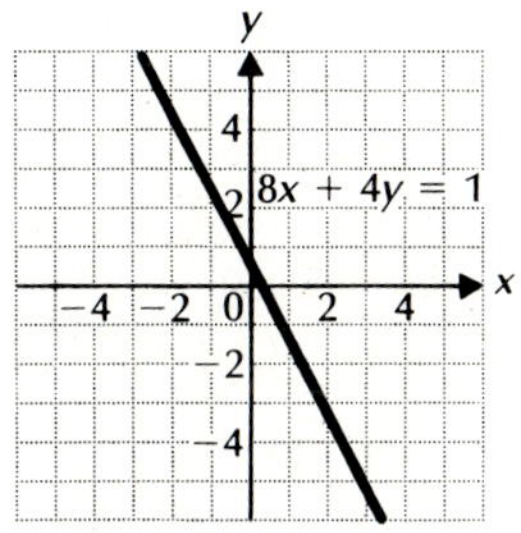

4. $\dfrac{7}{5}$ **5.** $m = \dfrac{\text{rise}}{\text{run}} = \dfrac{3}{4}$
6. $-\dfrac{13}{8}$ **7.** $\dfrac{1}{3}$
8. $-\dfrac{8.9}{12.5}$, or -0.712
9. 0 **10.** Does not exist
11. Zero slope: (a), (e); no slope: (b), (c)

Exercise Set 2.3, pp. 65–66

For problems 1–13, each graph is a line through the intercepts given below.
1. $(0, -2)$, $(2, 0)$ **3.** $(0, -1)$, $(\frac{1}{3}, 0)$ **5.** $(0, -5)$, $(4, 0)$
7. $(0, -5)$, $(-1, 0)$ **9.** $(0, -3)$, $(5, 0)$ **11.** $(0, 3)$, $(-5, 0)$
13. $(0, -7)$, $(2.8, 0)$ **15.** -2 **17.** 5 **19.** $-\frac{1}{3}$
21. No slope **23.** Slope is 3 **25.** No slope
27. Zero slope **29.** Zero slope **31.** No slope
33. Slope is $\frac{1}{2}$ **35.** No slope **37.** Zero slope
39. 4 miles **41.** $-\dfrac{b}{m}$
43. The slope of A is twice the slope of B
45. **(a)** $-\dfrac{5c}{4b}$; **(b)** no slope; **(c)** $\dfrac{a+d}{f}$ **47.** $-\frac{8}{3}$, $-\frac{2}{3}$, $\frac{4}{3}$

Margin Exercises, Section 2.4

1. $y + 2 = 3(x - 1)$, $y + 2 = 3x - 3$, or $y = 3x - 5$
2. $y - \frac{3}{4} = -\frac{4}{3}(x + 5)$, $y - \frac{3}{4} = -\frac{4}{3}x - \frac{20}{3}$, or $y = -\frac{4}{3}x - \frac{71}{12}$
3. $y + 3 = -\frac{5}{3}x + \frac{20}{3}$, $y - 2 = -\frac{5}{3}x + \frac{5}{3}$, or $y = -\frac{5}{3}x + \frac{11}{3}$
4. $y + 5 = -17x - 51$, $y - 12 = -17x - 68$, or $y = -17x - 56$ **5.** $y + 1 = \frac{8}{7}x - \frac{16}{7}$, or $y = \frac{8}{7}x - \frac{23}{7}$
6. $y - 4 = -2x + 10$, or $y = -2x + 14$

Exercise Set 2.4, pp. 69–71

1. $y = 4x - 10$ **3.** $y = -2x + 15$ **5.** $y = 3x + 2$
7. $y = -2x + 16$ **9.** $y = -7$ **11.** $y = \frac{1}{2}x + \frac{7}{2}$ **13.** $y = x$
15. $y = \frac{5}{2}x + 5$ **17.** $y = \frac{1}{4}x + \frac{17}{4}$ **19.** $y = \frac{2}{5}x$
21. $y = 3x + 5$ **23.** Yes **25.** No **27.** Yes
29. $y = -\frac{1}{2}x + \frac{17}{2}$ **31.** $y = \frac{5}{7}x - \frac{17}{7}$ **33.** $y = \frac{1}{3}x + 4$
35. Yes **37.** No **39.** $y = \frac{1}{2}x + 4$ **41.** $y = \frac{4}{3}x - 6$
43. $y = \frac{5}{2}x + 9$ **45.** \$35 **47.** \$136,000

49. **(a)** Grade = 4%, $y = 4\%x$; **(b)** grade = 6.7%, $y = 6.7\%x$
51. $y = -\frac{7}{3}x + \frac{22}{3}$
53. The side that contains (3, 4) and (6, 9) has slope $\frac{5}{3}$. The side that contains (−2, 7) and (3, 4) has slope $-\frac{3}{5}$. Since $\frac{5}{3} \cdot (-\frac{3}{5}) = -1$, these two sides are perpendicular.
55. (1), (4) **57.** 7 **59.** $y = -0.5x - 0.6$

Margin Exercises, Section 2.5

1. (a) and (b) are functions
2. Domain: items in the store; range: a set of values (prices)
3. 17 **4.** 15 **5.** 2 **6.** \$15 **7.** 523.3 in^3
8.

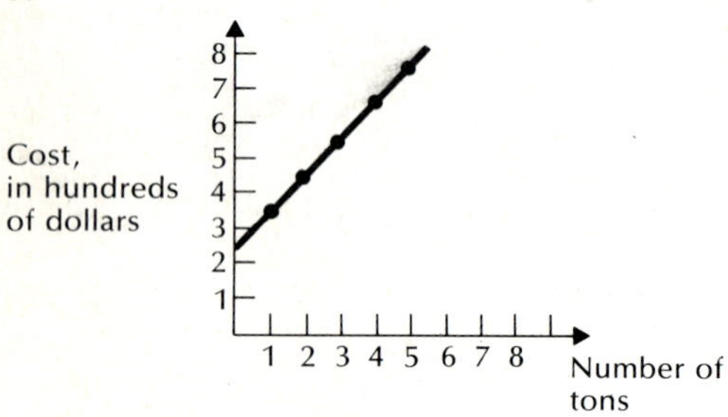

9. **(a)** \$480; **(b)** \$300 **10.** A function **11.** Not a function
12. Not a function **13.** A function

Exercise Set 2.5, pp. 77–79

1. No **3.** No **5.** Yes **7.** Yes **9.** **(a)** 1; **(b)** −3; **(c)** −6; **(d)** 9 **11.** **(a)** 0; **(b)** 1; **(c)** 57; **(d)** $5t^2 + 4t$
13. **(a)** 15; **(b)** 32; **(c)** 20; **(d)** 4
15. **(a)** $\frac{2}{3}$; **(b)** $\frac{10}{9}$; **(c)** 0; **(d)** not possible
17. −2 **19.** −2156 **21.** \$23.55 **23.** $4\sqrt{3}$ **25.** 36π
27. \$1400 **29.** Yes **31.** Yes **33.** 18, 20, 22
35. **(a)** 3.1497708; **(b)** 55.731468; **(c)** 3178.20675; **(d)** 1166.70323 **37.** Yes **39.** $y = \frac{15}{4}x - \frac{13}{4}$
41. $g(0) = 1 - 0^2 = 1$; $f(1) = 2(1) + 1 = 3$ **43.** −8
45. −48

Margin Exercises, Section 2.6

1. **(a)** $R = -0.01t + 10.43$; **(b)** 9.81, 9.73 seconds; **(c)** 2063

Exercise Set 2.6, pp. 81–82

1. **(a)** $E = \frac{3}{20}t + 72$; **(b)** 77.9 years, 78.3 years
3. **(a)** $D = \frac{1}{5}t + 20$; **(b)** 27.4 quadrillion joules, 30 quadrillion joules **5.** **(a)** $R = -0.075t + 46.8$; **(b)** 42.3 seconds, 41.6 seconds; **(c)** 2021
7. **(a)** $C = 0.3\text{ m} + 10$; **(b)** \$70 **9.** 21.1°C
11. 100.03916 cm, 99.96796 cm
13. **(a)** Plan A; $E = 600 + 0.04x$, plan B: $E = 100 + 0.06x$; **(b)** $x > 25{,}000$ **15.** \$151,000, \$201,000

Summary and Review: Chapter 2, pp. 82–83

1. [2.1] Yes **2.** [2.1] No **3.** [2.1] No **4.** [2.1] Yes

5. [2.1]

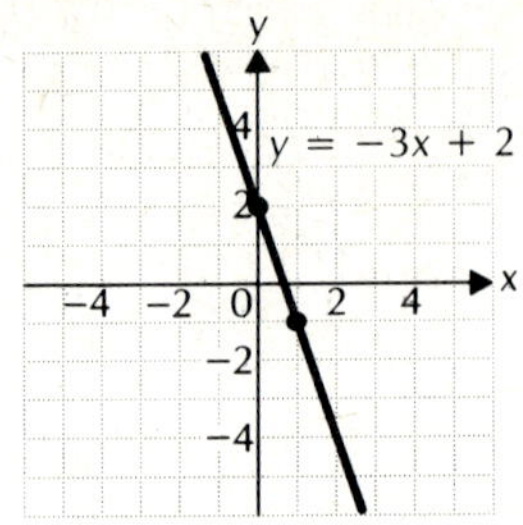

6. [2.1]

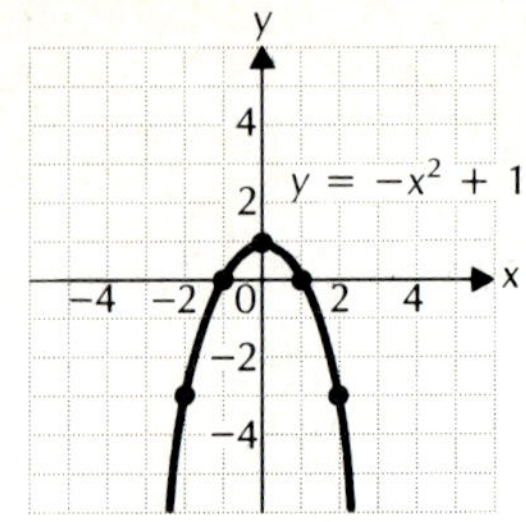

7. [2.2] Slope is −4, y-intercept is −9
8. [2.2] Slope is $\frac{1}{3}$, y-intercept is $-\frac{7}{6}$
9. [2.2]

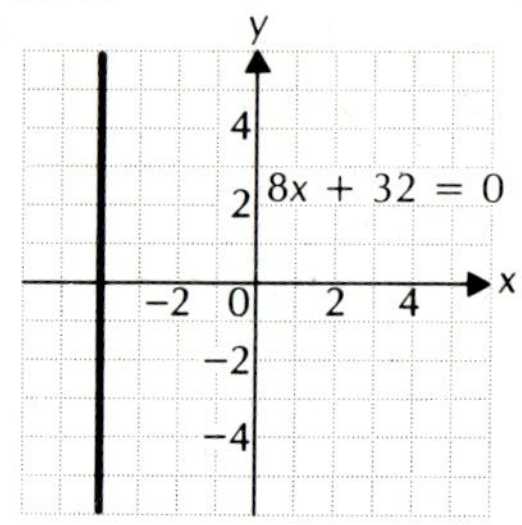

10. [2.2] Yes **11.** [2.2] Yes
12. [2.2] No **13.** [2.2] No

14. [2.3]

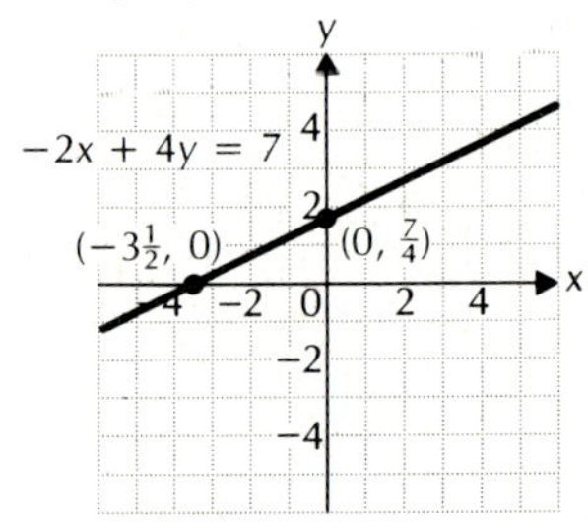

15. [2.3] $\frac{4}{7}$
16. [2.3] No slope
17. [2.4] $y = -2x - 2$
18. [2.4] $y = \frac{5}{6}x + \frac{10}{3}$
19. [2.4] Perpendicular
20. [2.4] Parallel
21. [2.4] Parallel
22. [2.4] $y = \frac{3}{5}x - \frac{31}{5}$
23. [2.4] $y = -\frac{5}{3}x - \frac{5}{3}$
24. [2.5] **(a)** −5; **(b)** −1; **(c)** −11; **(d)** 9
25. [2.5] **(a)** 1; **(b)** 4; **(c)** 1; **(d)** 16
26. [2.6] **(a)** $y = 1.5x - 0.75$; **(b)** \$6.75
27. Slope = 1.8; y-intercept is 32 **28.** −4.5

CHAPTER 3

Margin Exercises, Section 3.1

1. Suppose Jones's salary is x and Smith's salary is y. Then $x = 100 + 2y$ and $x = 3y - 900$.
2. Yes **3.** No
4. (0, 1)

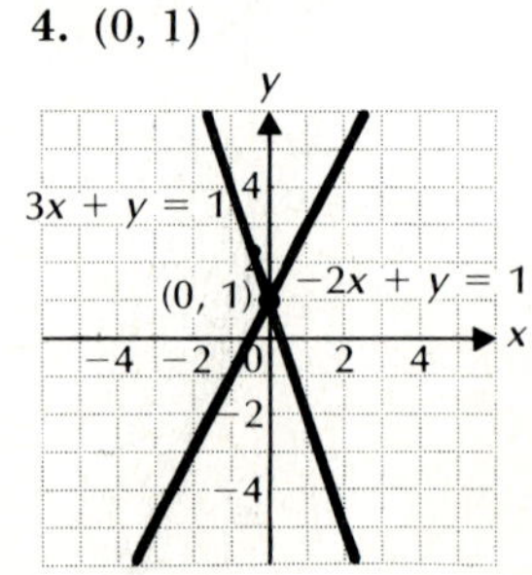

5. (2, 1)

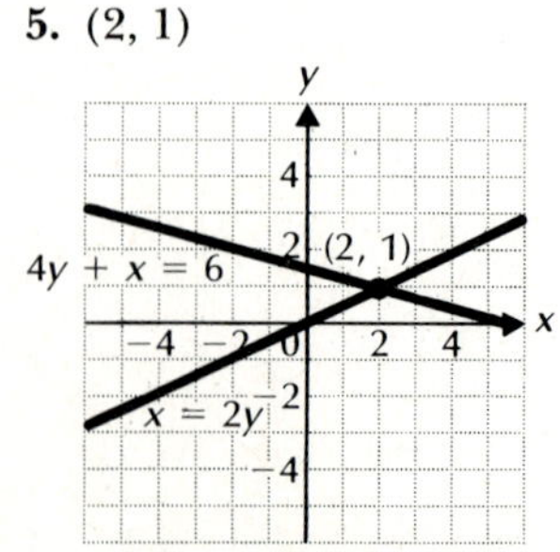

Exercise Set 3.1, p. 88

1. $x + y = -42$, $x - y = 52$ **3.** $2l + 2w = 628$, $l = w + 6$ **5.** $x = y - 9$, $x + y = 43$

7. $x + y = 250$, $1.5x + 2y = 441$ **9.** $x + y = 152$, $x = 6y + 5$ **11.** Yes
13. No **15.** Yes **17.** No **19.** Yes **21.** Yes
23. (3, 1) **25.** (3, 2) **27.** (1, −5) **29.** (2, 1)
31. $(\frac{5}{2}, -2)$ **33.** (3, −2) **35.** $\varnothing$ **37.** (1, 1)
39. (−2, 2) **41.** $x = 2y$, $x + 20 = 3y$
43. $2l + 2w = 156$, $l = 4(w - 6)$ **45.** $d + q = 58$, $q + 7 = 4d$

Margin Exercises, Section 3.2

1. (2, 4) **2.** (5, 2) **3.** (−3, 2) **4.** (13, 16) **5.** (1, 4)
6. $(\frac{1}{3}, \frac{1}{2})$ **7.** (2, 3) **8.** (5, 3) **9.** $132x + 912y = 10$, $42x - 6y = 12$
10. $9x + 10y = 5$, $9x - 4y = 3$ **11.** (−3, 2) **12.** $(-\frac{1}{3}, \frac{1}{2})$
13. (2, −1)

Exercise Set 3.2, pp. 93–94

1. (−4, 3) **3.** (−3, −15) **5.** (2, −2) **7.** (−2, 1)
9. $(\frac{1}{2}, \frac{1}{2})$ **11.** $(\frac{19}{8}, \frac{1}{8})$ **13.** (1, 2) **15.** (3, 0)
17. (−1, 2) **19.** $(\frac{128}{31}, -\frac{17}{31})$ **21.** (6, 2) **23.** $(\frac{140}{13}, -\frac{50}{13})$
25. (4, 6) **27.** $(\frac{43}{12}, \frac{5}{4})$ **29.** $(\frac{110}{19}, -\frac{12}{19})$ **31.** $(\frac{1}{2}, -\frac{1}{2})$
33. $(-\frac{4}{3}, -\frac{19}{3})$ **35.** (90.91, −90.91) **37.** (140, 60)
39. (23.118878, −12.039964) **41.** $\left(\frac{a + 2b}{7}, \frac{a - 5b}{7}\right)$
43. $p = 2$, $q = -\frac{1}{3}$ **45.** $(-\frac{1}{5}, \frac{1}{10})$

Margin Exercises, Section 3.3

1. 22 white, 8 red **2.** 30 L of 5%, 70 L of 15%
3. \$1800 at 7%, \$1900 at 9% **4.** $l = 32.5$ ft, $w = 12.5$ ft
5. 280 km

Exercise Set 3.3, pp. 99–100

1. 5, −47 **3.** 24, 8
5. 150 lb soybean meal, 200 lb corn meal
7. 5 L of each **9.** \$4100 at 14%, \$4700 at 16%
11. \$725 at 12%, \$425 at 11% **13.** 375 km **15.** $1\frac{3}{4}$ hr
17. 8 white, 22 yellow **19.** 13 at \$9.75, 32 at \$8.50
21. Maria 20, Carlos 28 **23.** $l = 160$ m; $w = 154$ m
25. $l = 31$ cm; $w = 12$ cm **27.** Joan: 14 yrs, James: 32 years
29. 82 **31.** 137° **33.** $4\frac{4}{7}$ L
35. 261 at 18 mpg, 204 at 24 mpg

Margin Exercises, Section 3.4

1. **(a)** No; **(b)** yes **2.** (2, 1, −1) **3.** $(2, -2, \frac{1}{2})$
4. (20, 30, 50)

Exercise Set 3.4, p. 104

1. Yes **3.** (1, 2, 3) **5.** (−1, 5, −2) **7.** (3, 1, 2)
9. (−3, −4, 2) **11.** (2, 4, 1) **13.** (−3, 0, 4)
15. (2, 2, 4) **17.** $(\frac{1}{2}, 4, -6)$ **19.** $(\frac{1}{2}, \frac{1}{3}, \frac{1}{6})$
21. $(\frac{1}{2}, \frac{2}{3}, -\frac{5}{6})$ **23.** (1, −1, 2) **25.** (1, −2, 4, −1)

27. $(-1, \frac{1}{5}, -\frac{1}{2})$ **29.** $a = 2$, $b = -1$, $c = 3$
31. $z = 8 - 2x - 4y$

Margin Exercises, Section 3.5

1. 2, −3, 4 **2.** 64°, 32°, 84° **3.** A, 112; B, 90; C, 85

3

Exercise Set 3.5, pp. 107–108

1. 17, 9, 79 **3.** 4, 2, −1 **5.** A = 34°, B = 104°, C = 42°
7. A = 25°, B = 50°, C = 105°
9. \$21 on Thursday, \$18 on Friday, \$27 on Saturday
11. First score is 74.5, second score is 68.5, third score is 82.
13. A: 1500, B: 1900, C: 2300
15. A: 900 gal/hr, B: 1300 gal/hr, C: 1500 gal/hr
17. 20 **19.** 180° **21.** 1869
23. Adult: 2050, senior citizen: 845, children: 210
25. First: \$4.50, second: \$10.00, third: \$18.00

Margin Exercises, Section 3.6

1. (−8, 2) **2.** (−1, 2, 3)

Exercise Set 3.6, pp. 110–111

1. $(\frac{3}{2}, \frac{5}{2})$ **3.** (−1, 2, −2) **5.** $(\frac{1}{2}, \frac{3}{2})$ **7.** $(\frac{3}{2}, -4, 3)$
9. (2, −2, 1) **11.** (1, −3, −2, −1)
13. 4 dimes, 30 nickels **15.** 10 nickels, 4 dimes, 8 quarters
17. 5 lb of \$4.05, 10 lb of \$2.70

Margin Exercises, Section 3.7

1. **(a)** Inconsistent; **(b)** consistent
2. **(a)** Inconsistent; **(b)** consistent
3. **(a)** Dependent; **(b)** independent
4. **(a)** Dependent; **(b)** independent

Exercise Set 3.7, p. 113

1. Inconsistent **3.** Consistent **5.** Consistent
7. Inconsistent **9.** Consistent **11.** Dependent
13. Independent **15.** Dependent **17.** Independent
19. Dependent **21.** No solution
23. Jennifer is 8; Roland is 32. **25.** No solution
27. (0, −5), (1, −2), (−1, −8) **29.** No solution
31. No solution
33. 27: consistent, dependent; 29: inconsistent, independent; 31: inconsistent, dependent
35. 2 **37.** $\left(\frac{1 - 3y}{2}, y\right)$

Margin Exercises, Section 3.8

1. **(a)** $P = 16x - 80{,}000$; **(b)** 5000 units **2.** (\$12.96, 352)

Exercise Set 3.8, pp. 116–117

1. **(a)** $P = 20x - 600{,}000$; **(b)** 30,000 units
3. **(a)** $P = 50x - 120{,}000$; **(b)** 2400 units

5. **(a)** $P = 80x - 10{,}000$; **(b)** 125 units
7. **(a)** $P = 40x - 75{,}000$; **(b)** 1875 units
9. **(a)** $P = 75x - 195{,}000$; **(b)** 2600 units **11.** (\$10, 1400)
13. (\$22, 474) **15.** (\$50, 6250) **17.** (\$10, 1070)
19. (\$14, 3882) **21.** $l = 235$ m, $w = 120$ m
23. 4 years, 8 years
25. **(a)** $C = 750x + 125{,}000$; **(b)** $R = 1050x$;
(c) $P = 300x - 125{,}000$; **(d)** \$5000 loss; **(e)** 417 units
27. **(a)** $C = 20x + 10{,}000$; **(b)** $R = 100x$;
(c) $P = 80x - 10{,}000$; **(d)** \$150,000 profit; **(e)** 125 units

Summary and Review: Chapter 3, pp. 117–119

1. [3.1] $(\frac{13}{6}, -\frac{4}{3})$ **2.** [3.1] $(-2, 1)$ **3.** [3.2] $(\frac{2}{5}, -\frac{4}{5})$
4. [3.2] $(-\frac{8}{5}, \frac{2}{5})$ **5.** [3.2] $(-\frac{11}{24}, -\frac{49}{48})$ **6.** [3.2] $(\frac{37}{19}, \frac{53}{19})$
7. [3.2] $(\frac{76}{17}, -\frac{2}{119})$ **8.** [3.2] $(2, 2)$
9. [3.3] Record album, \$7; poster, \$6
10. [3.3] 25 pens, 23 pencils
11. [3.3] 10 liters of 30% alcohol, 30 liters of 50% alcohol
12. [3.4] $(10, 4, -8)$ **13.** [3.4] $(-\frac{7}{3}, \frac{125}{27}, \frac{20}{27})$
14. [3.4] $(2, 0, 4)$
15. [3.5] $m\angle A = 90°$, $m\angle B = 67.5°$, $m\angle C = 22.5°$
16. [3.5] 641 **17.** [3.5] 5 \$20 bills, 15 \$5 bills, 19 \$1 bills
18. [3.6] $(1, 2, 3)$ **19.** [3.6] $(-3, 0, 4)$
20. [3.7] Inconsistent **21.** [3.7] Consistent
22. [3.7] Dependent **23.** [3.7] Independent
24. [3.8] **(a)** $C = 175x + 35{,}000$; **(b)** $R = 225x$;
(c) $P = 50x - 35{,}000$; **(d)** profit of \$25,000; **(e)** 700 units
25. $(2, \frac{1}{2})$ or $(1, 1)$ **26.** $(5, 0)$ or $(3, 4)$

CUMULATIVE REVIEW, CHAPTERS 1–3, pp. 119–120

1. 14.87 **2.** -22 **3.** -42.9 **4.** 20 **5.** $-\frac{21}{4}$ **6.** 6
7. -5 **8.** $\frac{10}{9}$ **9.** $-\frac{32}{5}$ **10.** $\frac{18}{17}$ **11.** x^{11} **12.** $-\frac{40x}{y^5}$
13. $-12x^{2a}y^{2b+3}$ **14.** y^{10} **15.** $-\frac{2a^{11}}{5b^{33}}$ **16.** $\frac{81x^{36}}{256y^8}$
17. 1.118 **18.** 1.11735×10^6 **19.** 4×10^2 **20.** 4×10^6
21. $b_1 = \frac{2A - b_2h}{h}$ **22.** Yes **23.** No

24.

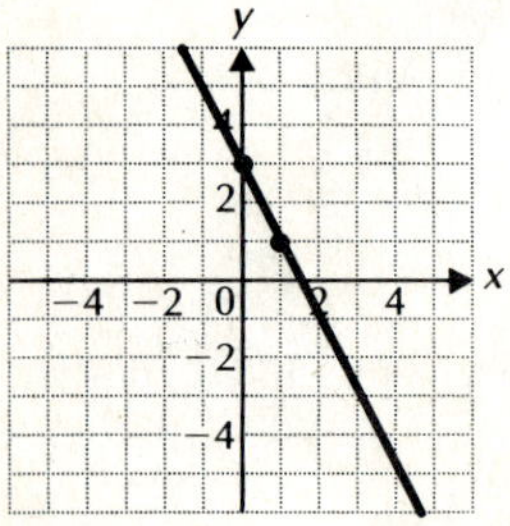

25.

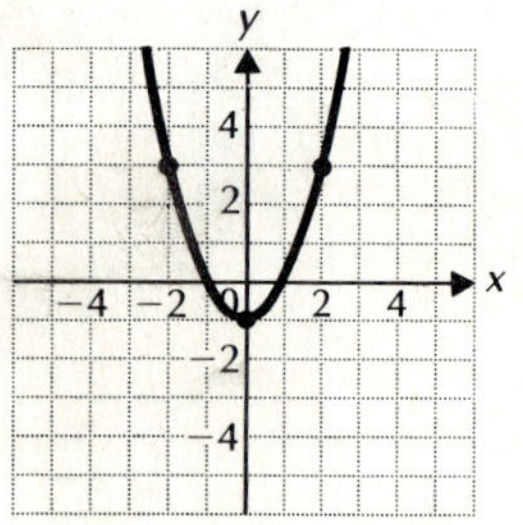

26.

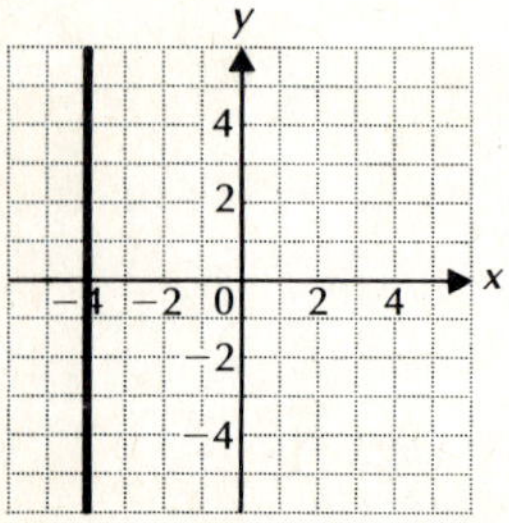

27.

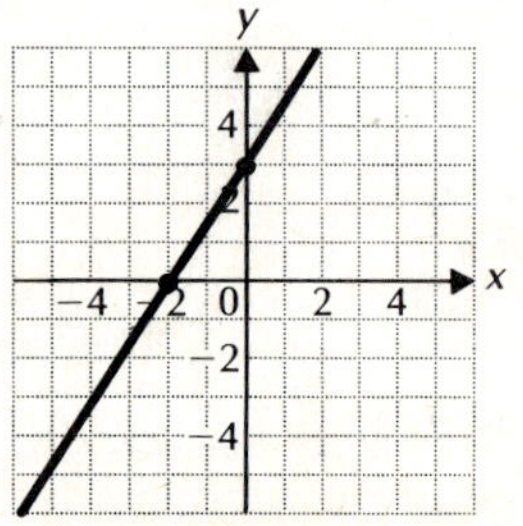

28. $\frac{9}{4}$, $(0, -3)$ **29.** $\frac{4}{3}$ **30.** $y = -3x - 5$
31. $x + 10y = 24$ **32.** -7 **33.** -31 **34.** 3 **35.** -2
36. -43 **37.** $8a^2 + 4a - 4$ **38.** $(1, 1)$ **39.** $(-2, 3)$
40. $(-3, \frac{2}{5})$ **41.** $(-3, 2, -4)$ **42.** $(0, -1, 2)$
43. 3, 5, 7 **44.** 90 **45.** $l = 10$ cm, $w = 6$ cm **46.** \$60

CHAPTER 4

Margin Exercises, Section 4.1

1. Terms: $-92x^5$, $-8x^4$, y^2, 5; leading term: $-92x^5$
2. 5, -4, -2, 1, -1, -5; leading coefficient: 5
3. Degrees of terms (left to right): 1, 0, 1, 7, 3; degree of polynomial: 7; leading term: $7x^2y^3z^2$; leading coefficient: 7
4. Monomials: $4x^2$, $-7y$, -3; binomials: $3x^2 - 2$, $8x^3 - 2x^2$, $5 - 3x$; trinomials: $5x^3 + 9x - 3$, $-4y^2 - 5y - 5$
5. Linear: -3, $-7y$, $5 - 3x$; quadratic: $3x^2 - 2$, $4x^2$, $-4y^2 - 5y - 5$; cubic: $5x^3 + 9x - 3$, $8x^3 - 2x^2$
6. $3x^2y^3 - 3y^2 + 5x^4y + x^3 - 5$ **7.** **(a)** 5; **(b)** 13; **(c)** 13
8. 11¢ per km **9.** $3y - 4x + 4xy^2$ **10.** $8xy^3 + 2x^3y - 9x^2y$
11. $-4x^3 + 2x^2 - 4x + 2$ **12.** $10y^5 - 4y^2 + 5$
13. $5p^2q^4 - 8p^2q^2 + 5$
14. **(a)** $-(4x^3 - 5x^2 + \frac{1}{4}x - 10)$; **(b)** $-4x^3 + 5x^2 - \frac{1}{4}x + 10$
15. **(a)** $-(8xy^2 - 4x^3y^2 - 9x - \frac{1}{5})$;
(b) $-8xy^2 + 4x^3y^2 + 9x + \frac{1}{5}$
16. **(a)** $-(-9y^5 - 8y^4 + \frac{1}{2}y^3 - y^2 + y - 1)$;
(b) $9y^5 + 8y^4 - \frac{1}{2}y^3 + y^2 - y + 1$ **17.** $3x^2 + 5$
18. $14y^3 - 2y + 4$ **19.** $p^2 - 6p - 2$
20. $3y^5 - 3y^4 + 5y^3 - 2y^2 - 3$
21. $9p^4q - 10p^3q^2 + 4p^2q^3 + 9q^4$ **22.** $y^3 - y^2 + \frac{4}{3}y + 0.1$

Exercise Set 4.1, pp. 126–127

1. 4, 3, 2, 1, 0; 4; $-11x^4$; -11 **3.** 3, 7, 6, 0; 7; $2y^7$; 2
5. 5, 6, 2, 1, 0; 6; $4a^2b^4$; 4 **7.** $-xy^3 + x^2y^2 + x^3y + 1$
9. $-5a^5y^5 - 4a^2y^3 + 3ay$ **11.** $3xy^3 + x^2y^2 - 9x^3y + 2x^4$
13. $-7ab + 4ax - 7ax^2 + 4x^6$ **15.** $P(4) = 54$, $P(0) = 2$
17. $P(-2) = -45$, $P(\frac{1}{3}) = -8\frac{19}{27}$
19. $-4a^3 + 2a^2 - 7a + 5$ **21.** $8a^3 - 16a^2 + 6a - 7$
23. $-3x^3 - 9x^2a - 9a^2x - 3a^3 + 6x^2 + 12xa + 6a^2 - 4$
25. $2x^2$ **27.** $3x + y$ **29.** $a + 6$ **31.** $-6a^2b - 2b^2$
33. $9x^2 + 2xy + 15y^2$ **35.** $-x^2y + 4y + 9xy^2$
37. $5x^2 + 2y^2 + 5$ **39.** $6a + b + c$ **41.** $-4a^2 - b^2 + 3c^2$
43. $-2x^2 + x - xy - 1$ **45.** $5x^2y - 4xy^2 + 5xy$
47. $9r^2 + 9r - 9$ **49.** $-3a^2 + 2a - 1$
51. $-2x^2y - 5xy + 3xy^2$
53. $-(5x^3 - 7x^2 + 3x - 6)$, $-5x^3 + 7x^2 - 3x + 6$
55. $-(-12y^5 + 4ay^4 - 7by^2)$, $12y^5 - 4ay^4 + 7by^2$
57. $13x - 6$ **59.** $-4x^2 - 3x + 13$ **61.** $2a - 4b + 3c$
63. $-2x^2 + 6x$ **65.** $-4a^2 + 8ab - 5b^2$ **67.** $-9y + 9$
69. $8a^2b + 16ab + 3ab^2$
71. $0.06y^4 + 0.032y^3 - 0.94y^2 + 0.93$
73. $27.94x^2 + 9.105y^2 + 8.15x$ **75.** 28; 190
77. $55x^5 - 59x^4 + 14x^3 + 12x^2 + 18x + 175$
79. $29x^5 - 15x^4 + 86x^3 - 68x^2 + 50x + 25$ **81.** $5x^2 - 8x$
83. $47x^{4a} + 40x^{3a} + 30x^{2a} + x^a + 4$
85. $x^{5b} + 4x^{4b} + x^{3b} - 6x^{2b} - 9x^b$

Margin Exercises, Section 4.2

1. $p^4 + p^3 - 12p^2 - 5p + 15$
2. $2x^4 - 8x^3 + 4x^2 - 21x + 20$
3. $8x^5 + 12x^4 - 20x^3 + 4x^2 - 15x + 6$ **4.** $y^2 + 6y - 40$
5. $2p^2 + 7pq - 15q^2$ **6.** $x^3y^3 + x^2y^3 + 2x^2y^2 + 2xy^2$
7. $a^2 - 2ab + b^2$ **8.** $x^2 + 16x + 64$ **9.** $9x^2 - 42x + 49$
10. $x^2 - 64$ **11.** $16y^2 - 49$ **12.** $49x^4y^2 - 4y^2$
13. $4x^2 + 12x + 9 - 25y^2$ **14.** $x^3 + 3x^2 + 3x + 1$
15. $x^3 - 3x^2 + 3x - 1$ **16.** $t^6 - 9t^4b + 27t^2b^2 - 27b^3$
17. $8a^6 - 60a^4b^2 + 150a^2b^4 - 125b^6$

Exercise Set 4.2, pp. 129–130

1. $10y^3$ **3.** $-20x^3y$ **5.** $-10x^5y^6$ **7.** $6x - 2x^2$
9. $3a^2b + 3ab^2$ **11.** $15c^3d^2 - 25c^2d^3$ **13.** $6x^2 + x - 12$
15. $s^2 - 9t^2$ **17.** $x^2 - 2xy + y^2$ **19.** $2y^2 + 9xy - 56x^2$
21. $a^4 - 5a^2b^2 + 6b^4$ **23.** $x^3 - 64$ **25.** $x^3 + y^3$
27. $a^4 + 5a^3 - 2a^2 - 9a + 5$
29. $4a^3b^2 + 4a^3b - 10a^2b^2 - 2a^2b + 3ab^3 + 7ab^2 - 6b^3$
31. $a^2 + 5a + 6$ **33.** $y^2 + y - 6$ **35.** $4a^2 + \frac{4}{3}a + \frac{1}{9}$
37. $x^2 - 4xy + 4y^2$ **39.** $b^2 - \frac{5}{6}b + \frac{1}{6}$ **41.** $2x^2 + 13x + 18$
43. $4x^2 - 4xy - 3y^2$ **45.** $x^2 + 6x + 9$
47. $4x^4 - 12x^2y^2 + 9y^4$ **49.** $a^4b^4 + 2a^2b^2 + 1$ **51.** $c^2 - 4$
53. $4a^2 - 1$ **55.** $9m^2 - 4n^2$ **57.** $x^4 - y^2z^2$
59. $y^3 + 15y^2 + 75y + 125$ **61.** $m^6 - 6m^4n + 12m^2n^2 - 8n^3$
63. $a^6 + 6a^4y + 12a^2y^2 + 8y^3$
65. $27a^3 - 54a^2b^2 + 36ab^4 - 8b^6$ **67.** 11 weekdays
69. $-\frac{4}{3}x^6y^{11}$ **71.** $-\frac{9}{4}r^{22}s^{14}$ **73.** $\frac{8}{21}x^8y^5$
75. $-16s^8t + 24s^6t^5 - 8s^4t^8 + 8s^3t^{11}$
77. $16x^4 - 32x^3 + 16x^2$
79. $-a^4 - 2a^3b + 25a^2 + 2ab^3 - 25b^2 + b^4$
81. $r^8 - 2r^4s^4 + s^8$ **83.** $a^2 + 2ac - b^2 - 2bd + c^2 - d^2$
85. $10y^2 - 38xy + 24x^2 - 212y + 306x + 930$
87. $16x^4 + 4x^2y^2 + y^4$ **89.** $x^{4a} - y^{4b}$
91. $a^3 - b^3 + 3b^2 - 3b + 1$ **93.** $\frac{1}{81}x^{12} - \frac{8}{81}x^6y^4 + \frac{16}{81}y^8$
95. $x^{a^2-b^2}$ **97.** \$1404.93

Margin Exercises, Section 4.3

1. $3(x^2 - 2)$ **2.** $4x^3(x^2 - 2)$ **3.** $3y^2(3y^2 - 5y + 1)$
4. $3x^2y(2 - 7xy + y^2)$ **5.** $(p + q)(2x + y + 2)$
6. $(y + 3)(2y - 11)$ **7.** $(x - 4)(x + 5)$ **8.** $(y^2 - 2)(5y + 2)$
9. $x(x(x(x - 2) + 2) - 5) - 4) + 6$; $P(6) = 5418$
10. $x(x(x(x(x + 0) + 5) + 4) + 0) + 12$; $P(-4) = -1268$

Exercise Set 4.3, pp. 132–133

1. $2a(2a + 1)$ **3.** $y(y - 5)$ **5.** $y^2(y + 9)$ **7.** $3x^2(2 - x^2)$
9. $4xy(x - 3y)$ **11.** $3(y^2 - y - 3)$ **13.** $2a(2b - 3c + 6d)$
15. $5(2a^4 + 3a^2 - 5a - 6)$ **17.** $(b - 2)(a + c)$
19. $(x - 2)(2x + 13)$ **21.** $2a^2(x - y)$ **23.** $(a + b)(c + d)$
25. $(b^2 + 2)(b - 1)$ **27.** $(y - 8)(y - 1)$
29. $12x(2x^2 - 3x + 6)$ **31.** $x^2(x^4 + x^3 - x + 1)$
33. $(2y^2 + 5)(y^2 + 3)$ **35.** $x(x(x(x - 3) - 5) + 4) + 2$
37. $x(x(x(x(5x + 0) - 3) + 4) - 5) + 1$
39. $x(x(x(2x - 3) + 5) + 6) - 4$; $P(5) = 1026$, $P(-2) = 60$, $P(10) = 17{,}556$
41. Between 19 words and 39 words inclusive
43. $2(2y^{2a} + 5)(y^{2a} + 3)$ **45.** $x^a(4x^b + 7x^{-b})$ **47.** 54

49. **(a)** $x(x(x(0.1x + 0) - 1) + 0) + 0.4$;
(b)

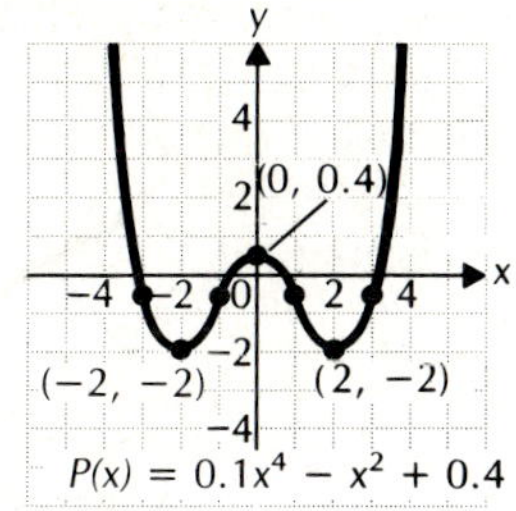

Margin Exercises, Section 4.4

1. (a), (b), (d) **2.** $(x + 7)^2$ **3.** $(3y - 5)^2$ **4.** $(4x + 9y)^2$
5. $(4x^2 - 5y^3)^2$ **6.** $-2(2a - 3b)^2$ **7.** $(x + 7)(x - 2)$
8. $(y - 2)(y + 1)$ **9.** $(3x + 2)(x + 1)$
10. $(2x + 3)(2x - 1)$ **11.** $2(4y - 1)(3y - 5)$
12. $4(2x - 1)(2x + 3)$ **13.** $(pq + 4)(pq + 3)$
14. $(2x^2y^3 + 5)(x^2y^3 - 4)$

Exercise Set 4.4, pp. 136–137

1. $(y - 3)^2$ **3.** $(x + 7)^2$ **5.** $(x + 1)^2$ **7.** $2(a + 2)^2$
9. $(y - 6)^2$ **11.** $y(y - 9)^2$ **13.** $3(2a + 3)^2$
15. $2(x - 10)^2$ **17.** $(1 - 4d)^2$ **19.** $(x + 5)(x + 4)$
21. $(y - 4)^2$ **23.** $(x - 9)(x + 3)$ **25.** $(m - 7)(m + 4)$
27. $(x + 9)(x + 5)$ **29.** $(y + 9)(y - 7)$ **31.** $(t - 7)(t - 4)$
33. $(x + 5)(x - 2)$ **35.** $(x + 2)(x + 3)$
37. $(8 - y)(4 + y)$ **39.** $(t + 5)(t + 3)$
41. $(3x + 1)(3x + 4)$ **43.** $(3a - 4)(a + 1)$
45. $(12z + 1)(z - 3)$ **47.** $(2t + 5)(2t - 3)$
49. $(3x + 5)(2x - 5)$ **51.** $(5y + 4)(2y - 3)$
53. $(4a - 1)(3a - 1)$ **55.** $(3a + 2)(3a + 4)$
57. $(5x + 3)(3x - 1)$ **59.** $6(3x - 4)(x + 1)$
61. $y(6y - 5)(3y + 2)$ **63.** $(5y + 2)(3y - 5)$
65. $(5y + 4)(2y + 3)$ **67.** $(6y + 5)(4y - 3)$
69. $(4a - 3)(5a - 2)$ **71.** $(x^2 + 16)(x^2 - 5)$
73. $(x + \frac{4}{5})(x - \frac{1}{5})$ **75.** $(y - 0.1)(y + 0.5)$
77. $(2x - 3y)(x + 2y)$ **79.** $(2x - 3y)(4x + 3y)$
81. $(7ab + 6)(ab + 1)$ **83.** $3(x - 11)(x + 15)$
85. $4y(y - 12)^2$ **87.** $7(x - 25y)(x + 25y)$
89. $15t(t - 7)(t + 3)$ **91.** $(x^a + 8)(x^a - 3)$
93. $(2x^a - 3)(2x^a + 1)$ **95.** $(\frac{1}{2}p - \frac{2}{3})^2$ **97.** 263.76 cm²

Margin Exercises, Section 4.5

1. $(y + 2)(y - 2)$ **2.** $(7x^2 + 5y^5)(7x^2 - 5y^5)$
3. $(5xy + 2a)(5xy - 2a)$ **4.** $(3x + 4y^3)(3x - 4y^3)$
5. $5(2x + y)(2x - y)$ **6.** $y^2(9x^2 + 4)(3x + 2)(3x - 2)$
7. $(x - 5 + 2y)(x - 5 - 2y)$ **8.** $x^2 + 14x + 49$
9. $y^2 - 10by + 25b^2$ **10.** $x^2 - \frac{2}{5}x + \frac{1}{25}$
11. $x^2 + 4.2x + 4.41$ **12.** $(x + 20)(x + 6)$
13. $2(x - 3)(x - 21)$ **14.** $(x + 6.2)(x + 2.2)$

Exercise Set 4.5, pp. 139–140

1. $(x + 4)(x - 4)$ **3.** $6(x + y)(x - y)$
5. $4x(y^2 + z^2)(y + z)(y - z)$ **7.** $a(2a + 7)(2a - 7)$
9. $3(x^4 + y^4)(x^2 + y^2)(x + y)(x - y)$

11. $a^2(3a + 5b^2)(3a - 5b^2)$ **13.** $(\frac{1}{5} - x)(\frac{1}{5} + x)$
15. $(0.2x + 0.3y)(0.2x - 0.3y)$ **17.** $(a + b + 3)(a + b - 3)$
19. $(r - 1 - 2s)(r - 1 + 2s)$
21. $2(m + n - 5b)(m + n + 5b)$ **23.** $(3 - a - b)(3 + a + b)$
25. $x^2 + 16x + 64$ **27.** $x^2 - 4.2x + 4.41$
29. $(x + 7)(x + 17)$ **31.** $(x - 21)(x - 5)$
33. $2(x + 9)(x + 19)$ **35.** $2(x - 12a)(x - 4a)$ **37.** 76
39. $(x + 5.766)(x - 1.284)$ **41.** $x^2 + \frac{2}{3}bx + \frac{b^2}{9}$
43. $ax^2 + bx + \frac{b^2}{4a}$ **45.** $(x - \frac{1}{2})(x - \frac{5}{2})$
47. $(x + 2.3)(x + 0.3)$ **49.** $(x + \frac{1}{5})(x + \frac{2}{5})$
51. $(x^2y^2 - 4)^2$ **53.** $(\frac{1}{6}x^4 + \frac{2}{3})^2$ **55.** $(x^a + y)(x^a - y)$
57. $2(2y^{2a} + 5)(y^{2a} + 3)$ **59.** $(x^a + 5)(x^a + 9)$

Margin Exercises, Section 4.6

1. $(x - 2)(x^2 + 2x + 4)$ **2.** $(4 - y)(16 + 4y + y^2)$
3. $(3x + y)(9x^2 - 3xy + y^2)$ **4.** $(2y + z)(4y^2 - 2yz + z^2)$
5. $(2xy)(2x^2 + 3y^2)(4x^4 - 6x^2y^2 + 9y^4)$
6. $(3x + 2y)(9x^2 - 6xy + 4y^2)(3x - 2y)(9x^2 + 6xy + 4y^2)$
7. $(x - 0.3)(x^2 + 0.3x + 0.09)$

Exercise Set 4.6, p. 142

1. $(x + 2)(x^2 - 2x + 4)$ **3.** $(y - 4)(y^2 + 4y + 16)$
5. $(w + 1)(w^2 - w + 1)$ **7.** $(2a + 1)(4a^2 - 2a + 1)$
9. $(y - 2)(y^2 + 2y + 4)$ **11.** $(2 - 3b)(4 + 6b + 9b^2)$
13. $(4y + 1)(16y^2 - 4y + 1)$ **15.** $(2x + 3)(4x^2 - 6x + 9)$
17. $(a - b)(a^2 + ab + b^2)$ **19.** $(a + \frac{1}{2})(a^2 - \frac{1}{2}a + \frac{1}{4})$
21. $2(y - 4)(y^2 + 4y + 16)$ **23.** $3(2a + 1)(4a^2 - 2a + 1)$
25. $r(s + 4)(s^2 - 4s + 16)$ **27.** $5(x - 2z)(x^2 + 2xz + 4z^2)$
29. $(x + 0.1)(x^2 - 0.1x + 0.01)$
31. $8(2x^2 - t^2)(4x^4 + 2x^2t^2 + t^4)$ **33.** 228 ft^2
35. $(x^{2a} + y^b)(x^{4a} - x^{2a}y^b + y^{2b})$
37. $3(x^a + 2y^b)(x^{2a} - 2x^ay^b + 4y^{2b})$
39. $\frac{1}{3}(\frac{1}{2}xy + z)(\frac{1}{4}x^2y^2 - \frac{1}{2}xyz + z^2)$ **41.** $x^2 - 16$

Margin Exercises, Section 4.7

1. $3y(y + 2x)(y - 2x)$ **2.** $7(a - 1)(a^2 + a + 1)$
3. $(2x + 3y)(4x^2 - 6xy + 9y^2)(2x - 3y)(4x^2 + 6xy + 9y^2)$
4. $(x - 2)(3 - bx)$ **5.** $5(y^4 + 4x^6)$ **6.** $3(x - 2)(2x + 3)$
7. $(a - b)^2(a + b)$ **8.** $3(x + 3a)^2$
9. $2(x - 5 + 3b)(x - 5 - 3b)$

Exercise Set 4.7, p. 145

1. $(x + 12)(x - 12)$ **3.** $(2x + 3)(x + 4)$
5. $3(x^2 + 2)(x^2 - 2)$ **7.** $(a + 5)^2$ **9.** $2(x - 11)(x + 6)$
11. $(3x + 5y)(3x - 5y)$
13. $(m + 1)(m^2 - m + 1)(m - 1)(m^2 + m + 1)$
15. $(x + y + 3)(x - y + 3)$
17. $2(5x - 4y)(25x^2 + 20xy + 16y^2)$ **19.** $(m^3 + 10)(m^3 - 2)$
21. $(c - b)(a + d)$ **23.** $(2c - d)^2$ **25.** $(x^2 + 2)(2x - 7)$
27. $2(x + 2)(x - 2)(x + 3)$
29. $2(2x + 3y)(4x^2 - 6xy + 9y^2)$ **31.** $(6y - 5)(6y + 7)$
33. $(a^4 + b^4)(a^2 + b^2)(a + b)(a - b)$ **35.** $ab(a + 4b)(a - 4b)$
37. 55 correct answers, 20 wrong answers

39. $(2x - 3)(10x + 13)$ **41.** $4(2a - 1)(9a + 40)$
43. $7(x - \frac{1}{2})(x^2 + \frac{1}{2}x + \frac{1}{4})$ **45.** $(c - 2d)^2(c^2 - cd + d^2)^2$
47. $(x + 1)(x - 1)(x + 7)(x - 7)$
49. $(s - 3t)(s + 3t)(s^2 + 3st + 9t^2)(s^2 - 3st + 9t^2)$
51. $(3x^{2s} + 4y^t)(9x^{4s} - 12x^{2s}y^t + 16y^{2t})$
53. $(2x + y - r + 3s)(2x + y + r - 3s)$
55. $(c^2d^2 + a^8)(cd - a^4)(cd + a^4)$ **57.** $c(c^w + 1)^2$
59. $y(y^4 + 1)(y^2 + 1)(y + 1)(y - 1)$
61. $3(a + b - c - d)(a + b + c + d)$ **63.** $4(3a^2 + 4)$ **65.** 0

Margin Exercises, Section 4.8

1. 4, 2 **2.** $\frac{1}{2}$, -3 **3.** 0, 2 **4.** -5 **5.** 0, 2, -3

Exercise Set 4.8, pp. 147–148

1. -7, 4 **3.** 4 **5.** 6 **7.** -5, -4 **9.** 0, -8
11. -3, 3 **13.** -6, 6 **15.** -5, -9 **17.** -9, 7
19. 7, 4 **21.** 8, -4 **23.** $-\frac{2}{3}$, -2 **25.** $\frac{3}{4}$, $\frac{1}{2}$ **27.** 0, 6
29. $-\frac{3}{4}$, $\frac{2}{3}$ **31.** -2, 2 **33.** $\frac{1}{2}$, 7 **35.** $-\frac{5}{7}$, $\frac{2}{3}$ **37.** 0, $\frac{1}{5}$
39. $-\frac{1}{5}$, $\frac{1}{5}$ **41.** 0, 3, -2 **43.** 0, -8, 8
45. 54 mph, 39 mph **47.** 4 **49.** -1, 11 **51.** $-\frac{5}{3}$, 5, 4
53. 2, -2

Margin Exercises, Section 4.9

1. 8, -6 **2.** 8 cm, 3 cm

Exercise Set 4.9, pp. 150–151

1. $\frac{7}{2}$, $-\frac{3}{2}$ **3.** -12, 11 **5.** Length is 12 cm; width is 7 cm
7. Length is 100 m; width is 75 m **9.** 9 and 11 **11.** 3 cm
13. Height is 7 cm; base is 16 cm **15.** 2
17. -10, -8, and -6; 6, 8, and 10
19. 11, 12, and 13; -2, -1, and 0 **21.** 3; 14
23. $x = \frac{14 - 2b}{5}$ **25.** Width is 30 in.; length is 40 in.; depth is 20 in. **27.** 54 cm^2

Summary and Review: Chapter 4, pp. 151–152

1. [4.1] $-x^3 + 2x^2 + 5x + 2$ **2.** [4.1] $-3x^4 + 3x^3 - x + 16$
3. [4.1] $x^3 + 6x^2 - x - 4$ **4.** [4.1] $-8xy^2 + 4xy + 13x^2y$
5. [4.1] $9x - 7$ **6.** [4.1] $-2a + 6b + 7c$
7. [4.1] $16p^2 - 8p$ **8.** [4.1] $6x^2 - 7xy + y^2$
9. [4.2] $-18x^3y^4$ **10.** [4.2] $8a^2b^2 + 2abc - 3c^2$
11. [4.2] $2x^3 + 5x^2y - 6xy - 8x - 20y - 15y^2$
12. [4.2] $20x^4 - 18x^3 - 47x^2 + 69x - 27$
13. [4.2] $36m^2 - 12mn + n^2$ **14.** [4.2] $x^3 + 6x^2 + 12x + 8$
15. [4.3] $x(6x + 5)$ **16.** [4.3] $3y^2(3y^2 - 1)$
17. [4.3] $3x(5x^3 - 6x^2 + 7x - 3)$ **18.** [4.4] $(a - 9)(a - 3)$
19. [4.4] $(3m + 2)(m + 4)$ **20.** [4.4] $(5x + 2)^2$
21. [4.5] $4(y + 2)(y - 2)$ **22.** [4.5] $(a^2 + 9)(a + 3)(a - 3)$
23. [4.3] $(y + 2)(3y^2 - 5)$ **24.** [4.5] $3(x - 3)(x + 15)$
25. [4.6] $(3x - 2)(9x^2 + 6x + 4)$
26. [4.6] $(0.4b - 0.5c)(0.16b^2 + 0.2bc + 0.25c^2)$
27. [4.7] $y(y^2 + 1)(y + 1)(y - 1)$
28. [4.7] $2(y^4 + 2z^4)(y^4 - 2z^4)$ **29.** [4.8] 10 **30.** [4.8] $\frac{2}{3}$, $\frac{3}{2}$

31. [4.8] $\frac{5}{4}, \frac{1}{2}$ **32.** [4.9] 5 **33.** [4.9] 3, 5, 7; $-7, -5, -3$
34. [4.9] Length is 8 inches; width is 5 inches
35. $2(2x + y)(2x - y)(4x^2 - 2xy + y^2)(4x^2 + 2xy + y^2)$
36. $\dfrac{(2y-3)(y-1)}{3(y+1)(3y-2)}$ **37.** $\frac{7}{4}$ or $\frac{4}{7}$

CHAPTER 5

Margin Exercises, Section 5.1

1. $1 < 8$ **2.** $5 > -2$ **3.** $-3 > -8$ **4.** True **5.** True
6. False **7.** Yes **8.** No **9.** Yes
10. $\{x|x < 3\}$

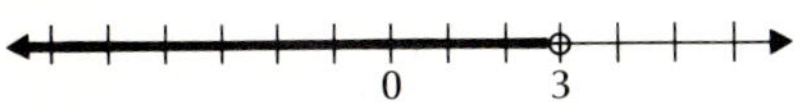

11. $\{x|x \geqslant -2\}$

12. $\{x|x > 3\}$

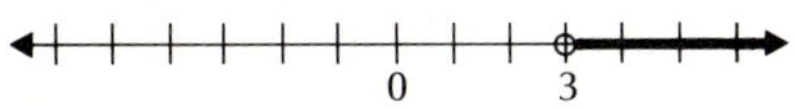

13. $\{x|x \leqslant 3\}$

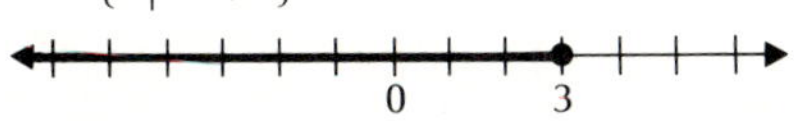

14. $\{x|x \geqslant -2\}$

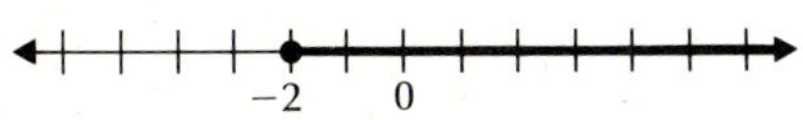

15. $\{y|y \leqslant \frac{3}{10}\}$ **16.** $\{y|y < \frac{-5}{12}\}$ **17.** $\{x|x \geqslant 12\}$
18. $\{y|\frac{-1}{5} \geqslant y\}$ **19.** $\{x|x < \frac{1}{2}\}$ **20.** $\{y|\frac{22}{13} \leqslant y\}$

Exercise Set 5.1, pp. 157–158

1. $\{x|x \leqslant 4\}$

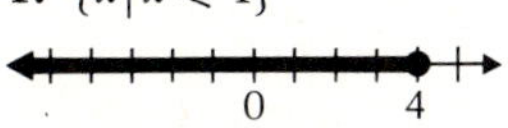

3. $\{x|x > 5\}$

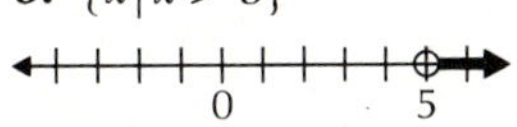

5. $\{x|x > -5\}$

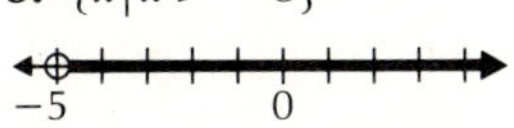

7. $\{y|y < 6\}$

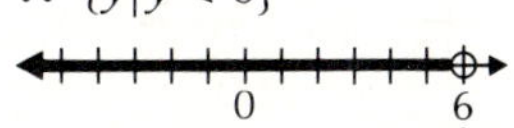

9. $\{a|a \leqslant -21\}$

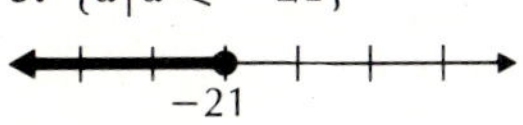

11. $\{t|t \geqslant -5\}$

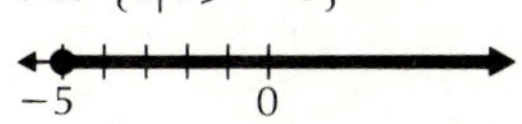

13. $\{y|y > -6\}$

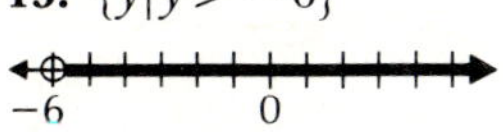

15. $\{x|x \leqslant 9\}$

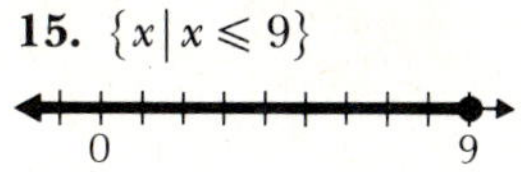

17. $\{x|x \geqslant 3\}$

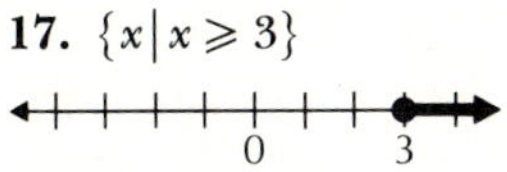

19. $\{x|x < -60\}$

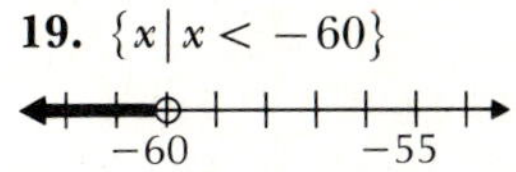

21. $\{x|x \leqslant 0.9\}$ **23.** $\{x|x \leqslant \frac{5}{6}\}$ **25.** $\{x|x < 6\}$
27. $\{y|y \leqslant -3\}$ **29.** $\{y|y > \frac{2}{3}\}$ **31.** $\{x|x \geqslant 11.25\}$
33. $\{x|x \leqslant \frac{1}{2}\}$
35. $36 < 4.5x < 63$; 9: least number of lawns, 13: number of lawns

37. $\{y|y \leqslant -8\frac{5}{6}\}$ **39.** $\{x|x > -\frac{2}{17}\}$ **41.** $\{m|m > \frac{7}{3}\}$
43. $\left\{x\middle|x \leqslant \dfrac{2}{a-1};\ a > 1\right\}$ **45.** $\left\{y\middle|y \geqslant \dfrac{5b+2a}{b(a-2)};\ a > 2\right\}$
47. $\left\{x\middle|x < \dfrac{2c-4m}{5c-d+2m}\right\}$ **49.** $\left\{k\middle|k \geqslant -\dfrac{9}{x}\right\}$
51. (a) False, $a = 2$, $b = 3$, $c = 4$, $d = 5$; $2 - 4 = 3 - 5$
(b) false, because $-3 < -2$, but $9 > 4$
53. Not equivalent, because $0 \cdot x < 0 \cdot 3 = 0 < 0$, which is not true.

Margin Exercises, Section 5.2

1. 71 or greater **2.** $n < 100$

Exercise Set 5.2, p. 160

1. 330 miles or less **3.** 84 or more, **5.** \$20,000
7. \$1.99 **9.** More than \$1450 **11.** More than \$15,000
13. Less than $166\frac{2}{3}$

Margin Exercises, Section 5.3

1. 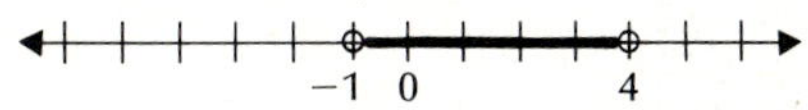

2. 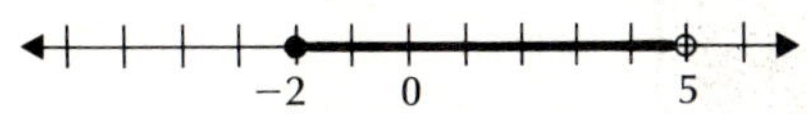

3. $\{x|-2 < x < 1\}$ **4.** 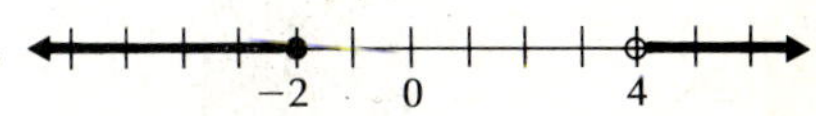

5. 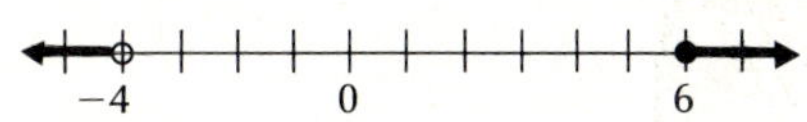

6. $\{x|x < 1 \text{ or } x \geq 6\}$ **7.** $\{x|x \geqslant \frac{7}{2} \text{ or } x < -2\}$

Exercise Set 5.3, pp. 163–164

1. 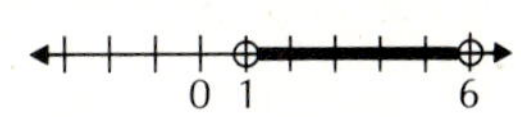
$1 < x < 6$

3. 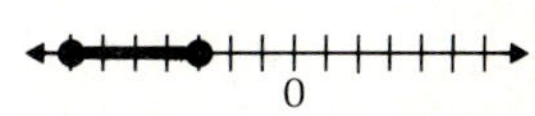
$-7 \leq y \leq -3$

5. 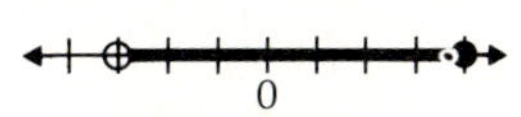
$-3 < x \leq 4$

7. 
$-6 < x \leq 2$

9. 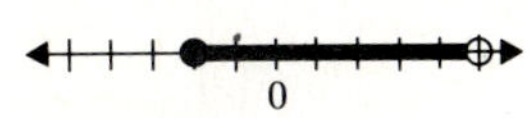
$-2 \leq x < 5$

11. 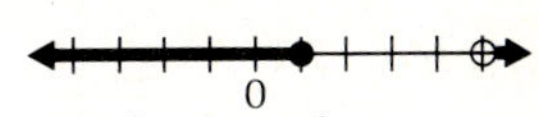
$x > 5$ or $x \leq 1$

13. $\{x|-4 < x < 6\}$ **15.** $\{y|-2 < y \leqslant 2\}$
17. $\{x|-\frac{5}{3} \leqslant x \leqslant \frac{4}{3}\}$ **19.** $\{x|-1 < x \leqslant 6\}$
21. $\{x|x \leqslant -9 \text{ or } x > -5\}$

23. 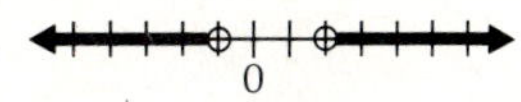
$x < -1$ or $x > 2$

25. 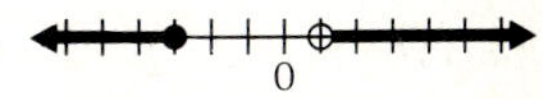
$x \leq -3$ or $x > 1$

27. 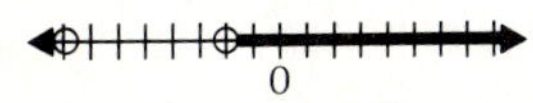
$x < -8$ or $x > -2$

29. $\{x|x < -9 \text{ or } x > -5\}$ **31.** $\{x|x \leqslant \frac{5}{2} \text{ or } x \geqslant 11\}$
33. $\{x|x < \frac{4}{3} \text{ or } x > 15\}$ **35.** $\{x|-\frac{3}{2} \leqslant x < \frac{9}{2}\}$
37. $\{m|-\frac{11}{3} < m \leqslant -3\}$ **39.** Between 23° and 113°
41. $\{a|-\frac{3}{2} \leqslant a \leqslant 1\}$
43. $\{x|-4 < x \leqslant 1\}$ **45.** $\{x|\frac{2}{5} \leqslant x \leqslant \frac{4}{5}\}$
47. $\{x|-\frac{1}{8} < x < \frac{1}{2}\}$ **49.** True **51.** False **53.** False
55. $\{x|-\frac{2}{5} \leqslant x \leqslant 2\}$ **57.** $\{x|x > \frac{12}{35}\}$

59. $\{x \mid x < -\sqrt{5} \text{ or } x > \sqrt{5}\}$ **61.** $\{x \mid -\frac{1}{3} < x < 2\}$
63. $\{x \mid -6.891 \leqslant x < 8.351\}$ **65.** $\{x \mid x > 1.8\}$
67. $\{x \mid x \leqslant -9 \text{ or } x \geqslant -7\}$ **69.** All real numbers
71. $\{x \mid x \leqslant -\frac{1}{5} \text{ or } x \geqslant 0\}$

Margin Exercises, Section 5.4

1. $7|x|$ **2.** x^8 **3.** $5a^2|b|$ **4.** $\frac{7|a|}{b^2}$ **5.** $9|x|$ **6.** 29 **7.** 5
8. 20 **9.** 6, −6

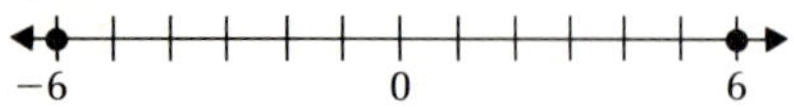

10. $\{x \mid -5 < x < 5\}$

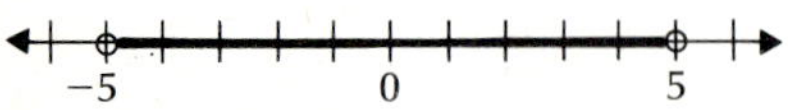

11. $\{y \mid y \leqslant -8 \text{ or } y \geqslant 8\}$

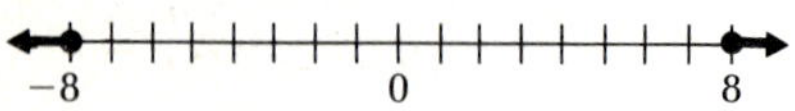

12. $\frac{5}{3}, \frac{-13}{3}$ **13.** $\{x \mid -2 < x < 5\}$
14. $\{x \mid x > \frac{11}{2} \text{ or } x < \frac{-3}{2}\}$

Exercise Set 5.4, p. 168

1. $3|x|$ **3.** $9x^2$ **5.** $4x^2$ **7.** $8|y|$ **9.** $\frac{4}{|x|}$ **11.** $\frac{x^2}{|y|}$
13. $4|x|$ **15.** 34 **17.** 11 **19.** 33 **21.** 5 **23.** $\{-3, 3\}$
25. $\{x \mid -3 < x < 3\}$ **27.** $\{x \mid x \leqslant -2 \text{ or } x \geqslant 2\}$
29. $\{t \mid t \leqslant -5.5 \text{ or } t \geqslant 5.5\}$ **31.** $\{-9, 15\}$
33. $\{x \mid -\frac{1}{2} \leqslant x \leqslant \frac{7}{2}\}$ **35.** $\{y \mid y < -\frac{3}{2} \text{ or } y > \frac{17}{2}\}$
37. $\{x \mid x \leqslant -\frac{5}{4} \text{ or } x \geqslant \frac{23}{4}\}$
39. Average greater than or equal to 2.8 **41.** $3x^4$
43. $3x^2|x|$ **45.** $\{m \mid -12 \leqslant m \leqslant 2\}$ **47.** No solution
49. $\{x \mid -1 \leqslant x \leqslant 2\}$ **51.** $\{x \mid 0.49705 \leqslant x \leqslant 0.50295\}$
53. $\{x \mid 0.794926 \leqslant x \leqslant 0.805074\}$ **55.** All real numbers
57. $\{x \mid -\frac{13}{54} < x < -\frac{7}{54}\}$ **59.** All real numbers
61. $\{-\frac{1}{4}, 1\}$ **63.** $\{x \mid x \leqslant 1\}$

Margin Exercises, Section 5.5

1. No

2.

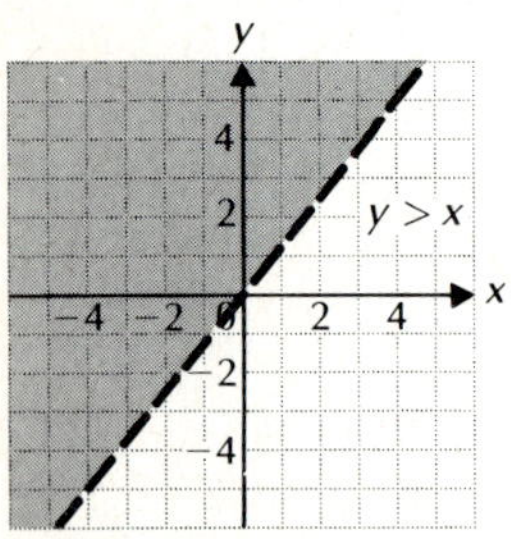

3.

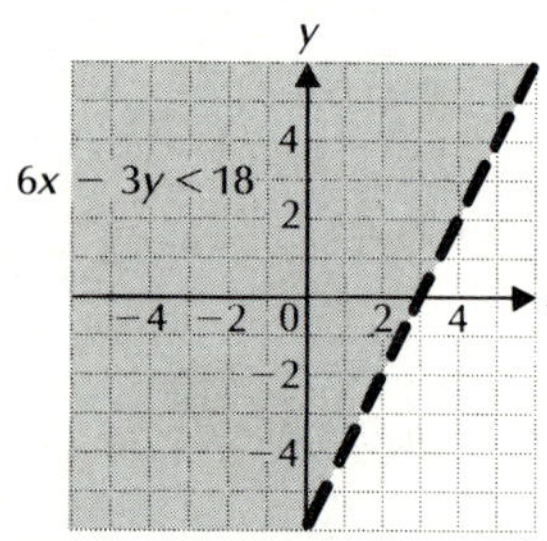

4.

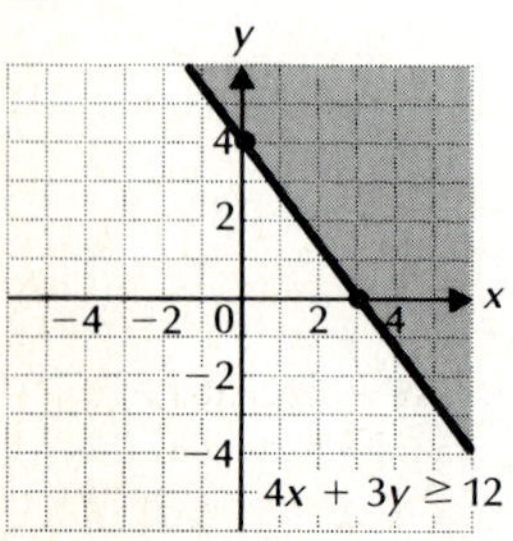

5.

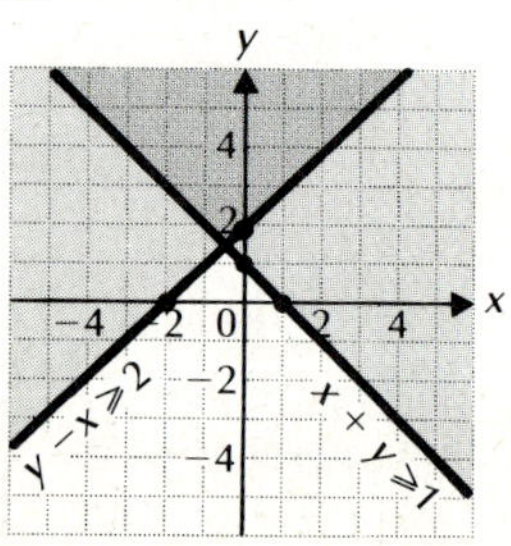

6.

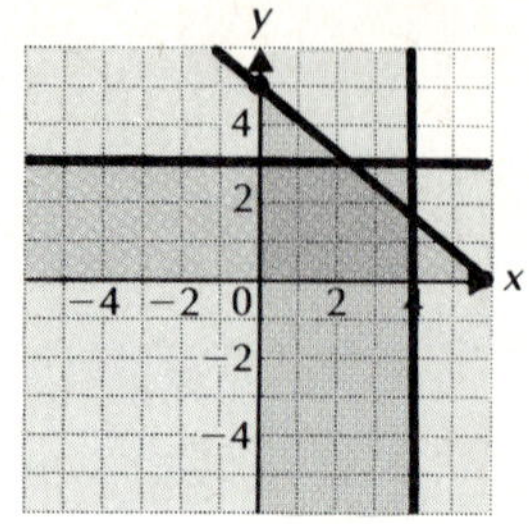

Exercise Set 5.5, pp. 172–173

1. Yes **3.** No

5. $y > 2x$

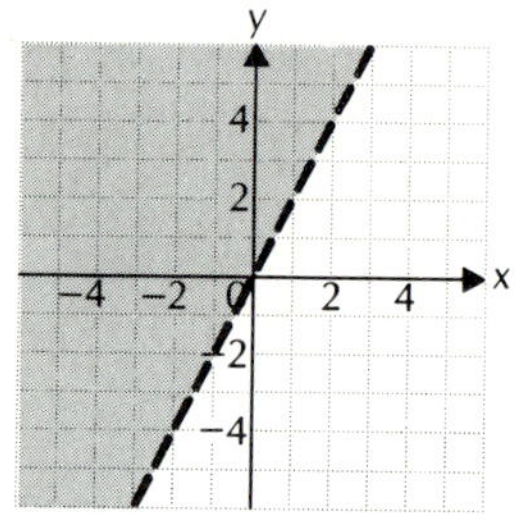

7. $y < x + 1$

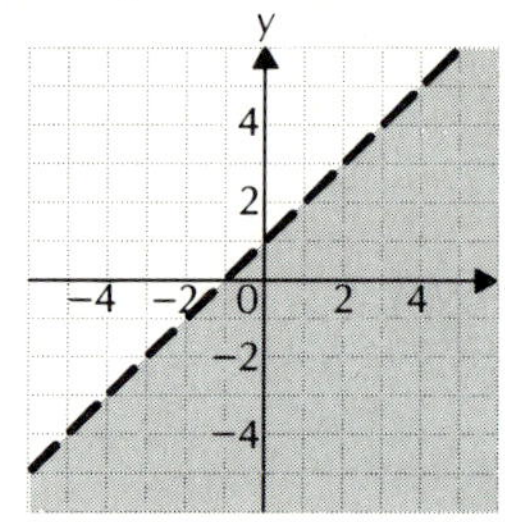

9. $y > x - 2$

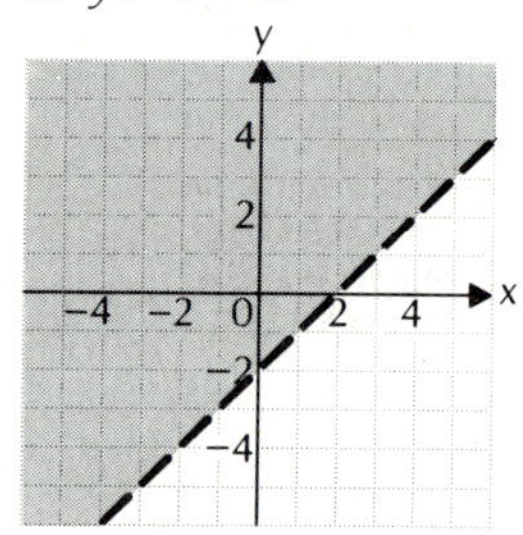

11. $x + y < 4$

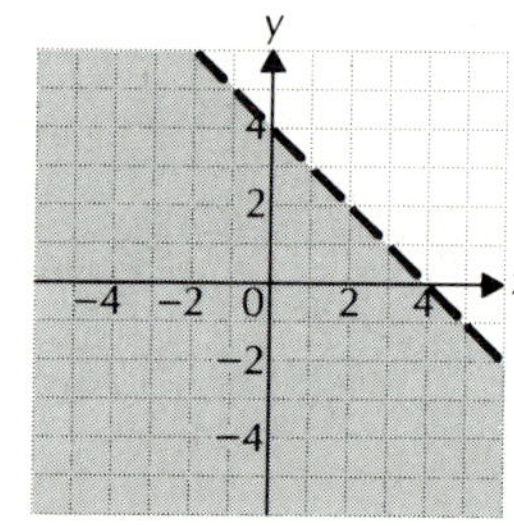

13. $3x + 4y \leqslant 12$

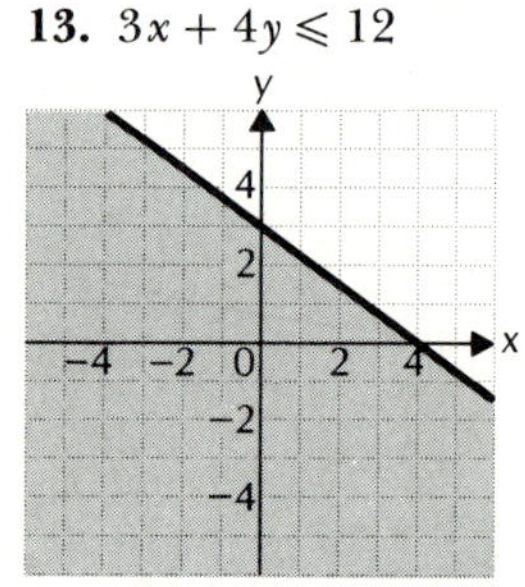

15. $2y - 3x > 6$

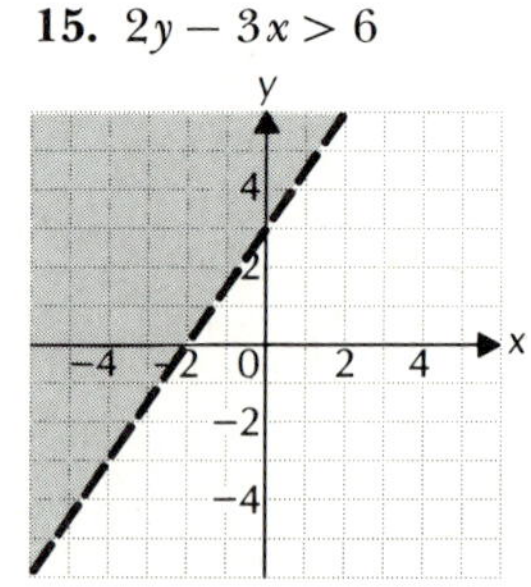

17. $3x - 2 \leqslant 5x + y$

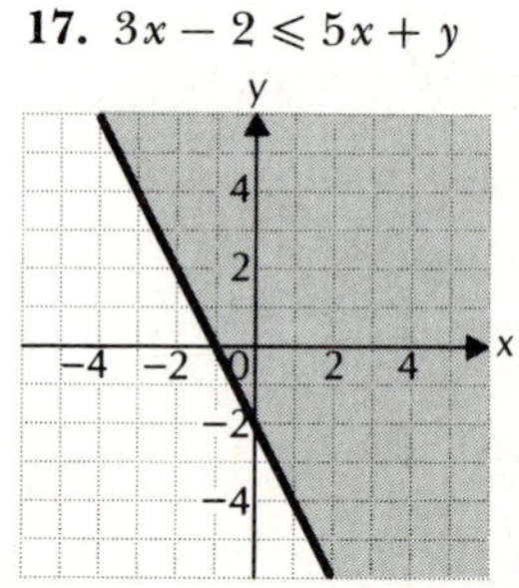

19. $x < -4$

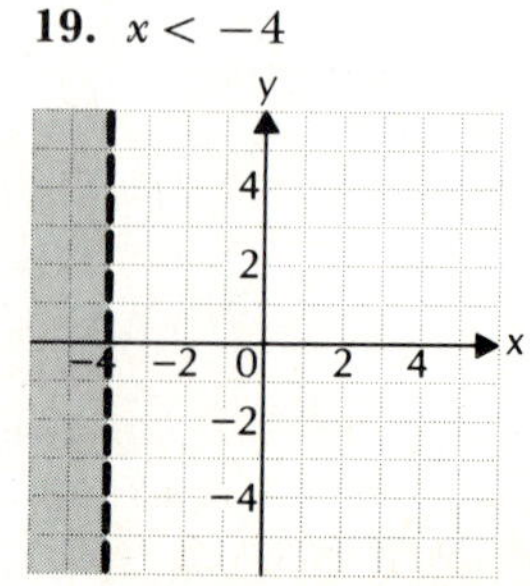

21. $x > -2$

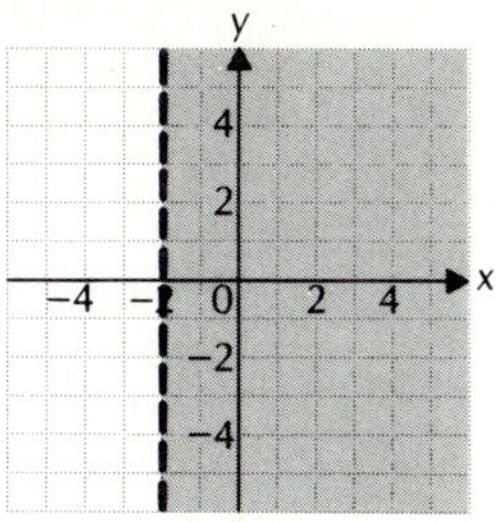

23. $-2 < y < 3$

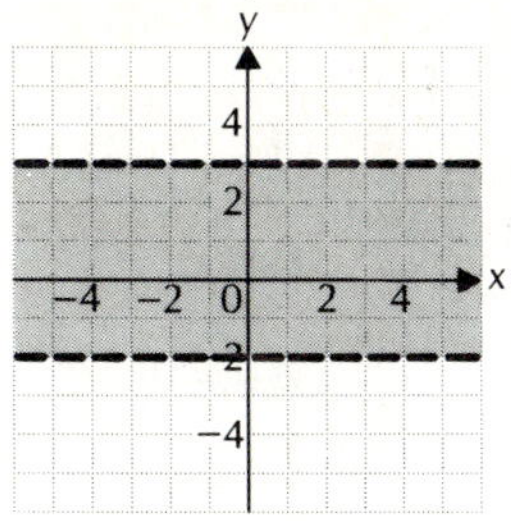

25. $-4 \leqslant x \leqslant 4$

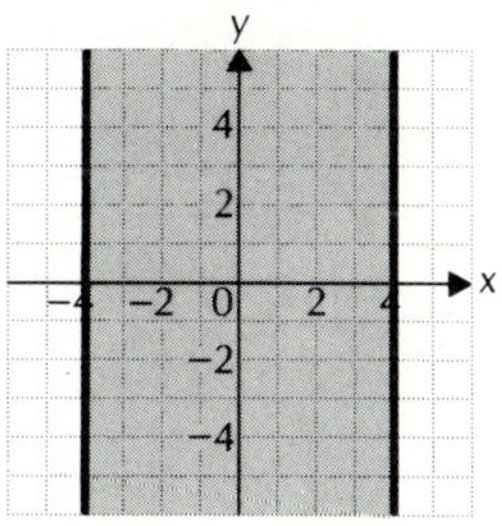

27.

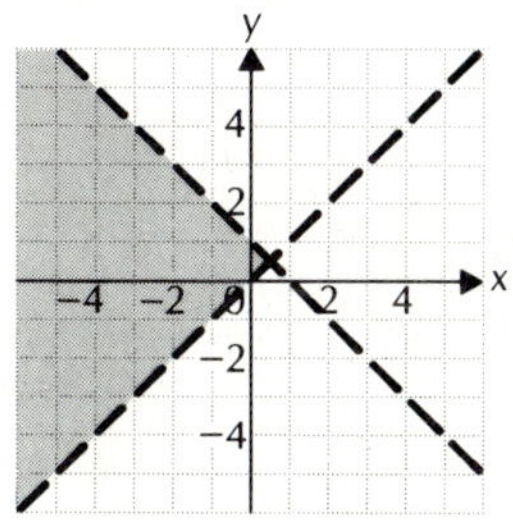

29.

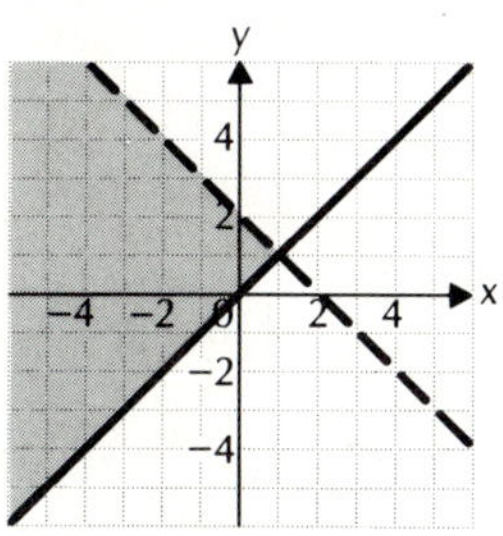

31.

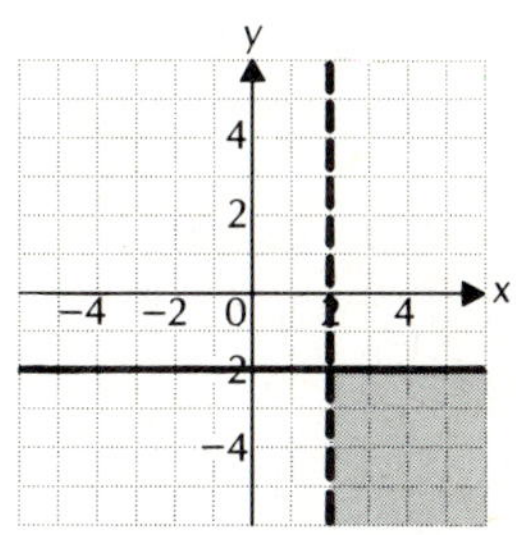

33.

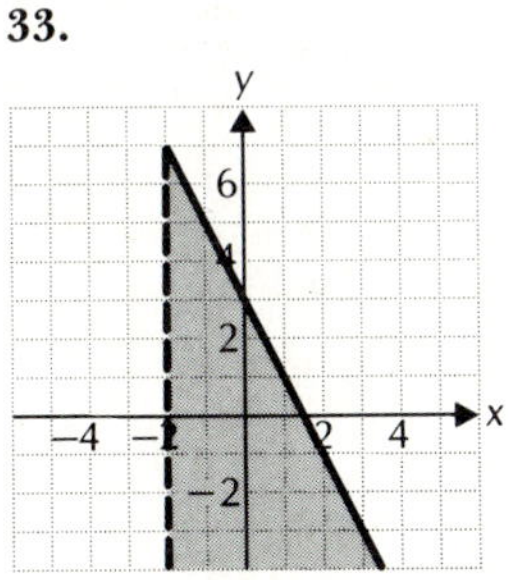

35.

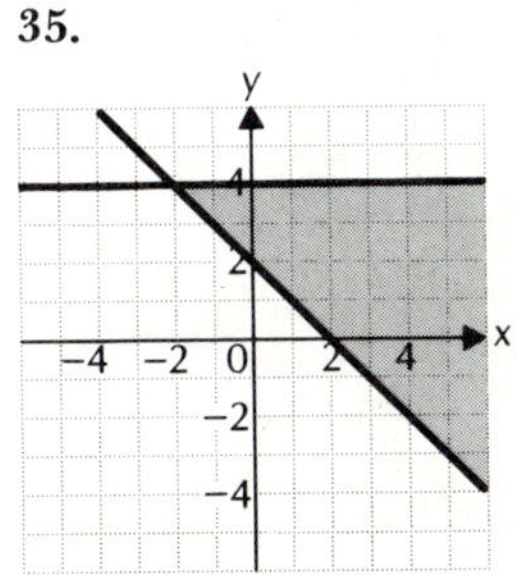

37.

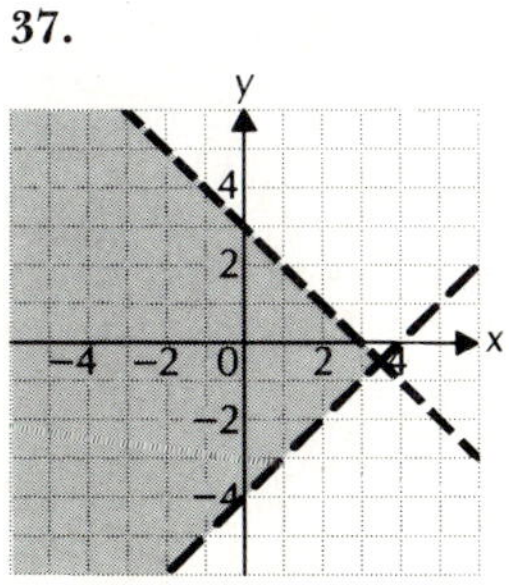

39.

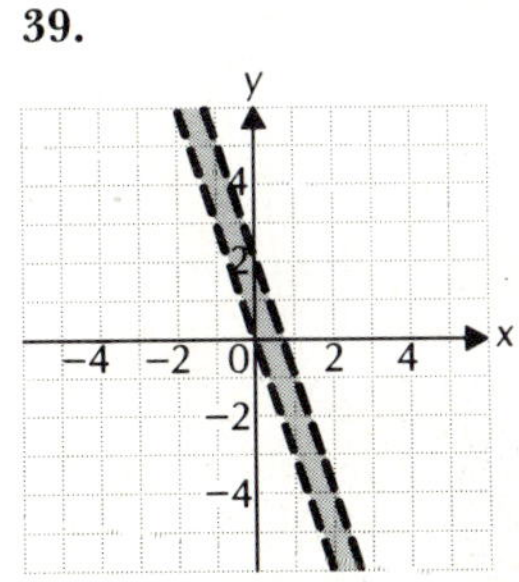

41. Vertices at (0, 1), (2, −3), (2, 5)
43. Vertices at (0, 0), (0, 6), (4, 4), (6, 0)
45. Vertices at (0, 0), (0, 4), $(\frac{40}{11}, \frac{24}{11})$, (5, 0) **47.** 15

49.

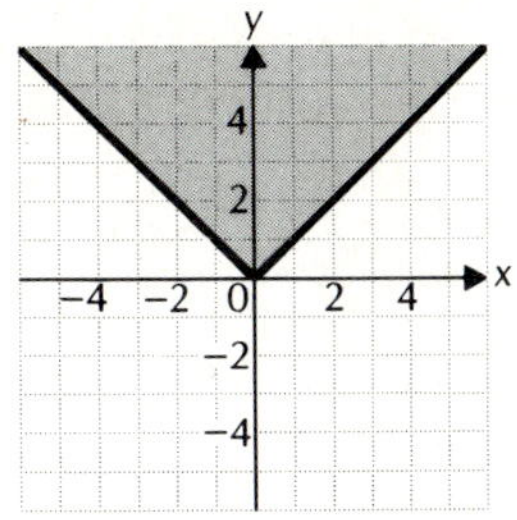

51

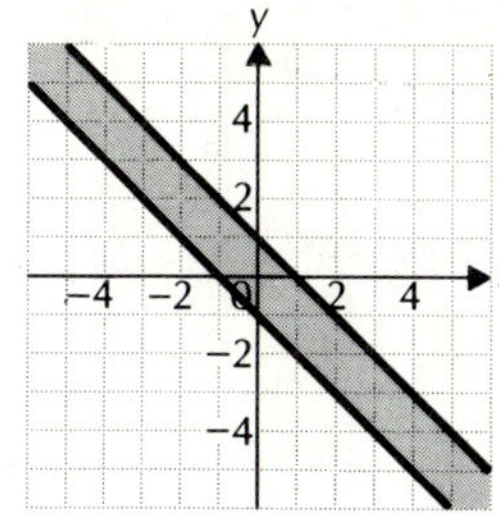

53.

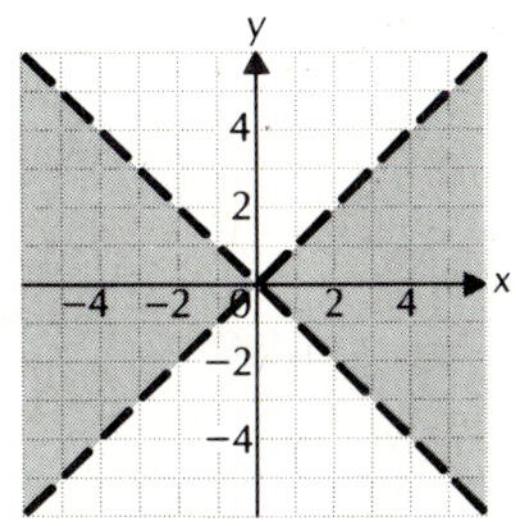

55.

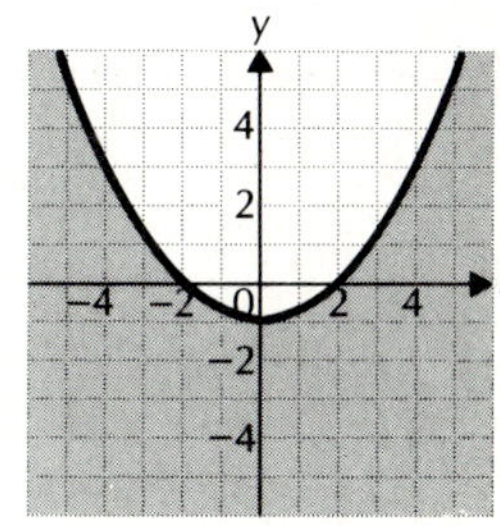

Margin Exercises, Section 5.6

1. Maximum value is 2, at (0, 1) and (2, 5); minimum value is −14, at (2, −3).
2. 40 hamburgers and 50 hotdogs. Profit will be $23.70.

Exercise Set 5.6, pp. 177–178

1. Maximum 168, when $x = 0$, $y = 6$; minimum 0, when $x = 0$, $y = 0$
3. Maximum 152, when $x = 7$, $y = 0$; minimum 32, when $x = 0$, $y = 4$ **5.** Maximum 102, 8 of A, 10 of B
7. $7000 bank X, $15,000 bank Y, maximum income $1395 **9.** Maximum $192, 2 knits, 4 worsteds
11. Minimum $460,000, 30 P-1 airplanes, 10 P-2 airplanes

Summary and Review: Chapter 5, pp. 178–179

1. [5.1] $\{x \mid x \leqslant -4\}$ **2.** [5.1] $\{x \mid x > 1\}$
3. [5.1] $\{a \mid a \leqslant -21\}$ **4.** [5.1] $\{y \mid y \geqslant -7\}$
5. [5.1] $\{y \mid y > -\frac{15}{4}\}$ **6.** [5.1] $\{y \mid y > -30\}$
7. [5.1] $\{x \mid x > -3\}$ **8.** [5.1] $\{y \mid y < -\frac{6}{5}\}$
9. [5.1] $\{x \mid x < -3\}$ **10.** [5.1] $\{y \mid y > -\frac{220}{23}\}$
11. [5.2] Scores higher than or equal to 92
12. [5.2] Amounts less than or equal to $10,000
13. [5.3] $\{x \mid -7 < x \leqslant 2\}$ **14.** [5.3] $\{x \mid -\frac{5}{4} < x < \frac{9}{4}\}$
15. [5.3] $\{x \mid x < -3 \text{ or } x > 1\}$
16. [5.3] $\{x \mid x < -11 \text{ or } x \geqslant -6\}$
17. [5.3] $\{x \mid x < -11 \text{ or } x > 1\}$
18. [5.3] $\{x \mid x < -6 \text{ or } x > 8\}$ **19.** [5.4] $\frac{3}{|x|}$
20. [5.4] $\frac{2|x|}{y^2}$ **21.** [5.4] $\frac{4}{|y|}$
22. [5.4] $x = 6$ or $x = -6$ **23.** [5.4] No solutions
24. [5.4] $\{x \mid x \leqslant -3.5 \text{ or } x \geqslant 3.5\}$ **25.** [5.4] $\{-5, 9\}$
26. [5.4] $\{x \mid -\frac{17}{2} < x < \frac{7}{2}\}$ **27.** [5.4] $\{x \mid -\frac{11}{3} \leqslant x \leqslant \frac{19}{3}\}$

28. [5.5]

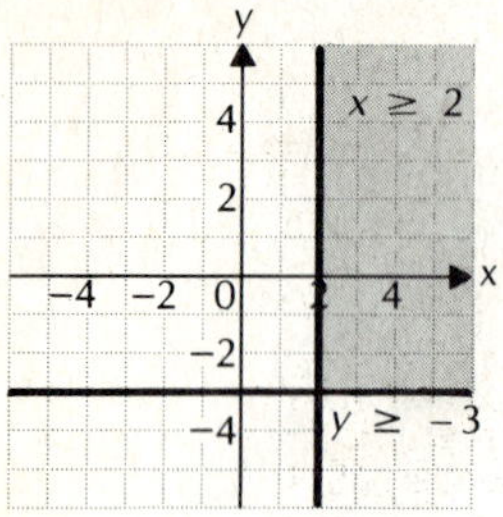

29. [5.5]

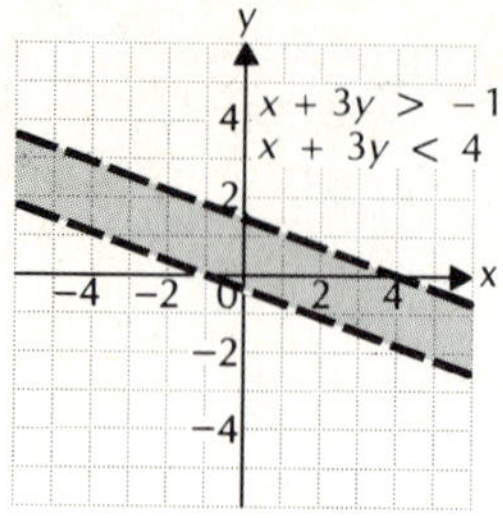

30. [5.5]

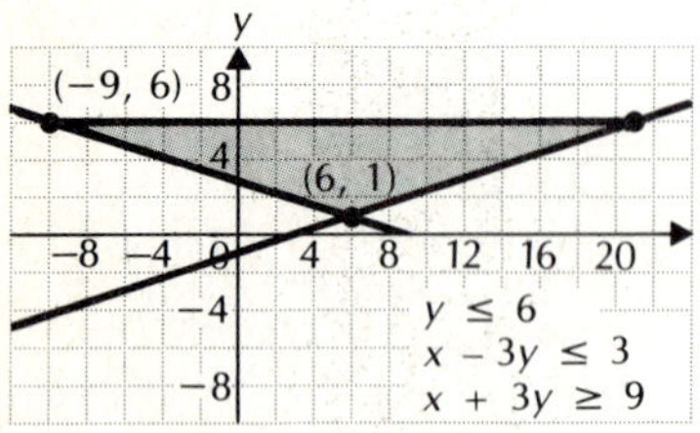

31. $\{x \mid -\frac{26}{15} \leqslant x \leqslant -\frac{14}{15}\}$ 32. $\{x \mid -\frac{8}{3} \leqslant x \leqslant -2\}$

CHAPTER 6

Margin Exercises, Section 6.1

1. (a) -1; (b) -4; (c) $\frac{2}{3}$; (d) does not exist, because the denominator is 0 2. $\dfrac{x^2-4}{5x+20}$ 3. $\dfrac{x^2+2xy+y^2}{x^2-9}$ 4. $\dfrac{3x^2+2xy}{5x^2+4xy}$ 5. $\dfrac{6x^3+4x^2-3xy-2y}{9x^2+18x+8}$ 6. $\dfrac{5-2a}{b-a}$ 7. $7x$ 8. $2a+3$ 9. $\dfrac{3x+2}{2x+4}$ 10. $\dfrac{2y+4}{y-1}$ 11. $\dfrac{3x^2-6xy+3y^2}{x+y}$ 12. $a-b$ 13. $\dfrac{x-5}{x+3}$ 14. $\dfrac{1}{x+7}$ 15. y^3-9 16. $\dfrac{x+5}{2x-10}$ 17. $\dfrac{2a^2b+2ab^2}{a-b}$

Exercise Set 6.1, pp. 186–187

1. $v(0)=\frac{2}{3}$, $v(3)=\frac{23}{6}$, $v(7)=\frac{163}{10}$ 3. $r(0)=0$, $r(4)=-184$, $r(5)$ does not exist 5. $\dfrac{3x(x+1)}{3x(x+3)}$ 7. $\dfrac{(t-3)(t+3)}{(t+2)(t+3)}$ 9. $\dfrac{3y}{5}$ 11. $a-3$ 13. $\dfrac{y-3}{y+3}$ 15. $\dfrac{t+4}{t-4}$ 17. $\dfrac{(x+4)(x-4)}{x(x+3)}$ 19. $\dfrac{y+4}{2}$ 21. $\dfrac{(x+5)(2x+3)}{7x}$ 23. $c-2$ 25. $\dfrac{1}{x+y}$ 27. 3 29. $\dfrac{(y-3)(y+2)}{y}$ 31. $\dfrac{2a+1}{a+2}$ 33. $\dfrac{(x+4)(x+2)}{3(x-5)}$ 35. $\dfrac{y(y^2+3)}{(y+3)(y-2)}$ 37. $\dfrac{x^2+4x+16}{(x+4)^2}$ 39. $\dfrac{2s}{r+2s}$ 41. $\dfrac{21{,}934.2x^2}{y^2-182{,}499.84}$ 43. $\dfrac{132.3t^2}{t^2-0.000049}$ 45. $\dfrac{x-3}{(x+1)(x+3)}$ 47. $\dfrac{m-t}{m+t+1}$ 49. $\dfrac{x^2+xy+y^2+x+y}{x-y}$ 51. $\dfrac{-2x}{x-1}$

Margin Exercises, Section 6.2

1. 90 2. 72 3. $\frac{47}{60}$ 4. $\frac{97}{72}$ 5. $5a^3b^2$ 6. $(y+3)(y+4)(y+4)$ 7. $2x^2(x-3)(x+3)(x+2)$ 8. $2(a+b)(a-b)$, or $2(a+b)(b-a)$

Exercise Set 6.2, p. 190

1. $2^2\cdot3^2$ 3. $2^4\cdot3^2$ 5. $2^3\cdot3^2$ 7. $3^2\cdot5$ 9. $2^3\cdot3^2\cdot7$ 11. $\frac{11}{10}$ 13. $\frac{43}{36}$ 15. $\frac{79}{60}$ 17. $24x^3$ 19. $12x^2y$ 21. $30a^3b^2$ 23. $(a+b)(a-b)$ 25. $6(y-2)$ or $6(2-y)$ 27. $3(y+3)(y-3)$ 29. $5(y-3)(y-3)$ 31. $(a+1)(a-1)(a-1)$ 33. $(x+2)(x-2)$ or $(x+2)(2-x)$ 35. $(x+5)(x+5)(x-3)$ 37. $(2r+3)(r-4)(3r-1)$ 39. 15 feet long, 12 feet wide 41. $(2x+1)(x-3)(x-3)(x-1)$ 43. $x^4(x^2+1)(x^2-1)(x^2+x+1)(x^2-x+1)$

Margin Exercises, Section 6.3

1. $\dfrac{12+y}{y}$ 2. $3x+1$ 3. $\dfrac{a-b}{b+2}$ 4. $\dfrac{y+12}{x^2+y^2}$ 5. $\dfrac{1-b^2}{3}$ 6. $\dfrac{2x^2+11}{x-5}$ 7. $\dfrac{3+7x}{4y}$ 8. $\dfrac{11x^2}{2x-y}$ 9. $\dfrac{9x^2+28y}{21x}$ 10. $\dfrac{3}{x+y}$ 11. $\dfrac{34y^2+21y-1}{3y(y-1)(y+1)}$ 12. $\dfrac{a+12}{a^2+3a}$ 13. $\dfrac{3y^2+12y+3}{(y-4)(y-3)(y+5)}$ 14. $\dfrac{2}{x-1}$

Exercise Set 6.3, pp. 194–195

1. 2 3. $\dfrac{3y+5}{y-2}$ 5. $a+b$ 7. $\dfrac{11}{x}$ 9. $\dfrac{1}{x+5}$ 11. $\dfrac{2y^2+22}{y^2-y-20}$ 13. $\dfrac{x+y}{x-y}$ 15. $\dfrac{3x-4}{x^2-3x+2}$ 17. $\dfrac{8x+1}{x^2-1}$ 19. $\dfrac{2x-14}{15x+75}$ 21. $\dfrac{-a^2+7ab-b^2}{a^2-b^2}$ 23. $\dfrac{y}{(y-2)(y-3)}$ 25. $\dfrac{3y-10}{y^2-y-20}$ 27. $\dfrac{3y^2-3y-29}{(y-3)(y+8)(y-4)}$ 29. $\dfrac{2x^2-13x+7}{(x+3)(x-1)(x-3)}$ 31. 0 33. $\dfrac{-3x^2-3x-4}{x^2-1}$ 35. 2 rolls of dimes, 5 rolls of nickels, 5 rolls of quarters 37. $\dfrac{2y^2+3-7x^3y}{x^2y^2}$ 39. $\dfrac{5y+23}{5-2y}$ 41. $\dfrac{x-y+x^2+xy}{x+y-x^2+xy}$

Margin Exercises, Section 6.4

1. $\dfrac{7(2y+1)}{2(7y-1)}$ 2. $\dfrac{x}{x+1}$ 3. $\dfrac{b+a}{b-a}$ 4. $\dfrac{a^2b^2}{b^2+ab+a^2}$

Exercise Set 6.4, pp. 197–198

1. $\frac{1+4x}{1-3x}$ **3.** $\frac{x^2-1}{x^2+1}$ **5.** $\frac{3y+4x}{4y-3x}$ **7.** $\frac{x+y}{x}$
9. $\frac{a^2(b-3)}{b^2(a-1)}$ **11.** $\frac{1}{a-b}$ **13.** $-\frac{1}{x(x+h)}$ **15.** $\frac{y-3}{y+5}$
17. Perimeter of the first rectangle is 15; perimeter of the second rectangle is 29
19. $\frac{5}{6}$ **21.** $\frac{11x+8}{4x+3}$ **23.** $\frac{b-a}{ab}$ **25.** a **27.** $\frac{3x+2}{5x+3}$
29. $\frac{-6x-3h}{x^2(x+h)^2}$ **31.** $\frac{1}{(1-x-h)(1-x)}$

Margin Exercises, Section 6.5

1. $\frac{2}{3}$ **2.** No solution **3.** 3 **4.** -3 **5.** 4, -3
6. 7 **7.** -13

Exercise Set 6.5, p. 203

1. $\frac{51}{2}$ **3.** -2 **5.** 144 **7.** -5, -1 **9.** 2 **11.** $\frac{17}{4}$
13. 11 **15.** No solution **17.** 2 **19.** $\frac{3}{5}$ **21.** 5
23. -145 **25.** $-\frac{10}{3}$ **27.** -3 **29.** -6, 5
31. No solution
33. 25 multiple choice, 30 true–false, 15 fill-ins
35. All reals except 1 and -1 **37.** 0.947
39. All real numbers except -2 **41.** Yes **43.** No
45. Yes

Margin Exercises, Section 6.6

1. 1, -7 **2.** $2\frac{2}{5}$ hr **3.** A, 32 hr; B, 96 hr
4. 280 mi **5.** 35.5 mph

Exercise Set 6.6, pp. 210–211

1. $\frac{35}{12}$ **3.** -3, -2 **5.** $3\frac{3}{14}$ hours **7.** $8\frac{4}{7}$ hours
9. $3\frac{9}{52}$ hours **11.** A: 10 days; B: 40 days **13.** 7 mph
15. 375 km **17.** Train A: 46 mph; train B: 58 mph
19. 9 km/h **21.** $21\frac{9}{11}$ min after 4:00
23. Boat: 14 km/h; stream: 10 km/h **25.** 3.75 km/h
27. 48 km/h **29.** $t=\frac{2}{3}$ hr

Margin Exercises, Section 6.7

1. $\frac{x^2}{2}+8x+3$ **2.** $5y^3-2y^2+6y$ **3.** $\frac{x^2}{2}+5x+8$
4. $x+5$ **5.** $3y^3-2y^2+6y-4$
6. $y^2-8y-24$, R -66; or $y^2-8y-24+\frac{-66}{y-3}$
7. $x^2+10x+10$, R 5; or $x^2+10x+10+\frac{5}{x-1}$
8. $y-11$, R $3y-27$; or $y-11+\frac{3y-27}{y^2-3}$

Exercise Set 6.7, p. 214

1. $6x^4-3x^2+8$ **3.** $-2a^2+4a-3$ **5.** y^3-2y^2+3y
7. $-6x^5+3x^3+2x$ **9.** $1-ab^2-a^3b^4$
11. $-2pq+3p-4q$ **13.** $x+7$ **15.** $a-12$, R 32
17. $y-5$ **19.** y^2-2y-1, R -8
21. $a^2+4a+15$, R 72 **23.** $4x^2-6x+9$ **25.** x^2+6
27. x^3+x^2-1, R 1 **29.** $2y^2+2y-1+\frac{8}{5y-2}$
31. $2x^2-x-9+\frac{3x+12}{x^2+2}$ **33.** 12, 13, 14
35. x^2+2y **37.** $x^3+x^2y+xy^2+y^3$ **39.** $\frac{14}{3}$

Margin Exercises, Section 6.8

1. $2x^2+2x+14$, R 34 **2.** $x^2-4x+13$, R -30
3. y^2-y+1

Exercise Set 6.8, pp. 216–217

1. x^2-x+1, R -4 **3.** $a+7$, R -47
5. $x^2-5x-23$, R -43 **7.** $3x^2-2x+2$, R -3
9. y^2+2y+1, R 12 **11.** $3x^3+9x^2+2x+6$
13. x^2+3x+9 **15.** y^3+y^2+y+1 **17.** -23°
19. $3.41x^3+8.2181x^2-4.444379x-10.710953$, R -39.283396
21. **(a)** 218; **(b)** 218; **(c)** they are the same.

Margin Exercises, Section 6.9

1. $T=\frac{PV}{k}$ **2.** $T=\frac{A+Vnt}{Vn}$ **3.** 6 cm

Exercise Set 6.9, p. 221

1. $d_1=\frac{d_2W_1}{W_2}$ **3.** $t=\frac{2s}{v_1+v_2}$ **5.** $r_1=\frac{Rr_2}{r_2-R}$
7. $s=\frac{Rg}{g-R}$ **9.** $r=\frac{2V-IR}{2I}$ **11.** $f=\frac{pq}{q+p}$
13. $r=\frac{nE-IR}{In}$ **15.** $H=m(t_1-t_2)S$ **17.** $e=\frac{Er}{R+r}$
19. $a=\frac{S-Sr}{1-r^n}$ **21.** 12 runs **23.** $5\frac{5}{9}$ ohms
25. **(a)** $t=\frac{ab}{b+a}$; **(b)** $a=\frac{tb}{b-t}$; **(c)** $b=\frac{ta}{a-t}$

Margin Exercises, Section 6.10

1. $\frac{2}{5}$; $y=\frac{2}{5}x$ **2.** 0.7; $y=0.7x$ **3.** 50 volts
4. 176, 250 tons **5.** 0.6; $y=\frac{0.6}{x}$ **6.** $7\frac{1}{2}$ hr

Exercise Set 6.10, pp. 226–227

1. $k=8$; $y=8x$ **3.** $k=36$; $y=3.6x$ **5.** $k=5$; $y=5x$
7. $k=\frac{15}{4}$; $y=\frac{15}{4}x$ **9.** $k=\frac{8}{5}$; $y=\frac{8}{5}x$ **11.** 6 amperes
13. 125,000 **15.** 532,500 tons **17.** 40 kg
19. $k=60$; $y=\frac{60}{x}$ **21.** $k=12$; $y=\frac{12}{x}$
23. $k=36$; $y=\frac{36}{x}$

25. $y = \frac{9}{x}$ **27.** $\frac{2}{9}$ ampere **29.** 160 cm^3 **31.** $6\frac{2}{3}$ hours
33. 10 cents per ounce **35.** 2074
37. $\frac{\text{Number of words in book}}{\text{Number of pages in book}} = \frac{\text{Number of words on } n \text{ pages}}{n}$; where n is some small number, say 1, 2, or 3.
39. Steve: 56; Stephanie: 42
41. **(a)** Inversely; **(b)** neither; **(c)** directly; **(d)** directly; **(e)** directly

Summary and Review: Chapter 6, pp. 227–228

1. [6.1] $\frac{y-8}{2}$ **2.** [6.1] $\frac{(x-2)(x+5)}{x-5}$ **3.** [6.1] $\frac{3a-1}{a-3}$
4. [6.1] $\frac{(x^3-64)}{(x+6)(x+2)}$ **5.** [6.2] $15y^2$ **6.** [6.2] $48x^3$
7. [6.2] $3(x^2-49)$ **8.** [6.2] $(x-2)(x+5)(x-4)$
9. [6.3] $\frac{x^2+11xy+y^2}{x^2-y^2}$ **10.** [6.3] $\frac{2(x^3+x^2y+xy^2-y^3)}{x^2-y^2}$
11. [6.3] $\frac{-y}{(y+4)(y-1)}$ **12.** [6.4] $\frac{1-x^2}{1+x^2}$
13. [6.4] $\frac{4x^2-3}{4x}$ **14.** [6.4] $\frac{(y-6)}{(y-9)}$ **15.** [6.5] 2
16. [6.5] $\frac{28}{11}$ **17.** [6.5] 6 **18.** [6.6] $5\frac{1}{7}$ hours
19. [6.6] 742.5 miles **20.** [6.7] $4s^2-3s-2rs^2$
21. [6.7] $y-3$ **22.** [6.7] y^2+3y+9
23. [6.7] $4x+3+\frac{-9x-5}{x^2+1}$ **24.** [6.8] $x^2+6x+20$, R 54
25. [6.8] $4x^2-6x+18$, R -59 **26.** [6.9] $g = \frac{Rs}{s-R}$
27. [6.9] $m = \frac{H}{S(t_1-t_2)}$ **28.** [6.10] 78
29. [6.10] 58 **30.** $\frac{2}{a}$ **31.** $\frac{-3x^2+9x-7}{(x-1)^2(x-2)}$
32. $3x^6-7x^5+14x^4+3x^3-6x^2+12x-3$, R 11

CUMULATIVE REVIEW: CHAPTERS 1–6, pp. 229-230

1. $-\frac{9}{5}$ **2.** 10 **3.** $\frac{13}{9}$ **4.** -3.11 **5.** -3 **6.** $\frac{1}{4}$
7. $\frac{7}{2}$ or -2 **8.** $\{x \mid x < 22\}$ **9.** $\{x \mid x > -\frac{8}{3}\}$
10. $\{y \mid y > -50\}$ **11.** $\{a \mid a > -7\}$ **12.** $\{x \mid x > -1\}$
13. $\{x \mid -10 < x < 13\}$ **14.** $\{x \mid x < -\frac{4}{3} \text{ or } x > 6\}$
15. $\{x \mid x < -6.4 \text{ or } x > 6.4\}$ **16.** $\frac{8}{3}$ or $\frac{4}{3}$
17. $\{x \mid -\frac{13}{4} < x < \frac{15}{4}\}$ **18.** $\{x \mid -3 \leqslant x \leqslant -1 \text{ or } 7 \leqslant x \leqslant 9\}$
19. $-\frac{5}{3}$ **20.** $-\frac{5}{8}$ **21.** $\frac{-192xz^{19}}{y}$ **22.** $-\frac{3}{8}y^{b+1}$
23. $4x = -7y + 5$ **24.** $-y = 2x^2 - 3$

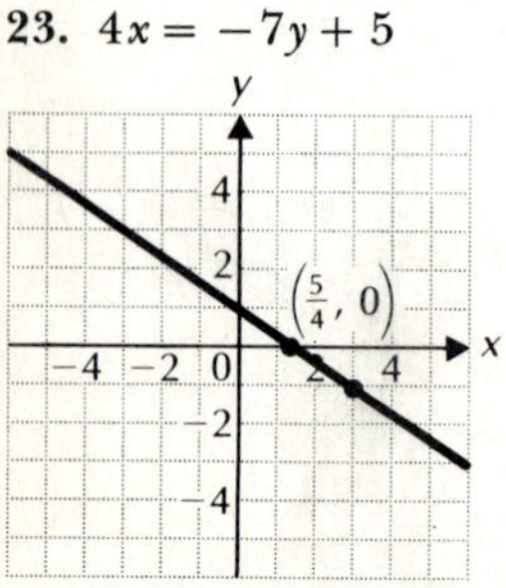

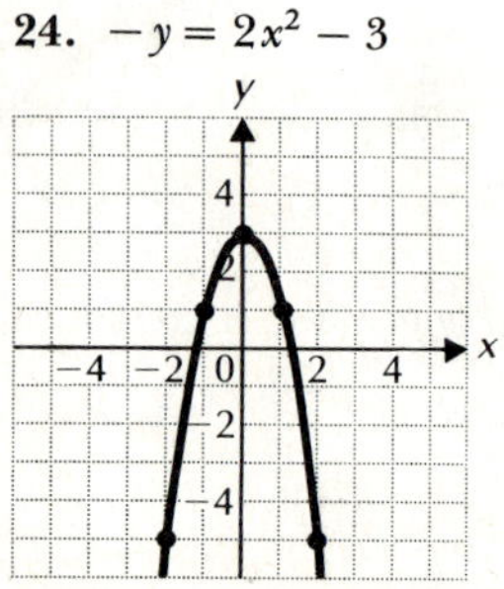

25.

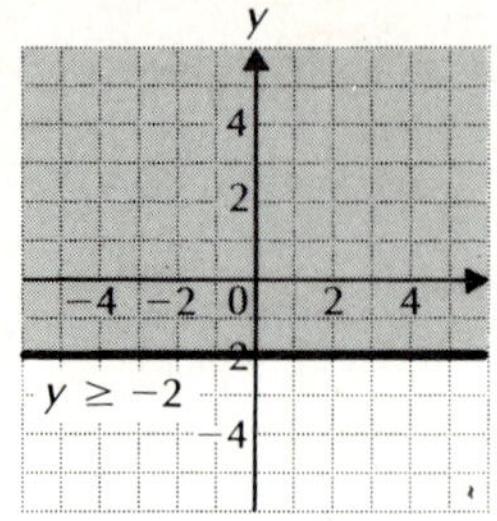

26.

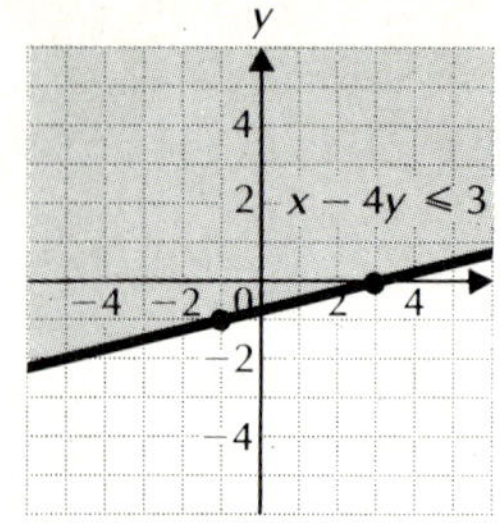

27. $x-2y=7$ **28.** $y=4x+10$
29. $-3x^3+9x^2+3x-3$ **30.** $-15x^4y^4$ **31.** $5a+5b-5c$
32. $x^3-12x^2+48x-64$ **33.** $15x^4-x^3-9x^2+5x-2$
34. $-m^3n^2-5mn^3-m^2n^2$ **35.** $\frac{y-6}{2}$ **36.** $x-1$
37. $\frac{a^2+7ab+b^2}{a^2-b^2}$ **38.** $\frac{m^2+5m-8}{(m-5)(m+1)}$ **39.** $\frac{3y^2-2}{3y}$
40. $\frac{y-x}{xy(x+y)}$ **41.** $9x^2-13x+26$, R (-50)
42. $2x^2(2x+9)$ **43.** $(x-6)(x+14)$
44. $(4y+9)(4y-9)$ **45.** $(3x-2)(2x+5)$
46. $8(2x+1)(4x^2-2x+1)$ **47.** $(x-8)^2$
48. $x^2(x^2+1)(x-1)(x+1)$
49. $(0.3b-0.2c)(0.09b^2+0.06bc+0.04c^2)$
50. $(4x+1)(5x+3)$ **51.** $(3x+7)(x-4)$
52. $\frac{P-2L}{2} = w$
53. Holds for any three consecutive even numbers
54. \$5.63 or more **55.** 6 mL **56.** 1 hr **57.** 56.8 yr

CHAPTER 7

Margin Exercises, Section 7.1

1. 3, -3 **2.** 6, -6 **3.** 11, -11 **4.** 1 **5.** 6 **6.** $\frac{9}{10}$
7. 0.08 **8.** **(a)** 4; **(b)** -4; **(c)** does not exist
9. **(a)** 7; **(b)** -7; **(c)** does not exist
10. **(a)** 12; **(b)** -12; **(c)** does not exist
11. **(a)** $f(1) = 1$; **(b)** $f(0)$ does not exist; 0 is not in the domain; **(c)** $f(4) = 4$
12. **(a)** $g(0) = -4$; **(b)** $g(10) = -6$; **(c)** $g(-6) = -2$; **(d)** $g(-10)$ does not exist; -10 is not in the domain
13. $|y|$ **14.** 24 **15.** $|5y|$ or $5|y|$ **16.** $|4y|$ or $4|y|$
17. $|x+7|$ **18.** $|2(x-2)|$ or $2|x-2|$
19. $|7(y+5)|$ or $7|y+5|$ **20.** $|x-3|$ **21.** z **22.** $3x+4$
23. $x-1$ **24.** 4 **25.** -3 **26.** $2(x+2)$ **27.** $-3(y-4)$
28. $-2x$ **29.** $3x+2$ **30.** **(a)** 3; **(b)** -3; **(c)** does not exist **31.** $2|x-2|$ **32.** $|x|$ **33.** $|x+3|$

Exercise Set 7.1, pp. 236–237

1. 4, -4 **3.** 12, -12 **5.** 20, -20 **7.** $-\frac{7}{6}$ **9.** 14
11. $-\frac{4}{9}$ **13.** 0.3 **15.** -0.07 **17.** p^2+4 **19.** $\frac{x}{y+4}$
21. $f(6) = \sqrt{20}$, $f(2) = 0$, $f(0)$ does not exist, $f(-3)$ does not exist
23. $t(4) = -3$, $t(-4)$ does not exist, $t(0) = -1$, $t(12) = -5$

25. $f(5) = \sqrt{26}$, $f(-5) = \sqrt{26}$, $f(0) = 1$, $f(10) = \sqrt{101}$, $f(-10) = \sqrt{101}$ **27.** $4|x|$ **29.** $7|c|$ **31.** $|a+1|$ **33.** $|x-2|$ **35.** $|2x+7|$ **37.** $4x$ **39.** $6b$ **41.** $a+1$ **43.** $2x+2$ **45.** 3 **47.** $-4x$ **49.** -6 **51.** $0.7(x+1)$ **53.** $f(7) = 2$, $f(26) = 3$, $f(-9) = -2$, $f(-65) = -4$ **55.** 5 **57.** -1 **59.** $-\frac{2}{3}$ **61.** $|x|$ **63.** $5|a|$ **65.** 6 **67.** $|a+b|$ **69.** y **71.** $x-2$ **73.** (a) 13; (b) 15; (c) 18; (d) 20 **75.**

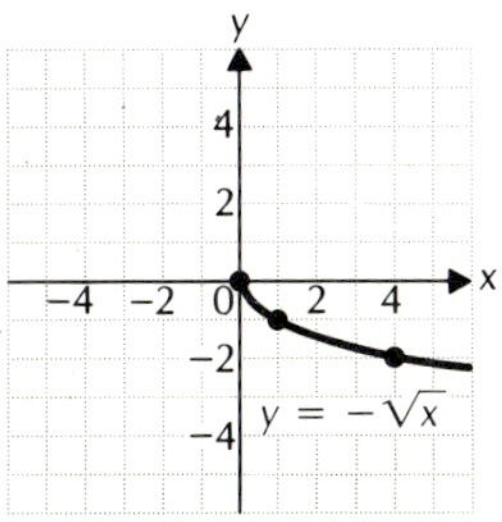

Margin Exercises, Section 7.2

1. $\sqrt{133}$ **2.** $\sqrt{x^2 - 4y^2}$ **3.** $\sqrt[4]{2821}$ **4.** $\sqrt[3]{8x^5 + 40x}$ **5.** $4\sqrt{2}$ **6.** $2\sqrt[3]{10}$ **7.** $10\sqrt{3}$ **8.** $6y$ **9.** $(x+2)\sqrt{3}$ **10.** $2bc\sqrt{3ab}$ **11.** $2\sqrt[3]{2}$ **12.** $3xy^2\sqrt[3]{3xy^2}$ **13.** $(a+b)\sqrt[3]{a+b}$ **14.** $3\sqrt{2}$ **15.** $6y\sqrt{7}$ **16.** $3x\sqrt[3]{4y}$ **17.** $7\sqrt{3ab}$ **18.** $2(y+5)\sqrt[3]{(y+5)^2}$ **19.** Rational **20.** Rational **21.** Rational **22.** Rational **23.** Irrational **24.** Irrational **25.** Rational **26.** 15.9 **27.** 20.1 **28.** 2.175×10^7 **29.** 1.773×10^{-7}

Exercise Set 7.2, pp. 240–241

1. $\sqrt{6}$ **3.** $\sqrt[3]{10}$ **5.** $\sqrt[4]{72}$ **7.** $\sqrt{30ab}$ **9.** $\sqrt[5]{18t^3}$ **11.** $\sqrt{x^2 - a^2}$ **13.** $\sqrt[3]{0.06x^2}$ **15.** $\sqrt[4]{x^3 - 1}$ **17.** $\sqrt{\dfrac{6y}{5x}}$ **19.** $2\sqrt{2}$ **21.** $2\sqrt{6}$ **23.** $6x^2\sqrt{5}$ **25.** $3x^2\sqrt[3]{2x^2}$ **27.** $2x^2\sqrt[3]{10x^2}$ **29.** $2\sqrt[4]{2}$ **31.** $3cd\sqrt[4]{2d^2}$ **33.** $(x+y)\sqrt[3]{x+y}$ **35.** $5\sqrt{2}$ **37.** 8 **39.** $6\sqrt{7}$ **41.** $30\sqrt{3}$ **43.** $2x^2y\sqrt{6}$ **45.** $25t^3\sqrt[3]{t}$ **47.** $(x+y)^2\sqrt[3]{(x+y)^2}$ **49.** Rational **51.** Rational **53.** Rational **55.** Irrational **57.** 15.9 **59.** 13.7 **61.** 18.7 **63.** 29.9 **65.** 15.5 **67.** 156,525 **69.** 684,251 **71.** 418,832 **73.** 0.000199 **75.** 0.000707 **77.** 8 **79.** (a) 20 mph; (b) 37.4 mph; (c) 42.4 mph

Margin Exercises, Section 7.3

1. $\frac{5}{6}$ **2.** $\dfrac{x}{10}$ **3.** $\dfrac{3x\sqrt[3]{2x^2}}{5}$ **4.** 5 **5.** $56\sqrt{xy}$ **6.** $5a$ **7.** $\frac{20}{7}$ **8.** $\dfrac{2ab}{3}$ **9.** 25 **10.** $6y\sqrt{6y}$

Exercise Set 7.3, pp. 243–244

1. $\frac{4}{5}$ **3.** $\frac{4}{3}$ **5.** $\dfrac{7}{y}$ **7.** $\dfrac{5y\sqrt{y}}{x^2}$ **9.** $\dfrac{2x\sqrt[3]{x^2}}{3y}$ **11.** $\sqrt{7}$ **13.** 3 **15.** $y\sqrt{5y}$ **17.** $2\sqrt[3]{a^2b}$ **19.** $3\sqrt{xy}$ **21.** $\sqrt{x^2 + xy + y^2}$ **23.** $6a\sqrt{6a}$ **25.** $4b\sqrt[3]{4b}$ **27.** $54a^3b\sqrt{2b}$ **29.** $2c\sqrt[3]{18cd^2}$ **31.** \$5.00 per hour **33.** (a) 1.62 sec; (b) 1.99 sec; (c) 2.20 sec **35.** $9\sqrt[3]{9n^2}$

Margin Exercises, Section 7.4

1. $13\sqrt{2}$ **2.** $10\sqrt[4]{5x} - \sqrt{7}$ **3.** $19\sqrt{5}$ **4.** $(3y+4)\sqrt[3]{y^2} + 2y^2$ **5.** $2\sqrt{x-1}$

Exercise Set 7.4, p. 245

1. $8\sqrt{3}$ **3.** $3\sqrt[3]{5}$ **5.** $13\sqrt[3]{y}$ **7.** $7\sqrt{2}$ **9.** $6\sqrt[3]{3}$ **11.** $21\sqrt{3}$ **13.** $38\sqrt{5}$ **15.** $122\sqrt{2}$ **17.** $9\sqrt[3]{2}$ **19.** $29\sqrt{2}$ **21.** $(1+6a)\sqrt{5a}$ **23.** $(2-x)\sqrt[3]{3x}$ **25.** $3\sqrt{2y-2}$ **27.** $(x+3)\sqrt{x-1}$ **29.** $5\sqrt{2}$ **31.** $15\sqrt[3]{4}$ **33.** $-9\sqrt{5}$ **35.** 0 **37.** 36 **39.** $a = 0$ and $b \geqslant 0$, or $a \geqslant 0$ and $b = 0$; any pair of numbers that does not satisfy the conditions stated above. **41.** $(25x - 30y - 9xy)\sqrt{2xy}$ **43.** $(7x^2 - 2y^2)\sqrt{x+y}$

7

Margin Exercises, Section 7.5

1. $5\sqrt{6} + 3\sqrt{14}$ **2.** $a\sqrt[3]{3} - \sqrt[3]{2a^2}$ **3.** $-4 - 9\sqrt{6}$ **4.** $3\sqrt{ab} - 4\sqrt{3a} + 6\sqrt{3b} - 24$ **5.** -3 **6.** $20 - 4y\sqrt{5} + y^2$ **7.** $58 + 12\sqrt{6}$

Exercise Set 7.5, pp. 246–247

1. $2\sqrt{6} - 18$ **3.** $\sqrt{6} - \sqrt{10}$ **5.** $2\sqrt{15} - 6\sqrt{3}$ **7.** -6 **9.** $3a\sqrt[3]{2}$ **11.** 1 **13.** -12 **15.** $a - b$ **17.** $1 + \sqrt{5}$ **19.** $7 + 3\sqrt{3}$ **21.** -6 **23.** $a + \sqrt{3a} + \sqrt{2a} + \sqrt{6}$ **25.** $2\sqrt[3]{9} - 3\sqrt[3]{6} - 2\sqrt[3]{4}$ **27.** $7 + 4\sqrt{3}$ **29.** 28, 82 **31.** (a) 19.485281; (b) 19.485281 **33.** $2x - 2\sqrt{x^2 - 4}$ **35.** $\sqrt[3]{y} - y$ **37.** $12 + 6\sqrt{3}$ **39.** $16(a+3) + 38a\sqrt{3(a+3)} - 15a^2$ **41.** $2\sqrt{34}$

Margin Exercises, Section 7.6

1. $\dfrac{\sqrt{6}}{3}$ **2.** $\dfrac{\sqrt{30}}{3}$ **3.** $\dfrac{\sqrt[3]{4}}{2}$ **4.** $\dfrac{2\sqrt{3ab}}{3b}$ **5.** $\dfrac{2x^2\sqrt{3xy}}{3y^2}$ **6.** $\dfrac{\sqrt[3]{28}}{2}$ **7.** $\dfrac{x\sqrt[3]{12x^2y^2}}{2y}$ **8.** $\dfrac{11}{\sqrt{66}}$ **9.** $\dfrac{7}{\sqrt[3]{196}}$ **10.** $\dfrac{\sqrt{3} + y}{\sqrt{3} + y}$ **11.** $\dfrac{\sqrt{2} - \sqrt{3}}{\sqrt{2} - \sqrt{3}}$ **12.** $-5(1 + \sqrt{2})$ **13.** $\dfrac{5 + \sqrt{7}}{5 + \sqrt{7}}$ **14.** $\dfrac{4}{3\sqrt{2} - \sqrt{10} - 3\sqrt{6} + \sqrt{30}}$

Exercise Set 7.6, pp. 251–252

1. $\dfrac{\sqrt{30}}{5}$ **3.** $\dfrac{\sqrt{30}}{3}$ **5.** $\dfrac{2\sqrt{15}}{5}$ **7.** $\dfrac{2\sqrt[3]{6}}{3}$ **9.** $\dfrac{\sqrt[3]{75ac^2}}{5c}$ **11.** $\dfrac{y\sqrt[3]{9yx^2}}{3x^2}$ **13.** $\dfrac{\sqrt[3]{x^2y^2}}{xy}$ **15.** $\dfrac{7}{\sqrt{21x}}$ **17.** $\dfrac{2}{\sqrt{6}}$ **19.** $\dfrac{52}{3\sqrt{91}}$ **21.** $\dfrac{7}{\sqrt[3]{98}}$ **23.** $\dfrac{7x}{\sqrt{21xy}}$ **25.** $\dfrac{5y^2}{x\sqrt[3]{150x^2y^2}}$

27. $\dfrac{ab}{3\sqrt{ab}}$ **29.** $\dfrac{5(8+\sqrt{6})}{58}$ **31.** $-2\sqrt{7}(\sqrt{5}+\sqrt{3})$
33. $\dfrac{\sqrt{15}+20-6\sqrt{2}-8\sqrt{30}}{-77}$ **35.** $\dfrac{x-2\sqrt{xy}+y}{x-y}$
37. $\dfrac{3\sqrt{6}+4}{2}$ **39.** $\dfrac{-11}{4(\sqrt{3}-5)}$
41. $\dfrac{-22}{\sqrt{6}+5\sqrt{2}+5\sqrt{3}+25}$ **43.** $\dfrac{x-y}{x+2\sqrt{xy}+y}$
45. $\dfrac{7}{43\sqrt{2}+66}$ **47.** $\dfrac{\sqrt{10}+2\sqrt{5}-2\sqrt{3}}{10}$ **49.** $\dfrac{b^2+\sqrt{b}}{b^2+b+1}$
51. $6a\sqrt[3]{36ab^2}$ **53.** $\dfrac{x-19}{x+31+10\sqrt{x+6}}$
55. $\left(\dfrac{5x+4y-3}{xy}\right)\sqrt{xy}$ **57.** $\dfrac{7\sqrt{3}}{39}$

Margin Exercises, Section 7.7

1. $\sqrt[4]{y}$ **2.** $\sqrt{3a}$ **3.** $\sqrt[4]{16}$, or 2 **4.** $\sqrt[3]{125}$, or 5
5. $\sqrt[5]{a^3b^2c}$ **6.** $19^{1/3}$ **7.** $(abc)^{1/2}$ **8.** $\left(\dfrac{x^2y}{16}\right)^{1/5}$
9. $\sqrt{x^3}$ or $(\sqrt{x})^3$, which simplifies to $x\sqrt{x}$
10. $\sqrt[3]{8^2}$ or $(\sqrt[3]{8})^2$, which simplifies to 4
11. $\sqrt{4^5}$ or $(\sqrt{4})^5$, which simplifies to 32 **12.** $(7abc)^{4/3}$
13. $6^{7/5}$ **14.** $\frac{1}{2}$ **15.** $\dfrac{1}{(3xy)^{7/8}}$ **16.** $7^{14/15}$
17. $5^{2/6}$ or $5^{1/3}$ **18.** $9^{2/5}$ **19.** $\sqrt{a}$ **20.** x **21.** $\sqrt{2}$
22. ab^2 **23.** xy^3 **24.** $\sqrt[4]{xy^2}$ **25.** $\sqrt[4]{63}$ **26.** $\sqrt[4]{a-b}$
27. $\sqrt[6]{x^4y^3z^5}$ **28.** $\sqrt[4]{ab}$

Exercise Set 7.7, p. 256

1. $\sqrt[4]{x}$ **3.** 2 **5.** $\sqrt[5]{a^2b^2}$ **7.** $\sqrt[3]{a^2}$ **9.** 8 **11.** $20^{1/3}$
13. $17^{1/2}$ **15.** $(cd)^{1/4}$ **17.** $(xy^2z)^{1/5}$ **19.** $(3mn)^{3/2}$
21. $(8x^2y)^{5/7}$ **23.** $\dfrac{1}{x^{1/3}}$ **25.** $\dfrac{1}{(2rs)^{3/4}}$ **27.** $10^{2/3}$
29. $x^{2/3}$ **31.** $5^{7/8}$ **33.** $7^{1/4}$ **35.** $8.3^{7/20}$ **37.** $10^{6/25}$
39. $\sqrt[3]{a^2}$ **41.** $2y^2$ **43.** $2\sqrt[4]{2}$ **45.** $\sqrt[3]{2x}$ **47.** $2c^2d^3$
49. $\dfrac{m^2n^4}{2}$ **51.** $\sqrt[4]{r^2s}$ **53.** $3ab^3$ **55.** $\sqrt[6]{392}$ **57.** $\sqrt[6]{4x^5}$
59. $\sqrt[6]{x^5-4x^4+4x^3}$ **61.** $\sqrt[6]{a+b}$ **63.** $\sqrt[12]{a^8b^9}$
65. $t\sqrt[4]{s}$ **67.** $\sqrt[10]{x^9y^7}$ **69.** $(a+b)\sqrt[12]{(a+b)^{11}}$
71. $343w^4$ **73.** $\sqrt[3]{9}+\sqrt[3]{6}+\sqrt[3]{4}$ **75.** $x^5y^7z\sqrt[p]{yz^3}$

Margin Exercises, Section 7.8

1. 100 **2.** No solution **3.** 4 **4.** 17 **5.** 9 **6.** 5
7. 13.038 **8.** 93.81 **9.** 22.2 miles

Exercise Set 7.8, pp. 261–262

1. 2 **3.** 168 **5.** 3 **7.** 19 **9.** $\frac{80}{3}$ **11.** 4
13. −27 **15.** 397 **17.** No solution **19.** $\frac{1}{64}$ **21.** −6
23. 5 **25.** $-\frac{1}{4}$ **27.** 3 **29.** 9 **31.** 7 **33.** $\frac{80}{9}$ **35.** −1
37. 6, 2 **39.** No solution **41.** 5 **43.** 12 **45.** 10
47. $\sqrt{58}\approx 7.62$ **49.** $\sqrt{2}\approx 1.41$ **51.** $\sqrt{12}\approx 3.46$
53. Both triangles have the same area. **55.** $l=2\sqrt{s^2-h^2}$
57. 208 miles **59.** $-\frac{8}{9}$ **61.** −8, 8 **63.** −4, 3
65. 1, 8 **67.** $\frac{1}{36}$, 36 **69.** 25 **71.** 3 **73.** 3 **75.** No

Margin Exercises, Section 7.9

1. $\sqrt{7}\,i$ **2.** $5i$ **3.** $-\sqrt{11}\,i$ **4.** $-6i$ **5.** $15-3i$
6. $-12+6i$ **7.** $3-5i$ **8.** $12-7i$ **9.** $-9-12i$
10. $-35-25i$ **11.** $-14+8i$ **12.** $11+10i$
13. i **14.** $-8i$ **15.** $5+12i$ **16.** $-21-20i$
17. $6-3i$ **18.** $-9+5i$ **19.** $\pi+\frac{1}{4}i$ **20.** 53 **21.** 10
22. $2i$ **23.** $\frac{10}{17}-\frac{11}{17}i$ **24.** Yes **25.** Yes

Exercise Set 7.9, pp. 266–267

1. $\sqrt{15}\,i$ **3.** $4i$ **5.** $-2\sqrt{3}\,i$ **7.** $8+i$ **9.** $9-5i$
11. $7+4i$ **13.** $-2-3i$ **15.** $-1+i$ **17.** $11+6i$
19. $1+5i$ **21.** $18+14i$ **23.** $38+9i$ **25.** $2-46i$
27. −1 **29.** $-125i$ **31.** $5-12i$ **33.** $-5+12i$
35. $-5-12i$ **37.** $\frac{8}{5}+\frac{1}{5}i$ **39.** $-i$
41. $-\frac{21}{49}-\frac{56}{49}i$, or $-\frac{3}{7}-\frac{8}{7}i$ **43.** Yes **45.** No
47. 412 legal pads **49.** $-4-8i$; $-2+4i$; $8-6i$
51. $-3-4i$ **53.** $-88i$ **55.** 8 **57.** $\frac{3}{5}+\frac{9}{5}i$ **59.** 1
61. $-\dfrac{7}{13}+\dfrac{2\sqrt{30}}{13}\,i$

Summary and Review: Chapter 7, pp. 267–268

1. [7.1] 48 **2.** [7.1] $5|x|$ **3.** [7.1] $\dfrac{-3}{4x}$ **4.** [7.1] −2
5. [7.1] $5|x|$ **6.** [7.2] $15\sqrt{2x}$ **7.** [7.2] m^2n^3
8. [7.2] $3ab^2\sqrt[3]{2a}$ **9.** [7.2] 15.7 **10.** [7.2] 25.3
11. [7.2] 250,998 **12.** [7.2] 0.00174 **13.** [7.3] 3
14. [7.3] $\dfrac{2\sqrt{5x^2}}{y^3}$ **15.** [7.3] $\dfrac{3a\sqrt{5a}}{5}$ **16.** [7.4] $36\sqrt{2}$
17. [7.4] $\sqrt[3]{2}$ **18.** [7.4] $(2x-2)\sqrt[3]{4x}$ **19.** [7.5] −41
20. [7.5] $105-24\sqrt{6}$ **21.** [7.5] $6+5\sqrt{x}+x$
22. [7.6] $\dfrac{\sqrt{21}}{6}$ **23.** [7.6] $\dfrac{48-8\sqrt{5}}{31}$
24. [7.6] $\sqrt{10}-5\sqrt{5}+2\sqrt{2}-10$ **25.** [7.7] $x^{5/8}y^{1/2}z^{3/8}$
26. [7.7] $5\sqrt[3]{5}$ **27.** [7.7] $\sqrt[5]{6}$ **28.** [7.7] $3x^4y^5$
29. [7.7] $\sqrt[4]{xy^2}$ **30.** [7.7] $\sqrt[6]{x^5}$ **31.** [7.7] $\sqrt[20]{(a-b)^7}$
32. [7.8] $\frac{7}{2}$ **33.** [7.8] 16 **34.** [7.9] $6+i$
35. [7.9] $-2-11i$ **36.** [7.9] $-11+15i$
37. [7.9] $20+4i$ **38.** [7.9] $53+9i$ **39.** [7.9] $-\frac{6}{13}+\frac{17}{13}i$
40. [7.9] $-\frac{43}{41}+\frac{23}{41}i$ **41.** $2x-3\sqrt{xy}-2y$
42. $\dfrac{x\sqrt{6y}+3y\sqrt{x}}{2x-3y}$ **43.** $12xy^{-1/2}$

CHAPTER 8

Margin Exercises, Section 8.1

1. $\sqrt{3}$ or $-\sqrt{3}$ **2.** 0 **3.** $\dfrac{2\sqrt{6}}{3}, \dfrac{-2\sqrt{6}}{3}$; or $\pm\dfrac{2\sqrt{6}}{3}$

4. $\frac{i\sqrt{2}}{2}, \frac{-i\sqrt{2}}{2}$; or $\pm\frac{i\sqrt{2}}{2}$ **5.** $0, \frac{3}{4}$ **6.** $\frac{3}{5}, 1$ **7.** 3, 8
8. 2 ft **9.** A to B is 6 mi, A to C is 8 mi

Exercise Set 8.1, pp. 276–277

1. $\pm\sqrt{5}$ **3.** 0 **5.** $\pm\frac{\sqrt{6}}{2}$ **7.** $\pm\frac{\sqrt{15}}{3}$ **9.** $\pm\frac{2}{5}i$
11. $\pm\frac{i\sqrt{3}}{3}$ **13.** $\pm i\sqrt{5}$ **15.** $\pm i\sqrt{7}$ **17.** $\pm\frac{3}{2}$ **19.** 0, 5
21. 0, -2 **23.** $0, \frac{2}{3}$ **25.** $0, -\frac{9}{14}$ **27.** 0, 5 **29.** 6, 1
31. 7, -1 **33.** $-7, -2$ **35.** $-\frac{3}{2}, -5$ **37.** $\frac{2}{3}, -4$
39. $\frac{5}{3}, -4$ **41.** $-\frac{7}{2}, -1$ **43.** 4 **45.** 10, 5 **47.** 3 cm
49. The length is 8 m; the width is 3 m.
51. The length is 26 km; the width is 13 km.
53. 12 m and 5 m **55.** 24 km and 7 km
57. $-\frac{5}{3}, 4, \frac{5}{2}$ **59.** $\frac{1}{3}, \frac{1}{18}$ **61.** $0, \frac{b}{a}$ **63.** 5, 6, and 7

Margin Exercises, Section 8.2

1. $\frac{-1+\sqrt{22}}{3}, \frac{-1-\sqrt{22}}{3}$; or $\frac{-1\pm\sqrt{22}}{3}$
2. $\frac{1+i\sqrt{7}}{2}, \frac{1-i\sqrt{7}}{2}$; or $\frac{1\pm i\sqrt{7}}{2}$ **3.** $2, \frac{-1}{3}$
4. 2.1, 0.2 **5.** 1.2, -1.9

Exercise Set 8.2, pp. 280–281

1. $-3\pm\sqrt{5}$ **3.** 1, -5 **5.** 3, -10 **7.** $2, -\frac{1}{2}$
9. $-1, -\frac{5}{3}$ **11.** $\frac{1\pm i\sqrt{3}}{2}$ **13.** $2\pm 3i$ **15.** $\pm i\sqrt{5}$
17. $\frac{-3\pm\sqrt{41}}{2}$ **19.** $-1\pm 2i$ **21.** $1\pm 2i$
23. $\pm\frac{\sqrt{10}}{2}$ **25.** 0, -1 **27.** $\frac{-1\pm 2i}{5}$ **29.** $\frac{3}{4}, -2$
31. $\frac{1\pm 3i}{2}$ **33.** $\frac{3}{2}, \frac{2}{3}$ **35.** 1.3, -5.3 **37.** 5.2, 0.8
39. 2.8, -1.3 **41.** 30 lb at $1.50; 20 lb at $2.50
43. 0.5700731, -0.7973458 **45.** 0.4567764, -0.6567764
47. 1.1753905, -0.4253905 **49.** $\frac{-1\pm\sqrt{1+4\sqrt{2}}}{2}$
51. $\frac{-\sqrt{5}\pm\sqrt{5+4\sqrt{3}}}{2}$ **53.** $\frac{-3\pm\sqrt{9-4i}}{2}$
55. $\frac{-y\pm\sqrt{-47y^2+108}}{6}$ **57.** $\frac{1\pm\sqrt{1+8\sqrt{5}}}{4}$
59. $\frac{-i\pm i\sqrt{1+4i}}{2}$ **61.** $\frac{1}{2}$

Margin Exercises, Section 8.3

1. $\frac{-2}{5}$, 10 **2.** 36 hr **3.** 20 knots, 30 knots

Exercise Set 8.3, pp. 284–285

1. 6, -4 **3.** $\pm\frac{3}{2}$ **5.** 25, -20 **7.** 4, -1 **9.** $\pm\sqrt{37}$
11. -5, 2 **13.** $\frac{1}{15}$, 2 **15.** $10, \frac{21}{2}$
17. First part, 40 mph; second part, 35 mph **19.** 35 mph
21. A: 350 km/h; B: 400 km/h **23.** 6 hr
25. 3.24 km/h **27.** A: 24.0 hr; B: 51.0 hr
29. $\frac{-3\pm\sqrt{57}}{2}$ **31.** $\frac{-i\pm\sqrt{23}}{4}$ **33.** $2.50

Margin Exercises, Section 8.4

1. Two real **2.** One real **3.** Two nonreal
4. $3x^2+7x-20=0$ **5.** $x^2+25=0$
6. $x^2+\sqrt{2}\,x-4=0$

Exercise Set 8.4, pp. 287–288

1. One real **3.** Two nonreal **5.** Two real **7.** One real
9. Two nonreal **11.** Two real **13.** Two real
15. One real **17.** $x^2+2x-99=0$
19. $x^2-14x+49=0$ **21.** $25x^2-20x-12=0$
23. $4x^2-2(c+d)x+cd=0$ **25.** $x^2-4\sqrt{2}\,x+6=0$
27. 2 quarts
29. The solutions of $ax^2+bx+c=0$ are given by
$x=\frac{-b\pm\sqrt{b^2-4ac}}{2a}$.
(a) $\frac{-b+\sqrt{b^2-4ac}}{2a}+\frac{-b-\sqrt{b^2-4ac}}{2a}=\frac{-2b}{2a}=\frac{-b}{a}$.
(b) $\frac{-b+\sqrt{b^2-4ac}}{2a}\cdot\frac{-b-\sqrt{b^2-4ac}}{2a}=$
$\frac{(-b)^2-(\sqrt{b^2-4ac})^2}{4a^2}=\frac{b^2-b^2+4ac}{4a^2}=\frac{4ac}{4a^2}=\frac{c}{a}$.
31. **(a)** $k=-\frac{3}{5}$; **(b)** $=-\frac{1}{3}$ **33.** **(a)** $k=9+9i$; **(b)** $3+3i$
35. $x^2-\sqrt{3}\,x+8=0$ **37.** $h=-36, k=15$
39. $|k|=2$ **41.** $k=6$

Margin Exercises, Section 8.5

1. $w_2=\frac{w_1}{A^2}$ **2.** $r=\sqrt{\frac{V}{\pi h}}$ **3.** $s=\frac{-R\pm\sqrt{R^2-\frac{4L}{C}}}{2L}$
4. $b=\sqrt{\frac{p^2a^2}{1+p^2}}$ **5.** 5 cm

Exercise Set 8.5, pp. 290–291

1. $s=\sqrt{\frac{A}{6}}$ **3.** $r=\sqrt{\frac{Gm_1m_2}{F}}$ **5.** $c=\sqrt{\frac{E}{m}}$
7. $b=\sqrt{c^2-a^2}$ **9.** $k=\frac{3+\sqrt{9+8N}}{2}$
11. $r=\frac{-\pi h+\sqrt{\pi^2h^2+2\pi A}}{2\pi}$ **13.** $g=\frac{4\pi^2 l}{T^2}$

15. $D = \sqrt{\dfrac{32LV}{g(P_1 - P_2)}}$ **17.** $v = \dfrac{c\sqrt{m^2 - m_0^2}}{m}$

19. **(a)** 3.91 sec; **(b)** 1.91 sec; **(c)** 79.6 m

21. **(a)** 18.75%; **(b)** 10%; **(c)** 11%; **(d)** 12.75%

23. 7 ft **25.** $90\sqrt{2} = 127.28$ ft

Margin Exercises, Section 8.6

1. $\pm 3, \pm 1$ **2.** 4 **3.** $4, 2, -1, -3,$ **4.** $\frac{1}{2}, \frac{-1}{3}$

Exercise Set 8.6, pp. 293–294

1. 81, 1 **3.** $\pm\sqrt{5}$ **5.** $-27, 8$ **7.** 16 **9.** $7, -1, 5, 1$

11. $4, 1, 6, -1$ **13.** $\pm\sqrt{2+\sqrt{6}}, \pm\sqrt{2-\sqrt{6}}$

15. $\frac{1}{3}, -\frac{1}{2}$ **17.** $2, -1$ **19.** \$18 **21.** 28.5

23. $3 \pm \sqrt{10}, -1 \pm \sqrt{2}$ **25.** $1 \pm \sqrt{2}, \dfrac{-1 \pm \sqrt{5}}{2}$

27. $\dfrac{5 \pm \sqrt{21}}{2}, \dfrac{3 \pm \sqrt{5}}{2}$ **29.** $\frac{1}{3}$ **31.** 1, 4 **33.** 19

35. $\dfrac{2}{51 \pm 7\sqrt{61}}$, or $\dfrac{7\sqrt{61} - 51}{194}$ and $\dfrac{-(7\sqrt{61} + 51)}{194}$

Margin Exercises, Section 8.7

1. $y = 7x^2$ **2.** $y = \dfrac{9}{x^2}$ **3.** $y = \frac{1}{2}xz$ **4.** $y = \dfrac{5xz^2}{w}$

5. 490 m **6.** 1 ohm

Exercise Set 8.7, pp. 298–299

1. $y = 15x^2$ **3.** $y = \dfrac{0.0015}{x^2}$ **5.** $y = xz$ **7.** $y = \frac{3}{10}xz^2$

9. $y = \dfrac{xz}{5wp}$ **11.** $355\frac{5}{9}$ ft **13.** 94.03 kg **15.** 97

17. y is tripled **19.** y is multiplied by $\dfrac{1}{n^2}$ **21.** $\dfrac{\pi}{4}$

23. y is multiplied by $\dfrac{1}{0.125}$, or 8

Margin Exercises, Section 8.8

1.

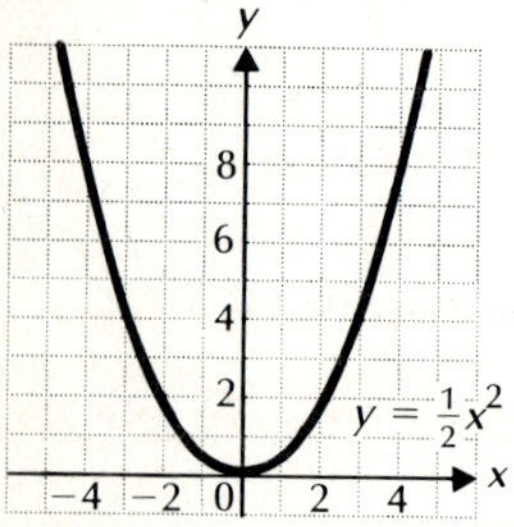

2.

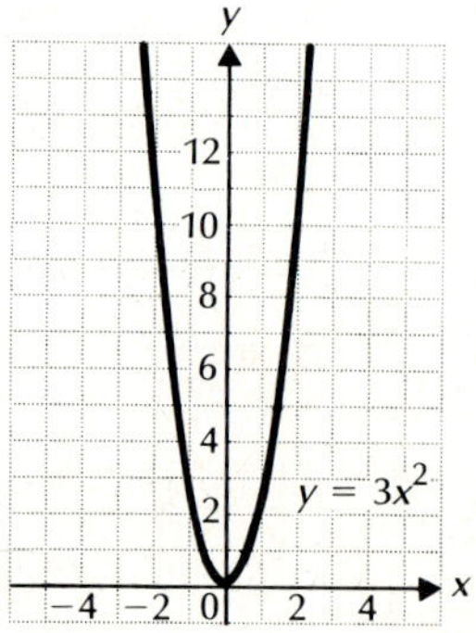

3.

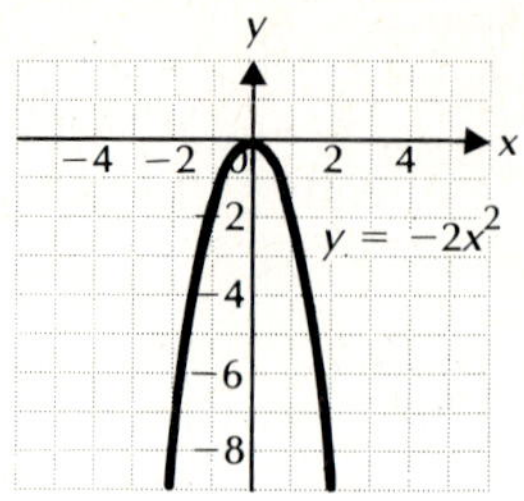

4. **(a)**

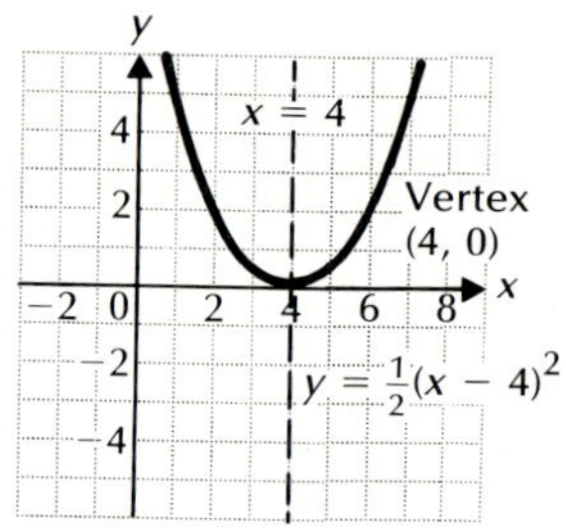

(b)

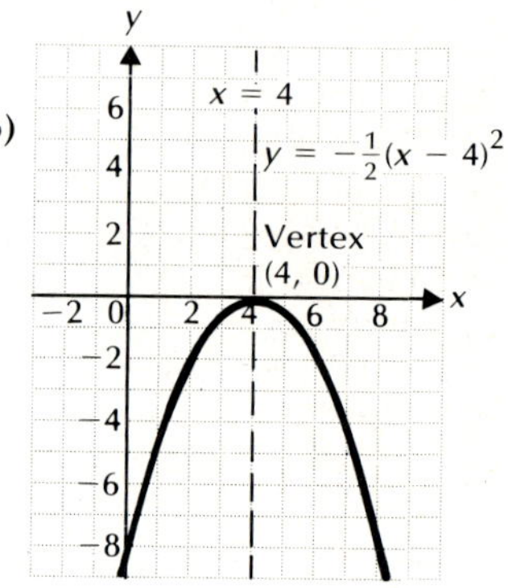

5. **(a)**, **(b)** **(c)** -4

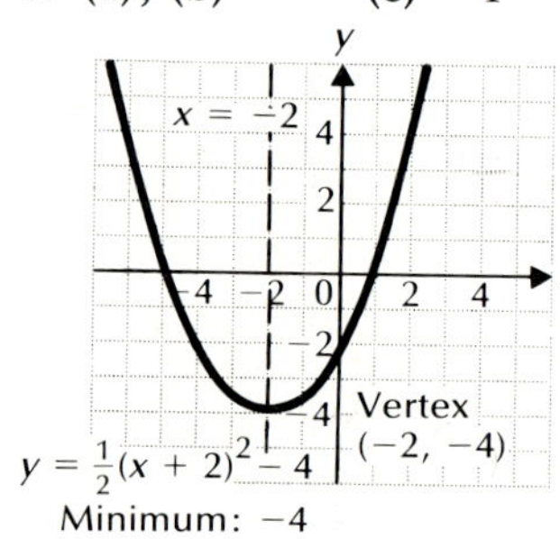

6. **(a)**, **(b)** **(c)** 3

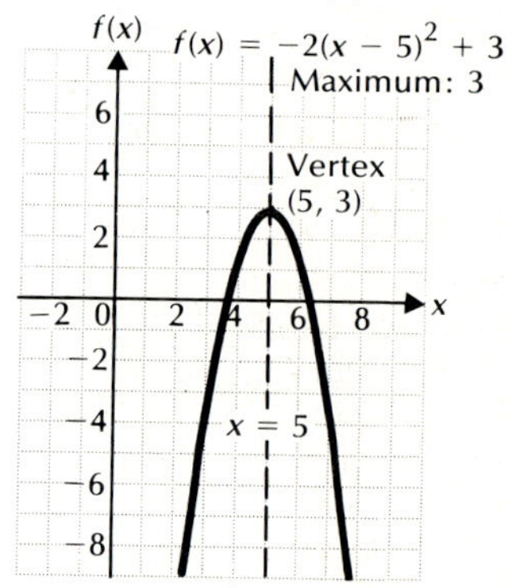

7. **(a)**

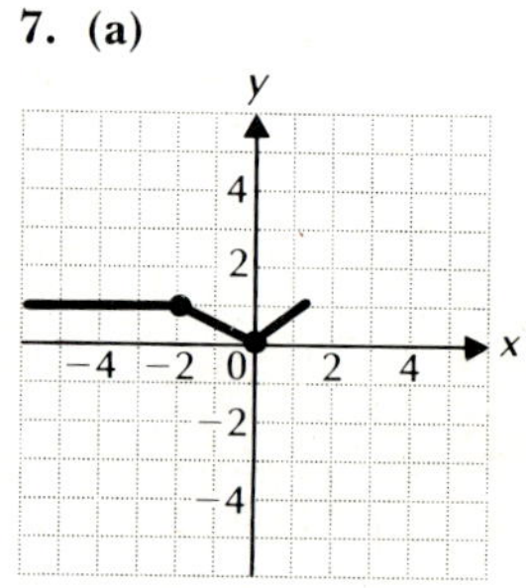

(b)

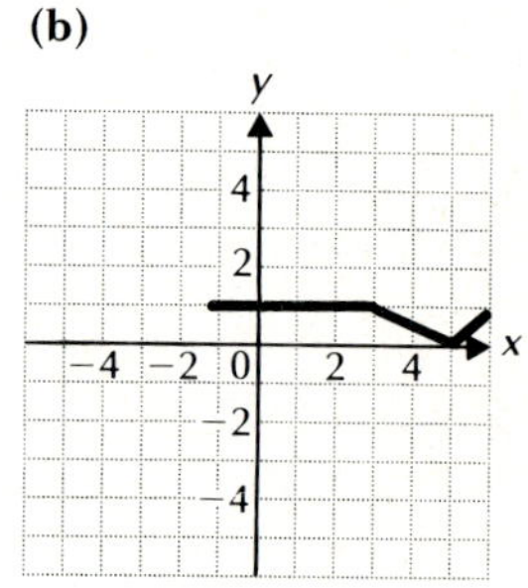

8. **(a)**

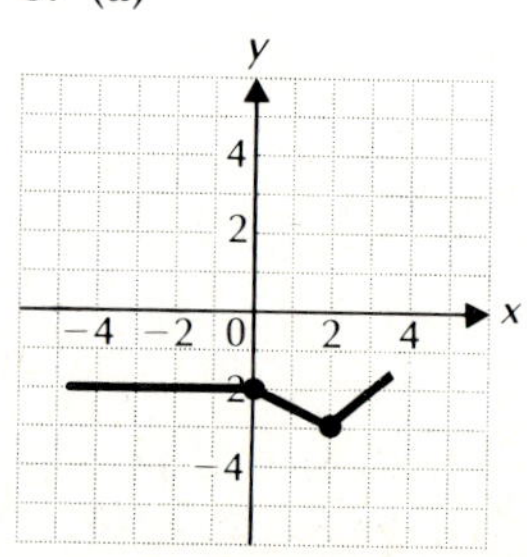

(b)

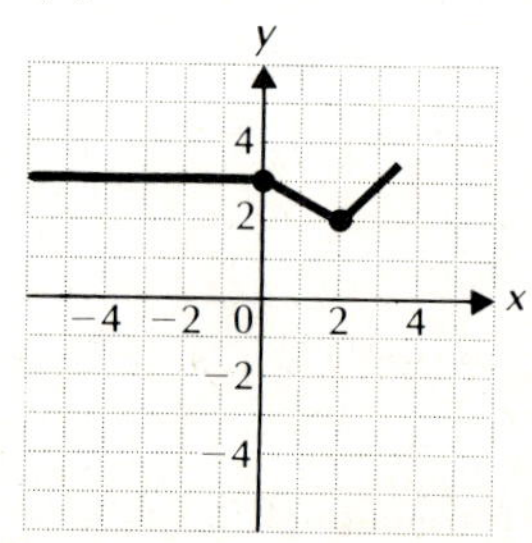

Exercise Set 8.8, p. 307

1. Parabola through $(-2, 4)$, $(0, 0)$, $(2, 4)$; $(0, 0)$; $x = 0$
3. Parabola through $(-1, -4)$, $(0, 0)$, $(1, -4)$; $(0, 0)$; $x = 0$
5. Parabola through $(1, 4)$, $(3, 0)$, $(5, 4)$; $(3, 0)$; $x = 3$
7. Parabola through $(-5, -1)$, $(-4, 0)$, $(-3, -1)$; $(-4, 0)$; $x = -4$
9. Parabola through $(2, 2)$, $(3, 0)$, $(4, 2)$; $(3, 0)$; $x = 3$
11. Parabola through $(-10, -2)$, $(-9, 0)$, $(-8, -2)$; $(-9, 0)$; $x = -9$
13. Parabola through $(0, 3)$, $(1, 0)$, $(2, 3)$; $(1, 0)$; $x = 1$
15. Parabola through $(0, -\frac{3}{4})$, $(\frac{1}{2}, 0)$, $(1, -\frac{3}{4})$; $(\frac{1}{2}, 0)$; $x = \frac{1}{2}$
17. Parabola through $(-2, \frac{1}{2})$, $(-1, 0)$, $(0, \frac{1}{2})$; $(-1, 0)$; $x = -1$
19. Parabola through $(1, 5)$, $(3, 1)$, $(5, 5)$; $(3, 1)$; $x = 3$; minimum $= 1$
21. Parabola through $(-3, 2)$, $(-1, -2)$, $(1, 2)$; $(-1, -2)$; $x = -1$; minimum $= -2$
23. Parabola through $(0, -1)$, $(1, -3)$, $(2, -1)$; $(1, -3)$; $x = 1$; minimum $= -3$
25. Parabola through $(-5, -2)$, $(-4, 1)$, $(-3, -2)$; $(-4, 1)$; $x = -4$; maximum $= 1$
27. $(9, 5)$; $x = 9$; minimum $= 5$
29. $(-\frac{1}{4}, -13)$; $x = -\frac{1}{4}$; minimum $= -13$
31. $(10, -20)$; $x = 10$; maximum $= -20$
33. $(-4.58, 65\pi)$; $x = -4.58$; minimum $= 65\pi$
35. Graph is moved horizontally 2 units in the negative direction.
37. Graph is moved vertically 3 units in the positive direction.
39. Graph is moved horizontally 4 units in the positive direction and vertically 2 units in the negative direction.
41. 4 12-oz sodas, 12 20-oz sodas, 11 32-oz sodas
43. $f(x) = -2x^2 + 4$ **45.** $f(x) = 2(x - 6)^2$
47. $f(x) = -2(x - 3)^2 + 8$ **49.** $f(x) = 2(x + 3)^2 + 6$
51. $f(x) = 2(x - 2)^2 - 3$ **53.** $y = -\frac{1}{2}(x - 1)^2 - 6$

Margin Exercises, Section 8.9

1.

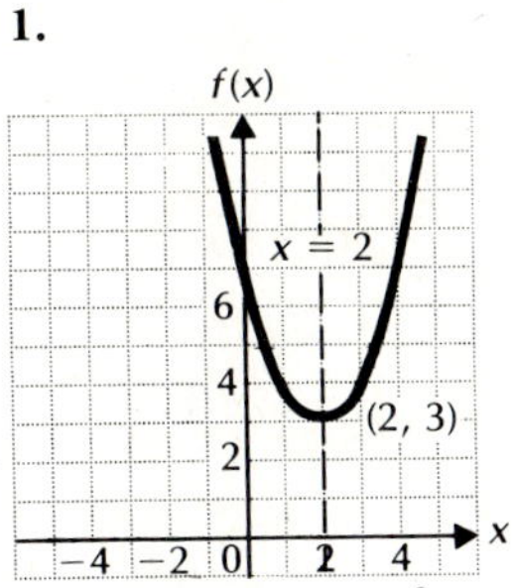

$f(x) = (x - 2)^2 + 3$

2.

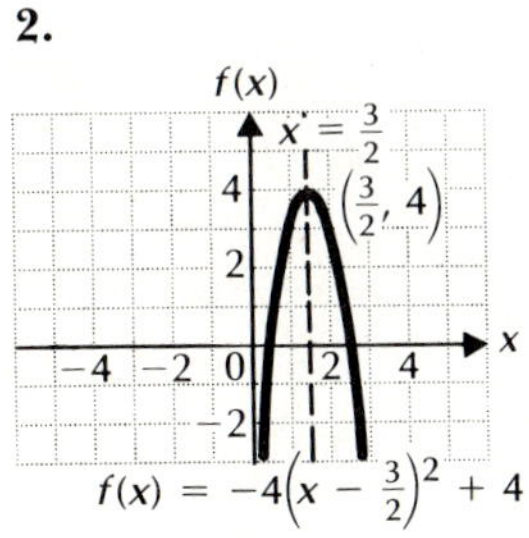

3. $(1 - \sqrt{6}, 0)$, $(1 + \sqrt{6}, 0)$ **4.** $(-1, 0)$, $(3, 0)$
5. $(-4, 0)$ **6.** None

Exercise Set 8.9, pp. 310–311

1. $f(x) = (x - 1)^2 - 4$; $(1, -4)$, $x = 1$, minimum $= -4$
3. $f(x) = -(x - 2)^2 + 10$; $(2, 10)$, $x = 2$, maximum $= 10$
5. $f(x) = (x + \frac{3}{2})^2 - \frac{49}{4}$; $(-\frac{3}{2}, -\frac{49}{4})$, $x = -\frac{3}{2}$, minimum $= -\frac{49}{4}$
7. $f(x) = (x - \frac{9}{2})^2 - \frac{81}{4}$; $(\frac{9}{2}, -\frac{81}{4})$, $x = \frac{9}{2}$, minimum $= -\frac{81}{4}$
9. $f(x) = 3(x - 4)^2 + 2$; $(4, 2)$, $x = 4$, minimum $= 2$
11. $f(x) = \frac{3}{4}(x + 6)^2 - 27$; $(-6, -27)$, $x = -6$, minimum $= -27$
13. $(2 + \sqrt{3}, 0)$, $(2 - \sqrt{3}, 0)$ **15.** $(3, 0)$, $(-1, 0)$
17. $(4, 0)$, $(-1, 0)$ **19.** $(4, 0)$, $(-1, 0)$
21. $\left(\frac{-2 + \sqrt{6}}{2}, 0\right)$, $\left(\frac{-2 - \sqrt{6}}{2}, 0\right)$ **23.** None
25. None **27.** The length is 12; the width is 5.
29. Minimum -6.95 **31.** $(62.75852, 0)$, $(31.941480, 0)$
33. None **35.** **(a)** $3.4, -2.4$; **(b)** $2.3, -1.3$
37. $f(x) = a\left[x - \left(-\frac{b}{2a}\right)\right]^2 + \frac{4ac - b^2}{4a}$

39.

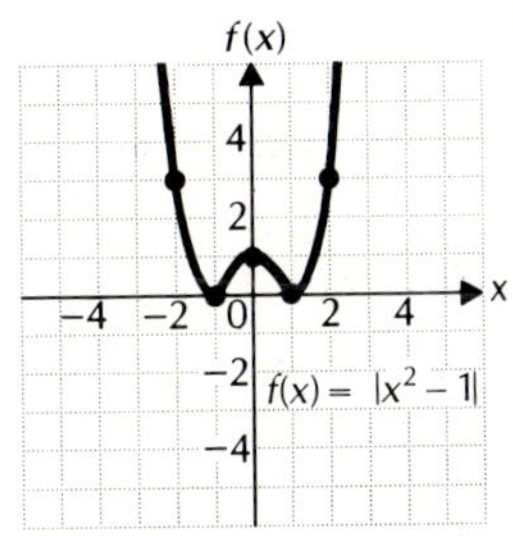

41. All points below parabola through $(1, -4)$, $(2, -5)$, $(3, -4)$.
43. All points below and including parabola through $(-2, 0)$, $(-\frac{5}{2}, -\frac{1}{4})$, $(-3, 0)$.
45. All points above parabola through $(-2, 2)$, $(-1, -1)$, $(0, 2)$.
47. All points above and including parabola through $(-2, 3)$, $(-1, -1)$, $(0, 3)$.

Margin Exercises, Section 8.10

1. 225 **2.** 25 m by 25 m **3.** $f(x) = x^2 - 2x + 1$
4. **(a)** $f(x) = 0.625x^2 - 50x + 1150$; **(b)** 510

Exercise Set 8.10, pp. 313–315

1. 19 ft by 19 ft; 361 ft^2 **3.** 121; 11 and 11
5. -4; 2 and -2 **7.** $-\frac{25}{4}$; $\frac{5}{2}$ and $-\frac{5}{2}$
9. $f(x) = 2x^2 + 3x - 1$ **11.** $f(x) = -3x^2 + 13x - 5$
13. **(a)** $f(x) = -4x^2 + 40x + 2$; **(b)** \$98
15. **(a)** $0.05x^2 - 5.5x + 250$; **(b)** 100
17. Minimum: -7.023400974
19. $f(x) = 0.0499218x^2 - 2.7573651x + 29.571379$
21. 19 cm by 19 cm; 180.5 cm^2 **23.** 4050 ft^2 **25.** \$6

Summary and Review: Chapter 8, pp. 315–316

1. [8.1] $\pm\frac{\sqrt{14}}{2}$ **2.** [8.1] $0, -\frac{5}{14}$ **3.** [8.1] 9, 3
4. [8.1] 7, -8 **5.** [8.1] 9 m, 40 m
6. [8.2] $-2 \pm \sqrt{2}$ **7.** [8.2] $\frac{1 \pm 2i\sqrt{2}}{-3}$

8. [8.2] $\frac{\pm\sqrt{14}}{2}$ **9.** [8.2] $2 \pm i$ **10.** [8.3] $\pm\sqrt{26}$
11. [8.3] $\frac{1 \pm \sqrt{481}}{15}$ **12.** [8.3] $4 \pm 4\sqrt{2}$
13. [8.3] 100 mph, 200 mph
14. [8.4] Two real solutions
15. [8.4] Two real solutions
16. [8.4] Two nonreal solutions
17. [8.4] Two nonreal solutions **18.** [8.5] $r = \frac{\sqrt{3A\pi h}}{\pi h}$
19. [8.5] 2.32 **20.** [8.6] $\pm\sqrt{6}$ **21.** [8.6] $-\frac{3}{2}$, 1
22. [8.6] $\pm\sqrt{2}$, $\pm\sqrt{7}$ **23.** [8.7] 23.71 in^2
24. [8.8] Parabola through $(-3, 2)$, $(-2, 0)$, $(-1, 2)$; minimum: $(-2, 0)$, $x = -2$
25. [8.8] Parabola through $(5, -2)$, $(3, 2)$, $(1, -2)$; maximum: $(3, 2)$; $x = 3$
26. [8.9], [8.10] 21.75 by 21.75 **27.** -6, $-2k$
28. $ax^4 - bx^3 - ac^2x^2 + bc^2x = 0$

CHAPTER 9

Margin Exercises, Section 9.1

1. **(a)** 1, 3, 9, $\frac{1}{3}$, $\frac{1}{9}$, $\frac{1}{27}$
(b)

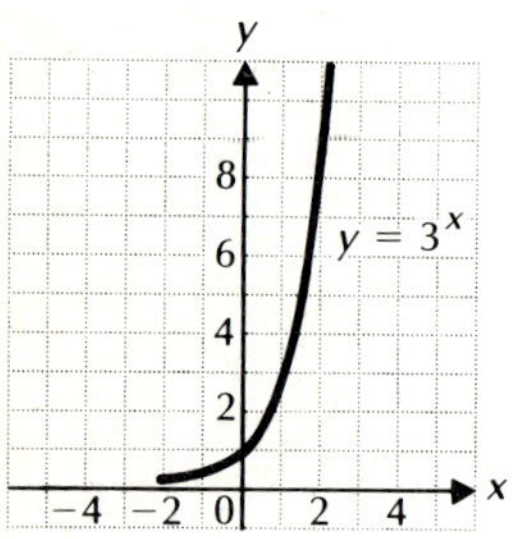

2.

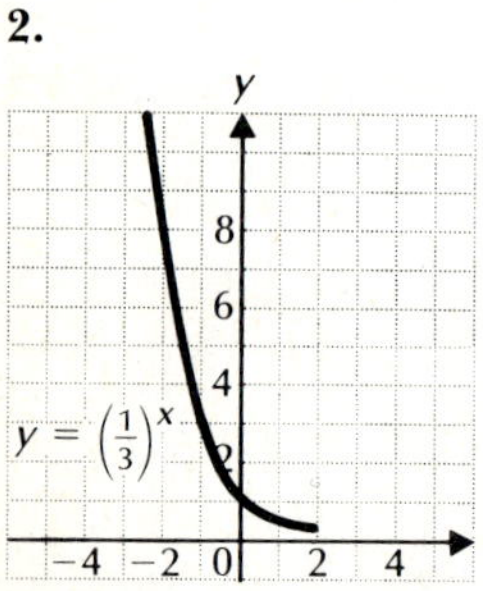

3.

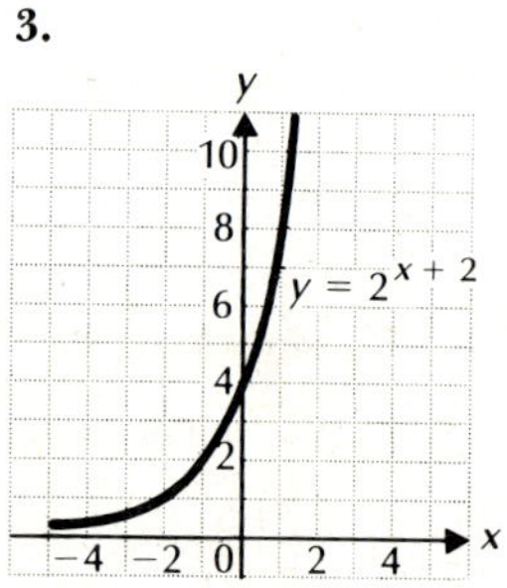

4.

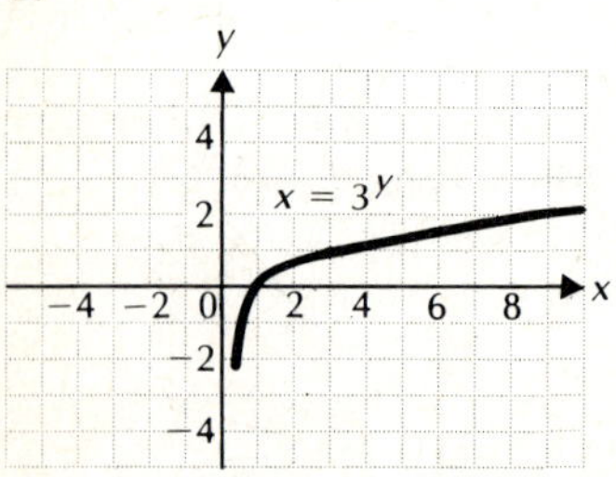

5. $f^{-1}(x) = x + 5$

6. **(a)** $q^{-1}(x) = 3x - 2$; **(b)** q adds 2 to each input, then divides by 3; q^{-1} multiplies by 3 and then subtracts 2. Yes, q and q^{-1} do opposite things.
7. $y = \pm\sqrt[4]{x}$. No.

Exercise Set 9.1, p. 323

1.

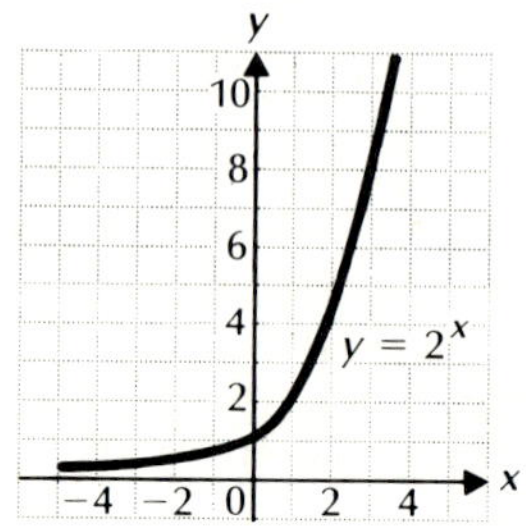

3.

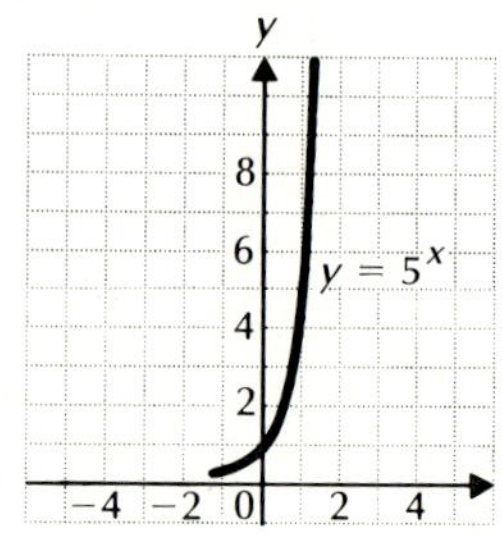

5.

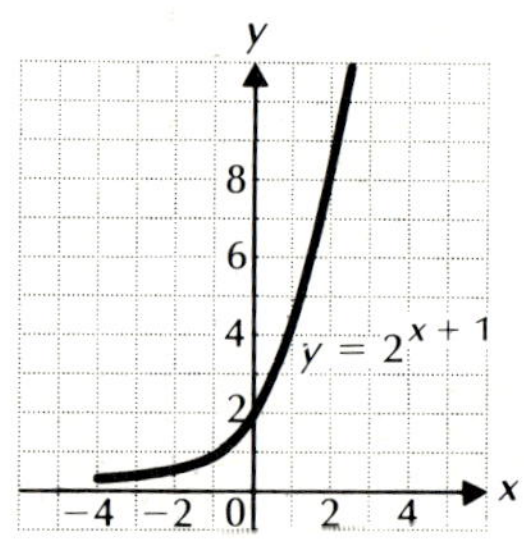

7.

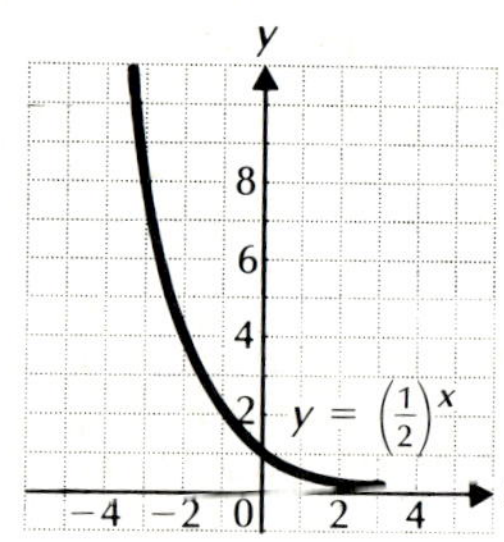

9.

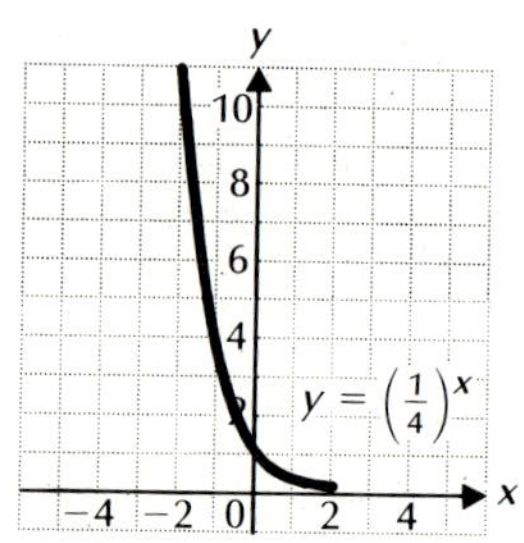

11.

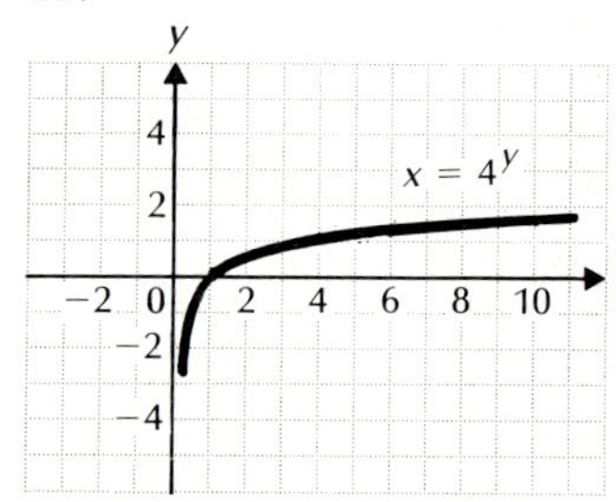

13.

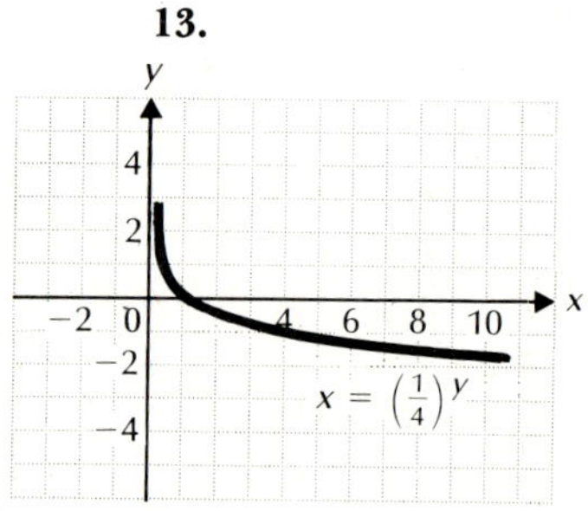

15. $f^{-1}(x) = x + 1$
17. $f^{-1}(x) = x - 4$
19. $f^{-1}(x) = x - 8$
21. $f^{-1}(x) = \frac{x - 5}{2}$
23. $g^{-1}(x) = \frac{3x - 1}{2}$
25. $g^{-1}(x) = \sqrt{\frac{x - 3}{2}}$
27. 30 ft, 55 ft, 91 ft

29.

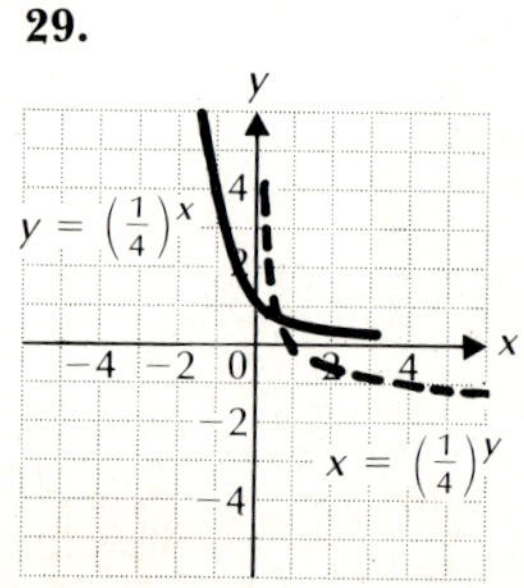

31.

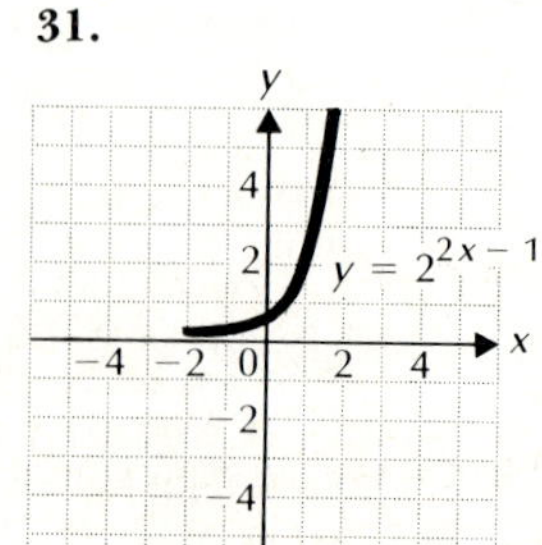

33.

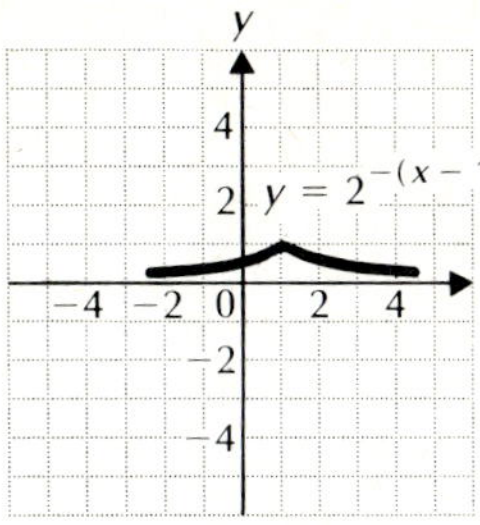

35.

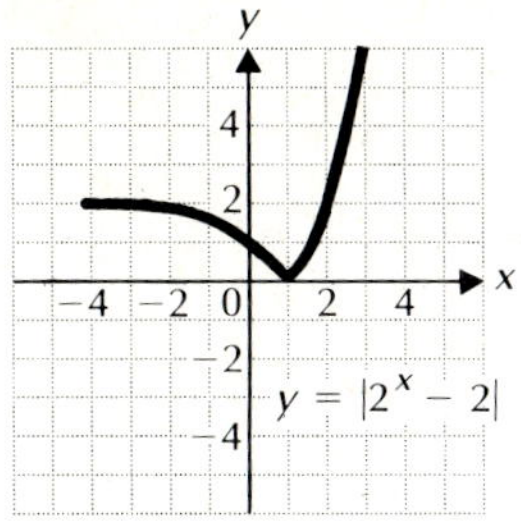

37. π^5

Margin Exercises, Section 9.2

1.

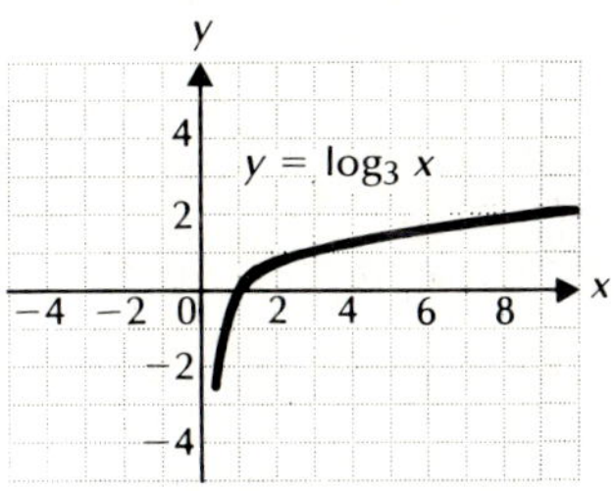

2. $0 = \log_6 1$
3. $-3 = \log_{10} 0.001$
4. $0.25 = \log_{16} 2$
5. $T = \log_m P$ **6.** $2^5 = 32$
7. $10^3 = 1000$ **8.** $a^7 = Q$
9. $t^x = M$ **10.** 10,000
11. 3 **12.** 4 **13.** -4
14. 0 **15.** 3 **16.** 17
17. 4.2 **18.** $3x^2 - 4$
19. 6 **20.** 3.2 **21.** 12 **22.** $5x - y^2$

Exercise Set 9.2, pp. 327–328

1.

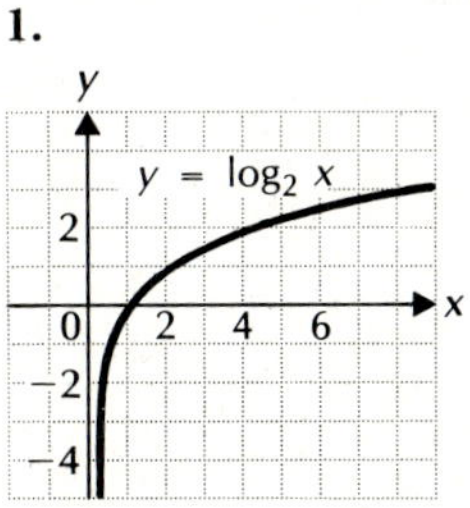

3.

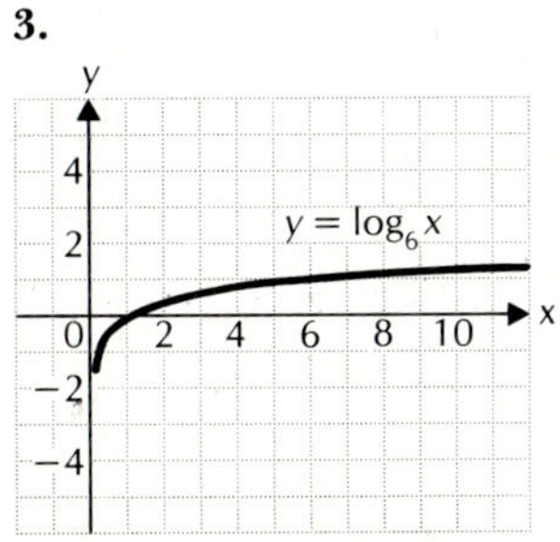

5. $3 = \log_{10} 1000$ **7.** $-3 = \log_5 \frac{1}{125}$ **9.** $\frac{1}{3} = \log_8 2$
11. $0.3010 = \log_{10} 2$ **13.** $3^t = 8$ **15.** $5^2 = 25$
17. $10^{-1} = 0.1$ **19.** $10^{0.845} = 7$ **21.** 9 **23.** 4
25. $\frac{1}{2}$ **27.** 2 **29.** 2 **31.** -1 **33.** 0 **35.** 4
37. -2 **39.** 0 **41.** 4 **43.** 9 **45.** $5x^2 + 7y^3$
47. 15°, 45°, 30°

49.

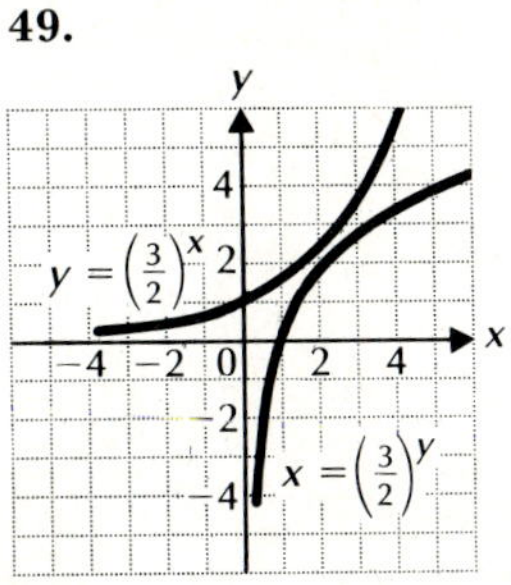

51.

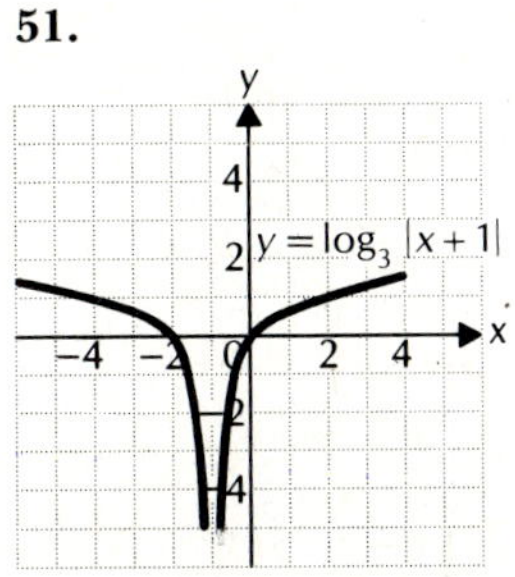

53. 25 **55.** $\frac{\sqrt{5}}{25}$ **57.** 6 **59.** 1296 **61.** $\frac{2}{3}$ **63.** x^2
65. 0 **67.** 1 **69.** -2

Margin Exercises, Section 9.3

1. $\log_5 25 + \log_5 5$ **2.** $\log_b P + \log_b Q$
3. $\log_3 (7 \cdot 5)$ or $\log_3 35$ **4.** $\log_a$ (CABIN) **5.** $5 \log_7 4$
6. $\frac{1}{2} \log_a 5$ **7.** $\log_b P - \log_b Q$ **8.** $\log_2 \frac{x}{25}$
9. $\frac{1}{2}(3 \log_a z - \log_a x - \log_a y)$ **10.** $\log_a \frac{x^5 z^{1/4}}{y}$

Exercise Set 9.3, pp. 331–332

1. $\log_2 32 + \log_2 8$ **3.** $\log_4 64 + \log_4 16$
5. $\log_c B + \log_c x$ **7.** $\log_a (6 \cdot 70)$ **9.** $\log_c (K \cdot y)$
11. $3 \log_a x$ **13.** $6 \log_c y$ **15.** $-3 \log_b C$
17. $\log_a 67 - \log_a 5$ **19.** $\log_b 3 - \log_b 4$
21. $\log_a \frac{15}{7}$ **23.** $2 \log_a x + 3 \log_a y + \log_a z$
25. $\log_b x + 2 \log_b y - 3 \log_b z$ **27.** $\log_a \frac{\sqrt[3]{x^2}\sqrt{y}}{y}$
29. $\log_a \frac{2x^4}{y^3}$ **31.** $\log_a \frac{\sqrt{a}}{x}$ **33.** 0.602 **35.** 1.699
37. 1.778 **39.** -0.088 **41.** 1.954 **43.** -0.602
45. $\frac{40}{11}$ m, $\frac{15}{11}$ m **47.** $\log_a (x^6 - x^4y^2 + x^2y^4 - y^6)$
49. $\frac{1}{2} \log_a (1 - s) + \frac{1}{2} \log_a (1 + s)$ **51.** $\frac{10}{3}$ **53.** $\frac{9}{8}$ **55.** 3
57. $\{x \mid x > 0\}$ **59.** -2

Margin Exercises, Section 9.4

1. 6.9042 **2.** -3.0482 **3.** 127,438 **4.** 1.0042
5. 11.083 **6.** -6.8666 **7.** 172.17 **8.** 1.0518
9. 0.6415 **10.** 0.9647 **11.** 7.73 **12.** 2.99 **13.** 3.8831
14. -2.9788 **15.** 838,000 **16.** 0.00472 **17.** 11.3645
18. -2.96805 **19.** 43,600 **20.** 1.750

Exercise Set 9.4, pp. 337–338

1. 0.8020893 **3.** 1.6532125 **5.** 1.7951846
7. 2.6404814 **9.** 4.1271048 **11.** -0.2441251
13. -1.2839967 **15.** -2.2069084 **17.** 1000
19. 501.18723 **21.** 2.9998532 **23.** 0.2841188
25. 0.0011161 **27.** 4.1271344 **29.** 8.3813735
31. -5.083206 **33.** 36.789041 **35.** 0.0022994
37. 0.7945 **39.** 0.9943 **41.** 0.5328 **43.** 0.4378
45. 5.01 **47.** 2.83 **49.** 6.06 **51.** 1.75 **53.** 1.6839
55. 2.7536 **57.** 4.6149 **59.** -2.1925 **61.** 99.3
63. 282 **65.** 0.357 **67.** 0.00638 **69.** 8.1491
71. -3.3496 **73.** 1640 **75.** 0.1310 **77.** 52.508353
79. 4.9855491 **81.** 48.586

Margin Exercises, Section 9.5

1. 3.6592 **2.** $0.3779 - 2$, or -1.6221 **3.** 5956
4. 0.0002856

Exercise Set 9.5, p. 341

1. 1.6194 **3.** 0.4689 **5.** 2.8130 **7.** -0.8462
9. -2.3709 **11.** -0.7506 **13.** 2.7786 **15.** 43.15
17. 4.444 **19.** 1.082 **21.** 0.7185 **23.** 0.002587
25. 0.01589 **27.** 609 cm^2 **29.** -0.3214 **31.** -0.0731

Margin Exercises, Section 9.6

1. 3 **2.** 2 **3.** 7 **4.** 5111

Exercise Set 9.6, pp. 345–346

1. 3 **3.** 3.3219 **5.** $\frac{5}{2}$ **7.** $-3, -1$ **9.** 1.4036

11. 2.7093 **13.** 3.6064 **15.** 5.6456 **17.** 10 **19.** $\frac{1}{3}$
21. 5 **23.** $1, 10^9$ **25.** $\pm 2\sqrt{6}$ **27.** 6.1 years
29. $k = 0.0201$; population = 1,242,763
31. **(a)** \$1.13; **(b)** 2002
33. **(a)** 2.37 feet per second; **(b)** 3.37 feet per second
35. \$0.57; \$1.49 **37.** -4 **39.** 4 **41.** 2 **43.** $\pm\sqrt{34}$
45. $10^{100,000}$ **47.** $\frac{9}{8}$ **49.** 11 **51.** $\frac{1}{100,000}$, 100,000
53. $-7, 3$ **55.** $\frac{1}{10}, \frac{1}{1000}$ **57.** $-\frac{1}{3}$ **59.** $\frac{3}{2}$ **61.** -3

Summary and Review: Chapter 9, pp. 346–347

1. [9.1]

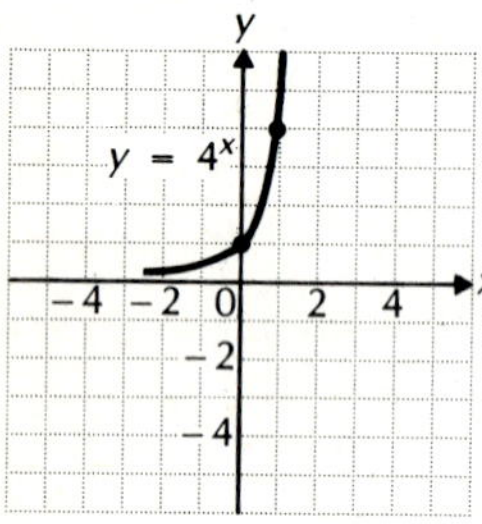

2. [9.1]

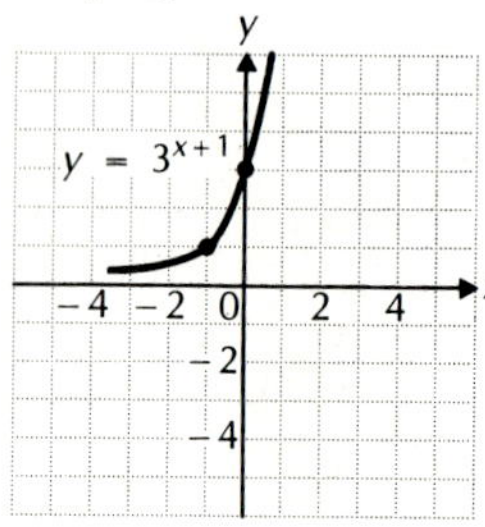

3. [9.1] $f^{-1}(x) = x - 2$ **4.** [9.1] $f^{-1}(x) = \dfrac{x+1}{3}$

5. [9.1] $f^{-1}(x) = x - 5$ **6.** [9.1] $f^{-1}(x) = \dfrac{x-3}{2}$

7. [9.2] $\log_{10} 10{,}000 = 4$ **8.** [9.2] $\log_6 216 = 3$
9. [9.2] $\log_{25} 5 = \frac{1}{2}$ **10.** [9.2] $\log_{81} 3 = \frac{1}{4}$
11. [9.2] $4^x = 16$ **12.** [9.2] $8^1 = 8$
13. [9.2] $10^{0.3010} = 2$ **14.** [9.2] $10^{-4} = 0.0001$
15. [9.3] $\log_a 120$ **16.** [9.3] $\log_x 6$
17. [9.3] $4 \log_a x + 2 \log_a y + 3 \log_a z$
18. [9.3] $\log_a 7 + \log_a x + \log_a y + 5 \log_a z$
19. [9.3] $\log_a x + \log_a y - 2 \log_a z$
20. [9.3] $\log_a 7 + 3 \log_a x - 2 \log_a y - \log_a z$
21. [9.4] 0.8482 **22.** [9.4] 0.5276 **23.** [9.4] 1.6561
24. [9.4] 1.9974 **25.** [9.4] -2.3279 **26.** [9.4] -1.2487
27. [9.5] 3.6593 **28.** [9.5] -0.2172 **29.** [9.4] 3.94
30. [9.4] 8.41 **31.** [9.4] 472 **32.** [9.4] 0.0545
33. [9.5] 4.316 **34.** [9.5] 0.004546 **35.** [9.4] 4.2047
36. [9.4] 6.0039 **37.** [9.4] -3.3668 **38.** [9.4] -1.4024
39. [9.5] 1.96 **40.** [9.4] 1.24 **41.** [9.4] 12.9
42. [9.5] 0.0302 **43.** [9.6] 4.585 **44.** [9.6] 1
45. [9.6] 20 **46.** [9.6] $\sqrt{43}$ **47.** $x = 2.06$, $y = 0.8$
48. $\{x \mid x > -0.317\}$ **49.** 4,294,967,295

CUMULATIVE REVIEW: CHAPTERS 1–9, pp. 348–349

1. $\frac{13}{7}$ **2.** $\frac{8}{5}$ **3.** $y > \frac{7}{2}$ **4.** $\{a \mid a > -1\}$
5. $\{x \mid \frac{2}{3} \leqslant x \leqslant \frac{8}{3}$, or $\frac{10}{3} \leqslant x \leqslant \frac{16}{3}\}$ **6.** 0, -1, or -3

7. $\frac{21}{5}$ **8.** $\frac{26}{3}$ **9.** -1 **10.** $\pm\sqrt{5}$ **11.** $\pm\dfrac{\sqrt{10}}{5}$

12. $\dfrac{-1 \pm 3i}{2}$ **13.** $-3 \pm \sqrt{6}$ **14.** 2 or 6

15. ± 1 or $\pm\sqrt{7}$ **16.** 5 **17.** $a = \pm\sqrt{c^2 - b^2}$ for $c^2 \geqslant b^2$

18. $h = \dfrac{A - 2r^2}{2r}$ **19.** $a^3 + b^3$ **20.** $-2xy^2 - 9xy$

21. $-xy$ **22.** $9x^2$ **23.** -5 **24.** $4mn^3(\sqrt[3]{m})$

25. $\dfrac{2a}{3}\sqrt{6a}$ **26.** $4\sqrt{14} + 4\sqrt{3} - 20\sqrt{6}$ **27.** $16 + 8\sqrt{3}$

28. $11 + 4i$ **29.** $-11 + 11i$ **30.** $-3 - 29i$
31. $-\frac{9}{20} - \frac{2}{5}i$

32. $y = (x+2)^2$

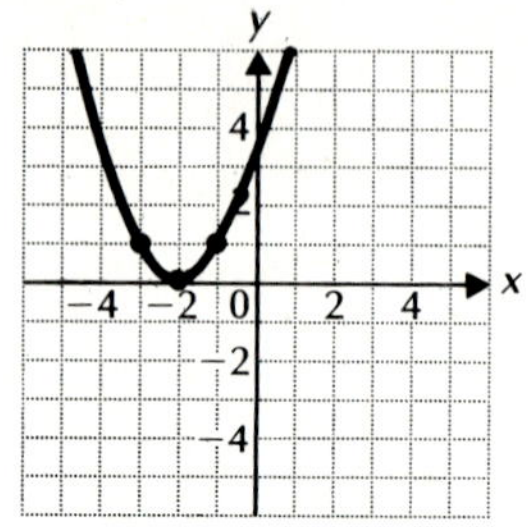

33. $x > y + 1$

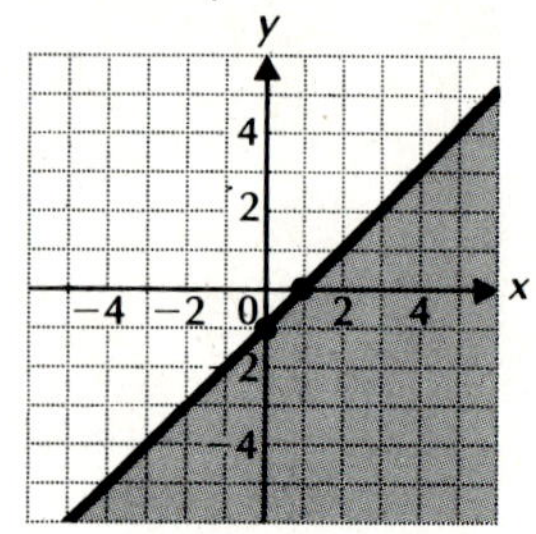

34. $y^2 = x^2 - 8x + 16$

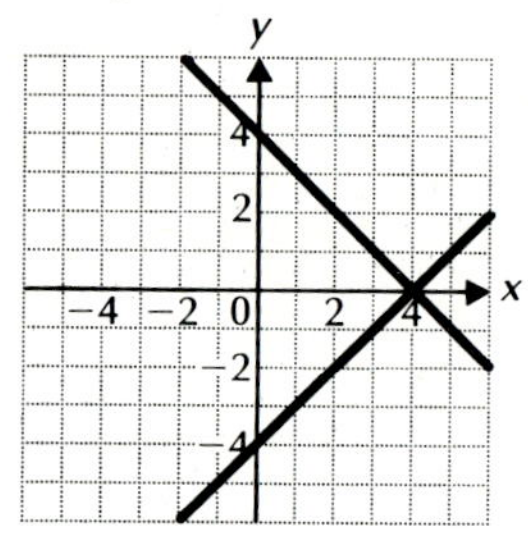

35. $f(x) = -(x+3)^2$

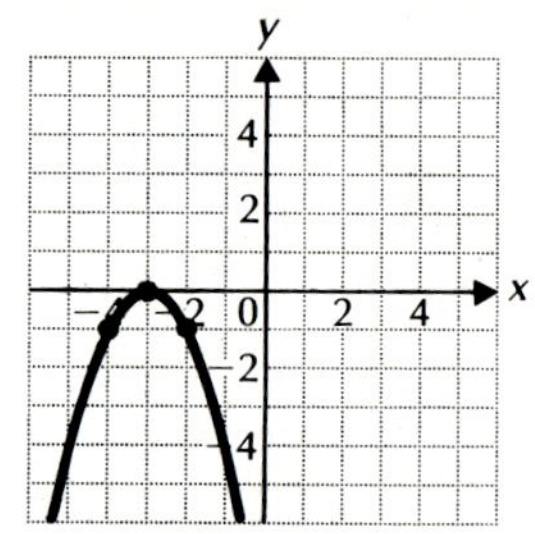

36. $f(x) = 2(x-5)^2$

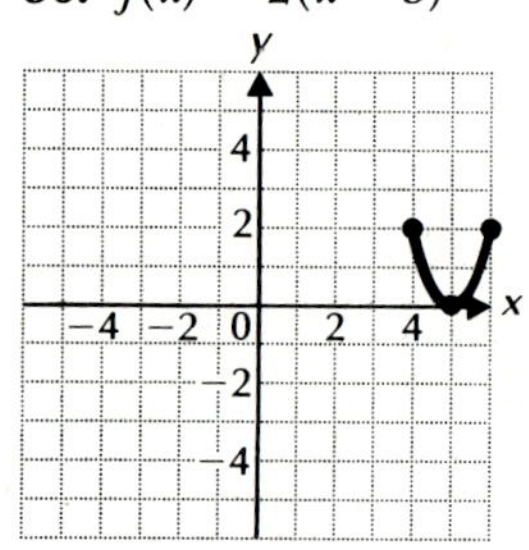

37. $f(x) = \frac{1}{4}(x-2)^2$

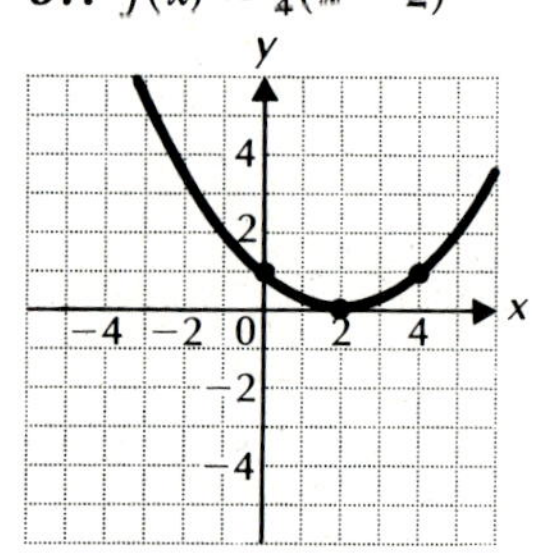

38. $f(x) = -3(x+1)^2 - 2$

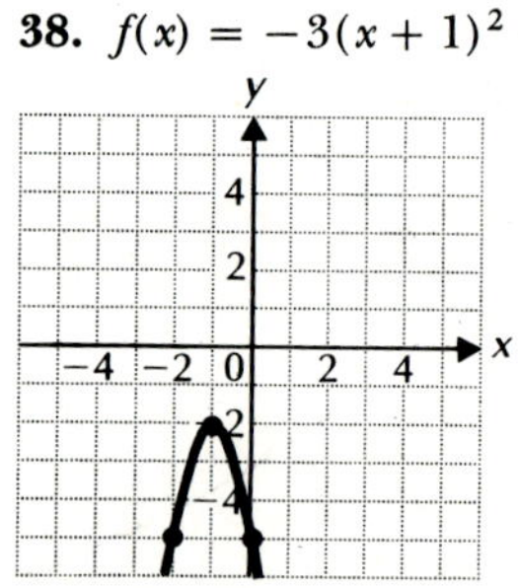

39. $f(x) = 4(x - \frac{1}{2})^2 - 10$

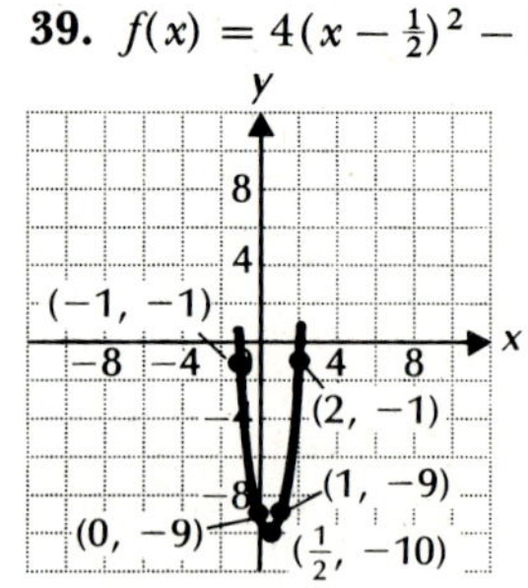

40. $f(x) = x^2 - 4x + 1$

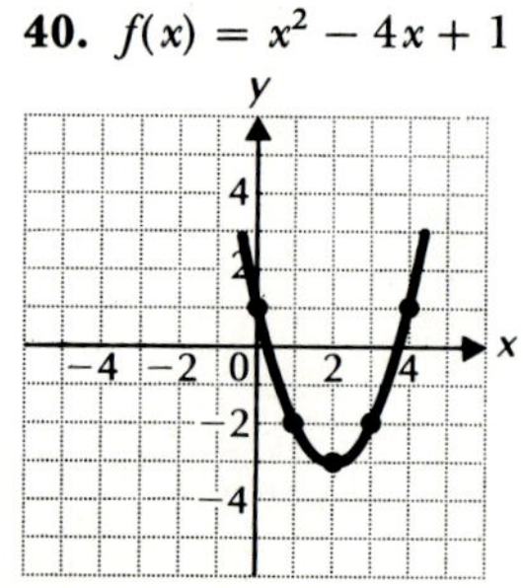

41. $y = 3^x$

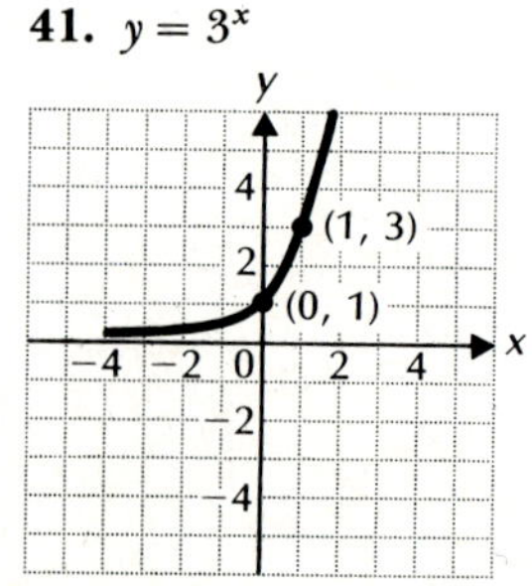

42. $f(x) = \dfrac{x+5}{2}$ **43.** $f(x) = 4 - x$

44. $\log_{10} 0.00001 = -5$ **45.** $\log_{169} 13 = \frac{1}{2}$ **46.** $3^x = 27$
47. $10^{0.4771} = 3$ **48.** 1.9566 **49.** 5.02 **50.** -1.2932
51. 5.9687 **52.** 1.97 **53.** -5.1396
54. $w = 14$ cm, $l = 19$ cm **55.** A: 100 mph and B: 150 mph or A: 250 mph and B: 300 mph
56. 27.5 ft × 27.5 ft; 756.25 ft^2 **57.** 39 ft

CHAPTER 10

Margin Exercises, Section 10.1

1. 14 **2.** 11 **3.** 10 **4.** $\sqrt{37} \approx 6.083$

Exercise Set 10.1, pp. 353–354

1. 5 **3.** $3\sqrt{2}$ **5.** 5 **7.** 8 **9.** $\sqrt{a^2 + 64}$
11. $\sqrt{a^2 + b^2}$ **13.** $2\sqrt{a}$ **15.** 80 **17.** Yes
19. If two points are on a vertical line, they have coordinates (a, y_1) and (a, y_2). Thus the distance from one to the other is $|y_2 - y_1|$. Now $\sqrt{(a - a)^2 + (y_2 - y_1)^2} = \sqrt{0 + (y_2 - y_1)^2} = \sqrt{(y_2 - y_1)^2} = |y_2 - y_1|$. The proof for a horizontal line is similar.
21. Let $P_1(x_1, y_1)$ and $P_2(x_2, y_2)$ be the endpoints of a segment and M be the point $\left(\frac{x_1 + x_2}{2}, \frac{y_1 + y_2}{2}\right)$. If $d(P_1M) = d(MP_2) = \frac{1}{2}d(P_1P_2)$, then M is the midpoint of P_1P_2. And
$d(P_1M) = \sqrt{\left(\frac{x_1 + x_2}{2} - x_1\right)^2 + \left(\frac{y_1 + y_2}{2} - y_1\right)^2} = \frac{1}{2}\sqrt{(x_2 - x_1)^2 + (y_2 - y_1)^2}$; $d(MP_2) = \sqrt{\left(x_2 - \frac{x_1 + x_2}{2}\right)^2 + \left(y_2 - \frac{y_1 + y_2}{2}\right)^2} = \frac{1}{2}\sqrt{(x_2 - x_1)^2 + (y_2 - y_1)^2}$; $d(P_1P_2) = \sqrt{(x_2 - x_1)^2 + (y_2 - y_1)^2}$.
23. Yes

Margin Exercises, Section 10.2

1.

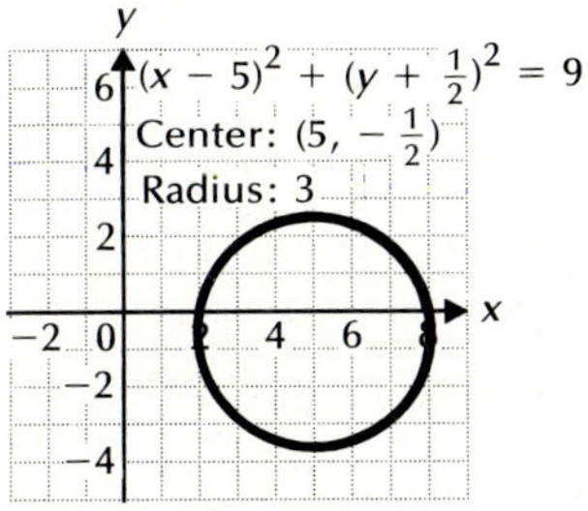

2. $(x + 3)^2 + (y - 1)^2 = 36$
3. $(0, 0)$, $r = 8$
4. $(x + 4)^2 + (y - 5)^2 = 49$; center: $(-4, 5)$; radius: 7

5.

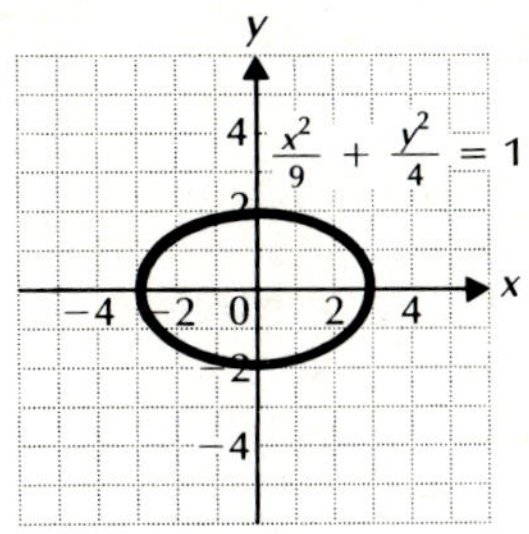

6.

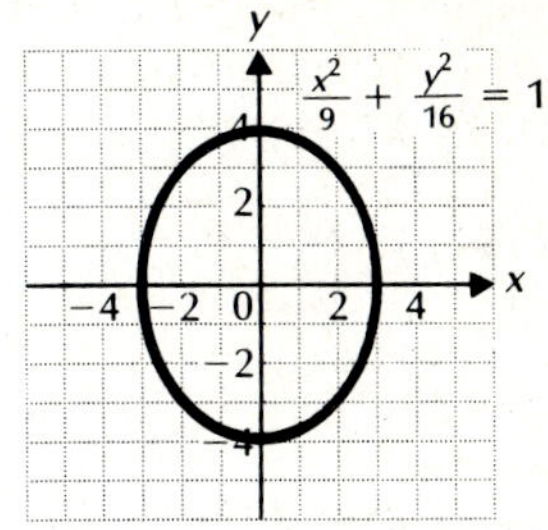

Exercise Set 10.2, pp. 358–359

1. $(-1, -3)$, 2 **3.** $(8, -3)$, $2\sqrt{10}$ **5.** $(0, 0)$, $\sqrt{2}$
7. $(5, 0)$, $\frac{1}{2}$ **9.** $(-4, 3)$, $2\sqrt{10}$ **11.** $(4, -1)$, 2
13. $(2, 0)$, 2 **15.** $x^2 + y^2 = 49$
17. $(x + 2)^2 + (y - 7)^2 = 5$
19. Vertices: $(-2, 0)$, $(2, 0)$, $(0, -1)$, $(0, 1)$
21. Vertices: $(-3, 0)$, $(3, 0)$, $(0, -4)$, $(0, 4)$
23. Vertices: $(-\sqrt{3}, 0)$, $(\sqrt{3}, 0)$, $(0, -\sqrt{2})$, $(0, \sqrt{2})$
25. Vertices: $(-\frac{1}{2}, 0)$, $(\frac{1}{2}, 0)$, $(0, -\frac{1}{3})$, $(0, \frac{1}{3})$
27. 4 inches
29. **(a)** $x^2 + y^2 = 2$; **(b)** $(x + 4)^2 + (y - 1)^2 = 72$; **(c)** $(x + 3)^2 + (y + 2)^2 = 9$; **(d)** $(x - 3)^2 + y^2 = 25$
31. **(a)** $\frac{x^2}{64} + \frac{y^2}{4} = 1$; **(b)** $\frac{(x - 5)^2}{25} + \frac{(y - 2)^2}{9} = 1$
33. The graph is a circle. **35.** Yes **37.** No
39. From the equation of a circle we have $b^2 + c^2 = a^2$; $c^2 = a^2 - b^2$; the slope of $\overline{AB}$ is $\frac{c}{b + a}$ and the slope of $\overline{BC}$ is $\frac{c}{b - a}$; since $\frac{c}{b + a} \cdot \frac{c}{b - a} = \frac{c^2}{b^2 - a^2} = \frac{a^2 - b^2}{b^2 - a^2} = -1$, $\overline{AB}$ and $\overline{BC}$ are perpendicular and angle ABC is a right angle.

Margin Exercises, Section 10.3

1.

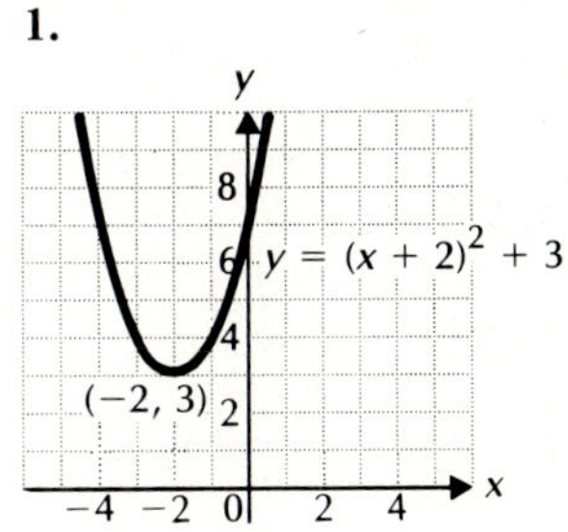

2.

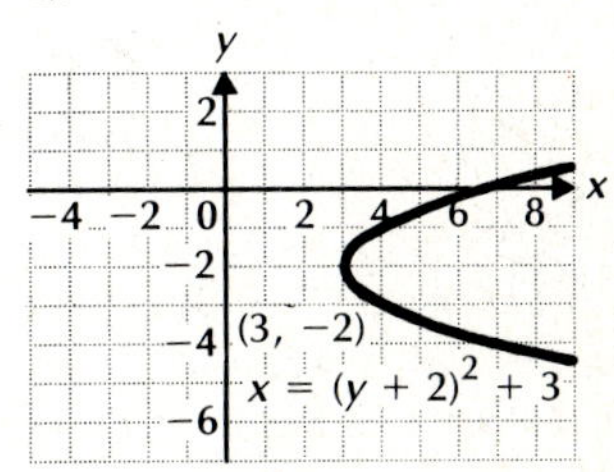

3.

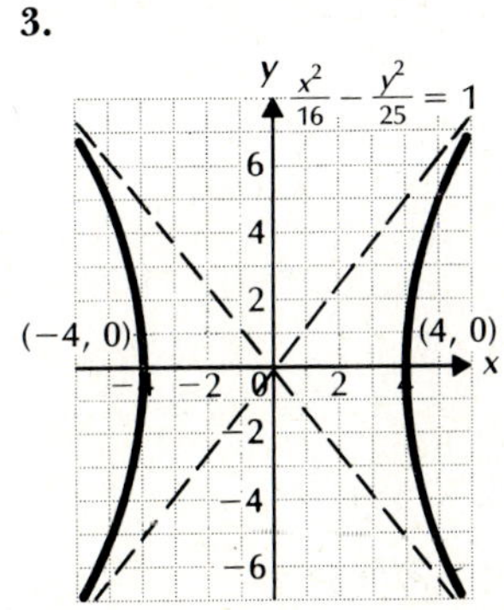

4.

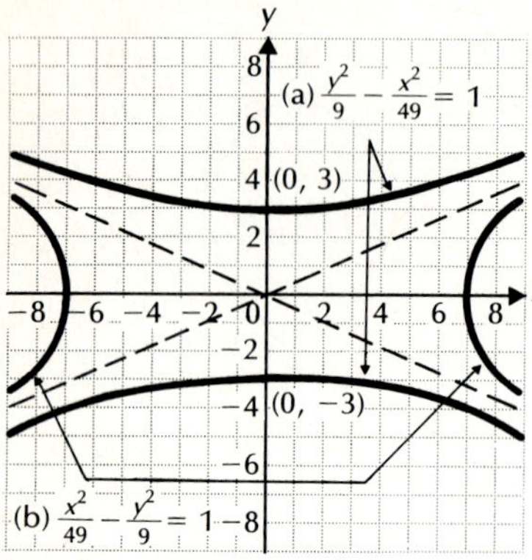

5.

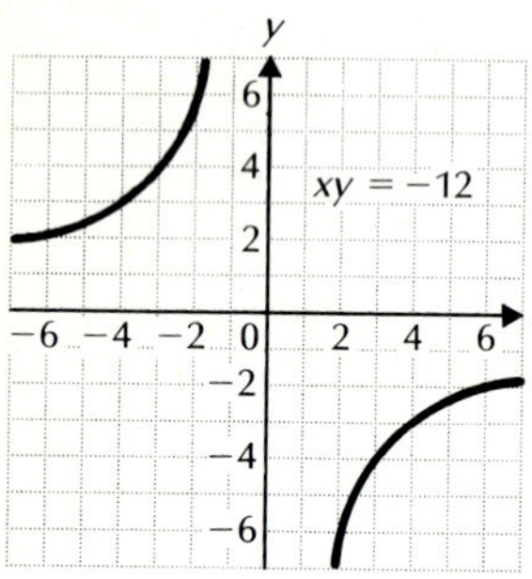

6.

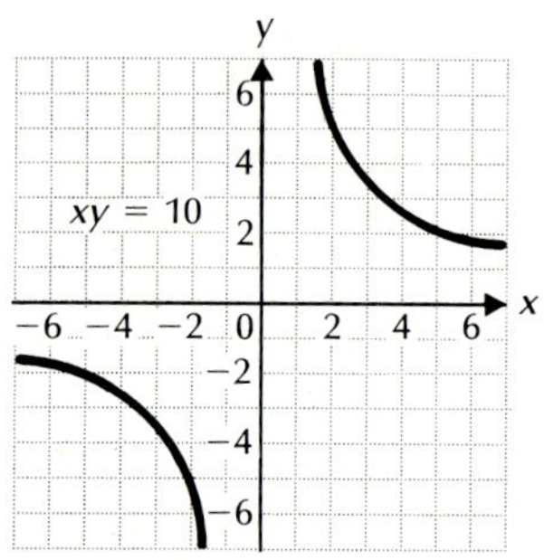

13.

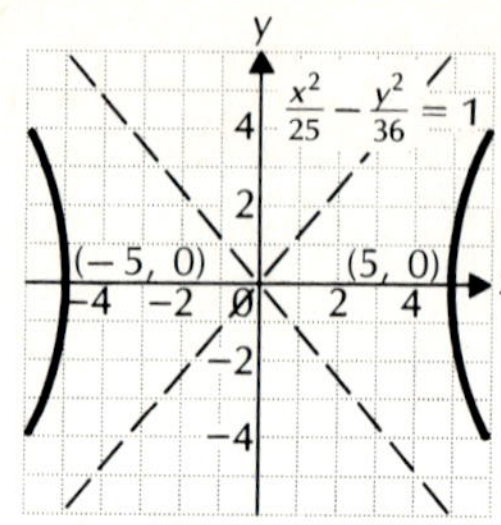

15.

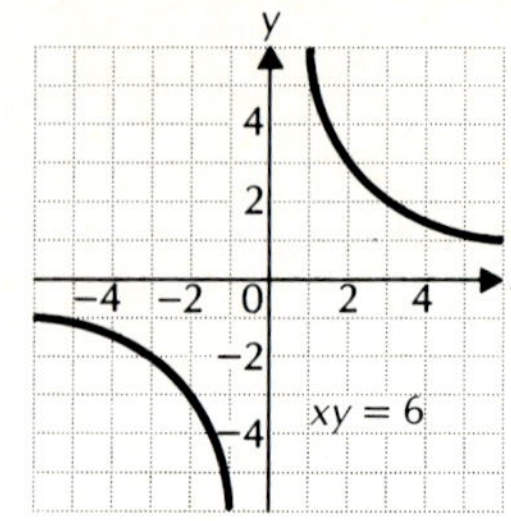

17.

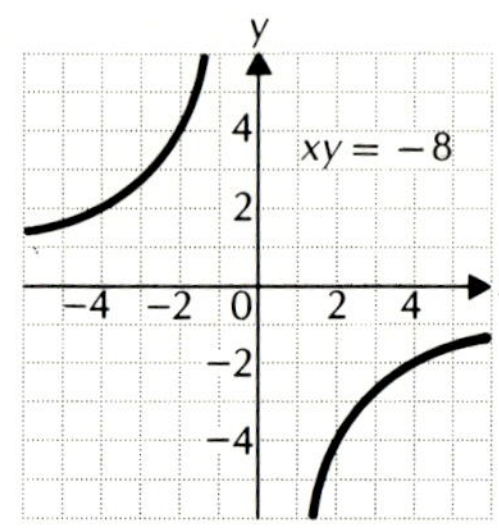

19. 17 oz

21.

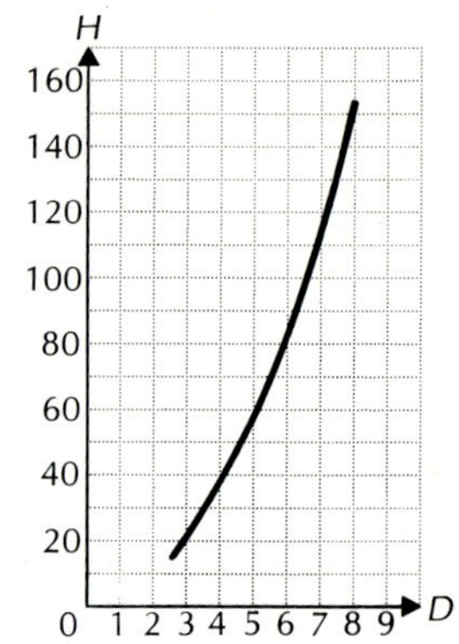

23. **(a)** Circle; **(b)** parabola; **(c)** hyperbola; **(d)** parabola; **(e)** ellipse; **(f)** circle.

Exercise Set 10.3, pp. 364–365

1.

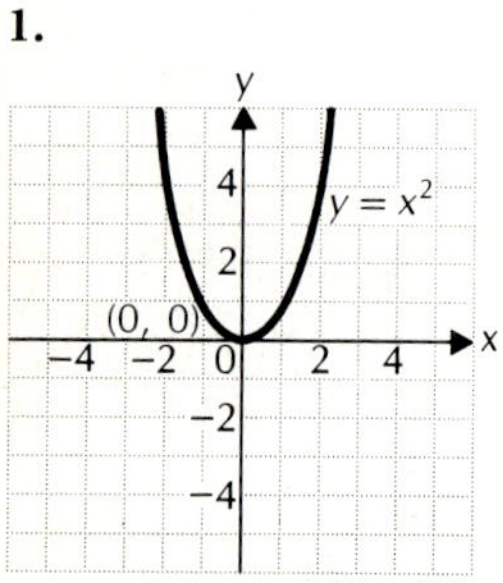

3.

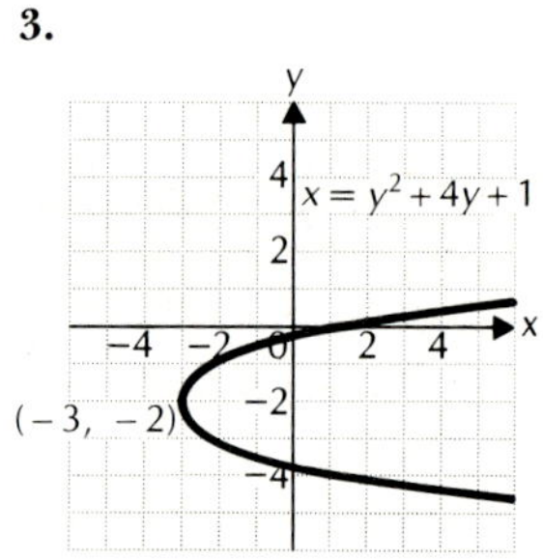

5.

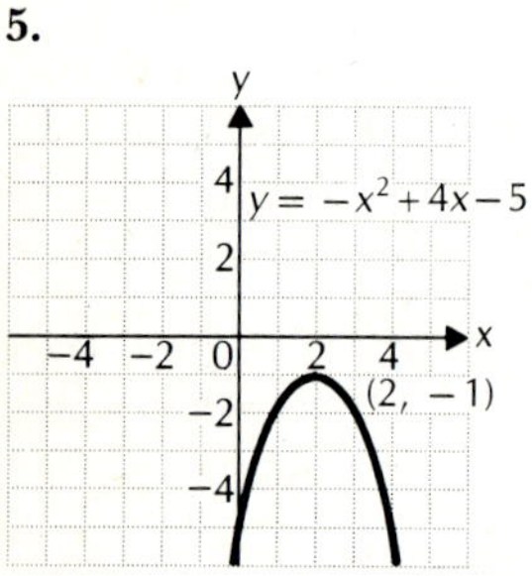

7.

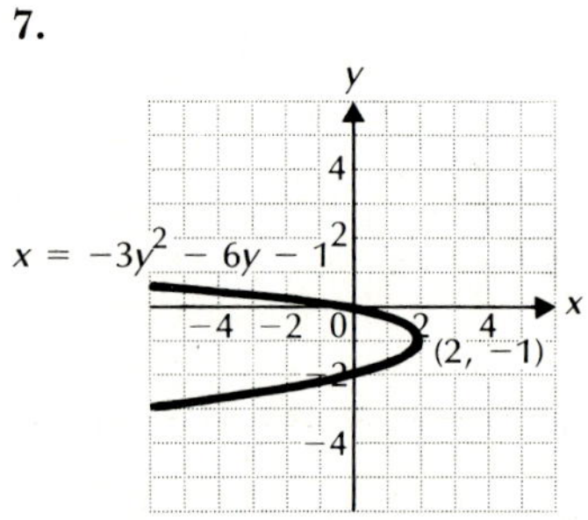

9.

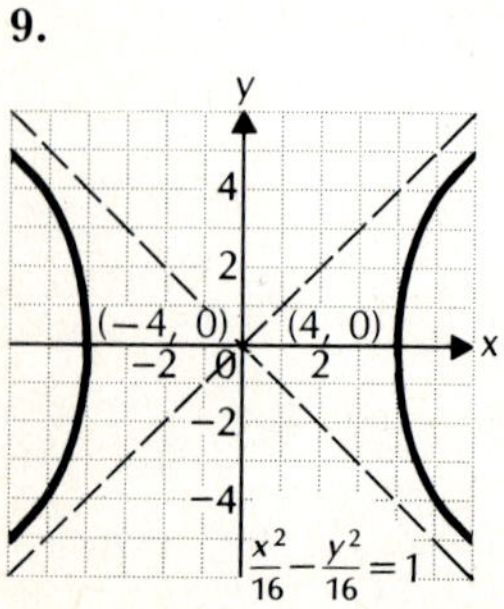

11.

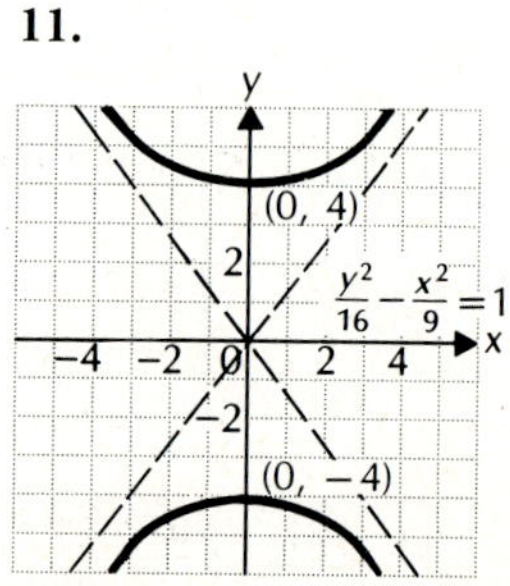

Margin Exercises, Section 10.4

1. (4, 3), (−3, −4) **2.** (4, 7), (−1, 2) **3.** $(-\frac{5}{7}, \frac{22}{7})$, (1, −2) **4.** 11 and 7 **5.** 12 m by 5 m

Exercise Set 10.4, pp. 368–369

1. (−4, −3), (3, 4) **3.** (0, 2), (3, 0) **5.** (−2, 1)
7. $\left(\frac{5+\sqrt{70}}{3}, \frac{-1+\sqrt{70}}{3}\right), \left(\frac{5-\sqrt{70}}{3}, \frac{-1-\sqrt{70}}{3}\right)$
9. $(3, 2), (4, \frac{3}{2})$ **11.** $(\frac{7}{3}, \frac{1}{3})$, (1, −1)
13. $(1, 4), (\frac{11}{4}, -\frac{5}{4})$ **15.** 3, 9 **17.** 6 cm, 8 cm
19. 4 in., 5 in. **21.** $(x + \frac{5}{13})^2 + (y - \frac{32}{13})^2 = \frac{5365}{169}$
23. $(x + 2)^2 + (y - 1)^2 = 4$

Margin Exercises, Section 10.5

1. (2, 3), (2, −3), (−2, 3), (−2, −3)
2. (3, 2), (−3, −2), (2, 3), (−2, −3)
3. Length is 2 in.; width is 1 in.

Exercise Set 10.5, pp. 371–372

1. (4, −3), (4, 3), (−5, 0) **3.** (−3, 0), (3, 0)
5. (−4, −3), (−3, −4), (3, 4), (4, 3)
7. No real solution
9. $(-\sqrt{2}, -\sqrt{14}), (-\sqrt{2}, \sqrt{14}), (\sqrt{2}, -\sqrt{14}), (\sqrt{2}, \sqrt{14})$
11. (−2, −1), (−1, −2), (1, 2), (2, 1)
13. (−3, −2), (−2, −3), (2, 3), (3, 2)
15. $\left(\frac{5-9\sqrt{15}}{20}, \frac{-45+3\sqrt{15}}{20}\right), \left(\frac{5+9\sqrt{15}}{20}, \frac{-45-3\sqrt{15}}{20}\right)$
17. 12, 13 and −12, −13 **19.** $l = \sqrt{3}$ m; $w = 1$ m
21. 16 ft, 24 ft **23.** $\pm\frac{1}{2}, \pm\frac{1}{4}$ **25.** 10 in. × 7 in. × 5 in.

Margin Exercises, Section 10.6

1. $\{x \mid x < -3 \text{ or } x > 1\}$ **2.** $\{x \mid -3 \leqslant x \leqslant 1\}$
3. $\{x \mid x < -4 \text{ or } x > 1\}$ **4.** $\{x \mid -4 \leqslant x \leqslant 1\}$
5. $\{x \mid x < -1 \text{ or } 0 < x < 1\}$ **6.** $\{x \mid 2 < x \leqslant \frac{7}{2}\}$
7. $\{x \mid x < 5 \text{ or } x > 10\}$

Exercise Set 10.6, p. 375

1. $\{x \mid x < -3 \text{ or } x > 5\}$ **3.** $\{x \mid -1 \leqslant x \leqslant 2\}$
5. $\{x \mid -1 < x < 2\}$ **7.** $\{x \mid x \leqslant -3 \text{ or } x \geqslant 3\}$
9. All real numbers **11.** $\{x \mid 2 < x < 4\}$
13. $\{x \mid x < -2 \text{ or } 0 < x < 2\}$
15. $\{x \mid -3 < x < -1 \text{ or } x > 2\}$
17. $\{x \mid x < -3 \text{ or } -2 < x < 1\}$ **19.** $\{x \mid x < 4\}$
21. $\{x \mid x < -1 \text{ or } x > 3\}$ **23.** $\{x \mid -\frac{2}{3} \leqslant x < 3\}$
25. $\{x \mid 2 < x < \frac{5}{2}\}$ **27.** $\{x \mid x < -1 \text{ or } 2 < x < 5\}$
29. $\{x \mid x \leqslant 0 \text{ or } x > 2\}$ **31.** $\{x \mid x > 0\}$
33. $\{x \mid x < -4 \text{ or } 1 < x < 3\}$ **35.** $\{x \mid 0 < x < \frac{1}{2}\}$
37. $\{x \mid 1 - \sqrt{3} \leqslant x \leqslant 1 + \sqrt{3}\}$

Summary and Review: Chapter 10, p. 376

1. [10.1] 4 **2.** [10.1] 5 **3.** [10.1] $\sqrt{130}$
4. [10.1] $\sqrt{9 + 4a^2}$ **5.** [10.2] $(-2, 3)$, $\sqrt{2}$
6. [10.2] $(5, 0)$, 7 **7.** [10.2] $(-3, 1)$, 3
8. [10.2] $(-4, 3)$, $\sqrt{35}$
9. [10.2] Circle with center $(0, 0)$ and radius 5
10. [10.2] Ellipse with vertices $(\sqrt{2}, 0)$, $(-\sqrt{2}, 0)$, $(0, 3)$, $(0, -3)$
11. [10.3]

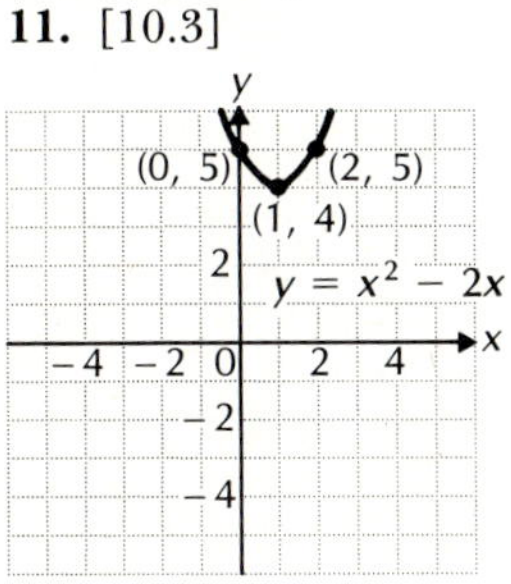

12. [10.3]

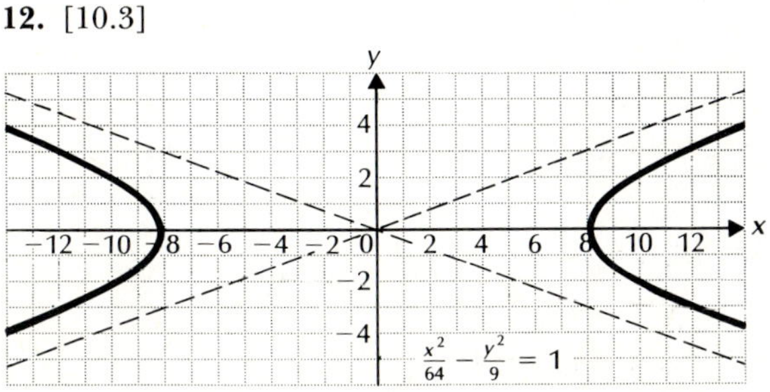

13. [10.4] $(7, 4)$ **14.** [10.4] $(2, 2)$, $(\frac{32}{9}, -\frac{10}{9})$
15. [10.4] 12 m by 7 m **16.** [10.4] 4, 8
17. [10.5] $(6, 8)$, $(-6, -8)$, $(-6, 8)$, $(6, -8)$
18. [10.5] $(-2, -2)$, $(2, 2)$, $(2\sqrt{2}, \sqrt{2})$, $(-2\sqrt{2}, \sqrt{2})$
19. [10.6] $\{x \mid -1 < x < 7\}$
20. [10.6] $\{x \mid -3 < x < 5\}$
21. $(2, i)$, $(-2, i)$, $(2, -i)$, $(-2, -i)$ **22.** $(0, 6)$, $(0, -6)$

CHAPTER 11

Margin Exercises, Section 11.1

1. **(a)** 1, 3, 5; **(b)** 67 **2.** **(a)** $1, -\frac{1}{2}, \frac{1}{3}, -\frac{1}{4}$; **(b)** $-\frac{1}{48}$
3. $2n$ **4.** n **5.** n^3 **6.** $\frac{x^n}{n}$ **7.** 2^{n-1} **8.** **(a)** 0; **(b)** 4
9. $3 + 2\frac{1}{2} + 2\frac{1}{3}$ **10.** $5^0 + 5^1 + 5^2 + 5^3 + 5^4$
11. $8^3 + 9^3 + 10^3 + 11^3$ **12.** $\sum_{k=1}^{5} 2k$ **13.** $\sum_{k=1}^{4} k^3$
14. $\sum_{k=1}^{6} \frac{x^k}{k}$ **15.** $\sum_{k=1}^{4} (k + 1)$

Exercise Set 11.1, pp. 382–383

1. 4, 7, 10, 13; 31; 46 **3.** $\frac{1}{2}, \frac{2}{3}, \frac{3}{4}, \frac{4}{5}; \frac{10}{11}; \frac{15}{16}$
5. $-1, 0, 3, 8; 80; 195$ **7.** $2, 2\frac{1}{2}, 3\frac{1}{3}, 4\frac{1}{4}; 10\frac{1}{10}; 15\frac{1}{15}$
9. $2n - 1$ **11.** $\frac{n+1}{n+2}$ **13.** $3^{n/2}$ **15.** $-(3n - 2)$
17. 28 **19.** 30 **21.** $\frac{1}{2} + \frac{1}{4} + \frac{1}{6} + \frac{1}{8} + \frac{1}{10} = \frac{137}{120}$
23. $2^0 + 2^1 + 2^2 + 2^3 + 2^4 + 2^5 = 63$
25. $\log 7 + \log 8 + \log 9 + \log 10$ **27.** $\sum_{k=1}^{6} \frac{k}{k+1}$
29. $\sum_{k=1}^{6} (-1)^k 2^k$ **31.** $\sum_{k=2}^{n} (-1)^k k^2$
33. \$7.67 **35.** $\frac{3}{2}, \frac{3}{2}, \frac{3}{2}, \frac{3}{2}, \frac{3}{2}; \frac{15}{2}$
37. 0, 0.693, 1.792, 3.178, 4.787; 10.45
39. 2, 2.25, 2.37037, 2.441406, 2.488320, 2.521626
41. $1 - \frac{1}{n+1}$

Margin Exercises, Section 11.2

1. $a_1 = 3.1$, $d = 0.8$ **2.** 50 **3.** 75
4. $a_1 = 7$, $d = 12$, 7, 19, 31, 43, . . . **5.** 20,100 **6.** 225 11
7. 5200

Exercise Set 11.2, p. 388

1. $a_1 = 2$, $d = 5$ **3.** $a_1 = 7$, $d = -4$ **5.** $a_1 = \frac{3}{2}$, $d = \frac{3}{4}$
7. 46 **9.** -41 **11.** 27th **13.** 102nd **15.** 101 **17.** 5
19. 28 **21.** 8, 5, 2, -1, -4; $d = -3$ **23.** 670
25. 2550 **27.** 735 **29.** 990 **31.** 3 **33.** 1275 **35.** n^2
37. $a_1 = p - 5q$, $d = 3p + 2q$ **39.** \$51,679.65

Margin Exercises, Section 11.3

1. 2^8, or 256 **2.** 11,718 **3.** $\frac{341}{256}$ **4.** $\frac{3}{2}$ **5.** $3\frac{1}{5}$
6. \$476.41

Exercise Set 11.3, pp. 394–395

1. 2 **3.** -1 **5.** $-\frac{1}{2}$ **7.** $\frac{1}{5}$ **9.** $\frac{1}{x}$ **11.** $a_n = 3^{n-1}$
13. $a_n = (-1)^{n-1}$ **15.** $a_n = \frac{1}{x^n}$ **17.** 64 **19.** 162
21. 648 **23.** $10\frac{1}{2}$ **25.** $\frac{33{,}333}{5000}$ **27.** $\frac{1 - x^{20}}{1 - x^2}$ **29.** $\frac{49}{4}$

31. 48 **33.** -4 **35.** 10 **37.** $5\frac{1}{3}$ ft **39.** 24 years
41. $645,826.93 **43.** 10,485.76 in. **45.** 3892 ft **47.** 10
49. $\frac{2}{3}$ **51.** $\frac{4}{33}$ **53.** $S_1 = 2$, $S_2 = 2\frac{1}{2}$, $S_3 = 2\frac{2}{3}$, $S_4 = 2\frac{17}{24}$, $S_5 = 2\frac{43}{60}$, $S_6 = 2\frac{517}{720}$, $S = 2\frac{3}{4}$

Margin Exercises, Section 11.4

1. $\frac{1}{2}$ **2.** $\frac{1}{6}$ **3.** (a) 6; (b) 27 **4.** $\frac{15}{91}$

Exercise Set 11.4, pp. 398–399

1. $\frac{2}{7}$ **3.** 0 **5.** $\frac{11}{4165}$ **7.** $\frac{15}{1001}$ **9.** $\frac{5}{36}$ **11.** $\frac{1}{36}$ **13.** $\frac{7}{1938}$
15. $\frac{1}{221}$ **17.** $\frac{9}{19}$ **19.** $\frac{18}{19}$ **21.** $\frac{1}{38}$ **23.** $\frac{9}{19}$

Margin Exercises, Section 11.5

1. 1 6 15 20 15 6 1; 1 7 21 35 35 21 7 1; 1 8 28 56 70 56 28 8 1
2. $a^9 + 9a^8b + 36a^7b^2 + 84a^6b^3 + 126a^5b^4 + 126a^4b^5 + 84a^3b^6 + 36a^2b^7 + 9ab^8 + b^9$
3. $x^4 - 4x^3y + 6x^2y^2 - 4xy^3 + y^4$
4. $128t^7 + 448t^5 + 672t^3 + 560t + 280t^{-1} + 84t^{-3} + 14t^{-5} + t^{-7}$ **5.** $x^9 - 9x^8y + 36x^7y^2 - 84x^6y^3 + 126x^5y^4 - 126x^4y^5 + 84x^3y^6 - 36x^2y^7 + 9xy^8 - y^9$
6. 40,320 **7.** 362,880 **8.** 126 **9.** 28 **10.** 7
11. 1 **12.** 56 **13.** 7 **14.** 9 **15.** 1
16. $x^5 + 25x^4b + 250x^3b^2 + 1250x^2b^3 + 3125xb^4 + 3125b^5$
17. $x^4 - 4x^3y^2 + 6x^2y^4 - 4xy^6 + y^8$ **18.** $181{,}440a^4b^3$
19. $103{,}680x^2$

Exercise Set 11.5, p. 406

1. $m^5 + 5m^4n + 10m^3n^2 + 10m^2n^3 + 5mn^4 + n^5$
3. $x^6 - 6x^5y + 15x^4y^2 - 20x^3y^3 + 15x^2y^4 - 6xy^5 + y^6$
5. $x^{10} - 15x^8y + 90x^6y^2 - 270x^4y^3 + 405x^2y^4 - 243y^5$
7. $729c^6 - 1458c^5d + 1215c^4d^2 - 540c^3d^3 + 135c^2d^4 - 18cd^5 + d^6$ **9.** $x^3 - 3x^2y + 3xy^2 - y^3$
11. $x^{-7} + 7x^{-6}y + 21x^{-5}y^2 + 35x^{-4}y^3 + 35x^{-3}y^4 + 21x^{-2}y^5 + 7x^{-1}y^6 + y^7$
13. $a^9 - 18a^7 + 144a^5 - 672a^3 + 2016a - 4032a^{-1} + 5376a^{-3} - 4608a^{-5} + 2304a^{-7} - 512a^{-9}$
15. $1 - n + \binom{n}{2} - \binom{n}{3} + \cdots + \binom{n}{n}(-1)^n$
17. $9 - 12\sqrt{3}t + 18t^2 - 4\sqrt{3}t^3 + t^4$ **19.** 720 **21.** 1
23. 1 **25.** 70 **27.** $15a^4b^2$ **29.** $-745{,}472a^3$
31. $1120x^{12}y^2$ **33.** $-1{,}959{,}552u^5v^{10}$
35. $x^{-8} + 4x^{-4} + 6 + 4x^4 + x^8$ **37.** $140\sqrt{2}$
39. 4368, 4368

Summary and Review: Chapter 11, pp. 407–408

1. [11.1] 1, 5, 9, 13; 29; 45 **2.** [11.1] 0, $\frac{1}{5}$, $\frac{1}{5}$, $\frac{3}{17}$; $\frac{7}{65}$; $\frac{11}{145}$
3. [11.1] $a_n = -2n$ **4.** [11.1] $a_n = n^2$
5. [11.1] $-2 + 4 + -8 + 16 + -32$
6. [11.1] $-3 + -5 + -7 + -9 + -11 + -13$
7. [11.1] $\sum_{k=1}^{5} 4k$ **8.** [11.1] $\sum_{k=1}^{5} \frac{1}{(-2)^k}$ **9.** [11.2] 85
10. [11.2] $2\frac{2}{3}$ **11.** [11.2] $d = 1.25$, $a_1 = 11.25$
12. [11.2] -544 **13.** [11.2] 8580 **14.** [11.2] 63
15. [11.2] 864 **16.** [11.3] $1024\sqrt{2}$ **17.** [11.3] $\frac{2}{3}$
18. [11.3] $a_n = 2(-1)^n$ **19.** [11.3] $4\frac{20}{27}$
20. [11.3] $-10\frac{4095}{6144}$ **21.** [11.3] $-4095x$ **22.** [11.3] 12
23. [11.3] 0.05 **24.** [11.3] 2446 **25.** [11.3] $9.27 per hour
26. [11.4] $\frac{1}{9}$ **27.** [11.4] $\frac{961}{133{,}225}$ **28.** [11.5] $21a^2b^5$
29. [11.5] $x^4 - 8x^3y + 24x^2y^2 - 32xy^3 + 16y^4$ **30.** $\frac{419}{990}$
31. (a) $\frac{1}{4}$; (b) $\frac{1}{2}$; (c) $\frac{1}{2}$

CUMULATIVE REVIEW, CHAPTERS 1–11, pp. 408–409

1. $\frac{17}{14}$ **2.** $3 \pm 3\sqrt{2}$ **3.** -1 or 5
4. $\{x \mid -\frac{6}{5} \leqslant x \leqslant \frac{2}{5}$ or $\frac{6}{5} \leqslant x \leqslant \frac{14}{5}\}$ **5.** 10 **6.** 0, 8, or 12
7. 69 **8.** $\pm\frac{9}{4}$ **9.** 0 or 1 **10.** $\frac{20}{3}$
11. $8m^3 + 12m^2n + 6mn^2 + n^3$ **12.** -3
13. $12\sqrt{2} + 6\sqrt{5} + 24\sqrt{3}$ **14.** $51 + 33i$
15. $-2a^4b^3 + 2a^3b^4 - 9$ **16.** $4(x - 7)(x + 7)$
17. $(4x + 1)(3x + 4)$ **18.** $8x(1 - 2x)(1 + 2x + 4x^2)$
19. $(25x - 9)(x + 1)$

20.

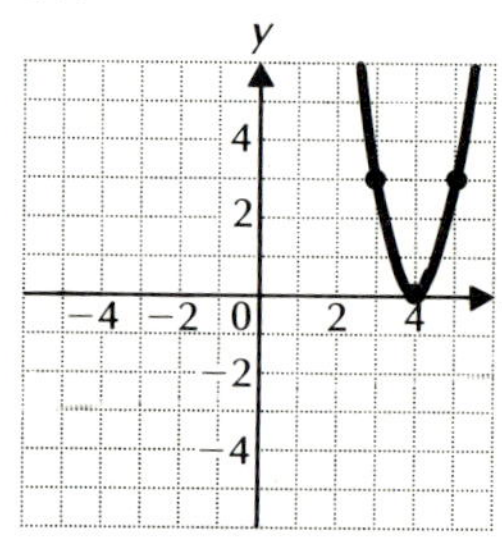

21.

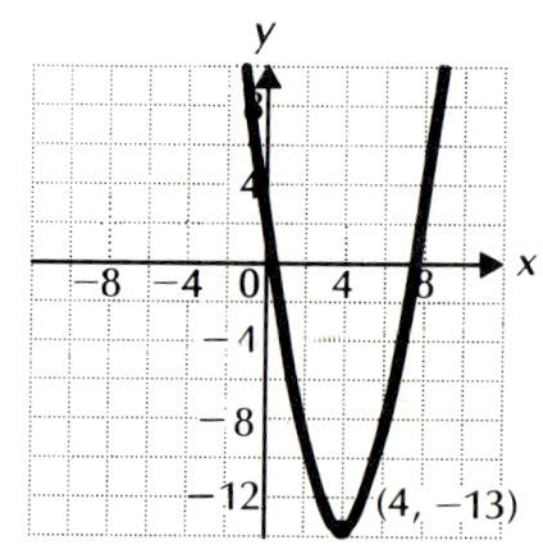

22.

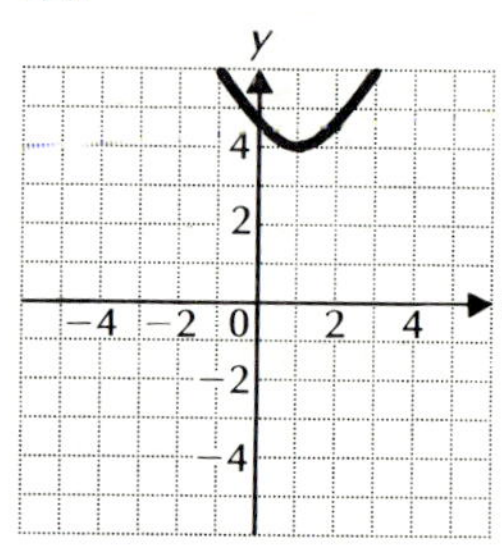

23.

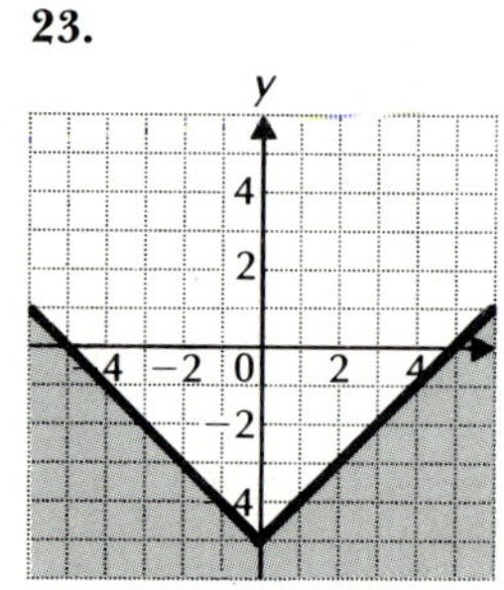

24.

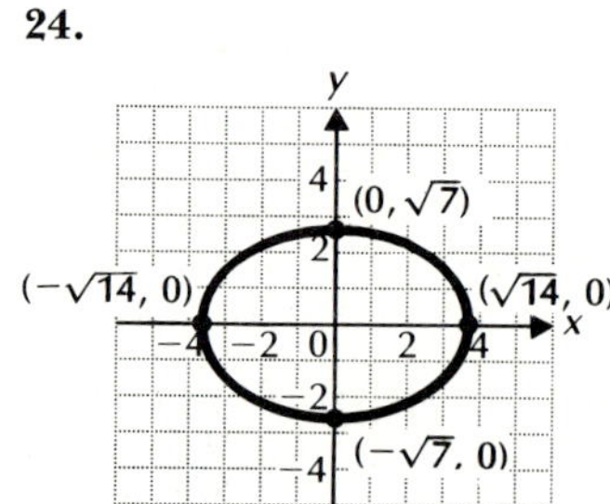

25.

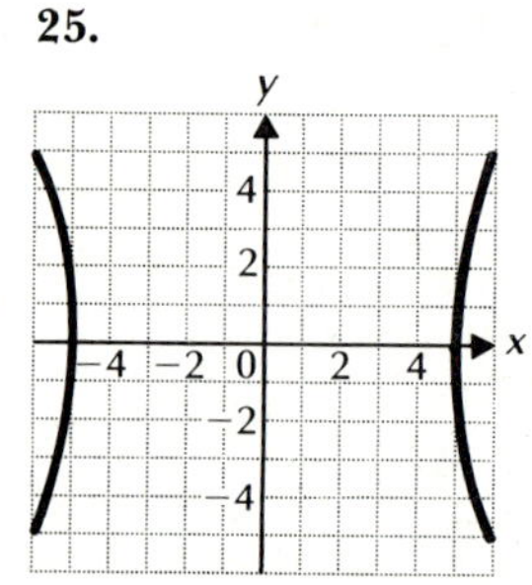

26. 4.7505 **27.** 8.01 **28.** 8.35 **29.** 8.3802 **30.** $5\sqrt{2}$
31. 7 **32.** $\sqrt{185}$ **33.** $2\sqrt{61}$ **34.** $c(-4, 2)$; $r = \sqrt{5}$
35. $c(1, -2)$, $V(6, -2)$ and $(-4, -2)$;
asymptotes: $2x + 5y + 8 = 0$ and $2x - 5y - 12 = 0$
36. (23, 22) **37.** $(\pm 9, \pm 6)$ **38.** 81 **39.** -121
40. 875 **41.** $16(\frac{1}{4})^{n-1}$ **42.** 40.96 **43.** $74.88671875x$
44. 15 cm × 17 cm **45.** 8 and 12 **46.** $13.15
47. $168.95 **48.** $\frac{1}{17}$ or 5.88%

INDEX

Miscellaneous Formulas of General Interest

Compound interest, 343
Cost of operating an automobile, 127
Distance an object falls, 289
Earned run average, 217
Games played in a league, 127
Height, estimating (anthropology), 82
Horizon, sighting to, 262
Horsepower of an engine, 365
Number of objects in a pile, 123
Parking, 237
Pendulum, period of, 244
Polygon, number of diagonals, 133
Population growth, 344
Road pavement messages, 256
Simple interest, 43
Speed of a skidding car, 241
Telephone calls between cities (gravity model), 299
Temperature conversion, Celsius to Fahrenheit, 164
Walking speed, 346
Wind chill temperature, 241

A
Abscissa, 51
Absolute value, 164
  and distance, 5
  equations with, 166
  and graphing, 165
  inequalities with, 166
  properties, 165
  and radical expressions, 234, 235, 236
Addition
  associative law, 23
  commutative law, 23
  of complex numbers, 263
  of exponents, 32
  of polynomials, 124
  of radical expressions, 244
  of rational expressions, 190–194
  of real numbers, 5
  signs, rules of, 5, 6
Addition method for systems of equations, 90, 101, 369
Addition principle
  for equations, 10
  for inequalities, 155
Additive inverse, 4
  and multiplying by -1, 25
  of polynomials, 125
  and subtraction, 5
Algebra notation, 27
Algebraic expressions, least common multiple, 189
Algorithm for solving systems of equations, 93
Annuity, 394
Antilogarithms, 333
  on a calculator, 333
  by interpolation, 340
  using tables, 335, 336
Applications, *see* Applied problems; Formulas; Miscellaneous formulas of general interest
Applied problems, 15–20, 35, 40, 79–81, 94–99, 105–107, 114–116, 148–150, 158–160, 175–177, 203–210, 222–224, 225, 261, 274–276, 282–284, 289, 296–298, 311–313, 343–345, 367, 370, 386–388, 392, 393
  break-even analysis, 114
  business and economics, 114–116
  distance an object falls, 289
  heights, estimating, 82
  interest, 43, 96, 343
  mixture, 95
  motion, 98, 207–210, 283
  population growth, 344
  supply and demand, 115
  work, 204, 282
Approximately equal to ($\approx$), 40
Approximating
  using scientific notation, 39
  solutions of quadratic equations, 279
  square roots, 233, 234, 239
Area, of a rectangle, 275
Arithmetic sequences, 383
  common difference, 383
  *n*th term, 384
  sum, 385
Ascending order, 122
Associative laws, 23
Asymptote
  exponential function, 319
  hyperbola, 361
Average speed, 211
Axes, 50
  x-, 52
  y-, 52
Axis of a hyperbola, 360

B
Base, 30
  of an exponential function, 318
  of a logarithmic function, 324
BASIC, 27
Binomial expansion
  using factorial notation, 403–406
  using Pascal's triangle, 399–401
  using preceding coefficient, 401–403
Binomial theorem, 401, 405

Binomials, 122
multiplication of (FOIL), 128
cubing, 129
products of sums, differences, 128
squares of, 128
Braces, 7, 26
Brackets, 26
Branches of a hyperbola, 364
Break-even analysis, 114
Break-even point, 114

C

Calculator
common logarithms on, 332
natural logarithms on, 333
square roots on, 233, 239
Catenary, 359
Center of a hyperbola, 360
Changing the sign, 5, 25, 125
Checking solutions
fractional equations, 200, 279, 281
inequalities, 156
logarithmic equations, 326
radical equations, 257
reducible to quadratic equations, 292
Circle, 354
equation
centered at origin, 355
standard form, 355
graph of, 355
unit, 359
Clearing of decimals, 12, 92
Clearing of fractions, 12, 92, 199
Coefficients, 122
Collecting like terms, 25, 124
Columns of a matrix, 109
Combining like terms, 25, 124
Common denominator, 191, 192
Common difference, 383
Common factor, 130
Common logarithms, 332
antilogarithm of, 333
on a calculator, 332
converting to natural logarithms, 336
using a table, 334
Common multiple, least, 188, 189
Common ratio, 389
Commutative laws, 23
Completing the square, 138
factoring by, 139
in graphing, 308, 355, 359
Complex conjugates, 264
Complex fractional expressions, 195
Complex numbers, 263
addition, 263
conjugate, 264
division, 265
multiplication, 264
as solutions of equations, 265, 273, 278
subtraction, 263
Composite function, 79
Compound inequalities, 161
Compound interest, 343
Conic sections, 354–357, 359–364
circle, 354
ellipse, 356
hyperbola, 359, 360
parabola, 359
Conjugate of complex numbers, 264
in division, 265
Conjunction, 161
Consistent system of equations, 87, 111
Constant, 42
variation, 222, 224
Constraints, 174
Converting
common logarithms to natural logarithms, 336
from decimal to scientific notation, 37
from exponential to logarithmic equations, 325
from logarithmic to exponential equations, 325
from scientific to decimal notation, 38
Convex polygon, 174
Coordinates
on a line, 165
in the plane, 51
Correspondence, *see* Function
Cost, 114
Counting principle, fundamental, 397
Cube roots, 235
Cubes, factoring sum and difference of, 141
Cubic polynomial, 122
Cubing binomials, 129

D

Data point, 80, 313
Decimal notation, 2, 3. *See also* Decimals.
converting to scientific notation, 37
Decimals, *see also* Decimal notation.
clearing of, 12, 92
ending, 239
repeating, 3, 239
terminating, 239
Degree of a polynomial, 122
Degree of a term, 122
Demand, 115
Denominators
least common, 191, 192
rationalizing, 247, 250
Dependent system of equations, 87, 112
Dependent variable, 75
Descending order, 122
Difference
common, 383
tabular, 339
Difference of cubes, factoring, 141
Difference of logarithms, 330
Difference of squares, factoring, 137
Direct variation, 221, 295
and problem solving, 222–224
Discriminant, 285
and intercepts, 310
Disjunction, 162
Distance
on a line, 165
between points in the plane, 353
Distance formula, 353
Distributive laws, 24
and factoring, 25
Division
of complex numbers, 265
with exponential notation, 33
by a monomial, 211
of polynomials, 211–216
with radical expressions, 242
of rational expressions, 186
of real numbers, 6
and reciprocals, 7, 186
remainder, 212
with scientific notation, 38
signs, rules of, 6
synthetic, 215
Domain of a function, 71

E

e, 333
Earned run average, 217
Edge of a half-plane, 171
Eliminating a variable, 90
Elimination method, 90
Ellipse, 356
equation of, 356, 358
foci, 356
graph of, 357
intercepts, 357
Empty set, 161
Ending decimals, 239
Equally likely outcomes, 396
Equations, 10. *See also* Formulas.
with absolute value, 166
of a circle, 355
consistent system of, 87, 111
dependent system of, 87, 112
of an ellipse, 356, 358
exponential, 325, 342
fractional, 199, 279, 281
graphs of
linear, 52, 55, 61
quadratic, 53, 299–304, 308–310
of a hyperbola, 360
inconsistent system of, 87
independent system of, 112
linear, 58, 59
logarithmic, 325, 342
with a missing variable, 58
with no solution, 13
of a parabola, 301, 302, 303, 359
containing parentheses, 13
point–slope, 67
polynomial, 145, 272
quadratic, 272
radical, 257
with any real number a solution, 13
reducible to quadratic, 292
slope–intercept, 58
solutions of, 10, 51
solving, *see* Solving equations
systems of, 86
two-point, 67
in two variables, 51
writing from solutions, 286
Equilibrium point, 115
Equivalent expressions, 21, 24
Estimating using scientific notation, 39
Evaluating expressions
by nested evaluation, 131
by substitution, 123
Even root, 235
Event, 396
Expansion, binomial, *see* Binomial expansion
Experimental probability, 395
Exponential equations, 342
converting to logarithmic equations, 325
solving, 342
Exponential functions, 318
base, 318
graphs of, 318–320
asymptotes, 319
inverse of, 320
Exponential notation, 30. *See also* Exponents.
base, 30
and division, 33
and multiplication, 32
and raising a power to a power, 34
and raising a product to a power, 34
and raising a quotient to a power, 35
and translating, 35

Exponentiation, 333
Exponents, 30
  addition of, 32
  fractional, 252–254
  integers as, 30
  irrational, 318
  laws of, 254
  multiplying, 34, 35
  negative, 30, 253
  of one, 30
  properties of, 31–35
  rational, 252–254
  subtracting, 33
  of zero, 30
Expressions
  equivalent, 21, 24
  factoring, 25
  fractional, 182
    complex, 195
  polynomial, 122
  radical, 232
  rational, 182

F

Factorial notation, 403
Factoring, 25, 130
  common factors, 130
  by completing the square, 139
  difference of cubes, 141
  difference of squares, 137
  by grouping, 131
  polynomials, general strategy, 143
  the radicand, 237
  solving equations by, 146, 273, 274
  sum of cubes, 141
  trinomial squares, 134
  trinomials, 134, 135
Familiarization, 15
First coordinate, 51
Fitting quadratic functions to data, 313
Five-step problem-solving strategy, 15
Fixed costs, 114
Foci of an ellipse, 356
FOIL, 128
Formulas, 42. *See also* Miscellaneous formulas of general interest.
  distance, 353
  fractional, 217
  and functions, 74
  for inverse functions, 321–323
  midpoint, 66, 354
  nth term
    arithmetic sequence, 384
    geometric sequence, 389
  quadratic, 277
  solving, 42, 217–220, 288, 289
  sum
    arithmetic sequence, 385
    geometric sequence, 390, 392
  in translating, 44
Fractional equations, 199
  formulas, 217
  quadratic, 279, 281
Fractional exponents, 252–254. *See also* Rational expressions.
Fractional expressions, 182
  complex, 195
  formulas, 217
Fractional notation, 2, 3, 239
Fractions, clearing of, 12, 92, 199
Function, 49, 71
  absolute value, 164
  composite, 79
  dependent variable, 75
  domain, 71
  exponential, 318
  and formulas, 74
  graphs of, 74
    vertical-line test, 76
  independent variable, 75
  input, 72
  inverse, 321, 327
  linear, 76
  logarithm, 324
  notation, 72
  output, 72
  polynomial, 122, 123
  quadratic, 299
  range, 71
  rational, 182
  sequence, 380
  square root, 233
  and tables, 73
  translations, 304–306
  of two variables, 173
Fundamental counting principle, 397

G

General strategy for factoring, 143
General term of a sequence, 380
Geometric sequences, 389
  common ratio, 389
  nth term, 389
  sum
    infinite, 392
    n terms, 390
Grade of a road, 70
Graphs
  of absolute value function, 165
  of circles, 355
  and completing the square, 308
  of ellipses, 357
  of equations, 52, 53
    systems of, 87
  of exponential functions, 318–320
  of functions, 74
    vertical-line test, 76
  horizontal line, 59
  of hyperbolas, 361–364
  of inequalities, 155
    in one variable, 155
    systems of, 172
    in two variables (linear), 169
  using intercepts, 61
  of inverse functions, 320
  of linear equations
    parallel to an axis, 58, 59
    procedure, 62
    slope, 56
    of the type $y = mx$, 52
    of the type $y = mx + b$, 55
    y-intercept, 56
  of logarithm functions, 324
  of parabolas, 300, 359
  of quadratic functions, 53, 299–304
    x-intercepts, 309
  of quadratic inequalities, 372
  solutions by, 87, 372
  translations, 304–306
  vertical line, 59
Gravity model, 299
Greater than (>), 154
Greater than or equal to ($\geqslant$), 154
Grouping, factoring by, 131

H

Half-plane, 171
Harmonic mean, 221
Horizon, sighting to, 262
Horizontal line, 59
  slope, 64
Horsepower, 365
Hyperbola, 359, 360
  asymptotes, 361
  axis, 360
  branches, 364
  center, 360
  equation of
    nonstandard form, 363
    standard form, 360
  graphs of, 361–364
  intercepts, 362
  vertices, 361
Hypotenuse, 259

I

i, 263
Identity, 14
Identity property
  of one, 21
  of zero, 21
Imaginary numbers, 263
Inconsistent system of equations, 87
Independent system of equations, 112
Independent variable, 75
Index of a radical expression, 235
Inequalities, 154
  with absolute value, 166
  addition principle for, 155
  with "and," 161
  checking solutions, 156
  compound, 161
  conjunction, 161
  disjunction, 162
  graphs of
    in one variable, 155
    quadratic, 372
    systems of, 172
    in two variables, 169
  linear, 169
  multiplication principle for, 156
  with "or," 163
  polynomial, 373
  quadratic, 372
  rational, 374
  solution, 154, 168
  solution set, 154
  solving, 154
    in one variable, 155–157, 162, 163, 166, 372–375
    systems of, 172
    in two variables, 168–171
  systems of, 171
Infinite sequence, 391
Infinite series, 391
Input, 72
Integers, 2
  as exponents, 30
Intercepts
  of ellipses, 357
  graphing using, 61
  of hyperbolas, 362
  x, 61, 62, 310
  y, 56, 62
Interest problems
  compound, 343
  simple, 43, 96

Interpolation, 338
  and antilogarithms, 340
  linear, 338
  and logarithms, 339
Intersection, 161
Inverse
  additive, 4
    and multiplying by -1, 25
    and subtraction, 5
  function, 321, 327
  multiplicative, 5. *See also* Reciprocals.
Inverse variation, 224, 295
  and problem solving, 225
Irrational numbers, 239
  as exponents, 318

J

Joint variation, 296

K

kth roots, 235

L

Laws
  associative, 23
  commutative, 23
  distributive, 24
  of exponents, 254
Leading coefficient, 122
Leading term, 122
Least common denominator, 191, 192
Least common multiple, 188, 189
Legs of a right triangle, 259
Less than ($<$), 154
Less than or equal to ($\leqslant$), 154
Like radicals, 244
Like terms, 124
  collecting (or combining), 25, 124
Line of symmetry, 300, 302, 303
Linear equations, 58, 59. *See also* Linear functions.
  graphs of, 52, 55
  point–slope, 67
  slope–intercept, 58
  systems of, 86
  two-point, 67
Linear functions. *See also* Linear equations.
  graphs, 76
  maximum, minimum value, 173
  and problem solving, 79–81
Linear inequality, 169
Linear interpolation, 338
Linear polynomial, 122
Linear programming, 174
Lines
  equations of, *see* Linear equations
  horizontal, 59
  parallel, 68
  perpendicular, 69
  slope of, 56
  vertical, 59
Logarithm functions, 324. *See also* Logarithms.
  base, 324
  graphs, 324
Logarithmic equations, 342
  converting to exponential equations, 325
  solving, 326, 342
Logarithmic functions, *see* Logarithm functions
Logarithms, *see also* Logarithm functions
  anti-, 333, 334, 336
  on a calculator, 332
  common, 332
  converting common to natural, 336
  difference of, 330
  and exponents, 324
  by interpolation, 339
  natural, 333
  of powers, 329
  of products, 328
  of quotients, 330
  sum of, 328
  using tables, 334, 336

M

Main diagonal, 110
Matrix, 109
  columns, 109
  main diagonal, 110
  row-equivalent, 110
  rows, 109
  and solving systems of equations, 109
  triangular form, 110
Maximum value
  of a linear function, 173
  of a quadratic function, 303, 311
Mean, harmonic, 221
Method of elimination, 90
Midpoint formula, 66, 354
Mil, 41
Minimum value
  of a linear function, 173
  of a quadratic function, 303, 311
Missing terms in dividends, 212
Missing variable in an equation, 58
Mixture problem, 95
Monomials, 122
  as divisors, 211
Motion problems, 98, 207–210, 283
Multiple, least common, 188, 189
Multiplication
  associative law, 23
  of binomials, 128, 129
  commutative law, 23
  of complex numbers, 264
  distributive laws, 24
  with exponential notation, 32
  of exponents, 34, 35
  by -1, 25
  by one, 7, 22, 37, 183
  of polynomials, 127
  of radical expressions, 237, 238, 246
  of rational expressions, 183
  of real numbers, 6
  with scientific notation, 38
  signs, rules of, 6
  and simplifying, 185
Multiplication principle
  for equations, 10
  for inequalities, 156
Multiplicative inverse, 5. *See also* Reciprocals.

N

nth term
  arithmetic sequence, 384
  geometric sequence, 389
Natural logarithms, 333
  antilogarithms, 333, 334, 336
  on a calculator, 333
  and tables, 336
Natural numbers, 2
Nature of solutions, quadratic equation, 285
Negative exponents, 30, 253
Negative square root, 232
Nested evaluation, 131
Nonsensible replacements, 21
Notation
  algebra, 27
  BASIC, 27
  decimal, 2, 3
  exponential, 30
  factorial, 403
  fractional, 2, 3, 239
  function, 72
  percent, 3, 4
  radical, 232
  for rational numbers, 2, 239
  scientific, 37
  sequence, terms of, 380
  set, 7
  set builder, 8
  sigma, 381
Numbers
  complex, 263
  imaginary, 263
  integers, 2
  irrational, 239
  natural, 2
  rational, 2, 239
  real, 2
  whole, 2
Numerators, rationalizing, 247, 249, 251

O

Odd root, 235
One
  as an exponent, 30
  identity property of, 21
  multiplying by, 7, 22, 37
  named with fractional notation, 7, 21
Operations
  order of, 27
  row-equivalent, 110
Order of operations, 27
Order of real numbers, 154
Ordered pair, 50
Ordered triple, 100
Ordinate, 51
Origin, 50
Outcomes, 396
Output, 72

P

Pairs, ordered, 50
Parabola, 300, 359
  equations of, 301, 302, 303
  graphs of, 300, 359
  line of symmetry, 300, 302, 303, 359
  vertex, 300, 302
Parallel lines, 68
Parentheses
  equations containing, 13
  omitting, 24
  within parentheses, 26
  removing, 26
Parking, 237
Pascal's triangle, 400
Partial sums, 381
Pendulum, 244
Percent notation, 3, 4
Percent symbol, 4
Perfect squares, 234
Perpendicular lines, 69
Pi (π), 239

Plane
  axes, 50
  ordered pairs in, 50
  origin, 50
  points, 50
  quadrants, 51
Plotting points, 50
Point, data, 80, 313
Points on a plane, 50
Point–slope equation, 67
Polygon, convex, 174
Polynomial equation, 145, 272
  solving, *see* Solving equations
Polynomial inequalities, 373
Polynomials, 122
  addition of, 124
  additive inverse, 125
  applications, 148–150
  in ascending order, 122
  binomials, 122
  coefficients, 122
  collecting (or combining) like terms, 124
  cubic, 122
  degree of, 122
  degree of a term, 122
  in descending order, 122
  division of, 211–216
    divisor a monomial, 211
    divisor not a monomial, 212
    synthetic, 215
  evaluating, 123, 131
  factoring, 130
    common factors, 130
    by completing the square, 139
    difference of cubes, 141
    difference of squares, 137
    general strategy, 143
    by grouping, 131
    sum of cubes, 141
    trinomial square, 134
    trinomials, 134, 135
  functions, 122, 123
  linear, 122
  monomial, 122
  multiplication, 127–129
  in one variable, 122
  quadratic, 122
  in several variables, 122
  subtraction, 125
  terms of, 122
  trinomial square, 133
  trinomials, 122
Population growth, 344
Powers
  logarithm of, 329
  principle of, 257
  and roots, 243
Power raised to a power, 34
Price axis, 116
Price, unit, 226
Principal square root, 232
Principle
  addition
    for equations, 10
    for inequalities, 155
  fundamental counting, 397
  multiplication
    for equations, 10
    for inequalities, 156
  P (probability), 396
  of powers, 257
  of zero products, 146
Probability, 395
  event, 396
  experimental, 395
  fundamental counting principle, 397
  outcomes, 396
  principle *P*, 396
  sample space, 396
  theoretical, 396
Problem solving, five-step strategy, 15
Problems, 14. *See also* Applied problems.
Product
  logarithm of, 328
  raising to a power, 34
Products of sums, differences, 128
Profit, 114
Programming, linear, 174
Properties
  of absolute value function, 165
  of exponents, 31–35
  identity, of one and zero, 21
  of logarithm functions, 328–330
Proportions, 222, 224. *See also* Direct variation; Inverse variation.
Pythagorean theorem, 259
  converse of, 353
  in three dimensions, 262

Q

Quadrants, 51
Quadratic equations, 272
  approximating solutions, 279
  discriminant, 285
  and fractional equations, 279, 281
  reducible to quadratic, 292
  solutions, nature of, 285
  solving, 272–274
    by factoring, 273, 274
    general strategy, 279
    by quadratic formula, 277
  standard form, 272
  writing from solutions, 286
Quadratic formula, 277
Quadratic functions, 299
  fitting to data, 313
  graphs of, 53, 299–304, 308–310
  maximum value of, 303, 311
  minimum value of, 303, 311
  and problem solving, 311
  translations, 304–306
Quadratic inequalities, 372
  graphing, 372
  solving, 372, 373
Quadratic polynomial, 122
Quadratic variation, 295
Quantity axis, 116
Quotient
  logarithm of, 330
  raising to a power, 35
  root of, 242

R

Radical equations, 257–259
Radical expressions, 232
  and absolute value, 234, 235, 236
  addition of, 244
  division with, 242
  and fractional exponents, 252
  index, 235
  like, 244
  multiplying with, 237, 238, 246
  and powers, 243
  quotients, 242
  radicand, 232
  rationalizing
    denominator, 247, 250
    numerator, 247, 249, 251
  simplifying, 237, 254, 255
  subtraction of, 244
Radical sign, 232
Radicals, *see* Radical expressions
Radicand, 232
  factoring, 237
Raising a power to a power, 34
Range of a function, 71
Ratio, common, 389
Rational exponents, 252–254
Rational expression, 182. *See also* Fractional expressions.
  addition, 190–194
  dividing, 186
  equivalent, 182
  least common multiple of denominators, 192
  multiplying, 183
  multiplying by one, 182
  reciprocals of, 185
  simplifying, 183
  subtraction, 190–194
Rational function, 182
Rational inequalities, 374
Rational numbers, 2, 239
  as exponents, 252–254
  notation, 2, 239
Rationalizing
  denominators, 247, 250
  numerators, 247, 249, 251
Real numbers, 2
  addition, 5
  division, 6
  multiplication, 6
  subtraction, 5
  order of, 154
  subsets, 2
Reciprocals, 5
  and division, 7
  and exponential notation, 30
  of rational expressions, 185
Reducible to quadratic equations, 292
Remainder, 212
Removing factors of 1, 183
Removing parentheses, 26
Repeating decimals, 3
Replacements, sensible, 21
Revenue, 114
Right triangle, 259
Rise, 62
Road grade, 70
Road pavement messages, 256
Roots
  cube, 235
  *k*th, odd and even, 235
  and powers, 243
  of quotients, 242
  square, 232
    approximating, 233
Row-equivalent matrices, 110
Row-equivalent operations, 110
Rows of a matrix, 109
Rules of signs
  addition, 5
  multiplication and division, 6
Run, 62

S
Sample space, 396
Scientific notation, 37
  approximating with, 39
  converting to, from decimal notation, 37
  converting to decimal notation, 38
  dividing with, 38
  estimating with, 39
  examples using, 40
  and logarithms, 335
  multiplying with, 38
  in problems, 40
Second coordinate, 51
Sensible replacements, 21
Sequence, 380
  arithmetic, 383–388
  geometric, 389–393
  infinite, 391
  of partial sums, 381
  terms, 380
Series, 381
  infinite, 391
  sigma notation, 381
Set-builder notation, 8
Sets
  empty, 161
  intersection, 161
  notation, 7, 8
  solution, 52
  union, 163
Sighting to the horizon, 262
Sigma notation, 381
Signs
  changing, 5
  rules of
    in addition, 5
    in division, 6
    in multiplication, 6
Similar terms, 124
  collecting (or combining), 25, 124
Simplifying
  complex fractional expressions, 195
  logarithmic expressions, 326, 328–331
  radical expressions, 237, 254, 255
  rational expressions, 183
Skidding car, speed of, 241
Slope, 56
  computing, 62–64
  lines without, 65
  of parallel lines, 68
  of perpendicular lines, 69
  zero, 64
Slope–intercept equation, 58
Solution sets
  of equations, 52
  of inequalities, 154
Solutions
  approximating, 279
  complex number, 265, 273, 278
  of equations, 10, 13, 51
  graphical, 87, 372
  of inequalities, 154, 168
  nature of, quadratic equations, 285
  of systems of equations, 86, 100
  writing equations from, 286
Solving equations
  with absolute value, 166
  addition principle, 10
  clearing
    of decimals, 12
    of fractions, 12
  exponential, 342
  by factoring, 146
  fractional, 199
  linear, 10–13
  logarithmic, 326
  multiplication principle, 10
  containing parentheses, 13
  polynomial, 145–147
  quadratic, 272–274
    by factoring, 146, 273, 274
    general strategy, 279
    by quadratic formula, 277
  radical, 257–259
  reducible to quadratic, 292
  systems, *see* Systems of equations
Solving formulas, *see* Formulas
Solving inequalities, 154
  in one variable, 155–157, 162, 163, 372–375
    with absolute value, 166
  systems of, 171
  in two variables, 168–171
Solving problems, five-step strategy, 15. *See also* Applied problems.
Speed, 98, 207
  average, 211
  of a skidding car, 241
Square
  of binomials, 128
  completing, 138
  trinomial, 133
Square root function, 233
Square roots, 232
  approximating, 233, 234, 239
  on a calculator, 233, 239
  negative, 232
  of negative numbers, 263
  perfect-square radicand, 234
  principal, 232
Squares, factoring difference of, 137. *See also* Square.
Standard form
  of circle equations, 355
  of ellipse equations, 356, 358
  of hyperbola equations, 360
  of quadratic equations, 272
Strategy
  factoring, 143
  problem-solving, 15
Subsets of real numbers, 2
Substitution method for solving systems, 89, 366, 370
Subtraction
  additive inverses and, 5
  of complex numbers, 263
  of exponents, 33
  of logarithms, 330
  of polynomials, 125
  of radical expressions, 244
  of rational expressions, 190–194
  of real numbers, 5
Sum
  of cubes, factoring, 141
  of logarithms, 328
  of polynomials, 124
  of sequences
    arithmetic, 385
    geometric, infinite, 392
    geometric, *n* terms, 390
    partial, 381
Sum and difference, product of, 128
Supply and demand, 115
Symmetry, line of, 300, 302, 303
Synthetic division, 215
Systems of equations, 86
  consistent, 87
  dependent, 87, 112
  of first and second degree, 365
  graphs of, 87
  inconsistent, 87
  independent, 112
  linear
    in three variables, 100
    in two variables, 86
  solution of, 86, 100
  solving
    addition method (method of elimination), 90, 101, 369
    by graphing, 87
    using matrices, 109
    substitution method, 89, 366, 370
  triangular form, 102
  two second-degree, 369
Systems of inequalities, 171
  graphing, 172

T
Tables
  antilogarithms using, 335
  common logarithms using, 334
  and functions, 73
  and natural logarithms, 336
  square roots using, 234, 239
Tabular difference, 339
Terminating decimals, 239
Terms
  coefficients of, 122
  collecting (or combining) like, 25, 124
  degree of, 122
  leading, 122
  like, 124
  missing, 212
  of a polynomial, 122
  of a sequence, 380
  similar, 124
Theorem
  binomial, 401, 405
  Pythagorean, 259
    in three dimensions, 262
Theoretical probability, 396
Total cost, 114
Translating to mathematical language, 15, 17–19
  with exponential notation, 35
  and formulas, 44
  using systems of equations, 86
Translations of graphs, 304–306
Triangular form
  of a matrix, 110
  of a system of equations, 102
Trinomials, 122
  factoring, 134
  squares, 133
    factoring, 134, 135
Two-point equation, 67

U
Union, 163
Unit circle, 359
Unit price, 226

V
Variable costs, 114
Variables
  dependent, 75
  eliminating, 90
  independent, 75

Variation
  direct, 221, 295
  inverse, 224, 295
  joint, 296
  and problem solving, 222–225, 296–298
  quadratic, 295
Variation constant, 222, 224
Vertex, 300, 302, 303
Vertical line, 59
  slope, 65
Vertical-line test for functions, 76

W

Walking speed, 346
Whole numbers, 2
Wind chill temperature, 241
Work problems, 204–207, 282

X

x-axis, 52
x-intercept, 61, 62, 309

Y

y-axis, 52
y-intercept, 56, 62
Yield, 97

Z

Zero
  as an exponent, 30
  identity property, 21
  named as a difference, 21
  slope, 64
Zero products, principle of, 146